T0215246

Lecture Notes in Computer Science   10409

Commenced Publication in 1973
Founding and Former Series Editors:
Gerhard Goos, Juris Hartmanis, and Jan van Leeuwen

More information about this series at http://www.springer.com/series/7407

Osvaldo Gervasi · Beniamino Murgante
Sanjay Misra · Giuseppe Borruso
Carmelo M. Torre · Ana Maria A.C. Rocha
David Taniar · Bernady O. Apduhan
Elena Stankova · Alfredo Cuzzocrea (Eds.)

Computational Science and Its Applications – ICCSA 2017

17th International Conference
Trieste, Italy, July 3–6, 2017
Proceedings, Part VI

 Springer

Editors

Osvaldo Gervasi ⓘ
University of Perugia
Perugia
Italy

Beniamino Murgante ⓘ
University of Basilicata
Potenza
Italy

Sanjay Misra ⓘ
Covenant University
Ota
Nigeria

Giuseppe Borruso ⓘ
University of Trieste
Trieste
Italy

Carmelo M. Torre ⓘ
Polytechnic University of Bari
Bari
Italy

Ana Maria A.C. Rocha ⓘ
University of Minho
Braga
Portugal

David Taniar ⓘ
Monash University
Clayton, VIC
Australia

Bernady O. Apduhan
Kyushu Sangyo University
Fukuoka
Japan

Elena Stankova ⓘ
Saint Petersburg State University
Saint Petersburg
Russia

Alfredo Cuzzocrea ⓘ
University of Trieste
Trieste
Italy

ISSN 0302-9743   ISSN 1611-3349 (electronic)
Lecture Notes in Computer Science
ISBN 978-3-319-62406-8   ISBN 978-3-319-62407-5 (eBook)
DOI 10.1007/978-3-319-62407-5

Library of Congress Control Number: 2017945283

LNCS Sublibrary: SL1 – Theoretical Computer Science and General Issues

Printed on acid-free paper

This Springer imprint is published by Springer Nature
The registered company is Springer International Publishing AG
The registered company address is: Gewerbestrasse 11, 6330 Cham, Switzerland

Preface

These multiple volumes (LNCS volumes 10404, 10405, 10406, 10407, 10408, and 10409) consist of the peer-reviewed papers from the 2017 International Conference on Computational Science and Its Applications (ICCSA 2017) held in Trieste, Italy, during July 3–6, 2017.

ICCSA 2017 was a successful event in the ICCSA conference series, previously held in Beijing, China (2016), Banff, Canada (2015), Guimarães, Portugal (2014), Ho Chi Minh City, Vietnam (2013), Salvador, Brazil (2012), Santander, Spain (2011), Fukuoka, Japan (2010), Suwon, South Korea (2009), Perugia, Italy (2008), Kuala Lumpur, Malaysia (2007), Glasgow, UK (2006), Singapore (2005), Assisi, Italy (2004), Montreal, Canada (2003), (as ICCS) Amsterdam, The Netherlands (2002), and San Francisco, USA (2001).

Computational science is a main pillar of most present research as well as industrial and commercial activities and plays a unique role in exploiting ICT innovative technologies. The ICCSA conference series have been providing a venue to researchers and industry practitioners to discuss new ideas, to share complex problems and their solutions, and to shape new trends in computational science.

Apart from the general tracks, ICCSA 2017 also include 43 international workshops, in various areas of computational sciences, ranging from computational science technologies to specific areas of computational sciences, such as computer graphics and virtual reality. Furthermore, this year ICCSA 2017 hosted the XIV International Workshop on Quantum Reactive Scattering. The program also features three keynote speeches and four tutorials.

The success of the ICCSA conference series in general, and ICCSA 2017 in particular, is due to the support of many people: authors, presenters, participants, keynote speakers, session chairs, Organizing Committee members, student volunteers, Program Committee members, international Advisory Committee members, international liaison chairs, and various people in other roles. We would like to thank them all.

We would also like to thank Springer for their continuous support in publishing the ICCSA conference proceedings.

July 2017

Giuseppe Borruso
Osvaldo Gervasi
Bernady O. Apduhan

Welcome to Trieste

We were honored and happy to have organized this extraordinary edition of the conference, with so many interesting contributions and participants coming from more than 46 countries around the world!

Trieste is a medium-size Italian city lying on the north-eastern border between Italy and Slovenia. It has a population of nearly 200,000 inhabitants and faces the Adriatic Sea, surrounded by the Karst plateau.

It is quite an atypical Italian city, with its history being very much influenced by belonging for several centuries to the Austro-Hungarian empire and having been through several foreign occupations in history: by French, Venetians, and the Allied Forces after the Second World War. Such events left several footprints on the structure of the city, on its buildings, as well as on culture and society!

During its history, Trieste hosted people coming from different countries and regions, making it a cosmopolitan and open city. This was also helped by the presence of a commercial port that made it an important trade center from the 18th century on. Trieste is known today as a 'City of Science' or, more proudly, presenting itself as the 'City of Knowledge', thanks to the presence of several universities and research centers, all of them working at an international level, as well as of cultural institutions and traditions. The city has a high presence of researchers, more than 35 per 1,000 employed people, much higher than the European average of 6 employed researchers per 1,000 people.

The University of Trieste, the origin of such a system of scientific institutions, dates back to 1924, although its roots go back to the end of the 19th century under the Austro-Hungarian Empire. The university today employs nearly 1,500 teaching, research, technical, and administrative staff with a population of more than 16,000 students.

The university currently has 10 departments: Economics, Business, Mathematical, and Statistical Sciences; Engineering and Architecture; Humanities; Legal, Language, Interpreting, and Translation Studies; Mathematics and Geosciences; Medicine, Surgery, and Health Sciences; Life Sciences; Pharmaceutical and Chemical Sciences; Physics; Political and Social Sciences.

We trust the participants enjoyed the cultural and scientific offerings of Trieste and will keep a special memory of the event.

Giuseppe Borruso

Organization

ICCSA 2017 was organized by the University of Trieste (Italy), University of Perugia (Italy), Monash University (Australia), Kyushu Sangyo University (Japan), University of Basilicata (Italy), and University of Minho, (Portugal).

Honorary General Chairs

Antonio Laganà	University of Perugia, Italy
Norio Shiratori	Tohoku University, Japan
Kenneth C.J. Tan	Sardina Systems, Estonia

General Chairs

Giuseppe Borruso	University of Trieste, Italy
Osvaldo Gervasi	University of Perugia, Italy
Bernady O. Apduhan	Kyushu Sangyo University, Japan

Program Committee Chairs

Alfredo Cuzzocrea	University of Trieste, Italy
Beniamino Murgante	University of Basilicata, Italy
Ana Maria A.C. Rocha	University of Minho, Portugal
David Taniar	Monash University, Australia

International Advisory Committee

Jemal Abawajy	Deakin University, Australia
Dharma P. Agrawal	University of Cincinnati, USA
Marina L. Gavrilova	University of Calgary, Canada
Claudia Bauzer Medeiros	University of Campinas, Brazil
Manfred M. Fisher	Vienna University of Economics and Business, Austria
Yee Leung	Chinese University of Hong Kong, SAR China

International Liaison Chairs

Ana Carla P. Bitencourt	Universidade Federal do Reconcavo da Bahia, Brazil
Maria Irene Falcão	University of Minho, Portugal
Robert C.H. Hsu	Chung Hua University, Taiwan
Tai-Hoon Kim	Hannam University, Korea
Sanjay Misra	University of Minna, Nigeria
Takashi Naka	Kyushu Sangyo University, Japan

| Rafael D.C. Santos | National Institute for Space Research, Brazil |
| Maribel Yasmina Santos | University of Minho, Portugal |

Workshop and Session Organizing Chairs

Beniamino Murgante	University of Basilicata, Italy
Sanjay Misra	Covenant University, Nigeria
Jorge Gustavo Rocha	University of Minho, Portugal

Award Chair

| Wenny Rahayu | La Trobe University, Australia |

Publicity Committee Chair

Stefano Cozzini	Democritos Center, National Research Council, Italy
Elmer Dadios	De La Salle University, Philippines
Hong Quang Nguyen	International University (VNU-HCM), Vietnam
Daisuke Takahashi	Tsukuba University, Japan
Shangwang Wang	Beijing University of Posts and Telecommunications, China

Workshop Organizers

Agricultural and Environmental Big Data Analytics (AEDBA 2017)

| Sandro Bimonte | IRSTEA, France |
| André Miralles | IRSTEA, France |

Advances in Data Mining for Applications (AMDMA 2017)

Carlo Cattani	University of Tuscia, Italy
Majaz Moonis	University of Massachusettes Medical School, USA
Yeliz Karaca	IEEE, Computer Society Association

Advances Smart Mobility and Transportation (ASMAT 2017)

| Mauro Mazzei | CNR, Italian National Research Council, Italy |

Advances in Information Systems and Technologies for Emergency Preparedness and Risk Assessment and Mitigation (ASTER 2017)

Maurizio Pollino	ENEA, Italy
Marco Vona	University of Basilicata, Italy
Beniamino Murgante	University of Basilicata, Italy

Advances in Web-Based Learning (AWBL 2017)

Mustafa Murat Inceoglu   Ege University, Turkey
Birol Ciloglugil   Ege University, Turkey

Big Data Warehousing and Analytics (BIGGS 2017)

Maribel Yasmina Santos   University of Minho, Portugal
Monica Wachowicz   University of New Brunswick, Canada
Joao Moura Pires   NOVA de Lisboa University, Portugal
Rafael Santos   National Institute for Space Research, Brazil

Bio-inspired Computing and Applications (BIONCA 2017)

Nadia Nedjah   State University of Rio de Janeiro, Brazil
Luiza de Macedo Mourell   State University of Rio de Janeiro, Brazil

Computational and Applied Mathematics (CAM 2017)

M. Irene Falcao   University of Minho, Portugal
Fernando Miranda   University of Minho, Portugal

Computer-Aided Modeling, Simulation, and Analysis (CAMSA 2017)

Jie Shen   University of Michigan, USA and Jilin University, China
Hao Chenina   Shanghai University of Engineering Science, China
Chaochun Yuan   Jiangsu University, China

Computational and Applied Statistics (CAS 2017)

Ana Cristina Braga   University of Minho, Portugal

Computational Geometry and Security Applications (CGSA 2017)

Marina L. Gavrilova   University of Calgary, Canada

Central Italy 2016 Earthquake: Computational Tools and Data Analysis for Emergency Response, Community Support, and Reconstruction Planning (CIEQ 2017)

Alessandro Rasulo   Università degli Studi di Cassino e del Lazio
                               Meridionale, Italy
Davide Lavorato   Università degli Studi di Roma Tre, Italy

Computational Methods for Business Analytics (CMBA 2017)

Telmo Pinto	University of Minho, Portugal
Claudio Alves	University of Minho, Portugal

Chemistry and Materials Sciences and Technologies (CMST 2017)

Antonio Laganà	University of Perugia, Italy
Noelia Faginas Lago	University of Perugia, Italy

Computational Optimization and Applications (COA 2017)

Ana Maria Rocha	University of Minho, Portugal
Humberto Rocha	University of Coimbra, Portugal

Cities, Technologies, and Planning (CTP 2017)

Giuseppe Borruso	University of Trieste, Italy
Beniamino Murgante	University of Basilicata, Italy

Data-Driven Modelling for Sustainability Assessment (DAMOST 2017)

Antonino Marvuglia	Luxembourg Institute of Science and Technology, LIST, Luxembourg
Mikhail Kanevski	University of Lausanne, Switzerland
Beniamino Murgante	University of Basilicata, Italy
Janusz Starczewski	Częstochowa University of Technology, Poland

Databases and Computerized Information Retrieval Systems (DCIRS 2017)

Sultan Alamri	College of Computing and Informatics, SEU, Saudi Arabia
Adil Fahad	Albaha University, Saudi Arabia
Abdullah Alamri	Jeddah University, Saudi Arabia

Data Science for Intelligent Decision Support (DS4IDS 2016)

Filipe Portela	University of Minho, Portugal
Manuel Filipe Santos	University of Minho, Portugal

Deep Cities: Intelligence and Interoperability (DEEP_CITY 2017)

Maurizio Pollino   ENEA, Italian National Agency for New Technologies,
                           Energy and Sustainable Economic Development, Italy
Grazia Fattoruso   ENEA, Italian National Agency for New Technologies,
                           Energy and Sustainable Economic Development, Italy

Emotion Recognition (EMORE 2017)

Valentina Franzoni   University of Rome La Sapienza, Italy
Alfredo Milani   University of Perugia, Italy

Future Computing Systems, Technologies, and Applications (FISTA 2017)

Bernady O. Apduhan   Kyushu Sangyo University, Japan
Rafael Santos   National Institute for Space Research, Brazil

Geographical Analysis, Urban Modeling, Spatial Statistics (Geo-and-Mod 2017)

Giuseppe Borruso   University of Trieste, Italy
Beniamino Murgante   University of Basilicata, Italy
Hartmut Asche   University of Potsdam, Germany

Geomatics and Remote Sensing Techniques for Resource Monitoring and Control (GRS-RMC 2017)

Eufemia Tarantino   Polytechnic of Bari, Italy
Rosa Lasaponara   Italian Research Council, IMAA-CNR, Italy
Antonio Novelli   Polytechnic of Bari, Italy

Interactively Presenting High-Quality Graphics in Cooperation with Various Computing Tools (IPHQG 2017)

Masataka Kaneko   Toho University, Japan
Setsuo Takato   Toho University, Japan
Satoshi Yamashita   Kisarazu National College of Technology, Italy

Web-Based Collective Evolutionary Systems: Models, Measures, Applications (IWCES 2017)

Alfredo Milani   University of Perugia, Italy
Rajdeep Nyogi   Institute of Technology, Roorkee, India
Valentina Franzoni   University of Rome La Sapienza, Italy

Computational Mathematics, and Statistics for Data Management and Software Engineering (IWCMSDMSE 2017)

M. Filomena Teodoro	Lisbon University and Portuguese Naval Academy, Portugal
Anacleto Correia	Portuguese Naval Academy, Portugal

Land Use Monitoring for Soil Consumption Reduction (LUMS 2017)

Carmelo M. Torre	Polytechnic of Bari, Italy
Beniamino Murgante	University of Basilicata, Italy
Alessandro Bonifazi	Polytechnic of Bari, Italy
Massimiliano Bencardino	University of Salerno, Italy

Mobile Communications (MC 2017)

Hyunseung Choo	Sungkyunkwan University, Korea

Mobile-Computing, Sensing, and Actuation - Fog Networking (MSA4FOG 2017)

Saad Qaisar	NUST School of Electrical Engineering and Computer Science, Pakistan
Moonseong Kim	Korean Intellectual Property Office, South Korea

Physiological and Affective Computing: Methods and Applications (PACMA 2017)

Robertas Damasevicius	Kaunas University of Technology, Lithuania
Christian Napoli	University of Catania, Italy
Marcin Wozniak	Silesian University of Technology, Poland

Quantum Mechanics: Computational Strategies and Applications (QMCSA 2017)

Mirco Ragni	Universidad Federal de Bahia, Brazil
Ana Carla Peixoto Bitencourt	Universidade Estadual de Feira de Santana, Brazil
Vincenzo Aquilanti	University of Perugia, Italy

Advances in Remote Sensing for Cultural Heritage (RS 2017)

Rosa Lasaponara   IRMMA, CNR, Italy
Nicola Masini   IBAM, CNR, Italy Zhengzhou Base, International
                         Center on Space Technologies for Natural and
                         Cultural Heritage, China

Scientific Computing Infrastructure (SCI 2017)

Elena Stankova   Saint Petersburg State University, Russia
Alexander Bodganov   Saint Petersburg State University, Russia
Vladimir Korkhov   Saint Petersburg State University, Russia

Software Engineering Processes and Applications (SEPA 2017)

Sanjay Misra   Covenant University, Nigeria

Sustainability Performance Assessment: Models, Approaches and Applications Toward Interdisciplinarity and Integrated Solutions (SPA 2017)

Francesco Scorza   University of Basilicata, Italy
Valentin Grecu   Lucia Blaga University on Sibiu, Romania
Jolanta Dvarioniene   Kaunas University, Lithuania
Sabrina Lai   Cagliari University, Italy

Software Quality (SQ 2017)

Sanjay Misra   Covenant University, Nigeria

Advances in Spatio-Temporal Analytics (ST-Analytics 2017)

Rafael Santos   Brazilian Space Research Agency, Brazil
Karine Reis Ferreira   Brazilian Space Research Agency, Brazil
Maribel Yasmina Santos   University of Minho, Portugal
Joao Moura Pires   New University of Lisbon, Portugal

Tools and Techniques in Software Development Processes (TTSDP 2017)

Sanjay Misra   Covenant University, Nigeria

Challenges, Trends, and Innovations in VGI (VGI 2017)

Claudia Ceppi	University of Basilicata, Italy
Beniamino Murgante	University of Basilicata, Italy
Lucia Tilio	University of Basilicata, Italy
Francesco Mancini	University of Modena and Reggio Emilia, Italy
Rodrigo Tapia-McClung	Centro de Investigación en Geografía y Geomática "Ing Jorge L. Tamayo", Mexico
Jorge Gustavo Rocha	University of Minho, Portugal

Virtual Reality and Applications (VRA 2017)

Osvaldo Gervasi	University of Perugia, Italy

Industrial Computational Applications (WICA 2017)

Eric Medvet	University of Trieste, Italy
Gianfranco Fenu	University of Trieste, Italy
Riccardo Ferrari	Delft University of Technology, The Netherlands

XIV International Workshop on Quantum Reactive Scattering (QRS 2017)

Niyazi Bulut	Fırat University, Turkey
Noelia Faginas Lago	University of Perugia, Italy
Andrea Lombardi	University of Perugia, Italy
Federico Palazzetti	University of Perugia, Italy

Program Committee

Jemal Abawajy	Deakin University, Australia
Kenny Adamson	University of Ulster, UK
Filipe Alvelos	University of Minho, Portugal
Paula Amaral	Universidade Nova de Lisboa, Portugal
Hartmut Asche	University of Potsdam, Germany
Md. Abul Kalam Azad	University of Minho, Portugal
Michela Bertolotto	University College Dublin, Ireland
Sandro Bimonte	CEMAGREF, TSCF, France
Rod Blais	University of Calgary, Canada
Ivan Blečić	University of Sassari, Italy
Giuseppe Borruso	University of Trieste, Italy
Yves Caniou	Lyon University, France
José A. Cardoso e Cunha	Universidade Nova de Lisboa, Portugal
Rui Cardoso	University of Beira Interior, Portugal
Leocadio G. Casado	University of Almeria, Spain
Carlo Cattani	University of Salerno, Italy

Mete Celik	Erciyes University, Turkey
Alexander Chemeris	National Technical University of Ukraine KPI, Ukraine
Min Young Chung	Sungkyunkwan University, Korea
Gilberto Corso Pereira	Federal University of Bahia, Brazil
M. Fernanda Costa	University of Minho, Portugal
Gaspar Cunha	University of Minho, Portugal
Alfredo Cuzzocrea	ICAR-CNR and University of Calabria, Italy
Carla Dal Sasso Freitas	Universidade Federal do Rio Grande do Sul, Brazil
Pradesh Debba	The Council for Scientific and Industrial Research (CSIR), South Africa
Hendrik Decker	Instituto Tecnológico de Informática, Spain
Frank Devai	London South Bank University, UK
Rodolphe Devillers	Memorial University of Newfoundland, Canada
Prabu Dorairaj	NetApp, India/USA
M. Irene Falcao	University of Minho, Portugal
Cherry Liu Fang	U.S. DOE Ames Laboratory, USA
Edite M.G.P. Fernandes	University of Minho, Portugal
Jose-Jesús Fernandez	National Centre for Biotechnology, CSIS, Spain
María Antonia Forjaz	University of Minho, Portugal
María Celia Furtado Rocha	PRODEB-Pós Cultura/UFBA, Brazil
Akemi Galvez	University of Cantabria, Spain
Paulino Jose Garcia Nieto	University of Oviedo, Spain
Marina Gavrilova	University of Calgary, Canada
Jerome Gensel	LSR-IMAG, France
María Giaoutzi	National Technical University, Athens, Greece
Andrzej M. Goscinski	Deakin University, Australia
Alex Hagen-Zanker	University of Cambridge, UK
Malgorzata Hanzl	Technical University of Lodz, Poland
Shanmugasundaram Hariharan	B.S. Abdur Rahman University, India
Eligius M.T. Hendrix	University of Malaga/Wageningen University, Spain/The Netherlands
Tutut Herawan	Universitas Teknologi Yogyakarta, Indonesia
Hisamoto Hiyoshi	Gunma University, Japan
Fermin Huarte	University of Barcelona, Spain
Andrés Iglesias	University of Cantabria, Spain
Mustafa Inceoglu	EGE University, Turkey
Peter Jimack	University of Leeds, UK
Qun Jin	Waseda University, Japan
Farid Karimipour	Vienna University of Technology, Austria
Baris Kazar	Oracle Corp., USA
Maulana Adhinugraha Kiki	Telkom University, Indonesia
DongSeong Kim	University of Canterbury, New Zealand
Taihoon Kim	Hannam University, Korea
Ivana Kolingerova	University of West Bohemia, Czech Republic

Dieter Kranzlmueller	LMU and LRZ Munich, Germany
Antonio Laganà	University of Perugia, Italy
Rosa Lasaponara	National Research Council, Italy
Maurizio Lazzari	National Research Council, Italy
Cheng Siong Lee	Monash University, Australia
Sangyoun Lee	Yonsei University, Korea
Jongchan Lee	Kunsan National University, Korea
Clement Leung	Hong Kong Baptist University, Hong Kong, SAR China
Chendong Li	University of Connecticut, USA
Gang Li	Deakin University, Australia
Ming Li	East China Normal University, China
Fang Liu	AMES Laboratories, USA
Xin Liu	University of Calgary, Canada
Savino Longo	University of Bari, Italy
Tinghuai Ma	NanJing University of Information Science and Technology, China
Sergio Maffioletti	University of Zurich, Switzerland
Ernesto Marcheggiani	Katholieke Universiteit Leuven, Belgium
Antonino Marvuglia	Research Centre Henri Tudor, Luxembourg
Nicola Masini	National Research Council, Italy
Nirvana Meratnia	University of Twente, The Netherlands
Alfredo Milani	University of Perugia, Italy
Sanjay Misra	Federal University of Technology Minna, Nigeria
Giuseppe Modica	University of Reggio Calabria, Italy
José Luis Montaña	University of Cantabria, Spain
Beniamino Murgante	University of Basilicata, Italy
Jiri Nedoma	Academy of Sciences of the Czech Republic, Czech Republic
Laszlo Neumann	University of Girona, Spain
Kok-Leong Ong	Deakin University, Australia
Belen Palop	Universidad de Valladolid, Spain
Marcin Paprzycki	Polish Academy of Sciences, Poland
Eric Pardede	La Trobe University, Australia
Kwangjin Park	Wonkwang University, Korea
Ana Isabel Pereira	Polytechnic Institute of Braganca, Portugal
Maurizio Pollino	Italian National Agency for New Technologies, Energy and Sustainable Economic Development, Italy
Alenka Poplin	University of Hamburg, Germany
Vidyasagar Potdar	Curtin University of Technology, Australia
David C. Prosperi	Florida Atlantic University, USA
Wenny Rahayu	La Trobe University, Australia
Jerzy Respondek	Silesian University of Technology Poland
Ana Maria A.C. Rocha	University of Minho, Portugal
Maria Clara Rocha	ESTES Coimbra, Portugal
Humberto Rocha	INESC-Coimbra, Portugal

Additional Reviewers

A. Alwan Al-Juboori Ali	School of Computer Science and Technology, China
Aceto Lidia	University of Pisa, Italy
Acharjee Shukla	Dibrugarh University, India
Afreixo Vera	University of Aveiro, Portugal
Agra Agostinho	University of Aveiro, Portugal
Aguilar Antonio	University of Barcelona, Spain
Aguilar José Alfonso	Universidad Autónoma de Sinaloa, Mexico
Aicardi Irene	Politecnico di Torino, Italy
Alberti Margarita	University of Barcelona, Spain
Alberto Rui	University of Lisbon, Portugal
Ali Salman	University of Magna Graecia, Italy
Alvanides Seraphim	University at Newcastle, UK
Alvelos Filipe	Universidade do Minho, Portugal
Amato Alba	Seconda Università degli Studi di Napoli, Italy
Amorim Paulo	Instituto de Matemática da UFRJ (IM-UFRJ), Brazil
Anderson Roger	University of California Santa Cruz, USA
Andrianov Serge	Saint Petersburg State University, Russia
Andrienko Gennady	Fraunhofer-Institut für Intelligente Analyse- und Informationssysteme, Germany
Apduhan Bernady	Kyushu Sangyo University, Japan
Aquilanti Vincenzo	University of Perugia, Italy
Asche Hartmut	Potsdam University, Germany
Azam Samiul	United International University, Bangladesh
Azevedo Ana	Athabasca University, USA
Bae Ihn-Han	Catholic University of Daegu, South Korea
Balacco Gabriella	Polytechnic of Bari, Italy
Balena Pasquale	Polytechnic of Bari, Italy
Barroca Filho Itamir	Universidade Federal do Rio Grande do Norte, Brazil
Behera Ranjan Kumar	Indian Institute of Technology Patna, India
Belpassi Leonardo	National Research Council, Italy
Bentayeb Fadila	Université Lyon, France
Bernardino Raquel	Universidade da Beira Interiore, Portugal
Bertolotto Michela	University Collegue Dublin, UK
Bhatta Bijaya	Utkal University, India
Bimonte Sandro	IRSTEA, France
Blecic Ivan	University of Cagliari, Italy
Bo Carles	ICIQ, Spain
Bogdanov Alexander	Saint Petersburg State University, Russia
Bollini Letizia	University of Milano-Bicocca, Italy
Bonifazi Alessandro	Polytechnic of Bari, Italy
Bonnet Claude-Laurent	Université de Bordeaux, France
Borgogno Mondino Enrico Corrado	University of Turin, Italy
Borruso Giuseppe	University of Trieste, Italy

Bostenaru Maria	Ion Mincu University of Architecture and Urbanism, Romania
Boussaid Omar	Université Lyon 2, France
Braga Ana Cristina	University of Minho, Portugal
Braga Nuno	University of Minho, Portugal
Brasil Luciana	Instituto Federal Sao Paolo, Brazil
Cabral Pedro	Universidade NOVA de Lisboa, Portugal
Cacao Isabel	University of Aveiro, Portugal
Caiaffa Emanuela	Enea, Italy
Campagna Michele	University of Cagliari, Italy
Caniato Renhe Marcelo	Universidade Federal de Juiz de Fora, Brazil
Canora Filomena	University of Basilicata, Italy
Caradonna Grazia	Polytechnic of Bari, Italy
Cardoso Rui	Beira Interior University, Portugal
Caroti Gabriella	University of Pisa, Italy
Carravilla Maria Antonia	Universidade do Porto, Portugal
Cattani Carlo	University of Salerno, Italy
Cefalo Raffaela	University of Trieste, Italy
Ceppi Claudia	Polytechnic of Bari, Italy
Cerreta Maria	University Federico II of Naples, Italy
Chanet Jean-Pierre	UR TSCF Irstea, France
Chaturvedi Krishna Kumar	University of Delhi, India
Chiancone Andrea	University of Perugia, Italy
Choo Hyunseung	Sungkyunkwan University, South Korea
Ciabo Serena	University of l'Aquila, Italy
Coletti Cecilia	University of Chieti, Italy
Correia Aldina	Porto Polytechnic, Portugal
Correia Anacleto	CINAV, Portugal
Correia Elisete	University of Trás-Os-Montes e Alto Douro, Portugal
Correia Florbela Maria da Cruz Domingues	Instituto Politécnico de Viana do Castelo, Portugal
Cosido Oscar	University of Cantabria, Spain
Costa e Silva Eliana	University of Minho, Portugal
Costa Graça	Instituto Politécnico de Setúbal, Portugal
Costantini Alessandro	INFN, Italy
Crispim José	University of Minho, Portugal
Cuzzocrea Alfredo	University of Trieste, Italy
Danese Maria	IBAM, CNR, Italy
Daneshpajouh Shervin	University of Western Ontario, USA
De Fazio Dario	IMIP-CNR, Italy
De Runz Cyril	University of Reims Champagne-Ardenne, France
Deffuant Guillaume	Institut national de recherche en sciences et technologies pour l'environnement et l'agriculture, France
Degtyarev Alexander	Saint Petersburg State University, Russia
Devai Frank	London South Bank University, UK
Di Leo Margherita	JRC, European Commission, Belgium

Dias Joana	University of Coimbra, Portugal
Dilo Arta	University of Twente, The Netherlands
Dvarioniene Jolanta	Kaunas University of Technology, Lithuania
El-Zawawy Mohamed A.	Cairo University, Egypt
Escalona Maria-Jose	University of Seville, Spain
Faginas-Lago, Noelia	University of Perugia, Italy
Falcinelli Stefano	University of Perugia, Italy
Falcão M. Irene	University of Minho, Portugal
Faria Susana	University of Minho, Portugal
Fattoruso Grazia	ENEA, Italy
Fenu Gianfranco	University of Trieste, Italy
Fernandes Edite	University of Minho, Portugal
Fernandes Florbela	Escola Superior de Tecnologia e Gest ão de Bragancca, Portugal
Fernandes Rosario	USP/ESALQ, Brazil
Ferrari Riccardo	Delft University of Technology, The Netherlands
Figueiredo Manuel Carlos	University of Minho, Portugal
Florence Le Ber	ENGEES, France
Flouvat Frederic	University of New Caledonia, France
Fontes Dalila	Universidade do Porto, Portugal
Franzoni Valentina	University of Perugia, Italy
Freitas Adelaide de Fátima Baptista Valente	University of Aveiro, Portugal
Fusco Giovanni	Università di Bari, Italy
Gabrani Goldie	Tecpro Syst. Ltd., India
Gaido Luciano	INFN, Italy
Gallo Crescenzio	University of Foggia, Italy
Garaba Shungu	University of Connecticut, USA
Garau Chiara	University of Cagliari, Italy
Garcia Ernesto	University of the Basque Country, Spain
Gargano Ricardo	Universidade Brasilia, Brazil
Gavrilova Marina	University of Calgary, Canada
Gensel Jerome	IMAG, France
Gervasi Osvaldo	University of Perugia, Italy
Gioia Andrea	Polytechnic University of Bari, Italy
Giovinazzi Sonia	University of Canterbury, New Zealand
Gizzi Fabrizio	National Research Council, Italy
Gomes dos Anjos Eudisley	Universidade Federal da Paraíba, Brazil
Gonzaga de Oliveira Sanderson Lincohn	Universidade Federal de Lavras, Brazil
Gonçalves Arminda Manuela	University of Minho, Braga, Portugal
Gorbachev Yuriy	Geolink Technologies, Russia
Grecu Valentin	University of Sibiu, Romania
Gupta Brij	Cancer Biology Research Center, USA
Hagen-Zanker Alex	University of Surrey, UK

Hamaguchi Naoki	Tokyo Kyoiku University, Japan
Hanazumi Simone	University of Sao Paulo, Brazil
Hanzl Malgorzata	University of Lodz, Poland
Hayashi Masaki	University of Calgary, Canada
Hendrix Eligius M.T.	Operations Research and Logistics Group, The Netherlands
Henriques Carla	Inst. Politécnico de Viseu, Portugal
Herawan Tutut	State Polytechnic of Malang, Indonesia
Hsu Hui-Huang	National Chiao Tung University, Taiwan
Ienco Dino	La Maison de la télédétection de Montpellier, France
Iglesias Andres	Universidad de Cantabria, Spain
Imran Rabeea	NUST Islamabad, Pakistan
Inoue Kentaro	National Technical University of Athens, Greece
Josselin Didier	Université d'Avignon et des Pays de Vaucluse, France
Kaneko Masataka	Kisarazu National College of Technology, Japan
Kang Myoung-Ah	Blaise Pascal University, France
Karampiperis Pythagoras	National Center of Scientific Research, Athens, Greece
Kavouras Marinos	University of Athens, Greece
Kolingerova Ivana	University of West Bohemia, Czech Republic
Korkhov Vladimir	Saint Petersburg State University, Russia
Kotzinos Dimitrios	University of Cergy Pontoise, France
Kulabukhova Nataliia	Saint Petersburg State University, Russia
Kumar Dileep	SR Engineering College, India
Kumar Lov	National Institute of Technology, Rourkela, India
Kumar Pawan	Institute for Advanced Study, Princeton, USA
Laganà Antonio	University of Perugia, Italy
Lai Sabrina	Università di Cagliari, Italy
Lanza Viviana	Lombardy Regional Institute for Research, Italy
Lasala Piermichele	Università di Foggia, Italy
Laurent Anne	Laboratoire d'Informatique, de Robotique et de Microélectronique de Montpellier, France
Lavorato Davide	University of Rome, Italy
Le Duc Tai	Sungkyunkwan University, South Korea
Legatiuk Dmitrii	Bauhaus University, Germany
Li Ming	University of Waterloo, Canada
Lima Ana	University of São Paulo (UNIFESP), Brazil
Liu Xin	École polytechnique fédérale de Lausanne, Switzerland
Lombardi Andrea	University of Perugia, Italy
Lopes Cristina	Instituto Superior de Contabilidade e Administracao do Porto, Portugal
Lopes Maria João	Instituto Universitário de Lisboa, Portugal
Lourenço Vanda Marisa	Universidade NOVA de Lisboa, Portugal
Machado Jose	University of Minho, Portugal
Maeda Yoichi	Tokai University, Japan
Majcen Nineta	Euchems, Belgium
Malonek Helmuth	Universidade de Aveiro, Portugal

Mancini Francesco	University of Modena and Reggio Emilia, Italy
Mandanici Emanuele	Università di Bologna, Italy
Manganelli Benedetto	Università degli studi della Basilicata, Italy
Manso Callejo Miguel Angel	Universidad Politécnica de Madrid, Spain
Margalef Tomas	Autonomous University of Barcelona, Spain
Marques Jorge	University of Coimbra, Portugal
Martins Bruno	Universidade de Lisboa, Portugal
Marvuglia Antonino	Public Research Centre Henri Tudor, Luxembourg
Mateos Cristian	Universidad Nacional del Centro, Argentina
Mauro Giovanni	University of Trieste, Italy
McGuire Michael	Towson University, USA
Medvet Eric	University of Trieste, Italy
Milani Alfredo	University of Perugia, Italy
Millham Richard	Durban University of Technoloy, South Africa
Minghini Marco	Polytechnic University of Milan, Italy
Minhas Umar	University of Waterloo, Ontario, Canada
Miralles André	La Maison de la télédétection de Montpellier, France
Miranda Fernando	Universidade do Minho, Portugal
Misra Sanjay	Covenant University, Nigeria
Modica Giuseppe	Università Mediterranea di Reggio Calabria, Italy
Molaei Qelichi Mohamad	University of Tehran, Iran
Monteiro Ana Margarida	University of Coimbra, Portugal
Morano Pierluigi	Polytechnic University of Bari, Italy
Moura Ana	Universidade de Aveiro, Portugal
Moura Pires João	Universidade NOVA de Lisboa, Portugal
Mourão Maria	ESTG-IPVC, Portugal
Murgante Beniamino	University of Basilicata, Italy
Nagy Csaba	University of Szeged, Hungary
Nakamura Yasuyuki	Nagoya University, Japan
Natário Isabel Cristina Maciel	University Nova de Lisboa, Portugal
Nemmaoui Abderrahim	Universidad de Almeria (UAL), Spain
Nguyen Tien Dzung	Sungkyunkwan University, South Korea
Niyogi Rajdeep	Indian Institute of Technology Roorkee, India
Novelli Antonio	University of Bari, Italy
Oliveira Irene	University of Trás-Os-Montes e Alto Douro, Portugal
Oliveira José A.	Universidade do Minho, Portugal
Ottomanelli Michele	University of Bari, Italy
Ouchi Shunji	Shimonoseki City University, Japan
Ozturk Savas	Scientific and Technological Research Council of Turkey, Turkey
P. Costa M. Fernanda	Universidade do Minho, Portugal
Painho Marco	NOVA Information Management School, Portugal
Panetta J.B.	Tecnologia Geofísica Petróleo Brasileiro SA, PETROBRAS, Brazil

Pantazis Dimos	Otenet, Greece
Papa Enrica	University of Amsterdam, The Netherlands
Pardede Eric	La Trobe University, Australia
Parente Claudio	Università degli Studi di Napoli Parthenope, Italy
Pathan Al-Sakib Khan	Islamic University of Technology, Bangladesh
Paul Prantosh K.	EIILM University, Jorethang, Sikkim, India
Pengő Edit	University of Szeged, Hungary
Pereira Ana	IPB, Portugal
Pereira José Luís	Universidade do Minho, Portugal
Peschechera Giuseppe	Università di Bologna, Italy
Pham Quoc Trung	HCMC University of Technology, Vietnam
Piemonte Andreaa	University of Pisa, Italy
Pimentel Carina	Universidade de Aveiro, Portugal
Pinet Francois	IRSTEA, France
Pinto Livio	Polytechnic University of Milan, Italy
Pinto Telmo	Universidade do Minho, Portugal
Pinet Francois	IRSTEA, France
Poli Giuliano	Université Pierre et Marie Curie, France
Pollino Maurizio	ENEA, Italy
Portela Carlos Filipe	Universidade do Minho, Portugal
Prata Paula	Universidade Federal de Sergipe, Brazil
Previl Carlo	University of Quebec in Abitibi-Témiscamingue (UQAT), Canada
Prezioso Giuseppina	Università degli Studi di Napoli Parthenope, Italy
Pusatli Tolga	Cankaya University, Turkey
Quan Tho	Ho Chi Minh, University of Technology, Vietnam
Ragni Mirco	Universidade Estadual de Feira de Santana, Brazil
Rahman Nazreena	Biotechnology Research Centre, Malaysia
Rahman Wasiur	Technical University Darmstadt, Germany
Rashid Sidra	National University of Sciences and Technology (NUST) Islamabad, Pakistan
Rasulo Alessandro	Università degli studi di Cassino e del Lazio Meridionale, Italy
Raza Syed Muhammad	Sungkyunkwan University, South Korea
Reis Ferreira Gomes Karine	Instituto Nacional de Pesquisas Espaciais, Brazil
Requejo Cristina	Universidade de Aveiro, Portugal
Rocha Ana Maria	University of Minho, Portugal
Rocha Humberto	University of Coimbra, Portugal
Rocha Jorge	University of Minho, Portugal
Rodriguez Daniel	University of Berkeley, USA
Saeki Koichi	Graduate University for Advanced Studies, Japan
Samela Caterina	University of Basilicata, Italy
Sannicandro Valentina	Polytechnic of Bari, Italy
Santiago Júnior Valdivino	Instituto Nacional de Pesquisas Espaciais, Brazil
Sarafian Haiduke	Pennsylvania State University, USA

Santos Daniel	Universidade Federal de Minas Gerais, Portugal
Santos Dorabella	Instituto de Telecomunicações, Portugal
Santos Eulália	SAPO, Portugal
Santos Maribel Yasmina	Universidade de Minho, Portugal
Santos Rafael	University of Toronto, Canada
Santucci Valentinoi	University of Perugia, Italy
Sautot Lucil	MR TETIS, AgroParisTech, France
Scaioni Marco	Polytechnic University of Milan, Italy
Schernthanner Harald	University of Potsdam, Germany
Schneider Michel	ISIMA, France
Schoier Gabriella	University of Trieste, Italy
Scorza Francesco	University of Basilicata, Italy
Sebillo Monica	University of Salerno, Italy
Severino Ricardo Jose	Universidade de Minho, Portugal
Shakhov Vladimir	Russian Academy of Sciences (Siberian Branch), Russia
Sheeren David	Toulouse Institute of Technology, France
Shen Jie	University of Michigan, USA
Silva Elsa	INESC Tec, Porto, Portugal
Sipos Gergely	MTA SZTAKI Computer and Automation Research Institute, Hungary
Skarga-Bandurova Inna	Technological Institute of East Ukrainian National University, Ukraine
Skoković Dražen	University of Valencia, Spain
Skouteris Dimitrios	SNS, Italy
Soares Inês Soares Maria Joana	Universidade de Minho, Portugal
Soares Michel	Federal University of Sergipe, Brazil
Sokolovski Dmitri	Ikerbasque, Basque Foundation for Science, Spain
Sousa Lisete	Research, FCUL, CEAUL, Lisboa, Portugal
Stener Mauro	Università di Trieste, Italy
Sumida Yasuaki	Center for Digestive and Liver Diseases, Nara City Hospital, Japan
Suri Bharti	Guru Gobind Singh Indraprastha University, India
Sørensen Claus Aage Grøn	University of Aarhus, Denmark
Tajani Francesco	University of Rome, Italy
Takato Setsuo	Kisarazu National College of Technology, Japan
Tanaka Kazuaki	Hasanuddin University, Indonesia
Taniar David	Monash University, Australia
Tapia-McClung Rodrigo	The Center for Research in Geography and Geomatics, Mexico
Tarantino Eufemia	Polytechnic of Bari, Italy
Teixeira Ana Paula	Federal University of Ceará, Fortaleza, Brazil
Teixeira Senhorinha	Universidade do Minho, Portugal
Teodoro M. Filomena	Instituto Politécnico de Setúbal, Portugal
Thill Jean-Claude	University at Buffalo, USA
Thorat Pankaj	Sungkyunkwan University, South Korea

Tilio Lucia	University of Basilicata, Italy
Tomaz Graça	Instituto Politécnico da Guarda, Portugal
Torre Carmelo Maria	Polytechnic of Bari, Italy
Totaro Vincenzo	Polytechnic University of Bari, Italy
Tran Manh Hung	University of Danang, Vietnam
Tripathi Ashish	MNNIT Allahabad, India
Tripp Barba Carolina	Universidad Autónoma de Sinaloa, Mexico
Tut Zohra Fatema	University of Calgary, Canada
Upadhyay Ashish	Indian Institute of Public Health-Gandhinagar, India
Vallverdu Jordi	Autonomous University of Barcelona, Spain
Valuev Ilya	Russian Academy of Sciences, Russia
Varela Leonilde	University of Minho, Portugal
Varela Tania	Universidade de Lisboa, Portugal
Vasconcelos Paulo	Queensland University, Brisbane, Australia
Vasyunin Dmitry	University of Amsterdam, The Netherlands
Vella Flavio	University of Rome, Italy
Vijaykumar Nandamudi	INPE, Brazil
Vidacs Laszlo	University of Szeged, Hungary
Viqueira José R.R.	Agricultural University of Athens, Greece
Vizzari Marco	University of Perugia, Italy
Vohra Varun	Japan Advanced Institute of Science and Technology (JAIST), Japan
Voit Nikolay	Ulyanovsk State Technical University Ulyanovsk, Russia
Walkowiak Krzysztof	Wroclaw University of Technology, Poland
Wallace Richard J.	University College Cork, Ireland
Waluyo Agustinus Borgy	Monash University, Melbourne, Australia
Wanderley Fernando	FCT/UNL, Portugal
Wei Hoo Chong	Motorola, USA
Yamashita Satoshi	National Research Institute for Child Health and Development, Tokyo, Japan
Yamauchi Toshihiro	Okayama University, Japan
Yao Fenghui	Tennessee State University, USA
Yeoum Sanggil	Sungkyunkwan University, South Korea
Zaza Claudio	University of Foggia, Italy
Zeile Peter	Technische Universität Kaiserslautern, Germany
Zenha-Rela Mario	University of Coimbra, Portugal
Zoppi Corrado	Università di Cagliari, Italy
Zullo Francesco	University of l'Aquila, Italy
Zunino Alejandro	Universidad Nacional del Centro, Argentina
Žemlička Michal	Univerzita Karlova, Czech Republic
Živković Ljiljana	University of Belgrade, Serbia

Sponsoring Organizations

ICCSA 2017 would not have been possible without the tremendous support of many organizations and institutions, for which all organizers and participants of ICCSA 2017 express their sincere gratitude:

University of Trieste, Trieste, Italy
(http://www.units.it/)

University of Perugia, Italy
(http://www.unipg.it)

University of Basilicata, Italy
(http://www.unibas.it)

Monash University, Australia
(http://monash.edu)

Kyushu Sangyo University, Japan
(www.kyusan-u.ac.jp)

Universidade do Minho, Portugal
(http://www.uminho.pt)

Contents – Part VI

**Workshop on Sustainability Performance Assessment: Models,
Approaches and Applications Toward Interdisciplinary and Integrated
Solutions (SPA 2017)**

Workshop on Advances in Spatio-Temporal Analytics (ST-Analytics 2017)

Short Papers

Workshop on Software Engineering Processes and Applications (SEPA 2017)

MC-DMN: Meeting MCDM with DMN Involving Multi-criteria Decision-Making in Business Process

Riadh Ghlala[1,2(✉)], Zahra Kodia Aouina[1,3], and Lamjed Ben Said[1,3]

[1] Université de Tunis, ISG, LR11ES03 SMART, 2000,
Cit Bouchoucha Le Bardo, Tunis, Tunisia
riadh.ghlala@isetr.rnu.tn, {zahra.kodia,lamjed.bensaid}@isg.rnu.tn
[2] Institut Supérieur des études Technologiques de Radès (ISETR), Radès, Tunisia
[3] Institut Supérieur de Gestion (ISG), Tunis, Tunisia

Abstract. The modelling of business processes and in particular decision-making in these processes takes an important place in the quality and reliability of IT solutions. In order to define a modelling standards in this domain, the Open Management Group (OMG) has developed the Business Process Model and Notation (BPMN) and Decision Model and Notation (DMN). Currently, these two standards are a pillar of several business architecture Frameworks to support Business-IT alignment and minimize the gap between the managers expectations and delivered technical solution. In this paper, we propose the Multi-Criteria DMN (MC-DMN) which is a DMN enrichment. It allows covering the preference to criteria in decision-making using Technique for Order Preference by Similarity to Ideal Solution (TOPSIS) as a Multi-Criteria Decision-Making (MCDM) method and therefore it gives more faithfulness to the real world and further agility face the business layer changes.

Keywords: BPMN · DMN · MCDM · MC-DMN · Preference to criteria · TOPSIS

1 Introduction

Nowadays, companies are operating in a very dynamic business environment, due, on one side, to the continuous changes in customer requirements and constraints of competition [1] and on the other side to the spectacular evolution of information technology and the need for Business-IT alignment [2]. This new setting has imposed a need for improvement and innovation in business processes. Business Process Model and Notation (BPMN) which is a standard in its domain, has been announced by the Object Management Group (OMG) in 2006 and is currently in version BPMN 2.0 since 2011 [3]. Several new features are introduced in the last version of BPMN in order to improve the modelling of business processes by covering aspects such as:

© Springer International Publishing AG 2017
O. Gervasi et al. (Eds.): ICCSA 2017, Part VI, LNCS 10409, pp. 3–16, 2017.
DOI: 10.1007/978-3-319-62407-5_1

- Integrating new elements in the specification (events, gateways, activities, etc.) [4].
- Adding new diagrams for better modelling interactions (collaboration and choreography diagrams) [5].
- Strengthening automation of business process execution by improving the Business Process Execution Language (BPEL) [6].

Decision-making represents another field of investigation in order to improve the business process modelling. It was also an OMG center of interest and has led to the invention of the Decision Model and Notation (DMN) in 2013 [7]. This new model is a BPMN add-in. It is structured in two distinct parts: (i) Decision Requirements Diagram containing the decision to study, business knowledge models, input data and Knowledge Source. (ii) Decision Logic which is represented by a decision table that can be converted into FEEL scripting language (Friendly Enough Expression Language) [8]. This table contains the rules with their input and output in addition to other technical details such as the hit policy. We can classify decisions handled by the DMN in several categories such as eligibility, validation, calculation, risk, fraud, etc. Decision-making by using DMN pushes towards defining roles and calls for specialization in construction of business processes. It also promotes agility and encourages the involvement of stakeholders in the project [9].

DMN is still in its preliminary versions (DMN 1.0 and DMN 1.1). At this stage, we can say that he has excelled as a graphical formalism for modelling decision-making, but has not yet encompassed many factors related to the decision. For example, DMN cannot deal with un-certainty and is limited to predefined decisions made from known criteria [10]. Unfortunately, this situation does not meet the expectations of managers who consider such decision-making process is often case-dependent and carried out under uncertainty [11].

In this paper, we propose our novel model Multi-Criteria DMN (MC-DMN) which is summarized in three topics:

- The proposal of a new structure of the DMN decision table in which we add the inputs' weights. We base the table on possible outputs rather than business rules and we introduce a new column that will contain the closeness computation of each output compared to the ideal alternative.
- The use of a function for transforming FEEL expression in numeric values compatible with the inputs of the MCDM Data Matrix.
- The application of TOPSIS method steps through an algorithm for the determination of the optimal rule.

This paper is structured as follows. Section 2 overviews the related work about decision-making and its relationship with business processes. It also presents the BPMN and DMN with samples based on Business Intelligence (BI) project and the focus on the lack of support for preference to criteria in decision-making. Section 3 is devoted to the presentation of MCDM methods which model preference to criteria in decision-making and especially the TOPSIS method.

Section 4 presents our proposal MC-DMN to enrich DMN by preference to criteria feature through the TOPSIS method. Finally, we summarize the presented work and outline its extensions.

2 Decision-Making in Business Process: Related Works

A literature review on decision-making and its relationship to business processes showed a progression in dealing with decision. Indeed, the first preoccupation was concentrated on the separation between decision-making modelling and process modelling [9,12]. Research in this field are based on the collection, the modelling and the integration of business rules in business processes [13,14]. This work is crowned in the industrial world by DMN [7] which has become a standard in modelling decision-making in business process [8]. The second focus is the serialization of business rules and automation of its processing and its exchange [15,16]. Several open-source and proprietary software [17–20] appeared and are in competition to implement both of BPMN and DMN standards based on BPEL and FEEL languages [6,8]. The current challenge is to make decision-making modelling closer to real world. This aim is ensured by giving importance to various aspects attached to decision-making. In this paper, we consider the decision-making covering the preference to criteria. According to OMG, business processes are modelled through two standards [8]: Business Process Model and Notation (BPMN) and Decision Model and Notation (DMN). The first is used to represent the various tasks and their relationships. The second supports decision-making in the business process. Figure 1 shows an extract from a BI process in which we use these standards to model decision-making at the data extraction task.

2.1 The BPMN Standard

BPMN is the substrate for modelling business processes. It allows to represent graphically the company's activities to ensure better collaboration between managers and IT engineers [3]. The example given in Fig. 1 describes the ETL software selection task in a BI process. The decision-making is delegated to DMN to work around the issue.

2.2 The DMN Standard

DMN is a BPMN add-in. It is structured in two distinct parts: (i) Decision Requirements Diagram containing the decision to study, business knowledge models, input data and knowledge source; (ii) Decision Logic which is represented by a decision table that can be converted into FEEL scripting language (Friendly Enough Expression Language) [8]. Delegating the decision-making task to a separate model of the BPMN allows a good mastery of this task by increasing the readability of the model and improving its reliability by the high grasp of the business rules.

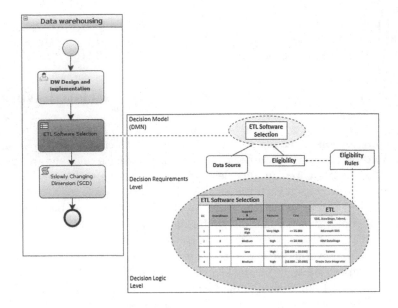

Fig. 1. Using BPMN and DMN in BI process

Decision Requirements Diagram (DRD)

It is a decision model which uses seven types of construct divided into four elements (Decision, Business Knowledge Model, Knowledge Source and Input Data) and three requirements (Information Requirement, Knowledge Requirement and Authority Requirement). This is a view of Decision Requirements Graph (DRG) representing requirements decision level in a global way.

DMN Decision Table

This table contains the rules with their input and output, in addition to other technical details such as the hit policy and Completeness indicator. We can classify decisions handled by the DMN in several categories such as eligibility, validation, calculation, risk, fraud, etc. Business rules can be represented in the DMN Decision Table in three ways: rows, columns or cross table. Figure 2 shows a business rule relative to ETL software selection with a representation with rules as rows.

1. Hit policies and Completeness indicator:
   Hit policies specify how to handle the case where multiple rules are triggered by an input configuration [7]. There are two hit policies categories: The first one is the single hit policies which could be one of these following kinds:
   - Unique: it is the default indicator designed by 'U'. It indicates that at most one rule can be triggered by a given input configuration. Thereby, there is no overlap between rules.
   - Any: with this hit designed by 'A', overlap between rules is possible, but when multiple rules are triggered, they must agree on the output objects. Otherwise, the result is undefined.

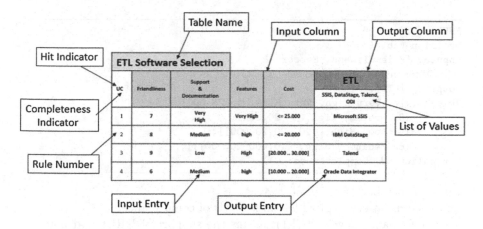

Fig. 2. DMN decision table related to ETL software selection

- Priority: with this hit designed by 'P', rules may overlap, with different outputs, but whenever multiple rules trigger, then the out-put is unambiguously computed by only considering the contribution of the triggered rule that has the highest priority.
- First: this hit designed by 'F' is a variant of the priority hit, in which priority is implicitly obtained from the ordering in which rules appear in the decision table.

The second one is the multiple hit policies which could be one of these following kinds:

- Output order: all hits are returned in a decreasing order of output priorities. An ordered list of output values represents output priorities.
- Rule order: all hits are returned in rule order.
- Collect: returns all hits in the arbitrary order. Operators, such as sum, min, max and count, can be added to the outputs.
- Rules missing and rules overlapping can lead to an ambiguous state. For this, completeness indicator is used to indicate that each possible input configuration must trigger at least one rule.

2. FEEL (Friendly Enough Expression Language):
FEEL is a special expression language included as a part of the DMN specification [7]. It is designed to write expressions and conditions in a simple way that could be easily understood by business professionals and developers. It is a simple language with inspiration drawn from Java, Java script, Xpath, SQL, Predictive Model Markup Language (PMML), Lisp, and others. In particular, FEEL extends JavaScript Object Notation (JSON) objects. We note that JSON object is a number, a string, a context (JSON calls them maps) or a list of JSON objects; FEEL adds date, time, and duration objects, functions, friendlier syntax for literal values, and does not require the context keys to be quoted [8]. The business rule, graphically represented in DMN decision table as shown in Figure may be transformed into FEEL as follows:

```
FEEL(
Decision table(
Inputs: [Friendliness, Support &
    Documentation, Features , Cost],
Outputs: [ETL],
Rules: [ [7, Very High, Very High, <= 25.000],
         [8, Medium, High, <= 20.000],
         [9, Low, High, [20.000 30.000]],
         [6, Medium , High, [10.000 20.000]]],
Hit policy: U, Completeness: C))
```

3. Conformance Levels:
   DMN specification [7] provides three levels of conformance:
   - Conformance Level 1 (CL1): means the support of DRDs and non executable decision tables. Decision logic can be represented informally, in natural language.
   - Conformance Level 2 (CL2) is the most common level adopted by software that implements the standard DMN as Camundar [17] Segnavio [18] Firstmodeler [19] and Trisotech [20]. CL2 tools support DRDs, decision tables, and literal expressions based on Simplified FEEL (S-FEEL). These software are able to verify the completeness and consistency of decision table through completeness indicator.
   - Conformance Level 3 (CL3): this level is not yet implemented. We expect it to support FEEL as a whole and even add new features such as the support of preference to criteria.

2.3 DMN Limitations Relative to Criteria Preference

Obviously, we notice the absence of preference to criteria in the DMN throughout its different mechanism. Indeed, we note that:

- Decision table deals only with predefined decisions made from known criteria and it does not provide recourse to the concept of weight to handle business rules priority.
- FEEL language, although it is extensible, does not discuss possible features like support of criteria preference.
- Compliance levels, according to DMN specification, does not reach an automation level in the selection of a business rule from a proposed list, it is satisfied merely with its graphic representation. The designer must comply with one of predefined hit policies in order to force a priority manually.

3 Multi-criteria Decision-Making

The decision-making becomes a difficult task in one of the following cases:

- A large number of possibilities to compare, in this case, it is a combinatorial optimization.

- A significant number of decision makers thus, the recourse to social choice and game theory.
- The consequences of actions are not safe, so we deal with decision with criteria preference.
- Several criteria to be taken into consideration, therefore we are facing to Multiple Criteria Decision-Making (MCDM).

Decision-making in business process is generally confronted with the last situation of difficulty which is multicriteria decision-making [11,21].

3.1   MCDM Methods

In literature [22], many terms have been used for MCDM such as: Multi-Criteria Decision Analysis (MCDA), Multi-Objective Decision-Making (MODM), Multi-Attributes Decision-Making (MADM) and Multi-Dimensions Decision-Making (MDDM). Literature [23] is rich with different types of multi-criteria decision-making methods. Several MCDM methods have been used by researchers to solve real-world problems in which multiple criteria are used in decision-making. Among these methods, we can cite as an example: Analytic Hierarchy Process (AHP), Analytic Network Process (ANP), Elimination and Choice Translating Reality (ELECTRE), Technique for Order Preference by Similarity to Ideal Solution (TOPSIS), etc.

3.2   TOPSIS

TOPSIS, developed by Hwang and Yoon in 1981, is a simple ranking method in conception and application. The choice of this method in our study is explained by two main arguments. Firstly, TOPSIS is known by its efficient modelling. This method is based at the same time on several promoting and inhibiting factors which suits our field of study that is decision-making in business processes. Secondly, we notice its wide use in many real-world applications [23] and therefore maturity of the method and its rich documentation.

TOPSIS method stepwise, as shown in Fig. 3, is composed of five stages, namely: (i) the construction of the normalized decision matrix, (ii) the construction of the weighted normalized decision matrix, (iii) the determination of the positive ideal and the negative ideal solutions, (iv) the calculation of the separation measures for each alternative and (v) the calculation of the relative closeness to the ideal solution.

4   DMN Under Multi-criteria Analysis

DMN's contribution in improving the decision-making in business process has become an established fact, but evolution's wheel is never stopped. The focus is now around upgrading DMN to cover likewise important aspects such as preference to criteria. Making the DMN able to handle preference to criteria can help

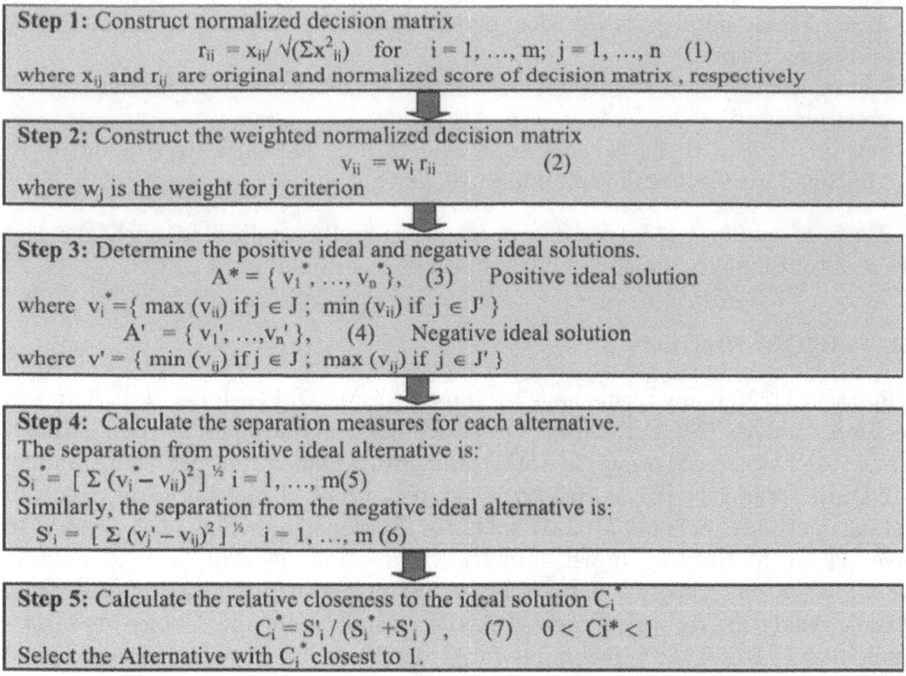

Step 1: Construct normalized decision matrix
$r_{ij} = x_{ij}/\sqrt{(\Sigma x^2_{ij})}$ for $i = 1, ..., m;\ j = 1, ..., n$ (1)
where x_{ij} and r_{ij} are original and normalized score of decision matrix , respectively

Step 2: Construct the weighted normalized decision matrix
$v_{ij} = w_j\, r_{ij}$ (2)
where w_j is the weight for j criterion

Step 3: Determine the positive ideal and negative ideal solutions.
$A^* = \{\, v_1^*, ..., v_n^*\,\}$,   (3)   Positive ideal solution
where $v_i^* = \{\ \max\,(v_{ij})$ if $j \in J$; $\min\,(v_{ij})$ if $j \in J'\ \}$
$A' = \{\, v_1', ..., v_n'\,\}$,   (4)   Negative ideal solution
where $v' = \{\ \min\,(v_{ij})$ if $j \in J$; $\max\,(v_{ij})$ if $j \in J'\ \}$

Step 4: Calculate the separation measures for each alternative.
The separation from positive ideal alternative is:
$S_i^* = [\ \Sigma\,(v_i^* - v_{ij})^2\,]^{\frac{1}{2}}\ i = 1, ..., m$(5)
Similarly, the separation from the negative ideal alternative is:
$S_i' = [\ \Sigma\,(v_j' - v_{ij})^2\,]^{\frac{1}{2}}\   i = 1, ..., m$ (6)

Step 5: Calculate the relative closeness to the ideal solution C_i^*
$C_i^* = S_i' / (S_i^* + S_i')$,   (7) $0 < C_i^* < 1$
Select the Alternative with C_i^* closest to 1.

Fig. 3. Stepwise procedure for performing TOPSIS methodology [21]

decision makers in situations where the choice is not obvious and automatic processing must be done to achieve a better result. Our guiding thread in this track is the striking similarity between the DMN reasoning and functioning of MCDM especially TOPSIS methods.

4.1   Analogy Between DMN Decision Table and TOPSIS Data Matrix

In this section, we focus on the analogy between the TOPSIS data matrix and the DMN decision table. Indeed, Fig. 4 shows that DMN decision table is structured as follows:

- The different inputs I_i ($i = 1, 2, ..., n$) where n is the number of inputs.
- The different business rules represented as row R_j ($j = 1, 2, ..., m$) where m is the number of business rules.
- The different expressions formulated in FEEL. exp_{ij} designates the ith input for the jth rule.
- The different outputs O_j associated to each business rule.

Analogously, Fig. 5 shows that MCDM data matrix is structured as follows:

- The different criteria c_i and their associated weight w_i ($i = 1, 2, ..., n$) where n is the number of criteria.

- The different alternatives represented as row A_j (j = 1, 2, ..., m) where m is the number of alternatives.
- The different numeric values. x_{ij} designates the ith input for the jth alternative.
- The closeness calculated for each alternative.

$$
\begin{bmatrix} R_1 \\ ... \\ ... \\ ... \\ R_m \end{bmatrix}
\begin{matrix} [\ I_1 & I_2 & I_3 & \cdots & I_n\ ] \\ \begin{bmatrix} exp_{11} & exp_{12} & exp_{13} & ... & exp_{1n} \\ ... & ... & ... & ... & ... \\ ... & ... & ... & ... & ... \\ ... & ... & ... & ... & ... \\ exp_{m1} & exp_{m2} & exp_{m3} & & exp_{mn} \end{bmatrix} \end{matrix}
\begin{bmatrix} O_1 \\ ... \\ ... \\ ... \\ O_m \end{bmatrix}
$$

Fig. 4. Schematization of DMN decision table

$$
\begin{bmatrix} A_1 \\ ... \\ ... \\ ... \\ A_m \end{bmatrix}
\begin{matrix} \begin{bmatrix} c_1 & c_2 & c_3 & \cdots & c_n \\ w_1 & w_2 & w_3 & ... & w_n \end{bmatrix} \\ \begin{bmatrix} x_{11} & x_{12} & x_{13} & ... & x_{1n} \\ ... & ... & ... & ... & ... \\ ... & ... & ... & ... & ... \\ ... & ... & ... & ... & ... \\ x_{m1} & x_{m2} & x_{m3} & ... & x_{mn} \end{bmatrix} \end{matrix}
\begin{bmatrix} Cl_1 \\ ... \\ ... \\ ... \\ Cl_m \end{bmatrix}
$$

Fig. 5. Schematization of TOPSIS data matrix

The TOPSIS method, although it is not a recent method, adapts well to change the preference to the criteria in DMN. We use the TOPSIS method due to its simplicity and ability to consider a non-limited number of alternatives and criteria in the decision-making task using both positive and negative impacts.

4.2   MC-DMN

MC-DMN is a novel notation based on the standard DMN. It can be implemented to make the DMN able to cover preference to criteria. The stepwise of the MC-DMN is to: (i) be based on a user interface representing the standard DMN decision table. (ii) Transform the decision table, on one hand, by integrating weights, and on the other hand by switching the table on the different outputs as rows. (iii) Apply a transform function on inputs for converting the FEEL expressions into numeric values. (vi) Run the TOPSIS method algorithm to calculate closeness.

New MC-DMN Decision Table

MC-DMN decision table is automatically generated from the standard DMN table decision. Figure 6 shows the structure of this new table with new informational elements such as weight that must be entered by the user and closeness to be calculated pursuant to TOPSIS method algorithm. The new layout of MC-DMN table decision is imposed according on TOPSIS Data Matrix.

$$\begin{bmatrix} c_1 & c_2 & c_3 & \cdots & c_n \\ w_1' & w_2' & w_3' & \cdots & w_n' \end{bmatrix}$$

$$\begin{bmatrix} O_1 \\ \cdots \\ \cdots \\ \cdots \\ O_m \end{bmatrix} \begin{bmatrix} exp_{11} & exp_{12} & exp_{13} & \cdots & exp_{1n} \\ \cdots & \cdots & \cdots & \cdots & \cdots \\ \cdots & \cdots & \cdots & \cdots & \cdots \\ \cdots & \cdots & \cdots & \cdots & \cdots \\ exp_{m1} & exp_{m2} & exp_{m3} & & exp_{mn} \end{bmatrix} \begin{bmatrix} Cl_1 \\ \cdots \\ \cdots \\ \cdots \\ Cl_m \end{bmatrix}$$

Fig. 6. DMN-MCDM

Transformation Function

The standard version of DMN uses the expression language FEEL to allow the manager express friendly the inputs in the decision table. This asset is always valued in our MC-DMN proposal despite the TOPSIS method requires only numeric values for the data matrix inputs. Therefore, the idea is to use a transformation function to convert FEEL inputs in numeric values. The template of the transformation function can be as follows:

```
int function FEEL_Transformation(val)
{
    if IsNumeric (val) then return val
        elseif IsString(val) then return AssociativeValue(val)
            elseif IsExpression(val) the return ExpressionToVal(val)
                else return DefaultValue
    end if;
}
```

It is to highlight that:

- AssociativeValue is an associative table with all possible strings and their associative numeric values.
- DefaultValue is an environment variable used to be the result of a conversion of a wrong value.
- Both AssociativeValue and DefaultValue should be considered as initialization parameters of the Framework.
- Both IsExpression and ExpressionToVal functions will represent the subject of a future research.

New Conformance Level

Even the selection of the business rule that will be retained is treated as an action determined by the manager across the choice of a hit policy. We aim across our MC-DMN proposal to upgrade to a new level of conformance reflecting a trend of automation. This new level of conformance will be reached by applying the stepwise TOPSIS method. It contributes necessarily to improving decision-making in business process especially when the choice is difficult to do manually by the manager due to preference to criteria adopted.

4.3   MC-DMN Applying Example for ETL Software Selection

In this section, we present an applicative example of our MC-DMN model in a BI process and specifically the task of selection of an ETL software. Figure 7 shows the DMN decision table in its original version.

ETL Software Selection					ETL
UC	Friendliness	Support & Documentation	Features	Cost	SSIS, DataStage, Talend, ODI
1	7	Very High	Very High	<= 25.000	Microsoft SSIS
2	8	Medium	high	<= 20.000	IBM DataStage
3	9	Low	High	[20.000 .. 30.000]	Talend
4	6	Medium	high	[10.000 .. 20.000]	Oracle Data Integrator

Fig. 7. Initial DMN decision table for ETL software selection

It is noted, for reasons of simplification, the absence of rules overlapping and rules missing. The first step of our model is to transform the decision table in its new structure and introduce the weight criteria. Figure 8 shows the new decision table according to our MC-DMN. It can also be noted that the values of the criteria are of different forms: quantitative numerical values, qualitative values, intervals and even expressions.

The next step consists on applying the transformation function to convert FEEL expression into numeric values. Figure 9 shows the result of this transformation. We clearly note that all the criteria are now represented with quantitative numerical values and therefore we can apply the TOPSIS method in this case.

Finally, the execution of TOPSIS method algorithm defines the best and the worst alternatives. Figure 10 illustrates that SQL Server Integration Services (SSIS) is the best ETL software and Talend is the worst.

ETL Software Selection						
UC	Weight	0.1	0.4	0.3	0.2	**ETL**
	Criteria	Friendliness	Support & Documentation	Features	Cost	Closeness
Microsoft SSIS		7	Very High	Very High	<= 25.000	
IBM DataStage		8	Medium	High	<= 20.000	
Talend		9	Low	High	[20.000 .. 30.000]	
Oracle Data Integrator		6	Medium	High	[10.000 .. 20.000]	

Fig. 8. New structure for MC-DMN decision table

ETL Software Selection						
UC	Weight	0.1	0.4	0.3	0.2	**ETL**
	Criteria	Friendliness	Support & Documentation	Features	Cost	Closeness
Microsoft SSIS		7	9	9	8	
IBM DataStage		8	7	8	7	
Talend		9	6	8	9	
Oracle Oata Integrator		6	7	8	6	

Fig. 9. MC-DMN decision table after transformation function

ETL Software Selection							
UC	Weight	0.1	0.4	0.3	0.2	**ETL**	
	Criteria	Friendliness	Support & Documentation	Features	Cost	Closeness	
Microsoft SSIS		7	9	9	8	0.74	⇐ Best
IBM DataStage		8	7	8	7	0.41	
Talend		9	6	8	9	0.17	⇐ Worst
Oracle Oata Integrator		6	7	8	6	0.45	

Fig. 10. MC-DMN decision table after stepwise procedure for performing TOPSIS methodology.

5   Conclusion and Future Works

Since decision-making is based on gathering, modelling and integrating business rules in business processes, this practice is not so stagnant but rather influenced by changes that can be made on these business rules. Associating weights to criteria and using a method to develop a priority between business rules allows to further increase the faithfulness and agility of the decision-making in business process modelling.

In this paper, we proposed a merger of two formalisms dealing with decision-making: DMN and MCDM to improve decision-making in business processes modelling.

Our future work will be scheduled on three axes. First, the implementation of a Framework to apply the MC-DMN approach. The second step is to validate this approach with case studies in the industrial environment. The third axis is the improvement of this approach in a cooperative, collaborative and distributed environment.

References

1. Harmon, P.: Business Process Change: A Guide for Business Managers and BPM and Six Sigma Professionals, 2nd edn. Morgan Kaufmann Publishers, Burlington (2007). ISBN 978-0-12-374152-3
2. Brian, E., Dima, P., Sven, J., Christian, H.: Aligning business and IT models in service-oriented architectures using BPMN and SoaML. In: Proceedings of the First International Workshop on Model-Driven Interoperability, Oslo, Norway, pp. 61–68. ACM, October 2010. doi:10.1145/1866272.1866281
3. OMG: Business Process Modeling Notation Specification 2.0 (2011). http://www.omg.org/spec/BPMN/2.0/PDF/
4. Nazaruka, E., Oviikova, V.: Specification of decision-making and control flow branching in topological functioning models of systems. In: Proceedings of the 10th International Conference on Evaluation of Novel Approaches to Software Engineering, ENASE, Barcelona, Spain, 29–30 April 2015, pp. 364–373 (2015)
5. Nie, H., Lu, X., Duan, H.: Supporting BPMN choreography with system integration artefacts for enterprise process collaboration. Enterp. Inf. Syst. 8(4), 512–529 (2014)
6. Nguyen, B.T., Nguyen, D.H., Nguyen, T.T.: Translation from BPMN to BPEL, current techniques and limitations. In: SoICT 2014, pp. 21–30 (2014)
7. OMG: Decision Model and Notation 1.0 (2015). http://www.omg.org/spec/DMN/1.0/PDF
8. Taylor, J., Fish, A., Vincent, P.: Emerging Standards in Decision Modeling - An Introduction to Decision Model & Notation in iBPMS Intelligent BPM Systems: Impact and Opportunity. Future Strategies Inc., Brampton (2013). ISBN 978-0-9849764-6-1
9. Batoulis, K., Meyer, A., Bazhenova, E., Decker, G., Weske, M.: Extracting decision logic from process models. In: Zdravkovic, J., Kirikova, M., Johannesson, P. (eds.) CAiSE 2015. LNCS, vol. 9097, pp. 349–366. Springer, Cham (2015). doi:10.1007/978-3-319-19069-3_22

10. Le Mauff, A., Bigand, M., Bourey, J.-P.: Separation of decision modeling from BPM using DMN for automating operational decision making. In: PRO-VE 2015 IFIP Working Conference on Virtual Enterprises, 05 October 2015
11. Batoulis, K., Baumgraß, A., Herzberg, N., Weske, M.: Enabling dynamic decision making in business processes with DMN. In: Reichert, M., Reijers, H.A. (eds.) BPM 2015. LNBIP, vol. 256, pp. 418–431. Springer, Cham (2016). doi:10.1007/978-3-319-42887-1_34
12. Biard, T., Mauff, A., Bigand, M., Bourey, J.-P.: Separation of decision modeling from business process modeling using new decision model and notation (DMN) for automating operational decision-making. In: Camarinha-Matos, L.M., Bénaben, F., Picard, W. (eds.) PRO-VE 2015. IAICT, vol. 463, pp. 489–496. Springer, Cham (2015). doi:10.1007/978-3-319-24141-8_45
13. Kluza, K., Nalepa, G.J.: Towards rule-oriented business process model generation. In: Proceedings of the Federated Conference on Computer Science and Information Systems, FedCSIS, Krakw, Poland, 8–11 September 2013, pp. 939–946 (2013)
14. Bajwa, I.S., Lee, M.G., Bordbar, B.: SBVR business rules generation from natural language specification. In: Artificial Intelligence for Business Agility - AAAI Spring Symposium Series, (SS-11-03), pp. 2–8 (2011)
15. Benson, T., Grieve, G.: UML, BPMN, XML and JSON. In: Benson, T., Grieve, G. (eds.) Principles of Health Interoperability. Health Information Technology Standards, pp. 55–81. Springer, Heidelberg (2016). doi:10.1007/978-3-319-30370-3_4
16. Ouyang, C., Dumas, M., ter Hofstede, A., van der Aalst, W.: From BPMN process models to BPEL web services. In: ICWS, Los Alamitos, pp. 285–292. IEEE, New York (2006)
17. Camunda Workflow and Business Process Management. https://camunda.org/
18. Signavio Decision Manager. http://www.signavio.com/products/decision-manager/
19. Decision Management Solutions: DecisionsFirst Modeler. http://decisionsfirst.com/
20. The Digital Enterprise Suite. http://www.trisotech.com/
21. Janssens, L., Smedt, J., Vanthienen, J.: Modeling and enacting enterprise decisions. In: Krogstie, J., Mouratidis, H., Su, J. (eds.) CAiSE 2016. LNBIP, vol. 249, pp. 169–180. Springer, Cham (2016). doi:10.1007/978-3-319-39564-7_17
22. Zardari, N.H., Ahmed, K., Shirazi, S.M., Yusop, Z.B.: Weighting Methods and their Effects on Multi-Criteria Decision Making Model Outcomes in Water Resources Management. SWST. Springer, Cham (2015). doi:10.1007/978-3-319-12586-2. ISBN 978-3-319-12585-5 (Print) 978-3-319-12586-2 (Online)
23. Behzadian, M., Otaghsara, S.K., Yazdani, M., Ignatius, J.: A state-of the-art survey of TOPSIS applications. Expert Syst. Appl. **39**(17), 13051–13069 (2012). doi:10.1016/j.eswa.2012.05.056

Context Sensitive Query Correction Method for Query-Based Text Summarization

Nazreena Rahman[✉] and Bhogeswar Borah

Department of Computer Science and Engineering, Tezpur University,
Sonitpur 784028, Assam, India
{naz1912,bgb}@tezu.ernet.in

Abstract. Contextual spell correction is very important for real word error correction. It gives the correct word for an incorrect word in a particular sentence. The traditional spell checker can correct those misspelled words which are not present in dictionary but here we try to develop a spell checker which can give appropriate word on the basis of the contextual meaning of the sentence. This spell checker is specially applied for error correction in query-based text summarization. Here, we try to combine both semantic based measure and lexical character matching to find the appropriate word for a particular sentence.

Keywords: Contextual spell correction · Real word error · Query-based text summarization · Semantic based measure and lexical character matching

1 Introduction

Errors in words always play a major issue while retrieving information in the field of natural language processing. Spelling errors can be of two types. One is non-word errors where words are not present in dictionary and do not have meaning. For example, if we do not know the correct spelling of *zebra*, we can write *jebra* in place of *zebra*. From the literature survey it is found that research is done extensively on non-word spell correction. Many traditional spell checking softwares are present through which we can correct these non-word errors easily. Some examples are Microsoft Word, GNU Aspell, Ispell, LibreOffice Writer etc. Microsoft word is a word processing software developed by Microsoft, GNU Aspell is a free and open source spell checker, Ispell is mainly used in Unix that supports many western languages word correction and LibreOffice is also a free and open source office suite. The other type of spelling error is real-word error. Here, spelling error in a word accidentally gives another actual word which is present in dictionary. For example, we write *piece* in place of *peace*. This can be possible by typographical errors or writer might get confused with homophone or near homophones. Homophone is a word, whose pronunciation is same with another word but meaning and spelling is different. For example: here and hear. These type of words are said as confused words. Real-word errors are found more

O. Gervasi et al. (Eds.): ICCSA 2017, Part VI, LNCS 10409, pp. 17–30, 2017.
DOI: 10.1007/978-3-319-62407-5_2

in Dyslexic text [1]. Dyslexia is a disorder that finds difficulty in reading, spelling and writing. From the studies, it is found that from 25% to 40% spelling errors are actually a valid english word [2].

Spell correction in real-word errors is much difficult compared to non-word errors. To deal with real-word error correction, syntactic and semantic analysis should be considered along with pragmatic knowledge about the language. In general, many spell checker give the suggestion of many probable words and user has to choose the correct word according to the context.

The rapid and continuous growth of text information makes it difficult to retrieve the exact information. Therefore, query based text summarization can be used for finding the summarized answer according to user's need. In query-based text summarization, we have a query with single or multiple text documents as input texts. In question answering system or information retrieval system, we need to extract those sentences which have similar meaning with the query. Therefore, it is very much important that the query should not contain any error. When the query is inserted, there is possibility that words in the query might be spelled wrongly. Here, we assume that these incorrect words can be of real-word type of errors. Therefore, to correct this type of real-word errors, we have to consider semantic measures. Semantic measures are always considerd to be an important feature to find the similarity and relatedness among words. Semantics is the study of meaning of word that is used to understand human expression through language. Hence, semantic measure will help us in finding appropriate word for incorrect words.

Word-net (Started by Miller in 1985) relations are applied widely in text analysis and artificial intelligence applications. This lexical database, Word-net is used for english language and was created by Cognitive Science Laboratory of Princeton University. Words with same lexical category are organized into synonym sets called synset. Different kind of synsets are related by different semantic relations in Word-net. Concepts in Word-net are linked together in a hierarchical structure. In Fig. 1, an example of words present in Word-net is shown.

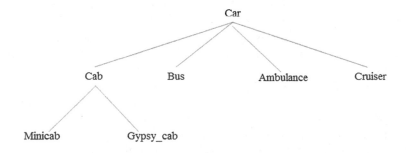

Fig. 1. An example of relationship of words present in Word-net hierarchy

In this paper, a semantics approach (CSQ Method) along with lexical character matching is proposed to find out the correct word for a confused word based on contextually appropriate for the specific sentence. This CSQ method can be applied particularly in query-based text summarization purpose for real-word error correction process. The remainder of the paper is as follows. Section 2 presents various works which are mainly done on real-word context sensitive spell correction, Sect. 3 gives the overview of the proposed CSQ method. Section 4 describes the experimental results. Finally, Sect. 5 covers the conclusion and future plans.

2 Literature Survey

Mays et al. [3] uses maximum likelihood based statistical techniques to find contextually correct word. Their method uses word trigram model. This trigram model computes the conditional probability of a word given by two prior words. This statistical technique models the spelling correction as a speech recognition process. A word string w is generated from text generator. Speller and typist will perform the transformation and a word string y will be produced which might not be similar with w. Finally, linguistic decoder will choose $\hat{w}$ which gives the maximum conditional probability value of w given y.

Golding et al. [4] introduces a new method which is based on trigram and Bayes. Trigram method is based on parts-of-speech trigrams. But this trigram model works only when part-of-speech of the words in the confusion set are different. Here, tagging probability depends on the previous two tags. Moreover, to deal with same parts-of-speech words in the confusion set, their method uses Bayes. Context-word feature based Bayes method uses two types of features: one is context-word and other is collocation. Context-word feature checks if a particular word is present within a certain range of the confused target word and collocation searches for adjacent words of certain length and/or parts-of-speech tagging of confused target word. They combine both the method and named it as Tribayes. This Tribayes uses trigram method for different parts-of-speech tagging and use Bayes method for same parts-of-speech tagging.

Another hybrid Bayesian method [5] is proposed by Golding. This method combines two complementary methods: context word and collocation. Earlier by Yarowsky [6] use decision lists where theses two context word and collocation methods are combined. Decision lists solve the problem by transforming the collected evidence into a single strongest piece of information. Golding initially uses the decision lists hybrid approach to solve the context-sensitive spelling correction problem. However, performance of error correction can be improved by using Bayesian classifier. This classifier does not take only the strongest single piece of evidence but also all available evidences to get better performance. Finally, a further combination is added by using trigram approach when the words in confusion set have different parts of speech and Bayesian approach for same parts of speech.

Hirst and Budanitsky [7] uses Jiang and Conrath semantic similarity measure [8] to find semantically unrelated words according to the context and variation of spelling of words that can be related to the context of the sentence. Their system finds the set of all possible words by insertion, deletion, substitution of single character and transposition of two nearby words. For each word, semantically related words are found from the whole text document and replace the word with highest similarity value.

Fossati and Di Eugenio [9] used Hidden Markov Model (HMM) framework to apply it on mixed trigram model. Each state of the HMM is represented as a pair of parts-of-speech (POS) tagged and a pair of POS tag with a valid dictionary word. The checked word (central word) is considered as a confused word having a set of confusion set. For each confused word present in confusion set, HMM matched the entire sentence. Here, Viterbi algorithm is used to find the most probable sequence of hidden sequence state. If the probability of label of state for a particular central word is higher, then that central word can be taken as a correct word for that particular sentence.

Samanta and Chaudhuri [10] uses bigram and trigram approch to detect and correct real-word errors. Bigram score is calculated by taking left and right neighbor of the candidate key of the sentence and trigram score is generated by the these three words. They consider single error in word detection and correction. Since this correction method depends on the immediate left and right of candidate word, hence their approach can correct errors appearing in alternate words. Initially, they find confusion set for corresponding candidate word using Levenshtein distance [11] from the dictionary. Their model calculates bigram and trigram probability score for each word in confusion set by using Markov chain rule. Here, BYU corpus is used to find n-gram probability. By using Maximum Likehood Estimation, bigram and trigram probabilities are obtained. Sometimes many proper bigrams and trigrams are not found in the corpus, hence they stem the words to increase the accuracy of their model.

Sharma et al. [12] proposes a model for real-word error correction. They use collocation feature by finding the presence of neighboring words. Trigram probability is calculated for each word in the confusion set and highest probability is considered as right one. They also use Bayesian approach to find context features. Bayesian approach finds all the words surrounded by the target words and calls it as features. It finds all nearby words of the target word and calculates the probability of textual information using a training corpus. Their method also uses synonyms of the contextual word if the exact word is not present in corpus. Thus, finally the highest score word is considered as a correct word. Sorokin [13] presents an automatic spelling correction algorithm. The algorithm uses noisy channel model and re-ranking of hypothesis based on features. This language independent model is applied for Russian language. The word-level and the sentence-level features are integrated here. There are three steps to find out the correct word: first step is candidate generation, second step is extraction of n-best list and final step is the feature-based ranking of hypothesis.

From the above study, it is seen that there is no context based spell checker which is used for spell correction particularly in query-based text summarization. However, existing spell checkers can be used for spelling correction in query-based text summarization, but efficiency is quite low while applying this spell checkers. Hence, we try to propose one context sensitive query correction method for query-based text summarization.

It is observed from the extensive analysis and survey of literature that n-gram matching similarity always plays a vital role while dealing with context-sensitive real-word errors. Additionally, we can strengthen the spell checker by using Word-net. Word-net gives different semantic relations which will eventually help us to find out the exact word for a specific sentence.

3   Overview of the Proposed Method: CSQ

3.1   Proposed Framework

The proposed method is a real-word error correction method. We try to correct those real words which are normally considered as confused words. These words can be obtained with the help of the commonly confused word list from the Random House Unabridged Dictionary [14].

Contextual spelling correction method always finds appropriate word which will be suitable for that specific sentence. In general, meaning of a word depends on the immediate left and immediate right words of that particular word. In fact, it is seen that strong semantic relation is there between the previous word (left word) and next word (right word) with the target word. Semantic measure gives contextually similar words which will be appropriate for the incorrect word present in a sentence. Semantic measure can be calculated by using semantic similarity and semantic relatedness. We try to find out the semantic measure by using neighboring contextual information with the target confused word. Semantic similarity always finds similar meaning words but semantic relatedness does not mean that two words or concepts are similar. Semantically related two words are said to be related words by considering their likeliness. For example, in bank, a bank account and a customer are related, hence there is a strong semantic similarity in between the two concepts. With the help of a lexical relation, different concepts are semantically related: like meronymy (hand-finger), antonymy (good-bad). It is also seen that any kind of functional relationship like frequent association or co-occurrence of ideas (tea-India) can be considered as semantically similar concepts.

Two concepts can be related by various ways by not considering only the similarity of two words. To find semantic relatedness, the follows relations are used:

1. Synonymy: Words that sound different but have the same or identical meaning as another word; example: 'achiever' is the synonym of 'success'.
2. Hyponymy: Word or phrase whose semantic field is more specific; example: 'domestic_cat' is the hyponym of 'cat'.
3. Hypernymy: It is also known as a superordinate, is broader than that of a hyponym; example: 'feline' is the hypernym of 'cat'.

4. Meronymy: Denotes a constituent part of, or a member of something; example: 'heart' is a meronym of 'body'.
5. Holonymy: Defines the relationship between a term denoting the whole and a term denoting a part of, or a member of, the whole; example: 'body' is a holonym of 'heart'.
6. Troponymy: Troponymy is the presence of a 'manner' relation; example: 'smile' is troponym of 'laugh'. It is used for verb.
7. Entailment: Any verb A entails verb B, if the truth of B follows logically from the truth of A; example: 'morning walk' is the entailment of 'early rising'.
8. Antonymy: A word having opposite meaning; example: 'bad' is a antonym of 'good'.

Now, to find out the semantic relatedness, we use hso similarity (Hirst and St-Onge) [15] using Word-net. The main idea of hso measures is that the similarity between two concepts is a function of the length of the path linking the concepts and the number of directions between the two concepts in the taxonomy. This semantic relatedness measure includes has-part, is-made-of, is-an-attribute-of type of relations. It is more generalized concept than semantic similarity concept. We can find similarity between words of different parts-of-speech. This path based measure classifies relation in Word-net in terms of direction; for example upward direction *is-a*, horizontal direction *has-part*. The following Fig. 2 shows the path lengths and Fig. 3 shows the number of changes of directions between two concepts or words that are present in Word-net. Here, we consider *is-a* relation between two nouns.

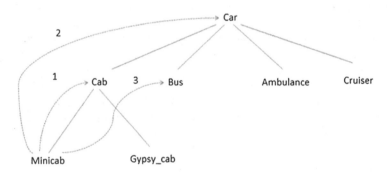

Fig. 2. A fragment showing path lengths in Word-net hypernym hierarchy

In Fig. 2 it is shown that path lengths between Minicab to Cab is 1, Minicab to Car is 2 (path is from Minicab to Cab and Cab to Car) and Minicab to Bus is 3 (path is from Minicab to Cab, Cab to Car and Car to Bus). Shortest path similarity says that if the shortest path distance between two sense or words in a graph is short, it signifies that two words are more similar. In fact, shortest path similarity calculates the number of edges between two concepts or words in the

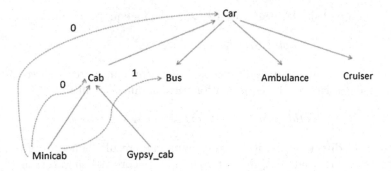

Fig. 3. A fragment showing changes of path directions in Word-net hypernym hierarchy

thesaurus graph. Here, two words having same parents are considered as more similar and words that are far away in the network are considered less similar.

From the Fig. 3, it is clear that the number of changes of direction between Minicab to Cab and Minicab to Car is 0. But, for Minicab to Bus, number of changes of direction is 1, as to go from Minicab to Bus, we have to go first to upward direction (from Minicab to Car) and again have to traverse to downward direction (from Car to Bus). Therefore, semantic relatedness always makes an effort to find a path which is not too long and also number of changes of direction is less. The required hso equation is

$$path\,weight(c_1, c_2) = 2*c - path\,length\,(c_1, c_2) - (k * direction\,changes\,(c_1, c_2)) \tag{1}$$

Here, c and k are the constants and values are $c = 8$ and $k = 1$. We have to normalize the semantic relatedness value as we get 16 as highest score if two words are completely similar. For, normalization, we use the following equation:

$$\frac{x - x_{min}}{x_{max} - x_{min}} \tag{2}$$

Moreover, we try to find similarity of confused words based on lexical character matching. Here, we assume that the appropriate word in the confused wordset may be present in the title of the input text document. Therefore, we take each word in the input text title and match with the words present in commonly confused wordset. Scores are given based on the same number of characters present between input text title words and words in confused wordset. Here, longest common contiguous subsequence is considered in sequence matching. The equation for sequence matching between two words or strings is as follows:

$$Sequence\,Matching\,Score = 2.0 * \frac{M}{T} \tag{3}$$

Here,
M = Matching characters between two strings.
T = Total number of characters present in both the strings.

Now, these two similarity scores are added. The equation will be

$$Total\ Similarity\ Score = A + B \tag{4}$$

To get more accuracy in result, we use weighting parameter for two different similarity values. Hence, the equation for total similarity will be:

$$Total\ Similarity\ Score = \alpha * A + \beta * B \tag{5}$$

Presence of direct word always gives significant result, hence we give higher priority to sequence matching character similarity score (B) than semantic relatedness scores (A). We tested the method on training set and optimize the accuracy by giving following values of α and β: $\alpha = 0.40$, $\beta = 0.60$; where $\alpha + \beta = 1$. Range of total similarity score is $0 \le$ Total Similarity Score ≤ 1.

The method chooses that word having the highest score. Finally, replacement of the correct spelled word with the incorrect word is done to get the correct query.

3.2   Description of Proposed Method (CSQ)

In query-based text summarization, user has to enter the query. It is quite possible that a user may enter incorrect words. Hence, we assume that the query contains wrong words.

The pseudo code of Context Sensitive Query Correction Method for Query Based Text Summarization (CSQ) method is as follows:

Data: Query (Q_i) and Title of the Input Text (T_i)
Result: Correct Query ($Q_{correct}$)
Do the stop word removal and stemming of the query
for *each confused word w in Q_i* **do**
  Find out the confused wordset (C_w) using dictionary
  for *each word c in C_w* **do**
    Find out the average semantic measure score of the word c with the previous word P_w and the next word N_w of Q_i using Eq. 1
    Find out the sequence matching of the target word with the title of the input text document (T_i) using Eq. 3
    Sum up all scores (*score*) using Eq. 5
    Replace w with c having highest *score*
  end
end

Algorithm 1: Steps of CSQ Method

3.3   Example Computation

To experiment the proposed method, DUC 2005 datasets are used (http://duc. nist.gov). Each dataset has 50 queries with 50 different topics. We take those queries where confused words are present. CSQ method is applied on DUC 2005 datasets. Here, we try to show it by taking a simple example. We take the query example as, "Identify and describe types of organized crime that crosses borders or involves more than one country". When the query is entered, it is written with spelling mistake as "Identify and describe types of organized crime that crosses *boarders* or involves more than one country" Now, we apply the CSQ method and results are as follows:

1. Pre-process the query by removing the stop words and stemming the query.
2. Initially, CSQ Method checks the presence of confused words in the query. Here, it finds the confused word 'boarder' and for 'boarder' the method gets this confusion wordset {boarder, border} accordingly. The score of semantic measure score (A) of confused word with next and previous word is shown in Table 1.

Table 1. A: Score

Confused words	Score1
Boarder	0.05
Border	0.08

3. Sequence matching of the target word from the confusion set is calculated with the title of the input text document and score (B) is shown in Table 2.

Table 2. B: Score

Confused words	Score3
Boarder	0.61
Border	0.67

4. Finally, we calculate the total similarity value by adding two similarity scores (Table 3):

Table 3. Total similarity scores of confused words

Confused words	Total score
Boarder	0.386
Border	0.434

Here highest score is 0.434 for "border". Hence "boarder" will be replaced by "border" in the query and it will be written as, "Identify and describe types of organized crime that crosses borders or involves more than one country".

4   Experimental Results and Discussion

4.1   Comparison of CSQ Method with Real-Word Spell Checkers

We try to evaluate CSQ method with other existing real-word error correction methods. Here, Baseline method, Hidden Markov Model Tagger and popular Ginger Software are used to compare with CSQ method. We use DUC 2005 and 2006 datasets. Each dataset contains 50 queries with 50 different topics. We take those queries where confused words are present.

The baseline method is based on frequency of occurrence of a word in the training corpus. For example, if the confused set is {piece, peace}, then the baseline method finds most occurred word between these two confused words from the training corpus and it suggests to change (or remain same) the word to the most common word in the test corpus.

The hidden markov model (HMM) is mainly used for parts of speech tagging. This probabilistic tagger chooses the tag sequence with highest probability. Hidden markov model is a special case of Bayesian interference for noisy-channel models. It helps to find out the appropriate word according to the context of the sentence. HMM tagger only works for different parts-of-speech words. The ginger spell checker checks every word depending on the context of the sentence.

To find out the accuracy of the spell checker, we use the following equation:

$$Accuracy = \frac{words\,correctly\,recognized\,by\,spell\,checker}{Total\,no\,of\,confused\,words} \tag{6}$$

To find out the sentence level precision and recall, we use the following measures:

- true positive (TP): correct word which is recognized as correct word by the spell checker.
- false positive (FP): incorrect word which is recognized as correct word by the spell checker.
- false negative (FN): correct word which is recognized as incorrect word by the spell checker.
- true negative (TN): incorrect word which is recognized as incorrect word by the spell checker.

To know about the capacity to detect correct sentences, we can use following recall and precision equations.

$$R_c = \frac{TP}{TP + FN} \tag{7}$$

$$P_c = \frac{TP}{TP + FP} \tag{8}$$

F-measure gives the harmonic mean of recall and precision values. Hence, the equation will be:

$$F = \frac{2 * R_c * P_c}{R_c + P_c} \tag{9}$$

Table 4. Performance measure with baseline systems for real-word errors

Method name	Accuracy	Recall	Precision	F
CSQ	76.92%	83.33%	90.90%	86.95%
Baseline method	38.46%	41.66%	83.33%	55.54%
HMM tagger	15.38%	16.66%	66.66%	26.65%
Ginger	46.15%	50%	85.71%	63.15%

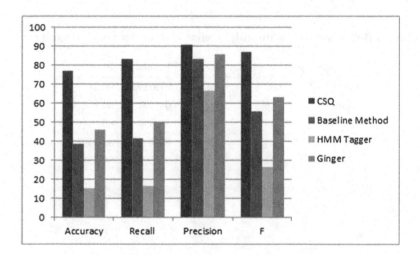

Fig. 4. Comparison of performance between CSQ and other existing methods for real-word errors

By using the above equations, CSQ method is compared with other existing systems. Results are shown in following Table 4 and Fig. 4.

From the above comparison, it is clear that CSQ method performs better in terms of all accuracy, precision, recall and F-measure values.

4.2   Comparison of CSQ Method with Non-word Spell Checkers

This CSQ Method also works well for non-word error detection and correction. Hence, we can also apply it for non-word error correction. To apply this method, we just need to find out the suggested wordset using dictionary. We apply CSQ method with every word in suggested wordset. Here, 50 queries are taken for doing the evaluation. We use TAC (Text Analytics Conference) 2009 datsets (http://tac.nist.gov). There are 44 documents each having 2 topics. For each topic, there are ten text documents.

We compare the CSQ method with some baseline methods like Longest Common Substring, Character Similarity, Microsoftword Spell Corrector and Minimum Edit Distance. Longest common substring (LCS) finds the longest string or strings between two or more strings. Character similarity (CS) finds the maximum common characters between two strings.

In a word document, Microsoft Spell Corrector (MSC) gives many suggested words for an incorrect word. Minimum Edit Distance (MED) gives minimum number of operations required to transfer from one word to another word. Insertion, deletion and substitution operations are used here for transformation from one string to other string. Detailed results are shown in following Table 5 and Fig. 5.

Table 5. Performance measure with baseline systems for non-real word errors

Method name	Accuracy	Recall	Precision	F
CSQ	82%	83.67%	97.61%	90.10%
LCS	18%	69%	64.3%	66.6%
CS	48%	85%	82.8%	83.9%
MSC	64%	89.2%	89%	89.1%
MED	42%	84%	80.8%	82.4%

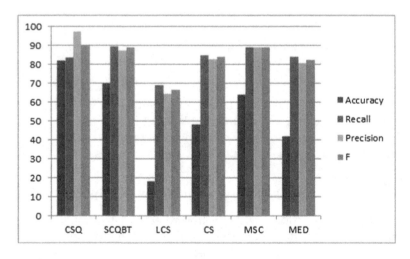

Fig. 5. Comparison of performance between CSQ and other existing methods for nonword errors

From the above comparison, it also proved that this CSQ method performs well in terms of accuracy. We also get high precision and F-measure value for all other existing systems except for recall values. Hence, we can use CSQ method both for real word and non-real word errors.

5   Conclusion

An effective context sensitive spell checker is suggested here which is based on both semantic measure and lexical character matching. Lexical character matching can be achieved by finding similar characters between two words and semantic measure can be found by using different semantic relations. CSQ method is helpful for real-word as well as non-word spelling correction particularly in query-based text summarization. This spell checker can correct more than one incorrect word present in a sentence. Semantic measure is calculated by using hso similarity. This new spell checker outperforms many existing contextual real -word and non-word spell checkers.

Though CSQ method performs substantially high compared to other existing spell checkers, but this method is only implemented on small confusion wordset. We can improve the scalability of identification and correction of wrong words by applying on large confusion wordset. This CSQ method also depends on word bi-gram. Therefore, if the incorrect word is the first or last word of the sentence, then bi-gram semantic measure can not find due to absence of previous or next word of the target word. In addition, we can improve this CSQ spell checker by introducing new similarity measure which can find semantic measure between words of different parts of speech.

References

1. Pedler, J.: Computer correction of real-word spelling errors in dyslexic text. Ph.D. thesis, University of London (2007)
2. Kukich, K.: Techniques for automatically correcting words in text. ACM Comput. Surv. (CSUR) **24**(4), 377–439 (1992)
3. Mays, E., Damerau, F.J., Mercer, R.L.: Context based spelling correction. Inf. Process. Manag. **27**(5), 517–522 (1991)
4. Golding, A.R., Schabes, Y.: Combining trigram-based and feature-based methods for context-sensitive spelling correction. In: Proceedings of the 34th Annual Meeting on Association for Computational Linguistics, Association for Computational Linguistics, pp. 71–78 (1996)
5. Golding, A.R.: A Bayesian hybrid method for context-sensitive spelling correction. arXiv preprint arXiv:cmp-lg/9606001 (1996)
6. Yarowsky, D.: A comparison of corpus-based techniques for restoring accents in Spanish and French text. In: Armstrong, S., Church, K., Isabelle, P., Manzi, S., Tzoukermann, E., Yarowsky, D. (eds.) Natural Language Processing Using Very Large Corpora, pp. 99–120. Springer, Heidelberg (1999)
7. Hirst, G., Budanitsky, A.: Correcting real-word spelling errors by restoring lexical cohesion. Nat. Lang. Eng. **11**(1), 87 (2005)
8. Jiang, J.J., Conrath, D.W.: Semantic similarity based on corpus statistics and lexical taxonomy. arXiv preprint arXiv:cmp-lg/9709008 (1997)
9. Fossati, D., Di Eugenio, B.: I saw tree trees in the park: how to correct real-word spelling mistakes. In: LREC (2008)
10. Samanta, P., Chaudhuri, B.B.: A simple real-word error detection and correction using local word bigram and trigram. In: ROCLING (2013)

11. Levenshtein, V.I.: Binary codes capable of correcting deletions, insertions, and reversals. In: Soviet Physics Doklady, vol. 10, pp. 707–710 (1966)
12. Gupta, S., et al.: A correction model for real-word errors. Procedia Comput. Sci. **70**, 99–106 (2015)
13. Sorokin, A.: Spelling correction for morphologically rich language: a case study of Russian. In: BSNLP 2017, p. 45 (2017)
14. Flexner, S.B., Hauck, L.C., et al.: Random House Unabridged Dictionary. Random House, New York (1993)
15. Hirst, G., St-Onge, D., et al.: Lexical chains as representations of context for the detection and correction of malapropisms. WordNet: Electron. Lexical Database **305**, 305–332 (1998)

Ontological Controlling the Lexical Items in Conceptual Solution of Project Tasks

P. Sosnin[(⊠)] and A. Pushkareva

Ulyanovsk State Technical University, ul. Severny Venets, str. 32,
432027 Ulyanovsk, Russia
sosnin@ulstu.ru, a.pushl206@gmail.com

Abstract. The paper presents a way of an ontological control of lexical items in a conceptual solution of project tasks. The proposed way is aimed at finding errors, correcting and preventing them, as well as supporting the processes of understanding in a real-time work of designers developing the software systems. The basis of the proposed way lays in the use of a precedent-oriented approach to work with project tasks, stepwise refinement in analysis and structuration of a designed project. The way is materialized in the instrumental environment WIQA (Working in Questions and Answers) supporting the conceptual designing. The chosen toolkit includes a number of components, one of which provides a creation and use of necessary ontologies.

Keywords: Conceptual designing · Lexical control · Mental imagery · Ontology · Question-answering · Software intensive system · Visual modeling

1 Introduction

Among the stages of designing the software intensive systems (SIS), the conceptual stage occupies a very important place. As known [1], this stage is a source of expensive semantic errors that essentially influence on the success in developing the SISs.

At this stage, the team of designers working in the certain instrumentally technological environment must solve a complex of domain tasks that specify a conceptual essence of the SIS that must be created. Furthermore, conceptual specifications of the SIS must be sufficient for programming the software components at the next stages.

At the conceptual stage, the work of the team and obtained results have the following features:

1. Members of the team are involved in corresponding workflows where designers solve appointed tasks in collaboration and coordination with each other.
2. Almost all artifacts that are created by designers are conceptual constructs or, in other words, they consist of components that are expressed using the necessary means of a professional natural language and appropriate diagrams.

Marked features point out that this activity suggests using reasonings based on a certain volume of appropriate lexical items that support the necessary level of understanding among designers. For achieving this, designers share the lexis in conditions when its usage is controlled.

O. Gervasi et al. (Eds.): ICCSA 2017, Part VI, LNCS 10409, pp. 31–46, 2017.
DOI: 10.1007/978-3-319-62407-5_3

In this paper, we propose the way of ontological control of lexical items in the conceptual solution of project tasks. The way is based on actions that support creating a project ontology, and a simultaneous usage of this ontology for controlling the used lexical items in the work of designers with appointed tasks. Since the interests of the paper are limited by the conceptual stage, when designers register the achieved states of the task solutions, these states of tasks are qualified as conceptual solutions of corresponding tasks. We implemented the proposed way in the instrumental environment WIQA [2] intended for conceptual designing the SISs.

The remainder of the article is structured as follows. Preliminary bases are presented in Sect. 2. Section 3 focuses on related works. The toolkit that supports the use of ontology is described in Sect. 4. Section 5 discloses the used workflows, and Sect. 6 contains a brief conclusion.

2 Preliminary Bases

2.1 Conceptual Solution of the Task

In the general case, a project task that will be solved by a designer can occur in the form of a new task Z, a statement of which (a description of the task) has a view of an intention of achieving the certain useful aim. From this point of time, the task starts its life cycle, during which the task will be to move from a state to state, enriching the content until a state in which it will receive the necessary decision and its material realization. At the conceptual stage of the life cycle, the task Z achieves the state of its conceptual solution.

Before the statement $St(Z, t_0)$ will acquire the observable shape of a text written in the language of the project, this statement has been created in the mind of a designer as a reaction to the corresponding task situation. This process includes mental imaginations, the activity of consciousness and other requested intellectual activities.

Let us suppose, that statement $St(Z, t_0)$ has got the textual shape, and this text is observable on the monitor screen. To move forward, the designer should check the lexical items used in the text and its understandability. These checks can be fulfilled with the use of the following actions:

1. Comparison of the used words with the current state of the ontology that is created in the process of project designing from its beginning.
2. Reflection of $St(Z, t_0)$ on an appropriate visual image or images that will activate the mental imagery and other intellectual activities in the mind of a designer who will register the result of the reflection beyond brains (in our case on the screen of the monitor).

Both kinds of checks lead to service tasks, the first of which is disclosed below. Our version of the second service task was described in publication [3]. Here we also mark, that a reflection of the text on graphics supports the achievement and expression of understanding, without which either personal or collective work of designers are impossible.

Processes of checking and finding errors will lead to the expansion of the information content of $St(Z, t_0)$, which the designer should analyze to create and register the next state $St(Z, t_1)$ of the task Z. Thus, step by step, the designer will build the conceptual solution of the task, and this process is presented in Fig. 1 in a general way.

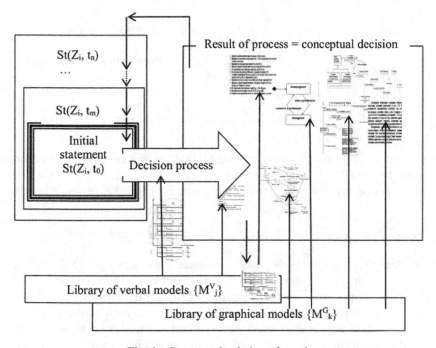

Fig. 1. Conceptual solution of a task

The scheme also includes two sources of conceptual models (verbal and graphical, for example, appropriate UML-diagrams) that are usefully applied to the solution process. They are conditionally called libraries, but this only emphasizes that such models are useful for gathering them in a whole for the possible reuse.

In the described research, the way of coordination was materialized in the instrumental environment WIQA where a designer creates and evolves the statement $St(Z_i, t)$ by the use of its stepwise refinement based on question-answering analysis and modeling [2]. Dynamic of this process is reflected in Fig. 2 by the question-answer tree (QA-tree).

The left column of the scheme indicates that the designer creates the statement of the task with the use of the question-answer analysis and modeling, the iterative process of which is implemented by a stepwise refinement. In its turn, the states $St(Z_i, t)$ reflects the process of the conceptual solution, which is oriented at its reuse as a precedent.

In creating the task statement, the central place occupies designers' reasoning that is based on thinking and doing, results of which are registered in the hierarchical structure of questions and answers written in the project language. For controlling this work, we

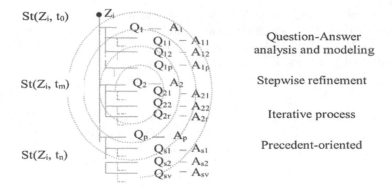

Fig. 2. Developing the statement of the task

suggest using mechanisms of design thinking [4] that concern as constructive actions of generating the textual units of the statement $St(Z_i, t)$ so testing these units at the level of their understandability, for example with the use of conceptual experimenting [5].

2.2 Project Ontology

The complex WIQA that has been created for conceptual designing the SISs consists of a kernel and a number of components. The kernel includes a semantic memory of the question-answer type, and a system of operations support the interactive works of the designer with cells of this memory and their content. One of such works is an iterative creation and use of an ontology by designers, the framework of which is shown in Fig. 3. In its applications, this framework has the generative potential that implicitly expresses the formal grammar of the project ontology.

Components extend the potential of the kernel. They include subsystem for pseudocode programming defined above the semantic memory, graphical editor supported

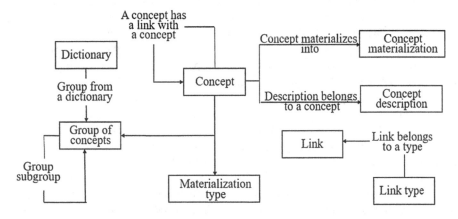

Fig. 3. Developing the statement of the task

creation and use graphical models $M_r^{Gy}(Z_i, t_k)$ embedded into their library, subsystem of an organizational structure of the team and subsystem of an agile management combined Kanban, Scrum and multitasking means.

Indicated functionalities are open for extending the potential of any of them. For example, a designer can use a program access to any ontology components that are shown in the ontology framework presented in Fig. 3. This access is one of our novelties that provides implementing the generative potential of the ontology framework.

3 Related Works

The first group of related works combines the papers focused on the subject area "Design Thinking." The informational source [6] discloses basic features of this subject area from the viewpoint of its use in the real practice of developing the SISs. Brown in [7] describes how design thinking transforms organizations and inspires innovation. The becoming of the DT-methodology and the state of affairs of this subject area are described in the paper [8].

The interesting analysis of the nature of design thinking is conducted in [9] where Dorst connects this nature with two types of abductive reasoning of the designer in interactions with situated problems. By the use of two model of mind, the paper [10] specifies the influence of reflections that are intertwined with language actions of the designer in DT-processes.

The next group of related works combines the papers focused on the domain "Concept Development and Experimentation (CD&E)". In [11] CD&E is defined as "A method which allows us to explore and predict, by way of experimentation, whether new concepts that may impact people, organization, process and systems will contribute to transformation objectives and will fit in a larger context". This document specifies the processes in this subject area where the role and place of conceptual experimentation in military applications are indicated in details. The following publication [12] defines the version of the occupational maturity of the CD&E-process. The paper also demonstrates some specialized solutions that are focused on a behavioral side of experimenting.

A set of typical kinds of ontologies (according to their level of dependence on a particular task or point of view) includes the top-level ontologies, domain ontologies, tasks ontologies and applied ontologies. All these types of ontologies are defined in [13, 14] as means that are used in different systems.

For the SISs, an adequate type of ontologies is the applied type that usually is expanded using the other ontologies types. By the publication [15], the theory and practice of applied ontologies "will require many more experiences yet to be made."

It is necessary to notice that the project ontology as a subtype of applied ontologies is essentially important for SISs. Project ontologies in the greater measure are aimed at the process of designing, but after refining, they can be embedded into implemented SISs.

The specificity of project ontologies is indicated in some publications. In the technical report [13] the main attention is paid to "people, process and product" and collaborative understanding in interactions. Investigating the possibility of the ontology-based project management is discussed in the paper [15].

As for developing software systems and considering ontological problems of software products, the use of ontology possibilities is investigated in the paper [9]. In means suggested in this article, the experience of task ontologies is taking into account also and, first of all, in the role of different kinds of knowledge. The place of knowledge in the task ontologies is reflected and discussed in the publication [16]. The role of knowledge connected with problem-solving models is opened in papers [17].

All papers indicated in this section were used as sources of requirements in developing the set of instrumental means provided the creation and use of the proposed ontology.

4   Toolset for Creating a Project Ontology

4.1   Basic Workflows

As told above, this paper is aimed at controlling (real-time checking) the lexis used by designers in a conceptual solution of project tasks with the use of stepwise refinement by the way presented in Fig. 3. To achieve this aim, we are trying to resolve the following problem:

1. To develop a toolset of forming a controlled dictionary aimed at finding and preventing lexical errors when processing project documentation texts for software systems.
2. To represent the dictionary in the ontological form and to introduce automation into the work of a designer due to the usage of software agents and program access to the ontology in its current state.
3. To develop a toolset within the toolkit WIQA.

Our main purpose is to raise the efficiency of designers' interaction (in the real-time) with experience bases when they develop software systems with the help of ontologies implemented into experience bases to systemize their content.

For this, we developed a text-processing toolset allowing to exercise ontological control over project documentation. The toolset can solve the following tasks:

1. **T1.** To create a new ontological dictionary in the toolkit WIQA based on a list of contender words derived from project documentation texts after filtering out stop words from their word forms.
2. **T2.** To complement a working vocabulary of concepts in the WIQA ontology using the mechanism of controlled lexis.
3. **T3.** To fetch out the surrounding of concepts in project documentation texts and to add them into the ontology as concepts definitions.

Each of these tasks has several workflows (**WF**) inside. Some of them are carried out by a software agent only (**WF^A**), and some of them involve a designer (**WF^D**).

Let us go into detail on each workflow of the task **T1**.

1. **T1.WF^D1.** Select a text to process.
2. **T1.WF^A1.** To split the text into word forms.
3. **T1.WF^A2.** To filter out stop words from the word forms and get a list of contender words that can be added to the ontology.
4. **T1.WF^D2.** To select items from the list of contender words, which are consistent with the selected domain and the project task being solved, and add them to the ontology as concepts.

The workflows of the task **T2** are partly equal to the ones of the task **T1**. However, there is one more workflow, which is carried out when adding concepts to the ontology simultaneously with the **T2.WF^D1** one. Note that the task **T2** can be repeated an unlimited number of times but only after finishing the task **T1**.

1. $T2.WF^D1 = T1.WF^D1$
2. $T2.WF^A1 = T1.WF^A1$
3. $T2.WF^A2 = T1.WF^A2$
4. $T2.WF^D2 = T1.WF^D2 \parallel T2.WF^A3$. To exercise control over dictionary (check if the concept has already been added to the ontology).

Two agents are involved in solving the tasks **T1** and **T2**. Agent A splits the source text into the word forms and Agent B filters out the stop words. The workflows of the tasks **T1** and **T2** are presented in Fig. 4.

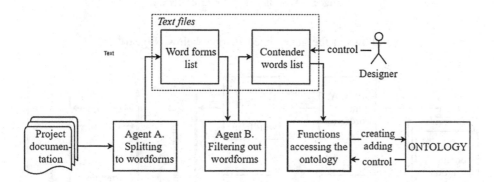

Fig. 4. Workflows of the tasks T1 and T2

Let us go into detail on each workflow of the task **T3**. This task relates to the working dictionary of a project ontology (or a task ontology) in its current state:

1. **T3.WF^D1.** Select a working dictionary to analyze.
2. **T3.WF^D2.** Select a text to process.
3. **T3.WF^A1.** To fetch out the ambiance of concepts in the ontological dictionary from the selected text and get a list of ambients.
4. **T3.WF^A2.** To filter the list of ambients using word collocation syntactic models and some other methods stated below and get a list of concepts "use cases."

5. **T3.WF^{D}3.** To pick up the concept use cases consistent with the natural language rules and the project task being solved. ‖ **T3.WF^{A}3.** To add the selected use cases into the ontology as concept definitions.

Two agents are involved in solving the task **T3**. **Agent C** fetches out the ambiance of each concept (2–3 words before and after each of the concepts) presented in the ontology at its current state and **Agent D** filters the ambiance using some syntactic models and semantic tokens described below. The workflows of the tasks **T3** are presented in Fig. 5.

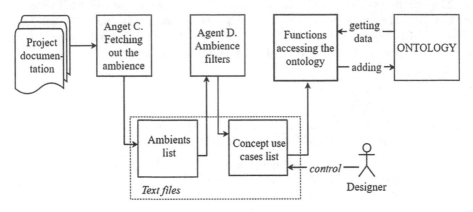

Fig. 5. Workflows of the tasks T3.

See all the workflows at the generalized scheme below (Fig. 6).

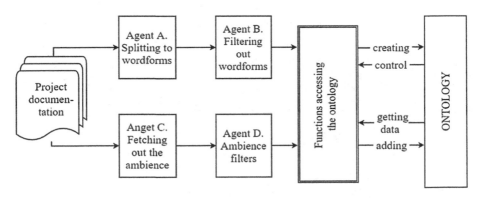

Fig. 6. Generalized scheme of project documentation texts processing to form an ontology in the WIQA system

Below we consider the working mechanisms of the Agents B and D as soon as they use some sophisticated linguistic mechanisms inside.

4.2   Agent B. Filtering Out Stop Words

Stop words are the words that do not have any important semantic meaning in the text and can be ignored in the process of creating a domain ontology which is a semantic model of a text.

We developed the following classification of English stop words (see Table 1):

Table 1. Classification of stop words

No	Type	Examples
1	Auxiliary parts of speech (articles, conjunctions, particles, prepositions, interjections, pronouns, etc.)	*a, the, to, since, that, in, under, oh, whose, mine*
2	Abbreviations, acronyms	*i.e., etc., PhD*
3	Words with general meaning	*have, think, become, the universe*
4	Words expressing assessment and modality	*probably, exactly, can, should, could, perhaps*
5	Introductory words	*nevertheless, anyway*
6	Foreign words	*Deutsche, vive*

We developed the classification after having analyzed a number of open stop word lists used by MySQL, Google and various NLP software systems [18, 19].

4.3   Agent D. Fetching Out the Ambiance

This agent works together with a specific software system – English POS-tagger [14]. It tags a text – the tags define the part of speech (noun, verb, adjective, particle, etc.) of each word form in a text. See Table 2 to discover the tags and their values of the POS-tagger that we use.

We discovered several syntactic models of English collocations (see Table 3) with the help of which the agent fetches out the ambiance of concepts and forms a list of concept use cases (Table 4).

Apart from the collocation models we use the following semantic tokens which help to fetch out the ambiance of a concept in a text to specify or expand its meaning:

- *That-token:* **<concept> that ... .**
  If "that" goes after a concept, the rest part (up to the full stop) of a sentence may relate to its definition.
- *Participle-token:* **<concept> V-ing ... ./,**
  If a Participle I goes after a concept, the rest of the sentence or its part (up to the full stop or a comma) may relate to the concept definition.
- *Enumeration-token:* **..., ..., ... <concept> ..., ..., ...**
  Any enumerations which go before or after a concept may relate to its definition.

To sum it all up, both the models and the tokens allow us to get a list of concept ambiance which a designer can add to the ontology.

Table 2. English POS-tagger tags

Tag	Explanation	Examples
NNP	Proper noun, singular	*Pierre, England*
CD	Cardinal number	*61, three*
NNS	Noun, plural	*researcher, men*
JJ	Adjective	*British, new*
MD	Modal	*can, ought*
VB	Verb, base form	*be, introduce*
DT	Determiner	*the, neither*
NN	Noun, singular or mass	*The spokeswoman, year*
IN	Preposition or subordinating conjunction	*in, because*
VBZ	Verb, 3rd person, singular, present	*has, says*
VBG	Verb, gerund or present participle	*according, rising*
CC	Coordinating conjunction	*so, versus*
VBD	Verb, past tense	*learnt, studied*
VBN	Verb, past participle	*improved, retired*
RB	Adverb	*mostly, far*
TO	To	*to*
PRP	Personal pronoun	*they, we*
RBR	Adverb, comparative	*higher, easier*
WDT	Wh-determiner	*why, when*
VBP	Verb, non 3rd person, singular, present	*include, calculate*
RP	Particle	*on, away*
PRP$	Possessive pronoun	*their, my*
JJS	Adjective, superlative	*highest, longest*
POS	Possessive ending	*'s*
EX	Existential there	*there*
WP	Wh-pronoun	*whom, whoever*
JJR	Adjective, comparative	*less, closer*
WRB	Wh-adverb	*where, whenever*
NNPS	Proper noun, plural	*angels, motors*
WP$	Possessive pronouns, plural	*whose*
PDT	Predeterminer	*quite, such*
RBS	Adverb, superlative	*best, hardest*
FW	Foreign word	*Deutsche, vive*
UH	Interjection	*alas, wow*
SYM	Symbol	*@, &*
LS	List item marker	*a, b, first, second*

4.4 Programmed Access to the Ontology

The ontology in the WIQA system can be accessed via the specific pseudocode functions that request the ontological database in the controlled pseudo code interpretation mode. There are two types of such functions:

Table 3. Classification of collocation models

Model	Explanation	Examples
Co-ordinating collocations		
N&N	Noun + conjunction + noun	*sun and moon*
A&A	Adjective + conjunction + adjective	*red and blue*
V&V	Verb + conjunction + verb	*read and write*

Table 4. Classification of collocation models (continuation)

Model	Explanation	Examples
Subordinate collocations		
AN	Adjective + noun	*semantic meaning*
VN	Noun + verb	*filter word forms*
VD	Verb + adverb	*analyze thoroughly*
DA	Adverb + adjective	*very useful*
NN	Noun + noun	*domain ontology*
PN	Participle + noun	*working dictionary*

1. Recording-data functions.
2. Getting-data functions.

The functions work with the following ontology entities:

- dictionary;
- group (dictionary section);
- subgroup (group section);
- word (an ontology concept or a dictionary item that in some cases can be presented as a collocation);
- definition (explanation of the concept meaning);
- relation (a link of a certain type that exists between two words).

The Figs. 7 and 8 below show the general layout of an ontology in the WIQA system with a part of our task ontology as an example.

Note that the pseudocode functions mentioned above also allow working with concept relations. A set of relation types is divided into the following groups (see these groups in Fig. 8):

1. Basic (part-and-whole, inheritance, instance, attribute);
2. Cause-and-effect (cause-and-effect);
3. Associations (by resemblance, by adjacency, by contrast, time-based, space-based, synonyms);
4. Pragmatic (participate, do, use (as an instrument)).
5. Pragmatic (participate, do, use (as an instrument)).

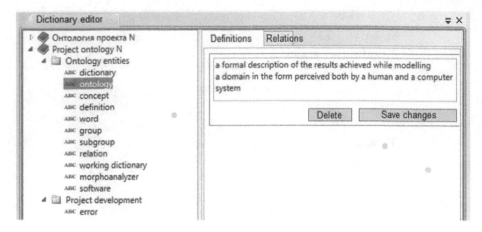

Fig. 7. Ontology layout in the WIQA system

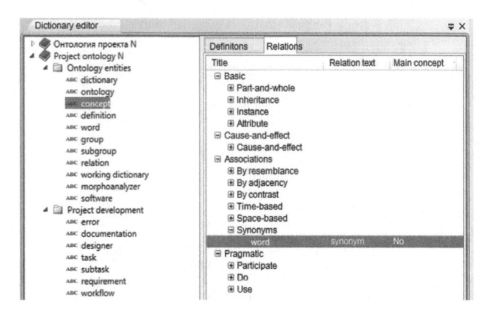

Fig. 8. Relation types structure in the WIQA system

5 Understanding of Textual Increments and Their Combinations

For the support of understanding, we developed two versions of using the ontology. The first version is based on the iterative coordination of the investigated textual unit with a corresponding semantic net that, in its turn, is reflected in a Prolog-like description. This description transforms to an executable prolog-prototype. Discovered errors are corrected after which the iterative process will continue until the state when verbal and graphical will be coordinated. This state signalizes about achieving the necessary understanding. In this process, interactions with the ontology provide the controlled use of lexis and demonstrate relations between nodes of the semantic net for its correcting and enriching. This version described in details in our paper [3].

The second version allows to a designer enter any textual increment ΔT of the task statement St(Z, t) and visualize the definitions of all the concepts, presented in the text, and all the concepts related to them with any type of relations. That helps to gain more information about the task and, therefore, get a better understanding of what has to be done. Figure 9 shows the interface providing this function.

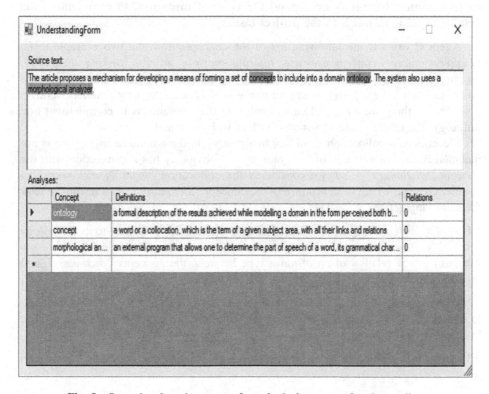

Fig. 9. Operational environment of ontological support of understanding

Extracted definitions and relators can be additionally used with the specialized utility for more fully their presentation and for their use in pseudo-code programs written by the designer.

6 Example of Applying for the Ontological Support

Let us consider the main opportunities of our toolset on the example of our project task. We have selected two subtasks for the demonstration purposes:

Z1. To fetch out the ontology concept use-cases with the help of mature mechanisms of logic and linguistic text processing at the level of lexical items and groups of lexical items, such as filtering out stop words, applying syntactic models and semantic tokens as well as using specific NLP tools, namely a POS-tagger.
Z2. To use the pseudocode functions allowing the program access to the project ontology in its current state, as well as to its working dictionary, to exercise lexical control namely to specify or expand the concept meanings, to complement the ontology and to adjust it to the project task.

Agent C extracts the following list of the concepts from the two example texts: *ontology concept, concept use-case, filtering out stop words, working dictionary, lexical control, concept meaning, project task, filtering out stop words, pseudocode functions allowing the program access to the project ontology in its current state.*

The first thing we have to do after analyzing the ambiance is to **complement our ontology**, the current state of which is shown in Figs. 7 and 8.

We noticed a collocation *"working dictionary"* in the ambiance list, which is not included in the current state of the ontology but obviously has a connection with the concept *"dictionary"* – so, we considered the collocation *"working dictionary"* as another concept, which is an *instance* of the concept *"dictionary."* As a result, we took the following actions:

A1. Add a new concept *"working dictionary"* to the ontology, namely to its *"Ontology entities"* group.
A2. Add a new relation of an *instance* type between the concepts *"dictionary"* and *"working dictionary."*

When considering the *"filtering out stop words"* phrase and the fact that there are to concepts *"filter"* and *"stop word"* in our ontology, we added a new relation of an instrument type:

A3. Add a new relation of an *instrument* type between the concepts *"filter"* and *"stop words."*

The phrase *"pseudocode functions allowing the program access to the project ontology in its current state"* in a list allows us to add a definition to the concept *"pseudocode function"* presented in our ontology:

A4. Add a new definition to the concept *"pseudocode function"* stated as follows: *"A function which allows the program access to the project ontology in its current state."*

After taking actions to complement the ontology, let us consider if there are any **changes** that we have to make **in the source text** based on the analyses.

In the first text, there is a concept *"use case,"* which has a relation of a *part-and-whole* type with the concept *"documentation"* in our ontology. Thus, we have to <u>edit the text</u> a little and formulate it as follows (the changes are underlined):

Z1^E. To fetch out the ontology concept use-cases <u>from the documentation</u> with the help of mature mechanisms of logic and linguistic text processing at the level of lexical items and groups of lexical items, such as filtering out stop words, applying syntactic models and semantic tokens as well as using specific NLP tools, namely a POS-tagger.

7   Conclusion

In conclusion, we would like to put emphasis on the fact that using a project ontology positively influences on designing the systems with software. This not only allows designers to more accurately develop the project documentation more accurate, but it also helps to reveal the facts, which are not obvious when a designer is thinking out solutions for tasks being solved.

The proposed way of ontological controlling focuses on conceptual solving the project tasks in conditions when, with using the stepwise refinement, designers formulate statements of tasks combining them in a coordinated system. Simultaneous creating the project ontology provides such coordination at the level of the used Lexis.

Basic features of the proposed way are caused by the use of the set of program agents that process textual units entering from actions of question-answer analysis and modeling in their applications to the statements of tasks. Implementing the way is managed by designers when they apply mechanisms of design thinking embedded into realizing the precedent-oriented approach to conceptual solutions of tasks. Our version of the automated design thinking includes the figuratively semantic support of mental imagination and conceptual experimentation that help to designers in achieving the necessary degree of understanding.

So, this way combines a number of lines of designers' activity including mental actions and actions beyond of brain. These actions are intertwined in processes of analysis and synthesis in their interactions with the accessible experience. All indicated features help to decrease the negative influence of human factors on the design process. We implemented the proposed way in the instrumental environment WIQA [1] intended for conceptual designing the SISs.

Acknowledgement. This work was supported by the Russian Fund for Basic Research (RFBR), Grant #15-07- 04809a and the State Contract № 2.1534.2017/ПЧ

References

1. Chaos reports 1994–2016 (2017). http://www.standishgroup.com
2. Sosnin, P.: A scientifically experimental approach to the simulation of designer activity in the conceptual designing of software intensive systems. IEEE Access **1**, 488–504 (2013)

3. Sosnin, P.: Figuratively semantic support in precedent-oriented solving the project tasks. In: Proceedings of the 10th International Conference on Application of Information and Communication Technologies (AICT-2016), pp. 479–483 (2016)
4. Introduction to Design Thinking (2017). https://experience.sap.com/skillup/introduction-to-design-thinking/
5. Sosnin, P.: Conceptual experiments in automated designing. In: Zuanon, R. (ed.) Projective Processes and Neuroscience in Art and Design, pp. 155–181. IG-Global (2016)
6. Häger, F., Kowark, T., Krüger, J., Vetterli, C., Übernickel, F., Uflacker, M.: DT@Scrum: integrating design thinking with software development processes. In: Plattner, H., Meinel, C., Leifer, L. (eds.) Design Thinking. Understanding Innovation, pp. 263–289. Springer, Heidelberg (2014)
7. Brown, T.: Change by Design: How Design Thinking Transforms Organizations and Inspires Innovation. HarperBusiness, New York (2009)
8. Leifer, L., Meinel, C.: Manifesto: design thinking becomes foundational. In: Plattner, H., Meinel, C., Leifer, L. (eds.) Design Thinking Research: Making Design Thinking Foundational, pp. 1–4. Springer, Heidelberg (2015)
9. Dorst, K.: The nature of design thinking. In: DTRS8 Interpreting Design Thinking: Design Thinking Research Symposium Proceedings, pp. 131–139 (2010). http://epress.lib.uts.edu.au/research/handle/10453/16590
10. Razavian, M., Tang, A., Capilla, R., Lago, P.: Reflective approach for software design decision making. In: Proceedings of the First Symposium on Qualitative Reasoning about Software Architectures, pp. 19–26 (2016)
11. MCM-0056-2010: NATO Concept Development and Experimentation (CD&E) Process. NATO HQ, Brussels (2010)
12. Wiel, W.M., Hasberg, M.P., Weima, I., Huiskamp, W.: Concept maturity levels bringing structure to the CD&E process. In: Proceedings I/ITSEC 2010. Interservice Industry Training, Simulation, and Education Conference, Orlando, pp. 2547–2555 (2010)
13. Garcia, A.C.B., Kunz, J., Ekstrom, M., Kiviniemi, A.: Building a project ontology with extreme collaboration and virtual design & construction. CIFE Technical Report #152, Stanford University (2003)
14. Guarino, N., Oberle, D., Staab, S.: What is an ontology? In: Staab, S., Studer, R. (eds.) Handbook on Ontologies. International Handbooks on Information Systems, vol. 2, pp. 1–17. Springer, Heidelberg (2009)
15. Fitsilis, P., Gerogiannis, V., Anthopoulos, L.: Ontologies for software project management: a review. J. Softw. Eng. Appl. 7(13), 1096–1110 (2014)
16. Martins, A.F., De Almeida F.R.: Models for representing task ontologies. In: Proceeding of the 3rd Workshops on Ontologies and their Application, Brazil (2008)
17. Eden, A.H., Turner, R.: Problems in the ontology of computer programs. Appl. Ontol. 2(1), 13–36 (2007). IOS Press, Amsterdam
18. Stopwords: http://www.ranks.nl/stopwords
19. Lande, D.V., Snarskii, A.A., Yagunova, E.V., Pronoza, E.V.: The use of horizontal visibility graphs to identify the words that define the informational structure of a text. In: 12th Mexican International Conference on Artificial Intelligence, pp. 209–215 (2013)

IoT-Based Healthcare Applications: A Review

Itamir de Morais Barroca Filho[1]([✉]) and Gibeon Soares de Aquino Junior[2]

[1] Metropole Digital Institute, Natal, Brazil
itamir.filho@imd.ufrn.br
[2] Department of Informatics and Applied Mathematics,
Federal University of Rio Grande do Norte, Natal, Brazil
gibeon@dimap.ufrn.br
http://www.dimap.ufrn.br, http://www.imd.ufrn.br

Abstract. The high cost of healthcare services, the aging population and the increase of chronic disease are becoming a global concern. Several studies have indicated the need for strategies to minimize the process of hospitalization and the effects related to the high cost of patient care. A promising trend in healthcare is to move the routines of medical checks from a hospital (hospital-centric) to the patient's home (home-centric). Moreover, recent advances in microelectronics, wireless, sensing and information have boosted the advent of a revolutionary model involving systems and communication technology, enabling smarter ways to "make things happen". This new paradigm, known as the Internet of Things (IoT), has a broad applicability in several areas, including healthcare. The full application of this paradigm in healthcare area is a common hope because it will allow hospitals to operate more efficiently and patients to receive better treatment. With the use of this technology-based healthcare approach, there is an unprecedented opportunity to improve the quality and efficiency of the medical treatment and consequently improve the wellness of the patients, as well as a better application of government financial resources. Based on this context, this paper aims to describe a review to comprehend the current state and future trends for healthcare applications based on IoT infrastructure, and also to find areas for further investigations.

Keywords: Healthcare · Internet of Things (IoT) · Patients · Technology · Review

1 Introduction

The high cost of healthcare services, the aging population and the increase of chronic disease are becoming a global concern. Many studies have indicated the need for strategies to minimize the institutionalization process and the effects of the high cost of patient care [19]. With the aim to minimize this concern, a promising trend in health treatments is to move the routines of hospital medical checks to the patient's home. But, to effectively achieve this, we need systems and communication technology to allow the patient's remote health monitoring.

© Springer International Publishing AG 2017
O. Gervasi et al. (Eds.): ICCSA 2017, Part VI, LNCS 10409, pp. 47–62, 2017.
DOI: 10.1007/978-3-319-62407-5_4

On the other hand, recent advances in microelectronics, wireless networks, sensing devices and information techonologies have fueled the advent of a revolutionary model involving systems and communication technology, enabling smarter ways to "make things happen". This new paradigm, known as the Internet of Things (IoT), has a broad applicability in several areas, including health. In this trend, it is estimated that by 2020 there will be around 20 billion "things" connected [14] and uniquely identifiable [17]. These "things" promote the basic idea of IoT that is pervasive computing around this kind of devices, such as RFID tags, sensors, actuators, mobile phones, etc. [5].

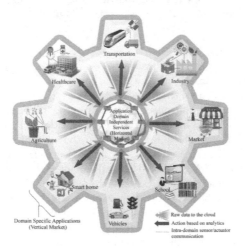

Fig. 1. IoT applications markets [2].

As presented in Fig. 1, the IoT will enable the development of applications in many markets, such as agriculture, industry, smart cities and healthcare. In particular in the healthcare market it is expected to see the development and application of this trend as part of its future, because it has the ability to allow hospitals to operate more efficiently and patients to receive better treatment. A type of healthcare application which will be focused on in conjunction with this new paradigm is the application of mobile health. The main objective of mobile health is to allow for the remote monitoring of the health status and the treatment of patients from anywhere [23].

In this context, the potential for change in the quality of life that can be promoted by IoT is unquestionable. Creating integrated utilities will lead to a qualitative change in the services to integrate information systems, computing and communication with extensive control [8]. Therefore, there is an urgent need for the development of technologies and applications related to IoT infrastructure for healthcare.

Thus, before the proposal of new platforms and solutions IoT-based for healthcare, it is essential to understand the state-of-art of the applications of this area and to realize that we performed a review based on Systematic literature reviews (SLR) method. According to Wohlin et al. [40], SLRs are conducted to identify, analyze and interpret all available evidence related to a specific research question, as it aims to give a complete, comprehensive and valid picture of the existing evidence, both the identification, analysis and interpretation must be conducted in a scientifically and rigorous way. So, this paper presents a review based on SLR method that was performed aiming to comprehend the current state and future trends for healthcare applications IoT-based, and also in order to find areas for further investigations.

Finally, in Sect. 2, we present the method for this review, focusing on the research questions, search process, inclusion and exclusion criteria, quality assessment and data collection. Continuing, in Sect. 3, we present the results for this method, regarding search results, quality evaluations and factors. In Sect. 4, we present discussion about the results, and in Sect. 5, we present the conclusions and future works of this research.

2   Method

This study has been undertaken as a systematic literature review based on the original guidelines as proposed by Kitchenham and Charters [26]. In this case, the goal of the review is to comprehend the current state and future trends in IoT-based healthcare applications. The steps in the systematic literature review method are documented in the following subsections.

2.1   Research Questions

Considering the context of this review, the research questions addressed by this study are:

RQ1. What are the main characteristics of healthcare applications based on IoT infrastructure?

RQ2. What are the patterns and protocols used in healthcare applications based on IoT infrastructure?

RQ3. What are the challenges and opportunities related to healthcare applications based on IoT infrastructure?

Regarding RQ1, about the characteristics of healthcare applications, we intend to analyze the functional and nonfunctional requirements, and for which area of healthcare the applications are intended.

2.2   Search Process

The studies selection was made on Scopus from Elsevier[1], as it indexes the main sources of computing in the academic area. The example of sources indexed by Scopus are presented in Table 1.

To define the search string we used terms related to health and Internet of Things (IoT). The main goal was to obtain the major number of researches of this particular applications. Thus, the defined search string was: ("Internet of Things" OR "IoT") AND health.

---

[1] http://scopus.com.

Table 1. Example of sources indexed by Scopus.

Source	Link
ACM digital library	http://dl.acm.org
IEEExplorer	http://ieeexplore.ieee.org
Science direct	http://www.sciencedirect.com
Springer link	http://link.springer.com

2.3   Inclusion and Exclusion Criteria

This review included works published at any year because we intended to find the biggest number of researches regarding the development of healthcare applications based on IoT infrastructure.

2.4   Quality Assessment

Each selected study was evaluated according to the following quality assessment (QA) questions:

QA1. Is the paper based on research (or is it merely a "lessons learned" report based on expert opinion)?
QA2. Is there a clear statement of the aims of the research?
QA3. Is there an adequate description of the context in which the research was carried out?
QA4. Is the study of value for research or practice?
QA5. Is there a clear statement of findings?

These criteria were based in Dybå and Dingsøyr [11] and they are grounded in three points that need to be addressed in the appreciation of the studies of the literature review:

– Rigour. Has a thorough and appropriate approach been applied to key research methods in the study?
– Credibility. Are the findings well-presented and meaningful?
– Relevance. How useful are the findings to the software industry and the research community?

These five quality assessment questions were scored as follows: 0 - in case of not attend the criteria; 0.5 - in case of partial attend of the criteria; and 1 - in case of fully attend of the criteria.

2.5   Data Collection

The data extracted from each study were: authors country, publication year, venue (journal of conference), goal, app characteristics, functional requirements, nonfunctional requirements, transfer protocols, formatting pattern, IoT platform, define ontologies?, communication protocols, application domain, hardware, interoperability with other systems?, application deployment, challenges and opportunities and additional comments.

3   Results

This section summarizes the results of this review. It specifies each stage of its execution and also presents an overview of the studies that were useful for answer the research questions. Finally, it describes the quality evaluation results of the read studies.

3.1   Search Results

We began to obtain the results by the execution of the search with the string described in Sect. 2.2 at Scopus (stage 1). This search returned 1355 studies, and then, we performed the analysis of the titles and abstracts of each one of them (stage 2). After this analysis, only the 46 studies presented in Table 2 remained. Finally, we performed a carefully read of these 46 studies and 33 of them were useful to answer the proposed research questions (stage 3). The Fig. 2 presents these stages of the study selection process. The results of the extraction of the 46 studies are presented in https://goo.gl/skZmns.

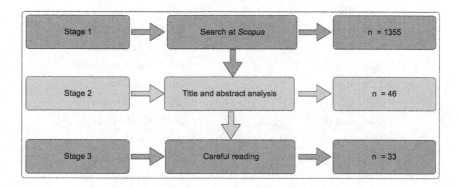

Fig. 2. Stages of the study selection process.

3.2   Overview of Studies

Considering the venue (journal or conference) of the selected studies, 72.7% are from conferences and 27.3% are from journals. Moreover, 6.1% of these papers are from 2017, 12.1% are from 2016, and we believe that in 2017 there will have more publications about healthcare application IoT-based than 2016, once we are at the beginning of 2017 and the number of publications of this year is

Table 2. The 46 carefully read studies.

Id	Authors	Year	Venue
S1	Yang et al. [43]	2014	Conference
S2	Jara et al. [23]	2013	Journal
S3	Fan et al. [12]	2014	Conference
S4	Castillejo et al. [6]	2013	Journal
S6	Doukas and Maglogiannis [10]	2012	Conference
S7	Poenaru and Poenaru [32]	2013	Conference
S8	Yang et al. [42]	2013	Conference
S9	Swiatek and Rucinsky [36]	2013	Conference
S10	Lopez et al. [29]	2013	Conference
S11	Trcek and Brodnik [38]	2013	Journal
S12	Hu et al. [21]	2013	Conference
S13	Le Moullec et al. [28]	2014	Conference
S14	Mohammed et al. [31]	2014	Conference
S18	Kevin et al. [24]	2014	Conference
S19	Sebestyen et al. [35]	2014	Conference
S20	Hassan et al. [18]	2014	Conference
S21	Hassan et al. [18]	2014	Conference
S24	Khattak et al. [25]	2014	Conference
S25	Jara et al. [22]	2014	Journal
S26	Tabish et al. [37]	2014	Conference
S27	Gia et al. [16]	2014	Conference
S28	Ray [34]	2015	Conference
S29	Raad [33]	2015	Conference
S30	Gao et al. [13]	2015	Conference
S34	van der Valk et al. [39]	2015	Conference
S35	Maksimović et al. [30]	2015	Conference
S37	Abdullah et al. [41]	2016	Conference
S38	Abawajy and Hassan [1]	2017	Journal
S39	Chen et al. [7]	2017	Journal
S40	Yang et al. [44]	2016	Journal
S41	Archip et al. [4]	2016	Conference
S42	Kodali et al. [27]	2015	Conference
S43	Gia et al. [15]	2015	Conference
S44	Al-Taee et al. [3]	2015	Conference
S45	Datta et al. [9]	2015	Conference
S46	Hossain and Muhammad [20]	2016	Journal

already about 50% of 2016. At the other side, only 3% are from 2012 and we did not find healthcare applications IoT-based before 2012. We believe that this is because of the maturity of IoT area. The Fig. 3 presents this distribution of the selected studies by year.

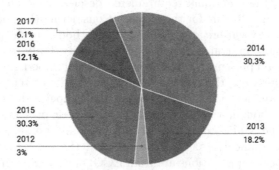

Fig. 3. Distribution of the studies by year.

Regarding the applications described in the studies, 63.7% of the papers do not specify where the healthcare application are deployed. Considering the other 36.3%, 12.1% presents solutions deployed at hospitals and 24.7% deployed at home. Moreover, these studies describe that main characteristics of these healthcare applications are the body and ambient monitoring. From the applications presented, only two studies, S6 and S34, presented the use of IoT Platforms, in this case, the ThingSpeak Platform[2]. Another observation is that seven of them define ontologies, they are S2, S3, S4, S10, S19, S25 and S45. One important point of these applications is that only S1 and S2 present interoperability with other systems, in the case of S1, with medical supply chain, emergency center and hospital, and S2 with clinical devices. So, the consequence of it is that the use of most of the presented healthcare applications in 93% of the selected studies would demand the change of existing systems of the hospitals.

3.3   Quality Evaluation Results

We evaluate using the criteria described in Sect. 2.4. The score of each study is presented in https://goo.gl/skZmns. All disagreements with scores were discussed and resolved. The results show that all studies scored more than 1, and only 7 of them had the maximum score (5): S1, S2, S18, S34, S35, S40 and S41.

4   Discussion

In this section, we discuss the answers to our research questions and then, we present the limitations and conclusions of this review.

---

[2] https://thingspeak.com/.

4.1 What are the Main Characteristics of Healthcare Applications Based on IoT Infrastructure?

Regarding the main characteristics of healthcare application based on IoT infrastructure, we collected their functional and nonfunctional requirements from the studies. Thus, the functional requirements described in the studies are the body and ambient monitoring for the patient. Considering the body monitoring, the data monitored by sensors attached to patient's body are the pulse oximeter, heart rate, galvanic skin, transpiration, muscle activity, body temperature, oxygen saturation, blood pressure, airflow, body movement, blood glucose, breathing rate and ECG. Moreover, the ambient monitoring is about sensors deployed in the patient's environment that capture data from temperature, light, humidity, location, body position, motion data, SPO2, atmospheric pressure and CO2. The Table 3 presents the studies and the patient's body and ambient data captured by the healthcare applications. We can note that the most frequent captured data of the healthcare applications are from ECG, body temperature, heart rate and blood pressure.

About the features of healthcare applications, there are some important nonfunctional requirements that represent a concern in this kind of applications. Thus the nonfunctional requirements cited by the studies are scalability, reliability, ubiquity, portability, interoperability, robustness, performance, availability, privacy, integrity, authentication and security. The Table 4 specify the nonfunctional requirements and the studies. We can note that the most cited nonfunctional requirements are security, interoperability, reliability and privacy.

Finally, we conclude that the main characteristics of healthcare applications in terms of functional requirements are the patient's body and ambient monitoring, with the mainly capture of data from ECG, body temperature, heart rate and blood pressure. With respect to nonfunctional requirements, the most important are security, interoperability, reliability and privacy.

The Fig. 4 presents the word clouds regarding the requirements of IoT-based healthcare applications.

4.2 What are the Patterns and Protocols Used in Healthcare Applications Based on IoT Infrastructure?

With respect to protocols, the collected data of the studies showed that there are two categories of protocols: communication, regarding network protocols, and application, regarding data transfer protocols. Thus, the communication protocols cited by the studies on the healthcare applications are 6LoWPAN, IEEE 802.15.4, Zigbee, Bluetooth, RFID, WIFI, Ethernet, GPRS, IEEE 802.15.6, 3G/4G, NFC and IrDA. Regarding the applications protocols, the studies cited: REST, YOAPY, HTTP, CoAP, XML-RPC and Web Services. The Table 5 presents the communication protocols and the studies. We can note that the most used communication protocols are Bluetooth, WIFI, 6LoWPAN, Zigbee and 3G/4G.

Table 3. Patient's body and ambient data captured by healthcare applications and the studies.

Data	Freq.	Studies
ECG	22	S1, S2, S6, S7, S10, S14, S19, S20, S26, S27, S28, S29, S30, S35, S37, S38, S39, S40, S41, S42, S43, S46
Body temperature	17	S1, S2, S4, S6, S7, S19, S20, S21, S24, S26, S28, S34, S35, S39, S41, S42, S45
Heart rate	11	S2, S4, S6, S18, S21, S24, S27, S29, S34, S39, S46
Blood pressure	9	S2, S8, S19, S20, S22, S28, S29, S35, S44
Oxygen saturation	6	S6, S19, S21, S24, S29, S39
Ambient temperature	5	S6, S8, S21, S30, S34
Body movement	4	S28, S30, S34, S37
SPO2	4	S27, S35, S41, S42
Pulse oximeter	3	S24, S28, S35
Breathing rate	3	S4, S21, S27
Muscle activity	2	S19, S35
Galvanic skin	2	S34, S35
Blood glucose	2	S7, S36
Ambient humidity	2	S8, S21
Airflow	2	S19, S35
Body position	2	S19, S35
Motion data	2	S6, S21
CO2	1	S8
Transpiration	1	S19
Ambient light	1	S30
Location	1	S6
Atmospheric pressure	1	S21

The Table 6 presents the application protocols and the studies. We can note that the most used application protocols are REST, HTTP and CoAP.

About the data format, the studies presented that the healthcare applications use HL7, XML, EHR, CSV, JSON and PHR. The Table 7 presents the data format and the studies. We can note that the most used are JSON, XML, HL7 and EHR.

The Fig. 5 presents word clouds regarding the technologies related to the patterns and protocols of IoT-based healthcare applications.

Table 4. Nonfunctional requirements of healthcare applications and the studies.

NFR	Freq.	Studies
Security	13	S2, S3, S6, S9, S10, S14, S19, S20, S28, S30, S38, S45, S46
Interoperability	10	S2, S3, S4, S6, S24, S27, S35, S38, S43, S45
Reliability	8	S2, S9, S20, S27, S30, S35, S37, S40
Privacy	8	S2, S6, S14, S19, S20, S28, S35, S38
Scalability	6	S2, S6, S14, S39, S44, S46
Availability	4	S2, S6, S9, S45
Performance	2	S14, S20
Authentication	2	S35, S46
Ubiquity	1	S10
Portability	1	S14
Robustness	1	S2
Integrity	1	S35

Fig. 4. Word clouds of the requirements of healthcare applications IoT-based.

4.3   What are the Challenges and Opportunities Related to Healthcare Applications Based on IoT Infrastructure?

The studies presented that are many challenges related to healthcare applications based on IoT infrastructure. In S6, the authors presented that health information management through mobile devices introduces several challenges: data storage and management (e.g., physical storage issues, availability and maintenance), interoperability and availability of heterogeneous resources, security and privacy (e.g., permission control, data anonymity, etc.), unified and ubiquitous access are a few to mention. Thus, according to S6, the vast amount of sensor data generated by the capture of these applications need to be managed properly for further analysis and processing. Another challenge regarding the data is the unstructured format of it, that, according to S14, the huge volume of data produced by the sensors is in an unstructured format, which is very complex to understand and requires different data storage mechanisms than the typical database management system (DBMS).

Table 5. Communication protocols of healthcare applications and the studies.

Com. protocols	Freq.	Studies
Bluetooth	19	S1, S2, S4, S6, S10, S14, S18, S19, S20, S28, S34, S35, S38, S39, S40, S43, S44, S45, S46
WIFI	17	S1, S3, S6, S19, S21, S29, S31, S34, S35, S37, S38, S39, S40, S43, S44, S45, S46
6LoWPAN	11	S2, S10, S17, S21, S24, S25, S26, S27, S28, S30, S43
Zigbee	11	S1, S2, S4, S8, S18, S28, S35, S40, S42, S43, S45
3G/4G	10	S1, S7, S20, S26, S31, S35, S37, S38, S40, S44
RFID	7	S1, S2, S3, S18, S25, S29, S37
IEEE 802.15.4	6	S4, S7, S9, S26, S35, S41
GPRS	3	S21, S35, S40
NFC	2	S10, S25
Ethernet	2	S1, S21
IEEE 802.15.6	1	S7
IrDA	1	S25

Table 6. Application protocols of healthcare applications and the studies.

Com. protocols	Freq.	Studies
REST	7	S4, S6, S14, S20, S35, S41, S45
HTTP	5	S34, S35, S40, S44, S45
CoAP	4	S2, S24, S30, S45
Web services	2	S27, S46
YOAPY	1	S2
XML-RPC	1	S9

Table 7. Data format and the studies.

Data format	Freq.	Studies
JSON	9	S4, S6, S18, S24, S34, S35, S41, S44, S45
XML	6	S6, S8, S18, S19, S27, S45
HL7	3	S2, S8, S24
EHR	3	S2, S25, S45
CSV	2	S6, S34
PHR	1	S25
HTML	1	S40

Fig. 5. Word clouds of technologies related to the patterns and protocols of IoT-based healthcare applications.

Still about challenges, in S18, the authors highlight that the existing home healthcare systems have drawbacks such as simple and few functionalities, weak interaction and poor mobility, and IoT is considered an effective method for healthcare monitoring system of the disabled and elderly people by the people-object interaction. Moreover, the authors in S18, describe that their future work is focused on the wireless body area networks combined with social networks, exploring the mobility impaired healthcare services based on social networking, and sharing the information of smart object more security and accuracy.

The authors in S19 describe an interoperability, political and administrative challenges, since the communication protocol of the devices is not open and a given device cannot be integrated in other (or multiple) applications. Moreover, according to S19, the implementation of these applications is a technical as well as a political and administrative challenge, once it implies not only in a technical infrastructure but also a number of regulatory measures, such as standards, regulations and institutional reorganization. Any regional or national implementation of such system must fulfill not only quality and safety requirements but also economical efficiency conditions.

In S20, the authors present the need for the development of new protocols that are reliable and energy efficient in data transmission, since routing protocols are critical for the system to work efficiently. In addition, they say that even though several protocols have been proposed for various domains, none of them has been accepted as a standard, and with the growing number of things, further research is required. Still, in S20, the authors also describe the need for the development of efficient data mining techniques for extracting useful knowledge from IoT data. Moreover, sometimes IoT-generated data are not always ready for direct consumption using visualization platforms and, therefore, new visualization schemes need to be developed. Another key challenge described by the authors in S20 regards the need to protect privacy information. They say that more innovative solutions need to be developed in privacy and security aspects.

The authors in S24 highlight the interoperability challenge, once there have been different studies and proposals for patient monitoring at hospital or at home for personal monitoring, a shared goal to produce an interoperable system adopting open standards for healthcare, for example HL7, and a seamless framework to be easily deployed in any given scenario for healthcare is still missing.

4.4   Limitations of This Review

The main limitation of this review is on the bias in the selection of publications and inaccuracy in data extraction. We strict follow the defined protocol, described in Sect. 2, to ensure that the selection process was unbiased. Another limitation is the used search string, described in Sect. 2.2, that although it was defined guided by the research questions, that there is a risk that some studies were omitted. The final limitation of this review is that we used Scopus from Elsevier to proceed with the search of the studies and, although it indexes others scientific repositories, inclusion in Scopus once a paper has been published takes some time, and so there is a risk that some already published studies were not yet included.

5   Conclusion and Future Works

This review was made aiming to comprehend the current state and future trends of healthcare applications based on IoT infrastructure, and also in order to find areas regarding it for further investigations. We started this review defining the method with research questions, search process, quality assessments and data collection. Then, we performed the search using the defined search string at Scopus from Elsevier (stage 1), resulting in 1355 studies. After this search, we performed the analyses of the titles and abstract of the studies (stage 2). Then, 46 studies remained in this review and they were carefully read (stage 3). For these 46 selected studies, we evaluate them according the quality assessment and 7 of them had the maximum score. From these 46 selected studies, 33 studies were useful to answer the research questions.

Using the extraction data, we were able to answer the research questions and provide the characteristics of healthcare applications based on IoT infrastructure (Sect. 4.1). We also described the protocols and data formats used in the studies (Sect. 4.2). Moreover, using the extract data from studies, we were able to find some challenges and opportunities for healthcare applications (Sect. 4.3). The challenges are related to the development of new solutions to resolve interoperability problems, data mining techniques for extraction of knowledge for IoT data, privacy and security problems. Regarding opportunities, there is an industry opportunity for companies that develop IoT-based healthcare applications, since healthcare industry is estimated to be more than $2 trillion by 2020 with an annual consumer market for remote/mobile monitoring devices at $40 billion globally [32].

Finally, with this review we were able to define a layered architecture for healthcare applications based on IoT Infrastructure. It considers the characteristics of these applications, functional and nonfunctional requirements, used protocols and patterns, and is composed of a layer of patients, monitoring, requirements, communication, middleware, systems and services, and users. As future works, we will present and improve this architecture, and it will be used for the development of a platform for remote health monitoring that will address issues like security and interoperability.

References

1. Abawajy, J.H., Hassan, M.M.: Federated internet of things and cloud computing pervasive patient health monitoring system. IEEE Commun. Mag. **55**(1), 48–53 (2017)
2. Al-Fuqaha, A., Guizani, M., Mohammadi, M., Aledhari, M., Ayyash, M.: Internet of things: a survey on enabling technologies, protocols, and applications. IEEE Commun. Surv. Tutor. **17**(4), 2347–2376 (2015)
3. Al-Taee, M.A., Al-Nuaimy, W., Al-Ataby, A., Muhsin, Z.J., Abood, S.N.: Mobile health platform for diabetes management based on the internet-of-things. In: 2015 IEEE Jordan Conference on Applied Electrical Engineering and Computing Technologies (AEECT), pp. 1–5. IEEE (2015)
4. Archip, A., Botezatu, N., Şerban, E., Herghelegiu, P.-C., Zală, A.: An IoT based system for remote patient monitoring. In: 2016 17th International Carpathian Control Conference (ICCC), pp. 1–6. IEEE (2016)
5. Atzori, L., Iera, A., Morabito, G.: The internet of things: a survey. Comput. Netw. **54**(15), 2787–2805 (2010)
6. Castillejo, P., Martinez, J.-F., Rodriguez-Molina, J., Cuerva, A.: Integration of wearable devices in a wireless sensor network for an e-health application. IEEE Wirel. Commun. **20**(4), 38–49 (2013)
7. Chen, M., Ma, Y., Li, Y., Wu, D., Zhang, Y., Youn, C.-H.: Wearable 2.0: enabling human-cloud integration in next generation healthcare systems. IEEE Commun. Mag. **55**(1), 54–61 (2017)
8. Chen, Y.: Analyzing and visual programming internet of things and autonomous decentralized systems (2016)
9. Datta, S.K., Bonnet, C., Gyrard, A., Da Costa, R.P.F., Boudaoud, K.: Applying internet of things for personalized healthcare in smart homes. In: 2015 24th Wireless and Optical Communication Conference (WOCC), pp. 164–169. IEEE (2015)
10. Doukas, C., Maglogiannis, I.: Bringing IoT and cloud computing towards pervasive healthcare. In: 2012 Sixth International Conference on Innovative Mobile and Internet Services in Ubiquitous Computing (IMIS), pp. 922–926. IEEE (2012)
11. Dybå, T., Dingsøyr, T.: Empirical studies of agile software development: a systematic review. Inf. Softw. Technol. **50**(9–10), 833–859 (2008)
12. Fan, Y.J., Yin, Y.H., Da Xu, L., Zeng, Y., Wu, F.: IoT-based smart rehabilitation system. IEEE Trans. Ind. Inf. **10**(2), 1568–1577 (2014)
13. Gao, R., Zhao, M., Qiu, Z., Yu, Y., Chang, C.H.: Web-based motion detection system for health care. In: 2015 IEEE/ACIS 14th International Conference on Computer and Information Science (ICIS), pp. 65–70. IEEE (2015)
14. Gartner, I.: Gartner says 6.4 billion connected "things" will be in use in 2016, up. 30 percent from 2015 (2015)
15. Gia, T.N., Jiang, M., Rahmani, A.M., Westerlund, T., Liljeberg, P., Tenhunen, H.: Fog computing in healthcare internet of things: a case study on ECG feature extraction. In: 2015 IEEE International Conference on Computer and Information Technology; Ubiquitous Computing and Communications; Dependable, Autonomic and Secure Computing; Pervasive Intelligence and Computing (CIT/IUCC/DASC/PICOM), pp. 356–363. IEEE (2015)
16. Gia, T.N., Thanigaivelan, N.K., Rahmani, A.M., Westerlund, T., Liljeberg, P., Tenhunen, H.: Customizing 6LoWPAN networks towards internet-of-things based ubiquitous healthcare systems. In: NORCHIP 2014, pp. 1–6. IEEE (2014)

17. Gubbi, J., Buyya, R., Marusic, S., Palaniswami, M.: Internet of things (IoT): a vision, architectural elements, and future directions. Future Gener. Comput. Syst. **29**(7), 1645–1660 (2013)
18. Hassan, M.M., Albakr, H.S., Al-Dossari, H.: A cloud-assisted internet of things framework for pervasive healthcare in smart city environment. In: Proceedings of the 1st International Workshop on Emerging Multimedia Applications and Services for Smart Cities, pp. 9–13. ACM (2014)
19. Hochron, S., Goldberg, P.: Driving physician adoption of mheath solutions. Healthc. Financ. Manag. **69**(2), 36–40 (2015)
20. Hossain, M.S., Muhammad, G.: Cloud-assisted industrial internet of things (IIoT)-enabled framework for health monitoring. Comput. Netw. **101**, 192–202 (2016)
21. Hu, F., Xie, D., Shen, S.: On the application of the internet of things in the field of medical and health care. In: Green Computing and Communications (Green-Com), 2013 IEEE and Internet of Things (iThings/CPSCom), IEEE International Conference on and IEEE Cyber, Physical and Social Computing, pp. 2053–2058. IEEE (2013)
22. Jara, A.J., Zamora, M.A., Skarmeta, A.F.: Drug identification and interaction checker based on IoT to minimize adverse drug reactions and improve drug compliance. Pers. Ubiquit. Comput. **18**(1), 5–17 (2014)
23. Jara, A.J., Zamora-Izquierdo, M.A., Skarmeta, A.F.: Interconnection framework for mHealth and remote monitoring based on the internet of things. IEEE J. Sel. Areas Commun. **31**(9), 47–65 (2013)
24. Kevin, I., Wang, K., Rajamohan, A., Dubey, S., Catapang, S.A., Salcic, Z.: A wearable internet of things mote with bare metal 6LoWPAN protocol for pervasive healthcare. In: Ubiquitous Intelligence and Computing, 2014 IEEE 11th International Conference on and Autonomic and Trusted Computing, and IEEE 14th International Conference on Scalable Computing and Communications and Its Associated Workshops (UTC-ATC-ScalCom), pp. 750–756. IEEE (2014)
25. Khattak, H.A., Ruta, M., Di Sciascio, E.: CoAP-based healthcare sensor networks: a survey. In: Proceedings of 2014 11th International Bhurban Conference on Applied Sciences and Technology (IBCAST), Islamabad, Pakistan, 14th–18th January 2014, pp. 499–503. IEEE (2014)
26. Kitchenham, B., Charters, S.: Guidelines for performing systematic literature reviews in software engineering. Technical report EBSE 2007–001, Keele University and Durham University Joint Report (2007)
27. Kodali, R.K., Swamy, G., Lakshmi, B.: An implementation of IoT for healthcare. In: 2015 IEEE Recent Advances in Intelligent Computational Systems (RAICS), pp. 411–416. IEEE (2015)
28. Le Moullec, Y., Lecat, Y., Annus, P., Land, R., Kuusik, A., Reidla, M., Hollstein, T., Reinsalu, U., Tammemäe, K., Ruberg, P.: A modular 6LoWPAN-based wireless sensor body area network for health-monitoring applications. In: 2014 Annual Summit and Conference Asia-Pacific Signal and Information Processing Association, (APSIPA), pp. 1–4. IEEE (2014)
29. López, P., Fernández, D., Jara, A.J., Skarmeta, A.F.: Survey of internet of things technologies for clinical environments. In: 2013 27th International Conference on Advanced Information Networking and Applications Workshops (WAINA), pp. 1349–1354. IEEE (2013)
30. Maksimović, M., Vujović, V., Perišić, B.: A custom internet of things healthcare system. In: 2015 10th Iberian Conference on Information Systems and Technologies (CISTI), pp. 1–6. IEEE (2015)

31. Mohammed, J., Lung, C.-H., Ocneanu, A., Thakral, A., Jones, C., Adler, A.: Internet of things: remote patient monitoring using web services and cloud computing. In: 2014 IEEE International Conference on Internet of Things (iThings), pp. 256–263. IEEE (2014)

32. Poenaru, E., Poenaru, C.: A structured approach of the Internet-of-Things eHealth use cases. In: E-Health and Bioengineering Conference (EHB), pp. 1–4. IEEE (2013)

33. Raad, M.W., Sheltami, T., Shakshuki, E.: Ubiquitous tele-health system for elderly patients with Alzheimer's. Procedia Comput. Sci. **52**, 685–689 (2015)

34. Ray, P.P.: Internet of things for sports (IpTSport): an architectural framework for sports and recreational activity. In: 2015 International Conference on Electrical, Electronics, Signals, Communication and Optimization (EESCO), pp. 1–4. IEEE (2015)

35. Sebestyen, G., Hangan, A., Oniga, S., Gál, Z.: eHealth solutions in the context of internet of things. In: Proceedings of IEEE International Conference on Automation, Quality and Testing, Robotics (AQTR 2014), Cluj-Napoca, Romania, pp. 261–267 (2014)

36. Swiatek, P., Rucinski, A.: IoT as a service system for eHealth. In: 2013 IEEE 15th International Conference on e-Health Networking, Applications & Services (Healthcom), pp. 81–84. IEEE (2013)

37. Tabish, R., Ghaleb, A.M., Hussein, R., Touati, F., Mnaouer, A.B., Khriji, L., Rasid, M.F.A.: A 3G/WiFi-enabled 6LoWPAN-based u-healthcare system for ubiquitous real-time monitoring and data logging. In: 2nd Middle East Conference on Biomedical Engineering, pp. 277–280. IEEE (2014)

38. Trcek, D., Brodnik, A.: Hard and soft security provisioning for computationally weak pervasive computing systems in e-health. IEEE Wirel. Commun. **20**(4), 22–29 (2013)

39. van der Valk, S., Myers, T., Atkinson, I., Mohring, K.: Sensor networks in workplaces: correlating comfort and productivity. In: 2015 IEEE Tenth International Conference on Intelligent Sensors, Sensor Networks and Information Processing (ISSNIP), pp. 1–6. IEEE (2015)

40. Wohlin, C., Runeson, P., Höst, M., Ohlsson, M.C., Regnell, B., Wesslén, A.: Experimentation in Software Engineering. Springer Science & Business Media, Heidelberg (2012). doi:10.1007/978-3-642-29044-2

41. Yaakob, N., Badlishah, R., Amir, A., binti Yah, S.A., et al.: On the effectiveness of congestion control mechanisms for remote healthcare monitoring system in IoT environmenta review. In: 2016 3rd International Conference on Electronic Design (ICED), pp. 348–353. IEEE (2016)

42. Yang, C.T., Liu, J.C., Liao, C.J., Wu, C.C., Le, F.Y.: On construction of an intelligent environmental monitoring system for healthcare. In: 2013 International Conference on Parallel and Distributed Computing, Applications and Technologies, pp. 246–253. IEEE (2013)

43. Yang, G., Xie, L., Mantysalo, M., Zhou, X., Pang, Z., Da Li, X., Kao-Walter, S., Chen, Q., Zheng, L.-R.: A health-IoT platform based on the integration of intelligent packaging, unobtrusive bio-sensor, and intelligent medicine box. IEEE Trans. Ind. Inf. **10**(4), 2180–2191 (2014)

44. Yang, Z., Zhou, Q., Lei, L., Zheng, K., Xiang, W.: An IoT-cloud based wearable ECG monitoring system for smart healthcare. J. Med. Syst. **40**(12), 286 (2016)

Fast Semi-blind Color Image Watermarking Scheme Using DWT and Extreme Learning Machine

Anurag Mishra[1], Ankit Rajpal[2(✉)], and Rajni Bala[3]

[1] Department of Electronics, Deen Dayal Upadhyaya College,
University of Delhi, Delhi, India
anurag_cse2003@yahoo.com
[2] Department of Computer Science, University of Delhi, Delhi, India
ankit.cs.du@gmail.com
[3] Department of Computer Science, Deen Dayal Upadhyaya College,
University of Delhi, Delhi, India
r_dagar@yahoo.com

Abstract. In this paper, a semi-blind watermarking of digital image based on Extreme Learning Machine (ELM) in DWT domain is proposed. The fourth level (LL4) low frequency sub-band coefficients are used for embedding the watermark. The machine is tuned iteratively and used for training and predicting the sub-band coefficients. The target fourth level sub-band coefficients are augmented by the quantized fourth level sub-band coefficients which are set as an input data-set to train the machine. A random key determines the starting position of the coefficients where the watermark is embedded. A binary watermark is embedded in the blue channel of four colored host images. This watermarking scheme strengthen the robustness towards popular interferences on images. The results of simulation clearly proves that the recovered watermark from signed and attacked images are similar to the embedded watermark. The time spans for training, embedding and extraction are computed and they show real time behavior (millisecond time spans) so that the proposed scheme is suitable for developing real time watermarking applications.

1 Introduction

Digital Image watermarking is mainly known to prevent copyright violation and piracy. The accelerated proliferation of internet and multimedia transmission has increased the attention paid to protecting copyright on multimedia content [1].

The watermark is broadly detected in three ways: Non-blind, Blind and Semi-blind. Non-blind/informed watermarking technique requires the host image to detect the watermark. A blind method does not require the host image to detect watermark. Semi-blind watermarking technique requires the key or the evolved model of the machine used and the signed image for detection [2,3].

© Springer International Publishing AG 2017
O. Gervasi et al. (Eds.): ICCSA 2017, Part VI, LNCS 10409, pp. 63–78, 2017.
DOI: 10.1007/978-3-319-62407-5_5

The present research in the area of image watermarking is focused towards optimization of twin criteria:- (1) The visual quality of the signed and attacked images, and (2) The robustness of the embedding scheme which is evaluated after implementing certain image processing attacks over the signed content. The visual quality is usually determined by computing the Peak Signal-to-Noise (PSNR) and Structural Similarity Index (SSIM) [4] parameters while semi-blind or blind extraction is carried out to examine the issue of robustness. The robustness assumes even more importance because this is the parameter which is used to ascertain the ownership verification and content authentication. It is found that if any watermarking scheme is biased towards robustness then it shall be at the expense of visual quality and vice-versa. Therefore, it becomes even more important to balance out these parameters. Historically, these two parameters are being optimized by extensive use of soft computing techniques. This is particularly so because these techniques are conventionally used to train the dataset pertaining to different problems and applications. The training usually yields optimal or sub-optimal solutions which are found to be helpful to reach specific conclusions. These techniques include different types of neural networks, Fuzzy Inference Systems (FIS) [5], Support Vector Machines (SVMs) [6] of different configurations, and meta-heuristic techniques [7]. The neural networks which are used for this purpose include Back-propagation Networks (BPN) [8], Radial-Basis Function Networks (RBF) [9] and several variants of Single-Layer Feed-Forward Neural Networks (SLFNs). Among these configurations, the BPN are found to be very slow especially to large datasets. Sometimes, they are found to be trapped within the local minima and its training does not complete. Therefore, these networks are rarely suitable for the design of security applications such as image watermarking. On the contrary, the RBF networks are fast and can be trained in an efficient manner as compared to their BPN counterparts. This is generally true for all applications including those based on image processing techniques. The Fuzzy Inference System (FIS) based techniques are not adaptive in nature. But, the use of fuzzy inference rules by these systems makes it more close to natural imitation of the processes involved in embedding of watermark and its extraction from images. In this case, more specifically, the visual quality of the signed and attacked images is found to be better as compared to their neural network counterparts. The issue of robustness using different FIS has also been examined. Both the neural network and FIS based watermarking schemes have their inherent limitations. These limitations are minimized by amalgamating neural networks and FIS to obtain it as a neuro-fuzzy architecture based scheme expected to give the best of both [10]. In any case, all these watermarking schemes are found to be intensive in terms of cost and time.

The third set of techniques is based on Support Vector Regression (SVR). These techniques are powerful techniques which show good generalization capabilities. The basic architecture of the SVM is found to be slow for image watermarking applications. It consumes a time span of the order of few minutes to carryout embedding of a small size watermark in a regular gray scale or color image. Their faster variants are also available. These include Finite Newton

Support Vector Regression (FNSVR) [11] and Least Square Support Vector Regression (LSSVR) [12]. The FNSVR is very fast and stable algorithm which converges in few iteration while the LSSVR is not as fast as FNSVR but it is found to be faster than the regular SVR.

Extreme Learning Machine (ELM) proposed by Huang et al. [13,14] is a fast algorithm for training Artificial Neural Network (ANN). ELM uses Single hidden Layer Feed forward Neural Network (SLFN) architecture in which the number of hidden neurons is to be set by human. The training of this machine is reported to have been completed within milliseconds with a reasonably good accuracy [14]. The conventional neural networks based on Back Propagation (BP) may take more time in manually tune the control parameters such as like stopping criteria, learning rate, learning epochs etc. and they sometimes get stuck in local minimum. The basic idea of ELM is that the input weights (connecting the input layer to the hidden layer) and the input bias of the hidden layer are randomly chosen and the output weights (connecting the hidden layer to the output layer) are then analytically computed based on Moore-Penrose generalized pseudo inverse [15]. Unlike other learning methods, ELM is very well suited for both differential and non-differential activation functions.

ELM and its variants like Online Sequential ELM (OS-ELM) and Weighted ELM (W-ELM) have been used for image watermarking as well [16–22]. However, a detailed analysis on color images, especially to examine the issue of robustness of the embedding scheme is lacking. The proposed work intends to fill this gap.

This research paper is organized as follows. Section 2 gives research contribution and motivation. Section 3 provides the mathematical modeling of ELM. The experimental details of the proposed watermarking algorithm is given in Sect. 4. The observed results and their analysis are discussed in Sect. 5. Finally, the paper is concluded in Sect. 6.

2 Research Contribution and Motivation

Different applications such as time series prediction, fault section identification and few other regression problems have proved that the Extreme Learning machine (ELM) is quite suitable for faster training and processing of the data-set. Usually, it finishes the training in millisecond range. Motivated by these outcomes, it was felt that this algorithm must find good image processing applications as well. Moreover, the commercial security related application such as image watermarking is a good choice on which the performance of this neural network can be tested. The proposed research work is oriented towards optimizing the trade-off between visual quality of signed/attacked images and robustness of the embedding scheme. In this paper, we propose a novel semi-blind color image watermarking scheme using the hybrid DWT-ELM architecture. One binary watermark is embedded and extracted from the blue channel of four different color host images of size 512×512. These images are Airplane, Lena, Peppers and Tiffany. It has now been established that a DWT or other wavelet transform based watermarking scheme yields better results in terms of embedding parameters such as visual quality of watermarks within the watermarked images. Moreover, the ELM algorithm has been used to eliminate time lapses to train the neural network. The performance of the ELM is well stated in the literature in comparison to that of other gradient descent based neural architectures such as BPN or RBF networks. However, this is true for few applications other than image watermarking. In the present case, we obtain the processing time spans in the order of millisecond. This makes this scheme particularly suitable for developing real time image processing applications. In our opinion, the natural corollary of the proposed work should be to extend this scheme to moving multimedia data such as audio and video both in compressed and uncompressed form. We further carry out multiple image processing attacks over signed host signals to examine the issue of robustness of the embedding scheme. These attacks are described in detail in Sect. 4. The image quality assessment is carried out by using two full reference metrics - PSNR and SSIM. The robustness of the embedding scheme is evaluated by computing the Normalized Correlation (NC) and the Bit-Error-Rate (BER). The embedding strength or the scaling factor is also optimized with respect to all four testing parameters - PSNR, SSIM, NC and BER. The watermarks have been extracted by using the original ELM model evolved during the embedding process. But, only the signed or the attacked images are used to recover the binary watermark without any reference to the original host image. Thus, the watermark recovery is carried out in a semi-blind manner. It is found that in the proposed DWT-ELM based watermarking scheme the embedding and extraction processes are well optimized and it is robust enough against the selected attacks. The time computation results clearly indicate that the scheme is quite suitable for real time running multimedia processing applications in general and the video watermarking in particular.

3   Extreme Learning Machine

Given any data-set with N training samples $(x_i, t_i)_{i=1,2,...,N}$, where $x_i \in \mathbf{R}^d$ and $t_i \in \mathbf{R}^m$, the neural network with single hidden layers having L hidden nodes generates an output which can be represented as:

$$o_i = \sum_{j=1}^{L} \beta_j g(a_j, b_j, x_i) = \sum_{j=1}^{L} \beta_j h_{ij}, i = 1, 2, \ldots, N \tag{1}$$

where $a_j \in \mathbf{R}^d$ and $b_j \in \mathbf{R}$ (j = 1, 2, ..., L) are learning parameters of the jth hidden node, respectively. $\beta_j \in \mathbf{R}^m$ is the link connecting the jth hidden node to the output node. In Eq. (1), $g(a_j, b_j, x_i)$ is the output of the jth hidden node corresponding to the input sample x_i. Huang et al. proves that SLFNs with L hidden nodes using an activation function g(.) is capable to approximate these N examples with zero error. This means that $\sum_{i=1}^{N} \|o_i - t_i\| = 0$, i.e. there exist β_j, a_j and b_j such that

$$H\beta = T \tag{2}$$

where $H = \{h_{ij}\}$ is the hidden-layer output matrix and $h_{ij} = g(a_j, b_j, x_i).\beta = (\beta_1\beta_2\ldots\beta_L)$ is the output weight matrix and $T = (t_1t_2\ldots t_N)$ is the desired output.

In case of ELM, the input weights and biases need not to be tuned by human and due to this feature, the computation of hidden-layer output matrix and the output weights becomes much easier. The computational cost of the construction of the network is very low which is possible because of the requirement of very few epochs by ELM.

Given the constraint of minimum norm least square, i.e. $\min\|\beta\|$ and $\min\|H\beta - T\|$, Huang et al. [14] has proposed a simple representation of the solution of the system of equations (Eq. 2) as:

$$\hat{\beta} = H^\dagger T \qquad (3)$$

where $H^\dagger$ is the Moore-Penrose generalized inverse [15] of the hidden-layer output matrix H. The brief algorithm of ELM algorithm is as follows.

Algorithm. ELM

Given the training set $(x_i, t_i), x_i \in \mathbf{R}^d, t_i \in \mathbf{R}^m$, an activation function
g: $\mathbf{R} \to \mathbf{R}$, and the number of hidden nodes L:
1. The hidden node parameters $(a_j, b_j), j = 1, 2, \ldots, L$ are assigned randomly;
2. Calculation of the hidden-layer output matrix H;
3. Calculation of output weight vector $\beta = H^\dagger T$

---

4   Experimental Details

The blue channels of four different standard host images of size 512×512 Airplane, Lena, Peppers and Tiffany are used to carry out watermark embedding in the proposed scheme in a semi-blind manner. For this purpose, quantized coefficients of the LL4 sub-band of the blue channel of the concerned image are used to carry out embedding and extraction processes. Different research groups [1,16,21] have concluded that the robust watermark embedding is practically possible if the embedding process is carried out either primarily in the low frequency band coefficients or to some extent in mid-frequency band coefficients in the transform domain. In the present work, the ELM is trained with the LL4 sub-band coefficients. A random key is used to determine the initial location of watermark embedding. The input data-set is of size 1024×2 while it produces an output sequence of size 1024×1 whose coefficients are close to the desired LL4 sub-band coefficients. A binary image of size 32×32 as watermark is tested in this experimental work.

4.1   Watermark Embedding Algorithm

The pseudo-code of the watermark embedding is given in Listing 1 below.

Listing 1. Embedding

Step 1. Apply 4-level DWT to transform the blue channel of the host image (512×512 size). Reshape LL4 (32×32 size matrix) to 1024×1 and set C_i to it.

Step 2. Perform quantization on C_i using Q. Augment C_i with its quantized values to construct a data-set of size 1024×2 to be supplied to the ELM. ELM constructs the model and predicts the output of size 1024×1 as given by Eq. (4):

$$P_i^{'} = ELM \left(Round \left(\frac{C_i}{Q} \right) \right) \quad (4)$$

This output is close to the desired output included in the data-set used to train the ELM. The optimized numerical value of Q is 32 for all practical computations.

Step 3. A random key is chosen to select the starting location of watermark embedding.

Step 4. Embed the watermark according to the Eq. (5) which uses the output generated after Step 2 ($P_i^{'}$):

$$C_{i+key}^{'} = P_{i+key}^{'} + K \times W_i \quad (5)$$

where W_i is the binary watermark, K is the scaling factor and $C_i^{'}$ are the modified LL4 sub-band coefficients obtained using Eq. (5).

Step 5. The signed image is generated by performing Inverse DWT.

The signed image is tested for visual quality by measuring the two full reference metrics:- Peak Signal-to-Noise Ratio (PSNR) and Structural Similarity (SSIM) [4]. Peak Signal-to-Noise Ratio (PSNR) in dB is one full reference metric to evaluate image quality of the signed image. This is given by Eq. (6).

$$PSNR = 10 log_{10} \frac{S^2}{MSE} \quad (6)$$

Among this, S is the luminance of image and MSE is the mean square error between signed and the host image.

Structural Similarity Index (SSIM) is another full reference image quality assessment metric which measures the structural distortion. The formulation is given in Eq. (7).

$$SSIM(x, y) = \frac{(2\mu_x \mu_y + C_1)(2\sigma_{xy} + C_2)}{(\mu_x^2 + \mu_y^2 + C_1)(\sigma_x^2 + \sigma_y^2 + C_2)} \quad (7)$$

In Eq. (7), μ_x and μ_y are mean intensity or luminance component of image signals x and y respectively. C_1 and C_2 are constants. Similarly, σ_x^2, σ_y^2 are variance of image signals x and y respectively, and σ_{xy}^2 is cross-covariance for images x, y.

4.2   Watermark Extraction Algorithm

The semi-blind extraction procedure to extract the watermark from the signed and attacked images is given in Listing 2 below.

Listing 2. Extraction

Step 1. Apply 4-level DWT Transform to the signed image of 512×512 size. Reshape the obtained LL4 (32×32 size) sub-band coefficients to 1024×1 and set C_i'' to it.

Step 2. Using Q, perform the quantization of C_i'', and use ELM prediction module to obtain the output from the already trained ELM model:

$$P_i'' = ELM\left(Round\left(\frac{C_i''}{Q}\right)\right) \quad (8)$$

Step 3. Recover the watermark W_i' using the Eq. (9) below, based on the output of the ELM in Eq. (8) and C_i'' and the secret key.

$$W_i' = \left(P_i'' - C_i''\right) \times (1/K) \quad (9)$$

where K is the scaling factor or embedding strength.

The normalized correlation NC (W,W') is used to compute the degree of similarity between the embedded and the recovered watermarks. The formulation of NC is given in Eq. (10).

$$NC(W, W') = \frac{\sum_{i=1}^{n} W(i)W'(i)}{\sqrt{\sum_{i=1}^{n} W(i)^2}\sqrt{\sum_{i=1}^{n} W'(i)^2}} \quad (10)$$

In Eq. (10), W is the watermark with dimensions $n \times n$. The value of NC varies between 0 and 1. If NC = 1, it indicates complete match between the embedded and recovered watermarks. Another unit-less parameter known as Bit Error Rate (BER) or Bit-Error Ratio which approximately estimates the bit error probability between two watermarks. The formulation is depicted in Eq. (11):

$$BER = \sum_{t=1}^{n \times n} \frac{W_t \oplus W_t'}{n \times n} \quad (11)$$

Where W denotes the embedded watermark and W' denotes the extracted watermark. Both W and W' are of size $n \times n$.

The signed images are also subject to selected image processing interferences to verify the issue of robustness. These attacks are:- (1) Median Filter (aperture = 3.0 and 5.0), (2) Scaling (re-sized to half and then restored to original size), (3) Cropping (Quarter of the watermarked image is cropped and the missing portion is filled with host image) (4) Salt and Pepper Noise (noise amount = 0.1%), (5) 5% Gaussian Noise addition, (6) JPEG compression (Q = 50, 75, 90) and (7) SPIHT Compression (rate = 0.6 bpp, 0.8 bpp and 1.0 bpp). The semi-blind extraction is performed in order to extract watermarks from the signed images before and after executing image processing interferences. The same key and the same ELM model are used during embedding and extraction. The semi-blind extraction is performed which means only the signed or the attacked image (as the case may be) is utilized to recover the watermark by predicting output using the already trained ELM. A comprehensive analysis of the results obtained in this simulation is given in Sect. 5.

5   Results and Discussion

5.1   Embedding and Extraction

Figure 1(a–d) shows four color host images respectively Airplane, Lena, Peppers and Tiffany. Figure 2 shows the binary watermark of size 32×32. This signed images depicted in Fig. 3(a–d) are obtained after embedding the binary watermark within host images based on Eq. (5). Their respective PSNR and SSIM values are PSNR $= 47.03$ dB and SSIM $= 0.9945$, PSNR $= 47.15$ dB and SSIM $= 0.9995$, PSNR $= 47.18$ dB and SSIM $= 0.9995$, PSNR $= 47.12$ dB and SSIM $= 0.9992$. High computed PSNR and SSIM values near unity indicate that the visual quality of these images is very good. Figure 1(a–d) depicts the binary watermarks extracted from color signed images.

(a) Airplane   (b) Lena   (c) Peppers   (d) Tiffany

Fig. 1. Original color host images - (a) Airplane (b) Lena (c) Peppers (d) Tiffany (Color figure online)

Fig. 2. Binary watermark

(a) Airplane   (b) Lena   (c) Peppers   (d) Tiffany

Fig. 3. Signed colored images: (a) Airplane (b) Lena (c) Peppers (d) Tiffany (Color figure online)

(a)   (b)   (c)   (d)

Fig. 4. Extracted binary watermarks (NC = 1.00 and BER = 0.0938 in each case)

Several numerical values of K are tested in our simulation. The best results which optimize visual quality and robustness are obtained for $K = 0.17$ (Fig. 5). Therefore, $K = 0.17$ is used throughout the experiment. The PSNR and SSIM values computed for the signed images are sufficiently high which indicates that the visual quality of these images is good. Moreover, the NC and BER values are respectively close to 1 and 0 which indicates that the watermark recovery from signed images is also successful. It is very clear from the plot depicted in Fig. 1 that PSNR continues to fall with an increase in numerical value of K. The SSIM remains constant throughout.

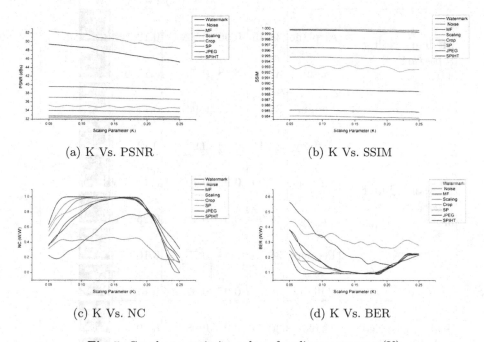

Fig. 5. Graphs to optimize value of scaling parameter (K)

However, the most glaring behavior with respect to K is shown by Normalized correlation coefficient (NC) and the Bit Error Rate (BER) shown in part (c) and (d) respectively. Both these parameters tend to change at around $K = 0.175$. Therefore, these plots taken in conjunction with those in (a) and (b) suggest that $K = 0.17$ is the optimum value of the scaling parameter.

5.2 Executing Image Processing Attacks

The robustness of our watermarking technique is verified using seven different image processing interferences are carried out on the four signed images of Fig. 3(a–d). These attacks are namely: (1) Median Filter (aperture = 3.0 and 5.0), (2) Scaling (re-sized to half and then restored to original size), (3) Cropping (Quarter of the signed image is cropped and the missing portion is filled with host image, (4) Salt and Pepper Noise (noise amount = 0.1%), (5) 5% Gaussian Noise addition, (6) JPEG compression (Q = 90, 75, 50) and (7) SPIHT Compression (rate = 0.6 bpp, 0.8 bpp and 1.0 bpp). The computed values of PSNR, SSIM and NC (W, W') and BER (W, W') for attacked images are compiled in Table 1.

Table 1. PSNR, SSIM, NC(W, W') and BER(W, W') values for attacked colored images.

Attack	Image	PSNR	SSIM	NC (W,W')	BER (W,W')	Extracted Watermark
Median Filter (*aperture* = 3.0)	Airplane	34.45	0.96	0.91	0.12	
	Lena	32.66	0.98	0.75	0.15	
	Peppers	32.80	0.99	0.94	0.11	
	Tiffany	33.41	0.98	0.93	0.12	
Median Filter (*aperture* = 5.0)	Airplane	29.90	0.92	0.70	0.19	
	Lena	29.96	0.97	0.49	0.25	
	Peppers	30.89	0.98	0.77	0.16	
	Tiffany	30.77	0.97	0.82	0.14	
Scaling (re-sized to half and then restored to original size)	Airplane	30.75	0.95	0.98	0.09	
	Lena	32.26	0.98	1.00	0.09	
	Peppers	30.43	0.98	0.99	0.09	
	Tiffany	30.42	0.97	0.98	0.09	

(*continued*)

Table 1. (*continued*)

Attack	Image	PSNR	SSIM	NC (W,W')	BER (W,W')	Extracted Watermark
Crop (quarter of the signed image and fill the portion with host image)	Airplane	48.43	0.99	0.74	0.18	
	Lena	48.63	0.99	0.72	0.19	
	Peppers	48.62	0.99	0.71	0.18	
	Tiffany	48.61	0.99	0.65	0.20	
Salt and Pepper Noise 0.1%	Airplane	34.34	0.95	0.99	0.09	
	Lena	34.72	0.99	0.99	0.09	
	Peppers	34.87	0.99	0.95	0.10	
	Tiffany	34.23	0.98	0.98	0.09	
Gaussian Noise 5%	Airplane	36.87	0.92	0.99	0.09	
	Lena	36.86	0.99	1.00	0.09	
	Peppers	36.94	0.99	0.97	0.10	
	Tiffany	37.33	0.99	0.99	0.09	
JPEG(Q.F=50)	Airplane	30.95	0.92	0.33	0.34	
	Lena	31.42	0.98	0.31	0.34	
	Peppers	29.18	0.97	0.24	0.35	
	Tiffany	29.80	0.96	0.28	0.35	
JPEG(Q.F=75)	Airplane	32.45	0.93	0.81	0.14	
	Lena	32.50	0.98	0.79	0.15	

(*continued*)

Table 1. (*continued*)

Attack	Image	PSNR	SSIM	NC (W,W')	BER (W,W')	Extracted Watermark
	Peppers	30.19	0.97	0.67	0.19	
	Tiffany	30.90	0.97	0.81	0.15	
JPEG(Q.F=90)	Airplane	33.85	0.95	0.94	0.11	
	Lena	33.89	0.98	0.98	0.09	
	Peppers	31.21	0.98	0.87	0.12	
	Tiffany	31.74	0.97	0.97	0.09	
SPIHT(bpp=0.6)	Airplane	41.21	0.97	0.96	0.10	
	Lena	37.97	0.99	0.98	0.10	
	Peppers	38.44	0.99	0.96	0.10	
	Tiffany	40.49	0.99	0.98	0.09	
SPIHT(bpp=0.8)	Airplane	42.29	0.98	0.99	0.09	
	Lena	38.63	0.99	0.98	0.09	
	Peppers	39.13	0.99	0.98	0.10	
	Tiffany	41.12	0.99	0.98	0.09	
SPIHT(bpp=1.0)	Airplane	43.05	0.98	0.99	0.09	
	Lena	39.27	0.99	0.99	0.09	
	Peppers	39.71	0.99	0.98	0.09	
	Tiffany	41.85	0.99	0.99	0.09	

Table 2 compiles the comparative results for NC (W, W') for different attacks over the Baboon color image between our work and that reported by Piao et al. [8].

Table 2. Comparison of NC (W, W') for our method with those reported by Piao et al. [8]

Attacks	Piao et al. [8] (PSNR = 39.02 dB)	Our method (PSNR = 47.11 dB)
No attack	0.9993	**1**
Gaussian noise (10%)	0.8732	**0.9234**
Gaussian low pass filter	0.9775	**0.9972**
Resizing (384 × 384)	0.9160	**1**
JPEG (Q = 50)	0.9841	0.3483
JPEG (Q = 50)	0.9872	0.4881
JPEG (Q = 50)	0.9931	0.7473
JPEG (Q = 50)	0.9851	0.8869
JPEG (Q = 50)	0.9970	0.9701
SPIHT (bpp = 1.0)	0.8928	0.7548

It is very clear from the data compiled in Tables 1 and 2 that the recovery process in the present case is very successful. The optimized scaling parameter (K) is primarily responsible for better results obtained in this work. The extracted binary watermarks are recognizable visually and show a high value of NC and low value of BER. This is particularly true for all seven attacks carried out in the present work. Therefore, it can be concluded that the proposed semi-blind watermarking scheme using ELM in the DWT domain for color images fulfills all required criteria for image watermarking. Moreover, as the ELM algorithm is fast, the second phase of this work relates to computing the time span consumed in different processes involved. Table 3 compiles the time spans (in milliseconds) consumed by various processes of the proposed watermarking scheme. There are very few works which are reported in literature to have embedded watermarks in the colored images. Moreover, none of them have compiled time computation results. Therefore, we are unable to compare our time computation results with those of other research papers.

From the results compiled in Table 3, it is quite clear that all the reported time spans for different processes involved in this scheme are in the millisecond range. Therefore, the proposed semi-blind watermarking scheme is fast due to quick training of the ELM. The training time span reported incorporates the embedding time also. The total embedding time is in the range of 234–312 ms which is very small. Moreover, these time spans are achieved with very promising results observed for visual quality and robustness criteria. Overall, this makes

Table 3. Computed time spans (in milliseconds) for different attacks used in proposed scheme.

Procedure	Airplane	Lena	Peppers	Tiffany
Training of ELM followed by embedding	312.50	250.00	234.37	312.50
Extraction from the signed image	140.62	140.62	125.00	203.12
Extraction from the signed image after median filter (aperture = 3.0) attack	125.00	140.62	187.50	187.50
Extraction from the signed image after scaling attack (re-sized to half and then restored to original size)	203.12	171.87	140.62	125.00
Extraction from signed image after crop (quarter of the watermarked image and fill the missing portion with host image) attack	187.50	156.25	140.62	187.50
Extraction from signed image after 0.1% Salt and Pepper noise attack	187.50	187.50	125.00	171.87
Extraction from signed image after 5% Gaussian noise attack	125.00	125.00	156.25	203.12
Extraction from signed image after JPEG (Q.F = 90) image compression attack	171.87	125.00	125.00	187.50
Extraction from signed image after SPIHT (bpp = 1.0) compression attack	125.00	140.62	140.62	156.25

the proposed scheme a very good candidate for faster implementation of watermarking of images.

We, therefore, conclude that real time watermark embedding and extraction is definitely possible as a result of fast training carried out by ELM technique. As in the case of few other applications such as time series prediction, fault section identification and other regression problems, it is possible to realize real time watermarking of colored images with a good optimization of visual quality of signed and attacked images and the robustness of embedding scheme. The capacity of the host signal to hold the watermark is assumed to be constant as the size of the watermark is extremely small in comparison to the size of the host signal. Thus, it can be certainly said that the proposed ELM based watermarking scheme is suitable for carrying out watermark embedding in color standard images with the fulfillment of real time constraints.

6   Conclusions

A semi-blind color image watermarking scheme using DWT and Extreme Learning Machine (ELM) is proposed. Blue channels are first segregated from color images. Low frequency LL4 sub-band coefficients of the segregated blue channel is used for watermark embedding. For this purpose, these coefficients are quantized and augmented with the desired LL4 sub-band coefficients. This constitutes the input data-set to train the ELM. A binary watermark of size 32×32 is embedded in the blue channel of four colored host images. The numerical value of the scaling parameter is optimized for the embedding. Watermarks are recovered from the signed images and are examined for robustness. The Normalized Correlation and Bit Error Rate are within the expected range. Visual quality of the signed images is also found to be high. The signed images are subject to seven different image interferences to examine the issue of robustness. The recovered watermarks from the attacked images are quite similar to their embedded counterparts and their NC and BER values are also in the expected range. It is therefore concluded that the proposed DWT-ELM color image watermarking scheme is quite robust to the selected attacks. In the last part of this work, time spans for different processes involved are computed. These time spans show real time behavior (millisecond time spans) so that the proposed scheme is suitable for developing real time watermarking applications.

References

1. Cox, I.J., Kilian, J., Leighton, F.T., Shamoon, T.: Secure spread spectrum watermarking for multimedia. IEEE Trans. Image Process. **6**(12), 1673–1687 (1997)
2. Jane, O., Elbaşi, E., İlk, H.: Hybrid non-blind watermarking based on DWT and SVD. J. Appl. Res. Technol. **12**(4), 750–761 (2014)
3. Mishra, A., Agarwal, C., Sharma, A., Bedi, P.: Optimized gray-scale image watermarking using DWT-SVD and Firefly Algorithm. Expert Syst. Appl. **41**(17), 7858–7867 (2014)
4. Wang, Z., Bovik, A.C., Sheikh, H.R., Simoncelli, E.P.: Image quality assessment: from error visibility to structural similarity. IEEE Trans. Image Process. **13**(4), 600–612 (2004)
5. Mishra, A., Goel, A.: A novel HVS based gray scale image watermarking scheme using fast fuzzy-ELM hybrid architecture. In: Cao, J., Mao, K., Cambria, E., Man, Z., Toh, K.-A. (eds.) Proceedings of ELM-2014 Volume 2. PALO, vol. 4, pp. 145–159. Springer, Cham (2015). doi:10.1007/978-3-319-14066-7_15
6. Shen, R.-M., Fu, Y.-G., Lu, H.-T.: A novel image watermarking scheme based on support vector regression. J. Syst. Softw. **78**(1), 1–8 (2005)
7. Agarwal, C., Mishra, A., Sharma, A.: Gray-scale image watermarking using GA-BPN hybrid network. J. Vis. Commun. Image Represent. **24**(7), 1135–1146 (2013)
8. Piao, C.-R., Fan, W., Woo, D.-M., Han, S.-S.: Robust digital image watermarking algorithm using BPN neural networks. In: Wang, J., Yi, Z., Zurada, J.M., Lu, B.-L., Yin, H. (eds.) ISNN 2006. LNCS, vol. 3973, pp. 285–292. Springer, Heidelberg (2006). doi:10.1007/11760191_42
9. Liu, Q., Jiang, X.: Design and realization of a meaningful digital watermarking algorithm based on RBF neural network. In: 2005 International Conference on Neural Networks and Brain, vol. 1, pp. 214–218. IEEE (2005)
10. Agarwal, C., Mishra, A., Sharma, A.: A novel gray-scale image watermarking using hybrid fuzzy-BPN architecture. Egypt. Inform. J. **16**(1), 83–102 (2015)
11. Mehta, R., Mishra, A., Singh, R., Rajpal, N.: Digital image watermarking in DCT domain using finite Newton support vector regression. In: 2010 Sixth International Conference on Intelligent Information Hiding and Multimedia Signal Processing (IIH-MSP), pp. 123–126. IEEE (2010)

12. Chaudhary, V., Mishra, A., Mehta, R., Verma, M., Singh, R.P., Rajpal, N.: Watermarking of grayscale images in DCT domain using least-squares support vector regression. Int. J. Mach. Learn. Comput. **2**(6), 725 (2012)
13. Huang, G.: The Matlab code for ELM (2004)
14. Huang, G.-B., Zhu, Q.-Y., Siew, C.-K.: Extreme learning machine: theory and applications. Neurocomputing **70**(1), 489–501 (2006)
15. Serre, D.: Matrices: Theory and Applications. Graduate Texts in Mathematics, vol. 216. Springer, New York (2002). (Translated from the 2001 french original)
16. Rajpal, A., Mishra, A., Bala, R.: Extreme learning machine for semi-blind grayscale image watermarking in DWT domain. In: Mueller, P., Thampi, S.M., Alam Bhuiyan, M.Z., Ko, R., Doss, R., Alcaraz Calero, J.M. (eds.) SSCC 2016. CCIS, vol. 625, pp. 305–317. Springer, Singapore (2016). doi:10.1007/978-981-10-2738-3_26
17. Rajpal, A., Mishra, A., Bala, R.: Multiple scaling factors based semi-blind watermarking of grayscale images using OS-ELM neural network. In: 2016 IEEE International Conference on Signal Processing, Communications and Computing (ICSPCC), pp. 1–6. IEEE (2016)
18. Singh, R.P., Dabas, N., Chaudhary, V., et al.: Online sequential extreme learning machine for watermarking in DWT domain. Neurocomputing **174**, 238–249 (2016)
19. Rajpal, A., Mishra, A., Bala, R.: Fast digital watermarking of uncompressed colored images using bidirectional extreme learning machine. IEEE (2017, in press)
20. Rajpal, A., Mishra, A., Bala, R.: Robust blind watermarking technique for color images using Online Sequential Extreme Learning Machine. In: 2015 International Conference on Computing, Communication and Security (ICCCS), pp. 1–7. IEEE (2015)
21. Mishra, A., Goel, A., Singh, R., Chetty, G., Singh, L.: A novel image watermarking scheme using extreme learning machine. In: The 2012 International Joint Conference on Neural Networks (IJCNN), pp. 1–6. IEEE (2012)
22. Singh, R.P., Dabas, N., Chaudhary, V., et al.: Weighted extreme learning machine for digital watermarking in DWT domain. In: 2015 International Conference on Intelligent Information Hiding and Multimedia Signal Processing (IIH-MSP), pp. 393–396. IEEE (2015)

Quality Enhancement of Location Based Services Through Real Time Context Aware Obfuscation Using Crowd Sourcing

Priti Jagwani[1(✉)] and Saroj Kaushik[2]

[1] School of Information Technology, Indian Institute of Technology,
Hauz Khas, New Delhi 110016, India
jagwani.priti@gmail.com
[2] Department of Computer Science and Engineering,
Indian Institute of Technology, Hauz Khas, New Delhi 110016, India
saroj@cse.iitd.ac.in

Abstract. Widespread usage of Location based services (LBS) has eventually raised the concern for user's privacy. Various privacy preserving techniques are based on the idea of forwarding cloaking area to service provider who might be untrusted party, instead of actual location of query issuer/client. For such scenarios, in which cloaking area is exploited for privacy, results of the query request are generally based on nearest distance between client and service requested. Such techniques do not include real time context which is important in determining security, accessibility, etc. of the service and enhancing service quality. In this work, a novel method, based on crowd-sourcing concept has been proposed which takes into account the real time context for determining results of query. A system consisting of real time context-aware component is coined. Real time context has been obtained through crowd-resources available in cloaking area of client. A fuzzy inference system (FIS) has been proposed which takes nearest distance and real time context parameters as input. Based on these parameters FIS generates a new rank for the service requested. This rank is the new position on the answer list for the service requested. A prototype of the proposed system is implemented. Evaluation of prototype has been done by taking feedback of 112 users about their satisfaction in the range (0–10). User feedback for the prototype is compared with feedback of other similar systems using Kruskal Wallis test for significant differences. It has been discovered that user satisfaction for proposed system stochastically dominates other prevalent systems.

Keywords: Location based services · Security · Quality enhancement · Fuzzy inference system · Cloaking area

1 Introduction

Location Based Services (LBSs) are important and useful in almost all applications in mobile systems. Wide usage of smart mobile devices and advancement in positioning technologies have open up sky limiting possibilities in the domain of LBSs. The LBSs

© Springer International Publishing AG 2017
O. Gervasi et al. (Eds.): ICCSA 2017, Part VI, LNCS 10409, pp. 79–94, 2017.
DOI: 10.1007/978-3-319-62407-5_6

are information services that provide users with required personalized contents, such as the nearest restaurants/hotels/clinics, which are retrieved from spatial databases. Potential LBSs applications can be countless ranging from Point of interest (POI) to complex navigation systems including proximity based marketing, fraud prevention, etc. These applications utilize the positioning capabilities of the device to determine the current location of the user. The location serves as an integral part of input for the location based services. The advantages of LBSs are very obvious and make life easier but they open up various vulnerabilities because of location knowledge. Location is often perceived as personal and sensitive information which can put a user's privacy in danger if mishandled by malicious actors. Thus, location disclosure can create threats to user's privacy. Recent research is being conducted on protecting location data [8, 11, 12, 23]. Approaches to safeguard location data can be based on policies and regulations, data transformations or location obfuscation.

Location obfuscation is a widely acknowledged category of location privacy mechanism in a ubiquitous computing environment. Obfuscation can be defined as deliberately degrading the quality of information in order to protect the privacy of the individual to whom that information refers. This process of degrading can be based on slightly altering, substituting or generalizing the location in order to avoid the real location of the user. A well focused technique of obfuscation is location cloaking in which exact location of user is blurred into a cloaked area based on some number K (K anonymity). After cloaking, a user is considered as K-anonymous if exact location of user cannot be distinguished among K-1 other users. The cloaked area contains at least K users including client. In a commonly used setting, client sends its service request to middleware which is actually a trusted party and is aware of user's location. The location of user can be determined using any of the available positioning techniques. Client gives query to service provider routed through middleware who generates clocking area around the client and sends it to location service provider (LSP). Assumption is that LSP may be considered as an untrusted party and hence should not know the actual location of user seeking service through him. Since a location service provider (LSP/Location Server) does not know the exact location of client thus providing client a security. LSP returns an answer set which are further filtered/arranged by middleware according to user's location. (Query and service request both refers to the request asked by client hence now onwards these terms can be used interchangeably).

These results of a service request are based on nearest distance method. Such methods lack in terms of inclusion of real time context. Results of such requests without inclusion of real time context may not be very effective for the client. For example, by getting a service or POI in the results which is not working presently, or is unreachable, etc. user may have a disappointing experience. To tackle such scenarios, real time context should be included with the results. Some context information can be collected through web also but here the requirement is of real time context for which the best sources are the human beings currently available at that place commonly known as crowd resources. Thus collection of real time context is based on crowd-sourcing application model where crowd resources are used.

According to Jeff Howe crowdsourcing refers to a distributed problem-solving model in which a crowd of undefined size is engaged to solve a complex problem through an open call. Crowdsourcing is also defined as the act in which a company or institution take a task and outsource it to an undefined network of people in the form of an open call [1]. The task solving by distributed large group of people, who belongs to different disciplines is an example of crowdsourcing approach. An organization identifies tasks and releases those tasks online to a crowd of outsiders who are interested in performing these tasks on behalf the organization for a stipulated fee or any other incentives. A vast number of individuals then offer to undertake the tasks individually or in a collaborative way. Researchers have investigated the opportunities of using crowdsourcing for mobile applications thus making use of the real world context. Overall crowdsourcing system has emerged as a new problem-solving paradigm [2].

A novel location obfuscation system is proposed in the paper to include real time context, referred as real time context-aware obfuscation system. The system works for determining results of a service request within a cloaking area. The paper sets out a formal system, in which not only obfuscation of location-based services for the purpose of security is implemented but service quality is enhanced by inclusion of real time context. The proposed system is a generic LBS system and point of interest (POI) query has been taken as an example service so henceforth discussion presented will be with reference to the POI services.

Validation of proposed system is done by implementing prototypes for the proposed system and other two systems namely nearest distance method (most prevalent one) and security accessibility system. Security accessibility is the system in which rank of a POI is determined by its ranking of security and accessibility. For this system result set may contain candidate answers outside the cloaking area. Feedbacks for all the three systems have been collected by 112 users in terms of a score in the range (1–10). Feedback score given by the users are compared using Kruskal Wallis test. This test proved to be significant which indicates that feedback scores for the proposed system stochastically dominate the values of feedback score for other systems. The main contributions of this work are as follows:

- Inclusion of real time context component with well known obfuscation framework.
- Blending of real time context with the results of nearest distance method using FIS.
- Use of Kruskal Wallis test for the stochastic establishment of the proposed system has been explored. The test verifies the value of user feedback for proposed and some prevalent systems for significant differences.

Remainder of this paper is organized as follows. Section 2 presents related work along with necessary background. Overall system design is presented in Sect. 3, followed by Sect. 4 containing experimental settings, implementation and performance analysis. Section 5 contains statistical establishment of the system while conclusion and future directions are discussed in Sect. 6.

2 Related Work

Spatial K-anonymity paradigm has been widely studied in [3–6]. The client sends its *service request* to a trusted anonymizer, which constructs an anonymizing spatial region (ASR)/cloaking area that contains the location of client along with other (K-1) client locations. The anonymizer then sends the cloaking area to the LBS. The latter executes the service request with respect to the cloaking area, and returns a superset of the results to the anonymizer, which filters out the false positives.

Context aware computing offers the promise of considerable improvement in service quality. Context awareness is the ability of systems to adapt more readily to user needs, models, and goals. Context awareness of privacy protection mechanism is a prominent research area. Pingley et al. introduced a Context-Aware Privacy-preserving LBS system (CAP) [7]. They devised a mechanism for protecting location privacy and achieving LBS accuracy on the basis of context. Challenges of violation of location privacy based on (i) user's location information contained in the LBS query payload, and (ii) by inferring a user's geographical location based on its device's IP address were addressed.

Zhang et al. in [8] focused on context aware location privacy protection (CLPP) for location based social networks (LBSN) where the privacy requirements of users are not constant and isolated. They designed algorithms to evaluate whether the users' published geo-content meet the user's privacy requirement.

Pournajaf et al. examined the problem of spatial task assignment in crowd sensing when participants utilized spatial cloaking to obfuscate their locations [9]. They investigated methods for assigning sensing tasks to participants, efficiently managing location uncertainty and resource constraints. They proposed an optimization approach which consisted of global optimization using cloaked locations followed by a local optimization using participants' precise locations without breaching privacy. Interaction between application and crowd resources is being optimized for spatial task assignment along with addressing privacy concerns of users.

Damiani et al. coined probe framework for personalized cloaking of private locations which combines privacy personalization and location privacy [10]. The framework can be integrated with K-anonymity techniques and with policy-based approaches to provide stronger privacy protection especially in novel geosocial applications. They presented the idea to safeguard sensitive locations comprehensive of a privacy model and an algorithm for the computation of obfuscated locations.

Fawaz et al. proposed 'LP-Doctor', a light-weight user-level tool that allows Android users to effectively utilize the OS's location access controls while maintaining the required app's functionality [11].

Ju and Shin presented EMP2 tool for Location Privacy Protection for Smartphone users using Quadtree Entropy Maps. EMP2 accurately estimates the uncertainty of users' intended destinations and dynamically adjusts the protection level to defend against sophisticated inference attacks based on query correlation. Effectiveness of EMP2 has been demonstrated for effective protection of users' location privacy with reasonable computation time and resource consumption [12].

All the above research works were based on protection of location privacy using various versions of obfuscation techniques. In the past, researchers in different fields have used crowdsourcing concept. So far, the crowdsourcing work included Jana [13], a mobile crowdsourcing system, which publish simple tasks such as translation, transcription, and filling out surveys for mobile phones users. Google also uses crowdsourcing model to collect the road traffic data and provide the real time traffic conditions. Other location aware crowdsourcing services include mCrowd [17], and

GigWalk. mCrowd is a mobile crowdsourcing system in which micro sensing tasks are performed by crowd using their mobile devices. Gigwalk is a crowd-sourced service pays users for writing reviews and posting pictures. Some of other popular crowd-sourcing systems are-Wikipedia, Yahoo! Suggestion Board, threadless, iStockphoto, InnoCentive, Sheep Market, Yahoo's flickr, MobiMission, Gopher game and CityExplorer, etc.

Crowd sourcing concept has been used in various research works. Erickson used geocentric crowdsourcing to enable cities and regions to more effectively address issues ranging from infrastructure to governance [14]. Liu et al. proposed location privacy recommender using user-user collaborative approach [15]. Yang et al. used Smart-phone based crowdsourcing approach for indoor localization problem [16]. Nghiem et al. proposed a knn p2p query processing approach for mobile adhoc system [18]. They used the concept of data sharing from peers. Despite of lots of research about crowdsourcing model, usage of crowdsourcing in protection of location privacy and in enhancement of service quality is not fully explored yet. To et al. described idea to offer location privacy guarantees to crowdsourcing worker, based on differential privacy and geocasting [19]. They investigated analytical models and task assignment strategies that balance multiple crucial aspects of spatial crowdsourcing functionality, such as task completion rate, worker travel distance and system overhead.

Hu et al. employed peer to peer spatial K anonymity to protect worker's location privacy [20]. They also coined optimized schemes for spatial task assignment without compromising worker's location privacy. Toch presented a crowdsourcing framework for privacy management of location information in ubiquitous environment named 'Super-Ego' [21]. Crowdsourcing has been used to predict the user's privacy prefer-ences for different location on the basis of the general user population.

Ubiquitous crowdsourcing is a little explored domain, in which the smart-phone users contribute information about their surrounding, thus providing a collective knowledge about the physical world. By applying ubiquitous crowdsourcing, the task of collecting information is performed continuously and in real-time. Mashhadi and Capra have explored the challenges for quality control in ubiquitous crowdsorucing and propose a technique that reasons on users mobility patterns and quality of their past contributions to estimate user's credibility [22]. In all the above works crowd resources are used in order to provide the service but in our work crowdsourcing is used to enhance the security and service quality of an already existing service by inclusion of real time context.

2.1 Problem Definition

In proposed work, the problem of service quality and security enhancement through inclusion of real time context obtained by crowdsourcing is explored. To explain the significance of inclusion of context in LBS, let us assume a scenario where, a user is asking for ATMs near his current location. Based on distance/geometry method, the user gets a list of 3 nearest ATMs; out of which the top/nearest two ATMs are not working or may be situated in isolated corners. The user may have a very embittered experience by these types of results. All these problems arise because the query results do not consider the just in time information, namely the security, accessibility, approachability and other elements of real time context while populating the reference space. Therefore those techniques are unable to provide user an upright service. So along with the protection of location privacy, need is for those techniques which are able to take into account the real time qualitative context in which users are located. For this the various parameters have been taken from crowd sources available at that place and these parameters are supplied as input to FIS. In this paper combination of server generated content (results of request generated by LSP) along with user-generated content (real time parameters) is being achieved using a fuzzy inference system (FIS). This is done in order to extend service quality of a task in the real world.

3   Overall System Design

This section presents the overall design and working flow of the proposed system. The system contains the following components: mobile client, middleware, crowd resources, location service provider (LSP). The middleware consist of various modules namely cloaking area generator module, real time context module, context based FIS module. Individual components of the proposed system are shown in Fig. 1 and described in detail in the following subsections.

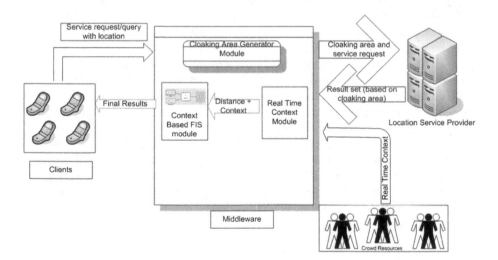

Fig. 1. Architecture of the proposed system

3.1   Mobile Client with Web Interface

Client is a mobile user who sends the service request about a particular service to middleware. Client is registered with the middleware. Current position of this registered user can be automatically obtained by GPS enabled device. Location of the client is known to middleware. Further, service request is given to middleware. Client does not want to reveal its location to the untrusted parties like location service provider (LSP) or Location server. In the proposed approach, after completion of service request, client gets the response to his service request based on real time context blended with nearest distance. Client can also see the rank of POI requested which shows the reachability, accessibility and security of the queried point/object.

3.2   Middleware

Middleware is the central and trusted entity which operates between client and LSP. In general scenarios a middleware takes the location and request of client. This location is obfuscated to maintain the privacy. Middleware comprises of 3 main modules. These are:

- Cloaking area generator module.
- Real time context module.
- Context based FIS module.

Cloaking Area Generator Module: Given exact location of the client, the module generates a cloaking area around that location. It is ensured that at least K-1 users must be present in that area apart from the user. Cloaking area will be generated enclosing those K-1 users and client. The value of K is determined by this module using the techniques available in literature [23]. Regular shaped cloaking regions can be circular or rectangular. It has been proved by research that circular cloaking regions incurs a higher processing cost [24]. So rectangular cloaking areas are preferable over circular. There can be a number of methods for generating the rectangular cloaking area like Hilbert curve, NNC (Nearest neighbor cloaking), etc. For the purpose of our experiments we have used nearest neighbor cloaking area method. In this method cloaking area is a minimum bounding box which encloses all K users. This cloaking area along with the service request is given to LSP.

Real Time Context Module: This module receives the parameters of real time context from crowd resources and result set obtained by LSP. Finally, it supplies rating of distance between service and client (nearest the distance, higher is the ranking); and aggregated values of real time context parameters (of POI) to context based FIS module.

A set of environmental states and settings that are important in determining behavior of an application or service can be termed as Context. The proposed system deals with some specific elements of context namely security, and density of place, accessibility/reachability. Values of all these elements are in the range of 1–5 where 5 represents the highest and 1 represents the lowest.

Security of the location depicts the well-being of the place. In general scenarios, some places are considered less/more safe than others depending upon their sensitivity and well being. If one is unable to decide upon safety of a place using intuition/perception/already available information about the place, it can be easily obtained by many readily available applications. Example of one of such application is "safetypin" [www.safetypin.com].

Density of a place can easily be observed by users present in the area. It is an observation of footfall in that area. For the proposed system, another factor of context is reachability/approachability that basically determines how accessible/functional is the POI? Ratings of all these elements are asked from user present in the cloaking area through a predefined set of questions. These questions contains a rank/rating about.

- Security.
- Density of the nearby place.
- Accessibility/reachability of the POI.

Rating for all the above mentioned parameters is given by users/crowd resources present in the cloaking area. This rating is in the form of crisp numbers as shown in Fig. 2. Rating given by all users is aggregated/averaged for each element. Question interface is designed in such a way that these questions can be easily answered by the crowd resources within few clicks. Set of questions and sample interface is shown in the figure below.

Sector 23 ICICI ATM
1. Security rating of the above place?

2. Approachability/accessibility rating of the above place?

3. Density of its surroundings?

Fig. 2. Questions and sample interface

The request for rating is based on a timing constraint i.e. system waits for 20 s for rating after displaying the rating request on the crowd resource's mobile screen. Crowd resource has to provide feedback within 20 s. Real time context module stores this rating given by crowd resources for future use also.

The module also receives the result set from the LSP. The result set contains list of POIs in the cloaking area. These POIs are arranged in the increasing order of distance between client and POI/service. In this way, every POI is now having a rating corresponding to nearest distance also. Finally, ratings of distance, security, approachability and density are supplied to context based FIS module for further processing.

Context Based FIS Module: This module collects the rating of distance and real time context parameters from real time context module and after processing it generates new rank of POIs to be sent to the client.

The main component of context based FIS module is Fuzzy inference system. Fuzzy inference is the process of transforming a given input into required output using fuzzy logic. This process involves membership functions, fuzzy if then else rules and fuzzy operators. This module obtain the aggregated values of ratings of all real time context parameters and distance between client and requested POI, from real time context module. These parameters serve as input to FIS. Firstly, fuzzification of these input values takes place based on membership functions. Then fuzzy inference engine converts these inputs into output on the basis of rules. Based on the rules written in rule base, output value of FIS is determined which is the new rank of the POI. The whole process is shown in Fig. 3

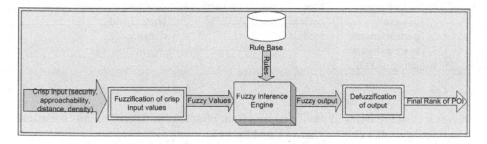

Fig. 3. Fuzzy inference system

Around 350 rules have been coined intuitively based on different values of inputs and output. Some rules of FIS are shown below:

- If security/well being of the place is low, approachability/accessibility is very low, density is moderate and distance is far then rating is very low.
- If security/well being is moderate, approachability/accessibility is low, density is moderate and distance is near then rating is medium.

Output of this module is a new crisp rating of POI. In this way final ranked list of recommended POIs is obtained. This final list of POIs is sent to client as a result. Collectively, input values, processing done and output values of all the three modules of middleware are given in the table below (Table 1).

3.3 Location Service Provider (LSP)/Location Server

LSP is the service and content providing entity which is considered as untrusted. Some common examples of LSPs are Navteq, Tele Atlas, Google and many others. Location service provider offers a number of different services like finding a route or searching specific information on objects of user interest and many others. Typically, LSP

Table 1. Input, processing and output of various modules of middleware

Module name	Input	Input received from	Processing done	Output	Output given to
Cloaking area generator	Service request and client location	Client	Generation of cloaking area (CA)	Service request and cloaking area	LSP
Real time context	1. Real time context parameters 2. Service response with respect to CA	1. Crowd resources 2. LSP	Aggregation of each real time context parameter obtained by all users and determining Rating of POI according to nearest distance	Aggregated values of real time context parameters and nearest distance rating of POI	Context based FIS module
Context based FIS	Aggregated values of real time context parameters and nearest distance rating of POI	Real time context module	Fuzzy inferencing based on fuzzy rules	Final rank of POIs based on real time context and distance	Client

receives a cloaking area and service request from middleware. Then it returns the result set pertaining to that cloaking area to middleware.

4   Implementation and Experiments

In this section, we describe implementation details followed by experimental settings.

Implementation: In order to show the feasibility of the proposed system, prototype of the client side application and middleware are implemented. Middleware resides on server side while client side module is for the mobile handset of client. The whole application is developed using Android Software Development kit and PHP. For appropriate comparison of the system we have developed three prototype applications. One that returns the result solely on the basis of nearest distance (systems which are already prevalent, now onwards called as nearest distance method), another one which retrieves results on the basis of security and accessibility ranking only, and the third one (our proposed idea) which provides the result on the basis of combination of nearest distance and security-accessibility ranking. Some snapshots (login screen and rating generation snapshot) of the developed prototype are shown below in Fig. 4.

Experimental Settings: For assessment and evaluation of the proposed system we have taken crowd resources to work on it. These are volunteers registered in the system. There are total 112 such volunteers. Every volunteer install all the three systems on their smart phones and used them for three weeks. Volunteers provide a feedback score for all the prototypes. Their feedback given in terms of score about all

Fig. 4. Sample snapshots of prototype

the three systems have been taken and recorded for further analysis. This feedback is on a scale of (1–10) where 1 represents the lowest score (lowest satisfaction) and 10 represents the highest.

Table 2 represents the feedback given by volunteers for all the three systems. First column shows the score while other three columns represent number of users who have given the score displayed in the left most column for the respective systems. For example, 10.6% users have given score in the range of 1–3 to the nearest distance prototype while 24% users have given score as 4–6 to the same. This data is shown in the following graph also (Fig. 5).

Table 2. Feedback scores given by users

Score	Security accessibility prototype	Nearest distance prototype	Nearest distance + security accessibility prototype
1–3	19.6%	10.6%	5.3%
4–6	80.4%	24%	35.6%
7–8	0%	65.4%	32.0%
9–10	0%	0%	22.3%

From the given data in above table, one can easily interpret that for security accessibility prototype performs poor on satisfaction (0%) on higher score (7–10), nearest distance prototype performs good on satisfaction at level (7–8) but no user has given score between (9–10). The proposed system has overall higher satisfaction with minimum score of dissatisfaction (1–3), and with 22.3% of users have given score in the range (9–10). Further statistical evaluation of all the three systems is done by Kruskal Wallis test results which prove that data on which experiment is done is significant. Following section describe statistical test on the feedback values. Data presented in this section proves our claim that proposed system performs better in terms of satisfaction.

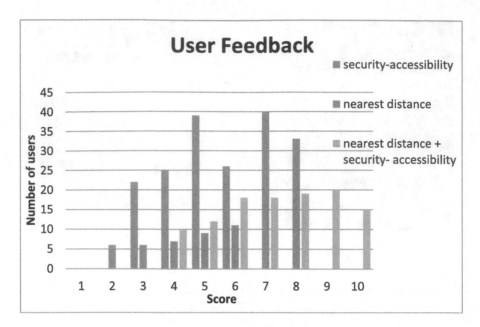

Fig. 5. Graphical representation of feedback scores

Performance Analysis: In the earlier subsection, performance of the proposed system has been evaluated in terms of feedback. The system has also been evaluated for response time as a metric. Results for response time taken as compared to the prevalent systems are shown in figure below.

Although the response time taken by our proposed system is greater than the prevalent system but this extra cost (in terms of time in sec) is paid in order to get real time security and current accessibility status. One may not want to go to an insecure place or an inaccessible/unapproachable, rather wants to invest some extra seconds in

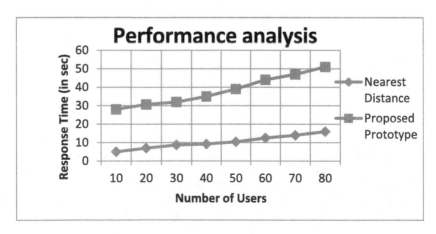

Fig. 6. Performance analysis in terms of response time

order to get actual real time status of POI. This fact has been proved by the better feedback given by users. This greater response time is because of an increased step of obtaining real time context from crowd resources and reranking the response on its basis. This step is not available in nearest distance method. Further, response time of a service is dependent on many design patterns like storage data structures, use of caching layers, archiving aging data to reduce table sizes, separate databases for read only and write only nodes, etc. Response time can certainly be improved by taking care of the above mentioned design patterns (Fig. 6).

5 Statistical Establishment of the System

In this section, for establishment of claims of proposed system, feedback values given by users are evaluated statistically. This evaluation is done to ascertain the fact that the difference between feedback values is significant and proposed system is performing better. For this Kruskal Wallis test has been applied on the feedback values. The Kruskal-Wallis H test (sometimes also called the "one-way ANOVA") is a nonparametric test that can be used to determine if there are statistically significant differences between the data samples. Also the Kruskal-Wallis H test does not assume normality in the data. That is why it is best suited for our data. Shown below are the results obtained from the kruskal's wallis test (with 5% significance level).

Here,

- var1 represents feedback ratings of the system with only distance as a metric
- var 2 represents the feedback of system with only security and accessibility as a metric
- var3 represents feedback of the system with distance and security-accessibility both as a metric.

Table 3 shows various summary statistics while Table 4 shows that there are significant differences between Var1 and Var3.

Table 3. Summary statistics

Variables	No. of observations	Observations with missing data	Observations without missing data	Min value	Max value	Mean	Std. deviation
Var1	10	3	7	6.00	40.0	16.0	14.3
Var2	10	6	4	22.0	39.0	28.0	7.5
Var3	10	0	10	1.00	18.0	11.2	6.2

Table 5 represents the null hypothesis and alternative hypothesis of the test performed and Table 6 contains sample mean scores, p value and significance level. Result interpretations are shown in Table 7.

Results clearly show that we can safely reject the null hypothesis which implies that the data is not drawn from the same population means the data is significantly distinct.

Table 4. Significant differences

	Var1	Var2	Var3
Var1		No	Yes
Var2	No		No
Var3	Yes	No	

Table 5. Test interpretation

H_0: The samples come from the same population. (Hypothesis)

H_a: The samples do not come from the same population. (Alternate Hypothesis)

Table 6. Results of Kruskal Wallis test

Sample mean score (Var 1)	4.616071429
Sample mean score (Var 2)	6.366071429
Sample mean score (Var 3)	6.598214
p-value (two-tailed)	0.0478
Alpha	0.05

Table 7. Results interpretation

As computed p-value is lower than the significance level alpha = 0.05, one should reject null hypothesis H_0, and accept the alternative hypothesis H_a

The risk to reject the null hypothesis H0 while it is true is lower than 4.78% (shown by p-value)

Also, highest sample mean score of feedback values for var 3 (proposed system) depicts outperformance the other systems

It is clear from the above results that there are significant differences between the var1 (feedback of the system with only distance as a metric) and var3 (feedback of the system with distance and security-accessibility both as a metric). Moreover, looking at the feedback ratings given by users, one can easily spot that proposed system has got higher scores of feedback. Highest sample mean score of feedback values for the proposed system depicts that our proposed system is performing better than the most prevalent nearest distance system.

6 Conclusions and Future Directions

In the work presented, a system has been proposed to enhance the security and accessibility of the location based services through inclusion of real time context. Experimental prototype of the proposed system is implemented by taking point of interest query as the example service. Crowdsourcing model has been used in obtaining the real time context which is collected from the crowd resources currently present in that area. Although a price is being paid in terms of increased response time (few seconds) but our system provides considerable satisfaction. Measures have been suggested to improve the response time. To evaluate the success of our prototype, feedback has been obtained from the users. Stochastic evaluation of the feedback showed that the proposed system outperforms the other prevalent systems.

The proposed system is in a prototype stage which can witness certain improvements in future. One can also device questions to be asked from crowd resources based on the service request type/content. Also to improve the service time, real time responses of previous queries (say of 15 min prior) can be saved in a separate database structure and the similar service request for the same area can be served on the basis of that stored data instead of asking crowd resources again. Moreover, credibility of crowd resources can be another area for improvement and further research based on which weightage to be given to the response of a particular crowd resource can be determined.

References

1. Howe, J.: Crowdsourcing: a definition, crowdsourcing: tracking the rise of the amateur. In: Crowdsourcing: Why the Power of the Crowd is Driving the Future of Business (2006)
2. Alt, F., et al.: Location-based crowdsourcing: extending crowdsourcing to the real world. In: Proceedings of the 6th Nordic Conference on Human-Computer Interaction: Extending Boundaries. ACM (2010)
3. Chow, C.-Y., Mokbel, M.F., Liu, X.: A peer-to-peer spatial cloaking algorithm for anonymous location-based service. In: Proceedings of the 14th Annual ACM International Symposium on Advances in Geographic Information Systems. ACM (2006)
4. Kalnis, P., et al.: Preventing location-based identity inference in anonymous spatial queries. IEEE Trans. Knowl. Data Eng. 19(12), 1719–1733 (2007)
5. Gedik, B., Liu, L.: Location privacy in mobile systems: a personalized anonymization model. In: 25th IEEE International Conference on Distributed Computing Systems (ICDCS 2005). IEEE (2005)
6. Yiu, M.L., et al.: Spacetwist: managing the trade-offs among location privacy, query performance, and query accuracy in mobile services. In: 2008 IEEE 24th International Conference on Data Engineering. IEEE (2008)
7. Pingley, A., et al.: CAP: a context-aware privacy protection system for location-based services. In: 29th IEEE International Conference on Distributed Computing Systems, ICDCS 2009. IEEE (2009)
8. Zhang, H., et al.: CLPP: context-aware location privacy protection for location-based social network. In: 2015 IEEE International Conference on Communications (ICC). IEEE (2015)
9. Pournajaf, L., et al.: Spatial task assignment for crowd sensing with cloaked locations. In: 2014 IEEE 15th International Conference on Mobile Data Management, vol. 1. IEEE (2014)
10. Damiani, M.L., Bertino, E., Silvestri, C.: The PROBE framework for the personalized cloaking of private locations. Trans. Data Priv. 3(2), 123–148 (2010)
11. Fawaz, K., Feng, H., Shin, K.G.: Anatomization and protection of mobile apps' location privacy threats. In: 24th USENIX Security Symposium (USENIX Security 2015) (2015)
12. Ju, X., Shin, K.G.: Location privacy protection for smartphone users using quadtree entropy maps. J. Inf. Priv. Secur. 11(2), 62–79 (2015)

13. Eagle, N.: txteagle: mobile crowdsourcing. In: Aykin, N. (ed.) IDGD 2009. LNCS, vol. 5623, pp. 447–456. Springer, Heidelberg (2009). doi:10.1007/978-3-642-02767-3_50
14. Erickson, T.: Some thoughts on a framework for crowdsourcing. In: Workshop on Crowdsourcing and Human Computation, pp. 1–4 (2011)
15. Liu, N.N., Zhao, M., Yang, Q.: Probabilistic latent preference analysis for collaborative filtering. In: Proceedings of the 18th ACM Conference on Information and Knowledge Management, CIKM 2009, pp. 759–766. ACM, New York (2009)
16. Yang, Z., Wu, C., Liu, Y.: Locating in fingerprint space: wireless indoor localization with little human intervention. In: Proceedings of the 18th Annual International Conference on Mobile Computing and Networking. ACM (2012)
17. Yan, T., et al.: mCrowd: a platform for mobile crowdsourcing. In: Proceedings of the 7th ACM Conference on Embedded Networked Sensor Systems. ACM (2009)
18. Nghiem, T.P., Waluyo, A.B., Taniar, D.: A pure peer-to-peer approach for kNN query processing in mobile ad hoc networks. Pers. Ubiquit. Comput. 17(5), 973–985 (2013)
19. To, H., Ghinita, G., Shahabi, C.: A framework for protecting worker location privacy in spatial crowdsourcing. Proc. VLDB Endow. 7(10), 919–930 (2014)
20. Hu, J., Huang, L., Li, L., Qi, M., Yang, W.: Protecting location privacy in spatial crowdsourcing. In: Cai, R., Chen, K., Hong, L., Yang, X., Zhang, R., Zou, L. (eds.) APWeb 2015. LNCS, vol. 9461, pp. 113–124. Springer, Cham (2015). doi:10.1007/978-3-319-28121-6_11
21. Toch, E.: Crowdsourcing privacy preferences in context-aware applications. Pers. Ubiquit. Comput. 18(1), 129–141 (2014)
22. Mashhadi, A.J., Capra, L.: Quality control for real-time ubiquitous crowdsourcing. In: Proceedings of the 2nd international workshop on Ubiquitous Crowdsouring. ACM (2011)
23. Jagwani, P., Kaushik, S.: K anonymity based on fuzzy spatio-temporal context. In: 2014 IEEE 15th International Conference on Mobile Data Management, vol. 2. IEEE (2014)
24. Kalnis, P., Ghinita, G., Mouratidis, K., Papadias, D.: Preventing location-based identity inference in anonymous spatial queries. TKDE 19(12), 1719–1733 (2007)

A Model-Based Testing Method for Dynamic Aspect-Oriented Software

Maria Laura Pires Souza and Fábio Fagundes Silveira[✉][iD]

Federal University of São Paulo – UNIFESP, São José dos Campos, Brazil
malaura.s1@gmail.com, fsilveira@unifesp.br

Abstract. Aspect-oriented programming (AOP) is used to implement crosscutting concerns such as persistence and safety in program units called aspects. To ensure that these concerns behave as specified and do not introduce faults into the application, rigorous software testing practices should be applied. Even though there are statements in the literature that the adoption of AOP takes a software to get better quality, it does not provide correctness by itself. Therefore, the test remains an important activity to ensure aspects are correctly integrated into the main system. Additionally, in a dynamic environment: new aspects may be incompatible with aspects already woven; and aspects to be removed can hold the system to an inconsistent state. Available approaches in the literature do not directly investigate the problem of testing dynamic aspects within the context of a target application. This paper presents a method to apply tests in dynamic aspects that verify the interactions between aspects and classes, as well as among aspects. Aiming to support the method, we also introduce a model to represent the dynamic behavior of aspects and a new strategy to derive testing cases. To evaluate the effectiveness of the test cases generated by the method, mutation operators were applied to the model and simulated with a model checker. Results showed that the approach is capable of detecting faults in dynamic aspects interactions into a target application.

Keywords: Dynamic aspect-oriented · Model-based testing · Mutation testing

1 Introduction

Aspect-oriented programming (AOP) consists in a programming technique that supports the implementation of cross-cutting concerns, such as persistence, security, and logging in units called aspects, where such concerns are implemented. To ensure that these additional concerns behave as specified and do not introduce defects in the application, rigorous and different test levels should be made in the application. Besides, these tests should be extended to the interaction between these aspects and also to the classes of the target application. The division of a system known as separation of concerns refers to the ability to identify,

© Springer International Publishing AG 2017
O. Gervasi et al. (Eds.): ICCSA 2017, Part VI, LNCS 10409, pp. 95–111, 2017.
DOI: 10.1007/978-3-319-62407-5_7

encapsulate, and manipulate concrete portions of software that are relevant to a particular concern [11].

Some difficulties can be verified when applying tests in aspects compositions [2,13] in a dynamic environment: loaded aspects may be incompatible with aspects already read and/or running; and removed aspects can lead the system to an inconsistent state.

Available approaches in the literature do not directly investigate the problem of testing dynamic aspects within the context of a target application. So, in order to find solutions to some of the problems mentioned above, this paper presents a testing method for dynamic AO applications. The method, called MESOADI, contemplates the verification of interactions between dynamic aspects, aiming to improve the behavioral tests between classes and aspects, and in composites of dynamic aspects, seeking to reduce the resources involved, such as effort and test time.

This paper is organized as follows: Sect. 2 presents AO definition, especially about dynamic aspects and model-based testing definition. Section 3 briefly summarizes related works on testing dynamic systems. Next, Sect. 4 describes the proposed method with their elements and a case study description. Section 5 highlights the main results and discussion are described. Finally, Sect. 6 concludes the paper and points out future works.

2   Dynamic AOP and Model-Based Testing

Aspect-oriented (AO) development was proposed due to the difficulties encountered in the treatment of code spreading and interlacing during maintenance and software development. Its purpose is to separate levels of concerns during development. The most common examples of crosscutting concerns, which are separated by AO, are those relating to non-functional requirements such as persistence, logging, authentication, security, fault tolerance, among others.

AOP introduces new concepts to the software development process. Among them is the aspect, which is the encapsulation unit of a crosscutting concern [2]. The aspects have structural and/or behavioral attributes that can be applied to various parts of the system. The process of insertion of the aspects, known as weaving, is responsible for injecting the aspects into the modules in which they should act. Weaver is the tool that combines object-oriented (OO) code with the AO code for the operation of the final system.

Regarding dynamic aspect-oriented programming (DAOP), it is possible to apply and remove aspects to a system at runtime, without the need to restart it, which is very useful when designing real world applications. Dynamic aspects are a necessary mechanism, especially if one aspect implements a crosscutting concerns at one point and the requirement for functionality changes dynamically, on the fly. One of the advantages of DAOP is that it removes AOP overhead when aspects are not required [1]. Also, it allows dynamic configuration of aspect behavior and aspect reconfiguration depending on the state of the base system.

Testing plays a critical role in any software development project. However, it is often overlooked as an expensive activity and hampered by the wide variety of programming languages, operating systems and hardware platforms that constantly evolve [12]. This creates serious problems in the software production phase, leading to high costs, poor business reputation and even killing human beings.

Model-Based Testing (MBT) consists of a technique for automatically generating a set of test cases (TC) by using models extracted from the software requirements [3]. Before any software testing activity, one has to validate the model to be sure that it will not cause any errors in the software or vice versa, so this model also needs to be tested. This way, there are rules for modeling software, as it is necessary for everyone involved in software development to have a standard model and know how to interpret them. There are several techniques for specifying systems that are used to add more stringency to the MBT, such as Finite State Machines (FSMs), which was used in this paper, Statecharts, and Petri nets.

The strategies TCs generation aim to verify if an implementation is correct with its specification, through the execution of activities of test and validation in systems described by models [8]. Although the strategies have a common goal, the difference between them is the cost, the size of the set and the effectiveness in finding defects in the system. For this paper, an extension of the Binder's *Round-Trip Path* strategy (RTP) was used. The RTP strategy traverses the FSM graph through an algorithm and generates a tree called *State Transition Tree* (STT) corresponding to that path, where the initial state is the root node.

3 Related Works

Several researches on AO systems and dynamically adaptive systems show up important works in these areas. Zhang and Cheng [15] separated the adaptive behavior and specifications of non-adaptive behaviors into dynamic programs. For this, a process was introduced for the construction of adaptive models, automatically generating adaptive programs of the models, besides checking and validating the models. For the authors, the main tasks for the adaptation of a point are to identify the states that are suitable for adaptation and to define adaptive transitions from these states. Zhao et al. [16] propose a definition for non-adaptive program and adaptive program through finite state machines. For them, adaptation is an action that changes the behavior of a state in one FSM to a state into another FSM.

Fuentes and Sánches [7] published a paper with the objective of presenting an extension of the Unified Modeling Language (UML) for the construction of aspect-oriented models. Silveira et al. [13] proposed the METEORA, a state-based testing method for AO programs, which provides classaspect and, more specifically, aspectaspect faults detecting capabilities. Ferrari et al. [6] describe an approach based on mutation testing for AspectJ programs. Lindström et al. [10] propose the use of AOM (Aspect-Oriented Models) mutation to test crosscutting concerns.

4   The MESOADI Method

The MESOADI method aims to apply state-based tests in dynamic aspects, through a model of behavioral representation of interactions between dynamic aspects. The model, called MEADI, consists of states, pseudo-states, and transitions and are based on models proposed by Zhang and Cheng [15] and Zhao et al. [16].

MEADI is composed of several FSMs with transitions between them. Each time an aspect is added or removed, the model represents an adaptation to the time and system at runtime, switching to another FSM, which represents the new behavior. This adaptation corresponds to an action that changes the behavior of a state in one FSM to a state in another FSM.

States that are added or removed by aspects have $<<aspect>>$ stereotypes and are yellow in color, while states added by classes are white in color. The change from one FSM to another happens through special transitions called *priority transitions*. They have colors in these transitions that indicate whether the aspect has been added (red) or removed (blue). In addition, the MEADI has an element that indicates when a pointcut is encountered and the type of advice that acts on it. Table 1 shows the advice representation and the possible priority transitions for a joinpoint in a transition t and state S, where t comes from. From the table, we see that the priority transition always shows the advice contained in the aspect that was added or removed and the transition in which it is applied. The state S appears in the priority transition because there are times when the transition is executed and when it goes to the other FSM needs to return to the state before the transition in which the aspect affects. An example is when a security aspect is added with advice before for password verification before moving on to the next state. The added aspect leads to the state that checks the password and if the password is incorrect, it must go back to the previous state and not continue to the next state (as would happen if the password were correct).

Table 1. Advices and priority transitions representations.

Advice	Symbol	Added Aspect	Removed Aspect
before	⟵	b{t, S}	-b{t}
around	⟷	ar{t, S}	-ar{t}
after	⟶	a{t, S}	-a{t}

When the priority transitions are found, they take priority over the other transitions, and the FSM moves to the state they point, already on another machine. This occurs in specific states, called *quiescent states* [15], only where the aspect is added or removed. The reason for the adaptations to occur only in the quiescent states is because before reaching these states, the program has no change in its behavior. Even if an aspect is added or removed at any time, the behavior of the program will change when this aspect is sensitized, causing the program to find the priority transition, which will lead to the transition to the FSM that models the new behavior.

The great limitation of the FSMs as a visual formalism for the description of complex systems is due to the problem of the explosion of the number of states and transitions that occur as the system becomes more complex [9]. To deal with this problem, the application under testing is modeled and tested by submodules.

4.1   Case Study: An Intelligent Transportation System

MESOADI was applied to two different case studies. The one chosen to be reported here refers to an intelligent transportation system [14]. The system, called ITS, is made up of a set of context-sensitive vehicles, and it was adapted to the dynamic OA context by Fuentes and Sánches [7].

In the ITS, vehicles navigate autonomously from a given origin to a predetermined destination. Each vehicle travels along a "virtual circuit", which has to be previously calculated with the aid of a GPS for a given target point [7]. This circuit can be any outdoor arena, and the vehicle can travel inside it. The vehicle receives the information from the GPS periodically, the time interval being dependent on the vehicle speed.

When driven autonomously, the vehicle needs to build a real-time perception of its surrounding environment so that, if an error occurs, it makes decisions about its next move. Before a trip, vehicles are notified with the information and guidelines of the "virtual circuit".

As the vehicle receives information from your GPS periodically, an error-handling module (an aspect) should monitor that the response time of the GPS is never exceeded, and react when this constraint is violated. If this constraint is violated, an error handling strategy must be applied. One possible solution is the temporary use of GPS data from a nearby vehicle. However, this is only possible if the vehicle is in circulation on a highway where the neighboring vehicles are going in the same direction and with an almost constant speed. If the vehicle is in the city, where the behavior of the vehicles is less predictable, information about other vehicles is out of use, and the human driver would be forced to manually control the vehicle until the GPS recovers.

A UML class diagram is shown in Fig. 1a to describe the ITS classes. This diagram shows two classes: (1) the Context class that store information about the current context of the vehicle, as speed mode (fast or slow) and type of path (city or highway); and (2) the Coordinator class which ensures that the foregoing elements cooperate properly so that the car is driven safely. This last class implements the behavior of the vehicle on or off, the driving mode is chosen (manual - when the driver drives the vehicle - or automatic - when the vehicle sails autonomously) and, in addition, there is verification of the next position it should follow if it is in automatic mode using GPS. Figure 1b shows a class diagram for the aspects of error handling and the change of context. Class names contain the $<<aspect>>$ stereotype indicating that class is one aspect.

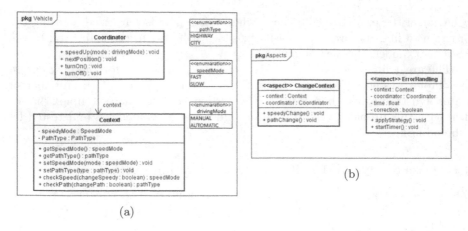

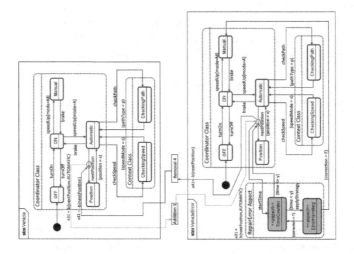

(a)

(b)

Fig. 1. Class diagram of (a) Coordinator and Context classes; and (b) ErrorHandling and ChangeContext aspects.

Fig. 2. MEADI with the ErrorHandling aspect added and removed from the application.

Using the MESOADI, Fig. 2 presents the MEADI with the ErrorHandling aspect that contains the FSM Vehicle and the FSM VehicleError. The aspect checks the length of time the vehicle has to receive information from your GPS. If an error occurs that is not resolved with respect to time, it needs to assume the manual mode. The modeling will not consider the various situations that may cause the error, only that an error has occurred that may or may not be corrected, because in this way the main objective, which is to show the iteration with the dynamic aspect, will be reached. This aspect is added to the system whenever the vehicle is navigating intelligently, so when the vehicle is navigating manually, the aspect can be removed. When the priority transition is found in the FSM Vehicle, one aspect is added, and this transition is followed up to the FSM Vehicle. The same happens when a priority transition is found in the FSM VehicleError, it will follow to FSM Vehicle, where the aspect does not exist.

The initial state of the FSM Vehicle is called OFF because the vehicle is off. When the vehicle is switched on, the FSM moves to the ON state, where the vehicle is stationary, waiting for the driving mode to be activated to accelerate the vehicle. This driving mode can be manual, leading to the MANUAL state, or automatic, leading to AUTOMATIC state. In this last state, the vehicle starts driving without a driver, so you need to check the speed mode (CHEKINGSPEED state) and the type of the path (CHEKINGPATH) to adapt to the environment. In addition, a "virtual circuit" is plotted and the POSITION state contains the precise information about the next point at which the vehicle will be driven. When the priority transition is found in FSM Vehicle, one aspect is added, and this transition is followed up to FSM VehicleError. The added aspect is responsible for verifying that the GPS data is received at the correct time and adds two new states to the model: TIMECOUNTER and ERRORHANDLING.

This aspect has a before advice with pointcut in the transition nextPosition because it does a time count before checking the next position. This time count is done in the state TIMECOUNTER. If the time count ends and the GPS has not yet returned any position, that is the time is longer than expected, a transition to apply a strategy to this error leads for the ERRORHANDLING state, where a treatment strategy error message is applied. If the strategy is efficient and corrects the error, the system continues in automatic mode. Otherwise, the system switches to manual mode.

The ChangeContext aspect is responsible for detecting the message that may result in changes to the contextual information, requiring an appropriate update of the context. It notices changes in speed mode and path type. The aspect contains a pointcuts in the checkSpeed transition and checkPath transition. When the speed mode check transition is called, if the speed value exceeds a certain constant value, the vehicle is considered to be in the fast mode. Otherwise, it is in the slow mode. If it goes from slow to fast or vice versa, the ChangeContext aspect will make this change. The same occurs when the vehicle is in the city and goes to a highway or vice versa.

Figure 3 shows the MEADI that contain the states added by this aspect. In this example, the FSM Vehicle starts the system without any aspect and when a priority transition is found (e11 or e12), it is followed, because the ChangeContext aspect has been added. At this point, the transition takes the system to the FSM ChangeContextVehicle, where two states are added, CHANGESPEED and CHANGEPATH. The advice is of the type around, because these states are substitutions of the states CHECKINGSPEED and CHECKINGPATH, respectively, given the condition that the speed or path has been changed. When the aspect is removed from the system, the transition e21 or e22 will be found, going to the FSM Vehicle.

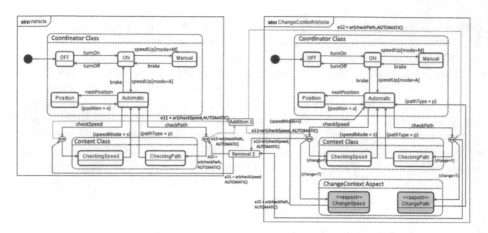

Fig. 3. MEADI with the ChangeContext aspect added and removed from the application.

4.2 Dynamic Combined Reacheability Tree

A state transition tree (STT) is the result of applying the RTP strategy to an FSM, and each path is a set of arcs from which a TC is derived. However, strategies developed for the OO paradigm, in general, can not be directly applied to the AO paradigm, taking into account the different specificities that exist for the treatment of cross-cutting concerns [13]. Furthermore, in a dynamic context, these strategies do not address the addition and removal of states and transitions at runtime.

In this work, STTs were modified and named Dynamic Combined Reachability Tree (DCRT) so that they can be applied to dynamic AO systems. DCRTs have priority transitions that take them to the other DCRTs. Each time a priority transition is found, it leads to the root of another tree, from which the TCs are derived. That is, the number of existing trees for the model is equal to the number of adaptations that the program may suffer, both when an aspect is added or removed. When there are static aspects or aspects already combined in the application, the states that represent them will be in the built DCRTs and, therefore, the number of trees will remain the same.

When an aspect is added, a pseudo-state will be the root of the new DCRT. This is due to the fact that a DCRT or TCs can not start with a transition and then a root state. In this case, it is necessary to have an initial state as root, which will be the pseudo-state. For the removed aspects, as the priority transition only indicates that it has been removed, it does not appear in the DCRT and therefore does not need to have a pseudo-state as root. The pseudo-states have the same priority transition nomenclature, but with the e uppercase, for example, for an e11 transition, the pseudo-state will be E11.

In DCRTs, states are represented by their names in capital letters, transitions by their names in lowercase letters, conditions are in square brackets and the returns of functions in parentheses, also in lowercase letters. The details of the construction of these DCRTs are presented below.

4.3   A New Strategy to Derive Test Sequences – RTP_{MESOADI}

In order to construct the DCRT from the dynamic modeling of the aspects, an adaptation of the original *Round-Trip Path* (RTP) method [3] was performed, named RTP_{MESOADI}. For each change in the MEADI, a new DCRT is constructed. The main change in the RTP relates to its stopping criterion, established as follows:

- When an aspect adds a state that is applied before, around, or after a transition, the transition t that precedes it and indicates the type of advice is saved. When the state appears again:
  - If the transition that precedes it is equal to t, then the DCRT repeats the state and stop the path;
  - Otherwise, the tree passes through the state and continues its generation obeying the original RTP stop criterion [3], that is, stop the path when a state is repeated or when it is a final state.

For the MEADI of Fig. 3, six DCRTs are derived using the RTP_{MESOADI}. The roots of these DCRTs are: initial state of the FSMs (1) Vehicle; (2) ChangeContextVehicle; quiescent states (3) e11; (4) e12; (5) e21; and (6) e22. For example, when the ChangeContext aspect is added, if the e11 priority transition from FSM

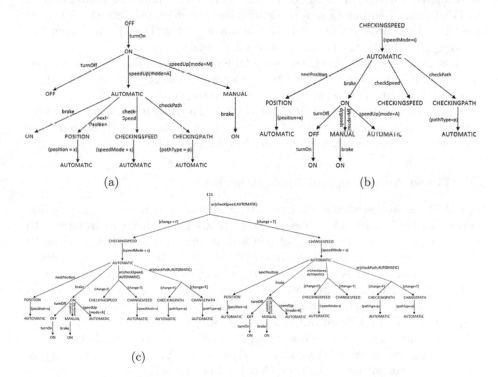

Fig. 4. DCRTs: (a) initial state of the FSM Vehicle; (b) quiescent state e21; and (c) quiescent state e11. The red nodes are leaf nodes. (Color figure online)

Vehicle to the FSM ChangeContextVehicle is found, a pseudo-state E11 is the root of a new DCRT. The same happens with the e12 priority transition. The e21 and e22 priority transitions also lead to the creation of two new DCRTs, but the root of these DCRTs corresponds to the states that these transitions point to. Figure 4 shows the DCRTs with the root trees in an initial state Vehicle, a quiescent state with the addition of the aspect (e11) and a quiescent state with the aspect removal (e21).

The same occurs for the MEADI of Fig. 2 that four DCRTs are derived using the algorithm. The roots of these DCRTs are: initial state of the FSMs (1) Vehicle; (2) VehicleError; quiescent states (3) $e31$; and (4) $e41$.

With the TC derivation strategy used ($RTP_{MESOADI}$), DCRTs were created where each path corresponds to an abstract TC. For TCs to have coverage of 100%, all paths of all DCRTs must be traversed.

5   Results and Discussion

To evaluate MESOADI and its elements, we measured the effectiveness of testing sequences generated by this method through the use of mutation testing. In this section, Mutation Operators (MOs) are described for aspect-oriented models to be evaluated through fault-based testing. Then, we report the translation of the ITS to timed automata in a model-checker called UPAAL. So, the MOs are applied to the automata for their simulation with the TCs derived from the $RTP_{MESOADI}$. Finally, the mutation scores are obtained.

MOs used here were based on and adapted from [10]. The work developed a mutation-based search for static AOP models. Therefore, for each of the proposed MOs, adaptations occurred in relation to the cited work. The MOs focus on the aspects and elements of AO (pointcuts and advice) because they are syntactic structures present only in AO models. Table 2 shows the MOs and their descriptions related to pointcuts and advices, respectively.

5.1   Timed Automata and Model-Checker

MESOADI was also analyzed using timed automata through the use of a model-checking tool called UPPAAL[1]. Timed automata are FSMs that are extended with clocks. The main focus in the use of UPPAAL in this work is the use of a model checker algorithm to simulate the result of the application of the TCs derived from the algorithm $RTP_{MESOADI}$, not the verification of time or performance.

Timed automata are finite state machines with timing constraints associated with their edges and states, and are intended to model the behavior of real-time systems [5]. In a timed automata clocks are represented by a finite set of real-valued variables C and events are represented by a finite alphabet Σ.

A network of timed automata $\mathcal{A}_1 || \ldots || \mathcal{A}_n$ over (Σ, C) is the parallel composition of n timed automata over (Σ, C), where components are required to

---

[1] Available at: http://www.uppaal.org/.

Table 2. Mutant operators to pointcuts and advices.

Mutantion operator – pointcuts	Description
Pointcut weakening (*PCW*)	Adds the pointcuts that should be in the FSM, but also adds new pointcuts in transitions that are not affected by the aspect
Incorrect pointcuts (*IPC*)	Add in the FSMs only the pointcuts that are not contained in the original FSMs
Pointcut strengthening (*PCS*)	Does not selecting one or more pointcuts, selecting only a subset of correct pointcuts
Mutantion operator – advices	Description
Advice on incorrect pointcut (*AIP*)	Associating existing advice with pointcuts that are also existent but incorrect
Advice replacement at pointcut (*ARP*)	Change the types of advices in the FSMs

synchronize on delay transitions and discrete transitions are required to be synchronized on complementary actions [10]. An action a? is complementary to a!.

UPPAAL is a tool for validation (through graphic simulation) and verification (through automatic model checking) of systems in real time. The idea is to model a system using timed automata, to simulate them, and then to check their properties. The simulation step is to run the system interactively to see if it works as planned. In the verification step, the accessibility properties are checked, that is, whether a particular state is accessible or not. This is called model-checking and is basically an exhaustive search that covers all possible dynamic behaviors of the system. Only the graphic simulation step is used for the analysis of MESOADI. In UPPAAL, the double-circle state indicates that it is the initial state. The system consists of a network of processes that are made up of a set of locations. The transitions between locations define how it behaves.

Through synchronizations, it is possible to invoke or activate one or more transitions using a previously defined synchronization channel. The use of the "!" tag can be seen as a send and the "?" tag as a reception. When a process is in a state from which there is a transition with c! synchronization, the only form of this transition is activated is if there is another process in another transition marked with c? or contrariwise.

The ITS was manually translated into automata for UPPAAL. Because the tool does not support AOP, some adaptations were necessary. The states that have been added by aspects are represented by the yellow color for a better understanding of the model. When an advice is found, the automata were modeled to behave as expected, bearing in mind whether the advice is before, after or around. All transitions in the model will be represented by synchronizations. The mutation operators that were used for the analysis do not affect variables, so even the transitions that in the original model exist variables will be represented by synchronizations.

The MEADI contains several FSMs, so for the simulations all the automata corresponding to each of these FSMs, and all the automata that have a quiescent state as the initial state, are modeled. In ITS there are nine automata with initial states in: (1) FSM Vehicle; (2) FSM ChangeContextVehicle; (3) FSM VehicleError; quiescent states (4) e11; (5) e12; (6) e13; (7) e21; (8) e22; and (9) e23. Automata 1, 2 and 3 will have the names of the FSMs to which they correspond and the automata 4, 5, 6, 7, 8 and 9 will have the name of their quiescent states, changing the e for E. Automats that do not contain states added by aspects will not be tested, because MOs only affect aspects. Thus, the MOs will be applied in five automata (ChangeContextVehicle, VehicleError, E11, E12 and E31).

The mutants generated from the original models are classified as: (1) killed mutants – when the mutant shows a behavior different from the original model; (2) equivalent mutants – when for each possible entry of the original model, the mutant version will show the same behavior (such mutants can not be distinguished from the original model by any test); and (3) stillborn mutants – when they are syntactically illegal and therefore are not accepted by the model verifier (for this reason, the latter are not considered in the analysis).

For the application of TCs, another process, called "TC", is created next to each of the automata. This process contains the TC using the synchronizations for the simulation in the automata. Figure 5 shows TC derived from the DCRT of Fig. 4a, which uses the FSM Vehicle of the MEADI, being applied to the original automaton no UPPAAL. In the lower right window, the vertical arrows show the transitions and the horizontal arrows show the synchronization between automata. The last state of the TC is called "TC_passed" (in this case it would be the "ON" state), because every time the simulator is deadlocked (shown in the upper left window), if the last state of the "TC" process is this, having applied a MO, it means that the mutant remains alive. If the simulator is blocked and has not yet reached this state, it means that the TC has failed. Therefore, the mutant is considered killed. However, if for any possible simulation, the result with mutant or no mutant is the same, the mutants are called equivalent mutants. Thus, all mutants that remain alive need to be analyzed in order to determine whether or not they are equivalent.

In the case of stillborn mutants, the tool points to a syntax error and the templates are not sent to the simulation.

The analysis performed with the obtained results refers to the metrics of confidence or adequacy of the test cases generated by the MESOADI, being used as reference the mutation scores. The automata used show all possible scenarios for the modeling of ITS with MESOADI.

The presented case study contains two aspects that can be added or removed, where one of them contains two pointcuts. The MOs were applied at these pointcuts and their advices and all the possibilities modeled. In the examples that will be presented, the transitions, states, and synchronizations represented by the red color show the elements added to the model. The states, transitions,

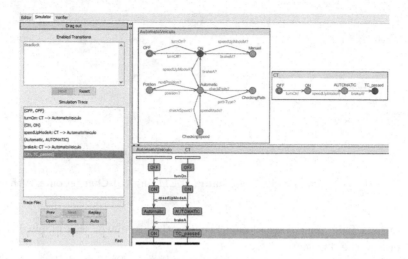

Fig. 5. Application of a TC in the automaton Vehicle

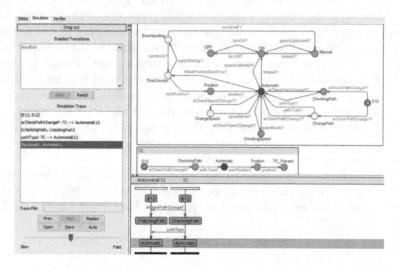

Fig. 6. Application of a TC to the automaton E12 mutated with the MO *PCW*

and synchronizations represented by gray color, and which contains two crossed traces, show the elements removed. The following are the MOs analyzed from this case study.

1. *PCW*: 7 mutants were generated for this operator, all killed by some TC. Figure 6 shows an example of this MO being applied to the E12 automaton and compared to a TC derived from RTP_{MESOADI}. This automaton should contain only the pointcuts of the ChangeContext aspect, however, the *PCW* operator also adds another pointcut of the ErrorHandling aspect in the "nextPosition"

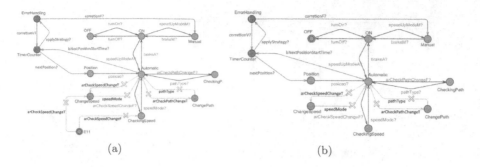

(a)   (b)

Fig. 7. *PCI* operator applied to automata (a) E11 and (b) ChangeContextVehicle.

transition. This mutant was killed because the simulator was blocked before the end of the TC scan.

2. *PCI*: 6 mutants were generated by this operator. Only 2 of them could be used since 4 mutants contained syntactic errors. The two mutants analyzed were killed by TCs. Figure 7a shows the two pointcuts of the ChangeContext aspect being deleted from the automaton and the pointcut of the ErrorHandling aspect being added. That is, the operator adds in the automaton E11 only a pointcut that is not contained in the original automaton. In the case of the E11 automaton, the initial state is removed and, for this reason, the mutant is then considered to be stillborn. Figure 7b shows how the automated ChangeContextVehicle was after the application of the *PCI* operator.

3. *PCS*: 6 mutants were generated by this operator. Of this total, 4 of them were used and killed, and 2 were characterized as stillborns, for the same reason as above.

4. *API*: 10 mutants were generated for this operator. For all of them, at least one TC derived from the RTP_{MESOADI} killed the mutants.

5. *ARP*: 10 mutants were generated by this operator. All of them were killed by the TCs applied.

5.2   Effectiveness Level of the Generated Test Cases

DeMillo [4] provides an objective measure for the adequacy of the TCs of the *P* program by defining a mutation score, which consists in evaluating the suitability of *T* in relation to the test. This score ranges from 0 to 1 and the higher, the greater the effectiveness of the test suite generated.

The generation of the mutants was done carefully and manually. None of the mutants generated by the *PCW*, *API*, and *ARP* operators were considered to be stillborn. That is, all were used in the evaluation. As no equivalent mutant was generated by these operators, the obtained mutation scores have resulted in 1. The mutation operators *PCI* and *PCS* were the only ones that generated stillborn mutants.

With the simulation of the application of the TCs to the automata temporized with MOs in UPPAAL, it was observed that the TCs generated by $RTP_{MESOADI}$ to the ITS presented a high degree of adequability (effectiveness). Table 3 summarizes the mutation scores obtained by the mutation operator used.

Table 3. Mutation scores for the ITS using the MESOADI.

Mutation operator	Generated mutants	Killed mutants	Equivalent mutants	Stillborn mutants	Mutation score
PCW	7	7	0	0	1.00
PCI	6	2	0	4	1.00
PCS	6	4	0	2	1.00
API	10	10	0	0	1.00
ARP	10	10	0	0	1.00
Total	39	33	0	6	

The high mutation score obtained can be explained by the fact that it is a small system and that it generated a low number of mutants. For this reason, no equivalent mutants were generated. To obtain a more accurate result and to generate equivalent mutants, the MESOADI and the generation of the mutants need to be automated for the application in a more complex system and empirically validated in a future work.

6   Conclusion and Future Work

The test activity of dynamic aspects is by no means a trivial task. In addition, despite its importance in real-world applications, the development and testing of dynamic aspects are still under-explored areas. In dynamic aspect-orientated, with the addition or removal of runtime aspects, the loaded aspects may be incompatible with aspects already read and/or running, and removed aspects may lead the system to an inconsistent state. This work described the difficulties encountered for modeling and applying the test activity in this type of system.

The MESOADI constitutes a proposal for the application of state-based tests for dynamic aspects. The behavior represented by MEADI is described through several Finite State Machines (FSMs), which have transitions (adaptations) between them, allowing to represent how the system can adapt as the context change occurs, in this case, adding or removing aspects dynamically. For the derivation of the test cases (TCs) several Dynamic Combined Reachability Tree (DCRT) are constructed, by means of the $RTP_{MESOADI}$, where each one represents an adaptation suffered by the test application in a dynamic way. From the constructed trees, we obtain the TCs, derived from the sequences of transitions of these trees. The results obtained by the application of MESOADI showed that the generated test sequences presented a high degree of suitability considering the mutation scores obtained in the evaluation.

Future work includes the development of a tool to support the MESOADI and carrying out an experimental study to evaluate in a more rigorous way the proposed approach.

Acknowledgments. The authors would like to thank CNPq (grant 455080/2014-3) and FAPESP for financial support.

References

1. Alam, F.E., Evermann, J., Fiech, A.: Modeling for dynamic aspect-oriented development. In: Proceedings of the 2nd Canadian Conference on Computer Science and Software Engineering, pp. 143–147. ACM (2009)
2. Alexander, R.T., Bieman, J.M., Andrews, A.A.: Towards the systematic testing of aspect-oriented programs. Rapport technique, Colorado State University (2004)
3. Binder, R.V.: Testing Object-Oriented Systems: Models, Patterns, and Tools. Addison-Wesley Professional, Boston (2001)
4. DeMillo, R.A.: Mutation analysis as a tool for software quality assurance. Technical report, DTIC Document (1980)
5. Dong, J.S., Hao, P., Qin, S., Sun, J., Yi, W.: Timed automata patterns. IEEE Trans. Softw. Eng. **34**(6), 844–859 (2008)
6. Ferrari, F.C., Rashid, A., Maldonado, J.C.: Towards the practical mutation testing of AspectJ programs. Sci. Comput. Program. **78**(9), 1639–1662 (2013)
7. Fuentes, L., Sánchez, P.: Dynamic weaving of aspect-oriented executable UML models. In: Katz, S., Ossher, H., France, R., Jézéquel, J.-M. (eds.) Transactions on Aspect-Oriented Software Development VI. LNCS, vol. 5560, pp. 1–38. Springer, Heidelberg (2009). doi:10.1007/978-3-642-03764-1_1
8. Fujiwara, S., Bochmann, G.V., Khendek, F., Amalou, M., Ghedamsi, A.: Test selection based on finite state models. IEEE Trans. Softw. Eng. **17**(6), 591–603 (1991)
9. Harel, D.: Statecharts: a visual formalism for complex systems. Sci. Comput. Program. **8**(3), 231–274 (1987)
10. Lindström, B., Offutt, J., Sundmark, D., Andler, S.F., Pettersson, P.: Using mutation to design tests for aspect-oriented models. Inf. Softw. Technol. **81**, 112–130 (2016)
11. Moreira, R.M., Paiva, A.C., Aguiar, A.: Testing aspect-oriented programs. In: 2010 5th Iberian Conference on Information Systems and Technologies (CISTI), pp. 1–6. IEEE (2010)
12. Myers, G.J., Sandler, C.: The Art of Software Testing. Wiley, Hoboken (2004)
13. Silveira, F.F., da Cunha, A.M., Lisbôa, M.L.: A state-based testing method for detecting aspect composition faults. In: Murgante, B., et al. (eds.) ICCSA 2014. LNCS, vol. 8583, pp. 418–433. Springer, Cham (2014). doi:10.1007/978-3-319-09156-3_30
14. Sivaharan, T., Blair, G.S., Friday, A., Wu, M., Duran-Limon, H., Okanda, P., Sørensen, C.F., EU FET: Cooperating sentient vehicles for next generation automobiles. In: ACM/USENIX MobiSys 2004 International Workshop on Applications of Mobile Embedded Systems (WAMES 2004 Online Proceedings) (2004)

15. Zhang, J., Cheng, B.H.C.: Model-based development of dynamically adaptive software. In: Proceedings of the 28th International Conference on Software Engineering, ICSE 2006, pp. 371–380. ACM, New York (2006)
16. Zhao, Y., Li, Z., Shen, H., Ma, D.: Development of global specification for dynamically adaptive software. Computing **95**(9), 785–816 (2013)

The Impacts of Using SNSs on e-WOM and Knowledge Sharing Through Social Capital: An Empirical Study in Vietnam

Quoc Trung Pham[✉] and Vi Khiet Huynh

School of Industrial Management, Bach Khoa University (VNU-HCM),
268 Ly Thuong Kiet, District 10, Ho Chi Minh City, Vietnam
pqtrung@hcmut.edu.vn

Abstract. Nowadays, Social Networking Sites (SNSs) have attracted billions of users all around the world. In Vietnam, SNSs become more and more familiar with Vietnamese people and Facebook is the most popular SNS. Social Capital is an important term in the study and social behavior, an important sector in the major universities of the world. However, Social Capital is still very vague concept, not only for citizens but also for policy makers in Vietnam. There are a few of studies have examined the impact of SNSs to Social Capital and Knowledge Sharing through word of mouth in environment of SNSs (e-WOM) in Vietnam. Based on previous research model, this research examines whether the intensity of using SNSs has a positive impact on users' Social Capital, Knowledge Sharing and e-WOM. A survey was conducted with Facebook users in Vietnam. The regression analysis was utilized to examine the nine main hypotheses through a questionnaire designed which is based on the Likert five-point scale. The results show that the intensity of using SNSs raise the Trust and Identification dimension of Social Capital, especially the Trust, and these dimensions have positive effects to e-WOM Quality on SNSs. It also finds out that the e-WOM Quality impacts positively on Knowledge Sharing through SNSs environment in the context of Vietnam.

Keywords: e-WOM · SNS · Knowledge sharing · Social capital · Vietnam

1 Introduction

The explosion of Information Technology has brought the tremendous changes of the socio-economic in the world. Today, web 2.0 creates new method of approaches for marketing or business industry and being a tool which supports knowledge sharing management (Lee 2011). Based on Web 2.0, the one of most utilized services is social media, which is defined as a network of Internet-based applications that create based on the ideological and technological foundations of Web 2.0, and that enables the creation and exchange of user-generated content. It includes social network sites (SNSs) (Chu and Kim 2011). The development of SNSs has formatted the way of connecting and interacting. The SNSs are defined as Websites that helps users to connect with other people and share their statuses and activities (Donath and Boyd 2004).

© Springer International Publishing AG 2017
O. Gervasi et al. (Eds.): ICCSA 2017, Part VI, LNCS 10409, pp. 112–127, 2017.
DOI: 10.1007/978-3-319-62407-5_8

Social capital is "the resource available to actors as a function of their location in the structure of their social relations" (Adler and Kwon 2002). The word of mouth (WOM) is one of the most important methods for giving and receiving information in social (Godes and Mayzlin 2004). The WOM has many forms, such as traditional WOM (offline form) and via Internet WOM (e-WOM) (Steffes and Burgee 2009). There are differences between traditional WOM and e-WOM although they are expected to be relevant (Hennig-Thurau et al. 2004). Traditional WOM is type of an immediate conversation, while e-WOM is a type of asynchronous interactions among people which is separated by time and space (Steffes and Burgee 2009). This research examines e-WOM, because it works in social networking environments, and SNSs are a perfect tool for checking e-WOM (Chu and Kim 2011).

Organizations have utilized SNSs for networking and collaboration. IBM and Microsoft use social networking tools to have strengthened weak link among colleagues (Huang et al. 2010). IBM has also used SNSs for keeping their employees contact, fulfill generation gaps, and encourage the innovating and collaborating (Majchrzak et al. 2009). Social networking tools are the good media to allow people interface together anywhere, any time (Majchrzak et al. 2009). Company can improve their innovation processes via their customer (Leimeister et al. 2009). Social networking tools also support the firms to improve interactions with their customers for crowdsourcing. Example Starbucks, SNSs have been utilized for contacting with customers, informing them about promotions, getting the suggestions and tracking consumer-to-consumer dialog (Gallaugher and Ransbotham 2010).

Many researchers have studied SNSs, their influence and practical usage. However, a few of paper research have examined the influence between SNSs, social capital, knowledge sharing, and e-WOM. Although there are some previous researches found out that WOM quality has a positive impacts on online Trust (Awad and Ragowsky 2008), the scenario has not been on SNSs. Furthermore, there is little research that has examined the influence of social capital, knowledge sharing to e-WOM on SNSs. Therefore, this paper's purpose is closing these research gaps by examining the impact of Intensity of Use of SNSs to users' Social Capital, Knowledge Sharing and e-WOM.

In Vietnam, Facebook is one of the most popular social network sites. The operation of "share" and "tagging" has increased the connection among the users in Facebook. In Vietnam, 44% users have more than 400 friends and 13% more than 1000 friends. This is a big surprise for researcher because the average number of friends on Facebook is about 100 friends in Japan. Half of Vietnamese user "will accept" if they get friend requests from strangers. Unlike the eastern countries of Southeast Asia, the Vietnamese generally open to share personal information and find the information other than managing it in private.

According Hofstede (1991), the individualism of Vietnam is different with South Korean and United States. It means the spirit of collectivism in South Korean and United States is different with Vietnam. Furthermore, the collectivism culture involves to the significant of emotional support and defined norms obtained through network social that are highlight in bonding social capital. Therefore, there must be a different to conduct this research in comparision with previous ones in Korea and US.

Specific research objectives of this research could be summarized as follows: (1) Measuring the impact of the SNS Usage Intensity to user's Social Capital, the impact of Social capital to Knowledge sharing and e-WOM, the impact of e-WOM to Knowledge sharing; and (2) Give the recommendation for promoting social development via SSNs by using its impact to e-WOM and Knowledge sharing. The structure of this paper includes: (2) definitions, (3) research model and hypotheses, (4) research process, (5) analysis results, and (6) conclusion and implications.

2 Definitions

2.1 Web 2.0

Web 2.0 is defined as new version of WWW sites which focused on dynamic content and interaction of participants. Web 2.0 allows firms to enhance existing capabilities by integrating multiple functions including knowledge management, project management, and social networks that connect people together (Bayus 2012). In addition, Web 2.0 also enhances the capability in many activities inside of business which includes internal marketing.

2.2 Social Network Sites

SNSs (belongs to Web 2.0) are web-based services that enable individuals to: (a) create a profile within a bounded system, (b) link a list of their friends, and (c) review and track their list of connections and those made by others within the system (Boyd and Ellison 2008). SNSs are considered the main source of creating social capital in modern organizations. However, the real impact of SNSs using on social capital of an organization is not known completely.

2.3 Social Capital

Social capital refers to intangible capital inside of an organization, which relating to the relationship between employees and customers. Social capital mentions to the collected resources through the relationships among people (Coleman 1988). Social capital must be created from community (Nahapiet and Ghoshal 1998). The Social capital enhances the knowledge transferring process through the social interaction (Huang et al. 2010). Social capital has three dimensions: (1) the structural dimension (network bonds, network configuration, and appropriable organization), (2) the cognitive dimension (shared codes and language, and shared descriptions), and (3), the relational dimension (trust, norms, obligations and identification) (Nahapiet and Ghoshal 1998).

2.4 e-WOM

The e-WOM is any statement made by customers about a product or a company, which is made available to many people through the Internet (Hennig-Thurau et al. 2004). e-WOM effects users' experiences in both positive and negative side (Sweeney et al. 2012). e-WOM quality has largely been researched in marketing method, especially on retailer websites. Nevertheless, there has been little research on e-WOM quality, while there are many researches about e-WOM. In e-business, e-WOM is considered a best strategy for attracting customers and conducting viral marketing. SNS members have trend to get valuable information about products or organizations from others' feedbacks, comments, sharings… which are very helpful for them in making the decision about online purchasing.

2.5   Knowledge

Knowledge is defined as "a justified belief" that increases one's capacity for doing something effectively (Alavi and Leidner 2001). Knowledge is a familiarity, understanding or awareness about somebody or something, such as object, information, descriptions, facts or skills, which is collected through education or experience by learning, discovering or perceiving. Knowledge is also said to be involved in the capacity of acknowledgment in human beings.

2.6   Knowledge Management

Knowledge management is defined as the system of process in organizing and managing knowledge processes, such as identifying gaps of knowledge, acquiring, developing, storing, distributing, sharing and applying the knowledge. The processes of managing knowledge has become to be critical in enhancing the performance of organizations, that can either be directed to more creation or more effectiveness. Knowledge also supplies the foundation for innovation and enhancement in organizations. (Verburg and Andriessen 2011).

2.7   Knowledge Sharing

Knowledge sharing is an important step of knowledge management process, in which one or both parties seeking and giving their knowledge, especially their tacit knowledge (know how, attitude, experience, ideas…). SNS is a good place for social interaction and sharing knowledge between members. Via SNSs, individuals can raise their opinion and also show their knowledge just in time. SNSs also enable users to communicate, exchange idea, interact and share their knowledge with other ones in naturally and friendly. SNSs create a good platform for keeping user's engagement by collect the feature, such as pages or groups.

3   Research Model

This research has referred to the previous study, Choi and Scott (2013) examined the model of impacts among Using SNSs, Social capital, e-WOM and Knowledge sharing in context of United State and Korean. Besides, Adhi (2014) explored whether the intensity of use of SNSs is related to users' social capital dimensions including structural, relational, and cognitive, and furthermore their relation with knowledge sharing behavior. Moreover, Hsu et al. (2013) examined how social relationship factors influence on e-WOM behaviors in SNSs. This study indicated that social capital positively impact on e-WOM behaviors, and there could be a difference in these impacts between Taiwan and Vietnam context.

In purpose to draw a full picture about these items in context of Vietnam, the research model of Choi and Scott (2013) is chosen. In Vietnam, due to specific of human, economic, social and political, this model research may not work or lead to new results which are different from other countries. This model could be summarized in the following figure (Fig. 1).

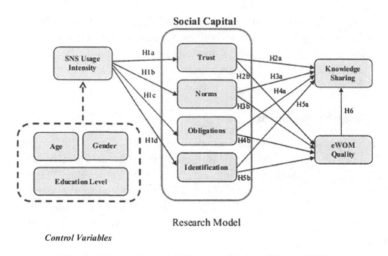

Research Model

Control Variables

Fig. 1. Research model (Source: Choi and Scott 2013)

Trust has been examined in various industries, and this leads to occurs many definitions of trust (Hsu et al. 2007). In this research, the trust is defined as "an expectation that others one chooses to trust will not behave opportunistically by taking advantage of the situation" (Gefen et al. 2003). From previous study, SNSs enable social networks to be visible to people, and therefore contribute to the trust between social network members. So we hypothesize the following:

Hypothesis H1a: The SNS usage intensity positively affects to trust in SNSs.

A norm occurs when people kept the socially defined right to control their action (Nahapiet et al. 1998). A norm creates a form of social capital when it occurs and is valid (Coleman 1988). Collaborative norms enable coordination and cooperation for mutual benefit (Putnam 2000). A norm is a reasonable way in a community, which is supported by SNSs to overcome geographic distance. Thus, we hypothesize the following:

Hypothesis H1b: The SNS usage intensity positively affects to norms in SNSs.

Obligations are known as credit and delegate "a commitment or duty to undertake some activity in the future" (Nahapiet and Ghoshal 1998). Obligations are different with norms because it relates to more personal relationships. Besides, previous research found out that virtual community is very important for developing obligations (Lesser and Storck 2001). So we hypothesize the following:

Hypothesis H1c: The SNS usage positively affects to obligations in SNSs.

Identification is defined as the process that each members see themselves as one with other members in their group (Nahapiet and Ghoshal 1998). As a member in the community, they build up a common identity for the community (Sherif et al. 2006). In the opposition, identification promotes members to join in virtual communities (Hung and Li 2007). Previous research shows that usage of SNSs has positive impacts on identification of the group. Therefore, we hypothesize the following:

Hypothesis H1d: The SNS usage intensity positively affects to identification in SNSs.

Based on Choi and Scott (2013), social capital including trust, norms, obligations and identification has a positive influence on knowledge sharing behavior of people using SNSs. Accordingly, we suggest the following hypotheses:

Hypothesis H2a: Trust positively affects to knowledge sharing in SNSs.
Hypothesis H3a: Norms positively affects to knowledge sharing in SNSs.
Hypothesis H4a: Obligations positively affects to knowledge sharing in SNSs.
Hypothesis H5a: Identifications positively affects to knowledge sharing in SNSs.

Based on Choi and Scott (2013), social capital including trust, norms, obligations and identification has a positive influence on e-WOM quality of people who use SNSs. Accordingly, we suggest the following hypotheses:

Hypothesis H2b: Trust positively affects to e-WOM quality in SNSs.
Hypothesis H3b: Norms positively affects to e-WOM quality in SNSs.
Hypothesis H4b: Obligations positively affects to e-WOM quality in SNSs.
Hypothesis H5b: Identifications positively affects to c-WOM quality in SNSs.

There is a relationship between e-WOM quality and online purchasing behavior, but there is a lack of proof about the relationship between e-WOM quality and knowledge sharing. However, knowledge sharing depends on trust and e-WOM could be a good input for building trust between members and lead them to knowledge sharing behavior (Chang et al. 2010). We realized that if SNS users get good e-WOM from other members, then they may be encouraged to share their knowledge and want to bring benefits to their friends. As a result, we hypothesize the following:

Hypothesis H6: e-WOM quality positively affects to knowledge sharing in SNSs.

4 Research Process

The research will pass through two stages, Pilot (n = 50) for primary test and finalize the questionnaire, and then, Quantitative phase (a survey is conducted with n = 357) for testing reliability of scales and analyzing the results. Below diagram presents the

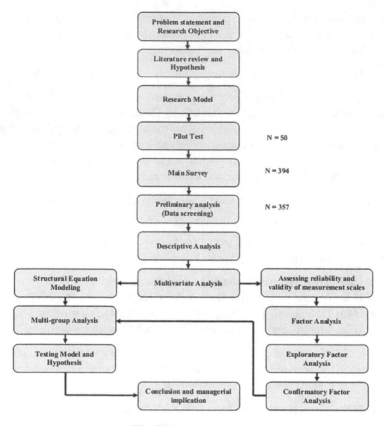

Fig. 2. Research process

entire research process with the main stages from survey which for data collection to data analysis (Cronbach Alpha test, EFA, CFA, SEM…) (Fig. 2).

4.1 Measurement Scales

There are seven constructs in this research model which are SNS usage intensity, Social capital (Trust, Norms, Obligations and Identification), Knowledge sharing, and e-WOM quality. Most of the elements in these constructs are built based on Choi and Scott (2013) as discussed earlier, some findings which are not suitable from the pilot stage which are removed out of the original questionnaire. The scales which are based on English are translated to Vietnamese and major meanings of the scales are keeping. A pilot test is firstly executed with 50 respondents to check if they have any difficulty in answering them. Adjusting of scales is active after this phase.

4.2 Sample Size and Breakdown

Indicated that we need at least 300 cases for a rational factor analysis to be successfully conducted. For the specific demographic group, the size of collected sample should be larger than 30. The larger size of collected sample, the more rational and typical the results and the lower the errors. So it would consider being an acceptable sample with 357 collected ones.

4.3 Data Analysis Methods

After collecting the data, the questionnaires are audited before being processed by the SPSS software version 20.0. Data analysis is executed by both SPSS and IBM's Analysis of Moment Structures (AMOS) version 21.0 on Windows OS. The structural process of analysis data in this research includes the below stages: Descriptive Statistics, Assessment of Measurement Scales, Exploratory Factor Analysis, Confirmatory Factor Analysis, Structural Equation Modelling for test model and hypothesis, and Bootstrap analysis.

5 Analysis Results

5.1 Sample Description

A total of 394 complete samples were collected including 200 responses (50%) from my Facebook friends, 127 responses (32%) from my company, and 57 responses (18%) from HCM Technology University. Based on the two ways of collecting data, the respondent is 279 (73%) for the paper survey, and 105 (27%) for the online (link sharing) survey.

However, only 357 questionnaires (90.61%) passed through the audited step. The number of qualified questionnaires met the sample size requirement suggested by Hair et al. (2006) to generate valid fit measurement and to avoid drawing inaccurate inference. The sample of 357 valid respondents was grouped based on their ages, genders and educational levels. Details are presented in the Table 1.

Table 1. Demographic profile on age, gender and educational

Demographic profile		Frequency	Percentage
Age (years of age)	Below 21	81	22.7%
	21–30	115	32.2%
	31–40	156	43.7%
	Above 40	5	1.4%
Gender	Male	221	61.9%
	Female	136	38.1%
Educational degree	High school	69	19.3%
	College/University	143	40.1%
	Master/PhD	103	28.9%
	Others	42	11.8%

We can see that the major group of respondents was between 21–40 years old (total 76%); followed by the youngest group with 22.7% and the oldest group is only 1.4% - this group maybe inconsiderable. Concerning the respondent ratio between males and females, the data shows that males joined mainly the survey about 1.5 times the number of females. About the educational levels, most of the participants had university or college degrees (66.4%). Especially, Master or PhD level is fairly high in this survey (28.9%). Besides, there is not much different in the number of participants from the lower and higher in the other education groups.

5.2 Preliminary Assessment of Measurement Scales

Uni-dimensionality Test
The total of factors are 10 constructs, namely, SNSs Usage Intensity, Trust, Norm, Obligation, Identifications, e-WOM Quality and Knowledge Sharing with total 32 measurement items. All these constructs must be uni-dimensional that means the indicators have only one underlying construct. Analysis result shows that all constructs have matched these requirements: Eigen value greater than 1; KMO index must be above 0.5 and less than 1; Total variance explained criteria >50%. So, the specific scales for each of these constructs are uni-dimensional.

Reliability Analysis
Cronbach's alpha coefficient of seven constructs was estimated, which ranged from 0.776 to $\alpha = 0.887$ (before deleting unfitting items) and from 0.776 to $\alpha = 0.935$ (after deleting unfitting items). There were three items which have corrected item-total correlation <0.3, which demonstrates the item is presenting different thing which do not belong to the scale. They were: TRUST4 (0.294), IDENTI5 (0.174) and E-WOMQL1 (0.188).

After deleting 3 above items, the alpha value of perceived Identification, e-WOM Quality was very high, over 0.90. The other constructs – Trust, Obligations, Norms and Knowledge Sharing had also values that is higher than 0.8. The construct of SNS Usage Intensity had lowest alpha value (0.776), and the highest one (0.935) belongs to construct of Identification. In addition, there was no item had corrected item-total correlation <0.3, which presents different thing which does not belong to the scale.

5.3 Exploratory Factor Analysis

Results of EFA
A next step of EFA with the same configuration as in previous step (example the principal axis factoring, Eigen-value must be ≥ 1 and method of PROMAX rotation) is executed to evaluate convergent validity and discriminant validity. The below rules must be followed for checking the outputs: Total value of cumulative variances is >60%; The factors are kept for further analysis only if it has eigenvalues of at least >=1.0; The

absolute value of difference between its max loading value and any other loading value must be >0.2; Minimum factor loading value for each of indicator must be >0.35.

Revised Research Model After EFA

The result of above EFA step also support a revision to previous research model for better reflection of these significant factors and their effects.

According to above criteria, after above steps, seven factors have been grouped as follows: e-WOM (E-WOMQL7, E-WOMQL5, E-WOMQL6, E-WOMQL2, E-WOMQL4 and E-WOMQL3); Trust (TRUST2, TRUST6, TRUST5, TRUST3 and TRUST1); Identifications (IDENTI2, IDENTI1, IDENTI3 and IDENTI4); Obligations (OBLIGA5, OBLIGA1, OBLIGA4, OBLIGA2 and OBLIGA3); Knowledge Sharing (KNOWSH1, KNOWSH5, KNOWSH4, KNOWSH2 and KNOWSH3); Norms (NORMSC3, NORMSC1 and NORMSC2); SNS Usage Intensity (SNSUIN2, SNSUIN1 and SNSUIN3).

5.4   Final Assessment of Measurement Scales Using CFA

The CFA Model analysis result showed that all criteria are satisfied (Chi-square/df < 2, GFI, TLI, CFI > 0.9, RMSEA < 0.05). Therefore, the model is fit with data.

Composite Reliability, Convergent Validity and Discriminant Validity

Table below shows the results of CR, AVE scores each pair of the constructs (Table 2).

Table 2. CR, AVE scores and correlation scores

	CR	AVE
SNSUIN	0.778	0.540
E-WOMQL	0.930	0.690
KNOWSH	0.807	0.458
OBLIGA	0.875	0.585
TRUST	0.890	0.620
IDENTI	0.935	0.784
NORMSC	0.865	0.683

Assessment of Model Fit

The overview of the model fit measured by AMOS showed that the full measurement model has reached a good fit with the data. Measurement model consistent with market data and no case of errors of the observed variables are correlated with each other, thus turning the pair agency monitoring achieving unitary (Steenkamp and van Trijp 1991).

5.5   Test Model and Hypothesis Using SEM

The SEM model contained the regressions of the relation between six potential constructs, they are: TRUST, NORMMSC, IDENTI, OBLIGA, E-WOMQL and KNOWSH and the dependent potential construct SNSUIN, which is formed based on the final output from CFA analysis with the paired arrows from determined latent constructs. These potential constructs are caculated by multi-item scales and each of items may have its own error limit as shown in the figure and table below (Fig. 3 and Table 3).

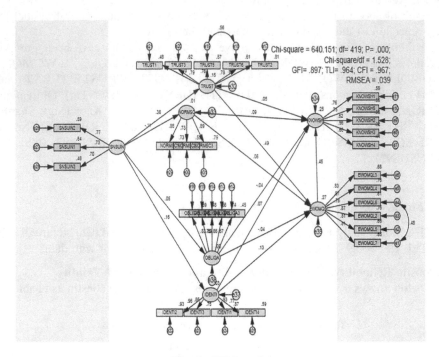

Fig. 3. SEM model

Table 3. Regression weights

Parameter			Estimate	S.E.	C.R	P label
TRUST	<—	SNSUIN	.427	.072	5.956	***
NORMSC	<—	SNSUIN	−.129	.075	−1.725	.085
OBLIGA	<—	SNSUIN	.040	0.47	.860	.390
IDENTI	<—	SNSUIN	.206	.079	2.598	.009
EWOMQL	<—	TRUST	.408	.046	8.777	***
EWOMQL	<—	NORMSC	.044	.041	1.086	.277
EWOMQL	<—	OBLIGA	−.056	.065	−.863	.388
EWOMQL	<—	IDENTI	.093	.036	2.616	.009
KNOWSH	<—	TRUST	.027	.038	.695	.487
KNOWSH	<—	NORMSC	.047	.031	1.507	.132
KNOWSH	<—	OBLIGA	−.035	.050	−.698	.485
KNOWSH	<—	IDENTI	.036	.027	1.326	.185
KNOWSH	<—	EWOMQL	.315	.052	6.087	***

Estimated Test of the Model by Boostrap Analysis

After executing SEM analysis, Boostrap analysis method is used to evaluate the reliability of the estimators. Number of repeat sampling in this research was chosen B = 500. Bootstrap estimation result is showed in below Table (Table 4).

Table 4. Bootstrap analysis

Parameter			SE	SE-SE	Mean	Bias	SE-Bias	CR
TRUST	<—	SNSUIN	0.058	0.002	0.378	−0.003	0.003	−1
NORMSC	<—	SNSUIN	0.073	0.002	−0.111	−0.001	0.003	−0.33333
OBLIGA	<—	SNSUIN	0.08	0.003	0.052	−0.002	0.004	−0.5
IDENTI	<—	SNSUIN	0.066	0.002	0.153	−0.005	0.003	−1.66667
E-WOMQL	<—	TRUST	0.049	0.002	0.489	−0.004	0.002	−2
E-WOMQL	<—	NORMSC	0.055	0.002	0.054	−0.002	0.002	−1
E-WOMQL	<—	OBLIGA	0.056	0.002	−0.045	0	0.003	0
E-WOMQL	<—	IDENTI	0.057	0.002	0.126	−0.004	0.003	−1.33333
KNOWSH	<—	TRUST	0.068	0.002	0.047	0.001	0.003	0.333333
KNOWSH	<—	NORMSC	0.057	0.002	0.089	0.003	0.003	1
KNOWSH	<—	OBLIGA	0.072	0.002	−0.034	0.005	0.003	1.666667
KNOWSH	<—	IDENTI	0.058	0.002	0.071	−0.001	0.003	−0.33333
KNOWSH	<—	E-WOMQL	0.066	0.002	0.444	−0.005	0.003	−1.66667

In summary, these measures show that the model has fits with the collected data and it could be used to present for the above hypotheses.

Testing of Hypotheses

The research hypotheses are continued to test and the result is as follows (Table 5):

Table 5. Hypothesis testing results

Hypothesis	Supported	Estimates	t-statistics
SNS usage intensity –> Trust	**Yes**	**0.427**	**5.956**
SNS usage intensity –> Norms	No	−0.129	−1.725
SNS usage intensity –> Obligations	No	0.04	0.860
SNS usage intensity –> Identification	**Yes**	**0.206**	**2.598**
Trust –> Knowledge sharing	No	0.27	0.695
Trust –> e-WOM quality	**Yes**	**0.408**	**8.777**
Norms –> Knowledge sharing	No	0.47	1.507
Norms –> e-WOM quality	No	0.044	1.086
Obligations –> Knowledge sharing	No	−0.035	−0.698
Obligations –> e-WOM quality	No	−.056	−0.863
Identification –> Knowledge sharing	No	0.036	1.326
Identification –> e-WOM quality	**Yes**	**0.093**	**2.616**
e-WOM quality –> Knowledge sharing	**Yes**	**0.315**	**6.087**

Multi-group Analysis

Chi-square test showed that there is no difference between Unconstrained Model and Constrained Model (P-value > 0.05). Therefore Constrained Model would be selected for analysis (due to higher degrees of freedom).

As a result, we conclude that there is no significant difference between Male and Female in regard to their perception of how different factors affect their using SNSs, e-WOM and Knowledge Sharing behaviors.

6 Discussion

Basically, this study has the similar result with previous study (Choi and Scott 2013). The SNS usage intensity has positive impact to both of dimension of Social Capital: Trust and Identification. The next finding is two these dimensions have main effects to e-WOM on SNSs space in Vietnam. The final result shows that Vietnamese people also think that e-WOM have good impact for Knowledge Sharing through SNSs environment. So, based on this result, some managerial implications for Vietnamese businesses, government, and SNS providers could be made as follows:

6.1 For Vietnamese Businesses

- First, it is a good foundation for organization: building the Trust and Identification among employees in the organization. The organization could use SNSs as a tool to build the Trust among employees and enlarge Identification in the organization. The organization could increase social interaction and shared vision in order to encourage knowledge sharing through SNSs.
- Second, this result could also provide the foundation for practitioners to use e-WOM on SNSs internally for building network and community to support knowledge management in organization.
- Third, users could use SNSs for supporting knowledge sharing in problem solving activities, such as software coding, technical supporting, or customer helping.
- Fourth, marketers also use the e-WOM as a marketing tool on SNSs. SNS could be good channel for firm to collect the feedbacks from customers. In positive side, these feedbacks are the priceless information for firm in improving or creating its product or service. On the other sides, many challenges that the firm could be faced. For example, in Vietnam, the organizations may face the risk which is dirty tricks from its competitors through SNSs or they cannot handle and adapt customers' opinions or complaint (Gallaugher and Ransbotham 2010).

6.2 For the Vietnamese Government

Some point of times, the Vietnamese authorities want to restrict using Facebook, they give some reasons: such as wasting working time, the risk of spread unhealthy information, Religious or regions conflicts, phishing or security online, crowd psychology... But Facebook reflects the real social; there have two sides of the object: positive side and negative side and the task of social managers should be restricting the negative side and enhancing the positive side of Facebook. There are many advantages and valuable effect for social:

- Connecting communities
- Orienting the information which is transmitted to citizen via SNSs based on their demographic: gender, age, kind of job, level of education...
- Building the Trust and Identifications dimensions in social.
- Improving the social policy.
- Spread legal Knowledge and social Knowledge.

6.3   For Vietnamese Providers of Social Networking Services

Creating virtual space with strong belief of members which is based on specific feature of group: age, hobby, interests... Based on this, they can build the Trust from the using SNSs.

Using network to collaborate among members for solving issue, sharing useful information...

The SNS management should focus more on how to create suitable policies for building the Trust dimension on online environment.

SNSs management also builds the platform that allows the member can share/public their skills/knowledge easily.

7   Conclusion and Future Research

In summary, this study finds out that the SNS usage intensity has a positive impact on Social capital (including: trust and identification), which will have a positive impact on e-WOM quality. The more people use their SNSs, the more social capital they feel obtain. Especially, this study finds that the level of trust between SNS members plays an important role in determining an individual's decision to share information/ knowledge in SNSs. Moreover, this research confirmed that e-WOM quality has a positive impact on knowledge sharing behavior in Vietnam.

The research provides the framework of social capital to test the relationship between knowledge sharing and e-WOM through SNSs in context of Vietnam. This framework is very significant because it indicates the theoretical foundation of the modeled impacts between e-WOM and Knowledge sharing through SNSs from the perspective of Social capital.

It would be a fine foundation for: Organizations, Social Managers improve their performance on managing; building a marketing of strategy; planning for Knowledge Sharing and Knowledge Management; disseminating educational content and human development as well as social issues are identified.

However, there are some of limitations in this study as below:

- Convenience sampling method.
- Just focuses on Facebook, most popular SNS in Vietnam.
- Only studies one aspect of norms (collaborative norms).

Some implications for future research could be summarized as follows:

- Examine the results with other users and in other environments.
- Conduct research on other SNSs, such as Twitter, Instagram, Myspace…
- Focus on other aspects of social capital and norms.
- Conduct in other country, in the new context of social or culture.

References

Adler, P.S., Kwon, S.W.: Social capital: prospects for a new concept. Acad. Manag. Rev. **27**(3), 17–40 (2002)

Adhi, P.: Understanding knowledge sharing and social capital in social network sites. Int. J. Sci. Res. **3**(3), 750–761 (2014)

Alavi, M., Leidner, D.E.: Review: knowledge management and knowledge management systems: conceptual foundations and research issues. MIS Q. **25**(1), 107–136 (2001)

Awad, N.F., Ragowsky, A.: Establishing trust in electronic commerce through online word of mouth: an examination across genders. J. Manag. Inf. Syst. **24**(4), 101–121 (2008)

Bayus, B.L.: Crowdsourcing new product ideas over time: an analysis of Dell's IdeaStorm community. Manage. Sci. **4**(1), 15–24 (2012)

Boyd, D.M., Ellison, N.B.: Social network sites: definition, history, and scholarship. J. Comput.-Med. Commun. **13**(1), 210–230 (2008)

Chang, C.-M., Hsu, M.-H., Cheng, H.-L., Lo, C.-H.: Exploring the antecedents of trust in virtual communities: a theoretical model for facilitating knowledge sharing. In: Proceedings of the 2010 International Conference on Business and Information, Kitakyushu, pp. 1–20 (2010)

Choi, J.H., Scott, J.E.: Electronic word of mouth and knowledge sharing on social network sites: a social capital perspective. J. Theor. Appl. Electron. Commer. Res. **8**(1), 69–82 (2013)

Chu, S.-C., Kim, Y.: Determinants of consumer engagement in electronic word-of-mouth (e-WOM) in social networking sites. Int. J. Advert. **30**(1), 47–75 (2011)

Coleman, J.S.: Social capital in the creation of human capital. Am. J. Sociol. **94**(1), 95–120 (1988)

Donath, J., Boyd, D.: Public displays of connection. BT Technol. J. **22**(4), 71–82 (2004)

Gallaugher, J., Ransbotham, S.: Social media and customer dialog management at Starbucks. MIS Q. Exec. **9**(4), 197–212 (2010)

Gefen, D., Karahanna, E., Straub, D.W.: Trust and TAM in online shopping: an integrated model. MIS Q. **27**(1), 51–90 (2003)

Godes, D., Mayzlin, D.: Using online conversations to study word-of-mouth communication. Mark. Sci. **23**(4), 545–560 (2004)

Hair, J.F., Anderson, R.E., Tatham, R.L., Black, W.C.: Multivariate Data Analysis. Prentice-Hall, New Jersey (2006)

Hennig-Thurau, T., Gwinner, K.P., Walsh, G., Gremler, D.D.: Electronic word-of-mouth via consumer-opinion platforms: what motivates consumers to articulate themselves on the internet? J. Interact. Mark. **18**(1), 38–52 (2004)

Hofstede, G.: Culture's Consequences. Sage Publications, Newbury Park (1991)

Hsu, M.-H., Ju, T.L., Yen, C.-H., Chang, C.-M.: Knowledge sharing behavior in virtual communities: the relationship between trust, self-efficacy, and outcome expectations. Int. J. Hum. Comput. Stud. **65**(2), 153–169 (2007)

Hsu, Y., Tran, T.-H.-C.: Social relationship factors influence on e-WOM behaviors in social networking sites: empirical study: Taiwan and Vietnam. Int. J. Bus. Humanit. Technol. **3**(3), 22–31 (2013)

Huang, K., Choi, N., Horowitz, L.: Web 2.0 use and organizational innovation: a knowledge transfer enabling perspective. In: Proceedings of the 16th Americas Conference on Information Systems, pp. 1–14 (2010)

Hung, K.H., Li, S.Y.: The influence of e-WOM on virtual consumer communities: social capital, consumer learning, & behavioral outcomes. J. Advert. Res. **47**(4), 485–495 (2007)

Lee, I.: Overview of emerging web 2.0-based business models and web 2.0 applications in businesses: an ecological perspective. Int. J. E-Bus. Res. **7**(4), 1–16 (2011)

Leimeister, J.M., Huber, M., Bretschneider, U., Krcmar, H.: Leveraging crowdsourcing: activation-supporting components for IT-based ideas competition. J. Manag. Inf. Syst. **26**(1), 197–224 (2009)

Lesser, E.L., Storck, J.: Communities of practice and organizational performance. IBM Syst. J. **40**(4), 831–841 (2001)

Majchrzak, A., Cherbakov, L., Ives, B.: Harnessing the power of the crowds with corporate social networking tools: how IBM does it? MIS Q. Exec. **8**(2), 103–108 (2009)

Nahapiet, J., Ghoshal, S.: Social capital, intellectual capital and the organizational advantage. Acad. Manag. Rev. **23**(2), 242–268 (1998)

Putnam, R.D.: Bowling Alone. Simon and Schuster, New York (2000)

Sherif, K., Hoffman, J., Thomas, B.: Can technology build organizational social capital? The case of a global IT consulting firm. Inf. Manag. **43**(7), 795–804 (2006)

Steenkamp, J.-B.E.M., van Trijp, H.C.M.: The use of LISREL in validating marketing constructs. Int. J. Res. Mark. **8**, 283–299 (1991)

Steffes, E.M., Burgee, L.E.: Social ties and online word of mouth. Internet Res. **19**(1), 42–59 (2009)

Sweeney, J.C., Soutar, G.N., Mazzarol, T.: Word of mouth: measuring the power of individual messages. Eur. J. Mark. **46**(1), 237–257 (2012)

Verburg, R., Andriessen, E.: A typology of knowledge sharing networks in practice. Knowl. Process Manag. **18**(1), 34–44 (2011)

An Experiment to Evaluate Software Development Teams by Using Object-Oriented Metrics

Jamille S. Madureira, Anderson S. Barroso, Rogerio P.C. do Nascimento, and Michel S. Soares$^{(\boxtimes)}$

Federal University of Sergipe, São Cristóvão, SE, Brazil
jamillemadureira@gmail.com, giga.anderson@gmail.com,
{rogerio,michel}@dcomp.ufs.br

Abstract. Managing a software project is a task that becomes increasingly difficult as software complexity increases. Gathering and interpreting software metrics is a mean to help the project team to achieve its goals. The objective of this work is to use software metrics to evaluate teams and individuals by analyzing current performance of developers. An experiment was realized in four organizations, two universities and two companies. Information about participants was collected and the object-oriented metrics of software produced by them were calculated. As a result, evidence was found that gathering software metrics is useful in activities of project management and to evaluate software development team members. In this way, software metrics can contribute during activities of software development, and also can advise managers with decisions that cause changes in the team.

Keywords: Experiment · Object-oriented software metrics · CK metrics · MOOD metrics · Project management

1 Introduction

Software that meets the needs of users is considered as one of the main goals of quality management [17]. Understanding the effort required by a development team is one of the factors that influence an organization to achieve better quality of the software development process. Individual performance tends to grow with increased knowledge of the application domain and its skill sets. Evaluation of individual and team effort by the project manager influences quality of the project [11,22].

Software measurement is the process of representation in quantitative numbers of software entities, such as processes, products and resources [26]. Software metrics are functions whose inputs are the software data and the output is a single numerical value that can be interpreted as the degree to which software has a certain attribute that affects its quality. Software metrics are intended to measure and evaluate performance characteristics of software and can be used to identify product defects and evaluate software quality [13].

© Springer International Publishing AG 2017
O. Gervasi et al. (Eds.): ICCSA 2017, Part VI, LNCS 10409, pp. 128–144, 2017.
DOI: 10.1007/978-3-319-62407-5_9

When a set of metrics is available, managers can assess whether the project is progressing according to project plans. Metrics are also useful for determining the current status of a project and evaluating its performance. Early identification of risks associated with different project activities provides an opportunity to increase efforts on the most critical tasks [13].

The main objective of this research is to use software metrics to evaluate teams and individuals by analyzing the current performance from a management point of view in a software development project. Within this research, software was evaluated by using the CK (Chidamber and Kemerer) [5] and MOOD (Metrics for Object Oriented Design) [1] metrics sets applied to softwares developed in Java. Results were analyzed to identify existence of evidence that using metrics is useful for management of projects and for evaluation of team members.

2   Related Works

Software metrics are often applied to evaluate characteristics of the software development process [12, 16, 21, 23, 32]. However, most frequently the focus of managers is on product's quality and meeting deadlines and costs. In this work, the application of software metrics will contribute to the decision making for project managers considering the technical qualification of developers. Within this paper, metrics are gathered to evaluate the development team by means of the produced software.

In the research described in [29], the authors state that measurement plays an important role in all phases of the software development process. They emphasize the measurement of different quality attributes, including reusability, ease of maintenance, testability, reliability and efficiency. The CK metrics were used to measure these attributes. The authors concluded that to develop high quality software, coupling between objects and lack of cohesion of methods should be kept low. A moderate level of inheritance tree depth and number of children are useful in building high quality software.

In the work reported in [33], the authors explain that the need for measurement to allow effective control in software projects is well recognized. It is necessary to understand the technical evolution of projects, especially where the projects have high levels of complexity. Measurement management would be useful to organizations as it would help to control the management of roles in a team.

3   Metrics for Object-Oriented Design

Object-oriented metrics have become important because of the growing popularity of this standard as the basis for most software systems developed in past years [24, 27]. Therefore, we have also considered object-oriented metrics in this work. More specifically, the CK and the MOOD sets of metric were applied to gather metrics from software. CK and MOOD are briefly introduced in the following subsections.

3.1   CK Metrics

The set of CK metrics was defined by Chidamber and Kemerer in 1994 [5]. The goal of these metrics is to measure the complexity of the project in relation to its impact on quality attributes such as usability, ease of maintenance, functionality and reliability. Six metrics were defined:

1. Depth of Inheritance Tree (DIT): is defined as the maximum length from the node to the root of the tree.
2. Number of Children (NOC): this metric determines the number of direct subclasses of a given class by evaluating the breadth of its hierarchy.
3. Coupling between Object Classes (CBO): is defined as the number of couplings between a given class with all other classes.
4. Response For a Class (RFC): is defined as a set of methods that can be performed in response to a message received by an object of that class.
5. Lack of Cohesion in Methods (LCOM): indicates the lack of method cohesion.
6. Weighted Methods Per Class (WMC): evaluates the complexity of a class, not considering the inherited methods.

3.2   MOOD Metrics

The MOOD metric set was proposed by Abreu and Carapuça in 1994 to measure the use of object-oriented design methods [1]. Six metrics were defined, and are briefly described as follows:

1. Method Inheritance Factor (MIF): is defined as a quotient between the sum of the inherited methods in all software classes and the total number of available methods for all classes.
2. Attribute Inheritance Factor (AIF): is defined as a quotient between the sum of the inherited attributes in all software classes and the total number of available attributes for all classes.
3. Method Hiding Factor (MHF): is calculated by the ratio of the number of hidden methods in all software classes and the total number of methods defined in all software classes.
4. Attribute Hiding Factor (AHF): is calculated by dividing the number of hidden attributes in all software classes and the total number of attributes defined in all software classes.
5. Polymorphism Factor (PF): is defined as the quotient between the real number of possible different polymorphic situations and the maximum number of possible distinct polymorphic situations per class.
6. Coupling Factor (CF): is defined as the ratio of the maximum possible number of couplings in the system to the real number of couplings not assigned to the inheritance.

4   Methodology

In order for metrics to be used effectively, software tools were used that collected the metrics selected and generated the result automatically. Software development team from two companies were set for the case study. In addition, the experiment was also executed with undergraduate students from two universities performing activities in a course. Metrics described in Sect. 3 were gathered from developed softwares. Results were analyzed and compared with the parameters already described in the literature.

In this context, the team was evaluated technically by means of the developed software, allowing project managers to take decisions about the team based on the technical performance of its members. Examples of decisions include the need for specific training, salary increases, suitability of project to work and job changes.

Planning and operation of the experiment are described in Sects. 6 and 7, respectively.

5   Reference Values

5.1   CK Metrics

In the work described in [25], the authors describe a statistical survey of the values of CK metrics and suggested values considered as low, high and anomalies. Several scientific journals and conferences were searched for studies that described the results in software analysis using the CK metrics. Considered programming languages were Java and C++. As a result, a metric rank was created based on its values.

In research described in this paper, only software products coded in Java were analyzed. Therefore, only the search results shown in [25] for this language will be considered. Table 1 shows the values found in [25] considered low, high, and anomaly for CK metrics.

Table 1. Reference values for CK metrics

Metric	Low	Normal	High	Anomaly
DIT	0	1	2	3
NOC	0	1–5	6–7	8
CBO	2	3–14	15–19	20
RFC	0	1–64	65–83	84
LCOM	0	1–297	298–387	388
WMC	1	2–21	22–28	29

5.2   MOOD Metrics

The objective of this study was to identify inappropriate behavior patterns of object-oriented software development according to the set of MOOD metrics. Values for the applied metrics were searched in the literature, in order to obtain criteria that serve as a basis for comparison with other developed software. When categorizing the values, four intervals were defined: low, normal, high and anomaly. In order to establish values that are considered anomalies for MOOD metrics, a search was performed in articles that describe results gathered from the application of the MOOD set of metrics. Data were extracted from the chosen articles, tabulated and analyzed using statistical methods. Finally, a metrics ranking was created based on these values.

Table 2. MOOD metrics obtained

Paper	Sw	Metric					
		MIF	AIF	MHF	AHF	PF	CF
[10]	S1	0.15	0.17	0.10	0.46	0.04	0.04
	S2	0.14	0.11	0.08	0.67	0.05	0.04
	S3	0.21	0.15	0.16	0.66	0.09	0.04
	S4	0.27	0.31	0.10	0.44	0.03	0.06
	S5	0.46	0.47	0.25	0.63	0.07	0.03
	S6	0.34	0.26	0.15	0.68	0.05	0.05
	S7	0.23	0.20	0.16	0.52	0.07	0.05
	S8	0.37	0.37	0.16	0.49	0.06	0.05
	S9	0.27	0.32	0.15	0.51	0.06	0.05
[19]	S10	0.34	0.11	0.18	0.62	0.05	0.10
[23]	S11	0.21	0.24	0.68	0.07	∅	∅
	S12	0.20	0.24	0.64	0.06	∅	∅
	S13	0.17	0.17	0.62	0.05	∅	∅
	S14	0.17	0.17	0.61	0.05	∅	∅
	S15	0.17	0.17	0.57	0.05	∅	∅
	S16	0.20	0.18	0.54	0.06	∅	∅
[7]	S17	0.14	∅	∅	∅	0.07	∅
	S18	0.12	∅	∅	∅	0.07	∅
	S19	0.11	∅	∅	∅	0.07	∅
	S20	0.11	∅	∅	∅	0.07	∅
	S21	0.21	∅	∅	∅	0.07	∅
[6]	S22	0.18	0.19	0.21	0.10	0.25	0.08
[2]	S23	0.02	0.06	0.89	0.95	0.10	∅
AVG		0.20	0.22	0.35	0.39	0.07	0.05
STDEV		0.10	0.10	0.26	0.29	0.05	0.02

The search was performed in the following databases: ACM Digital Library, IEEE Xplore, ScienceDirect and SpringerLink. To limit the scope of the search, initially only studies conducted from year 2006 to 2016 were selected. This criterion was chosen to leave the present study with information of current articles. Due to the low number of works found, the search was expanded for another ten years. We started from 23 articles found. After a manual filter, based on the abstracts, articles that did not fit the proposal of this study were rejected. After this task, 06 articles were chosen.

Table 2 shows metric values, mean, and standard deviation obtained from the analyzed articles. Unreported values are represented by the symbol $\varnothing$.

For each MOOD metric, four classes of values were defined: low, normal, high and anomaly. Values below or equal to the result of the subtraction between average (AVG) and the standard deviation (STDEV) are classified as low. Normal values are between low and high values. Values that are greater than or equal to the result of the sum between average (AVG) and standard deviation (STDEV) and below the anomaly are considered as high. Finally, an anomaly is classified according to the proposal of [18], where it is calculated by the sum between average (AVG) and standard deviation (STDEV) plus 50%. Values found are shown in Table 3.

Table 3. Reference values for MOOD metrics

Metrics	Values			
	Low	Normal	High	Anomaly
MIF	≤0.10	0.11–0.29	0.30–0.34	≥0.35
AIF	≤0.12	0.13–0.31	0.32–0.36	≥0.37
MHF	≤0.09	0.10–0.60	0.61–0.73	≥0.74
AHF	≤0.10	0.11–0.67	0.68–0.82	≥0.83
PF	≤0.02	0.03–0.11	0.12–0.13	≥0.14
CF	≤0.03	0.04–0.06	0.07	≥0.08

6 Experiment Definition and Planning

In this section we will present the definition and planning of the experiment, following the guidelines described in [34].

6.1 Goal Definition

The goal of this research was to use metrics to evaluate software products developed according to the object oriented paradigm. Results were used to evaluate performance of the development team. Thus, the experiment was realized with programmers of two companies and undergraduate students from two universities. The goal of the experiment was formalized using the GQM model originally proposed in [3]:

1. Analyze the applicability of metrics to evaluate software quality from the point of view of project management;
2. In order to technically evaluate the software development team;
3. From the point of view of programmers and software development managers;
4. In the context of programmers in companies and undergraduate students.

6.2 Planning

6.2.1 Hypotheses Formulation

Questions to be answered in this research are: (1) Can object-oriented software metrics be used to evaluate software quality from the point of view of project management?; and (2) Is it possible to technically evaluate software development teams by means of software metrics?

In order to answer these questions, Software Engineering metrics have been applied for measuring software products. Chosen metrics were described in Sect. 3.

After defining the goals and metrics, the following hypotheses were considered:

1. Hypothesis 1
   (a) H0GP: Software metrics are not useful in activities of project management.
   (b) H1GP: Software metrics are useful in activities of project management.
2. Hypothesis 2
   (a) H0AE: Software development teams can not be technically evaluated by means of metrics gathered from their software products.
   (b) H1AE: Software development teams can be technically evaluated by means of metrics gathered from their software products.

6.2.2 Independent Variable

Independent variable of this experiment was the application of object-oriented metrics. In order to collect the metrics, a search was performed to find a tool that would perform this task automatically and obey some criteria, such as: being simple to use, non-commercial software, and that could be able to collect metrics using the two sets of metrics proposed in this research, CK and MOOD. After this search, the chosen tool was IntelliJ Idea [14]. Figures 1 and 2 shows examples of the metrics CK and MOOD, respectively.

6.2.3 Dependent Variables

Concerning MOOD metrics, the dependent variables are the results found, since they are unique values for the whole project. In relation to CK metrics:

1. Average metric values: for each metric, it was calculated by the ratio of the sum of all values to the total quantity of values.
2. Maximum metric values: for each metric, the highest among all values was observed.
3. Minimum value of metrics: for each metric, the lowest of all values was observed.

Metrics Chidamber-Kemerer metrics for Project 'untitled' from Sex, 30 set 2016 15:07:47 BRT

Class metrics

class	CBO	DIT	LCOM	NOC	RFC	WMC
Siren	2	0	3	0	4	5
Synthesizer	2	0	3	0	4	5
TestAlarmSys	2	0	1	0	3	2
TestBuildingMonitor	1	0	1	0	2	1
TestEquipment	3	0	1	0	7	3
WindowSensors	3	0	1	0	5	8
Total						86
Average	2,84	0,05	1,16	0,05	4,11	4,53

Fig. 1. Result of CK metrics

Metrics MOOD metrics for Project 'untitled' from Sex, 30 set 2016 15:06:59 BRT

Project metrics

project	AHF	AIF	CF	MHF	MIF	PF
project	2,33%	4,44%	15,79%	0,00%	2,08%	0,00%

Fig. 2. Result of MOOD metrics

6.2.4 Interfering Variables

Variables that can influence the experiment are: the number of participants, the experiment in the companies was only realized in public institutions, and the experiment was realized with undergraduate students from different institutions.

6.2.5 Selection of Participants

The experiment was realized with two different groups: undergraduate students and programmers in their job environment. Two universities allowed the participation of their students in this research. To ensure the effectiveness of this research, in each university the professor requested to all students to develop the project based on a set of common requirements. To select the participants, only those whose projects solved the proposed problem with completeness were chosen.

Two public companies allowed the case study to be applied in their software development environment. Each company allowed three developers to participate in the experiment. To ensure the effectiveness of the experiment, the analyzed software products were coded by only one programmer, or at most one pair. The process of selecting the participants followed the criterion of finding programmers who had coded alone or in pair the software selected for the research.

6.2.6 Experiment Design

The experiment was designed to evaluate the development team by means of the quality of the software produced. For its execution, two activities were realized: on-line questionnaire to collect data from the team (applied only in the companies) and gathering of object-oriented software metrics.

6.2.7 Instrumentation

Tools used in the experiment were Google Forms, to conduct online surveys, which results can be stored in spreadsheets and graphics; and IntelliJ Idea, to automatically collect software metrics and displays the results in tables.

7 Operation of the Experiment

7.1 Preparation and Execution

To prepare the experiment, the following steps were followed:

1. Form: it was defined that the participants would answer a questionnaire about their education, programming experience, working time in the company, among other information.
2. IntelliJ Idea tool training: a training about the tool with programmers and professors, from installation to how to calculate metrics.

At the end of the previous steps, the experiment started, followed according to the planning described in Sect. 6.2.

7.1.1 Data Collection

After the instructions were explained, all participants stated that they had no doubts about the experiment. Then, they agreed to answer the questionnaire and collect the proposed metrics.

For the selected softwares, all metrics from CK and MOOD sets were collected. The results were compared with values described in Sect. 5. The analysis of the collected data is presented in Sect. 8 of this paper.

7.2 Data Validation

For the realization of the experiment, the sets of metrics CK and MOOD were considered. In this context, the averages, the largest and the smallest values of the results of each CK metric and the result of the MOOD metrics were computed.

8 Results

8.1 Data Collected from Participating Companies

Research was realized in the development sector of two public companies. Table 4 depicts the relationship between the companies, the programmers, and their software product.

Table 5 shows the age of each programmer, as well as the training level and whether they have any certification in the area.

Table 6 presents data about professional programming experience.

Table 4. Company × Software × Programmer

Company	Software	Programmer
C1	S1	P1, P2
	S2	P3
C2	S3	P4, P5
	S4	P6

Table 5. Scholarity

Programmer	Age	Scholarity	Certification	Capacity in progress
P1	25	Bachelor		Master's degree
P2	27	Bachelor	X	
P3	28	Bachelor		
P4	26	Bachelor		
P5	38	Bachelor		
P6	30	Bachelor	X	Master's degree

Table 6. Programming experience

Time (years)	P1	P2	P3	P4	P5	P6
Up to 1						
Between 2 and 4	X			X		
Between 5 and 7		X	X			
Between 8 and 10					X	X
Over 10						

It was also asked to each programmer how long do they have been working in the company, and it was found that only one participant (P6) had been working between 8 and 10 years at the same company. The others affirmed their working period varied between 2 and 4 years. Experience of programming, in years, is as follows: P1–4, P2–7, P3–5, P4–4, P5–10, and P6–2.

8.2   Data Collected on Undergraduate Students

In total, 09 undergraduate students from two different universities participated in the experiment. This information is shown in Table 7.

8.3   Metrics Results

Tables 8 and 9 shows the values of metrics collected from software developed by the students and programmers, where abbreviations mean respectively:

Table 7. Students' information

University	Course	Student	Semester	Age
U1	Information Systems	ST1	9	24
		ST2	4	23
		ST3	9	23
		ST4	9	23
	Computer Science	ST5	9	22
U2	Information Systems	ST6	9	24
		ST7	9	23
		ST8	8	23
		ST9	6	23

Table 8. Result of students metrics

SW	NC	SM	CK metrics						MOOD metrics					
			DIT	NOC	CBO	RFC	LCOM	WMC	MIF	AIF	MHF	AHF	PF	CF
S1	04	Ag	2.00	1.33	2.00	67.50	0.75	67.00	0.00	0.71	0.66	0.93	1.00	1.00
		Mx	6.00	1.00	4.00	148.00	1.00	190.00						
		Mn	0.00	0.00	1.00	0.00	0.00	0.00						
S2	06	Ag	1.00	1.00	3.66	10.00	1.83	6.16	0.00	0.00	0.15	0.90	1.00	0.87
		Mx	1.00	1.00	5.00	25.00	5.00	13.00						
		Mn	1.00	1.00	2.00	2.00	1.00	2.00						
S3	07	Ag	1.00	1.00	1.71	7.29	1.86	9.43	0.00	0.26	0.63	0.73	1.00	0.50
		Mx	1.00	1.00	3.00	20.00	3.00	22.00						
		Mn	1.00	1.00	0.00	1.00	1.00	1.00						
S4	04	Ag	1.00	1.00	3.00	22.5	3.25	57.50	0.00	0.00	0.14	1.00	1.00	1.00
		Mx	1.00	1.00	3.00	26.00	6.00	95.00						
		Mn	1.00	1.00	3.00	19.00	1.00	36.00						
S5	25	Ag	1.13	0.13	1.33	7.58	1.96	8.08	0.00	0.00	0.35	0.78	1.00	0.09
		Mx	2.00	3.00	3.00	38.00	5.00	38.00						
		Mn	1.00	0.00	0.00	0.00	0.00	0.00						
S6	11	Ag	1.91	0.18	2.91	10.91	2.45	10.18	0.13	0.13	0.00	1.00	0.14	0.29
		Mx	3.00	2.00	9.00	42.00	5.00	40.00						
		Mn	1.00	0.00	1.00	2.00	0.00	1.00						
S7	12	Ag	2.31	0.23	2.46	7.62	1.62	5.23	0.24	0.28	0.00	0.94	0.24	0.21
		Mx	3.00	2.00	10.00	25.00	6.00	18.00						
		Mn	1.00	0.00	1.00	3.00	0.00	2.00						
S8	05	Ag	1.60	1.20	3.20	9.80	1.80	6.60	0.80	0.78	0.00	0.00	0.02	0.80
		Mx	2.00	2.00	4.00	12.00	5.00	9.00						
		Mn	1.00	0.00	3.00	6.00	1.00	1.00						
S9	05	Ag	1.80	0.60	2.40	10.80	2.20	7.80	0.48	0.35	0.00	0.82	0.06	0.60
		Mx	3.00	2.00	3.00	15.00	4.00	12.00						
		Mn	1.00	0.00	2.00	6.00	1.00	1.00						

Table 9. Result of companies metrics

SW	NC	SM	CK Metrics						MOOD Metrics					
			DIT	NOC	CBO	RFC	LCOM	WMC	MIF	AIF	MHF	AHF	PF	CF
C1	18	Ag	1.00	0.56	3.11	19.72	3.06	19.56	0.00	0.00	0.16	0.90	1.00	0.18
		Mx	1.00	3.00	8.00	48.00	8.00	46.00						
		Mn	1.00	0.00	0.00	1.00	1.00	1.00						
C2	06	Ag	1.00	0.17	1.17	56.17	3.00	79.00	0.00	0.00	0.29	0.89	1.00	0.50
		Mx	1.00	1.00	4.00	137.00	7.00	176.00						
		Mn	1.00	0.00	0.00	2.00	1.00	3.00						
C3	1053	Ag	1.00	0.57	7.52	14.43	2.40	13.59	0.65	0.50	0.09	0.97	0.02	0.01
		Mx	4.00	172.00	296.00	132.00	51.00	183.00						
		Mn	0.00	0.00	0.00	0.00	0.00	0.00						
C4	40	Ag	0.00	0.00	2.37	11.40	2.57	12.23	0.00	0.00	0.02	0.88	1.00	0.07
		Mx	0.00	0.00	8.00	61.00	18.00	95.00						
		Mn	0.00	0.00	0.00	0.00	0.00	0.00						

SW = software, NC = total number of classes, SM = statistical method of evaluation (Ag = average, Mx = maximum, Mn = minimum). Acronyms for the CK and MOOD metrics were explained in Sects. 3.1 and 3.2.

8.4 Data Analysis and Interpretation

Results of metrics collected are analyzed based on the researches shown in Sect. 5. The acronyms of software between C1 to C4 correspond to companies, whereas S1 to S9 refer to students.

8.4.1 CK Metrics

Results of the CK metrics are analyzed and compared to values presented in Table 1.

DIT: metric deep inheritance tree constitutes greater complexity in the project, as a high number of methods and classes are involved [5]. Softwares C1, C2 and C4 meet the standards, whereas software C3 show a DIT = 4 which is considered an anomaly. From the students' softwares, S6, S7 and S9 have classes at the limit of the value considered as anomaly, and S1 has doubled that value.

NOC: this metric evaluates efficiency, reuse, and testability [8]. In [35] it was observed that the NOC metric is inversely proportional to error propensity. The more children a class have, the more tested it was. In software C3 value was detected far above the limit considered as anomaly.

CBO: excessive coupling between classes prevents reuse. The higher the coupling number, the maintenance will be more difficult and the more accurate testing needs to be performed [5]. In software C3 found value was much higher than that of the anomaly. In this way, it is possible to affirm that the maintenance of these software is a difficult task to realize.

RFC: the greater the number of methods that can be invoked from a class, the greater its complexity is, which makes testing more difficult [5]. All the software produced in the companies, along with S1, obtained values superior to the anomaly.

LCOM: the cohesion of methods of a class is desirable since it promotes encapsulation. The lack of cohesion implies that the class should be divided into two or more subclasses. With low cohesion, complexity increases as well as the probability of errors during the development process [5]. All analyzed softwares were below the value considered as anomaly.

WMC: the number and complexity of methods involved makes it possible to predict how much time and effort it will take to develop and maintain the class. The greater the number of methods in a class, the greater the potential impact on the child classes, since these will inherit all methods defined in the class. Classes with a large number of methods are likely to be a specific application, limiting the possibility of reuse [5,8]. All softwares from companies obtained low values for this metric. Among students, S1, S4, S5 and S6 have classes at the limit of the value considered as anomaly.

8.4.2 MOOD Metrics

Results of the MOOD metrics are analyzed and compared to values presented in Table 3.

MIF: high values indicate excessive inheritance, leading to greater coupling and reducing the possibility of reuse [9]. In this context, C3, S8 and S9 reached the value of the anomaly, in softwares C1, C2, C4, S1, S2, S3, S4, S5 the zero value was identified for this metric, indicating a lack of inheritance.

AIF: high values indicates the overuse of inheritance [9]. For this metric, anomalous values were found in C3, S1 and S8. However, in softwares C1, C2, C4, S2, S4 and S5 the results were null.

MHF: high values indicates the software has mostly private methods, which compromises functionality and reuse. On the other hand, low values indicate that most methods are public, which makes them unprotected [28]. Of the analyzed softwares, none presented an anomalous result, but S1 obtained a value considered high, C3 and C4 presented low values and S6, S7, S8, S9 obtained zero results.

AHF: high values reduces functionality and reuse and low values indicate that most attributes are vulnerable [28]. Of the analyzed softwares, C1, C2, C3, C4, S1, S2, S4, S6 and S7 reached the value considered as anomaly, S8 obtained null value, while the others resulted in high values.

PF: polymorphism suggests that, in some cases, the predominant methods could contribute to reduce complexity and therefore to make the system more understandable and easier to maintain [15]. In softwares C1, C2, C4, S1, S2, S3, S4, S5, S6 and S7 anomalous values were detected for this metric. With the exception of S6 and S7, all other software reached the maximum possible value (1.00).

CF: as the coupling between classes increases, the density of defects and rework must also increase. In this way, excessive coupling in software systems has a negative impact on software quality and should be avoided [15]. Softwares S1, S4 and all developed by students reached the value considered as an anomaly for this metric.

9   Software Metrics vs. Development Team

Metrics help in the management of software projects, since they are able to evaluate reuse (DIT, NOC, CBO, WMC, MIF, AIF), efficiency (NOC), test capability (NOC, CBO, RFC), maintenance (CBO, WMC, CF), failure in class design (LCOM) and development time (WMC). By applying the metrics during project development, it is possible to evaluate in which points software should be improved and not only when the product is finalized. In this way, the H1GP hypothesis is confirmed, evidencing that software product metrics are useful in project management.

Concerning the CK metrics, it was observed that the worst results were found in C3 software. Anomalous values were detected for metrics DIT, NOC, CBO and RFC. Information reported in Sect. 8.1 shows that P4 and P5 programmers (C3 software developers) have only undergraduate degrees, with no certification or other training in progress. It has also been observed that there is a difference between the experience of these programmers. While P5 has an experience between 8 to 10 years in programming, P4 has between 2 and 4 years. Another difference is in the language experience. While P4 has 4 years of experience, P5 has been working with this language for 10 years. On the other hand, C1 and C4 softwares obtained the best results. When observing the training level of programmers, it is observed that P1 and P6 are studying toward their master's degree.

About the MOOD metrics, anomalous values were detected in all evaluated softwares. In the S1 software, anomalous values were not found only for metrics MIF and MHF. For the CF metric, anomalous values were found in all softwares developed by students. The best performance was achieved by the student S9, anomalous values were found only in two metrics (MIF and CF).

According to the results, the H1AE hypothesis is confirmed, meaning that software development teams can be evaluated technically by means of software product metrics, rejecting the H0AE hypothesis.

Although the results of the experiment were satisfactory, it presents threats to its validity that can not be disregarded.

As for internal threats to validity, the selected softwares were chosen following the criterion of being coded in maximum in pair, but there were doubts on whether other programmers participated in the development. This threat was mitigated with the commitment of the managers of the sectors and professors that the information was true. As for external threats to validity, we can mention that the low number of participants can be a threat as it may influence the results of the experiment. This threat was reduced by performing the experiment in four different locals, which means that the environmental culture had little influence on the experiment.

Finally, as for construction threats to validity, the reduced number of metrics can be considered a threat to the experiment. This threat was mitigated using the most commonly used object-oriented metric sets in the literature.

10   Conclusion and Future Works

The goal of this work was to use software metrics to evaluate development teams by analyzing the current performance of developers. An experiment was realized with eleven students from two universities and six programmers from two companies. Information about participants was collected and metrics of software produced by them were gathered.

After analyzing the results, evidences were found that gathering metrics is useful in project management and for evaluation of members of software development teams. By means of metrics a number of characteristics were analyzed, including reuse, efficiency, testability, ease of maintenance, among others. It was observed that the worst results were obtained by programmers with only undergraduate training. Another factor that influenced software was the choice of developers for the projects. Pairs with different experience produced software with unwanted results. In this sense, software metrics can help to monitor project development in relation to the quality of the final product, in the selection of new contractors, and even in decisions that cause changes to the team, such as choosing a new project manager, increasing salaries and also how to invest in training.

For future work, it is suggested to collect information that may contribute to a better understanding of the system, such as deadline of the project, periodicity of meetings about the project, if the company cares about training, among other project management activities.

Acknowledgement. The authors would like to thank the Brazilian research agency CNPq (grant 445500/2014-0) for the financial support.

References

1. Abreu, F.B., Carapuça, R.: Object-oriented software engineering: measuring and controlling the development process. In: Proceedings of the 4th International Conference on Software Quality (1994)
2. Arora, D., Khanna, P., Tripathi, A., Sharma, S., Shukla, S.: Software quality estimation through object oriented design metrics. Int. J. Comput. Sci. Netw. Secur. 11(4), 100–104 (2011)
3. Basili, V.R., Weiss, D.M.: A methodology for collecting valid software engineering data. IEEE Trans. Softw. Eng. 10(6), 728–738 (1984)
4. Boehm, W., Brown, R., Lipow, M.: Quantitative evaluation of software quality, pp. 592–605. IEEE Computer Society Press (1976)

5. Chidamber, S.R., Kemerer, C.F.: A metrics suite for object oriented design. IEEE Trans. Softw. Eng. **20**(6), 476–493 (1994)
6. Elish, M.O., Al-Yafei, A.H., Al-Mulhem, M.: Empirical comparison of three metrics suites for fault prediction in packages of object-oriented systems: a case study of eclipse. Adv. Eng. Softw. **42**(10), 852–859 (2011)
7. El-lateef, T.A., Yousef, A.H., Ismail, M.F.: Object oriented design metrics framework based on code extraction. In: International Conference on Computer Engineering and Systems, pp. 291–295 (2008)
8. Gandhi, P., Bhatia, P.K.: Reusability metrics for object-oriented system: an alternative approach. Int. J. Softw. Eng. (IJSE) **1**(4), 63–72 (2010)
9. Gupta, A., Batra, G., et al.: Analyzing theoretical basis and inconsistencies of object oriented metrics. Int. J. Comput. Sci. Eng. **4**(5), 803 (2012)
10. Harrison, R., Counsell, S.J., Nithi, R.V.: An evaluation of the MOOD set of object-oriented software metrics. IEEE Trans. Softw. Eng. **24**(6), 491–496 (1998)
11. Heinemann, L., Hummel, B., Steidl, D.: Teamscale: software quality control in real-time. In: Companion Proceedings of the 36th International Conference on Software Engineering, pp. 592–595 (2014)
12. Hussain, S.: Threshold analysis of design metrics to detect design flaws: student research abstract. In: Proceedings of the 31st Annual ACM Symposium on Applied Computing, SAC 2016, pp. 1584–1585 (2016)
13. Iqbal, S., Khan, A., Naeem, M.: Yet another set of requirement metrics for software projects. Int. J. Softw. Eng. Appl. **6**(1), 19–28 (2012)
14. Jet Brains: IntelliJ Idea (2016). https://www.jetbrains.com/idea/
15. Jassim, F., Altaani, F.: Statistical approach for predicting factors of mood method for object oriented. Int. J. Comput. Sci. Issues (2013). (Citeseer)
16. Coscia, J.L.O., Crasso, M., Mateos, C., Zunino, A., Misra, S.: Predicting web service maintainability via object-oriented metrics: a statistics-based approach. In: Murgante, B., et al. (eds.) ICCSA 2012. LNCS, vol. 7336, pp. 29–39. Springer, Heidelberg (2012). doi:10.1007/978-3-642-31128-4_3
17. Jung, H.W.: Validating the external quality subcharacteristics of software products according to ISO/IEC 9126. Comput. Stand. Interfaces **29**(6), 653–661 (2007)
18. Lanza, M., Marinescu, R.: Object-Oriented Metrics in Practice: Using Software Metrics to Characterize, Evaluate, and Improve the Design of Object-Oriented Systems. Springer Science & Business Media, Berlin (2007)
19. Ma, Y., He, K., Du, D., Liu, J., Yan, Y.: A complexity metrics set for large-scale object-oriented software systems. In: Sixth IEEE International Conference on Computer and Information Technology, pp. 189–189 (2006)
20. Mao, M., Jiang, Y.: A coherent object-oriented software metric framework model: software engineering. In: International Conference on Computer Science and Software Engineering, vol. 2, pp. 68–72. IEEE (2008)
21. Mäurer, L., Hebecker, T., Stolte, T., Lipaczewski, M., Möhrstädt, U., Ortmeier, F.: On bringing object-oriented software metrics into the model-based world - verifying ISO 26262 compliance in simulink. In: Amyot, D., Fonseca i Casas, P., Mussbacher, G. (eds.) SAM 2014. LNCS, vol. 8769, pp. 207–222. Springer, Cham (2014). doi:10.1007/978-3-319-11743-0_15
22. Nair, T.R.G., Selvarani, R.: Defect proneness estimation and feedback approach for software design quality improvement. Inf. Softw. Technol. **54**(3), 274–285 (2012). Elsevier

23. Olague, H.M., Etzkorn, L.H., Gholston, S., Quattlebaum, S.: Empirical validation of three software metrics suites to predict fault-proneness of object-oriented classes developed using highly iterative or agile software development processes. IEEE Trans. Softw. Eng. **33**(6), 402–419 (2007)
24. Radjenović, D., Heričko, M., Torkar, R., Živkovič, A.: Software fault prediction metrics: a systematic literature review. Inf. Softw. Technol. **55**(8), 1397–1418 (2013)
25. Juliano, R.C., Travençolo, B.A.N., Soares, M.S.: Detection of software anomalies using object-oriented metrics. In: Proceedings of the 16th International Conference on Enterprise Information Systems ICEIS, vol. 2, pp. 241–248 (2014)
26. Runeson, P., Höst, M.: Guidelines for conducting and reporting case study research in software engineering. Empir. Softw. Eng. **14**(2), 131 (2009). Springer
27. Misra, S., Akman, I., Koyuncu, M.: An inheritance complexity metric for object-oriented code: a cognitive approach. Sadhana **36**(3), 317 (2011). (0973-7677)
28. Sastry, J.S.V.R.S., Ramesh, K.V., Padmaja, M.: Measuring object-oriented systems based on the experimental analysis of the complexity metrics. Int. J. Eng. Sci. Technol. **3**(1), 3726–3731 (2011)
29. Srivastava, S., Kumar, R.: Indirect method to measure software quality using CK-OO suite. In: International Conference on Intelligent Systems and Signal Processing, pp. 47–51 (2013)
30. Subramanyam, R., Krishnan, M.S.: Empirical analysis of CK metrics for object-oriented design complexity: implications for software defects. IEEE Trans. Softw. Eng. **29**(4), 297–310 (2003)
31. Suresh, Y., Pati, J., Rath, S.K.: Effectiveness of software metrics for object-oriented system. Procedia Technol. **6**, 420–427 (2012)
32. Wallace, L.G., Sheetz, S.D.: The adoption of software measures: a technology acceptance model (TAM) perspective. Inf. Manag. **51**(2), 249–259 (2014)
33. Walworth, T., Yearworth, M., Shrieves, L.: Knowledge management for metrics: enabling analysis and dissemination of metrics. In: 8th Annual IEEE Systems Conference (SysCon), pp. 199–205 (2014)
34. Wohlin, C., Runeson, P., Höst, M., Ohlsson, M., Regnell, B., Wesslén, A.: Experimentation in Software Engineering. Springer Science & Business Media, Heidelberg (2012)
35. Zhou, Y., Leung, H.: Empirical analysis of object-oriented design metrics for predicting high and low severity faults. IEEE Trans. Softw. Eng. **32**(10), 771–789 (2006)

An Event-Based Technique to Trace Requirements Modeled with SysML

Telmo Oliveira de Jesus and Michel S. Soares$^{(\boxtimes)}$

Department of Computing, Federal University of Sergipe, São Cristóvão, Brazil
jtelmooliveira@yahoo.com.br, mics.soares@gmail.com

Abstract. Changes in requirements occur throughout the software process, from requirements elicitation and analysis through the operation of software. Requirements traceability makes it possible to identify the origin and the dependence of software requirements. Studies show that the tools and current requirements traceability methods are inadequate and hamper the practical use of traceability. In this work, we propose an event-based technique for tracing requirements modeled with the Systems Modeling Language (SysML). A software tool is developed to support this approach. Whenever a requirement is changed, the tool provides automatic communication with stakeholders. The tool is evaluated by information technology professionals using the Technology Acceptance Model (TAM) model that encompasses the concepts of Perceived Usefulness (PU), Perceived Ease of Use (PEOU) and Perceived Usage (PUE). As a result, we observed a high acceptance of the technique and associated tool.

Keywords: Event based traceability · Systems Modeling Language · Technology Acceptance Model · Requirements Engineering

1 Introduction

Requirements Engineering (RE) can be classified into two groups of activities: requirements development, which includes the activities of eliciting, documenting, analyzing and validating requirements, and requirements management, which includes maintenance-related activities such as traceability and requirements change management [1]. Requirements traceability is a mechanism for the management, development and maintenance of the software development process [2].

Requirements traceability life cycle includes four tasks [3]: acquisition, use, maintenance and improvement. Acquisition comprises creating traceability links between requirements and other software artifacts such as source code and test cases. These links are created manually or with the help of software tools. Traceability assists in change impact analysis and requirement dependency analysis.

© Springer International Publishing AG 2017
O. Gervasi et al. (Eds.): ICCSA 2017, Part VI, LNCS 10409, pp. 145–159, 2017.
DOI: 10.1007/978-3-319-62407-5_10

Maintenance consists of continuous review and updating of traceability links. Finally, enhancement improves the quality of traceability links.

Effective traceability practice helps in understanding software, impact analysis, finding and eliminating software errors and defects, and communicating among team members [4]. The importance of traceability has been widely recognized [5–7], and has been considered in several patterns of development [8].

The importance of requirements traceability can also be measured by the number of methods, standards, and maturity models that explicitly address this issue [9]. There is empirical evidence that requirements traceability lowers the expected rate of defects in software [7]. Therefore, traceability has a great impact on the final quality of software. Despite its importance, traceability is perhaps one of the most difficult activities performed in the software development process. The cost, effort, and discipline required to create and maintain fast-tracking software tracing links can be high [6].

Torkar et al. [5] carried out a systematic review on requirements traceability. One of the objectives of this review was to identify requirements traceability techniques. The most referenced techniques (discussed in more than one article) are event-based traceability, information retrieval, rule-based traceability, traceability matrix, and value-based traceability. Manual traceability methods, such as traceability matrix, involve tracing data manually, which for large software projects requires a considerable amount of time and effort [10]. Software tools helps improving accuracy and reduces the time required to identify traceability relationships [11].

Another issue is that requirements are often described in natural language, which leads to issues such as lack of uniform notation, and challenges, including how to describe and maintain requirements changes. In order to help with this issue, SysML [12], a system modeling language, introduces a specific diagram to model requirements. Within SysML, the Requirements Diagram helps to better organize the requirements and explicitly shows the various types of relationships between the different requirements.

Developing software effectively considering traceability among the artifacts produced is still a major challenge in practice. The key challenges in deploying traceability, such as high cost, time constraints, difficulty in maintaining traceability due to changes in software, organizational problems, and the use of manual methods [10]. In this way, there are still several problems to be solved in this theme. This article proposes a technique supported by a tool that will help in the communication with the stakeholders in the requirements changes.

This work has the objective of creating a requirements traceability technique based on event-based traceability along with the criterion of automatic communication with stakeholders. Input to this technique is using SysML Requirements tables and relationships. A tool was developed and an evaluation of the tool was carried out by a group of professionals in the field of computer science using TAM (Technology Acceptance Model) [13].

2   Background and Literature Survey

2.1   Event Based Traceability

Event Based Traceability (EBT) [14], based upon the event-notifier design pattern [15], offers a solution to maintain accurate traceability by establishing loosely coupled traceability links through the use of publish-subscribe relationships between dependent objects [16]. This type of traceability scheme is designed to handle changes [17].

In EBT, objects use an event server to subscribe to the requirements on which they are dependent. When an important change occurs regarding requirements, including actions such as replacement, refinement, or abandonment involving one of those requirements, an event notification message is published. The message is received by the event server and forwarded to all dependent objects [14]. A subscriber manager handles all incoming event notifications for a set of similarly typed objects and follows predefined protocols to deal with specific types of event messages for that object. The event manager may automatically update the managed object, whilst in other cases the event message must be stored in an event log to await manual intervention [17].

2.2   Communication with Stakeholders

Communication between the client and the developer is often fraught with ambiguities. Defects in communication between software team, mostly the requirements engineer and the test engineer, can lead to defects in testing. It is common sense that it is cheaper to catch and remove faults as early in the development process as possible. For faults caused by communication failures, reduction of faults through improvements in communication is therefore an important tactic.

Requirements management involves communication among the members of the project team and stakeholders and adaptation to changing requirements throughout the project. Requirements traceability tools must allow the discussion of project issues automatically. As an example, electronic messages should be reported to stakeholders when a discussion is generated or a requirement is changed [18].

2.3   SysML

SysML [19] is a modeling language which supports the specification, analysis, design, verification and validation of a broad range of complex systems. As an evolution of UML 2.0, SysML can be applied to model systems that may include software, hardware, information, processes and personnel. The language is useful in specifying requirements, behavior, structure, allocations of elements to models, and constraints on system properties [20]. SysML has been often applied in a variety of important activities during the system life cycle, including communication with stakeholders, improving system knowledge, model execution and documentation for maintenance [21,22].

SysML Requirements Diagram. The SysML Requirements diagram helps in better organizing requirements, and also shows explicitly the various kinds of relationships between different requirements. Another advantage of using this diagram is to standardize the way of specifying requirements through a defined semantics. SysML allows the representation of requirements as model elements, which mean that requirements are part of the system architecture [23].

SysML Requirements diagram allows several ways to represent requirements relationships, including relationships for defining requirements hierarchy, satisfying requirements, deriving requirements, verifying requirements and refining requirements. The relationships can improve the specification of systems, as they can be used to model requirements [20]. Relationships hierarchy, satisfy, derive, verify, refine and trace are briefly explained below.

The concept of **hierarchy** permits reuse of requirements. In this case, a common requirement can be shared by other requirements. The **satisfy** requirement describes how a model satisfies one or more requirements. It represents a dependency relationship between a requirement and a model element, such as other SysML diagrams, that represents that requirement. The **derive** relationship relates a derived requirement to its source requirement. During the process of Requirements Engineering, new requirements are created from previous ones. Normally, the derived requirement is under a source requirement in the hi-erarchy. The **verify** relationship defines how a test case can verify a requirement. This includes standard verification methods for inspection, analysis, demonstration or test. The **refine** relationship describes how a model element (or set of elements) can be used to later refine a requirement [20].

The **trace** relationship provides a general purpose relationship between a requirement and any other model element. The semantics of the trace relationship is not well-defined as the other relationships. For instance, a generic trace dependency can be used to emphasize that a pair of requirements are related in a different way not defined by other SysML relationships [24].

Requirements Table. SysML allows the representation of requirements, their properties and relationships in a tabular format. One proposed table shows the hierarchical tree of requirements from a master one. The fields proposed for Table 1 are the requirement ID, its name and type. There is a table for each requirement that has child requirements related by the relationship hierarchy.

Table 1. A SysML hierarchy requirements table [20]

Id	Name	Type

Other information items can be represented, as shown in Table 2. For instance, the requirement Id, the name of the requirement, to which requirement it is related (if any), the type of relationship and the requirement type. Therefore, whenever a requirement is changed or deleted by a stakeholder, the tables are useful to show that this can affect other requirements.

Table 2. A SysML requirements relationship table [20]

Id	Name	RelatesTo	RelatesHow	Type

3   Criteria for Traceability Requirements Tool

In order to efficiently manage requirements, the members of a project need to use tools throughout the software development cycle [18]. Some attributes of a requirement can be mentioned, as for instance, status, priority, necessary effort to implement, and related stakeholders [25].

Status of a requirement may cover various aspects, such as content, validation and agreement. Content status specifies the current status of the requirement content (e.g., idea, concept, detailed content). Validation status specifies the current validation status (e.g., not valid, contains error, in correction process). Status agreement specifies the current status of the agreement negotiation (e.g., not agreed, agreed, in conflict) [25].

Priority of a requirement specifies the level of importance of the requirement regarding the prioritization of selected properties (e.g., order of implementation, importance for market acceptance, cost of lost opportunity in the event of non-implementation) [25]. In this work, priority will be considered regarding the order of implementation. Prioritization is important because users have expectations on the software being developed. Delivery of requirements as perceived by users, besides contributing to their satisfaction, may be essential for the development of the project.

When referring to the requirements prioritization, possible terms include essential, important and desirable. Essential priority refers to what is really necessary for the software, without which it cannot be given as full or fit for deployment. They are requirements that, if not implemented, hinder the software deployment. The important priority should be part of the scope, but it does not block the deployment. In this case, the software contains a pending scope that will be answered in due course. This type of requirement may be delayed after deployment. Finally, the desirable priority refers to requirements that are not necessary to the software completion and its deployment. Without a desirable requirement, software should work in a satisfactory manner.

Also known as effort, the difficulty of a requirement refers to estimated effort to implement the requirement [25]. This attribute is very important in software factories, as it needs to receive a specification of explicit and complete software requirements as the specification matches the expected final product.

4   A Proposed Approach for Requirements Tracing with Supporting Tool

Management of communications on a software project involves processes that ensure generation, collection, distribution, storage, retrieval and final destination of the project information in a timely and adequate manner. These processes provide the critical connections between people and information that are necessary for successful communication. Project managers may spend an excessive amount of time communicating with the project team, stakeholders and sponsor. Everyone involved in the project must understand how communications affect the project as a whole. Among these management processes, an important one is related to stakeholders involved in managing communications to meet the requirements of a project and resolve occurring issues.

Requirements are often related to each other [26,27]. These interactions affect various software development activities, such as release planning, change management and reuse [20]. Therefore, it is almost impossible to plan systems releases only based on the highest priority requirements, without considering which requirements are related to each others and the type of these relationships [20].

This work presents an approach to improve the communication between the stakeholders in the development of software based on EBT with the criterion of automatic communication. The input for the approach are SysML Requirements Tables, as discussed in Subsect. 2.3 of this paper. In order to support this approach, a tool has been developed including the attributes mentioned in Sect. 3.

The tool was developed using the C Sharp programming language and using the SQL Server database management system. The tool has three registrations

Fig. 1. Stakeholders registration

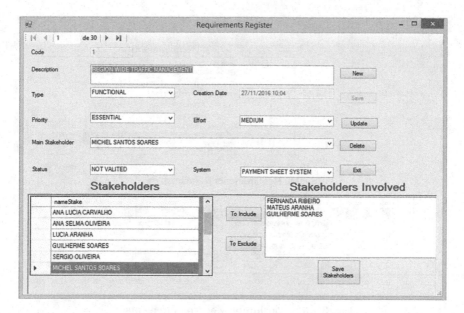

Fig. 2. Requirements registration

(stakeholders, requirements and relationships between requirements) and two queries (requirements and history of requirements changes).

The **stakeholders registration** (Fig. 1) aims to store all stakeholders in any system with emphasis on the email attribute that is fundamental to the main proposal of the tool.

The **requirements registration** (Fig. 2) is intended to control all system requirements with emphasis on email notifications to stakeholders (main and involved) in the requirement.

Each requirement has a main stakeholder (mandatory) with the possibility of registering other interested parties involved. **Relationship between requirements** has as purpose the connection of requirements and their type of relationship mentioned in Sect. 2.3. This is the functionality of the tool that makes use of SysML Requirements Tables. This registration is depicted in Fig. 3.

An application may have several links with requirements. The **requirements query** has the objective of visualizing all the requirements registered in the system.

Requirements change history tracks all changes made to the system. All inclusions and modifications in attributes are stored for further queries. This control is based on activated triggers in the database management system.

Figure 4 shows the class diagram of the tool.

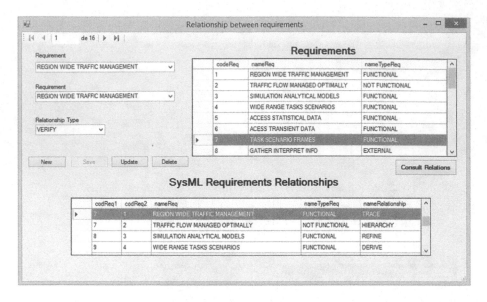

Fig. 3. Relationship between requirements

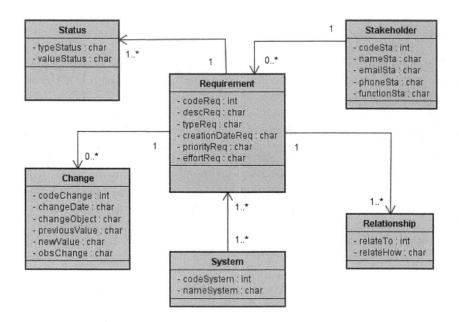

Fig. 4. UML class diagram of the tool

5   Evaluation of the Tool with TAM (Technology Acceptance Model)

Technology Acceptance Model (TAM) [13] is a model created specifically for the study of the usage behavior of computer technologies. TAM suggests that when users are presented with a new technology, a number of factors influence their decision about how and when they will use it. Originally, TAM proposes two main variables, Perceived Ease of Use - PEOU, and Perceived Usefulness - PU. A third variable is considered in this research: Perceived Usage - PUE, as introduced in [28].

PEOU variable can be defined as "how the person believes that learning to use a certain technology is without effort" [13]. It is important to know if there will be any resistance from the users to the new technology. This variable refers to the expectations of the individual in the exemption of physical or mental effort for the use of a certain system or technology. Sometimes softwares are difficult to use, frustrating the users, which may even lead to a decrease in their performance to finish their tasks.

PU can be defined as "how the person believes that using a certain technology will improve its performance on a certain task" [13]. This is important because it gives the right idea of how users think the software will help them to achieve their goals on their tasks at work.

PUE is the variable that seeks to point out the features that are really used in the software. This variable finds which are the features that have most important use in the tool. Given the right importance to the features, one can see how the software will be used, which features will be used, and the ones not considered by most users.

Evaluation is performed using two techniques: surveys, based on the TAM theory, and interviews. Participants answered a questionnaire composed of 15 statements in which they made their opinions explicit categorized in this follow way: Perceived Use (PU) - 7 questions, Perceived Easy of Use (PEOU) - 4 questions, and Perceived Usage (PUE) - 4 questions. A 5-point Likert scale was proposed to measure perceived attitudes of the subjects by providing a range of responses to each statement. The scale ranged from (1) strongly disagree, (2) disagree, (3) neutral, (4) agree and (5) strongly agree. All the evaluators are professionals in the area of information technology, working in companies as programmers, systems analysts, managers, directors, assistant professors and graduate students.

This study employed quantitative methods for analyzing data. Data collected was subjected to a descriptive and inferential analysis involving mean, standard deviation and frequency count. In addition, the Spearman correlation technique was used to measure the correlation between variables PU, PEOU, and PUE. In this way, we present the following research hypotheses:

– **H1**. Perceived Usefulness positively influences the Perceived Usage of the traceability requirements tool.

– **H2**. Perceived Ease of Use positively influences Perceived Usage of the traceability requirements tool.
– **H3**. Perceived Ease of Use positively influences Perceived Usefulness of the traceability requirements tool.

6   Results

The questionnaire was answered by 24 professionals. Answers regarding each question are shown in Tables 3, 4 and 5, in which "m" represents the mean, "s" represents the standard deviation, and "pos" indicates the number of positive answers, given in modulus because of negative statements. We arbitrarily considered as positive the answers "Agree" or "Strongly Agree" (values 4 or 5). For negative sentences we considered 1 and 2 for a positive response.

Table 3. Perceived usefulness (PU) - Statements 1 to 7 (N = 24)

Statement	1	2	3	4	5	m	s	pos
1 - Using the tool will improve performance of the requirements management activities	0	0	0	10	14	4.58	0.50	24
2 - The information provided in the tool is interesting for me	0	0	4	9	11	4.29	0.75	20
3 - I think the tool adds value	0	0	1	7	16	4.63	0.49	23
4 - I think the tool allows one to better control changes in requirements	0	1	1	2	20	4.63	0.58	22
5 - Requirements registration is important for controlling requirements	0	0	1	7	16	4.58	0.50	23
6 - Relationship between requirements in the tool is important for understanding the impact of changes in requirements	0	1	1	2	20	4.71	0.75	22
7 - History of requirements changes offered by the tool is important during software development	0	0	1	7	16	4.63	0.58	23

6.1   Responses to TAM Questionnaire

By analyzing averages presented in Tables 3, 4 and 5, it turns out that participants of the poll have shown high agreement in relation to most of the items from the survey. This mean they think the tool is easy to use, is useful and that they would use it often.

Besides, from data we can conclude that 93,6% of users have a positive opinion, (4 or higher) for most of the questionnaire. The positive assessment of respondents showed a 95,2% participation for PU, a 89,6% for PEOU and

Table 4. Perceived ease of use (PEOU) - Statements 1 to 4 (N = 24)

Statement	1	2	3	4	5	m	s	pos
1 - The tool is easy to navigate	0	0	2	4	18	4.67	0.64	22
2 - I think the tool is user-friendly	0	0	6	6	12	4.25	0.85	18
3 - I can quickly find the information I need in the tool	0	0	1	11	12	4.46	0.59	23
4 - Overall, it is easy for me to use the tool	0	1	0	7	16	4.58	0.72	23

Table 5. Perceived usage (PUE) - Statements 1 to 4 (N = 24)

Statement	1	2	3	4	5	m	s	pos
1 - Using the tool increases my performance at work	0	0	2	16	6	4.17	0.56	22
2 - Using tool provides improved communication between stakeholders	0	0	1	6	17	4.67	0.56	23
3 - E-mail notification improves communication regarding changes in software development	0	0	0	9	15	4.63	0.49	24
4 - I would use the tool in my daily requirements engineering activities	0	0	1	9	13	4.42	0.78	22

a 94,8% for PUE. With these results, it is noted that the technology is well-received, they hope for software that is easy to use and make their work performance better.

Items that showed an acceptance from the poll were sentence 6 (PU) and sentence 1 (PEOU). Items that showed a rejection from the poll were sentence 2 (PU) and sentence 2 (PEOU). The item that showed the best mean was sentence 6 (PEOU), while the lowest mean was identified in sentence 1 (PUE).

Regarding the standard deviation, the items that showed a low standard deviation, that is, the items that were closer to the mean were sentences 3 (PU) and sentence 3 (PUE). On the contrary, items that showed a high standard deviation, that is, items that were more scattered over a wide range of values, were sentences 2 (PEOU) and sentence 4 (PUE).

The Spearman coefficient measures the intensity of the relationship between ordinal variables. The Spearman coefficient varies between -1 and 1. The closer to these extremes, the greater the association between the variables. The statistical significance of a result is an estimated measure of the degree to which this result is "true" (in the sense that it is actually what occurs in the population, or in the sense of "population representativeness"). It represents a decreasing index of the reliability of a result. The higher the value of significance, the less one can believe that the observed relationship between variables in the sample is a reliable indicator of the relationship between the respective variables in the population. In many areas of research, the value of 0.05 is customarily treated as an "acceptable limit" of error [29].

In this way, all the hypotheses were confirmed, since they reached less than 0.05 in the value of statistical significance. Spearman correlation (Sc) coefficient showed a moderate correlation, showing a significant and positive correlation for all hypotheses, as presented in Table 6.

Table 6. Correlation between hypothesis and Spearman coefficient

Hypothesys	Sc	Significance
H1 - Perceived usefulness positively influences perceived usage	0.431	0.036
H2 - Perceived ease-of-use positively influences perceived usage	0.414	0.044
H3 - Perceived ease-of-use positively influences perceived usefulness	0.502	0.012

6.2 Interviews

With regard to the subjects, analyzing the number of years of experience in the area of information technology, 4,2% have with up to 5 years of experience, 12,5% between 6 and 10 years of experience, 37,5% between 11 and 15 years of experience, 37,5% between 16 and 20 years of experience and 8,3% have more than 21 years of experience.

Subjects suggested additional improvements to the tool. In this article, most important suggestions are: improvements in queries mainly regarding new filters, adding new options in the status attribute of requirements, adding the expected date of delivery of each requirement, adding new information (system to which the requirement belongs, description of the requirement) in the e-mail sent to stakeholders, create functionality to view the relationships of SysML, and make the tool available both for Web and mobile environments.

7   Threats to Validity

This study has endeavored to establish a research model for the successful evaluation of a software tool for managing traces between requirements realized with professionals in the area of information technology (programmers, systems analysts, managers, directors, assistant professors and graduate students). However, the study has some limitations that can be addressed in future research.

The study had 24 people surveyed, which is a relatively low number for a reliable estimate. However, the survey was answered by subjects working at various organizations, and most of them are experienced information technology professionals.

Another perceived threat is that notification messages of changes in requirements are only sent if at the time of modification the stakeholder has an Internet connection.

8   Conclusion

One of the activities related to requirements management is requirements traceability, which is an important mechanism for managing and auditing the entire software development process. Traceability influences the quality of software products making it easy to reuse.

Requirements traceability techniques, such as the event-based technique, combined with the SysML relations concept allow better control of requirements and improved communication with stakeholders. Such concepts supported by a tool helps to improve accuracy by reducing the time required for requirements engineering activities.

The research aimed to create a requirements traceability approach supported by a tool. This research used the evaluation criteria based on the TAM model, which is an empirical evaluation method applied to understand the use and behavior of users of a certain technology.

As for future work, it is suggested to improve the tool with suggestions given by the subjects mentioned in Sect. 6.2. In addition, new research regarding the use of other requirements traceability techniques combined with SysML can be performed.

Acknowledgement. The authors would like to thank the Brazilian research agency CNPq (grant 445500/2014-0) for the financial support.

References

1. Parviainen, P., Tihinen, M., Lormanms, M., van Solingen, R., Nasr, E., Coulin, C., Zowghi, D., Garda, J.A., Casa, J.A., Vázquez, R.G., et al.: Requirements Engineering for Sociotechnical Systems (2005)
2. Lago, P., Muccini, H., Van Vliet, H.: A scoped approach to traceability management. J. Syst. Softw. **82**(1), 168–182 (2009)
3. Egyed, A., Grünbacher, P., Heindl, M., Biffl, S.: Value-based requirements traceability: lessons learned. In: Lyytinen, K., Loucopoulos, P., Mylopoulos, J., Robinson, B. (eds.) Design Requirements Engineering: A Ten-Year Perspective. LNBIP, vol. 14, pp. 240–257. Springer, Heidelberg (2009). doi:10.1007/978-3-540-92966-6_14
4. Asuncion, H.U., Francois, F., Taylor, R.N.: An end-to-end industrial software traceability tool. In: Proceedings of the the 6th Joint Meeting of the European Software Engineering Conference and the ACM SIGSOFT Symposium on The Foundations of Software Engineering, ESEC-FSE 2007, New York, NY, USA, pp. 115–124. ACM (2007)
5. Torkar, R., Gorschek, T., Feldt, R., Svahnberg, M., Raja, U.A., Kamran, K.: Requirements traceability: a systematic review and industry case study. Int. J. Softw. Eng. Knowl. Eng. **22**(03), 385–433 (2012)
6. Cleland-Huang, J., Gotel, O.C., Huffman Hayes, J., Mäder, P., Zisman, A.: Software traceability: trends and future directions. In: Proceedings of the on Future of Software Engineering, Hyderabad, India, pp. 55–69. ACM (2014)
7. Rempel, P., Mader, P.: Preventing defects: the impact of requirements traceability completeness on software quality. IEEE Trans. Softw. Eng. (2016)

8. Cleland-Huang, J., Gotel, O., Zisman, A.: Software and Systems Traceability, vol. 2. Springer, Heidelberg (2012). GBR

9. Rempel, P., Mäder, P.: A quality model for the systematic assessment of requirements traceability. In: Proceedings of the 23rd IEEE International Requirements Engineering Conference, RE, pp. 176–185 (2015)

10. Kannenberg, A., Saiedian, H.: Why software requirements traceability remains a challenge. J. Defense Softw. Eng. **22**(5), 14–19 (2009)

11. Lucia, A.D., Fasano, F., Oliveto, R., Tortora, G.: Recovering traceability links in software artifact management systems using information retrieval methods. ACM Trans. Softw. Eng. Methodol. (TOSEM) **16**(4), 13 (2007)

12. Delligatti, L.: SysML Distilled: A Brief Guide to the Systems Modeling Language, 1st edn. Addison-Wesley Professional, Boston (2013)

13. Davis, F.D.: Perceived usefulness, perceived ease of use, and user acceptance of information technology. MIS Q. **13**, 319–340 (1989)

14. Buchgeher, G., Weinreich, R.: Automatic tracing of decisions to architecture and implementation. In: Proceedings of the 2011 Ninth Working IEEE/IFIP Conference on Software Architecture, WICSA 2011, pp. 46–55 (2011)

15. Gupta, S., Hartkopf, J.M., Ramaswamy, S.: Event notifier: a pattern for event notification. In: More Java gems, pp. 131–153. Cambridge University Press, Cambridge (2000)

16. Huang, J.: Robust Requirements Traceability for Handling Evolutionary and Speculative Change. University of Illinois at Chicago, Chicago (2002)

17. Cleland-Huang, J., Chang, C.K. Sethi, G., Javvaji, K., Hu, H., Xia, J.: Automating speculative queries through event-based requirements traceability. In: Proceedings of the IEEE Joint International Conference on Requirements Engineering, pp. 289–296. IEEE (2002)

18. Hong, Y., Kim, M., Lee, S.-W.: Requirements management tool with evolving traceability for heterogeneous artifacts in the entire life cycle. In: Proceedings of the 2010 Eighth ACIS International Conference on Software Engineering Research, Management and Applications (SERA), Washington, pp. 248–255. IEEE (2010)

19. OMG. OMG Systems Modeling Language (SysML), Version 1.4. Technical report (2015)

20. Soares, M.S., Vrancken, J.: Model-driven user requirements specification using SysML. J. Softw. **3**(6), 57–68 (2008)

21. Briand, L., Falessi, D., Nejati, S., Sabetzadeh, M., Yue, T.: Traceability and SysML design slices to support safety inspections: a controlled experiment. ACM Trans. Softw. Eng. Methodol. **23**, 9:1–9:43 (2014)

22. Ribeiro, F.G.C., Rettberg, A., Pereira, C.E., Soares, M.S.: A model-based engineering methodology for requirements and formal design of embedded and real-time systems. In: Proceedings of the Hawaii International Conference on System Sciences (HICSS), pp. 6131–6140 (2017)

23. Balmelli, L., Brown, D., Cantor, M., Mott, M.: Model-driven systems development. IBM Syst. J. **45**(3), 569–585 (2006)

24. Balmelli, L., et al.: An overview of the systems modeling language for products and systems development. J. Object Technol. **6**(6), 149–177 (2007)

25. Pohl, K., Rupp, C.: Requirements Engineering Fundamentals: A Study Guide for the Certified Professional for Requirements Engineering Exam-Foundation Level-IREB Compliant. Rocky Nook Inc., San Rafael (2011)

26. Nair, S., de la Vara, J.L., Sen, S.: A review of traceability research at the requirements engineering conferencere. In: 2013 21st IEEE International Requirements Engineering Conference (RE), pp. 222–229, July 2013

27. Mader, P., Egyed, A.: Do developers benefit from requirements traceability when evolving and maintaining a software system? Empir. Softw. Eng. **20**(2), 413–441 (2015)
28. Adams, D.A., Nelson, R.R., Todd, P.A.: Perceived usefulness, ease of use, and usage of information technology: a replication. MIS Q. **16**, 227–247 (1992)
29. Field, A.: Discovering Statistics Using SPSS. Sage Publications, Thousand Oaks (2009)

Test Case/Step Minimization for Visual Programming Language Models and Its Application to Space Systems

Paulo Nolberto dos Santos Alarcon$^{(\boxtimes)}$ and Valdivino Alexandre de Santiago Júnior

Instituto Nacional de Pesquisas Espaciais (INPE),
Av. dos Astronautas, 1758, São José dos Campos, São Paulo, SP, Brazil
paulonsalarcon@gmail.com, valdivino.santiago@inpe.br

Abstract. Visual Programming Languages have been widely used in the context of Model-Based Development, and they find a particular appeal for the design of satellite subsystems, such as the Attitude and Orbit Control Subsystem (AOCS) which is an extremely complex part of a spacecraft. The software testing community has been trying to ensure high quality products with as few defects as possible. Given that exhaustive generation and execution of software test cases are unfeasible in practice, one of the initiatives is to reduce the sets of test cases required to test a Software/System Under Test, while still maintaining the efficiency (ability to find product defects, code coverage). This paper presents a new methodology to generate test cases for Visual Programming Language models, aiming at minimizing the set of test cases/steps but maintaining efficiency. The approach, called specification Patterns, modified Condition/Decision coverage, and formal Verification to support Testing (PCDVT), combines the Modified Decision/Condition Coverage (MC/DC) criterion, Model Checking, specification patterns, and a minimization approach by identifying irreplaceable tests in a single method, taking advantage of the benefits of all these efforts in a unified strategy. Results showed that two instances of PCDVT presented a lower cost (smaller number of test steps) and, basically, the same efficiency (model coverage) if compared with a specialist ad hoc approach. We used the AOCS model of a Brazilian satellite in order to make the comparison between the methods.

Keywords: Model-Based Testing · Test case/step minimization · Model Checking · Specification patterns

1   Introduction

In space systems engineering, quality assurance represents an important role in the system development. Due to the critical and complex nature of a spacecraft, the development of methodologies, methods and techniques to assure its overall quality before launching is highly necessary [1].

© Springer International Publishing AG 2017
O. Gervasi et al. (Eds.): ICCSA 2017, Part VI, LNCS 10409, pp. 160–175, 2017.
DOI: 10.1007/978-3-319-62407-5_11

The Attitude and Orbit Control Subsystem (AOCS) [2] is one of the most important and complex subsystems of a satellite. It is the subsystem responsible for maintaining the spacecraft attitude and orbit ensuring the success of the mission. AOCS must present high quality and robustness. One way to obtain the quality of a system is through well defined processes such as testing. In the context of software development, testing [3,4] is one of the several processes related to Verification & Validation [5].

In order to bring quality assurance to earlier phases, several projects adopt the Model-Driven Development (MDD) [6] strategy where models are tested, verified and improved before the concrete implementation of a complex system/subsystem. As a consequence, models developed can be used to automatically generate test cases through Model-Based Testing (MBT) [7] where such test cases are partially or completely generated from a model describing some aspect (i.e. functionality, performance) of a software product.

A common issue in MBT is test case explosion. In other words, if very detailed models are considered, the amount of test cases to be executed is considerably great [7] making impossible the execution of such test cases withing feasible time. Hence, many studies aim to reduce the amount of test cases necessary to be executed via test case minimization [8–11]. However, test case minimization has been exploited more in the context of regression testing. Thus, it is interesting to investigate and propose new solutions for test cases which are going to be run for the first time.

This work presents a new methodology to generate a minimized set of test cases directly from certain types of Visual Programing Languages (VPLs) while maintaining efficiency of the test suite. Our approach, called specification Patterns, modified Condition/Decision coverage, and formal Verification to support Testing (PCDVT), combines the Modified Decision/Condition Coverage (MC/DC) criterion [12], Model Checking [13], specification patterns mappings for Linear Temporal Logic (LTL) [14], and a minimization approach by Identifying irreplaceable tests [11] in a single method, taking advantage of the benefits of all these efforts in a unified strategy. We present an empirical evaluation where we compare two instances of PCDVT against a set of test cases created by a specialist ad hoc approach [15]. All test cases were run over an AOCS model of a Brazilian satellite and the number of test steps and MC/DC coverage obtained for each test suites were compared. The two instances of PCDVT presented a lower cost (smaller number of test steps) and, basically, the same efficiency (model coverage) if compared with the specialist ad hoc approach.

This paper is structured as follows. Section 2 presents the main characteristics the VPLs must possess so that PCDVT can be used. Section 3 presents the PCDVT methodology. Section 4 shows an overview of the Brazilian satellite AOCS model we used as case study. Experimental assessment where we compared two instances of PCDVT with a specialist ad hoc test case generation approach is in Sect. 5. Section 6 presents related work. In Sect. 7, we show the conclusions and future directions of this research.

2   Visual Programing Languages: Required Features

The PCDVT methodology was designed to support block diagram and state-transition VPLs. Our methodology assumes that VPLs must have the following features: (i) the model is composed by blocks and sub-blocks; (ii) each block can present decisions impacting the outcome; and (iii) each possible outcome can lead to different blocks and execution paths. Figure 1 shows an example of VPL model presenting the features described above. Some VPLs which follow these characteristics are SciLab/Xcos [16], Yed [17], and Simulink [18].

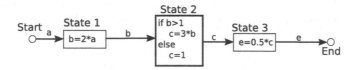

Fig. 1. Example of a general VPL code (model)

3   The PCDVT Methodology

The methodology was developed aiming to make it feasible the generation of test cases directly from some types of VPLs, by reducing the effort to execute the test suite. Furthermore, PCDVT relies on the MC/DC criterion to expedite the qualification of an MDD process since this is an important criterion considered in the DO-178C standard [12]. The PCDVT methodology is presented in Fig. 2. In the next subsections, we will detail the methods and principles of the methodology.

3.1   MC/DC Analysis and Derivation of LTL Properties

MC/DC is a coverage criterion developed in the context of structural (white box) testing aiming to assure that each condition, in a program's decision, must be tested for all possible results at least once, and every program's decision must be tested for all possible results at least once [19].

Figure 3 presents the algorithm to identify the conditions in accordance with MC/DC criterion. In other words, this algorithm takes as input the model and outputs all representations of transitions composed by the source state, s_i, the destination state, s_f, and the condition that independently affects the outcome or True otherwise. Such conditions or True are latter used to generate the LTL properties to support the Model Checking process. In Fig. 3, we have: $M = $ Model; $t = $ State transition in the model; $c_i = $ Condition in a given transition; $I = $ Set of conditions that independently affects the decision; C: Set of all conditions or True values; $s_i = $ Source state, $s_f = $ Destination state; Θ: Set of representations of transitions.

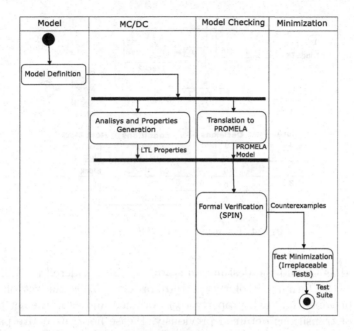

Fig. 2. The PCDVT methodology

Data: Model, M
Result: Set of representations of transitions, Θ
1 $\Theta \leftarrow \emptyset$
2 $C \leftarrow \emptyset$
3 **for** *each* $t \in M$ **do**
4   **for** *each* $c_i \in t$ **do**
5   **if** $c_i \in I$ **then**
6              $C \leftarrow C \cup c_i$
7   **end**
8   **else**
9              $C \leftarrow C \cup True$
10   **end**
11   **end**
12 **end**
13 $\Theta \leftarrow \Theta \cup \{(s_i, s_f, c_i)\}$
14 **return** Θ

Fig. 3. MC/DC analysis

Figure 4 shows how the algorithm presented works for a simple block. For Block 1, the algorithm will generate 2 conditions to be tested, while for the other blocks it will consider the condition True. Thus, the set of conditions (C) for all blocks, considering the algorithm shown in Fig. 3, becomes $\{u_1 = 1, u_1! = 1, True, True, True\}$.

For each condition obtained via the algorithm in Fig. 3, a LTL property is generated. The LTL properties are derived via two specification patterns for LTL. Considering the Absense Pattern with scope Between Q and R [14], Fig. 5 presents the algorithm, where the source state (s_i) of the transition of the model

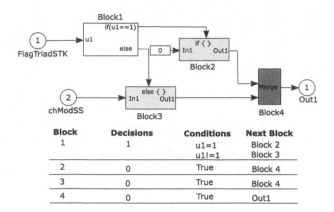

Block	Decisions	Conditions	Next Block
1	1	u1=1	Block 2
		u1!=1	Block 3
2	0	True	Block 4
3	0	True	Block 4
4	0	True	Out1

Fig. 4. Example of MC/DC analysis

is considered as Q, and the destination state (s_f) is considered as R. The condition (c) is the formula P of the pattern present. L_a is the set of all LTL properties obtained for each property (l_a) obtained and Θ is the set of representations of transitions obtained previously. Please note the entire pattern is completely negated (!) to force the counterexample generation. In these algorithms, we have the LTL temporal modalities Always ([]); Eventually (<>); Until($\cup$); and Next ($\circ$).

In Fig. 6, we see the Absence Pattern properties derivation for the model in Fig. 4. For each block, at least one LTL property is generated. For decision blocks, each condition identified via MC/DC analysis generates a property for the given transition. When there is no decision, the condition is always set as True and a property is generated.

Data: Set of representations of transitions, Θ
Result: LTL properties due to the Absence Pattern with scope between Q (s_i) and R (s_f)
1 $L_a \leftarrow \emptyset$
2 **for** *each* $c \in \Theta$ **do**
3 $\quad |\quad l_a \leftarrow !([]((s_i \& !s_f \& <> s_f) \rightarrow (!c \cup s_f))$
4 **end**
5 $L_a \leftarrow L_a \cup l_a$
6 **return** L_a

Fig. 5. Derivation of ABSENCE PATTERN property in LTL

The other specification pattern is the Chain Response with Global scope where the algorithm is presented in Fig. 7. Here, we considered the 2 stimulus-1 response chain, i.e. P responds to S, T. Hence, S is the source state (s_i) of the transition of the model, T is the condition (c) previously identified, and P is the destination state (s_f). L_r is the set of all properties (l_r) obtained and Θ is the set of representations of transitions obtained previously.

Again, the entire pattern is completely negated (!) to force the counterexample generation. An example can be seen in Fig. 8.

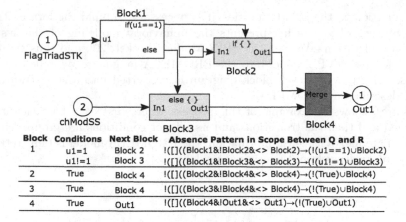

Fig. 6. Example: LTL properties for Absence Pattern

Block	Conditions	Next Block	Absence Pattern in Scope Between Q and R
1	u1=1	Block 2	!([]((Block1&!Block2&<> Block2)→(!(u1==1)∪Block2)
	u1!=1	Block 3	!([]((Block1&!Block3&<> Block3)→(!(u1!=1)∪Block3)
2	True	Block 4	!([]((Block2&!Block4&<> Block4)→(!(True)∪Block4)
3	True	Block 4	!([]((Block3&!Block4&<> Block4)→(!(True)∪Block4)
4	True	Out1	!([]((Block4&!Out1&<> Out1)→(!(True)∪Out1)

Fig. 6. Example: LTL properties for Absence Pattern

Data: Set of representations of transitions, Θ
Result: LTL Properties due to the Chain Response Pattern with Global scope
1 $L_r \leftarrow \emptyset$
2 **for** *each* $c \in \Theta$ **do**
3 | $l_r \leftarrow !([](s_i \& o <> c \to o(<> (c \& <> s_f))))$
4 **end**
5 $L_r \leftarrow L_r \cup l_r$
6 **return** L_r

Fig. 7. Derivation of CHAIN RESPONSE property in LTL

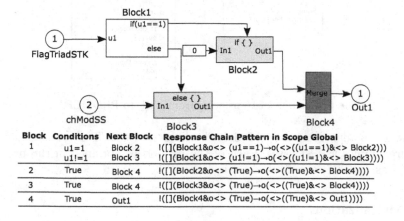

Block	Conditions	Next Block	Response Chain Pattern in Scope Global
1	u1=1	Block 2	!([](Block1&o<> (u1==1)→o(<>((u1==1)&<> Block2)))
	u1!=1	Block 3	!([](Block1&o<> (u1!=1)→o(<>((u1!=1)&<> Block3))))
2	True	Block 4	!([](Block2&o<> (True)→o(<>((True)&<> Block4))))
3	True	Block 4	!([](Block3&o<> (True)→o(<>((True)&<> Block4))))
4	True	Out1	!([](Block4&o<> (True)→o(<>((True)&<> Out1))))

Fig. 8. Example: LTL properties for Response Chain Pattern

3.2   Model Checking

Model Checking is a method used to verify if system behavioral models correspond to the specification [13]. Using a formal specification, typically described as temporal logic, the model checker (tool that has an implementation of Model Checking) explores the model searching for inconsistencies and returning a counterexample showing the flaw.

In addition to the LTL properties, it is necessary to build the formal Transition System (TS) which represents the behavioral modeling of the system under consideration. We used the SPIN Model Checker [20], and thus we need to transform the VPL model into the PROMELA language. We relied on the proposal of [21] where each block diagram is converted into a state-transition model described in PROMELA.

Figure 9 shows an example of this process where first each block identifier is labeled and then the transitions and its trigger conditions are grouped in an intermediate file. With such information, the PROMELA model is built retrieving the information from the original model.

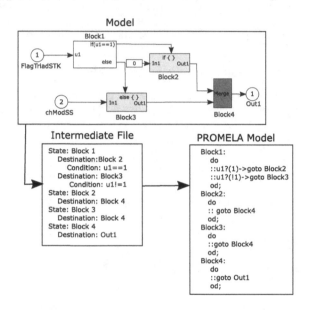

Fig. 9. Example of translation from VPL to PROMELA

The main idea related to Model Checking in PCDVT is to consider the counterexamples as test cases. Each of the properties obtained in the previous steps (Figs. 5 and 7) are checked against the TS.

3.3   Minimization

The last step of PCDVT is the minimization of the test suite previously obtained via Model Checking. Test case minimization is a process of test suite reduction with low impact in the ability to find defects according to this new test suite.

In PCDVT, the minimization process is made via the greedy algorithm combined with the irreplaceable tests evaluation strategy [11]. In this strategy, test cases are evaluated according to their requirement coverage and execution time. Tests with wide requirement coverage and low execution time receive higher weights. When a requirement is verified by only one test case, then it is automatically chosen to compose the optimized test suite. Since the irreplaceable test strategy was developed in the context of regression tests, it uses information not available during the test creation, such as execution time. The algorithm was slightly adapted in PCDVT. Each LTL property is treated as a requirement and the step count is treated as the execution time, since it directly affects the test duration.

A counterexample analysis was added before the minimization process, were the counterexample steps are analyzed and labeled in accordance with their similarity, based on the work of [10]. Figure 10 describes the redundancy analysis algorithm, where: CE: Set of all counterexamples obtained in the Model Checking step; ce_i: Counterexample in the set CE, being also considered as a set of test steps; η: Labeling operator. An example of redundancy analysis is shown in Fig. 11 where test case 1 is contained in test case 2, then, according with the

Data: Set of all counterexamples, CE
Result: Set of labeled counterexamples, CE_l

```
1  CE_l ← ∅
2  for each ce_i ∈ CE do
3      for each ce_j ∈ CE,   i < j do
4          if ce_i == ce_j then
5              ce_i ← η(ce_i, ce_j)
6              ce_j ← η(ce_j, ce_i)
7          end
8          else
9              if ce_i > ce_j ∧ ce_j ⊂ ce_i then
10                 ce_i ← η(ce_i, ce_j)
11             end
12             else
13                 if ce_i < ce_j ∧ ce_i ⊂ ce_j then
14                     ce_j ← η(ce_j, ce_i)
15                 end
16             end
17         end
18     end
19     CE_l ← CE_l ∪ ce_i ∪ ce_j
20 end
21 return CE_l
```

Fig. 10. Redundancy analysis algorithm

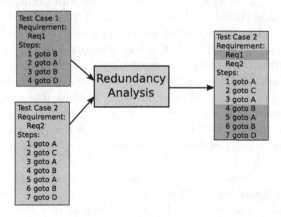

Fig. 11. Example of redundancy analysis

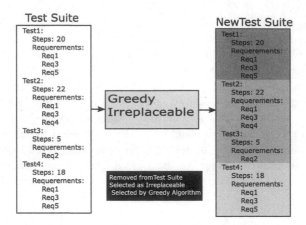

Fig. 12. Example of minimization run

algorithm of Fig. 10, test case 2 is associated with the requirements verified by test case 1. With all requirements mapped for all tests, the Greedy algorithm is executed combined with the irreplaceable tests strategy [11], reducing the test suite (see Fig. 12).

4   Brazilian Satellite AOCS Model

The AOCS [2] is a subsystem that provides information and maintain proper spacecraft attitude and orbit during all phases of the mission, since the separation from the launcher until the end of its operational life. In other words, it is responsible for maintaining the space application position and orientation in space.

The case study is a 2-dimension AOCS model of Lattes-1 satellite [22] presenting 3 operation modes (see Fig. 13), being them Sun Pointing Mode (SPM), Earth Pointing Mode (EPM) and Velocity Control Mode (VCM). Furthermore, it presents a Fault Detection, Isolation and Recovery Module (FDIR), which is able to trigger the SPM. The AOCS was developed to be tolerant to single and combined failure, and the FDIR must trigger the transition to mode SPM when occur composed failure, i.e. two or more components of the same faulted type simultaneously. We selected this case study because it comes from a critical domain (space application) and hence it is a good example to show the usefulness of PCDVT. Also, the use MC/DC criteria to drive the test case generation is important in order to improve the overall code coverage during quality assurance process.

In this model, the following sensors and actuators are considered:

Sun Sensor: Responsible for the sun direction determination. There are 7 sensors, being 3 always illuminated by the Sun independently the satellite orientation; Magnetometer: Responsible for measuring the Earth magnetic field. There

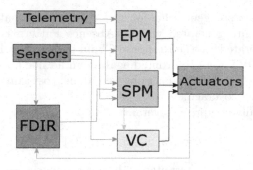

Fig. 13. AOCS block diagram

are 2 magnetometers in the satellite; Gyroscope: Responsible for measuring the satellite angular rate. There are 2 gyroscopes in the satellite; Star Tracker: Sensor responsible for the attitude and angular rate determination with high accuracy. This sensor is used only in the EPM mode and there are two of them; Reaction Wheels: Actuator responsible for the attitude maneuver execution. There are two reaction wheels for redundancy and both are used in the maneuver during normal operation.

4.1   Operation Modes

Sun Pointing Mode (SPM) is the initial mode of AOCS and also considered as contingency mode. It is triggered by telecommands or FDIR (see Fig. 13). It is the simplest mode, where the satellite locks its solar panes and follows the Sun direction. In the mode the mission does not operate and only telemetry and attitude control remain active. The satellite control is done using only sun sensors and magnetometers for attitude determination.

Earth Pointing Mode (EPM) is the mode where the satellite executes its mission, being always triggered by telecommands (see Fig. 13). In this mode all components are active and the satellite keep pointing to the center of Earth by default, but other points can be configured through telecomand.

Velocity Control Mode (VCM) is the mode where the satellite operates when it is in rotation and needs to stabilize (see Fig. 13). In the mode the stabilization process is done in way that no component is damaged. This is a mode that can only be used in the beginning of the mission, since once stabilized, it must keep pointing accordingly to the mission requirements.

5   Experimental Assessment

In order to verify the feasibility of the PCDVT methodology, an experimental evaluation was developed using partial translation of the AOCS model. In this model we took into account the EPM and SPM operation modes, and the FDIR for reaction wheels. Such model is composed by 33 blocks.

Two test suites were generated for the specification patterns previously defined. The test suite generated via the Absence Pattern is called PCDVT-A, and the one generated via the Response Chain pattern is called PCDVT-RC. Results due to PCDVT were confronted against an ad hoc approach created by a domain specialist [15]. Two comparisons were done: cost (amount of test steps) and efficiency (model coverage).

A test case usually comprises several test steps[1]. In other words:

$$tc = \{ts_i \quad | \quad i \in \mathbb{N} \backslash \{0\}\} \tag{1}$$

where tc = test case, and ts_i = test step i. However, one test case, tc_1, might have associated only 1 test step and, for instance, a second test case, tc_2, might be composed of 50 test steps. Thus, comparing the cost of two test suites considering the amount of test cases it is not adequate. Based on this, we defined the cost perspective of our evaluation as the total amount of test steps. Moreover, if we consider a uniform execution time for each test step (i.e. one test step takes α time to be executed), the amount of test steps is directly proportional to the test suite execution time.

In total, 34 LTL properties were generated via PCDVT using both patterns, resulting in 34 test cases in the test suite for each pattern. After the minimization activity, the test suite due to PCDVT-A was reduced to 1 test case while the test

Table 1. Test suite: PCDVT-A

Test case	Test steps	MC/DC coverage
1	25	53%
Total/Mean	25	53%

Table 2. Test suite: PCDVT-RC

Test case	Test steps	MC/DC coverage
1	7	53%
4	25	52%
7	6	53%
11	6	53%
12	10	50%
20	6	57%
23	6	52%
30	7	53%
31	9	59%
Total/Mean	88	53.4%

---

[1] A test step is an atomic activity to prepare or stimulate the System/Software Under Test. The stimulus can contain the test input data and the expected results.

suite due to PCDVT-RC was reduced to 10 test cases (see Tables 1, 2 and 3²). But, as our cost measure is the number of test steps, we presented in Fig. 14 the total test step count due to PCDVT-A, PCDVT-RC, and the specialist

Table 3. Manual test suite [15]

Test case	Test steps	MC/DC coverage
1	3	45%
2	5	59%
3	8	51%
4	8	51%
5	8	52%
6	8	51%
7	8	53%
8	8	52%
9	8	53%
10	8	52%
11	8	53%
12	8	52%
13	8	53%
14	8	53%
15	8	52%
16	8	53%
17	8	52%
18	8	51%
19	8	52%
20	8	53%
21	8	52%
22	8	52%
23	8	53%
24	8	53%
25	8	52%
26	8	53%
27	8	51%
28	8	53%
29	8	53%
30	8	52%
Total/Mean	232	52.23%

---

² Note that, in these tables, the last row implies the total number of test steps and the mean MC/DC coverage.

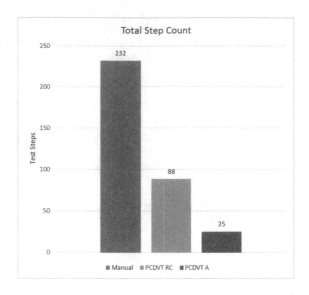

Fig. 14. Comparison of test step count for each test suite

ad doc strategy. As we note, PCDVT-A was the better solution of all with a total of 25 test steps (a reduction of 89% compared to the ad hoc approach). The PCDVT-RC was also better than the specialist approach (82 test steps; reduction of 62%).

Regarding efficiency (model coverage), the three test suites presented similar MC/DC coverage (see Fig. 15). Therefore, the results presented in this section

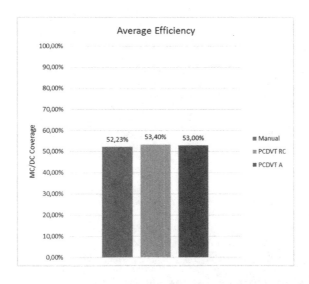

Fig. 15. Comparison of MC/DC coverage for each test suite

show that both PCDVT instances, PCDVT-A and PCDVT-RC, presented a smaller cost (amount of test steps) and basically the same efficiency (model coverage) when compared with a specialist manual and ad-hoc approach [15]. Particularly, the instance PCDVT-A was better than PCDVT-RC in terms of cost and was able to generate test cases for all properties obtained.

6   Related Work

For quality assurance applied to space systems engineering, several techniques have been proposed in areas like test case generation [5,23] and formal verification [24,25]. But no work deals with the use of model checking for test case generation, and does not consider the MC/DC coverage criteria in the space systems engineering context. Most of these MBT techniques are applied to functional test. On the other hand, this work uses these techniques and MC/DC coverage criteria which is associated to structural testing.

Ferrante [9] uses VPL to generate test cases, but Simulink is the only language supported and in this method the models are enriched with test objectives. This technique needs modifications in the models and does not consider coverage criteria. Furthermore, each test case considers only one test objective making the minimization impracticable.

Several approaches have been proposed where Model Checking helps test case generation [9,26–31], but no work considers MC/DC criterion, and only Fraser [26] deals with test case minimization.

Several algorithms and evaluation criteria have been proposed to guide the minimization process, among which stand out Genetic Algorithms [32], code coverage [10,26], requirement coverage [8,11], defects previously found [33], and test execution time [11]. However most of the techniques were developed to be applied in the context of regression testing by using information obtained after a history of executions [11,33,34]. But, such information are not available during the test case generation for the first time.

7   Conclusions

In this paper, we presented a new methodology, PCDVT, in order to provide a minimized test suite (regarding the number of test steps) for VPL models. PCDVT combines several concepts related to software testing (MC/DC, test minimization) and formal verification (Model Checking, specification patterns) to generate a set of test cases that in the end aims to derive a reduced set of test steps, demanding less effort to be executed.

Results have shown that 2 instances of PCDVT, PCDVT-A and PCDVT-RC, were better in terms of cost, and presented basically the same efficiency of an specialist ad-hoc approach. Case study was a non-trivial system, the AOCS of a Brazilian satellite. Particularly, the Absence Pattern with scope between Q and R was the best of all three approaches. This demonstrates the potential of this strategy for complex projects of space systems and other VPL applications.

The PCDVT methodology is partially automated. Hence, in the future we aim to complete the full automation of the tool that supports the methodology. We also intend to consider a more thorough model of the AOCS, with all sensors and actuators. Considering this expanded model, we will compare PCDVT with the specialist approach in terms of the cost and efficiency as defined in this research. This comparison will be done via a rigorous evaluation, e.g. a controlled experiment or quasiexperiment.

References

1. Pisacane, V.L.: Fundamentals of Space Systems. Oxford University Press, New York (2005)
2. Wertz, J.R., Larson, W.J.: Space Mission Analysis and Design. Microcosm Press, Hawthorne (1999). 976 p.
3. Mathur, A.P.: Foundations of Software Testing. Dorling Kindersley (India), Pearson Education in South Asia, Delhi (2008). 689 p
4. Delamaro, M.E., Maldonado, J.C., Jino, M.: Introdução ao teste de Software. Elsevier, Brasil (2007)
5. Santiago Júnior, V.A.: SOLIMVA: a methodology for generating model-based test cases from natural language requirements and detecting incompleteness in software specifications. Ph.D. thesis, Instituto Nacional de Pesquisas Espaciais (INPE) (2011)
6. Balasubramanian, K., Gokhale, A., Karsai, G., Sztipanovits, J., Neema, S.: Developing applications using model-driven design environments. Computer **39**(2), 33–40 (2006)
7. Utting, M., Pretschner, A., Legeard, B.: A taxonomy of model-based testing approaches. Softw. Test. Verif. Reliab. **22**(5), 297–312 (2012)
8. Campos, J., Abreu, R.: Encoding test requirements as constraints for test suite minimization. In: 2013 Tenth International Conference on Information Technology: New Generations (ITNG), pp. 317–322. IEEE (2013)
9. Ferrante, O., Ferrari, A., Marazza, M.: Model based generation of high coverage test suites for embedded systems. In: 19th IEEE European Test Symposium, vol. 99(2), pp. 335–337 (2014)
10. Fraser, G., Wotawa, F.: Mutant minimization for model-checker based test-case generation. In: Testing: Academic and Industrial Conference Practice and Research Techniques-MUTATION, TAICPART-MUTATION 2007, pp. 161–168. IEEE (2007)
11. Lin, C., Tang, K., Kapfhammer, G.M.: Test suite reduction methods that decrease regression testing costs by identifying irreplaceable tests. Inf. Softw. Technol. **56**, 1322–1344 (2014)
12. Holloway, C.M.: Towards understanding the do-178c/ed-12c assurance case. In: 7th IET International Conference on System Safety, Incorporating the Cyber Security Conference 2012, pp. 1–6 (2012)
13. Baier, C., Katoen, J.: Principles of Model Checking. MIT Press, Cambridge (2008)
14. Dwyer, M.B., Avrunin, G.S., Corbett, J.C.: Property specification patterns for finite-state verification. In: Proceedings of the Second Workshop on Formal Methods in Software Practice, pp. 7–15. ACM (1998)
15. Alarcon, P.N.S.: Test specification and procedures for AOCS. Technical report, São José dos Campos (2013)

16. Janík, Z., Žáková, K.: Online design of Matlab/Simulink and SciLab/Xcos block schemes. In: 2011 14th International Conference on Interactive Collaborative Learning (ICL), pp. 241–247 (2011)
17. Wiese, R., Eiglsperger, M., Kaufmann, M.: yFiles visualization and automatic layout of graphs. In: Jünger, M., Mutzel, P. (eds.) Graph Drawing Software, pp. 173–191. Springer, Heidelberg (2004). doi:10.1007/978-3-642-18638-7_8
18. Hanselman, D.C., Littlefield, B.: Mastering Matlab 7. Pearson/Prentice Hall, Upper Saddle River (2005)
19. Chilenski, J.J., Miller, S.P.: Applicability of modified condition/decision coverage to software testing. Softw. Eng. J. 9(5), 193–200 (1994)
20. Bartocci, E., DeFrancisco, R., Smolka, S.A.: Towards a GPGPU-parallel SPIN model checker. In: Proceedings of the 2014 International SPIN Symposium on Model Checking of Software, pp. 87–96 (2014)
21. Yamada, C., Miller, D.M.: Using spin to check simulink stateflow models. Int. J. Netw. Distrib. Comput. 4(1), 65–74 (2016)
22. Araujo, H.A.B.: AOCS design specification. Technical report, São José dos Campos (2012)
23. Alarcon, P.N.S., Carvalho, F.G.M., Simoes, A.R.: Geração automática de casos de teste aplicada ao projeto de aocs de satélites artificiais. In: XX Congresso Brasileiro de Automática (CBA 2014), pp. 1652–1659 (2014)
24. Gan, X., Dubrovin, J., Heljanko, K.: A symbolic model checking approach to verifying satellite onboard software. Sci. Comput. Program. 82, 44–55 (2014)
25. Nardone, V., Santone, A., Tipaldi, M., Glielmo, L.: Probabilistic model checking applied to autonomous spacecraft reconfiguration. In: 2016 IEEE Metrology for Aerospace (MetroAeroSpace), pp. 556–560 (2016)
26. Fraser, G., Wotawa, F., Ammann, P.E.: Testing with model checkers: a survey. Softw. Test. Verif. Reliab. 19(3), 215–261 (2009)
27. Ferrante, O., Marazza, M., Ferrari, A.: Formal specs verifier ATG: a tool for model-based generation of high coverage test suites. In: 19th IEEE European Test Symposium, vol. 99(2), pp. 335–337 (2014)
28. Zeng, H., Miao, H., Liu, J.: Specification-based test generation and optimization using model checking. In: Symposium on Theoretical Aspects of Software Engineering (TASE 2007) (2007)
29. Enoiu, E.P., Causevic, A., Ostrand, T.J., Weyuker, E.J., Sundmark, D., Pettersson, P.: Automated test generation using model checking: an industrial evaluation. Int. J. Softw. Tools Technol. Transf. 18, 335–353 (2014)
30. Yeolekar, A., Unadkat, D., Agarwal, V., Kumar, S., Venkatesh, R.: Scaling model checking for test generation using dynamic inference. In: 2013 IEEE Sixth International Conference on Software Testing, Verification and Validation, pp. 184–191 (2013)
31. Gent, K., Hsiao, M.S.: Functional test generation at the RTL using swarm intelligence and bounded model checking. In: 2013 22nd Asian Test Symposium, pp. 233–238 (2013)
32. Wang, S., Ali, S., Gotlieb, A.: Minimizing test suites in software product lines using weight-based genetic algorithms. In: Proceedings of the 15th Annual Conference on Genetic and Evolutionary Computation, pp. 1493–1500 (2013)
33. Dandan, G., Tiantian, W., Xiaohong, S., Peijun, M.: A test-suite reduction approach to improving fault-localization effectiveness. Comput. Lang. Syst. Struct. 39(3), 95–108 (2013)
34. Singh, R., Santosh, M.: Test case minimization techniques: a review. Int. J. Eng. Res. Technol. (IJERT) 2(12) (2013)

A System Based on Intelligent Documents
A Case Study

Jaroslav Král[1], Petr Novák[2], and Michal Žemlička[3,4(✉)]

[1] Faculty of Informatics, Masaryk University,
Botanická 68a, 602 00 Brno, Czech Republic
kral@fi.muni.cz
[2] Business Systems, a.s., Štěpánská 51, Praha 1, Czech Republic
petrn@bsys.cz
[3] Faculty of Mathematics and Physics, Charles University,
Malostranské nám. 25, 118 00 Praha 1, Czech Republic
zemlicka@sisal.mff.cuni.cz, zemlicka.michal@azd.cz
[4] AŽD Praha, a.s., Závod Technika, Žirovnická 2/3146, 106 17 Praha 10,
Czech Republic

Abstract. We discuss a collection of document-oriented software design patterns and tools used by a medium-sized software firm during implementation of an autonomous extension of already used business system. The attitudes are enabled by a document management system used as a data tier managing documents being similar to generalized spreadsheets. It enables construction of easily usable interfaces of software entities, typically software services. End users feel it as a transparent variant of their everyday activities. The built (sub)systems can have many further crucial quality characteristics. The system is transparent, modifiable, and has many further desirable properties. They are applicable in small as well as in large projects. We show that properly applied document-oriented philosophy enables many interesting and usable software engineering solutions.

Keywords: Document-oriented design patterns · Document systems in small firms · Document-oriented software engineering methods

1 Introduction

Document-oriented software architecture becomes a powerful software engineering paradigm. Tools for document management enable many useful practices during all stages of software systems life cycle. Document orientation simplifies the decomposition of software systems into coarse-grained autonomous components and transparent integration of autonomous software entities. It is important for large systems [19]. It can be applied in usefully in small-to-medium projects. The document management systems (DMS) and document-oriented philosophy enable many further design patterns [2,9,21] and tools in the design, development, and use of information systems.

© Springer International Publishing AG 2017
O. Gervasi et al. (Eds.): ICCSA 2017, Part VI, LNCS 10409, pp. 176–187, 2017.
DOI: 10.1007/978-3-319-62407-5_12

The solution provides systems cooperating using exchange of intelligent documents. Intelligent documents stand here for documents that provide part of the functionality of the system (testing, part of computation). At the same time the intelligent documents serve as a communication mean. The benefits of the solution could be important. It is surprising that the transparency, intuitiveness, seeming simplicity, and need for autonomy of individual entities (systems or components) are for many people difficult to accept and use.

We illustrate the use of the principles of a new coarse-grained document-oriented development patterns in action in a successful project called Requisitions. The aim of the project is to develop a material requirements management system RqS for a middle-sized Prague hospital. RqS was developed by a small Prague software company. The project is intuitively very simple but there are strong requirements on the modifiability, management, usability. There were several unsuccessful attempts to implement RqS as a part of hospital enterprise resource planning system (ERP, [10,23]). It follows that the RqS system has to be implemented as autonomous subsystem communicating with the ERP.

We show that service-oriented paradigm using document-oriented interfaces opens new opportunities and allows interesting solutions for small and medium-sized software companies even in the case of quite small systems. They open new opportunities for improvement of quality of their products in the sense of ISO 25010 [13]. Solutions discussed in this case study are usable also in very large software projects.

Structure of the paper is as follows: Sect. 2 introduces the problem, Sect. 3 reminds related work. Section 4 analysis original solution and requirements on the new one, Sect. 5 architecture of the described solution. Implementation details are in Sect. 6. Remarks on experience with implementation and deployment are in Sect. 7. Section 8 reveals currents state of the project, Sect. 9 its future development and further research in given direction. Finally, conclusions are in Sect. 10.

2 Material Control System

A middle-sized hospital in Prague decided to digitalize and enhance its system of consumption material control (RqC). The analysis of existing RqC system led to the following conclusion: The existing RqC system was based exclusively on the exchange of paper documents. It was operated almost independently on hospital enterprise resource planning system (ERP).

The individual requirements are defined using predefined paper forms on basic level typically by accredited nurses. The forms have a structure similar to list business documents like orders. The request must meet various limitations individually but also the collection of request must fit into various limitations too (like budget or quality).

The collection of requests from several nurses is passed to matrons analysing the request on the level of clinics. The matrons pass the collection of their requests consolidated for their clinics to a higher level. For example, matrons send their collections to higher level managers to be more globally analysed, summarized, and consolidated.

Finally, the requests all over the hospital are sent as an order to ERP for further processing. It implies that the process is even less formal than the second mode in bimodal systems [8].

Crucial requirement was that the requests should meet future needs. They are rather difficult to estimate exactly due to emergency situations or varying consumption of medical operations. The requests must be analysed, modified, and summarized at various levels of the organizational structure of the hospital. Finally, they were transformed into an order document sent to ERP.

The main properties of the system were acceptable but it was further necessary to increase accessibility of the data, to standardize and manage the processes, and to make the processes less laborious. New management tools were needed. Analysing the situation led to the observation that the material identification was still not sufficient.

It often happened that there is a requirement in a requisition that does not match any item in a company catalogue.

It was decided to order the modernization and digitalization of RqC by a small IT firm Business Systems Prague[1]. The firm has experience in similar areas – in the open source software for information systems.

2.1   Aims of the Project

The aims of the project are:

1. Gradual enhancement of the system.
2. Substantial enhancement of systems usability for nurses, physicians, and management.
3. Reduction of the irregular events in business processes (enhanced smoothness).
4. Simplifying and enhancement of management tools.
5. Some management tools should be available for nurses and other end-users.
6. Management tools should be available on various organizational levels.
7. Easy detection of unexpected or strange indications.
8. Gradual optimization of business processes.
9. Reduction of instores.
10. Openness for user involvement solving unexpected and rare cases or accidents.

2.2   Basic Critical System Requirements

– RqS should be designed and felt as a digitalization of current informal system. It must be operated almost independently on the existing ERP of the hospital[2].

---

[1] http://www.bsys.cz.

[2] This requirement is based on several years experience of the hospital. There were many attempts to implement RqS as a part of the existing ERP but all the attempts have failed.

- The individual material requests must be according to existing practices and policies defined (generated) on a basic organizational level by accredited people (typically chief nurse). They can use document templates. The requests are items of the document and must meet various limitations (budget, delivery time, quality, ...). The request can and should be optimized. The documents with the requests is confirmed by the nurse and made accessible to upper organizational level – to a matron. It is logically sent to the matron.
- The matrons collect dynamically the documents provided by subordinate nurses for checking and optimization of the requests. The collection of the documents are then opened (logically sent) to a supervisor (manager).
- The manager analysis the request documents of all subordinated clinics (matrons). The manager optimizes the collections of requirements as a whole. The manager finally confirms the modified requests and sends them to the management of the hospital.
- The management of the hospital performs optimization and checks at the hospital level (for example using summarized data over entire hospital). It is a two-step process. First the central accountant analysis the requirements according the budget limitations and the final confirmation is done by a responsible manager. Finally the request documents are transformed onto order documents and sent to ERP.
- ERP attempts to fulfil the request and generates documents allowing the matrons to get required material.

It is possible to modify the process so that at every level it is possible to consult experts. It can be possible to add or omit some steps.

2.3   Autonomy of ERP and Requisitions Systems

It is preferable to design the system as a transparent digitalization of the existing users' practice, i.e. as an autonomous component Requisitions cooperating with existing autonomous "component" – ERP system. Under *component* we understand here a clearly defined part of the reality communicating with its neighbourhood (with the real world as well as with other components) using digitalized business documents. In our case the Requisitions communicate with ERP using digitalized ordering process.

The document transfer can be implemented by sending them using a proper middleware or by setting organizational data (metadata) of the documents (and by proper handling of such (meta)data by the document management system).

The pair of autonomous entities (Requisition and ERP) cooperating as peer-to-peer is a very important architecture solution. It works like a coarse-grained pair of autonomous services known from SOA [14,19].

Let us summarize the main effects:

1. The software changes on requisitions side and in ERP are almost independent.
2. Both sides can be therefore modernized easily and autonomously and can be even replaced by different implementation of themselves.

3. It is crucial from the developers point of view that the solution can be used repeatedly for different enterprises (they are reusable).
4. Small enterprises can use simple ERP. The ERP can be outsourced.

3 Related Work

Described system is by its nature information system combining features from both classic information systems as well as form managerial ones (it handles both transactional data as well as aggregate data). Both groups of information systems are deeply studied and described in literature – let us remember e.g. [18,22]. As the system is at their border, we must be careful when trying to apply the recommendations from there.

Similar situation is with storing the data: It is necessary to store the requirements, their modifications during requirements concentration, and later real need of the requested commodity. It is reasonable to take into account various data stores: relational or object relational [11], document-based, or even something custom.

Important part of the solution will be handling of business processes. There are many methodologies and tools for modelling and execution of business processes (workflow [1,25], BPMN and BPEL [4,6], and others [3,26]. They are usually developed for stable and polished business processes. The business processes that should be supported accidentally do not fit in this class. They need to be time to time accommodated – e.g. when something expensive is necessary is necessary but current funds are low.

4 Summary of Analysis of Non-digitalized Material Demand Management Processes

The hospital used quite successfully non-digitalized systems (nRqS) having the following properties: nRqS uses paper requirement documents (in the form of requisition) RqD generated and modified by accredited persons. It is operated almost independently on a hospital ERP. RqD consists of metadata (head) and a list of individual material request items (Rqst). RqD's are created and filled by accredited head nurses using paper document templates.

RqD had to meet various limitations individually. Some collections of RqD must moreover fulfil various limitations too (budget, quality, . . .) and should be optimized (orchestrated).

Collections of RqD from a head nurses are sent to their matrons. The collection of RqD's sent to a matron is analysed and modified by the matron on the level of matron's clinic if necessary. The Rqst in RqD in the collection can be modified online or can be sent back to be updated. The matron finally accepts her collection of RqD and virtually sends it to her manager to be more globally analysed, summarized, and consolidated and sent to hospital management.

Finally, all the accepted/agreed RqD's of the hospital are analysed, modified, and accepted by hospital management. The collection of the accepted RqD is transformed into order document(s) to be sent to ERP for further processing.

4.1   Crucial Decisions

The main properties of nRqS are acceptable but it is necessary to increase accessibility of the data, to standardize the processes, and to make the processes less laborious. It is required to digitalize the processes (with the smallest possible changes from the users' point of view). For the sake of usability it is preferred to enable spreadsheet-like interface as much as possible.

The existing ERP remains unchanged except the policies of its use. It is therefore necessary to keep the digitalized requisition system (RqS) and the hospital ERP autonomous. Document-oriented tools enable the design of the system as a pair like a simple p2p network of two services – RqS and ERP – communicating via digitalized RqD.

The RqS should be designed and felt as a transparent digitalization of already existing non-digitalized RqS. RqS must be able to operate almost independently on the existing ERP of the hospital[3].

It was decided to order the development of RqS at the company Business Systems Prague. The company had experience in similar domain – in open source software for information systems.

4.2   RqS and ERP Can Be Used Like a Pair of Autonomous Services in SOA

RqS are designed as a transparent digitalization of the existing users' practice, i.e. as an autonomous component RqS cooperating with existing autonomous "component" – ERP system. The autonomy of components is an important characteristics of software system quality especially in service orientation [7,20]. Under *component* we understand here a clearly defined part of reality communicating with its neighbourhood (with the real world as well as with other components) using digitalized documents. In our case the RqS communicates with the hospital ERP using an order document. The order documents can have initially a paper form. It is good to enable both paper and digital formats.

The pair of autonomous entities is an important solution. Let us summarize the most important effects:

1. Enterprises need not change/modify ERP they use.
2. RqS can be reused easily at different enterprises.
3. The changes in RqS and in ERP are almost independent. Both sides can be modernized easily and autonomously and can be even replaced by different implementations from different vendors.
4. It is possible to integrate RqS with existing software systems.
5. The discussed solution enables construction of bimodal [5,24] systems.

---

[3] This requirement is based on several years experience of the hospital. There were many attempts in the hospital to implement RqS as a part of the existing ERP but all the attempts have failed.

5 Architecture of Digitalized Requisitions

Digitalized RqS can be viewed as a collection of digitalized Use Cases (UsC). An UsC is controlled/managed by its actors (accredited persons) and performs its event transforming collection of its input requisitions documents onto an output collection of requisitions documents. An UsC is identified by event type, actor profession, actor position and actor PID.

An UsC is activated by its actor. The actor must verify whether needed RqD are already available on input. UsC behaves like and application in classical sense. It has moreover some properties of services in the sense of SOA.

UsC have basic abilities of applications. They can be generated, activated, operated, and finished. It has some features of REA [12] structure or adaptive case management (ACM [16]).

Input requisitions documents can be generated manually and output requisitions documents can be printed. If used properly, it enables prototyping and incremental development. Individual UsC can be developed autonomously and then integrated.

6 Implementation of the System Based on Use Cases

UsC should be transparent and usable. The users have different professions. They are well trained to work with spreadsheets. It is preferable to design and implement UsC such that it behaves like a spreadsheet application. Digital format of RqD is a spreadsheet table.

It can be achieved if UsC is implemented as a pair of a spreadsheet (Excel) application and DMS (Alfresco[4]) serving as data tier and partially also as application logic tier of a 3-tier information systems architecture. DMS can be common for all UsC.

The system was developed and for some time successfully used by user organization. The system has been (using the already collected experience) successfully modernized to have a form of client-server application.

The client is here a spreadsheet equipped by a clever connector enabling collaboration with modernized DMS being on a server DMS use document database. Documents are as a rule encapsulated spreadsheet tables. Some turns known from management information systems [22] are used. The use of document database as data tier of the document management system brought another advantage: it is quite easy to get to the original documents if necessary.

6.1 Advantages for Users

- This solution enabled a very high usability of use cases and of Repositions (RqS) as a whole. Users feel the system easy to use and very transparent. They need not be aware about the implementation aspects. They feel it as a simple enhancement of existing practices.
- Many tools useful in management offered by Excel are available in UsC.

---

[4] https://www.alfresco.com/.

6.2   Implementation Issues

The orchestration of Excel and Alfresco was not easy. IT experts tend to use classical technologies like object databases. They dislike coarse-grained interfaces and are not ready to use document-oriented techniques, especially in a framework of a spreadsheet.

7   Experience with System Implementation and Deployment

The development and deployment of the system has been done in several stages.

The manual processes have been consolidated. The experience of both partners was properly used. The existing processes were updated, consolidated, and made more transparent. The necessary changes were designed to influence the involved users as less as possible. The changes have been therefore accepted as an improvement putting the process and paper documents into order.

The used documents have been digitized. The digitized documents looked as the paper original ones[5]. It was possible to print them out.

Checking and approving of requisitions. Originally it has been done by managers. Later it has been done also by head matrons. There has been generated virtual arrangements for given unit and a responsible person could modify it. The digital requisition directly checked if the requisition fits within limits. The changes have been made to the original (subunit) requisition forms and the impacts have been shown on both local and units' requisitions.
    The approvers can behave like by editing a document. It is a situation that they know and are familiar with it.

At the beginning the managers, physicians, and nurses felt the system as a screen version of the paper documents. It played a key role – they have not been afraid of it.

Observations

– It was always possible to print any requisition form (what has been used).
– The simplicity, transparency, and intuitiveness of the system interface have significant impact not only on the positive acceptance of the system by its users but also on the ease of learning of newcomers: There are often cases when they must quickly after their arrival start to work with the system. They usually have no problem with it.
– It seems that the system is only a small adaptation of a well-known approach. Practical effect was that even a small firm has been able to implement large system and modify current system for big companies.

---

[5] It is known as an important aspect of information systems [5].

– Although it looks simple, for the company having experienced programmers it took quite a long time to implement the system and deploy it in several steps (one of them is described here). It is common as it means change of the working and thinking habits.
– Mastering this approach the company got ahead of competitors.

The solution is relatively simple, although some complex tools and techniques (document management system, smart documents, service-oriented architecture) developed for larger systems have been used. The techniques usual for large systems have been used for small (probably middle-sized) project that could be developed by a small software vendor.

One of the main advantages of the requisitions system is that properly collected requisitions (expectations of future consumption) allow more precise handling of reserves (if the reserves can serve centrally, they can be significantly smaller).

8   Current State of the Project

The requisition generation and approval system has been incrementally developed. The system is welcome by all user levels (nurses, matrons, and several levels of management). Experts of various levels gradually took part in the process. It has been welcome that new services are easily added. It was, in fact, an incremental development and deployment. The system interface to the ERP has been for a long time based on paper, currently is digitized. It appears that the key role has been played by the following the rule that material orders cannot be planned precisely but the impreciseness can be lowered by putting the individual requests together and by proper business rules.

It was crucial for success that entire development process had the character of incremental development in the large what allowed requirements tuning and reduced It was further enhanced by the fact that paper documents were and are properly used (the introduction of electronic version can take significant advantage from that). The digitalization allowed further improvements of the system.

The path to a simple implementation of adaptive case management [16] has been found. It has been allowed by the fact that approving has similar features as flow control of batch systems. It is interesting that there was no need for organized education and training of the users – they were able to educate and train themselves.

9   Open Questions for Further Research

The use of DMS allows solution of many things just by proper setting of indicators in metadata or in waybills. An example is a virtual sending of documents for processing by individual responsible workers. It is not clear what the limits of such solution are when it is, from arbitrary reason, necessary to really send the documents somewhere. The issue can be in the fact that solution using state indicators can be more difficult to maintain and more error prone. It can complicate system decomposition and therefore also its flexibility and stability. The use of flags for flow control playing the role of triggers is a very strong tool. As many other very powerful tools also this technique can easily become hard to handle and easy to make mistakes. Roughly speaking: It is not clear how such solution influences individual aspects of quality of the solution [13,15,17,27].

10 Conclusions

Requisitions system is successfully in use up to generation orders for ERP. It is welcome. It is open for new use cases. The in fact incremental deployment is highly appreciated. The digitalized filling of requirement documents (requisitions) simplifies and optimizes business processes.

The chosen solution is for standard users (healthcare personnel, especially nurses) as well as for management highly transparent and requires almost no training. The main reasons are:

- Understandability and simple control of the interface.
- Well-defined requirement documents (requisitions).
- Simple on-line availability of managerial information and tools.
- The requirements better correspond to future needs. It makes the operation of the hospital smoother.

It was important that there were excellent contact people at the user side.

The system enabled several possibilities of supervision and analysis – either using tools of direct analysis of existing documents in the form of spreadsheet tables using tools of a proper spreadsheet, either using transformation tools of more spreadsheet documents into summary ones.

It was and it is crucial that the requisitions system has been developed as an autonomous add-on (software service in the sense of SOA) to the existing ERP system. The service Requisitions has been developed and afterwards integrated with the hospital's ERP. Software services therefore need not be developed using decomposition. They can be integrated, replaced, outsourced, or insourced as necessary.

Digitization of (business) documents offers important opportunities for management without harming common employees.

The way the project has been developed and used is not common. Although the task seems to be simple, its solution required application of several tricks that are not usual in classic situations. Some of the tricks are already common (cooperation of legacy parts with newly developed ones, gradual decomposition of the situation into a network of (pseudo)services in the sense of batch systems, and non-standard use of document-based interfaces in the form of smart complex documents. It is open whether it is better to send the documents or to share them using their metadata and rules for handling them.

The development process was quite hard for the developers but for the users it seemed that quite nothing happens. And if something happened then it was advantageous for them. It is especially important for the cases when the primary task of the users is hard and critical.

It has been the smoothest deployment of a project of that size the developers take part in or heard about.

A new trend of this time is that the experience of mankind (e.g. general structure of documents, business basics, ...) is more and more applied (i.e. the programmers do not always find their own approach as before, they started to apply existing approaches and processes). This trend leads to getting closer computer science to classic social and technical processes. It brings unexpected effects and gives to the ones that mastered the approach significant competitive advantage.

The described system has features interesting for many potential customers.

References

1. van der Aalst, W., van Hee, K.: Workflow Management: Models, Methods, and Systems (Cooperative Information Systems). The MIT Press, Cambridge (2004)
2. Alexander, C., Ishikawa, S., Silverstein, M.: A Pattern Language: Towns, Buildings, Construction. Oxford University Press, Oxford (1977)
3. André, É., Choppy, C., Reggio, G.: Activity Diagrams Patterns for Modeling Business Processes, pp. 197–213. Springer International Publishing, Heidelberg (2014). doi:10.1007/978-3-319-00948-3_13
4. Andrews, T., Curbera, F., Dholakia, H., Goland, Y., Klein, J., Leymann, F., Liu, K., Roller, D., Smith, D., Thatte, S., Trickovic, I., Weerawarana, S.: Specification: business process execution language for web services version 1.1 (2003)
5. Brandon, J.: Why paper still rules the enterprise. CIO Magazine, January 2016. http://www.cio.com/article/3025928/printers/why-paper-still-rules-the-enterprise.html
6. Business Process Management Initiative: Business process modelling notation (2004). http://www.bpmn.org/
7. Erl, T.: Service-Oriented Architecture: Concepts, Technology, and Design. Prentice Hall PTR, Upper Saddle River (2005)
8. Foster, M.: Case Management Part 1: An Introduction (2013). http://www.ateam-oracle.com/case-management-part-1-an-introduction/
9. Gamma, E., Helm, R., Johnson, R., Vlissides, J.: Design Patterns. Elements of Reusable Object-Orieneted Software. Addison-Wesley, Boston (1993)
10. Ganesh, K., Mohapatra, S., Anbuudayasankar, S.P., Sivakumar, P.: Enterprise Resource Planning. Management for Professionals. Springer International Publishing, Cham (2014)
11. Garcia-Molina, H., Ullman, J.D., Widom, J.: Database Systems, 2nd edn. Pearson, Upper Saddle River (2009)
12. Hruby, P.: Model-Driven Design Using Business Patterns. Springer, Heidelberg (2006)
13. International Organization for Standardization, International Electrotechnical Commission: ISO/IEC 25010: 2011 systems and software engineering – systems and software quality requirements and evaluation (SQuaRE) – system and software quality models (2011)

14. Jiao, W.: Using autonomous components to improve runtime qualities of software. IET Softw. **5**, 1–20 (2011)
15. Kan, S.H.: Metrics and Models in Software Quality Engineering, 2nd edn. Addison-Wesley Longman Publishing, Boston (2002)
16. Khanna, A.: Managing unpredicatbility using BPM for adaptive case management, July 2013. http://www.oracle.com/us/technologies/bpm/bpm-for-adative-case-mgmt-1972799.pdf
17. Kitchenham, B., Pfleeger, S.L.: Software quality: the elusive target. IEEE Softw. **13**(1), 12–21 (1996). http://dx.doi.org/10.1109/52.476281
18. Král, J.: Informační Systémy. Science, Veletiny, Czech Republic (1998)
19. Král, J., Žemlička, M.: Autonomous components. In: Hlaváč, V., Jeffery, K.G., Wiedermann, J. (eds.) SOFSEM 2000. LNCS, vol. 1963, pp. 375–383. Springer, Heidelberg (2000). doi:10.1007/3-540-44411-4_26
20. Král, J., Žemlička, M.: Software confederations - an architecture for global systems and global management. In: Kamel, S. (ed.) Managing Globally with Information Technology, pp. 57–81. Idea Group Publishing, Hershey (2003)
21. Král, J., Žemlička, M.: Crucial patterns in service-oriented architecture. In: Proceedings of ICDT 2007 Conference, p. 24. IEEE CS Press, Los Alamitos (2007). doi:10.1109/ICDT.2007.9
22. Laudon, K.C., Laudon, J.P.: Management Information Systems: Managing the Digital Firm, 13th edn. Pearson, Upper Saddle River (2014)
23. O'Leary, D.E.: Enetrprise Resource Planning Systems. Cambridge University Press, Cambridge (2000)
24. Pezzini, M.: Composite applications: where development and integration meet. Bus. Integr. J. 16–20 (2004)
25. Pokorný, J.: Workflow management systems: a survey of possibilities. In: Dias Cooelho, J., et al. (eds.) Proceedings of 4th European Conference on Information Systems (ECIS 1996), Lisbon, Portugal, pp. 253–263 (1996)
26. Schedlbauer, M.J.: The Art of Business Process Modeling. Cathris Group, Sudbury (2010)
27. Žemlička, M., Král, J.: Software architecture and software quality. In: Gervasi, O., et al. (eds.) ICCSA 2016. LNCS, pp. 139–155. Springer International Publishing, Cham (2016). doi:10.1007/978-3-319-42092-9_12

Proposing an IoT-Based Healthcare Platform to Integrate Patients, Physicians and Ambulance Services

Itamir de Morais Barroca Filho[1]([✉]) and Gibeon Soares de Aquino Junior[2]

[1] Metropole Digital Institute, Natal, Brazil
itamir.filho@imd.ufrn.br
[2] Department of Informatics and Applied Mathematics,
Federal University of Rio Grande Do Norte, Natal, Brazil
gibeon@dimap.ufrn.br
http://www.dimap.ufrn.br, http://www.imd.ufrn.br

Abstract. There is a global concern with the increasing number of patients at hospitals caused by population ageing and chronic diseases. Several studies indicated the need to minimize the process of hospitalization and the effects related to the high cost of patient care. In this context, a promising trend in healthcare is to move medical checks routines from a hospital (hospital-centric) to the patient's home (home-centric), but to make it possible we need Internet of Things (IoT)-based technologies and platforms to enable remote health monitoring. The application of IoT in healthcare area is a common hope because it will allow hospitals to operate more efficiently and patients to receive better treatment. With the use of IoT-based healthcare applications, there is an unprecedented opportunity to improve the quality and efficiency of the medical treatment and consequently improve patients' wellness, as well as a better application of the government's financial resources. Thus, based on this context, this paper aims to propose an IoT-based healthcare platform for the remote monitoring of patients in critical condition. It involves embedding sensors and actuators in patients, physicians, clinical staff, medical equipment and physical spaces in order to monitor, track and alert.

Keywords: Healthcare · Internet of Things (IoT) · Patients · Remote · Monitoring · Platform

1 Introduction

The population ageing and the increase of chronic diseases are becoming a global concern because it might result in more patients at hospitals. Moreover, several studies have indicated the need for strategies to minimize the institutionalization process and the effects of the high cost of patient care [15]. With the aim to minimize this concern, a promising trend in health treatments is to move the medical checks routines from the hospital (hospital-centric) to the patient's

© Springer International Publishing AG 2017
O. Gervasi et al. (Eds.): ICCSA 2017, Part VI, LNCS 10409, pp. 188–202, 2017.
DOI: 10.1007/978-3-319-62407-5_13

home (home-centric). However, to effectively achieve this, we need IoT-based healthcare applications to allow the patient's remote health monitoring.

On the other hand, recent advances in microelectronics, wireless, sensing and information have fueled the advent of a revolutionary model involving systems and communication technology, enabling smarter ways to "make things happen". This new paradigm, known as the Internet of Things (IoT), has a broad applicability in several areas, including health. In this trend, it is estimated that by 2020 there will be around 20 billion "things" connected [12] and uniquely identifiable [13]. These "things" promote the basic idea of IoT, which is pervasive computing around the range of devices, such as RFID tags, sensors, actuators, mobile phones, etc. [4].

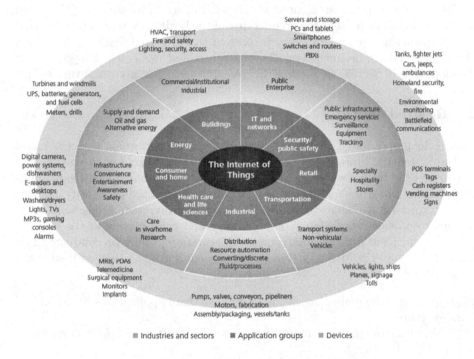

Fig. 1. IoT industries and sectores [29].

As presented in Fig. 1, the IoT will make possible the development of applications in many industries and sectors, such as transportation, buildings, energy, home, industrial, and healthcare. In healthcare market, it is expected to see the development of applications in this trend as part of the future, since it has the ability to allow hospitals to operate more efficiently and patients to receive better treatment. A type of healthcare application which will be focused in conjunction with this new paradigm is the application of mobile health. The main objective of mobile health is to allow remote monitoring of the patients' health status and treatment from anywhere [18].

In this context, the potential for change in the quality of life that can be promoted by IoT is unquestionable. Creating integrated utilities will lead to a qualitative change in the services to integrate information systems, computing and communication with extensive control [8]. Therefore, there is an urgent need for the development of technologies and applications related to IoT infrastructure for healthcare.

Therefore, we need IoT-based solutions in order to improve the healthcare of the patients in a critical situation, making their remote healthcare possible, and hence reducing the high number of patients in hospitals. Thus, this paper aims to propose an IoT-based healthcare platform for remote monitoring of patients in critical condition. It involves embedding sensors in patients, physicians, clinical staff, medical equipment and physical spaces in order to monitor, track and alert.

Finally, the paper is structured as follows: in Sect. 2, we present the related works; in Sect. 3 we present the architecture of the healthcare platform approaching technical aspects; and in Sect. 4 we present this proposed platform describing the design and requirements aspects, the actors and use cases that guided its development; in Sect. 5, we present this research's conclusions and future works.

2   Related Works

Before the proposal of a new healthcare platform, we performed a review based on the Systematic literature reviews (SLR) method. The goal of the review was to comprehend the current state and future trends in IoT-based healthcare applications. According to Wohlin et al. [30], SLRs are conducted to identify, analyze and interpret all available evidence related to a specific research question. Since it aims to provide a complete, comprehensive and valid picture of the existing evidence, the identification, analysis and interpretation must be conducted in a scientifically and rigorous way. Thus, the research questions that addressed the review were:

RQ1. What are the main characteristics of healthcare applications based on IoT infrastructure?
RQ2. What are the patterns and protocols used in healthcare applications based on IoT infrastructure?
RQ3. What are the challenges and opportunities related to healthcare applications based on IoT infrastructure?

Regarding the main characteristics of healthcare application based on IoT infrastructure (RQ1), we collected their functional and nonfunctional requirements from the studies. Thus, the functional requirements described in the studies are the patient's body and environment monitoring. Considering the body monitoring, the data monitored by sensors attached to patient's body are the pulse oximeter, heart rate, galvanic skin, transpiration, muscle activity, body temperature, oxygen saturation, blood pressure, airflow, body movement, blood glucose, breathing rate and ECG [1,6,11,18,25,31,32]. Moreover, the environment monitoring is related to sensors deployed in the patient's environment that capture data from temperature, light, humidity, location, body position, motion data, SPO2, atmospheric pressure and CO2 [3,7,16,33]. Moreover, when it comes to healthcare applications' features, there are some important nonfunctional requirements that represent a concern in this kind of application. Thus the nonfunctional requirements cited by the studies are scalability, reliability, ubiquity, portability, interoperability, robustness, performance, availability, privacy, integrity, authentication and security [3,7,11,16,18,33].

Regarding protocols (RQ2), the data collected from the studies [2,9,16,17, 26–28] showed that there are two protocols categories: communication, regarding network protocols, and application, regarding data transfer protocols. Thus, the communication protocols cited by the studies on the healthcare applications are 6LoWPAN, IEEE 802.15.4, Zigbee, Bluetooth, RFID, WIFI, Ethernet, GPRS, IEEE 802.15.6, 3G/4G, NFC and IrDA. Regarding the applications protocols, the studies cited: REST, YOAPY, HTTP, CoAP, XML-RPC and Web Services. When it comes to data format, the studies presented that the healthcare applications use HL7, XML, EHR, CSV, JSON and PHR.

The studies showed that there are many challenges related to healthcare applications based on IoT infrastructure (RQ3). The authors presented that health information management through mobile devices introduces several challenges: data storage and management (e.g., physical storage issues, availability and maintenance), interoperability and availability of heterogeneous resources, security and privacy (e.g., permission control, data anonymity, etc.), unified and ubiquitous access are a few to mention [10]. Moreover, the authors highlight the interoperability challenge, since there have been different studies and proposals for patient monitoring at the hospital or for personal monitoring at home, but a shared goal to produce an interoperable system adopting open standards for healthcare, such as HL7, and a seamless framework to be easily deployed in any given scenario for healthcare is still missing [20].

Thus, considering the presented challenges in the current solutions, we proposed an architecture, which will be described in Sect. 3, addressing the interoperability, security and privacy requirements. This architecture will be used in the development of the IoT-based platform proposed in Sect. 4.

3   The Solution's Architecture

With the performed review, described in Sect. 2, we were able to define an architecture to guide the development of the proposed healthcare platform. This architecture, presented in Fig. 2, is composed of 7 layers and was used in the development of the proposed IoT-based healthcare platform.

Thus, the layers are: requirements, users, systems and services, middleware, monitoring, communication, and patients:

– The requirements layer is responsible for features that make IoT-based healthcare applications secure and reliable. It is composed of components that assure scalability, reliability, ubiquity, portability, interoperability, robustness, performance, availability, privacy, integrity, authentication and security. It is

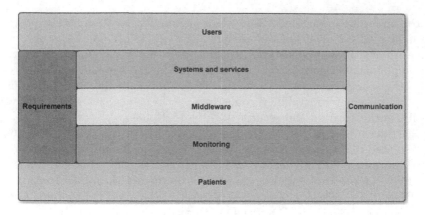

Fig. 2. Architecture for IoT-based healthcare platform.

important to emphasize that because of the responsibility of this layer, it is connected to the systems and services, middleware, and monitoring layers.

- The users layer is composed of part of the actors of healthcare applications: physicians, hospital administrator, nurses, family, pharmaceutical and clinical staff.
- The communication layer is responsible for making all components communicate with each other. Thus, it is composed by communication protocols: 6LoWPAN, IEEE 802.15.4, Zigbee, Bluetooth, RFID, WIFI, Ethernet, GPRS, IEEE 802.15.6, 3G/4G, NFC and IrDA.
- The monitoring layer is responsible for monitoring the patient's body and environment. Regarding the body monitoring, the data are captured from pulse oximeter, heart rate, galvanic skin response, transpiration, muscle activity, body temperature, oxygen saturation, blood pressure, airflow, body movement, blood glucose, breathing rate and ECG sensors. Regarding the environment monitoring, the data are captured from temperature, light, humidity, location, body position, motion, SPO2, pressure and CO2 sensors.
- The middleware layer is responsible for receiving the data from the myriad of sensors in the monitoring layer, processing them, and making them available for the systems and services layer. Thus, this layer is composed of IoT platforms such as ThingSpeak, FIWARE, Kaa, AzureIoT and AWS IoT.
- The systems and services layer is responsible for the services, formatting patterns and applications protocols. Thus, it is composed of the ambulance, hospital and pharmacy systems. Moreover, it uses formatting protocols such as HL7, XML, EHR, CSV, JSON and PHR. Finally, in this layer, we have applications protocols such as REST, CoAP, YOAPY, HTTP, XML-RPC and web services.
- The patients layer is composed of patients, and it defines healthcare areas where IoT-based applications would be useful: rehabilitation, respiratory diseases, elderly, obesity, arterial hypertension and diabetes. Patients who need healthcare for these problems are classified as patients in a critical situation.

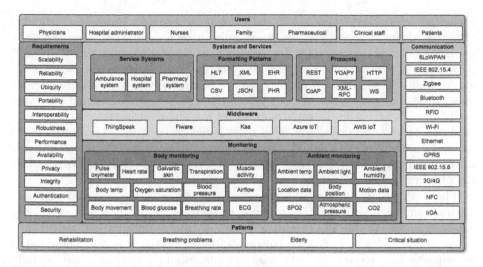

Fig. 3. Expanded architecture for the IoT-based healthcare platform.

Finally, considering the defined layers, Fig. 3 presents the healthcare architecture layers with the cited technologies. This architecture was used for the development of the proposed healthcare platform.

4   An IoT-Based Healthcare Platform

This section presents the IoT-based healthcare platform describing the design and requirements aspects, the actors and use cases that guided its development. It also explains the platform's architecture considering its modules, their relationship, components and protocols.

4.1   Design Issues and Requirements

The main goal of the proposed IoT-based healthcare platform is to provide remote monitoring for patients in a critical situation, and it was developed considering the need to transfer the healthcare from the hospital (hospital-centric) to the patient's home (home-centric). This platform is IoT-based and integrates patients, physicians and ambulance services in order to promote better care and fast preventive and reactive urgent actions. It address challenges like interoperability, security and privacy. Regarding requirements, this platform has *Remote Patient and Environment Monitoring, Patient Healthcare Data Management, Patient Health Condition Management and Emergency and Crisis Management*.

The *Remote Patient and Environment Monitoring* involves the acquisition of data from sensors attached to the patient's body and in the environment (patient's home). The data acquired from the sensors are used by clinical staff (physician and nurses) for healthcare treatment and emergency alert purposes.

Thus, the sensors attached to the patient's body provide information about: ECG, blood pressure and glucose, heart rate, oxygen saturation, temperature, breathing rate and capnography. The sensors from the environment provide information about environment temperature, location with latitude and longitude, and humidity. This is important because the control of the environment's temperature and humidity can directly affect the patient's treatment. Regarding the location, it assists in the rapid response of the ambulance service. Therefore, since the patient in critical condition is at home and not in a hospital, which is a more controlled environment, this ambient information is of greater importance for effective healthcare and enriches the remote monitoring provided by this platform.

The *Patient Healthcare Data Management* records the data about the patient: name, gender, date of birth, contacts, address, family information, physician information (name and contacts), health insurance information, health situation and the history of monitoring sensors and emergency alerts. These data are important for physicians and nurses to understand the current situation and history of patients, and also to facilitate the accurate monitoring of the health treatment.

The *Patient Health Condition Management* considers the patient's healthcare data, especially the health situation and history of the sensors' monitoring data, to allow the definition of critical level values for the sensors, which are important to enable the rapid response in case of emergency. It also defines rules to actions considering the settled critical levels for a patient and the related alerts.

Finally, the *Emergency and Crisis Management* address information about the patient's health condition and the services that should be alerted in case of emergency with a monitored patient in a critical situation. Since this patient is at home and not in a hospital, the efficiency of a rapid response in an emergency case can be the deciding factor between life and death.

To achieve the presented requirements, this platform is composed of ten use cases, presented by the use case diagram in Fig. 4, and four actors: the hospital operator, physician and nurses, ambulance service operator and the patient and family.

Considering the hospital operator actor, it interacts with the use cases related to patient, health insurance and clinical staff data, which are:

- Patient's data management: it allows to register data related to Patient Healthcare Data Management mentioned earlier;
- Clinical staff data management: it allows to register data related to the clinical staff (physician and nurse). These data include: name, contacts and specialty from physician and nurses;
- Health insurance data management: it allows to register data from the health insurance;
- Patient and physician association: it allows to register the responsibility of a physician with a patient.

Regarding the physician and nurse actors, they can use the patient's data management and interact with other use cases related to Remote Patient and

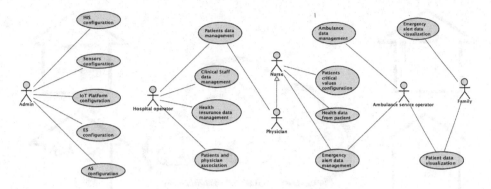

Fig. 4. Healthcare platform use case diagram.

Environment Monitoring, Patient Healthcare Data, Patient Health Condition Management and Emergency and Crisis Management, such as:

- Patient's critical values configuration: it allows the definition of critical level values for the sensors attached to the patient's body, which are considered in the alerts and notifications;
- Health data from patient: it allows the visualization of real-time health data from the sensors deployed in any patient's body and environment;
- Emergency alert data management: it allows the notification and alerts to be presented and managed by physicians and nurses.

The ambulance service operator actor uses the emergency alert data management use case and interacts with other use cases related to Emergency and Crisis Management, which are:

- Ambulance data management: it allows the management of data from the ambulances, such as real-time location and activation of an ambulance to an emergency;
- Patient data visualization: it allows the visualization of real-time health data from the sensors deployed in a single patient's body and environment, along with his/her home location.

Finally, the patient and family actor uses the patient's data visualization use case and interacts with the emergency alert data visualization, related to Emergency and Crisis Management. The emergency alerts and patient data are available only to the patient him/herself or to his/her family member. Thus, the proposed platform was developed considering the presented requirements, actors, and use cases. It provides integration between patients, physician and ambulance services, for the efficient healthcare of patients in critical condition. In the following subsection, the modules overview is presented with details of the platform's modules.

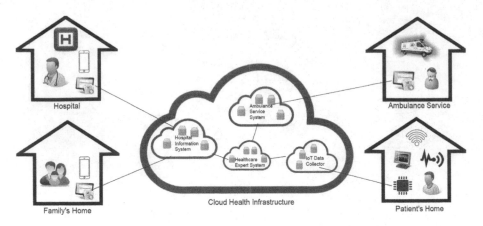

Fig. 5. IoT-based platform to integrate patients, physicians and ambulance services.

4.2 Modules Overview

The IoT-based healthcare platform, presented in Fig. 5, is composed of five modules: Patient's Home, Cloud Health Infrastructure, Hospital, Family's Home and Ambulance Service. These modules address the solution's functional requirements and work together to achieve the goal of remote monitoring and efficient healthcare for patients in critical condition.

Considering the Patient's Home module, it is mainly composed of sensors that provide body and environment remote monitoring. The sensors attached to the patient's body, described in the previous section, are part of a multi-parameter portable patient monitoring, which continuously measures his/her vital signs. This monitoring is configured by the clinical staff at the patient's home and does not require his intervention. Regarding security and essential performance issues, the monitor is in agreement with the IEC standard, 60601-1-11:2015 (IEC 60601, 2015), which defines the basic safety and essential performance of medical electrical equipment and medical electrical systems for use in the home healthcare environment. The environment monitoring sensors, described in the previous section, are also deployed at the patient's home.

Thus, this multi-parameter monitor is connected through the Internet to the Cloud Health Infrastructure module, as presented in Fig. 6. It uses HL7 [14] as a standard for data streaming. The environment sensors are connected to a gateway through 6LoWPAN protocol. Regarding 6LoWPAN, it is a protocol for Wireless Sensor Networks (WSNs) defined to enable IPv6 packets to be carried on top of low power wireless networks, specifically exploiting IEEE 802.15.4 protocol [20]. This gateway is also connected by the Internet to the IoT Platform of the Cloud Health Infrastructure module. The reason to use a gateway for the sensors is because they do not have interfaces for direct connection to the Internet.

Regarding the Cloud Health Infrastructure module, it is composed of the IoT Data Collector, the Healthcare Expert System, the Hospital Information

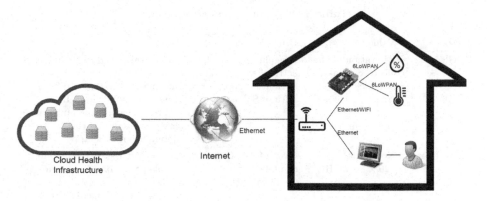

Fig. 6. The connections between the multi-parameter monitor and ambient sensors.

System and the Ambulance Service System. These systems implement the use cases described previously. The features of each system are:

- IoT Data Collector: responsible for receiving sensors' data. This is a challenge for achieve IoT due to the huge number of devices integrated into this component and their diversity in terms of data formats and protocols;
- Healthcare Expert System: configuration of the patient's critical values for alerts that are displayed in the Hospital Information and Ambulance System. These critical values are used in the rules defined by the physicians and nurses. This system provides standard mapped rules, presented in Table 1, and performs machine learning and analytics to assist the physician in defining the appropriate critical values for each patient. Regarding the rule, it is composed of a type of sensor, a value, a critical level and its action. This action is associated with a color, according to the Manchester Triage System, which is displayed on the related information systems.
- Hospital Information System: management of data from patients, clinical staff and health insurance, and the association between patients and physicians. It also provides mobile views for patient data and alerts visualization.
- Ambulance Service System: management of data from emergency alerts and ambulances. It also provides visualization of the patient's data.

The data received by the IoT Data Collector are used by the Healthcare Expert System, which contains rules and creates derived information to be used by the Hospital Information System and the Ambulance Service System. The Healthcare Expert System uses Machine Learning and Analytics techniques based on the huge amount of data received to produce knowledge about the patient's health behavior. This knowledge is then stored in the computer and users call upon the computer for specific advice as needed. The computer can make inferences and arrive at a specific conclusion [21]. For example, critically ill patients, particularly those with hemodynamic instability signals, need a diagnosis and immediate treatment. This condition presents itself with signs of tissue

Table 1. Rules of the expert system.

Sensor	Value	Critical level	Action
ECG	Alteration of the electrocardio-graphic trace	Irregular electrical activity, ventricular fibrillation supra or infra of ST asystole	Red alarm issued to the medical staff on the Hospital Information System and Physician mobile app. If asystole, it alerts the ambulance service system and the ambulance mobile app
Blood Pressure	Normal is 120 × 80 mmHG	Over 140 × 90 mmHg or under 90 × 60 mmHg	Green alert issued to the family mobile app. Considering the patient's history, a red alert can be issued to the medical staff in the Hospital Information System and Physician mobile app
Blood Glucose	Normal between 100 and 126 mgdL	Over 200 mg/dL or under 60 mg/dL	Green alert issued to the family mobile app. Considering the patient's history, a red alert can be issued to the medical staff in the Hospital Information System and Physician mobile app
Heart rate	Normal between 60 and 100 bpm	Under to 60 bpm or over a 100 bpm	Red alarm issued to the medical staff in the Hospital Information System and Physician mobile app. Considering the patient's history, it can alert the ambulance service system and the ambulance mobile app
Breathing rate	Normal between 16rpm and 20rpm	Under to 12rpm or over to 35rpm	Green alert issued to the family mobile app. Considering the patient's history and clinical staff programming, it can alert the ambulance service system and the ambulance mobile app
Temperature	Normal between 36 C and 37,5 C	Over 37.5 C	Green alert issued to the family mobile app. Considering the patient's history and in cases of desirable hypothermia, clinical staff can program the threshold
Capnography	Normal between 35 mmHg and 45 mmHg	Under to 35 mmHg or over to 55 mmHg	Red alarm issued to the medical staff in the Hospital Information System, Physician mobile app, the ambulance service system and the ambulance mobile app

perfusion and impaired tissue oxygenation, which is usually detected by macro-circulatory parameters or global hemodynamic measurements such as blood pressure and oxygen saturation in arterial blood [5]. When a critical value is captured by this healthcare expert system, it automatically generates an alert message to the Hospital Information System, the physician, the patient's family and ambulance service system providing support for a specific decision on when and how to intervene. Thus, the patient's state of the classification system is issued along with the monitoring values. Another example is if the data from the sensors show that the patient's heart rate is zero, it can translate this as a heart attack. Therefore, this Healthcare Expert System also notifies the Hospital Information System, the physician, the patient's family and Ambulance Service System.

Moreover, the Healthcare Expert System also provides an API to make the patient's information available to authorized third party systems, with the use of OAuth V2.0 [23], taking into account privacy and ethics. This API is composed of RESTFul Web Services [24] and uses JSON [19]. The purpose of this API is to make it easier to develop new solutions with the use of this data to promote innovation in the healthcare area. As a result, companies and researchers can benefit from this use.

The Hospital module is used by physicians, nurses and clinical staff, and it uses the Hospital Information System. This system contains the patient's records, including information about age, gender, name, contacts, family contact. It also provides real-time remote monitoring of patients in critical condition. Integrated with this Hospital Information System, there is a mobile app, which, in case of any problem with a patient, notifies the physician responsible. With this notification, the mobile app also presents the real-time situation and data from the sensors, such as ECG, blood pressure, blood glucose, heart rate, oxygen saturation, temperature, and breathing rate.

The Family's Home module is used by the patient's family and is connected to the Hospital Information System. However, the information presented is exclusive to their related patient. It provides a mobile app that displays real-time monitoring from the sensors connected to the patient. Finally, this mobile app displays less information than the version used by the physicians and clinical staff because some of the data from sensors require medical expertise to be understood.

Regarding the environment monitoring provided by this platform, all data from the sensors - temperature, location, and humidity - are presented at the Hospital Information System and its mobile apps and at the ambulance service system. This monitoring is important because the environment's temperature and humidity control can directly affect the patient's health treatment.

The Ambulance Service module is connected to the Ambulance Service System and is used by operators. The ambulances have a mobile app that is connected to the Ambulance Service System, using 3G/4G, to receive real-time information from the body and environment monitoring sensors, emergency alarms and patient's situation. It is important to emphasize that the monitored location data from the patient is a key point for an effective response from the ambulance service.

Considering the platform's presented modules and their purposes there are important requirements that need to be addressed: privacy, security, interoperability, scalability, reliability, robustness, ubiquity, portability, performance and availability. Since all the information transmitted from the sensors to the systems and mobile apps are sensible, there is a need for privacy in this communication. To assure privacy, this platform uses encryption. There is also the special need for security, which is mainly guaranteed by authentication [22]. The IoT platform from the Cloud Health Infrastructure module assures scalability, integrity, portability and interoperability between the different types of connected monitoring devices. This module's expert system addresses the need for ubiquity, considering the defined rules and the customization feature. Finally, the proposed platform and its modules organization aim to achieve good performance, robustness, reliability, and availability regarding the information from patient monitoring. Moreover, with the use of HL7 and an IoT data collector, we propose to solve interoperability issues, and with permissions controls and OAuth V2, we propose to solve security and privacy concerns.

5   Conclusion and Future Works

This paper presented an IoT-based healthcare platform for remote monitoring of patients in critical condition, promoting the process of moving the medical checks routines from the hospital (hospital-centric) to the patient's home (home-centric). It involves sensors in patients and integrates physicians, clinical staff, medical equipment and physical spaces in order to monitor, track and alert.

Before the proposal of this healthcare platform, we performed a review aiming to comprehend the current state and future trends of healthcare applications based on IoT infrastructure, and also in order to find areas regarding it for further investigations. Using the extraction data, we were able to answer the research questions and provide the characteristics of healthcare applications based on IoT infrastructure, as well as the protocols and data formats used in the studies, and we were also able to find some challenges and opportunities for healthcare applications (Sect. 2). The mentioned challenges are related to the development of new solutions to resolve interoperability, privacy and security problems.

In the possession of the knowledge obtained with the review, we were able to propose a layered architecture for healthcare applications based on IoT infrastructure presented in Sect. 3. It considers the characteristics of these applications, functional and nonfunctional requirements, used protocols and patterns, and is composed of the following layers: patients, monitoring, requirements, communication, middleware, systems and services, and users.

Thus, after the design of the platform architecture, we developed it considering the specified requirements and addressing issues like security and interoperability. This platform is composed of modules for Remote Patient and Environment Monitoring, Patient Healthcare Data Management, Patient Health Condition Management and Emergency and Crisis Management. In Sect. 4 we presented the design issues requirements and its modules overview. Finally, as future works, we will document and improve this architecture, present implementations details, and evaluate the proposed healthcare platform with real patients.

References

1. Abawajy, J.H., Hassan, M.M.: Federated internet of things and cloud computing pervasive patient health monitoring system. IEEE Commun. Mag. **55**(1), 48–53 (2017)
2. Al-Taee, M.A., Al-Nuaimy, W., Al-Ataby, A., Muhsin, Z.J., Abood, S.N.: Mobile health platform for diabetes management based on the internet-of-things. In: 2015 IEEE Jordan Conference on Applied Electrical Engineering and Computing Technologies (AEECT), pp. 1–5. IEEE (2015)
3. Archip, A., Botezatu, N., Şerban, E., Herghelegiu, P. C., Zală, A.: An IoT based system for remote patient monitoring. In: 2016 17th International Carpathian Control Conference (ICCC), pp. 1–6. IEEE (2016)
4. Atzori, L., Iera, A., Morabito, G.: The internet of things: a survey. Comput. Netw. **54**(15), 2787–2805 (2010)
5. Bazerbashi, H., Merriman, K.W., Toale, K.M., Chaftari, P., Carreras, M.T.C., Henderson, J.D., Yeung, S.C.J., Rice, T.W.: Low tissue oxygen saturation at emergency center triage is predictive of intensive care unit admission. J. Crit. Care **29**(5), 775–779 (2014)
6. Castillejo, P., Martinez, J.-F., Rodriguez-Molina, J., Cuerva, A.: Integration of wearable devices in a wireless sensor network for an e-health application. IEEE Wirel. Commun. **20**(4), 38–49 (2013)
7. Chen, M., Ma, Y., Li, Y., Wu, D., Zhang, Y., Youn, C.-H.: Wearable 2.0: enabling human-cloud integration in next generation healthcare systems. IEEE Commun. Mag. **55**(1), 54–61 (2017)
8. Chen, Y.: Analyzing and visual programming internet of things and autonomous decentralized systems (2016)
9. Datta, S.K., Bonnet, C., Gyrard, A., Da Costa, R.P.F., Boudaoud, K.: Applying internet of things for personalized healthcare in smart homes. In: 2015 24th Wireless and Optical Communication Conference (WOCC), pp. 164–169. IEEE (2015)
10. Doukas, C., Maglogiannis, I.: Bringing IoT and cloud computing towards pervasive healthcare. In: 2012 Sixth International Conference on Innovative Mobile and Internet Services in Ubiquitous Computing (IMIS), pp. 922–926. IEEE (2012)
11. Fan, Y.J., Yin, Y.H., Da Xu, L., Zeng, Y., Wu, F.: IoT-based smart rehabilitation system. IEEE Trans. Ind. Inf. **10**(2), 1568–1577 (2014)
12. Gartner: Gartner says 6.4 billion connected "things" will be in use in 2016, up. 30 percent from 2015 (2015)
13. Gubbi, J., Buyya, R., Marusic, S., Palaniswami, M.: Internet of things (IoT): a vision, architectural elements, and future directions. Future Gener. Comput. Syst. **29**(7), 1645–1660 (2013)
14. Health level seven (2017). http://www.hl7.org/
15. Hochron, S., Goldberg, P.: Driving physician adoption of mheath solutions. Healthc. Financ. Manage. **69**(2), 36–40 (2015)
16. Hossain, S.M., Muhammad, G.: Cloud-assisted industrial internet of things (IIoT)-enabled framework for health monitoring. Comput. Netw. **101**, 192–202 (2016)
17. Jara, A.J., Zamora, M.A., Skarmeta, A.F.: Drug identification and interaction checker based on iot to minimize adverse drug reactions and improve drug compliance. Pers. Ubiquit. Comput. **18**(1), 5–17 (2014)

18. Jara, A.J., Zamora-Izquierdo, M.A., Skarmeta, A.F.: Interconnection framework for mhealth and remote monitoring based on the internet of things. IEEE J. Sel. Areas Commun. **31**(9), 47–65 (2013)
19. JSON: Introducing json (2016). http://www.json.org/
20. Khattak, H.A., Ruta, M., Di Sciascio, E.: Coap-based healthcare sensor networks: a survey. In: Proceedings of 2014 11th International Bhurban Conference on Applied Sciences & Technology (IBCAST), Islamabad, Pakistan, 14th–18th January 2014, pp. 499–503. IEEE (2014)
21. Liao, S.-H.: Expert system methodologies and applicationsa decade review from 1995 to 2004. Expert Syst. Appl. **28**(1), 93–103 (2005)
22. Maksimović, M., Vujović, V., Perišić, B.: A custom internet of things healthcare system. In: 2015 10th Iberian Conference on Information Systems and Technologies (CISTI), pp. 1–6. IEEE (2015)
23. OAuth: Oauth 2.0 (2017)
24. Oracle: What are restful web services? (2016). http://docs.oracle.com/javaee/6/tutorial/doc/gijqy.html
25. Ray, P.P.: Internet of things for sports (IoTsport): an architectural framework for sports and recreational activity. In: 2015 International Conference on Electrical, Electronics, Signals, Communication and Optimization (EESCO), pp. 1–4. IEEE (2015)
26. Sebestyen, G., Hangan, A., Oniga, S., Gál, Z.: ehealth solutions in the context of internet of things. In: Proceedings of IEEE International Conference Automation, Quality and Testing, Robotics (AQTR 2014), Cluj-Napoca, Romania, pp. 261–267 (2014)
27. Swiatek, P., Rucinski, A.: IoT as a service system for ehealth. In: 2013 IEEE 15th International Conference on e-Health Networking, Applications & Services (Healthcom), pp. 81–84. IEEE (2013)
28. van der Valk, S., Myers, T., Atkinson, I., Mohring, K.: Sensor networks in workplaces: correlating comfort and productivity. In: 2015 IEEE Tenth International Conference on Intelligent Sensors, Sensor Networks and Information Processing (ISSNIP), pp. 1–6. IEEE (2015)
29. Wilson, S.: Rising tide: Exploring pathways to growth in the mobile semiconductor industry (2013)
30. Wohlin, C., Runeson, P., Höst, M., Ohlsson, M.C., Regnell, B., Wesslén, A.: Experimentation in Software Engineering. Springer Science & Business Media, Heidelberg (2012)
31. Yaakob, N., Badlishah, R., Amir, A., binti Yah, S.A., et al.: On the effectiveness of congestion control mechanisms for remote healthcare monitoring system in IoT environment a review. In: 2016 3rd International Conference on Electronic Design (ICED), pp. 348–353. IEEE (2016)
32. Yang, G., Xie, L., Mantysalo, M., Zhou, X., Pang, Z., Da Li, X., Kao-Walter, S., Chen, Q., Zheng, L.-R.: A health-iot platform based on the integration of intelligent packaging, unobtrusive bio-sensor, and intelligent medicine box. IEEE Trans. Ind. Inf. **10**(4), 2180–2191 (2014)
33. Yang, Z., Zhou, Q., Lei, L., Zheng, K., Xiang, W.: An iot-cloud based wearable ECG monitoring system for smart healthcare. J. Med. Syst. **40**(12), 286 (2016)

A Systematic Literature Review
on Microservices

Hulya Vural, Murat Koyuncu$^{(\boxtimes)}$, and Sinem Guney

Atilim University, Ankara, Turkey
hulya.vural.tr@gmail.com, sinemmguney@gmail.com,
mkoyuncu@atilim.edu.tr

Abstract. The cloud is an emerging paradigm which leads the way for different approaches and standards. The architectural styles are evolving based on the requirements of the cloud as well. In recent years microservices is seen as the architecture style for scalable, fast evolving cloud applications. As part of this paper, a systematic mapping study was carried out around microservices. It is aiming to find out the current trends around microservices, the motivation behind microservices research, emerging standards and the possible research gaps. The obtained results can help researchers and practitioner in software engineering domain who want to be aware of new trends about SOA and cloud computing.

Keywords: Cloud · SOA · Web services · Microservices · Systematic mapping

1 Introduction

Service-oriented architecture (SOA) has emerged as a means of developing distributed systems where the components are stand-alone services [37]. Services are basic units which are developed independently and made accessible over the Internet. Standard internet protocols are used for service communication among different computers. SOA provides many advantages to develop easy and economic distributed software systems and, therefore, it is the leading technology for interoperability on today's internet world. Service-oriented software engineering defines evolution of existing software engineering approaches to develop dependable and reusable services considering the requirements and characteristics of this technology [37]. Service-oriented computing (SOC) is the paradigm that utilizes services as the fundamental elements for developing applications. Therefore, service-oriented software engineering aims at designing and developing service-based applications consonant with SOC paradigm and SOA principles using software engineering methodologies.

After the popularity of cloud computing in recent years, new trends in the software engineering have emerged, such as going to market with minimal viable product and making small development teams autonomous. The architectural styles have also evolved based on the cloud environment needs [36]. One of those new architectural styles is microservices. The aim of the microservices is to divide the business behavior into small services which can run independent of each other. As mentioned by Martin

© Springer International Publishing AG 2017
O. Gervasi et al. (Eds.): ICCSA 2017, Part VI, LNCS 10409, pp. 203–217, 2017.
DOI: 10.1007/978-3-319-62407-5_14

Fowler, "While there is no precise definition of this architectural style, there are certain common characteristics around organization around business capability, automated deployment, intelligence in the endpoints, and decentralized control of languages and data" [1]. Another definition for microservices is "Microservices are small, autonomous services that work together" [2].

The characteristics of the microservices are listed as follows [1]:

- Componentization via Services
- Organized around Business Capabilities
- Products not Projects
- Smart endpoints and dumb pipes
- Decentralized Governance
- Decentralized Data Management
- Infrastructure Automation
- Design for failure
- Evolutionary Design

The microservices are developed, deployed and maintained separately. This allows the teams to be autonomous where they can decide on the technology to use which best addresses the current needs of the business behavior. The language and the database might be different from one microservice to another. They do not share data between each other, instead they use Representational State Transfer (REST) protocol to communicate to each other. The most important benefits of using microservices are agility, autonomy, scalability, resilience and easy continuous deployment.

Even though microservices were first mentioned at [60] in 2010, the definition of the microservice mentioned in that study does not totally map to the current microservice definition in literature. The study carried out in 2010 [60] defines microservices as light services using REST. It does not mention most of the characteristics listed at [1].

There has been another systematic mapping carried out on microservices in 2016 [40]. In that study, the research questions are around the architectural diagrams used for microservices' representation, the quality attributes and the challenges. However, the emerging standards and de facto tools are not mentioned.

In this paper, the aim is to not only analyze the emerging standards but also the types of research conducted and the practical motivations around carrying out the microservices architecture.

2 Method

This study is conducted a systematic mapping as defined in [3] with one modification (see Fig. 1). The modification is that, we carry out the keywording according to the whole paper instead of keywording according to the abstract. The reason for the modification to the original process is to enhance the classification criteria through adding new areas.

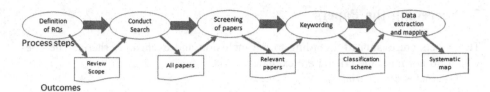

Fig. 1. Process steps and outcomes

2.1 Research Questions

Three research questions are determined as follows:

RQ1: What type of research is conducted on microservices?
RQ2: What are the main practical motivations behind microservices related research?
RQ3: What are the emerging standards and de facto tools on microservices solutions?

2.2 Search Sampling

The search is conducted using Web of Science (Thomson Reuters Web of Knowledge), which includes the following online databases:

- ACM (Association for Computing Machinery) Digital Library [4]
- CiteSeer [5]
- Computer Source [6]
- ebrary [7]
- Human-Computer Interaction Bibliography [8]
- IEEE Xplore [9]
- INSPEC [10]
- INSPEC Archive [11]
- Nature [12]
- Science [13]
- Science & Technology Collection [14]
- SciTech Connect [15]
- Springer LINK [16]

2.3 Search Iteration

The search is carried on with the following criteria:

- Keywords: microservice OR micro-service
- Research Area: Computer Science

The search iteration has returned 39 results [17–32, 38–60].

3   Screening the Papers

The papers are evaluated according to the inclusion and exclusion criteria. The ones which do not meet the criteria are excluded.

The inclusion criteria:

* All papers returned from the search criteria

Exclusion criteria:

* If the microservices is just mentioned in the research but the focus of the research is not directly on microservices.

Out of 39, 2 papers were excluded based on the exclusion criteria [30, 41]. As a result, 37 papers are included into the mapping process.

4   Keywording

As part of keywording four different categorization schemes are identified:

* Service models in cloud computing
* Operational areas
* Research types
* Emerging standards and tools.

4.1   Service Models in Cloud Computing

The service models in cloud computing are classified in three different types [33]:

* Infrastructure as a Service (IaaS): the infrastructure is supplied as a service (e.g. virtual machine, hard disk, load balancer etc.).
* Platform as a Service (PaaS): The platform is supplied as a service (e.g. Azure SQL, Tomcat etc.).
* Software as a Service (SaaS): The software itself is supplied as a service (e.g. Office 365, Gmail etc.).

Even though the microservice architecture style is shaped considering cloud needs, the research papers returned as part of the search criteria do not necessarily use cloud. As a result, on premise installations (OnPrem) are also included in the service models.

4.2   Operational Areas

In [34] several different operational areas are called out for cloud:

* Accounting and billing
* SLA management (Service Level Agreement)
* Service/resource provisioning

- Capacity planning
- Configuration management
- Security and privacy assurance
- Fault management

Some of the research papers included in the current study are focusing on cloud whereas some are not. As a result, the operational areas were modified to fit the needs as follows:

- Cost comparison
- Availability/Resiliency
- Performance
- Security
- Test technique
- Functionality/Design
- Analytics/Monitoring
- Scalability
- Deployment

The answer for the second research question (RQ2) will be based on the modified operational areas.

4.3 Research Types

In [35], 6 different research types are called out (See Table 1). The answer to the first research question (RQ1) will be based on these 6 research types.

4.4 Emerging Standards and Tools

The papers included in this systematic mapping study can be seen as a representation of the common tools used for microservices. Given that microservices is a new concept, the standards are not yet well formed. The current systematic study aims also to give an answer on emerging standards for microservices.

5 Data Extraction and Mapping

The results obtained from mapping are converted into different graphs and they are given below in a way to answer the defined research questions.

RQ1: What type of research is conducted on microservices?

The papers are mapped to the research types as seen in Fig. 2. The most widely used research type is Solution Proposal which is followed by Validation Research and Evaluation Research.

The papers are classified according to the service types as seen in Fig. 3. Almost half of the papers did not explicitly mention the service type they were targeting

Table 1. Research types

Class	Description
Validation research	Techniques investigated are novel and have not yet been implemented in practice. Techniques used are for example experiments, i.e., work done in the lab
Evaluation research	Techniques are implemented in practice and an evaluation of the technique is conducted. That means, it is shown how the technique is implemented in practice (solution implementation) and what are the consequences of the implementation in terms of benefits and drawbacks (implementation evaluation). This also includes identification of problems in industry
Solution proposal	A solution for a problem is proposed, the solution can be either novel or a significant extension of an existing technique. The potential benefits and the applicability of the solution is shown by a small example or a good line of argumentation
Philosophical papers	These papers sketch a new way of looking at existing things by structuring the field inform of a taxonomy or conceptual framework
Opinion papers	These papers express the personal opinion of somebody whether a certain technique is good or bad, or how things should have been done. They do not rely on related work and research methodologies
Experience papers	Experience papers explain what and how something has been done in practice. It has to be the personal experience of the author

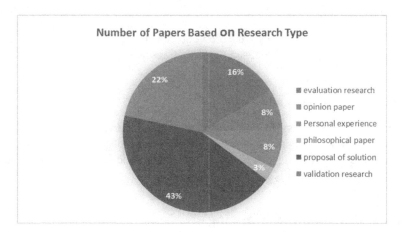

Fig. 2. Research types

(represented as NA in the figure). SaaS by far the most common service type being investigated. Also, some papers refer to more than one service type.

The bubble chart in Fig. 4 illustrates an analysis based on research types versus service types. The figure shows that there are only two studies on IaaS investigation regarding microservices. This is an expected outcome given that the microservices is a high level architectural style. On the other hand, there is only one official philosophical research papers on microservices. Most probably the reason is that the philosophical

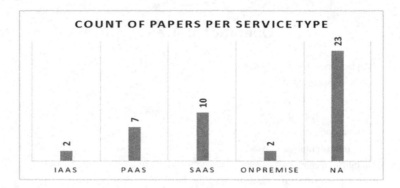

Fig. 3. Count of papers per service type

statement of microservices was laid out by Lewis and Fowler [1] on 2014. Mostly, the research is around Solution Proposal which do not explicitly call out the possible service types applicable for that solution.

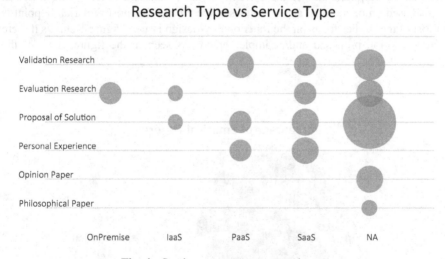

Fig. 4. Service types versus research types

RQ2: What are the main practical motivations behind microservices related research?

The papers are mapped to the operational areas and obtained results are shown in Fig. 5. The main motives are around functionality followed by performance and test techniques. Given that the microservices paradigm was first mentioned around 2014 and official research papers started to show up in 2015, it is natural to expect the functionality be main concerns of research.

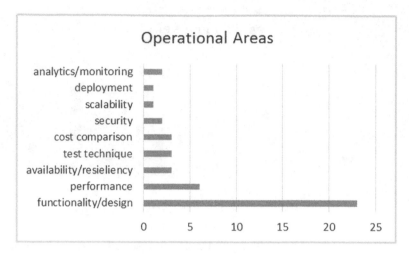

Fig. 5. Operational areas

Figure 6 aims to answer if the study has empirical results or not. Our analysis shows that the empirical studies are currently small in amount. Figure 7 illustrates an analysis based on operational areas versus service types. The most remarkable point is that most of the studies focus on the functionality/design issues. Figure 8 shows if there is a new solution proposed and/or implemented. As seen in the figure, most of the

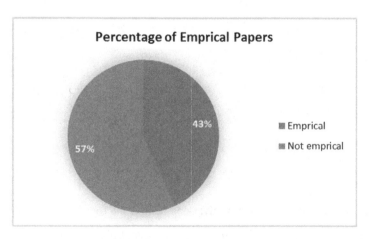

Fig. 6. Empirical results in research

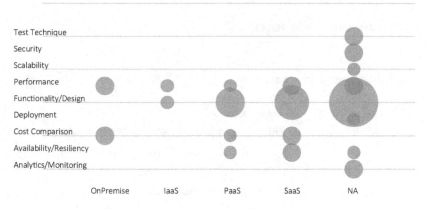

Fig. 7. Operational area vs service type.

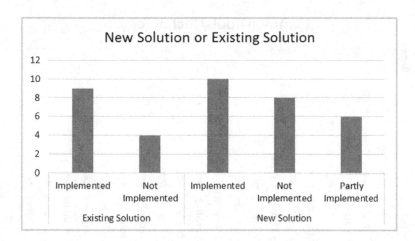

Fig. 8. Implementation of solutions.

research propose new solutions. Another noticeable point is that the implementation ratio of new solutions is higher than the implementation ratio of existing solutions.

RQ3: What are the emerging standards and de facto tools on microservices solutions?

The occurrence of standards proposed or implemented in the research papers included into the systematic mapping can be seen in Fig. 9. The figure includes all the standards either implemented or proposed in systematic mapping papers. As clearly seen in the figure, REST can be called out as the standard for Microservices, even though there is one outlier paper which used non-REST protocol in their study [28].

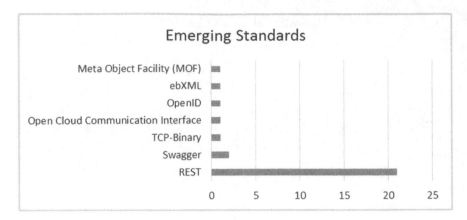

Fig. 9. Emerging standards

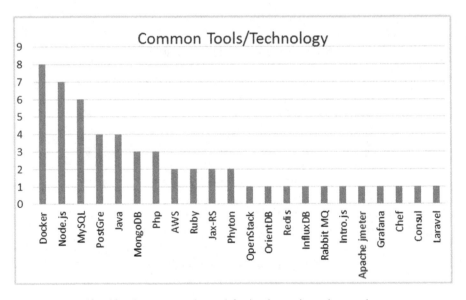

Fig. 10. Common tools used for implementing microservices

Only Swagger is used for microservice markup language. It is interesting to see that WADL or API Blueprint is not mentioned.

The occurrence of tools used in proposed or implemented solutions can be found in Fig. 10. Docker is seen as the most frequently used tool in studies.

The microservices topic is new and the official research started to show up in research papers in 2015. As a result, it is expected for the number of research on microservices to increase over time. Figure 11 shows publication numbers over time. The last search was carried out on the Web of Science in January 2017. On the figure,

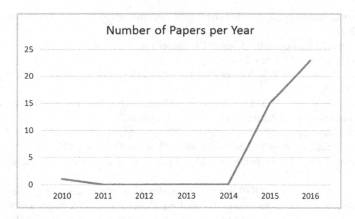

Fig. 11. Number of microservices papers over time without applying exclusion criteria (searched on January 20, 2017)

the line shows the trend. It is seen that the amount of papers increases radically and the trend line is going up.

6   Conclusions and Future Work

The term microservices was first appeared in 2014. All academic papers about microservices belong to 2015 and 2016. From that, we conclude that it is completely a new topic.

Considering the mapping results, we can conclude that microservices is a trending topic and our prediction is that we will see increasing trend in the near future.

Another important conclusion that we draw from the systematic mapping is that there are not enough empirical studies to clarify many issues under discussion related to microservices. Also, there is no research specifically targeting the fragile points of microservices such as distributed transactions.

References

1. Lewis, J., Fowler, M.: "Microservices" martinfowler.com. http://martinfowler.com/articles/microservices.html. Accessed 20 Dec 2016
2. Newman, S.: Building Microservices. O'Reilly Media, Inc., Sebastopol (2015)
3. Petersen, K., Feldt, R., Mujtaba, S., Mattsson, M.: Systematic mapping studies in software engineering. In: 12th International Conference on Evaluation and Assessment in Software Engineering, vol. 17, p. 1 (2008)
4. Dl.acm.org: ACM Digital Library (2016). http://dl.acm.org/dl.cfm. Accessed 05 Jan 2016
5. Citeseerx.ist.psu.edu: CiteSeerX (2016). http://citeseerx.ist.psu.edu/. Accessed 05 Jan 2016
6. Search.ebscohost.com: Computer Source (2016). http://search.ebscohost.com/login.aspx?authtype=ip,uid&profile=ehost&defaultdb=cph. Accessed 05 Jan 2016

7. Site.ebrary.com: ebrary: Server Message (2016). http://site.ebrary.com/lib/utexas. Accessed 05 Jan 2016
8. Hcibib.org: HCI Bibliography: Human-Computer Interaction Resources (2016). http://www.hcibib.org/. Accessed 05 Jan 2016
9. Ieeexplore.ieee.org: IEEE Xplore Digital Library (2016). http://ieeexplore.ieee.org/. Accessed 05 Jan 2016
10. Search.ebscohost.com: INSPEC (2016). http://search.ebscohost.com/login.aspx?authtype=ip,uid&profile=ehost&defaultdb=inh. Accessed 05 Jan 2016
11. Search.ebscohost.com: INSPEC Archive (2016). http://search.ebscohost.com/login.aspx?authtype=ip,uid&profile=ehost&defaultdb=ieh. Accessed 05 Jan 2016
12. Nature.com: Journal home: Nature (2015). http://www.nature.com/nature. Accessed 05 Jan 2016
13. Sciencemag.org: Science (2016). http://www.sciencemag.org/. Accessed 05 Jan 2016
14. Search.ebscohost.com: Science and Technology Collection (2016). http://search.ebscohost.com/login.aspx?authtype=ip,uid&profile=ehost&defaultdb=syh. Accessed 05 Jan 2016
15. Osti.gov: SciTech Connect: Your connection to science, technology, and engineering research information from the U.S. Department of Energy (2016). http://www.osti.gov/scitech/. Accessed 05 Jan 2016
16. Springerlink.com: Home - Springer (2016). http://www.springerlink.com. Accessed 05 Jan 2016
17. Lysne, O., Hole, K., Otterstad, C., Ytrehus, O., Aarseth, R., Tellnes, J.: Vendor malware: detection limits and mitigation. Computer **49**(8), 62–69 (2016)
18. Heorhiadi, V., Rajagopalan, S., Jamjoom, H., Reiter, M., Sekar, V.: Gremlin: systematic resilience testing of microservices. In: 2016 IEEE 36th International Conference on Distributed Computing Systems (ICDCS) (2016)
19. Villamizar, M., Garces, O., Ochoa, L., Castro, H., Salamanca, L., Verano, M., Casallas, R., Gil, S., Valencia, C., Zambrano, A., Lang, M.: Infrastructure cost comparison of running web applications in the cloud using AWS lambda and monolithic and microservice architectures. In: 2016 16th IEEE/ACM International Symposium on Cluster, Cloud and Grid Computing (CCGrid) (2016)
20. Villamizar, M., Garces, O., Castro, H., Verano, M., Salamanca, L., Casallas, R., Gil, S.: Evaluating the monolithic and the microservice architecture pattern to deploy web applications in the cloud. In: 2015 10th Computing Colombian Conference (10CCC) (2015)
21. Sun, Y., Nanda, S., Jaeger, T.: Security-as-a-service for microservices-based cloud applications. In: 2015 IEEE 7th International Conference on Cloud Computing Technology and Science (CloudCom) (2015)
22. Rahman, M., Gao, J.: A reusable automated acceptance testing architecture for microservices in behavior-driven development. In: 2015 IEEE Symposium on Service-Oriented System Engineering (2015)
23. Le, V., Neff, M., Stewart, R., Kelley, R., Fritzinger, E., Dascalu, S., Harris, F.: Microservice-based architecture for the NRDC. In: 2015 IEEE 13th International Conference on Industrial Informatics (INDIN) (2015)
24. Alpers, S., Becker, C., Oberweis, A., Schuster, T.: Microservice based tool support for business process modelling. In: 2015 IEEE 19th International Enterprise Distributed Object Computing Workshop (2015)
25. Bak, P., Melamed, R., Moshkovich, D., Nardi, Y., Ship, H., Yaeli, A.: Location and context-based microservices for mobile and internet of things workloads. In: 2015 IEEE International Conference on Mobile Services (2015)

26. Malavalli, D., Sathappan, S.: Scalable microservice based architecture for enabling DMTF profiles. In: 2015 11th International Conference on Network and Service Management (CNSM) (2015)
27. Krylovskiy, A., Jahn, M., Patti, E.: Designing a smart city internet of things platform with microservice architecture. In: 2015 3rd International Conference on Future Internet of Things and Cloud (2015)
28. Ciuffoletti, A.: Automated deployment of a microservice-based monitoring infrastructure. Procedia Comput. Sci. **68**, 163–172 (2015)
29. Meinke, K., Nycander, P.: Learning-based testing of distributed microservice architectures: correctness and fault injection. In: Bianculli, D., Calinescu, R., Rumpe, B. (eds.) SEFM 2015. LNCS, vol. 9509, pp. 3–10. Springer, Heidelberg (2015). doi:10.1007/978-3-662-49224-6_1
30. Pahl, C., Jamshidi, P.: Software architecture for the cloud – a roadmap towards control-theoretic, model-based cloud architecture. In: Weyns, D., Mirandola, R., Crnkovic, I. (eds.) ECSA 2015. LNCS, vol. 9278, pp. 212–220. Springer, Cham (2015). doi:10.1007/978-3-319-23727-5_17
31. Nicolaescu, P., Klamma, R.: A methodology and tool support for widget-based web application development. In: Cimiano, P., Frasincar, F., Houben, G.-J., Schwabe, D. (eds.) ICWE 2015. LNCS, vol. 9114, pp. 515–532. Springer, Cham (2015). doi:10.1007/978-3-319-19890-3_33
32. Koren, I., Nicolaescu, P., Klamma, R.: Collaborative drawing annotations on web videos. In: Cimiano, P., Frasincar, F., Houben, G.-J., Schwabe, D. (eds.) ICWE 2015. LNCS, vol. 9114, pp. 671–674. Springer, Cham (2015). doi:10.1007/978-3-319-19890-3_54
33. Wikipedia: Cloud computing (2016). https://en.wikipedia.org/wiki/Cloud_computing. Accessed 05 Jan 2016
34. Fatema, K., Emeakaroha, V., Healy, P., Morrison, J., Lynn, T.: A survey of cloud monitoring tools: taxonomy, capabilities and objectives. J. Parallel Distrib. Comput. **74**(10), 2918–2933 (2014)
35. Wieringa, R., Maiden, N., Mead, N., Rolland, C.: Requirements engineering paper classification and evaluation criteria: a proposal and a discussion. Requirements Eng. **11**(1), 102–107 (2005)
36. Are Gartner's Predictions on Track - Gartner's Top 10 Strategic Technology Trends for 2016: At a Glance from October 6, 2015 (2016). https://www.linkedin.com/pulse/how-well-did-gartner-do-prediction-gartners-top-10-strategic?trk=pulse-det-nav_art. Accessed 20 Dec 2016
37. Sommerville, I.: Software Engineering, 10th edn. Pearson, London (2016). (Chap. 18)
38. Braun, E., Düpmeier, C., Kimmig, D., Schillinger, W., Weissenbach, K.: Generic web framework for environmental data visualization. In: Wohlgemuth, V., Fuchs-Kittowski, F., Wittmann, J. (eds.) Advances and New Trends in Environmental Informatics. PI, pp. 289–299. Springer, Cham (2017). doi:10.1007/978-3-319-44711-7_23
39. Linthicum, D.: Practical use of microservices in moving workloads to the cloud. IEEE Cloud Comput. **3**(5), 6–9 (2016)
40. Alshuqayran, N., Ali, N., Evans, R.: A systematic mapping study in microservice architecture. In: 2016 IEEE 9th International Conference on Service-Oriented Computing and Applications (SOCA) (2016)
41. Inagaki, T., Ueda, Y., Ohara, M.: Container management as emerging workload for operating systems. In: 2016 IEEE International Symposium on Workload Characterization (IISWC) (2016)
42. Ueda, T., Nakaike, T., Ohara, M.: Workload characterization for microservices. In: 2016 IEEE International Symposium on Workload Characterization (IISWC) (2016)

43. Florio, L., Nitto, E.: Gru: an approach to introduce decentralized autonomic behavior in microservices architectures. In: 2016 IEEE International Conference on Autonomic Computing (ICAC) (2016)
44. Gadea, C., Trifan, M., Ionescu, D., Ionescu, B.: A reference architecture for real-time microservice API consumption. In: Proceedings of the 3rd Workshop on CrossCloud Infrastructures & Platforms - CrossCloud 2016 (2016)
45. Renz, J., Hoffmann, D., Staubitz, T., Meinel, C.: Using A/B testing in MOOC environments. In: Proceedings of the Sixth International Conference on Learning Analytics & Knowledge - LAK 2016 (2016)
46. Hasselbring, W.: Microservices for scalability. In: Proceedings of the 7th ACM/SPEC on International Conference on Performance Engineering - ICPE 2016 (2016)
47. Scarborough, W., Arnold, C., Dahan, M.: Case study. In: Proceedings of the XSEDE16 on Diversity, Big Data, and Science at Scale – XSEDE 2016 (2016)
48. Kecskemeti, G., Marosi, A., Kertesz, A.: The ENTICE approach to decompose monolithic services into microservices. In: 2016 International Conference on High Performance Computing and Simulation (HPCS) (2016)
49. Barais, O., Bourcier, J., Bromberg, Y., Dion, C.: Towards microservices architecture to transcode videos in the large at low costs. In: 2016 International Conference on Telecommunications and Multimedia (TEMU) (2016)
50. Kang, H., Le, M., Tao, S.: Container and microservice driven design for cloud infrastructure DevOps. In: 2016 IEEE International Conference on Cloud Engineering (IC2E) (2016)
51. Messina, A., Rizzo, R., Storniolo, P., Tripiciano, M., Urso, A.: The database-is-the-service pattern for microservice architectures. In: Renda, M.E., Bursa, M., Holzinger, A., Khuri, S. (eds.) ITBAM 2016. LNCS, vol. 9832, pp. 223–233. Springer, Cham (2016). doi:10.1007/978-3-319-43949-5_18
52. Hassan, S., Bahsoon, R.: Microservices and their design trade-offs: a self-adaptive roadmap. In: 2016 IEEE International Conference on Services Computing (SCC) (2016)
53. Bogner, J., Zimmermann, A.: Towards integrating microservices with adaptable enterprise architecture. In: 2016 IEEE 20th International Enterprise Distributed Object Computing Workshop (EDOCW) (2016)
54. Kratzke, N., Peinl, R.: ClouNS - a cloud-native application reference model for enterprise architects. In: 2016 IEEE 20th International Enterprise Distributed Object Computing Workshop (EDOCW) (2016)
55. Thiele, T., Sommer, T., Stiehm, S., Jeschke, S., Richert, A.: Exploring research networks with data science: a data-driven microservice architecture for synergy detection. In: 2016 IEEE 4th International Conference on Future Internet of Things and Cloud Workshops (FiCloudW) (2016)
56. Qanbari, S., Pezeshki, S., Raisi, R., Mahdizadeh, S., Rahimzadeh, R., Behinaein, N., Mahmoudi, F., Ayoubzadeh, S., Fazlali, P., Roshani, K., Yaghini, A., Amiri, M., Farivarmoheb, A., Zamani, A., Dustdar, S.: IoT design patterns: computational constructs to design, build and engineer edge applications. In: 2016 IEEE First International Conference on Internet-of-Things Design and Implementation (IoTDI) (2016)
57. Guo, D., Wang, W., Zeng, G., Wei, Z.: Microservices architecture based cloudware deployment platform for service computing. In: 2016 IEEE Symposium on Service-Oriented System Engineering (SOSE) (2016)
58. Safina, L., Mazzara, M., Montesi, F., Rivera, V.: Data-driven workflows for microservices: genericity in jolie. In: 2016 IEEE 30th International Conference on Advanced Information Networking and Applications (AINA) (2016)

59. Kratzke, N.: About microservices, containers and their underestimated impact on network performance. In: 6th International Conference on Cloud Computing, GRIDs, and Virtualization (CLOUD COMPUTING) (2015)
60. Fernandez-Villamor, J.I., Iglesias, C., Garijo, M.: MICROSERVICES lightweight service descriptions for rest architectural style. In: 2nd International Conference on Agents and Artificial Intelligence (ICAART 2010) (2010)

Separation Logic for States Dependencies in Life Cycles of Android Activities and Fragments

Mohamed A. El-Zawawy[1,2]([✉])

[1] College of Computer and Information Sciences, Al Imam Mohammad Ibn Saud
Islamic University (IMSIU), Riyadh, Kingdom of Saudi Arabia
[2] Department of Mathematics, Faculty of Science, Cairo University,
Giza 12613, Egypt
maelzawawy@cu.edu.eg

Abstract. Millions of smart phones are run by Android, the most common operating system for mobile devices. Main component of Android is applications providing most functionalities of Android smart devices. These applications include e-mail client, entertainment, educational, news, banking, maps, and contacts applications. Testing and verifying Android applications is an important direction of research.

This paper presents a separation logic for state transitions of activities and fragments (important components of applications) during their life cycles. The logic considers the necessary coordinations between the state transitions in the two life cycles. The logic is a good tool to verify and test applications against various issues including security ones.

Keywords: Android · Activities · Fragments · Axiomatic semantics · Separation logic · Memory logics · Static analysis

1 Introduction

Android is the most common operating system for mobile devices and it is running millions of smart phones. Android has managed to maintain an increasing share of the smart devices market. The popularity of mobile phones led to focus on the mobile phone and tablet computing rather than on desktop and laptop computing. Android applications are main tools to provide most functionalities of Android smart devices. There are a wide range of applications including e-mail client, entertainment, SMS program, educational, browser, news, calendar, banking, maps, and contacts applications. The great success of Android pointing out the need for effective methods for testing and verifying Android applications. Java is the main language to write Android applications which are compiled by Android SDK tools [1].

Activities, services, content providers, and broadcast receivers are the main building blocks used to develop applications. Each of these blocks may enable the application user or the system to treat the application. These components are not entirely independent and hence may affect each others. Activities are

O. Gervasi et al. (Eds.): ICCSA 2017, Part VI, LNCS 10409, pp. 218–232, 2017.
DOI: 10.1007/978-3-319-62407-5_15

the main tools for interacting with users. An activity provides a user interface in single frame. For example, an educational application would include one activity to present titles of topics and another activity to show details of a selected topic. Main communications between the system and the applications are enabled by the activities. These communications for example enables the system to be up-to-date with the activity content to react properly [1].

A life cycle built of various activity states is used to properly treat activities. The user iterations (of navigating through, stopping, and destroying) with activities are translated into activities transitions through the states of the life cycle. Each state corresponds to a callback method that is included in the activity class and used to indicate and admit a state change. Examples of the callback methods are *onCreate()*, *onPause()*, *onResume()*, and *onDestroy()* [1].

Fragments provide important techniques for utilizing activities as they act as components of activities interfaces. Fragments are modular entities that are reusable in various activities. Moreover, a single activity can have many fragments contributing to the design of a multi-pane user interface. Fragments have their own life cycles. Fragments react independently to input events. Also fragments may be added to or removed from the activates during the run time of activity.

The life cycle of a fragment is affected by the life cycle of the activity hosting the fragment; a fragment receives an equivalent life cycle callback to that received by the hosting activity. Therefore the reception of *onStop()* by the activity results in the reception of *onStop()* by each fragment hosted by the activity. Achieving smooth interaction between fragments and activities requires fragments life cycle to include more callbacks than that in the life cycle of activities. The extra callbacks of fragments include: *onAttach()*, *onActivityCreated()*, and *onDetach()*. It is the goal of this paper to develop formal means to model the effect of activity life cycle on that of fragments.

Separation logic [18] is a program logic to reasoning about shared mutable data structures. These structures are ones including fields that can be accessed from various points. This logic accounts for instructions for allocation, modifying, and deallocation shared structures. Assertions of separation logic use separating implication and separating conjunction which model the concept of sub-assertions satisfied in disjoint regions of memory.

This paper aims at extending separation logic to precisely model the effect of the life cycle of activities on that of fragments (focusing on memory regions). This includes extending the set of assertions of separation logic and describing the effects of various Dalvik instructions and life-cycle callbacks using the new set of assertions. The soundness of the extended logic presented in this paper is formalized. The main objective of designing the logic presented here is to use it as ground for building analysis techniques for Android application. The targeted analyses are those rely on precise modeling of the interaction between the life cycles of activates and fragments. These analyses typical work on *bundles* which are objects used for communications between activities and fragments.

Motivation
This paper is motivated by the need for a logic to reason about the coordination between activities and fragments during their life cycles transitions.

Paper Outline
The rest of the paper is organized as follows. Section 2 presents the logic in several subsections. The syntax and semantics of assertion language used in the propose logic is shown in Sect. 2.1. Inferences rules of the logic for Android procedures, activities, and fragments are shown in Sects. 2.2, 2.3, and 2.4, respectively. Section 3 reviews related work and proposes future work.

2 Activities and Fragments Logic for States Transitions

This section presents an extended version of separation logic for state transitions of activities and fragments during their life cycles. The logic considers the transitions of activities with consideration of the effects on the fragments hosted by the activities. The other type of effects is also considered when studying transitions of fragment states.

During their life cycles, activities can be in any of the following states:

constructor, onCreate, onStart, onRestart, onResume, running, onPause, onStop, onDestroy.

Then the following transitions between states are possible for activities:

(constructor, onCreate), (onCreate, onStart), (onRestart, onStart), (onStart, onResume), (onResume, running), (running, onPause), (onPause, onResume), (onPause, onStop), (onStop, onRestart), (onResume, onPause), (onStart, onStop), (onStop, onDestroy).

During their life cycles, fragments can be in any of the following states:

constructor, onCreate, onStart, onResume, running, onPause, onStop, onDestroy, onAttach, onCreateView, onActivityCreated, onDestroyView, onDetach.

Then the following transitions between states are possible for fragments:

(constructor, onAttach), (onAttach, onCreate), (onCreate, onCreateView), (onCreateView, onActivityCreated), (onActivityCreated, onStart), (onStart, onResume), (onResume, onPause), (onPause, onStop), (onStop, onDestroyView), (onDestroyView, onDestory), (onDestroyView, onCreateView), (onDestory, onDetach).

In this paper we focus on the set of Android Procedures (APs) most related to coordination between activities and fragments:

1. startActivity(Activity class object ref e); starting activity object at reference e.

2. fragmentTransaction.add(Activity class object e_1, Fragment class object e_2); adding fragment object at e_2 to activity object at e_1.
3. fragmentTransaction.commit(Activity class object e_1, Fragment class object e_2); running fragment at e_2 included in activity at e_1.
4. findViewById(view, i); looking for the address of the view whose *id* is in register i.
5. inflater.inflate(Activity class object ref e_1, Fragment class object ref e_2);
6. setContentView(Activity class object ref e, layout);
7. finish;

We define the activity as an extension class of "AppCompatActivity" class with the following properties:

- layout: structure of activity view.
- finished.
- backStack: list of hosted fragments by the activity.
- aFreg: currently active fragment of the class.
- status: holds the activity state.
- class: indicates the object class.

We define the a fragment as an extension of the "Fragment" class with the fields:

- layout.
- finished.
- parent: pointer to the hosting activity.
- status.
- class: indicates the object class.

2.1   Assertion Syntax and Semantics

The proposed logic has set of assertions including the typical ones of predicate calculus and the ones defined as follows:

Definition 1.

$$s_a \in Astates ::= constructor \mid onCreate \mid onStart \mid onRestart \mid onResume \mid running \mid$$
$$onPause \mid onStop \mid onDestroy$$
$$s_f \in Fstates ::= constructor \mid onCreate \mid onStart \mid onResume \mid running \mid onPause \mid$$
$$onStop \mid onDestroy \mid onAttach \mid onCreateView \mid onActivityCreated \mid$$
$$onDestroyView \mid onDetach$$
$$A \in Assert ::= emp \mid e \mapsto_a o \mid e \mapsto_f o \mid e_1 \mapsto_{aa} e_2 \mid e_a \mapsto_{af} e_f \mid e \mapsto_{as} s_a \mid$$
$$e \mapsto_{fs} s_f \mid Act_a\ e \mid !Act_a\ e \mid Act_f\ e \mid \mathcal{F}(e) \mid e \equiv e' \mid \mathcal{A}(e) \mid A * A \mid A -\!\!* A.$$

The rest of the paper uses the following abbreviations:

- $e \mapsto_{af} e_f^1, e_f^2, \dots, e_f^n$ abbreviates $e \mapsto_{af} e_f^1, e \mapsto_{af} e_f^2, \dots, e \mapsto_{af} e_f^n$.

$$\mathbf{emp} \propto (s, h, a_{[]}) \Longleftrightarrow^{def} h = \emptyset. \tag{1}$$

$$\mathcal{A}(e) \propto (s, h, a_{[]}) \Longleftrightarrow^{def} e \in a_{[]}. \tag{2}$$

$$e \mapsto_a o \propto (s, h, a_{[]}) \Longleftrightarrow^{def} h = \{(e, o)\} \wedge o.class \in \mathcal{A} \ (names \ of \ activity \ calsses) \tag{3}$$

$$e \mapsto_f o \propto (s, h, a_{[]}) \Longleftrightarrow^{def} h = \{(e, o)\} \wedge o.class \in \mathcal{F} \ (names \ of \ fragment \ calsses) \tag{4}$$

$$e_1 \mapsto_{aa} e_2 \propto (s, h, a_{[]}) \Longleftrightarrow^{def} \exists o_1, o_2. \ e_1 \mapsto_a o_1 \wedge e_2 \mapsto_a o_2 \wedge a_{[]} = a_{[]}^1; e_1; e_2; a_{[]}^2. \tag{5}$$

$$e_a \mapsto_{af} e_f \propto (s, h, a_{[]}) \Longleftrightarrow^{def} \exists o_a, o_f. \ e_a \mapsto_a o_a \wedge e_f \mapsto_f o_f \wedge e_a \in a_{[]} \wedge \tag{6}$$
$$h(e_f).parent = e_a \wedge e_f \in h(e_a).BackStack. \tag{7}$$

$$e \mapsto_{as} s_a \propto (s, h, a_{[]}) \Longleftrightarrow^{def} \exists o. \ e \mapsto_a o \wedge o(status) = s_a. \tag{8}$$

$$e \mapsto_{fs} s_f \propto (s, h, a_{[]}) \Longleftrightarrow^{def} \exists o. \ e \mapsto_f o \wedge o(status) = s_f. \tag{9}$$

$$Act_a \ e \propto (s, h, a_{[]}) \Longleftrightarrow^{def} \exists o. \ e \mapsto_a o \wedge a_{[]} = e; a'_{[]}. \tag{10}$$

$$!Act_a \ e \propto (s, h, a_{[]}) \Longleftrightarrow^{def} \exists o. \ e \mapsto_a o \wedge a_{[]} = [e]. \tag{11}$$

$$Act_f \ e \propto (s, h, a_{[]}) \Longleftrightarrow^{def} e' \mapsto_{af} e \wedge Act_a \ e' \wedge h(e')(aFrag) = e \wedge \tag{12}$$
$$h(e')(BackStack) = [e; -]. \tag{13}$$

$$\mathcal{F}(e) \propto (s, h, a_{[]}) \Longleftrightarrow^{def} \exists o. \ e \mapsto_a o \wedge o(finished) = true. \tag{14}$$

$$e \equiv e' \propto (s, h, a_{[]}) \Longleftrightarrow^{def} \exists o, o'. \ e \mapsto_a o \wedge e' \mapsto_a o' \wedge o(class) = o'(class) \wedge o = o' \tag{15}$$

Fig. 1. Assertion formal semantics for activities and fragments logic.

$- e, e_f^1, e_f^2, \ldots, e_f^n \mapsto_{as} onPause$ abbreviates $e \mapsto_{as} onPause, e_f^1 \mapsto_{fs}$
$onPause, e_f^2 \mapsto_{fs} onPause, \ldots, e_f^n \mapsto_{fs} onPause.$

Example 1. An example of assertion $A_1 \in \mathcal{A}$ is defined as

$$A_1 = ((v_0 \mapsto_{as} onCreate) * (v_1 \mapsto_{as} onDestory) * (v_2 \mapsto_{fs} constructor)) \wedge (v_0 \mapsto_{aa} v_1) \wedge \mathcal{F}(v_1).$$

The assertion describes a Dalvik state representing an application with two activities and one fragment.

Figure 1 presents the formal semantics of the set of assertions. The semantics assumes that the application semantics states has three components $(s, h, a_{[]})$. The first component (s) captures the statues of the Dalvik machine registers. The second component (h) captures the memory statues as a function from set of locations to set of class objects. The last component $(a_{[]})$ is the list of the activity references of the application on hand [5, 15].

$$\frac{\{e \mapsto_a o\}\, onCreate(e)\, \{e \mapsto_{as} onCreate\}}{\begin{array}{l} \{e \mapsto_a o * R\} \\ startActivity(Activity\ class\ object\ ref\ l) \\ \{e \mapsto_a o \wedge e \mapsto_{as} onCreate \wedge Act_a\ e * R\} \end{array}} \quad (16)$$

$$\frac{\begin{array}{l} \{e_3 \mapsto_f -\}\, onAttch(e_2)\, \{e_3 \mapsto_{fs} onAttch\} \\ \{e_3 \mapsto_{fs} onAttch\}\, onCreate(e_2)\, \{e_3 \mapsto_{fs} onCreate\} \end{array}}{\begin{array}{l} \{e_1 \mapsto_{as} onCreate * e_2 \mapsto_{fs} constructor * R\} \\ fragmentTransaction.add(Activity\ class\ object\ ref\ e_1, \\ Fragment\ class\ object\ ref\ e_2) \\ \{e_3 \mapsto_{fs} onCreate \wedge (e_1 \mapsto_{af} e_2 * R)\} \end{array}} \quad (17)$$

$$\frac{\{e_1 \mapsto_{af} e_2\}\, onCreateView(e_2)\, \{e_2 \mapsto_{fs} onCreateView\}}{\begin{array}{l} \{e_1 \mapsto_{af} e_2 * R\} \\ fragmentTransaction.commit(Activity\ class\ object\ ref\ e_1, \\ Fragment\ class\ object\ ref\ e_2) \\ \{e_1 \mapsto_{af} e_2 \wedge Act_f\ e_2 \wedge e_2 \mapsto_{fs} onCreateView * R\} \end{array}} \quad (18)$$

$$\overline{\{e \mapsto_a - * R\}\, finish(e)\, \{(\mathcal{F}(e) \wedge e \mapsto_a - \wedge (e \mapsto_{af} e_f \Rightarrow \mathcal{F}(e_f))) * R\}} \quad (19)$$

Fig. 2. Inference rules for logic of android procedures.

2.2   Android Procedures Inference Rules

Figure 2 presents inference rules of Android Procedures (APs). Rule 18 states that the command

$$fragment Transaction.commit(Activity\ class\ object\ ref\ e_1,\ Fragment\ class\ object\ ref\ e_2)$$

assumes a memory state in which $e_1 \mapsto_{af} e_2$ i.e. reference e_1 points to an activity object that is linked to a fragment object pointed to by reference e_2. In this case, running the callback method *onCreate View* of the fragment ensures a memory states in which the status of the fragment object is *onCreate View*. Then the ".commit" command ensures the creation and activation of the fragment within the activity.

Example 2. The semantics of executing the command

$$fragment Transaction.add(v_0, v_2);$$

on a Dalvik state satisfying the assertion A_1 of Example 1 results (by applying Rule 17 in a state satisfying the following assertion:

$$A_2 = ((v_0 \mapsto_{as} onCreate) * (v_1 \mapsto_{as} onDestory) * (v_2 \mapsto_{fs} onCreate)) \wedge (v_0 \mapsto_{aa} v_1) \wedge (v_0 \mapsto_{af} v_2) \wedge \mathcal{F}(v_1).$$

Example 3. Suppose that the class of the object referenced by v_2 (of Example 1) has the following definition for the callback method *onCreate View (object reference l)*:

```
onCreateView (object reference v₂: void
. registers 1
return−void ;
```

The semantics of executing the commands

$$fragment Transaction.commit(v_0, v_2);$$

on a state satisfying the assertion A_2 of Example 2 results (by applying Rule 18) in a state satisfying the following assertion:

$$A'_3 = ((v_0 \mapsto_{as} onCreate) * (v_1 \mapsto_{as} onDestory) * (v_2 \mapsto_{fs} onCreate View)) \wedge (v_0 \mapsto_{aa} v_1) \wedge (v_0 \mapsto_{af} v_2) \wedge \mathcal{F}(v_1) \wedge Act_f(v_2).$$

We now assume that the system runs the following callback methods

$$onActivity Created(v_2);\ onStart(v_0);\ onStart(v_2);\ onResume(v_0);\ onResume(v_2);$$

Then we will have a Dalvik state satisfying

$$A_3 = ((v_0 \mapsto_{as} onResume) * (v_1 \mapsto_{as} onDestory) * (v_2 \mapsto_{fs} onResume)) \wedge (v_0 \mapsto_{aa} v_1) \wedge (v_0 \mapsto_{af} v_2) \wedge \mathcal{F}(v_1) \wedge Act_f(v_2).$$

$$\frac{}{\{\mathcal{A}(e) * R\}\ aRemove\ \{\neg\mathcal{A}(e) * R\}} \quad (20)$$

$$\frac{}{\{\mathcal{A}(e) * R\}\ aActivation\ \{Act_a\ e * R\}} \quad (21)$$

$$\frac{}{\{Act_a\ e * R\}\ aDeActivation\ \{\mathcal{A}(e) \wedge \neg Act_a\ e * R\}} \quad (22)$$

$$\frac{\begin{array}{c}\text{the application user did an action } Act\\ \{Act_a\ e\}\ onActListner(e)\ \{q\}\end{array}}{\{Act_a\ e * R\}\ aListner\ \{q * R\}} \quad (23)$$

$$\frac{\begin{array}{l}\{!Act_a\ e \wedge e \mapsto_{af} e_f^1, e_f^2, \ldots, e_f^n \wedge e, e_f^1, e_f^2, \ldots, e_f^n \mapsto_{as} OnResume\}\\ onPause(e)\\ \{e \mapsto_{af} e_f^1, e_f^2, \ldots, e_f^n \wedge e \mapsto_{as} onPause \wedge e_f^1, e_f^2, \ldots, e_f^n \mapsto_{af} OnResume\}\\ onPause(e_f^1)\\ \{e \mapsto_{af} e_f^1, e_f^2, \ldots, e_f^n \wedge e, e_f^1 \mapsto_{as} onPause \wedge e_f^2, \ldots, e_f^n \mapsto_{af} OnResume\}\\ \vdots\\ onPause(e_f^n)\\ \{e \mapsto_{af} e_f^1, e_f^2, \ldots, e_f^n \wedge e, e_f^1, e_f^2, \ldots, e_f^n \mapsto_{as} onPause\}\end{array}}{\begin{array}{l}\{!Act_a\ e \wedge e, e_f^1, e_f^2, \ldots, e_f^n \mapsto_{as} OnResume \wedge e \mapsto_{af} e_f^1, e_f^2, \ldots, e_f^n * R\}\\ Stopping_1\\ \{e \mapsto_{af} e_f^1, e_f^2, \ldots, e_f^n \wedge e, e_f^1, e_f^2, \ldots, e_f^n \mapsto_{as} onPause * R\}\end{array}} \quad (24)$$

$$\frac{}{\{Act_a\ e \wedge e \mapsto_{as} onDestroy\}\ aFinished\ \{\neg\mathcal{A}(e)\}} \quad (25)$$

$$\frac{\begin{array}{l}(\xi, \xi') \in \{(running, onPause), (onPause, onStop)\}\\ \{\mathcal{F}(e) \wedge Act_a\ e \wedge e \mapsto_{af} e_f^1, e_f^2, \ldots, e_f^n \wedge e, e_f^1, e_f^2, \ldots, e_f^n \mapsto_{as} \xi\}\\ \xi'(e)\\ \{e \mapsto_{af} e_f^1, e_f^2, \ldots, e_f^n \wedge e \mapsto_{as} \xi' \wedge e_f^1, e_f^2, \ldots, e_f^n \mapsto_{af} \xi\}\\ \xi'(e_f^1)\\ \{e \mapsto_{af} e_f^1, e_f^2, \ldots, e_f^n \wedge e, e_f^1 \mapsto_{as} \xi' \wedge e_f^2, \ldots, e_f^n \mapsto_{af} \xi\}\\ \vdots\\ \xi'(e_f^n)\\ \{e \mapsto_{af} e_f^1, e_f^2, \ldots, e_f^n \wedge e, e_f^1, e_f^2, \ldots, e_f^n \mapsto_{as} \xi'\}\end{array}}{\begin{array}{l}\{\mathcal{F}(e) \wedge Act_a\ e \wedge e \mapsto_{af} e_f^1, e_f^2, \ldots, e_f^n \wedge e, e_f^1, e_f^2, \ldots, e_f^n \mapsto_{as} \xi * R\}\\ aStopping_2\\ \{e \mapsto_{af} e_f^1, e_f^2, \ldots, e_f^n \wedge e, e_f^1, e_f^2, \ldots, e_f^n \mapsto_{as} \xi' * R\}\end{array}} \quad (26)$$

$$\frac{}{\{Act_a\ e \wedge e \mapsto_{as} onResume \wedge e \mapsto_{af} e_f^1, e_f^2, \ldots, e_f^n * R\}\ aPressBack\ \{\mathcal{F}(e, e_f^1, e_f^2, \ldots, e_f^n) * R\}} \quad (27)$$

Fig. 3. Logic rules for activity state transitions (part 1).

2.3   Activities Inference Rules

Figures 3 and 4 present the inference rules for activities state transitions during their life cycles taking into consideration effects on hosted fragments. Rule 22 captures the situation when an active activity of an application becomes not active. The rule requires a memory state in a separate part of it, the concerned activity is active. In the resulting memory state, the concerned activity is still in the application list ($\mathcal{A}(e)$) but not active ($\neg Act_a\ e$).

Rule 24 expresses transition from the state *onResume* of an activity in a memory location referenced by e to the state *onPause*. The rule treats the case when the application has only one activity ($!Act_a\ e$). The rule assumes that the activity hosts n fragments pointed-to be references $e_f^1, \ldots, e_f^n$. This is denoted by $e \mapsto_{af} e_f^1, e_f^2, \ldots, e_f^n$. The rule requires that all the hosted fragments are in the same state as the activity ($e, e_f^1, e_f^2, \ldots, e_f^n \mapsto_{as} onResume$). The rule ensures the transitions of the hosted fragments states from *onResume* to *onPause*. This is done via running the callback methods *onPause* for the activity and all its hosted fragments.

Rule 26 models the state transition of an activity in a memory location referenced by e from the state $\xi = running$ to $\xi' = onPause$ or from the state $\xi = onPause$ to $\xi' = onStop$. This can only be done if the activity property *finished* is true. The rule assumes that the activity hosts n fragments pointed to be references $e_f^1, \ldots, e_f^n$ denoted by $e \mapsto_{af} e_f^1, e_f^2, \ldots, e_f^n$. The rule requires that all the hosted fragments are in the same state as the activity ($e, e_f^1, e_f^2, \ldots, e_f^n \mapsto_{as} \xi$). The rule ensures the transitions of the hosted fragments states from ξ to ξ'. This is done via running the callback methods ξ' for the activity and all its hosted fragments.

Example 4. Suppose that the class A_1 of the object v_0 (of Example 1) has a method *onPause (object reference l)* whose definition is as follows:

```
onPause (object reference v0): void
.registers 1
return-void;
```

Now suppose that another activity rather than v_0 comes into the foreground, hence the status of v_0 changes to *onPause*. This action can be modeled by running Rule 19 followed by Rule 24 on a state satisfying the assertion A_3 of Example 3. The result would be a state satisfying the following assertion:

$$A_4' = ((v_0 \mapsto_{as} onPause) * (v_1 \mapsto_{as} onDestory) * (v_2 \mapsto_{fs} onPause)) \wedge (v_0 \mapsto_{aa} v_1) \wedge (v_0 \mapsto_{af} v_2) \wedge \mathcal{F}(v_0) \wedge \mathcal{F}(v_1) \wedge \mathcal{F}(v_2).$$

We now assume that the system runs the following callback methods

$$onStop(v_0); onStop(v_2); onDestory(v_0); onDestroyView(v_2);$$

Then we will have a Dalvik state satisfying

$$A_4 = ((v_0 \mapsto_{as} onDestor) * (v_1 \mapsto_{as} onDestory) * (v_2 \mapsto_{fs} onDestroyView)) \wedge (v_0 \mapsto_{aa} v_1) \wedge (v_0 \mapsto_{af} v_2) \wedge \mathcal{F}(v_0) \wedge \mathcal{F}(v_1) \wedge \mathcal{F}(v_2).$$

$$\{Act_a\ e \wedge e \mapsto_{af} e_f^1, e_f^2, \ldots, e_f^n \wedge e \mapsto_{as} onDestroy \wedge e' \mapsto_{af} e_f^{1\prime}, e_f^{2\prime}, \ldots, e_f^{n\prime} \wedge$$
$$e \equiv e' \wedge \forall i.\ e_f^i \equiv e_f^{i\prime}\}$$
$$constructor(e')$$
$$\{Act_a\ e' \wedge e' \mapsto_{as} constructor \wedge e' \mapsto_{af} e_f^{1\prime}, e_f^{2\prime}, \ldots, e_f^{n\prime}\}$$
$$constructor(e_f^{1\prime})$$
$$\{Act_a\ e' \wedge e', e_f^{1\prime} \mapsto_{as} constructor \wedge e' \mapsto_{af} e_f^{1\prime}, e_f^{2\prime}, \ldots, e_f^{n\prime}\}$$
$$\vdots$$
$$constructor(e_f^{n\prime})$$
$$\{Act_a\ e' \wedge e', e_f^{1\prime}, e_f^{2\prime}, \ldots, e_f^{n\prime} \mapsto_{as} constructor \wedge e' \mapsto_{af} e_f^{1\prime}, e_f^{2\prime}, \ldots, e_f^{n\prime}\}$$

$$\rule{10cm}{0.4pt}$$

$$\{Act_a\ e \wedge e \mapsto_{af} e_f^1, e_f^2, \ldots, e_f^n \wedge e \mapsto_{as} onDestroy \wedge e' \mapsto_{af} e_f^{1\prime}, e_f^{2\prime}, \ldots, e_f^{n\prime} \wedge$$
$$e \equiv e' \wedge \forall i.\ e_f^i \equiv e_f^{i\prime} * R\}$$
$$aResize$$
$$\{Act_a\ e' \wedge e', e_f^{1\prime}, e_f^{2\prime}, \ldots, e_f^{n\prime} \mapsto_{as} constructor \wedge e' \mapsto_{af} e_f^{1\prime}, e_f^{2\prime}, \ldots, e_f^{n\prime} * R\}$$

(28)

$$\{Act_a\ e \wedge e \mapsto_{af} e_f^1, e_f^2, \ldots, e_f^n \wedge e, e_f^1, e_f^2, \ldots, e_f^n \mapsto_{as} running\}$$
$$onDestroy(e)$$
$$\{e \mapsto_{af} e_f^1, e_f^2, \ldots, e_f^n \wedge e \mapsto_{as} onDestroy \wedge e_f^1, e_f^2, \ldots, e_f^n \mapsto_{af} \xi\}$$
$$onDestroyView(e_f^1)$$
$$\{e \mapsto_{af} e_f^1, e_f^2, \ldots, e_f^n \wedge e \mapsto_{as} onDestroy \wedge$$
$$e_f^1 \mapsto_{as} onDestroyView \wedge e_f^2, \ldots, e_f^n \mapsto_{af} running\}$$
$$\vdots$$
$$onDestroyView(e_f^n)$$
$$\{e \mapsto_{af} e_f^1, e_f^2, \ldots, e_f^n \wedge e \mapsto_{us} onDestroy \wedge e_f^1, e_f^2, \ldots, e_f^n \mapsto_{as} onDestroyView\}$$

$$\rule{10cm}{0.4pt}$$

$$\{Act_a\ e \wedge e \mapsto_{af} e_f^1, e_f^2, \ldots, e_f^n \wedge e, e_f^1, e_f^2, \ldots, e_f^n \mapsto_{as} running * R\}$$
$$aStopping_3$$
$$\{e \mapsto_{af} e_f^1, e_f^2, \ldots, e_f^n \wedge e \mapsto_{us} onDestroy \wedge e_f^1, e_f^2, \ldots, e_f^n \mapsto_{us} onDestroyView * R\}$$

(29)

$$\rule{10cm}{0.4pt}$$

$$\{Act_a\ e \wedge e \mapsto_{as} onDestroy \wedge \mathcal{F}(e) * R\}\ aRemove\ \{\forall o. \neg(e \mapsto_a o) * R\}$$ (30)

Fig. 4. Logic rules for activity state transitions (part 2).

2.4   Fragments Inference Rules

Figures 5 and 6 present the logic rules for fragment state transitions. The rules considers coordination with the state of the activity hosting the fragment changing state.

Rule 31 formalize the scenario when an active fragment hosted by an activity of an application becomes not active. The rule obviously assumes a memory state in which a separate region has an active activity ($Act_a\ e$) and an active fragment

(Act_f e_f). The Rule also requires the fragment to be linked to the activity. In the resulting memory state, the concerned fragment becomes not active ($\neg Act_f$ e_f).

Rule 34 expresses fragment state transition from the state ξ to the state ξ' where

$$(\xi, \xi') \in \{(constructor, onAttach), (onAttach, onCreate),$$

$$(onCreate, onCreateView), (onCreateView, onActivityCreated)\}.$$

The precondition of the rule assumes a memory state including a separate region containing an active activity (Act_a e). The precondition assumes that the activity contains the fragment changing status ($e \mapsto_{af} e_f$). The activity status has to be *onCreate* for the fragment to change its status as is indicated above.

Rule 36 expresses fragment state transition from the state ξ to the state ξ' where

$$(\xi, \xi') \in \{(onStop, onDestroyView),$$

$$(onDestroyView, onDestory), (onDestory, onDetach)\}.$$

The rule specifies a precondition for the initial memory state requiring a separate region containing an active activity (Act_a e). The activity is assumed to host the fragment under transformation ($e \mapsto_{af} e_f$). The activity status has to be *onDestroy* for the fragment to change status from ξ to ξ'. The transformation amounts to executing the callback method of the fragment corresponding to the state ξ' ($\xi'(e_f)$).

Rule 40 formalizes the scenario where a fragment changes its state from *onResume* to the state *onPause*. The rules specifies a precondition for the initial memory state requiring a separate region containing an active activity (Act_a e). The activity is assumed to host the fragment under transformation ($e \mapsto_{af} e_f$). The activity status has to be *onPause* and the fragment status has to be *onResume* ($e \mapsto_{as} onPause \wedge e_f \mapsto_{fs} onResume$) for the fragment to change its state as indicated above. The transformation amounts to executing the callback method of the fragment corresponding to the state ($onPause(e_f)$).

Example 5. Suppose that the class F of the object v_2 (of Example 1) has the following definition for the callback method *onDestroy (object reference v_2):*

```
onDestroy (object reference $v_2$): void
.registers 2
return−void;
```

Now suppose that the system changes the status of the fragment v_2 of Example 4 to *onDestroy*. This action can be modeled by running Rule 36 on a state satisfying the assertion A_4 of Example 4. The result would be a state satisfying the following assertion:

$$A_5' = ((v_0 \mapsto_{as} onDestor) * (v_1 \mapsto_{as} onDestory) * (v_2 \mapsto_{fs} onDestroy))$$
$$\wedge (v_0 \mapsto_{aa} v_1) \wedge (v_0 \mapsto_{af} v_2) \wedge \mathcal{F}(v_0) \wedge \mathcal{F}(v_1) \wedge \mathcal{F}(v_2).$$

$$\frac{}{\{Act_a\ e \wedge Act_f\ e_f \wedge e \mapsto_{af} e_f * R\}\ fDeActivation\ \{\neg Act_f\ e_f * R\}} \quad (31)$$

$$\frac{}{\{Act_a\ e \wedge Act_f\ e_f \wedge e \mapsto_{af} e_f \wedge e \mapsto_{af} e'_f \wedge e'_f \neq e_f * R\}\ fActivation\ \{Act_f\ e'_f * R\}} \quad (32)$$

$$\frac{\text{the application user did an actionAct one}_f}{\{Act_a\ e \wedge Act_f\ e_f \wedge e \mapsto_{af} e_f\}\ onActionListner(e_f)\ \{q\}}{\{Act_a\ e \wedge Act_f\ e_f \wedge e \mapsto_{af} e_f * R\}\ fListner()\ \{q * R\}} \quad (33)$$

$$\frac{\begin{array}{c}(\xi, \xi') \in \{(constructor, onAttach), (onAttach, onCreate),\\ (onCreate, onCreateView), (onCreateView, onActivityCreated)\}\\ \{Act_a\ e \wedge e \mapsto_{af} e_f \wedge e \mapsto_{as} onCreate \wedge e_f \mapsto_{fs} \xi\}\ \xi'(e_f)\ \{e_f \mapsto_{fs} \xi'\}\end{array}}{\begin{array}{c}\{Act_a\ e \wedge e \mapsto_{af} e_f \wedge e \mapsto_{as} onCreate \wedge e_f \mapsto_{fs} \xi * R\}\\ fCreation\\ \{e_f \mapsto_{fs} \xi' * R\}\end{array}} \quad (34)$$

$$\frac{\begin{array}{c}\{Act_a\ e \wedge e \mapsto_{af} e_f \wedge e \mapsto_{as} onStart \wedge e_f \mapsto_{fs} onActivityCreated\}\\ onStart(e_f)\ \{e_f \mapsto_{fs}\ onStart\}\end{array}}{\begin{array}{c}\{Act_a\ e \wedge e \mapsto_{af} e_f \wedge e \mapsto_{as} onCreate \wedge e_f \mapsto_{fs} \gamma * R\}\\ fStart\\ \{e_f \mapsto_{fs}\ onStart * R\}\end{array}} \quad (35)$$

$$\frac{\begin{array}{c}(\xi, \xi') \in \{(onStop, onDestroyView),\\ (onDestroyView, onDestory), (onDestory, onDetach)\}\\ \{Act_a\ e \wedge e \mapsto_{af} e_f \wedge e \mapsto_{as} onDestroy \wedge e_f \mapsto_{fs} \xi\}\ \xi'(e_f)\ \{e_f \mapsto_{fs} \xi'\}\end{array}}{\begin{array}{c}\{Act_a\ e \wedge e \mapsto_{af} e_f \wedge e \mapsto_{as} onDestroy \wedge e_f \mapsto_{fs} \xi * R\}\\ fStopping\\ \{e_f \mapsto_{fs} \xi' * R\}\end{array}} \quad (36)$$

$$\frac{}{\begin{array}{c}\{Act_a\ e \wedge \neg Act_f\ e_f \wedge e \mapsto_{af} e_f \wedge e_f \mapsto_{fs} onDetach \wedge \mathcal{F}(e_f) * R\}\\ fRemove\\ \{\neg(e \mapsto_{af} e_f) * R\}\end{array}} \quad (37)$$

Fig. 5. Logic rules for fragment state transitions (part 1).

Now suppose that the system changes the status of the fragment v_2 of Example 4 to *onDetach*. This action can be modeled by running Rule 36 on a state satisfying the assertion A'_5. The result would be a state satisfying the following assertion:

$$A_5 = ((v_0 \mapsto_{as} onDestor) * (v_1 \mapsto_{as} onDestory) * (v_2 \mapsto_{fs} onDetach)) \wedge (v_0 \mapsto_{aa} v_1) \wedge (v_0 \mapsto_{af} v_2) \wedge \mathcal{F}(v_0) \wedge \mathcal{F}(v_1) \wedge \mathcal{F}(v_2).$$

Example 6. By applying Rule 37 once then Rule 30 twice on a state satisfying assertion A_5 of Example 5, we get an empty heap (satisfying the assertion *emp*).

Fig. 6. Logic rules for fragment state transitions (part 2).

A straightforward structural induction proves the following theorems.

Theorem 1 (*Preservation*). *If A is a well-formed assertion according to Definition 1 and $\{A\}\ X\ \{A'\}$ using any of the logic rules defined above, then A' is a well-formed assertion.*

Theorem 2 (*Progress*). *Suppose that A is a well-formed assertion. Then A is a stuck one (satisfied by stuck states) or there is a well-formed assertion A' such that $\{A\}\ X\ \{A'\}$ using inference rules defined above.*

Theorem 3 (*Soundness*). *Suppose that A is a well-formed assertion and $(s, h, a_{[]})$ and $(s', h', a'_{[]})$ are well-formed Dalvik states. Suppose also that $A \propto (s, h, a_{[]})$ and $X : (s, h, a_{[]}) \Rightarrow (s', h', a'_{[]})$ and $\{A\}\ X\ \{A'\}$. Then $A' \propto (s', h', a'_{[]})$.*

3   Related and Future Work

This section reviews the work on Android and Java systems most related to the technique presented in this paper. The related work belongs to one of two classes. The first class is logical and analyses methods [4–6,12,15,18]. The second class includes security methods that can benefit from the logics such as the one presented in the current paper [2,3,7,8,10,19,20].

For Java programs, symbolic execution is a technique for automatically constructing test cases. Android applications are executed on Davlik virtual machines and are event driven and vulnerable to path divergence. These two facts challenge applying symbolic execution to Android applications. The work in [14] proposed solutions to these issues. This was partially done by profile analysis methods to link handlers with events to build Android drivers capturing all correct events. Yet another interesting symbolic execution, Symdroid, for Dalvik bytecode is presented in [13].

In [16], a logic (Hoare-style) is proposed for the Java sequential kernel. Many Java concepts such as class and interface, recursive methods, dynamic binding, inheritance, and aliasing are considered in this logic. The paper discusses the logic soundness using an SOS semantics. Focusing on Java bytecode, [17] shows how to use Isabelle theorem prover to build a programming logic. The designed logic is shown useful to prove characteristics of programs with loops. This technique is convenient for Android programs as it was motivated by designing a procedure to assist Java Just-In-Time (JIT) compilers which are used in compiling Android applications.

In [11], the temporal logic (TL) is used to build a specification language for security of Android applications. In this work, the TL temporal operators are employed to enable recognizing timing-related security policies. The language is designed to enable dynamic application-monitoring in Android devices.

For future work extending the work presented in this paper, it is a good idea to study utilizing the logic presented in the paper to build precise security techniques [9]. Hence the goal would be to extend the logic presented her in a way that guarantees that Android applications with logical-judgment in the extended system are safe (concerning specific security issues).

References

1. Android developers. http://developer.android.com
2. Arzt, S., Rasthofer, S., Fritz, C., Bodden, E., Bartel, A., Klein, J., Le Traon, Y., Octeau, D., McDaniel, P.: Flowdroid: precise context, flow, field, object-sensitive and lifecycle-aware taint analysis for android apps. ACM Sigplan Not. 49(6), 259–269 (2014)
3. Barrera, D., Kayacik, H.G., van Oorschot, P.C., Somayaji, A.: A methodology for empirical analysis of permission-based security models and its application to android. In: Proceedings of the 17th ACM Conference on Computer and Communications Security, pp. 73–84. ACM (2010)

4. Bartel, A., Klein, J., Le Traon, Y., Monperrus, M.: Dexpler: converting android dalvik bytecode to jimple for static analysis with soot. In: Proceedings of the ACM SIGPLAN International Workshop on State of the Art in Java Program Analysis, pp. 27–38. ACM (2012)
5. El-Zawawy, M.A.: An operational semantics for android applications. In: Gervasi, O., et al. (eds.) ICCSA 2016. LNCS, vol. 9790, pp. 100–114. Springer, Cham (2016). doi:10.1007/978-3-319-42092-9_9
6. El-Zawawy, M.A.: A type system for android applications. In: Gervasi, O., et al. (eds.) ICCSA 2016. LNCS, vol. 9790, pp. 115–128. Springer, Cham (2016). doi:10.1007/978-3-319-42092-9_10
7. Enck, W., Gilbert, P., Han, S., Tendulkar, V., Chun, B.-G., Cox, L.P., Jung, J., McDaniel, P., Sheth, A.N.: Taintdroid: an information-flow tracking system for realtime privacy monitoring on smartphones. ACM Trans. Comput. Syst. (TOCS) 32(2), 5 (2014)
8. Enck, W., Octeau, D., McDaniel, P., Chaudhuri, S.: A study of android application security. In: USENIX security symposium, vol. 2, p. 2 (2011)
9. Enck, W., Ongtang, M., McDaniel, P.: Understanding android security. IEEE Secur. Priv. 7(1), 50–57 (2009)
10. Felt, A.P., Chin, E., Hanna, S., Song, D., Wagner, D.: Android permissions demystified. In: Proceedings of the 18th ACM Conference on Computer and Communications Security, pp. 627–638. ACM (2011)
11. Gunadi, H., Tiu, A.: Efficient runtime monitoring with metric temporal logic: a case study in the android operating system. In: Jones, C., Pihlajasaari, P., Sun, J. (eds.) FM 2014. LNCS, vol. 8442, pp. 296–311. Springer, Cham (2014). doi:10.1007/978-3-319-06410-9_21
12. Huisman, M., Jacobs, B.: Java program verification via a hoare logic with abrupt termination. In: Maibaum, T. (ed.) FASE 2000. LNCS, vol. 1783, pp. 284–303. Springer, Heidelberg (2000). doi:10.1007/3-540-46428-X_20
13. Jeon, J., Micinski, K.K., Foster, J.S.: Symbolic execution for dalvik bytecode. Technical report, Symdroid (2012)
14. Mirzaei, N., Malek, S., Păsăreanu, C.S., Esfahani, N., Mahmood, R.: Testing android apps through symbolic execution. ACM SIGSOFT Softw. Eng. Notes 37(6), 1–5 (2012)
15. Payet, E., Spoto, F.: An operational semantics for android activities. In: Proceedings of the ACM SIGPLAN 2014 Workshop on Partial Evaluation and Program Manipulation, pp. 121–132. ACM (2014)
16. Poetzsch-Heffter, A., Müller, P.: A programming logic for sequential Java. In: Swierstra, S.D. (ed.) ESOP 1999. LNCS, vol. 1576, pp. 162–176. Springer, Heidelberg (1999). doi:10.1007/3-540-49099-X_11
17. Quigley, C.L.: A programming logic for java bytecode programs. In: Basin, D., Wolff, B. (eds.) TPHOLs 2003. LNCS, vol. 2758, pp. 41–54. Springer, Heidelberg (2003). doi:10.1007/10930755_3
18. Reynolds, J.C.: Separation logic: a logic for shared mutable data structures. In: 2002 Proceedings of 17th Annual IEEE Symposium on Logic in Computer Science, pp. 55–74. IEEE (2002)
19. Schreckling, D., Köstler, J., Schaff, M.: Kynoid: real-time enforcement of fine-grained, user-defined, and data-centric security policies for android. Inf. Secur. Tech. Rep. 17(3), 71–80 (2013)
20. Yan, L.-K., Yin, H.: Droidscope: seamlessly reconstructing the OS and dalvik semantic views for dynamic android malware analysis. In: USENIX Security Symposium, pp. 569–584 (2012)

Posting Graphs for Finding Non-Terminating Executions in Asynchronous Programs

Mohamed A. El-Zawawy[1,2]([envelope])

[1] College of Computer and Information Sciences, Al Imam Mohammad Ibn Saud
Islamic University (IMSIU), Riyadh, Kingdom of Saudi Arabia
[2] Department of Mathematics, Faculty of Science,
Cairo University, Giza 12613, Egypt
maelzawawy@cu.edu.eg

Abstract. Asynchronous programming is one of the currently dominant programming techniques. The key idea of asynchronous programming is to use a queue to post computations as tasks (events). This queue is used then by the application to pick events asynchronously for processing. Asynchronous programming was found very convenient for achieving a massive percentage of software systems used today such as Gmail and Facebook which are Web 2.0 JavaScript ones.

This paper presents a novel technique for searching for nonterminating executions in asynchronous programming. The targeted cases of nontermination are those caused by the posting concept. The proposed technique is based on graphical representation for posting behaviours of asynchronous programs. Proofs for termination and correctness of the proposed method is outlined in the paper.

Keywords: Posting graphs · Asynchronous programs · Non-terminating · Multi-threaded programs · Program analysis

1 Introduction

The concept of asynchronous (event-driven) programming can be realized as the natural and convenient development of the thread concept [2,21]. Asynchronous Programming is today one of the dominant programming techniques. The key idea of asynchronous programming is to use a queue to post computations as tasks (events). This queue is used then by the application to pick events asynchronously for processing. This programming style was found very convenient for achieving a massive percentage of software systems used today. For example, asynchronous programming is used in applications such as Gmail, Facebook which are Web 2.0 JavaScript ones.

Asynchronous programming was also found useful to achieve large scale communications used by languages such as Scala (actor-based) and "node.js" [4] which is a platform. Such languages are main tools to build distributed servers such as LinkedIn, Twitter, and PayPal. Millions of used today Android (and

© Springer International Publishing AG 2017
O. Gervasi et al. (Eds.): ICCSA 2017, Part VI, LNCS 10409, pp. 233–245, 2017.
DOI: 10.1007/978-3-319-62407-5_16

other similar operating systems) mobile applications are accomplished using concepts of asynchronous programming. Modern concepts such as embedded systems, Internet-of-Things (IoT), sensor networks also benefit from the asynchronous programming style [5].

Techniques of program analyses for traditional programming methodologies (synchronous programming) are not typically and directly applicable for asynchronous programming [10]. This is so as the later have unique execution properties that are significantly differ from that of former techniques. The difference is caused by the fact that asynchronous programming languages combine short tasks from different methods in an accurate approach. Program analysis is very important for asynchronous programming because natural complexity of them makes it vulnerable to errors caused by programming mistakes and faulty designs [13,15].

One of the common errors in asynchronous programs is divergent (nonterminating) or erroneous behaviors. The bad news about this error is that it is tricky and not easy to reproduce during testing and simulation phases. This is due to nondeterminism in picking tasks from the queue and in processor scheduling. This makes techniques for formal verification of asynchronous programs ungovernable. This is also justified by the fact that these techniques are affected by the huge number of message-queues contents and processor context switches. There are many proposed solutions to the divergence problem where some of them are supported with correctness proofs. However most of these solutions either does not seem efficient enough or does not conveniently guarantee important features such as eventual quiescence [13,15].

This paper presents a new method to find nonterminating executions in asynchronous programs resulting from disobeying the important property of eventual quiescence. This property describes the situation where finite number of posting actions in asynchronous programs results in a nonterminating queue of posted methods where methods keep posting others infinitely.

The proposed technique is based on representing posting behaviour of asynchronous programs graphically. The graphical representation of the program is composed of components each of them represents a posting trace that begins with a posting command in the main method of the program. The technique removes the components that are guaranteed not to contribute to any nonterminating executions. These components are those with no loops. For each loop of the remaining components (with loops), a loop-start point is recognized. The technique determines for each loop-start points pairs of values for the program main global-variable. The values of each such pair represents the value at reaching the loop-start point and the value of returning to the same point. These values are then used as the search space to reproduce possible nonterminating executions.

Motivation

This paper is motivated by the need for an efficient yet simple enough algorithm for finding nonterminating executions in asynchronous programs like that of Example 1.

Contributions

The contribution of the paper is a novel, efficient and simple enough technique for reporting nonterminating execution in asynchronous programs. The technique is supported with mathematical proofs for its correctness and termination. The technique is demonstrated on an extended and rich example of an asynchronous program.

Paper Outline

The rest of the paper has the following sections. Section 2 presents the graphical resignations (partial posting graphs and complete posting graphs) of the posting behaviours of asynchronous programs. Section 3 introduces the main technique of the paper demonstrated in details on program of Example 1. A review of the work most related to the presented technique is shown in Sect. 4. Finally, Sect. 5 concludes the paper.

2   Posting Graphs

This section presents a graph representation for the posting relationship between program methods. The main idea of our technique is based on these graphical forms. We start by a formal definition for Asynchronous programs in The following definition.

Definition 1. *An asynchronous program P is a tuple (x, m, M) where x is the program's global variable, m is the main method of the program, and M is the set of methods posted by m or by other posted methods. The set of methods explicitly posted by m is denoted by M_m. Clearly $M_m \subseteq M$.*

We assume that the domain of possible valuations (denoted by V) for the global variable x has finite number of elements.

Example 1. Figure 1 shows the ping-family program for which $M = \{Ping, Pong, Dongo, Tango, Sango\}$ and $M_m = \{Ping, Pong, Dongo, Sango\}$.

A graph representing the posting relationship in an asynchronous program is composed of components (*partial posting graphs*) whose definition is introduced in Definition 2. The idea is that for each $\alpha \in M_m$, a partial posting graph is built to capture all the posting actions (and the relationship among them) that could happen upon executing α.

Definition 2. *Let $P = (x, m, M)$ be an asynchronous program. A partial posting graph of P (denoted by PPG^P) is a directed graph whose nodes include:*

- *a start node (labeled "Start"),*
- *an end node (labeled "Stop"),*
- *set of nodes (denoted by method nodes) each of them is labeled with method of M, and*
- *set of nodes (denoted by point nodes) representing potential program points.*

```
1   var x: bool;
2   proc Main ()
3       {x := false;
4       post Ping ();
5       post Pong ();
6       post Dongo ();
7       post Sango ();
8       return();}
9   proc Ping ()
10      {if (x == false)
11      then {post Ping();
12      x := true;}
13      return();}
14  proc Pong ()
15      {if (x == true)
16      then {post Pong();
17      x := false;}
18      return();}
19  proc Dongo()
20      {x := true;
21      post Tango();
22      return();}
23  proc Tango()
24      {x := false;
25      return();}
26  proc Sango()
27      {post Sango();
28      x := false;
29      return();}
```

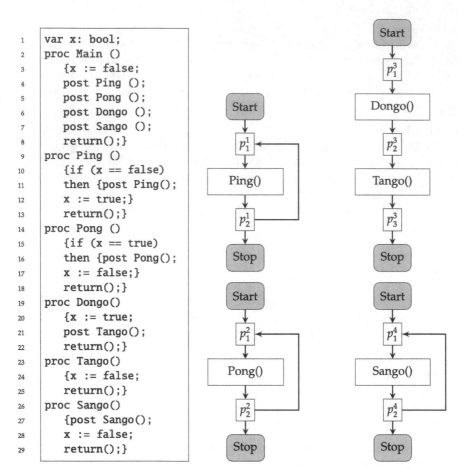

Fig. 1. The ping-family program.

Fig. 2. The Ping-PPG (PPG^1) and Pong-PPG (PPG^2) of ping-family program.

Fig. 3. The Dongo-PPG (PPG^3) and Sango-PPG (PPG^4) of ping-family program.

A directed path from method-node m_1 to method-node m_2 indicates that method m_1 may post method m_2. The first method node in the graph must correspond to a method in M_m. The nodes just before and after method nodes are point-nodes.

The set of all partial posting graphs of an asynchronous program composes the *complete posting graph* of the program introduced formally in the following definition.

Definition 3. *Let $P = (x, m, M)$ be an asynchronous program. A complete posting graph of P (denoted by CPG^P) is the graph whose components are all the partial posting graphs of P.*

The following lemma (Lemma 1) formalizes the idea of correspondence between elements of M_m of an asynchronous program $P = (x, m, M)$ and the set of partial posting graphs of P.

Lemma 1. *Let $P = (x, m, M)$ be an asynchronous program. The number of components of CPG^P equals the number of different methods posted by m ($|M_m|$).*

Proof. Let u be the number of components of CPG^P and v be the number of different methods posted by m. Suppose $\alpha \in M_m$. Then a partial posting graph PPG_α^P whose first method-node corresponds to α exists. Since $\{PPG_\alpha^P \mid \alpha \in M_m\} \subseteq CPG^P$, then $v \leq u$. Now because CPG^P includes only graphs corresponding to methods of M_m and each method α corresponds to only on partial graph PPG_α^P, then $u \leq v$. Hence $u = v$.

Certain point nodes of partial posting graphs play a crucial rule in our technique. These points are recognized in the following definition.

Definition 4. *Let PPG^P be a partial posting graph of the asynchronous program $P = (x, m, M)$. A point-node n^p of PPG^P is a loop-start if n^p has more than one in-edge.*

For simplicity, the proposed technique of this paper assumes that each partial posting graph has at most one loop-start node. For each loop-start node, pairs of valuations of the global variable of the asynchronous programs are of special importance in our technique and are detailed in the following definition.

Definition 5. *Suppose that $P = (x, m, M)$ is an asynchronous program and PPG_i^P is a partial posting graph of P that has exactly one loop-start node p. A pair of valuation $(v_1, v_2) \in V \times V$ is a return-pair of p, if when executing methods in PPG_i^P starting from p with x assigned the valuation v_1, it is possible to return to p again with x assigned the value v_2. The set of all such pairs for PPG_i^P is denoted by RV_i.*

Example 2. Figures 2 and 3 show the four PPGs (PPG^1, PPG^2, PPG^3, and PPG^4) of the ping-family program (of Example 1) which together form the CPG of the program. The number of different methods posted by the main method is 4 which equals the number of partial posting graphs. However the number of posted methods in the whole program is 5. The graphs PPG^1, PPG^2, PPG^3, and PPG^4 have the following sets of point nodes $\{p_1^1, p_2^1\}, \{p_1^2, p_2^2\}, \{p_1^3, p_2^3, p_2^3\}$, and $\{p_1^4, p_2^4\}$, respectively. Among all these points the points p_1^1, p_1^2, and p_1^4 are the only loop-start ones.

3 Finding the Nonterminating Executions Using Posting Graphs

This section presents our technique for searching for nonterminating behaviours of asynchronous programs using the posting graphs presented in the previous section. The main proposed algorithm for searching for nonterminating executions is *Asynchronous-Termination* which is presented in Algorithm 1. The algorithm starts (lines $2-4$) by removing from the set of *PPGs* elements that have no loops. This step is justified by the fact that such *PPGs* will not create repeated posting patterns (sequence of methods reposting themselves), hence will not contribute to the existence of nonterminating behaviours of the program. However these *PPGs* (with no loops) may help other *PPGs* with loops to create repeated posting patterns via assigning the global variable a convenient valuation. This is treated by studying the *PPGs* with loops on all possible valuation of the global variable (finite number of valuations by assumption).

Example 3. Implementing steps $2-4$ of *Asynchronous-Termination* on the program of Example 2 would result in removing PPG^3 whose execution will only change the value of x without having any affect on nontermination of the program. The effect of PPG^3 on x is considered by studying PPG^1, PPG^2 and PPG^4 on all possible values of x.

Algorithm 1. Asynchronous-Termination

Input: $P = (x, m, M)$: an asynchronous program, and $G = \{PPG_1^P, ..., PPG_n^P\}$: the set of partial posting graphs of P.

Output: Determines whether P has nonterminating executions with some details about the nontermination if exists; returning 0 for non-termination and 1 for termination.

1: **procedure** ASYNCHRONOUS-TERMINATION
2:     **for each** $(1 \leq i \leq n)$ **do**
3:         **if** $(PPG_i^P$ has no loop-start nodes) **then**
4:             $G \leftarrow G \setminus \{PPG_i^P\}$;
5:     **if** $(G = \emptyset)$ **then**
6:         **return** 1; – terminating program
7:     **for each** $(PPG_i^P \in G)$ **do**
8:         $RV_i \leftarrow$ **Calc-Return-Values**(P, PPG_i^P);
9:         **if** (for some $v \in V, (v, v) \in RV_i$); **then**
10:            **return** (0,v,i); – non-terminating execution at value v for graph PPG_i
11:    $r \leftarrow$ **Find-Non-Termination**$(RV_1, \ldots, RV_n)$;
12:    **return** r;

The halting of the algorithm in lines (6) is justified by The following theorem.

Theorem 1. *Let $P = (x, m, M)$ be an asynchronous program which has n partial posting graphs, $PPG_1^P, \ldots, PPG_n^P$. If none of the partial posting graphs has loop-start nodes, then the program has no nonterminating executions.*

Proof. The complete posting behavior of each method posted in the main method (and that also of their posted decedents) is represented by one of the PPG^P. If for all i, PPG_i^P has no loop-start nodes, then it also has no loops.

Therefore methods in PPG_i^P do not repost any of the methods in PPG_i^P. This guarantees that there will be no repetition patterns of posting methods to have a nonterminating executions.

For each PPG_i of G (line 7), the algorithm calls the method *Calc-Return-Values* (line 8) that calculates the set RV_i of *return-pairs* of the loop-start node of PPG_i. If RV_i has a pair (v, v) then the program has nonterminating executions as proved by Theorem 2.

Theorem 2. *Let $P = (x, m, M)$ be an asynchronous program which has PPG^P as one of its partial posting graphs. Suppose also that PPG^P has a loop-start point p. If (v, v) is a return-pair for p, then P has a nonterminating execution.*

Proof. Suppose that the methods in PPG^P constituting the loop starting at p are $m_1, m_2, \ldots, m_n$. Then the following is a potential nonterminating execution for P:

$$(\{m_1\} \cup M, v) \rightsquigarrow (\{m_2\} \cup M, v_1) \rightsquigarrow \ldots \rightsquigarrow (\{m_n\} \cup M, v_{n-1}) \rightsquigarrow (\{m_1\} \cup M, v) \rightsquigarrow \ldots$$

The set M denotes a set of posted methods. This completes the proof.

Example 4. Running Algorithm *Calc-Return-Values* (step 8 of *Asynchronous-Termination* algorithm) on PPG^1, PPG^2, and PPG^4 of Example 2 would result in the sets $RV_1 = \{(false, true)\}, RV_2 = \{(true, false)\}$ and $RV_4 = \{(true, false), (false, false)\}$, respectively. The *Asynchronous-Termination* algorithm does not call *Calc-Return-Values* for PPG^3 as it has no loops (loop-start nodes). The pair $(true, true)$ is not included in RV_1 as tracing the execution from the node p_1^1 of PPG^1 would not lead to reposting *Ping* and going back to the loop-start node p_1^1. Running steps $9 - 10$ (of the *Asynchronous-Termination* algorithm) on RV_4 would lead to terminating the algorithm with the result that the program has nonterminating executing. This is obviously correct from the content of the method *Sango()*.

If the program P is not found to be terminating or having nonterminating executions until step 11 of the *Asynchronous-Termination* algorithm, this step calls the *Find-Non-Termination* algorithm with the sets RV's. This later algorithm returns the final output of the main algorithm.

The *Calc-Return-Values* algorithm takes as input an asynchronous program $P = (x, m, M)$ together with one of its partial posting graph PPG^P that has exactly one loop whose loop-start node is p. For each value $v \in V$ (the domain of all possible values of x) the algorithm executes (step 5) the methods in PPG^P starting form p until returning again to the point p or reaching the stop node of PPG^P. In case the stop node is reached, the algorithm ignores (step 7) the current valuation in v and starts examining another valuation from the domain V. If the execution leads to returning back to the point p with valuation v' for the global variable x. Then the set RV is augmented with the pair (v, v') (step 9).

Algorithm 2. Calc-Return-Values

Input: $P = (x, m, M)$: an asynchronous program, and PPG^P: a partial posting graph
of P that has exactly one loop-start node (p).
Output: The set of return-pairs of the node p.
1: **procedure** CALC-RETURN-VALUE
2:      $RV \leftarrow \emptyset$
3:      **for each** $(v \in V)$ **do**
4:          $x \leftarrow v$
5:          Execute methods starting from p until reaching the stop node or p;
6:          **if** (the reached node is the stop node) **then**
7:              Continue;
8:          **if** (the reached node is p) **then**
9:              $RV \leftarrow RV \cup \{(v, x)\}$
10:     **return** RV

Lemma 2. *For inputs $P = (x, m, M)$ (an asynchronous program) and PPG_i^P (a partial posting graph of P that has exactly one loop-start node p), the algorithm Calc-Return-Values always terminates and returns the set RV_i of return-pairs of p (Definition 5).*

Proof. The termination of the algorithm is guaranteed by our assumption that the set of possible valuations V of global variable is finite. Correctness of the algorithm is clear.

The input of the algorithm *Find-Non-Termination* is a set of sets of return-pairs $\{RV_1, \ldots, RV_n\}$ (Definition 5) each of them represents a partial posting graph. The algorithm extracts nonterminating executions from the input return-pairs sets, if any. This is done as follows. For each set RV_i, the algorithm augments the set with its transitive closure (step 3). The algorithm then checks whether the augmented set RV_i has a pair (v, v) (step 4). If this is the case the algorithm concludes that the program P has a nonterminating execution caused by methods of graph PPG_i^P when x has the valuation v. This is justified by Theorem 3.

Theorem 3. *Supposed that RV_i' is the augmented set of return-pairs produced by step 4 of the Find-Non-Termination algorithm. If for some $v \in V, (v, v) \in RV_i'$, then the program has a nonterminating execution.*

Proof. As (v, v) is in the transitive closure of RV_i, then there exists $v_1, \ldots, v_m \in V$ such that $(v, v_1), (v_1, v_2), \ldots, (v_n, v) \in RV_i$. Suppose that PPG_i^P is the partial posting graph corresponding to RV_i. Suppose that the methods in PPG_i^P constituting the loop starting at the loop-start node of PPG_i^P are $m_1, m_2, \ldots, m_n$. Suppose that the following trace of execution is denoted by μ:

$$\mu = m_1; m_2; \ldots; m_n;$$

Then the following is a potential nonterminating execution for P:

$$(\mu, v) \rightsquigarrow (\mu, v_1) \rightsquigarrow (\mu, v_2), \ldots, (\mu, v_n), \rightsquigarrow (\mu, v) \ldots$$

This completes the proof.

Remark 1. Our techniques assumes that the global variable does not get modified before the loop-start node. Without this assumption, a backward analysis would be needed to find a suitable initial value for the global variable at the start node of the graph to arrive at the loop-start node with a certain value for the global variable.

If the pairs (v, v) of step 4 do not exist in any of the sets RV'_i, the algorithm continues after the loop (step 6) searching for a sequence of pairs $(v_{i_1}, v_{i_2}), (v_{i_2}, v_{i_3}), \ldots, (v_{i_k}, v_{i_1})$ in the sets RV'_i. The existence of such set of pairs guarantees the existence of a non-termination execution of P as Theorem 4 proves. Hence in this case the algorithm reports the details of the nontermination execution in step (7) and halts. In case the sequence (of step 6) can not be found the algorithm halts after reporting (in step 8) the program is a terminating one.

Theorem 4. *Supposed that the sets $\{RV'_i\}$ is the augmented sets of return values produced by step 4 of the Find-Non-Termination algorithm. If there are pairs $(v_{i_1}, v_{i_2}), (v_{i_2}, v_{i_3}), \ldots, (v_{i_k}, v_{i_1})$ such that $(v_{i_j}, v_{i_{j+1}}) \in RV'_{i_j}$ then the program has a nonterminating execution.*

Proof. Since $(v_{i_j}, v_{i_{j+1}}) \in RV'_{i_j}$ which is the transitive closure of RV_{i_j}. Then there exists $v_1, \ldots, v_r \in V$ such that $(v_{i_j}, v_1), (v_1, v_2), \ldots, (v_r, v_{i_{j+1}}) \in RV_{i_j}$. Suppose $PPG^P_{i_j}$ is the partial posting graph corresponds to RV_{i_j}. Suppose that the methods in $PPG^P_{i_j}$ constituting the loop starting at loop-start node of $PPG^P_{i_j}$ are $m^1_{i_j}, m^2_{i_j}, \ldots, m^{r+1}_{i_j}$. Suppose that the following trace of execution is denoted by γ_{i_j}:

$$\gamma_{i_j} = (m^1_{i_j}, v_{i_j}) \rightsquigarrow (m^2_{i_j}, v_{i_{j+1}}) \rightsquigarrow, \ldots, (m^{r+1}_{i_j}, v_r).$$

Now we define the following trace of execution:

$$\delta = \gamma_{i_1}; \gamma_{i_2}; \ldots; \gamma_{i_k}$$

Then the following is a potential nonterminating execution for P:

$$(\delta)_{x=v_{i_1}} \rightsquigarrow (\delta)_{x=v_{i_2}} \rightsquigarrow \ldots \rightsquigarrow (\delta)_{x=v_{i_k}} \ldots$$

This completes the proof.

Example 5. Consider the program P and its *PPGs* of Example 2. If the *Find-Non-Termination* algorithm is run for sets of return-pairs RV_1, RV_2, and RV_4 of Example 4. The test in Step 6 of the algorithm reports one of the following cases and then halts:

- $(0, (false, true), (1, 2))$. This a nontermination case as follows:

  $(Ping(), false) \rightsquigarrow (Pong(), true) \rightsquigarrow (Ping(), false) \rightsquigarrow (Pong(), true) \rightsquigarrow \ldots$

- $(0, (true, false), (2, 1))$. This a nontermination case as follows:

  $(Pong(), true) \rightsquigarrow (Ping(), false) \rightsquigarrow (Pong(), true) \rightsquigarrow (Ping(), false) \rightsquigarrow \ldots$

Algorithm 3. Find-Non-Termination

Input: $RV_1, \ldots, RV_n$.
Output: A non-termination sequence of execution if any.
1: **procedure** FIND-NON-TERMINATION
2:      **for each** $(1 \leq i \leq n)$ **do**
3:          $RV'_i \leftarrow$ ***Transitive-closure***(RV_i);
4:          **if** (for some $v \in V, (v, v) \in RV'_i$); **then**
5:              **return** (0, v, i); – non-terminating execution at value v for graph PPG_i
6:          **if** there are pairs $(v_{i_1}, v_{i_2}), (v_{i_2}, v_{i_3}), \ldots, (v_{i_k}, v_{i_1})$ such that $(v_{i_j}, v_{i_{j+1}}) \in RV'_{i_j}$
            then
7:              **return** $(0, (v_{i_1}, \ldots, v_{i_k}), (i_1, i_2, \ldots, i_k))$ – Non-termination execution at
            value v_{i_1} involving the determined graph indexes and values in the specified order.
8:      **return** 1; – Terminating program.

Lemma 3. *The algorithm Asynchronous-Termination always terminates.*

The proof of the following corollary follows from Theorems 2, 3, and 4.

Corollary 1. *For inputs* $P = (x, m, M)$ *(an asynchronous program) and* $PPG_1^P, \ldots, PPG_n^P$ *(the set of partial posting graphs of P), if the algorithm Asynchronous-Termination returns 0, then P has nonterminating executions.*

Theorem 5. *Suppose* $P = (x, m, M)$ *is an asynchronous program with a nonterminating execution. If the Asynchronous-Termination algorithm is run on P with its partial posting graphs, then the nonterminating execution will be reported by the algorithm.*

Proof. The nonterminating execution of the program is caused by a repeating pattern of posted methods. Such repetition would only occur with the existence of loop-start nodes in some partial-posting graphs and a reflexive return-pair (v, v) in the union of sets of return-Paris. Such pair is (as discussed above) discoverable by *Asynchronous-Termination* algorithm.

The following corollary results from Corollary 1 and Theorem 5.

Corollary 2. *Suppose* $P = (x, m, M)$ *is an asynchronous program. The program has a nonterminating execution if and only if such execution is discoverable by the Asynchronous-Termination algorithm.*

4   Related Work

This section reviews work most related to the technique presented in this paper. The review considers the following categories: program analysis techniques for asynchronous programs, concurrent programs, multi-threaded programs, and nonterminating executions.

The absence of synchronization between random delay of message-based communication, and parallel running methods in asynchronous programs cause non-terminating executions. The work in [15] proposes a method for reporting divergent behaviours in asynchronous programs. The method is composed of three steps. The first two steps are transformation ones transferring the problem to the world of sequential programs. The third step uses SMT-based verification tools for sequential programs to find the nontermination executions. The solution seems to be very involved in away that affect it efficiency. The technique presented in our paper is much simpler than that presented in [15]. Also the correctness of our work is supported by a set of theorems which is not the case for the work in [15].

Many techniques have been proposed for treating asynchronous programs [5, 12, 15–17, 22, 26]. It is possible that traditional architectures for processors designed for the properties of synchronous programs degenerate performance of asynchronous programs. This issue was addressed in [5] which proposes an architecture (Event Sneak Peek – ESP) that is transferring bottlenecks of micro-architectural in asynchronous programs. In ESP, the processor checks task queue to build future event-knowledge. This knowledge is used in this technique to avoid stalling due to latency of cache misses.

Program analysis techniques for concurrent programs is an active direction of research [1, 14, 20]. Determining the effects of running statements during context switches involving threads (where reordering is possible or not) is done via relaxed memory consistency models. Considering partial and total store ordering, acquire and release consistency, and relaxed memory ordering for models of relaxed memory consistency, the research in [1] suggests a concurrent program logic. Processing used by [1] is partially similar to ours (presented in the current paper) as it expresses a concurrent program in the form of a set of acyclic graphs. These graphs allow modeling commands dependencies. A serious weakness with this argument, however, is that it does not seem to consider nontermination cases and does no report robust experimental results.

Among research related to our work is that focusing on analyzing multi-threaded programs [11, 18, 19, 23–25]. For multi-threaded programs and using sparse analysis on the result of other analyses of thread interference, [25] suggests a scalable pointer analysis that is flow-Sensitive. Although the scalability of the technique is proved experimentally on relatively large programs, the soundless and the completeness against a relevant model (operational semantics for example) of the method are not established in the paper.

Program termination is a very important program analysis which receives research efforts [3, 6–9, 13]. Nontermination of heap treating programs was tackled in [3] which proposed a proof system (Hoare-style) that relies on separation logic and cyclic proofs. Termination of these programs are measured using judgements (in the system) built starting from a specific program point satisfying a certain (separation logic) precondition. Proofs in this system are cyclic derivations based on infinitely often unfolding of inductive predicates. This results in discarding infinite proof paths via an argument that is infinite descent. This technique is not directly applicable for studying or proving termination of asynchronous programs as it does not consider the idea of queuing tasks for future asynchronous execution. However, similar techniques can built to study nontermination of asynchronous programs.

The work in [13] presents a technique for testing termination of asynchronous programs using the idea of code augmentation towards capturing the traces of nonterminating executions. However the technique presented in the current paper is much simpler and more efficient than that of [13]. This is justified by the fact that the technique of this paper, recognizes and removes away posted tasks (methods) that are not possible to contribute to a nonterminating execution of the program. Also the current technique reports detailed information about possible nonterminating executions.

5   Conclusion

This paper presented a novel technique for finding nonterminating behaviours for asynchronous programming which is one of prominent programming styles today. This programming technique offers the concept of posting which allows queuing methods for later asynchronous picking and processing. The paper targeted nontermination cases caused by the posting concept. The proposed technique was built on graphical representation for posting behaviours of asynchronous programs. The paper proved termination and correctness of the algorithms proposed in the paper.

References

1. Abe, T., Maeda, T.: Concurrent program logic for relaxed memory consistency models with dependencies across loop iterations. J. Inf. Process. **25**, 244–255 (2017)
2. Andrews, G.R.: Concurrent Programming: Principles and Practice. Benjamin/Cummings Publishing Company, San Francisco (1991)
3. Brotherston, J., Bornat, R., Calcagno C.: Cyclic proofs of program termination in separation logic. In: ACM SIGPLAN Notices, vol. 43, pp. 101–112. ACM (2008)
4. Cantelon, M., Harter, M., Holowaychuk, T.J., Rajlich, N.: Node. js in Action. Manning, Greenwich (2014)
5. Chadha, G., Mahlke, S., Narayanasamy, S.: Accelerating asynchronous programs through event sneak peek. ACM SIGARCH Comput. Archit. News **43**(3), 642–654 (2016)
6. Colón, M.A., Sipma, H.B.: Practical methods for proving program termination. In: Brinksma, E., Larsen, K.G. (eds.) CAV 2002. LNCS, vol. 2404, pp. 442–454. Springer, Heidelberg (2002). doi:10.1007/3-540-45657-0_36
7. Cook, B., Podelski, A., Rybalchenko, A.: Termination proofs for systems code. In: ACM SIGPLAN Notices, vol. 41, pp. 415–426. ACM (2006)
8. Cook, B., Podelski, A., Rybalchenko, A.: Proving program termination. Commun. ACM **54**(5), 88–98 (2011)
9. Cousot, P.: Proving program invariance and termination by parametric abstraction, Lagrangian relaxation and semidefinite programming. In: Cousot, R. (ed.) VMCAI 2005. LNCS, vol. 3385, pp. 1–24. Springer, Heidelberg (2005). doi:10.1007/978-3-540-30579-8_1

10. Cristian, F.: Synchronous and asynchronous. Commun. ACM **39**(4), 88–97 (1996)
11. El-Zawawy, M.A.: Detection of probabilistic dangling references in multi-core programs using proof-supported tools. In: Murgante, B., et al. (eds.) ICCSA 2013. LNCS, vol. 7975, pp. 516–530. Springer, Heidelberg (2013). doi:10.1007/978-3-642-39640-3_38
12. El-Zawawy, M.A.: An efficient layer-aware technique for developing asynchronous context-oriented software (acos). In: 2015 15th International Conference on Computational Science and Its Applications (ICCSA), pp. 14–20. IEEE (2015)
13. ElZawawy, M.A.: Finding divergent executions in asynchronous programs. In: Gervasi, O., et al. (eds.) International Conference on Computational Science and Its Applications. LNCS, vol. 9790, pp. 410–421. Springer, Heidelberg (2016). doi:10.1007/978-3-319-42092-9_32
14. El-Zawawy, M.A., Alanazi, M.N.: An efficient binary technique for trace simplifications of concurrent programs. In: 2014 IEEE 6th International Conference on Adaptive Science & Technology (ICAST), pp. 1–8. IEEE (2014)
15. Emmi, M., Lal, A.: Finding non-terminating executions in distributed asynchronous programs. In: Miné, A., Schmidt, D. (eds.) SAS 2012. LNCS, vol. 7460, pp. 439–455. Springer, Heidelberg (2012). doi:10.1007/978-3-642-33125-1_29
16. Ganty, P., Majumdar, R.: Algorithmic verification of asynchronous programs. ACM Trans. Program. Lang. Syst. (TOPLAS) **34**(1), 6 (2012)
17. Jhala, R., Majumdar, R.: Interprocedural analysis of asynchronous programs. In: ACM SIGPLAN Notices, vol. 42, pp. 339–350. ACM (2007)
18. Lucia, B., Ceze, L.: Cooperative empirical failure avoidance for multithreaded programs. In: ACM SIGPLAN Notices, vol. 48, pp. 39–50. ACM (2013)
19. Musuvathi, M., Qadeer, S.: Iterative context bounding for systematic testing of multithreaded programs. In: ACM SIGPLAN Notices, vol. 42, pp. 446–455. ACM (2007)
20. Musuvathi, M., Qadeer, S., Ball, T., Basler, G., Nainar, P.A., Neamtiu, I.: Finding and reproducing Heisenbugs in concurrent programs. In: OSDI, vol. 8, pp. 267–280 (2008)
21. Ousterhout, J.: Why threads are a bad idea (for most purposes). In: Presentation Given at the 1996 Usenix Annual Technical Conference, vol. 5, San Diego, CA, USA (1996)
22. Pnueli, A., Rosner, R.: On the synthesis of an asynchronous reactive module. In: Ausiello, G., Dezani-Ciancaglini, M., Della Rocca, S.R. (eds.) ICALP 1989. LNCS, vol. 372, pp. 652–671. Springer, Heidelberg (1989). doi:10.1007/BFb0035790
23. Sabelfeld, A., Sands, D.: Probabilistic noninterference for multi-threaded programs. In: Proceedings of 13th IEEE Computer Security Foundations Workshop, CSFW-13, pp. 200–214. IEEE (2000)
24. Savage, S., Burrows, M., Nelson, G., Sobalvarro, P., Anderson, T.: Eraser: a dynamic data race detector for multithreaded programs. ACM Trans. Comput. Syst. (TOCS) **15**(4), 391–411 (1997)
25. Sui, Y., Di, P., Xue, J.: Sparse flow-sensitive pointer analysis for multithreaded programs. In: 2016 IEEE/ACM International Symposium on Code Generation and Optimization (CGO), pp. 160–170. IEEE (2016)
26. Syme, D., Petricek, T., Lomov, D.: The f# asynchronous programming model. In: Rocha, R., Launchbury, J. (eds.) PADL 2011. LNCS, vol. 6539, pp. 175–189. Springer, Heidelberg (2011). doi:10.1007/978-3-642-18378-2_15

Software Analytics for Web Usability: A Systematic Mapping

Lucas Henrique Pellizon[1], Joelma Choma[2], Tiago Silva da Silva[1(✉)], Eduardo Guerra[2], and Luciana Zaina[3]

[1] Universidade Federal de São Paulo - UNIFESP, São José dos Campos, SP, Brazil
silvadasilva@unifesp.br
[2] Instituto Nacional de Pesquisas Espaciais - INPE, São José dos Campos, SP, Brazil
[3] Universidade Federal de São Carlos - UFSCar, Sorocaba, SP, Brazil

Abstract. Software usability has become a key factor for the success or failure of web-based systems. However, traditional evaluation methods – user tests and field observations – are expensive and time consuming when applied to a large number of users. In order to deal with these inherent difficulties and costs and to propose a new method to automatically capture and analyze web usage data, we carried out a systematic mapping on web analytics and web usability. A total of 970 studies were identified, of which only 42 studies were selected for this mapping. We found out that most studies are focused on tools for capturing information on user's navigation, however, few tools have presented mechanisms for visualization of these user interaction data.

1 Introduction

In the last years, software usability has been seen as an important software quality attribute. For example, several ISO (International Organization for Standardization) standards [19–21] include usability as a component of the overall quality of a software product [31]. For web-based systems, in which the distance from a competitor is just one click, usability became a key factor in its success or failure.

In general, usability evaluations are carried out through inspection or observation methods. Traditionally, the observation of the user interaction with software systems is based on carrying out user tests and field observations [51]. Although these methods are popular and efficient, they are expensive and time consuming. User tests are considered expensive because of their costs to find users to perform a test, to move the users to a usability laboratory, to set up the infrastructure, conduct the test, collect and analyze the data generated during the tests.

Field observations are also an expensive method due to their necessity to one or more usability analysts observing users while they interact with the software systems in their work environment for long periods of time. Observing users in their real work environment, performing their own tasks, allow us to keep the

O. Gervasi et al. (Eds.): ICCSA 2017, Part VI, LNCS 10409, pp. 246–261, 2017.
DOI: 10.1007/978-3-319-62407-5_17

context of usage, an important element in a usability analysis. Nevertheless, analyzing video recording or observing users by video-conference are also extremely time consuming activities.

Due to these costs, a common practice is to analyze the behavior of a few users in user tests or field observations. Analyzing the behavior of a few users limits the evaluation to just a qualitative analysis. Moreover, some problems can be just highlighted in a quantitative analysis and also their impact can just be evaluated considering a large number of users [51].

In order to deal with the difficulties, risks and costs of carrying out user tests in laboratories and/or observing users in field studies to analyze the usability of web applications, and aiming at proposing a new method to automatically capture and analyze web usage data, we carried out a systematic mapping about the employment of software analytics to enable web usability.

The remainder of this paper is organized as follows: Sect. 2 presents a short background on software analytics usability. Section 3 introduces the research method and presents our research questions. Section 4 presents our results and the answers for the research questions. Section 5 highlights some limitations of this study, and presents our final considerations and future work.

2   Background

In this Section, we present some concepts related to software analytics and web analytics, particularly focusing on navigation behaviour and usage data.

2.1   Software Analytics

Web systems generate a huge amount of information, which includes not only the sending and receiving data requests, but also actions and decision making processes in order to reach a certain goal. The software analytics area is concerned with collecting, exploring and analyzing this large amount of data aiming to transform it into actions to improve the product and its usability [7].

This study refers to the software analytics geared towards web-based systems – an area also known as web analytics [25]. Within this context, we have mainly focused on the means of capturing navigation and usage data, since these data can provide valuable information on navigation behaviour and how software systems are actually being used. Real usage data analysis collected from web logs, for instance, can lead to better understanding of how users use software in practice.

We list below some of the most common means used to capture navigation behaviour and usage data.

- Server Logger: a web server automatically saves all requests received by users, such as pages, images, files, among others services. Usually, these records contain the client IP address, the time and date of the request, the requested resource, the status of the request, the HTTP method used, the size of the object returned to the client, and the referring web resource. Hence, of these logs can be extracted detailed information about the browsing behaviour of visitors to a website [47].

- Proxy Logger: a proxy-based system logs all communication between the client browser and the web server. However, the user's browser needs to be set up in order to direct web requests and captured events to the proxy-server. Basically, a proxy logger collects the same information type captured by a Server Log [16].
- Browser Plugin: a specific tool can be created for certain browsers in order to collect not only the requests made to the server, but also some behaviors, such as keyboard and mouse actions [37].
- JavaScript: a code is added to the website in order to inspect certain user actions, such as mouse and keyboard usage, for example. This JavaScript code must be added by the website developer himself, unless he chooses to take a integration with a proxy aiming to automatically add it to the page [44].
- Proxy + JavaScript: as mentioned in the previous item, the JavaScript code is automatically added through a proxy acting as an intermediary between the client and server. This approach has the same characteristics as the previous one and, in addition, it avoids the need to manually add the inspection code to the product. On the other hand, developers may be confronted with security issues [26].

2.2   Usability

Usability is a quality attribute that assesses how easy user interfaces are to use, in which the navigability is an important measure of how easily users can locate and access the necessary information to achieve their goals. According to [36], usability can be defined by five quality components: (i) Learnability: ease of learning; (ii) Efficiency: efficiency of use; (iii) Memorability: ease of remembering how to use; (iv) Errors: low rate of error during use; (v) Satisfaction: a pleasing user experience.

Online usability studies commonly use strategies for collecting both qualitative and quantitative data on either user attitudes or behaviors, and measuring the user experience. Quantitative metrics are useful, for example, to test the awareness of key features, or verify the intuitiveness of the navigation, while qualitative metrics can provide traceability into the nature of the problems that users encounter, or insight on possible design solutions [47].

3   Research Method

3.1   Systematic Mapping

Systematic mapping is a research method that provides an overview of a research area, and allows us to identify the quantity and type of research and results available within specific area, or a phenomenon of interest. This research method consists of classifying, conducting thematic analysis and identifying publications in order to identify research gaps. Also, it can provide indications for lack of evaluation or validation research in certain areas with limited effort. Usually, systematic mapping studies do not study articles in depth in order to identify best practices based on empirical evidence. Nevertheless, it can be a first step toward a systematic review for further investigation on a specific topic [41].

For this study, we follow a mapping process based on the guidelines provided by [42], in which at least two of the authors participated in each step of the analysis of publications in order to allow us to immediately resolve disagreements as to how publications should be keyworded through discussion. The mapping process consists of four steps: (i) definition of research questions, (ii) search for relevant publications, (iii) screening of papers and (iv) mapping of publications.

3.2 Research Questions

The goal of this systematic mapping study is to get an overview of existing research on software analytics approaches focusing on web-based software usability – encompassing techniques, methods and tools commonly employed to capture and analyze navigation behaviour and usage data. In order to narrow the mapping study scope, we defined six research questions as follows:

- RQ1: How are the studies on web analytics and usability distributed over time?
- RQ2: What are the most common types of studies, and in what environments are they conducted?
- RQ3: What is the research focus of these studies?
- RQ4: What types of technologies have been used to capture navigation behaviour and usage data?
- RQ5: How can the captured information be visualized?
- RQ6: What is the relationship between capture technologies and data visualization?

Having defined the research questions, we selected some search keywords related to the two interest subjects: analytics and usability. Table 1 presents the keywords used in the search. Once we are focusing our study on web-based software, the term "web" is common for both areas.

3.3 Search Strategy

Taking into account that this is a preliminary study on the two aforementioned areas of interest, we chose to conduct a search for publications relevant by performing an automated search from a single database. Thus, we chose to use the Scopus digital library[1] because this library is an online abstract and indexing

---

[1] http://www.scopus.com/.

Table 1. Keywords related to interest areas.

Search terms	
Subject	Keywords
Analytics	capture
	collect
	analyze
	web
Usability	user behavior
	usability
	user experience
	web

service provided through Elsevier[2], which also indexes content provides by other digital libraries such as IEEE Xplore[3], ACM [4] and Springer[5], as content coverage guide [45].

The search in the digital library was conducted in October 2016. According to our search terms, we used the following query expression in order to search in the electronic database: (capture **OR** collect **OR** analyze) **AND** (user behavior **OR** usability **OR** user experience) **AND** (web).

Furthermore, the Scopus search engine allowed us to specify the subject area (*Engineering or Computing*) and the document type (*Conference Paper, Article, Chapter or Article in Press*). This search resulted in 970 documents.

3.4   Screening of Papers

For the purpose of selecting only relevant studies aiming to answer our research questions, we applied the inclusion and exclusion criteria present in Table 2 to titles and abstracts.

Table 2. Inclusion and exclusion criteria

Screening criteria	
Inclusion	Exclusion
-Peer reviewed studies	-Non-English publications
-Addresses web analytics and usability	-Panel summaries, keynotes and posters
-Means for capture navigation behaviour	-Non-accessible in full-text
-Technologies for capture usage data	-Capture technologies based on Server Logs

[2] https://www.elsevier.com.br/.
[3] http://www.ieee.org/.
[4] http://dl.acm.org/.
[5] https://link.springer.com/.

Table 3. Distribution of publication per digital libraries.

Papers source	
Database Name	# Relevant Papers
ACM	14–33.33%
Springer	9–21.43%
IEEE Xplore	9–21.43%
Other	10–24.81%

Based on the screening criteria, we selected 42 relevant papers, which are listed in the Table 4. Table 3 presents the distribution of these publications per digital libraries. ACM, Springer and IEEEXplore stand out as the main sources of 76.1% of the publications (32 of 42). Despite we having selected only documents with title and abstract in English, we found out later a paper written in Portuguese by Brazilian researchers. However, due to the relevance of the study, we decided to keep this paper in our collection.

Table 4. Selected studies

ID	Paper	Ref.	ID	Paper	Ref.
1	Hong et al.	[16]	22	Peska	[40]
2	Paternó and Paganelli	[39]	23	Huang et al.	[17]
3	Waterson et al.	[53]	24	Deufemia et al.	[10]
4	Shahabi and Banaei-Kashani	[46]	25	Premchaiswadi and Romsaiyud	[43]
5	Niño et al.	[37]	26	De Vasconcelos et al.	[15]
6	Chui and Li	[8]	27	Li et al.	[27]
7	Cuddihy et al.	[9]	28	Apaolaza et al.	[2]
8	Arroyo et al.	[4]	29	Apaolaza	[1]
9	Atterer et al.	[5]	30	Nakano et al.	[34]
10	Ignatova and Brinkman	[18]	31	Uchida et al.	[48]
11	Mor et al.	[33]	32	Mao et al.	[30]
12	Zhu et al.	[55]	33	De Vasconcelos and Santos	[52]
13	Rivolli et al.	[44]	34	Breslav et al.	[6]
14	Kiura et al.	[24]	35	Papatheocharous et al.	[12]
15	Ding et al.	[11]	36	Zahoor et al.	[54]
16	Luna et al.	[28]	37	Apaolaza et al.	[3]
17	Vargas et al.	[49]	38	Arbelaitz et al.	[32]
18	Majji and Singh	[29]	39	Kalavri et al.	[23]
19	Vargas et al.	[50]	40	Gkantouna	[14]
20	Leal and Dias	[26]	41	Garcia and Paiva	[13]
21	Mayz et al.	[31]	42	Neelima and Rodda	[35]

4   Results

In this section, we present the results of the analysis and mapping of the selected papers answering our research questions.

4.1   Publications Distribution Over Time (RQ1)

In the set of relevant papers, the first papers to be published date from 2001 [16, 39]. Nonetheless, no paper was selected in 2003 or 2004. The number of publications per year in the studied field increased moderately from 2011 onwards, as can be seen in Fig. 1. Also, we note that there was a peak in 2014 in which six papers were published, however, the average in the period 2011–2016 is four papers per year.

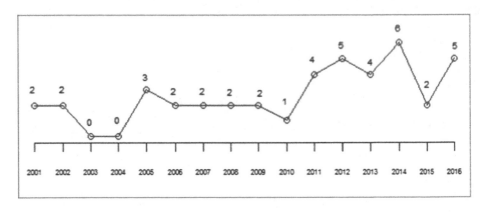

Fig. 1. Number of relevant papers on Web Analytics-Usability per year

4.2   Research Classification (RQ2)

First, we classified the papers into three categories of research according to the context in which it was developed: (i) academic environment, (ii) industrial case study, or (iii) academia-industry partnership. As shown in Fig. 2-a, there are a small share of studies developed in partnership with industry (7 of 42), while we find no case study involving practitioners of the software development industry in this research follow-up. Thus, the greater part of the studies was developed in an academic environment (35 of 42).

Secondly, we classified the studies according to the type of validation into two categories: (i) experimental or (ii) non-experimental. We identified that 30 studies (71%) were experimentally validated, as shown in Fig. 2-b. However we have not yet evaluated the experimental studies in relation to compliance with guidelines proposed for reporting experiments in software engineering [22].

Table 5 summarizes the number of studies classified according to the context and type of research. We noticed that most experimental studies have been performed in academic contexts (24 of 30).

Table 5. Research classification according to context and research type

Context	Type	
	Experimental	Non-experimental
Academia	24–69%	11–31%
Academia-Industry	6–86%	1–14%

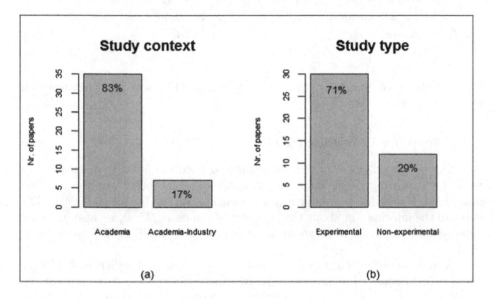

Fig. 2. Research classification

4.3   Research Focus (RQ3)

As for the research focus, when analyzing the selected set of papers, we identified three predominant type of research. Table 7 presents the papers distributed by these categories. More than half of the papers (52.4%) are focused on (i) "Tools for capturing navigation data" of which one part deals with both capture and data analysis (8 of 22), while another part (7 of 22) describes only about data capture mechanisms on user navigation. Kiura et al. [24], for instance, implements a tool that captures the user's navigation and reproduces the data captured for further analysis, which enables to product owners to improve the product usability. In another example, Zhu et al. [55] propose an user behavior analyzer, which can help the product owner find faults without having to reproduce the navigation.

Also, we identified that 41% of papers are divided between (ii) "Analysis of navigation from existing tools" and (iii) "Real-time page customization". With respect to the real-time page customization, we note that tools of capturing are as needed as the contextual analysis. Huang et al. [17] uses these user characteristics, for instance, to improve results of search engines. And finally, we found

Table 6. Research focus of the selected papers

Research focus		
#	Research focus	Papers
22	Tools for capturing navigation data	$[1,2,4,6,8,15-18,23,24,26,27,33,34,$ $37,39,44,46,48,53,55]$
10	Analysis of navigation from existing tools	$[3,5,9,12,13,35,43,49,50,52]$
7	Real-time page customization	$[10,28-32,40]$
3	Other matters	$[11,14,54]$

out that the three remaining papers refer to matters that are out of our research scope (Table 6).

4.4 Means for Capturing (RQ4)

As aforementioned, most of the studies are focused specifically on means and tools for capturing usage information. However, some other studies – whose main focus is not the means of capture – mention the type of technology used to capture the information about the behavior of the users. Thus, we also included these studies in our analysis about means of capture, totalling 32 papers examined.

We have identified four types of means for capturing: (i) JavaScript, (ii) Proxy Logger, (iii) Browser Plugin, and (iv) Java Applet. In the Fig. 3, we present the result of the mapping of such technologies distributed over time. Additionally, we listed the papers according means of capture mentioned, as showed in the Table 7.

Table 7. Means for capturing user navigation.

Types of means for capturing		
#	Means	Papers
21	JavaScript	$[1-6,9,10,15,17,18,24,26,30,34,39,44,49,50,52,55]$
8	Proxy/Server Logger	$[8,16,29,32,33,35,43,53]$
3	Browser Plugin	$[37,46,48]$

A large part of the studies (21 of 32–66%) uses the JavaScript technology as the main means to capture of client logs. Of these studies that use JavaScript, 13 studies (62%) insert JavaScript directly into the source code of the application, while the other 8 studies (38%) it is injected through a proxy (Table 4-a).

The second most commonly used means of capture is the Proxy Logger. We found four studies related to Server Logger $[32,33,35,43]$, which added to the Proxy Logger studies. Also, we found three studies about Browser Plugin,

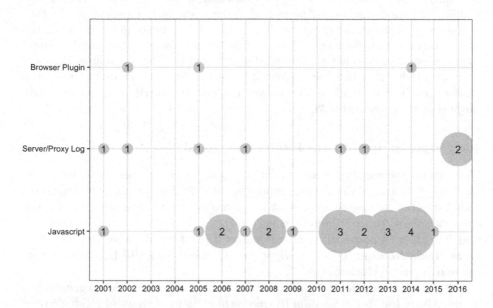

Fig. 3. Means for capturing user navigation over time

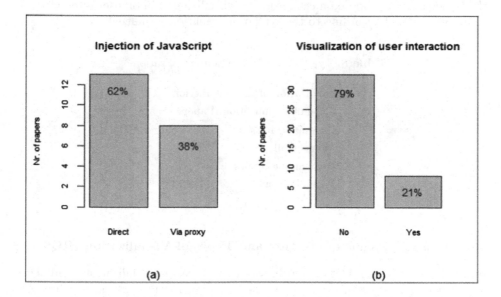

Fig. 4. Means of capture and visualization mechanisms

including a study about Java Applet published in 2001. Shahabi and Banaei-kashani [46] used the Java Applet technology to create a remote agent that transparently runs at the client machine, where is uploaded only once when the user enters the web-site. This agent tracks the user interactions only during a single session and does not store any information at the client machine. But over time, this technology has been replaced by technologies which do not rely on a browser plugin, once modern browser vendors decided restrict and reduce plugin support in their products [38].

4.5 Visualization of User Interaction (RQ5)

Mechanisms for viewing recorded information about the user interaction by tracking of screen activities – e.g., mouse movements, key presses, and log traces – are important to enable a posterior and more in-depth analysis on user experience with the software. However, when analyzing the set of studies, we observed that few papers report such mechanisms (see Table 4-b). We have identified four types of visualization: (i) page navigation, (ii) mouse trail, (iii) heat map graphics, and (iv) page elements.

Page navigation refers to the accessed pages by an user in order to accomplish a goal. Mouse trail is the result of the mouse movement of an user in a single page in order to identify it's path to reach a goal. Heat map graphics is a visual way to represent where are the users focusing in a certain page, by collecting mouse data. And, page elements refers to the HTML elements each user is accessing, which can be used for data mining techniques or others purposes. Table 8 presents the 9 studies that proposed mechanisms for visualizing data on user interaction, which we mapped according to the type of information visualized.

Table 8. Type of user interaction visualization.

Visualization of user interaction		
#	Type of visualization	Papers
3	Page navigation	[16,37,53]
1	Mouse trail	[5]
3	Heat map graphics	[4,6,24]
3	Page elements	[17,24,44]

4.6 Mapping Means of Capture and Types of Visualization (RQ6)

In order to verify the relationship between the types of visualizations and the means of capturing, we have analyzed the 9 studies that present mechanisms to visualize the data about the user interaction, as shown in Fig. 5. According to the means used to capture the information, we can notice that there are restrictions on the types of information visualized. In other words, proxy logger and

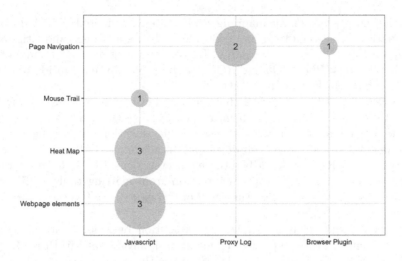

Fig. 5. Mapping means of capture and types of visualization.

browser plugin-based technologies offer limited resources, while *JavaScript* is a technology that allows us to implement more complex data capture functionality.

Niño et al. [37] proposed a client-side browser-embedded tool to capture and replay only user navigation sessions. While, Atterer et al. [5] use JavaScript tracking code in order to capture data about mouse movements, keyboard input and more, and then, mouse trails are combined to a screenshot of the website to show the user navigation.

5 Final Remarks and Limitations

In this paper we present the findings of a systematic mapping carried out to get an overview of existing research on software analytics for web usability, focusing on techniques, methods and tools commonly employed to capture and analyze navigation behaviour and usage data. We have selected and analyzed 42 relevant papers published between 2001 and 2016.

With regards to the threats to validity of our mapping study, we know that the selection of search terms and digital libraries can exclude some relevant studies in our search results. Furthermore, we searched only in a single library, and studies published in proceedings not indexed by this digital library may not appear in our analysis. Nonetheless, in this first stage, this was not our main concern since the intention was to have an overview about the research area.

Another limitation of this work is that we did not define a criteria for quality assessment of included studies based on the aims and research questions of our study. As regards the extraction and categorization process, it was carried out by the first author, while the first coauthor provided input to resolve ambiguities during the process.

Furthermore, a systematic mapping study can be considered less reliable than a systematic review when we usually consider only the abstracts and titles alone rather than the full text of each paper. However, aiming to answer some of our research questions (RQ4, RQ5 and RQ6), we had to analyze a considerable part of the text besides the title and abstract.

In response to RQ1, we found that the number of publications in this field increased moderately from 2011 onwards, and the average of the last six years is four papers per year. Regarding the research classification (RQ2), the majority of experimental studies have been developed in academic context.

To respond to RQ3, we identified three research focus: (i) "tools for capturing navigation data", (ii) "analysis of navigation from existing tools", (iii) "real-time page customization". More than half of the papers are focused on tools for capturing navigation data.

Concerning the means for capturing navigation behaviour and usage data (RQ4), we identified three types of technologies: (i) JavaScript, (ii) Proxy Logger, and (iii) Browser Plugin. However, JavaScript is the most used technology.

With respect to the visualization of user interaction (RQ5), there are few studies that presented some mechanism for visualization of these data. And, analysing the relationship between capture technologies and data visualization (RQ6), we can notice that there are restrictions on the types of information visualized depending on the technology employed.

As future work, we intend to carry out a systematic literature review to answer more specific issues. With this purpose and based on the results of this mapping, we will adjust our search terms as well as consider more research sources, and also look at the references of the relevant papers.

Acknowledgments. This work was supported by FAPESP [grant number 2014/25779-3].

References

1. Apaolaza, A.: Identifying emergent behaviours from longitudinal web use. In: Uist, pp. 53–56 (2013)
2. Apaolaza, A., Harper, S., Jay, C.: Understanding users in the wild, pp. 13:1–13:4 (2013)
3. Apaolaza, A., Harper, S., Jay, C.: Longitudinal analysis of low-level web interaction through micro behaviours. In: Proceedings of the 26th ACM Conference on Hypertext and Social Media - HT 2015, pp. 337–340 (2015)
4. Arroyo, E., Selker, T., Wei, W.: Usability tool for analysis of web designs using mouse tracks. In: CHI 2006 Extended Abstracts on Human Factors in Computing Systems CHI 2006, no. (3), p. 484 (2006)
5. Atterer, R., Wnuk, M., Schmidt, A.: Knowing the user's every move. In: Proceedings of the 15th International Conference on World Wide Web - WWW 2006, p. 203 (2006)
6. Breslav, S., Khan, A., Hornbæk, K.: Mimic: visual analytics of online micro-interactions. In: Proceedings of the 2014 International Working Conference on Advanced Visual Interfaces - AVI 2014, pp. 245–252 (2014)

7. Buse, R.P.L., Zimmermann, T.: Information needs for software development analytics, pp. 987–996 (2012)
8. Chui, C.K., Li, C.H.: Navigational structure mining for usability analysis, pp. 126–131 (2005)
9. Cuddihy, E., Wei, C., Barrick, J., Maust, B., Bartell, A.L., Spyridakis, J.H.: Methods for assessing web design through the internet. In: Proceedings of ACM CHI 2005 Conference on Human Factors in Computing Systems, vol. 2, pp. 1316–1319 (2005)
10. Deufemia, V., Giordano, M., Polese, G., Tortora, G.: Inferring web page relevance from human-computer interaction logging, pp. 653–662 (2006)
11. Ding, Z., Jiang, M., Pu, G., Sanders, J.W.: Modelling and verification of web navigation. In: Gaedke, M., Grossniklaus, M., Díaz, O. (eds.) ICWE 2009. LNCS, vol. 5648, pp. 181–188. Springer, Heidelberg (2009). doi:10.1007/978-3-642-02818-2_13
12. Papatheocharous, E., Belk, M., Germanakos, P., et al.: Towards implicit user modeling based on artificial intelligence, cognitive styles and web interaction data. Int. J. Artif. Intell. Tools **23**(2), 1–21 (2014)
13. Garcia, J.E., Paiva, A.C.R.: An automated approach for requirements specification maintenance. Adv. Intell. Syst. Comput. **444**(March), 827–833 (2016)
14. Gkantouna, V.: Mining interaction patterns in the design of web applications for improving user experience, pp. 219–224 (2016)
15. Guarino de Vasconcelos, L., Coelho dos Santos, R.D., Baldochi, L.A.: Exploring client logs towards characterizing the user behavior on web applications. In: Proceedings of SPIE - The International Society for Optical Engineering, p. 8758 (2013)
16. Hong, J.I., Heer, J., Waterson, S., Landay, J.A.: WebQuilt: a proxy-based approach to remote web usability testing. ACM Trans. Inf. Syst. **19**(3), 263–285 (2001)
17. Huang, J., White, R.W., Buscher, G., Wang, K.: Improving searcher models using mouse cursor activity. In: Proceedings of the 35th International ACM SIGIR Conference on Research and Development in Information Retrieval - SIGIR 2012, p. 195 (2012)
18. Ignatova, E.D., Brinkman, W.P.: Clever tracking user behaviour over the web: enabling researchers to respect the user. In: BCS-HCI 2007 Proceedings of the 21st British HCI Group Annual Conference on People and Computers: HCI...but not as We Know it - vol. 2, 2 September 2007, pp. 179–182 (2007)
19. ISO/IEC: Iso/iec std. 9241-11: Ergonomic requirements for office work with visual display terminals. Part11: "guidance on usability". Technical report, International Organization for Standardization (1998)
20. ISO/IEC: Iso/iec std. 9126-1: Software engineering - product quality. Technical report, International Organization for Standardization (2001)
21. ISO/IEC: Iso/iec std. 25010-3: Systems and software engineering: Software product quality and system quality in use models. Technical report, International Organization for Standardization (2009)
22. Jedlitschka, A., Ciolkowski, M., Pfahl, D.: Reporting experiments in software engineering. In: Shull, F., Singer, J., Sjøberg, D.I.K. (eds.) Guide to Advanced Empirical Software Engineering, pp. 201–228. Springer, Heidelberg (2008)
23. Kalavri, V., Blackburn, J., Varvello, M., Papagiannaki, K.: Like a pack of wolves: community structure of web trackers. In: Karagiannis, T., Dimitropoulos, X. (eds.) PAM 2016. LNCS, vol. 9631, pp. 42–54. Springer, Cham (2016). doi:10.1007/978-3-319-30505-9_4

24. Kiura, M., Ohira, M., Matsumoto, K.: Webjig: an automated user data collection system for website usability evaluation. In: Jacko, J.A. (ed.) HCI 2009. LNCS, vol. 5610, pp. 277–286. Springer, Heidelberg (2009). doi:10.1007/978-3-642-02574-7_31
25. Kumar, L., Singh, H., Kaur, R.: Web analytics and metrics: a survey. In: Proceedings of the International Conference on Advances in Computing, Communications and Informatics, pp. 966–971. ACM (2012)
26. Leal, J.P., Dias, H.: A framework to develop meta web interfaces. In: Proceedings of the IADIS International Conference WWW/Internet 2011, ICWI 2011, pp. 301–308 (2011)
27. Li, W., Cao, G., Qin, T., Cao, P.: A hierarchical method for user's behavior characteristics visualization and special user identification. In: IEEE International Conference on Networks, ICON (2013)
28. Luna, E.R., Garrigós, I., Rossi, G.: Capturing and validating personalization requirements in web applications. In: 2010 1st International Workshop on the Web and Requirements Engineering, WeRE 2010, pp. 13–20 (2010)
29. Majji, S., Singh, S.R.: Proxy server: integrating client side information and query (2011)
30. Mao, J., Liu, Y., Zhang, M., Ma, S.: Estimating credibility of user clicks with mouse movement and eye-tracking information. In: Zong, C., Nie, JY., Zhao, D., Feng, Y. (eds.) Natural Language Processing and Chinese Computing. Communications in Computer and Information Science, vol. 496, pp. 263–274. Springer, Heidelberg (2014)
31. Mayz, M.A., Curtino, D.M., De la Rosa, A.: Avoiding laboratories to collect usability data: two software applications. In: Sistemas Y Tecnologias De Informacion, vols. 1 and 2, pp. 145–149 (2012)
32. McKerlich, R., Ives, C., McGreal, R.: Measuring use and creation of open educational resources in higher education. Int. Rev. Res. Open Distance Learn. **14**(4), 90–103 (2013)
33. Mor, E., Minguillon, J., Santanach, F.: Capturing user behavior in e-learning environments. In: 3rd International Conference on Web Information Systems and Technologies (WEBIST 2007), pp. 464–469 (2007)
34. Nakano, A., Tanaka, A., Akiyoshi, M.: A preliminary study of relation induction between HTML tag set and user experience. In: Kurosu, M. (ed.) HCI 2014. LNCS, vol. 8512, pp. 49–56. Springer, Cham (2014). doi:10.1007/978-3-319-07227-2_6
35. Neelima, G., Rodda, S.: Predicting user behavior through sessions using the web log mining. In: 2016 International Conference on Advances in Human Machine Interaction (HMI 2016), pp. 1–5, March 2016
36. Nielsen, J., Loranger, H.: Prioritizing Web Usability. New Riders Publishing, Thousand Oaks (2006)
37. Niño, I.J., De La Ossa, B., Gil, J.A., Sahuquillo, J., Pont, A.: CARENA: a tool to capture and replay web navigation sessions. In: 3rd IEEE/IFIP Workshop on End-to-End Monitoring Techniques and Services, E2EMON 2005, pp. 127–141 (2005)
38. Oracle'blogs: Moving to a plugin-free web (2015). https://blogs.oracle.com/java-platform-group/entry/moving_to_a_plugin_free
39. Paternò, F., Paganelli, L.: Remote automatic evaluation of web sites based on task models and browser monitoring. In: CHI 2001 Extended Abstracts on Human Factors in Computing Systems, pp. 283–284 (2001)
40. Peska, L.: User feedback and preferences mining. In: Masthoff, J., Mobasher, B., Desmarais, M.C., Nkambou, R. (eds.) UMAP 2012. LNCS, vol. 7379, pp. 382–386. Springer, Heidelberg (2012). doi:10.1007/978-3-642-31454-4_41

41. Petersen, K., Feldt, R., Mujtaba, S., Mattsson, M.: Systematic mapping studies in software engineering. EASE **8**, 68–77 (2008)
42. Petersen, K., Vakkalanka, S., Kuzniarz, L.: Guidelines for conducting systematic mapping studies in software engineering: an update. Inf. Softw. Technol. **64**, 1–18 (2015)
43. Premchaiswadi, W., Romsaiyud, W.: Extracting weblog of Siam university for learning user behavior on mapreduce. In: 2012 4th International Conference on Intelligent and Advanced Systems (ICIAS 2012), vol. 1, pp. 149–154, June 2012
44. Rivolli, A., Marinho, D.A., Pansanato, L.T.E.: WAUTT: uma ferramenta para o rastreamento da interação do usuário com aplicações interativas web. In: Companion Proceedings of the XIV Brazilian Symposium on Multimedia and the Web, pp. 179–181 (2008)
45. Scopus, S.: Content coverage guide (2016)
46. Shahabi, C., Banaei-Kashani, F.: A framework for efficient and anonymous web usage mining based on client-side tracking. In: Kohavi, R., Masand, B.M., Spiliopoulou, M., Srivastava, J. (eds.) WebKDD 2001. LNCS, vol. 2356, pp. 113–144. Springer, Heidelberg (2002). doi:10.1007/3-540-45640-6_6
47. Tullis, T., Albert, W.: Measuring the User Experience, Second Edition: Collecting, Analyzing, and Presenting Usability Metrics, 2nd edn. Morgan Kaufmann Publishers Inc., San Francisco (2013)
48. Uchida, H., Swick, R., Sambra, A.: The web browser personalization with the client side triplestore. In: Mika, P., et al. (eds.) ISWC 2014. LNCS, vol. 8797, pp. 470–485. Springer, Cham (2014). doi:10.1007/978-3-319-11915-1_30
49. Vargas, A., Weffers, H., Da Rocha, H.V.: Discovering and analyzing patterns of usage to detect usability problems in web applications. In: International Conference on Intelligent Systems Design and Applications, ISDA (Ic), pp. 575–580 (2011)
50. Vargas, A., Weffers, H., da Rocha, H.V.: A method for remote and semi-automatic usability evaluation of web-based applications through users behavior analysis. In: ACM Proceedings of the 7th International Conference on Methods and Techniques in Behavioral Research - MB 2010, pp. 1–5 (2010)
51. Vargas, A., Weffers, H., da Rocha, H.V.: Analyzing user interaction logs to evaluate the usability of web applications. In: 2011 3rd Symposium on Web Society, pp. 61–67. IEEE (2011)
52. Vasconcelos, L.G.D., Santos, R.D.C.D.: Classifying user experience of web applications in real time using client logs (2014)
53. Waterson, S.J., Hong, J.I., Sohn, T., Landay, J.A., Heer, J., Matthews, T.: What did they do? Understanding clickstreams with the WebQuilt visualization system. In: Proceedings of the Working Conference on Advanced Visual Interfaces AVI 2002, p. 94 (2002)
54. Zahoor, S., Rajput, D., Bedekar, M., Kosamkar, P.: Inferring web page relevancy through keyboard and mouse usage. In: 2015 International Conference on Computing Communication Control and Automation, pp. 474–478 (2015)
55. Zhu, Z., Wang, Q., Xu, H., Wang, H., Bing, L.: Research on a method of Ajax-based web user behavior collection. In: 2008 3rd International Conference on Pervasive Computing and Applications, ICPCA 2008, vol. 2, pp. 965–969 (2008)

Representing Contexual Relations with Sanskrit Word Embeddings

Ishank Sharma[1], Shrey Anand[1], Rinkaj Goyal[1], and Sanjay Misra[2(⊠)]

[1] USICT, GGS Indraprastha University, New Delhi, India
[2] Covenant University, Ota, Nigeria
sanjay.misra@covenantuniversity.edu.ng

Abstract. Language processing of Sanskrit presents various challenges in the field of computational linguistics. Prosodical, orthographic and inflectional complexities encountered in Sanskrit texts makes it difficult to apply linguistic analysis methods relevant for western European languages. The inadequacy of contemporary computational approaches in the analysis of Sanskrit language is vivdly apparent. In this exposition, we focus on the challenge of learning syntactic and semantic similarities in a rich Sanskrit literature. We present a simple yet effective approach of representing Sanskrit words in a continuous vector space. We utilise word embeddings in similarity, compositionality and visualization tasks to test its efficacy. Experiments show that our method produces interpretable vector offsets exhibiting shared relationships.

1   Introduction

The Sanskrit language is believed to be originated in 1200-1000 BCE Rigvedic period. Since then, this ancient Indo-Aryan language gradually evolved over a period of 3000 years expanding its transitional vocabulary and rich morphological structures. Through the ages, Sanskrit has been recorded and preserved in the form of literary traditions. These vast text corpora are a primary source for understanding the political, social and cultural history of pre-modern India and the world. Scholars often distinguish Sanskrit into Vedic Sanskrit and Classical Sanskrit. Classical or Panini's Sanskrit developed out of earlier Vedic dialects (written in Vedic Sanskrit) in the first millennium BCE. Dakshiputra Panini's grammar Ashtadhyayi ostensibly deals with the Sanskrit language. Panni's grammar defines Classical Sanskrit by a set of rules, recursive and productive in nature [13]. Sanskrit has rich morphology compositions - both inflectional as well as derivational. This morphological ambiguity is introduced on three levels [10]- Inherent (isolated Sanskrit forms), Sandhi (words merged through euphonic rules) and bahuvrïhi (neither word of the compound refers to the actual idea that it describes). Hence, there exists a requirement to parse both the inflectional as well as derivational information to achieve the correct structure. In spite of the existence of a formally defined and well-described grammar, such challenges complicate the development of computational tools for the analysis of Sanskrit texts.

O. Gervasi et al. (Eds.): ICCSA 2017, Part VI, LNCS 10409, pp. 262–273, 2017.
DOI: 10.1007/978-3-319-62407-5_18

Sanskrit has a remarkably rich linguistic tradition of over more than three millennia, which remains under-investigated. Our work aims at contributing to the computational processing of the language and understanding contextual relationships in varied Sanskrit literature of India's Vedic and Classical eras.

We present a more detailed account of background work in NLP and Sanskrit literature in the next section: Related Work. Section 3 expounds our proposed methodology encompassing data preprocessing, word embeddings and visualisation techniques. Experiments conducted with the framework have been described in Sect. 4 and their results in Sect. 5. Finally, we summarise our analysis with the scope for future work in the Sect. 6.

2   Related Work

Natural language processing (NLP) systems have evolved significantly over the years. Conventionally, these systems used words as discrete atomic symbols ignoring the relationships that may exist between them. N-gram model, a statistical approach of modelling language is an example of such systems [19]. Although an effective and simple approach, it is usually dependent on the quality and size of the data to deliver reasonable results. On the contrary, vector space modeling is a more complex technique that outperforms N-gram models by exploiting the information of relationships between words. The vector space representation maps (embeds) semantically similar words to nearby points. An extensive computational study of the language English has resulted in several methods to represent words in a continuous vector space [2,6,20]. Mikolov et al. [22] in their pioneering work, presented two state-of-the-art models Continuous Bag of words (CBOW) and Continuous Skip-gram Model, to derive word vector embeddings[1] from large datasets. Both of these architectures follow a similar learning mechanism for modelling language to find syntactic and semantic word similarities. While training the CBOW architectural network, multiple words in the source context (in any order) are used to classify the target word. On the contrary, skip-gram model predicts the context words within a range of a given input word (see Fig. 1).

Tomas Mikolov and colleagues termed vector representations of the words as "word2vec" (w2v) in association to their software[2]. In their later work, several extensions including a negative sampling objective were used to enhance both the training speed and the quality of the vectors [7,21].

Although most work in computational language processing is concentrated on European languages [4], during the last two decades, Sanskrit has gained considerable attention from the linguistics community [3]. Researchers have developed multiple tools for text parsing using statistical as well as stochastic methods [8,11,23]. Recent works have explored Sanskrit word segmentation with path constrained random walks [16], statistical machine translation [25] and Adaptor Grammars combined with ensemble based approach for compound type identification in Sanskrit literary works [17]. Other studies highlight the challenges in

---

[1] A word represented as dense vector.
[2] https://code.google.com/archive/p/word2vec/.

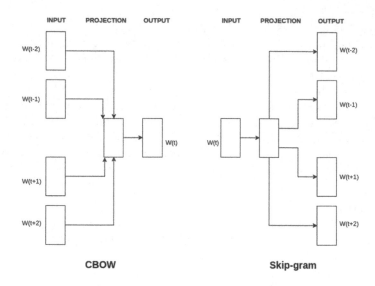

Fig. 1. CBOW and Skip-gram architectures

development and analysis of Hindi wordnet in association with other Indic languages including Sanskrit [14] and implicit processing of linear prediction (LP) residual signal for discriminating Sanskrit speech [24].

Word vector embeddings have been applied for morphological analysis [10] and detecting sentence boundaries in Sanskrit sentences using recurrent neural networks [9]. However, in this work, we use word embeddings to interpret contextual relations in Sanskrit literature.

3   Proposed Framework

3.1   Data Pre-processing

We use Unicode Devanagari characters in our dataset to preserve the original transcription of texts. After collecting Sanskrit literary texts (Table 1) from the web in Unicode format, we further perform basic text pre-processing such as removing HTML tags, English alphabets (a-zA-Z), Numerals (0–9), certain symbols(;, .!@\" < ~ >' $#%& * \n), Devanagari numerals (०१२३४५६७८९) and pūrṇavirāma[3] (॥ ।). Since the dataset contains both Vedic and classical Sanskrit writings, we do not apply saṃdhi viccheda[4] to break words at their corresponding word boundaries adhering to conserving lexical information of each word. Table 1 lists total number of words obtained after pre-processing.

---

[3] Period symbol in Sanskrit.
[4] Sandhi splitting.

Table 1. Literary works in our dataset.

Literature	Word tokens
amarakośa	13,584
bhagavad-gītā	6,774
brahma purāṇa	1,39,717
mahābhārata	6,82,582
Numbers descriptions	2212
pañcatantra	42,472
rāmāyaṇa	173937
skanda purāṇa	22,657
todala tantra	3746
Vedas	33,512

3.2 Word Embeddings

Computational language processing, initially developed for western European languages, does not encapsulate suitable rules required for analysing the complex and rich composition of Sanskrit. Morphological relationships dictate the underlying association of this language in contrast to western European languages where positional structure constrains the word order and in turn interpretability. Dependency parsing is more suitable for free word order languages like Sanskrit [1]. Moreover, paryāyavācī[5] are extensively used in Sanskrit depending upon the context and explication. For example, mahābhārata uses 35 synonyms to refer the warrior Arjuna, or 14 for the concept "mountain" [9]. Acknowledging these challenges, we extract low dimensional embeddings to examine overall similarities between word compositionality of sentences in early Sanskrit works. We experiment with distributed word embeddings using *genism*[6] library. The CBOW (Continuous Bag-of-Words) approach is suitable[7] for this task because of free word order nature of Sanskrit and, consequently to determine a word given its context. Thus, mitigating the before-mentioned problems even with small but lexically rich dataset. A CBOW model for 100 iterations with fixed window size 7 and minimum word frequency 5 (ignoring words less than this frequency) is trained using the dataset.

---

[5] Sanskrit word for synonym. For example- आरोहण (ArohaNa) can refer to words such as mount, climb, ride, depending upon the context.

[6] https://radimrehurek.com/gensim/models/word2vec.html.

[7] Although, we also applied Skip Gram model with same settings, the resultant word vectors were not of adequate quality.

3.3 Visualization

The visualization of the learned embeddings through dimensionality reduction provides a graphical interpretation of the vector space representation of words. This process reveals clusters suggesting that semantically similar words are proximately embedded. Our objective in visualizing word vectors is to preserve the relationship among high-dimensional data when mapped to low-dimensional space. This approach can be formulated as visualization of manifold embeddings. A state-of-the-art algorithm t-SNE [18] visualizes data by preserving local structures while showing global information. Barnes-Hut t-SNE (bhtsne), a tree-based variation of original t-SNE outperforms it with respect to both speed and visualization capacity [27]. We apply bhtsne[8] with different perplexities. Furthermore, we also experiment with K-means clustering to represent word vectors in continuous space as discussed in next section.

4 Experiments

The resultant vocabulary achieved from our CBOW model comprises of 23,407 unique word vectors. Using these embeddings, we experiment with following variants of our approach:

1. **w2v+PCA+K-means-** In this method, we first apply Principal Component Analysis (PCA) to reduce word embeddings to two dimensions. Then, we use K-means clustering with 50, 75 and 100 clusters. None of the settings performed well in identifying related word groups. Reducing sparse non-linear manifold data to a 2-dimensional linear subspace with PCA induces information loss [5,15,29]. Moreover, since K-means clustering utilises distance metrics and assumes similar data variance to compute clusters [12], local neighbourhood information may disappear in the global representation of data. Another limitation of its clustering mechanism is its dependency upon the choice of fixed number of clusters and initial cluster centres [28].
2. **w2v+PCA+bhtsne-** We use this combined approach with fixed theta value (0.5) and reducing the word embeddings size from 270 dimensions to 50 dimensions using PCA. We examined bhtsne with different perplexities- 30, 40 nd 50. The cost function becomes nearly constant after executing the algorithm for 50,000 iterations. In our experiments, although perplexity measures 40 and 50 gave similar visualizations, we choose perplexity = 40 (Refer Fig. 2) for further analysis as it offered more interpretability in visualizations of similar words from w2v.

---

[8] https://github.com/lvdmaaten/bhtsne.

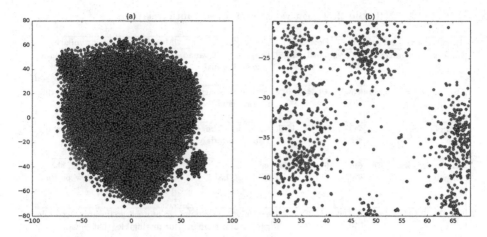

Fig. 2. Word vectors bhtsne representation with perplexity = 40 (a) All words in vocabulary. (b) Zoomed version of subspace.

5   Results

5.1   Similarity

Table 2 shows a target word and its top four closest words in similarity relationship (TOP4)[9]. The similarity between two words is calculated using cosine similarity (Eq. 1),

$$sim_{w2v}(w_t, w_j) = \frac{w_t \cdot w_j}{||w_t|| \cdot ||w_j||} \tag{1}$$

where w_t and w_j are corresponding word embeddings of t and j.

Semantic relationships for most words are inferred using TOP4. However, in some cases, we have to examine more than TOP4 since the Sanskrit language has several inflections and synonyms. For example in Table 2, for गच्छाम (go), गच्छामहे (gacchāmahe; first person plural present of go) with similarity value = 0.357 and गमिष्यामो (gamiṣyāmo; shall go or shall move) with similarity value = 0.3496, appears in TOP8. Although both of these words have the same root, their similarity values are low. For a few words, we cannot capture any direct conceptual or syntactical associations. The literal meaning of paśuḥ is an animal or organism which is associated with unrelated TOP4 words (Table 2).

Furthermore, an alternative approach to determining dissimilar word from a words group is to compute dot product with the unit mean vector of word group. The most dissimilar offset will have the minimum contribution to the relationship. Table 3 displays dissimilarity relationships for words in a word groups.

---

[9] From here on we will use TOPn to refer n closest similar words.

Table 2. Target words and its semantic relations

Target word	TOP4 (Devanagari)	TOP4 (IAST)	Meaning (semantic similarity)
Ganges: गङ्गा (gaṅgā)	भागीरथी, देवनदी, गोदावरी, गौरी	bhāgīrathī, devanadī, godāvarī, gaurī	river, divine river, river, goddess pārvatī (sister of gaṅgā)
Organism or animal: पशुः (paśuḥ)	वेदवित्, स्वर्विदः, मित्रा, विशुद्ध	vedavit, svarvidaḥ, mitrā, viśuddham	knower of the Vedas, celestial, friend, pure or clear
God: देवः (devaḥ)	देवदेवः, भगवान्, विष्णुः, वरदः	vadevaḥ, bhagavāñ, viṣṇuḥ, varadaḥ	god of gods, god, lord viṣṇuḥ, wish granter
Many arrows: बाणाः (bāṇāḥ)	सैनिका, नेमे, शिलाशिताः, बाणा	sainikā, neme, śilāśitāḥ, bāṇā	soldier, a part or divide, sharpened on stones, arrow or hind part of arrow
Go: गच्छाम (gacchāma)	गच्छावो, गच्छामो, गम्यतां, तन्नः	gacchāvo, gacchāmo, gamyatāṃ, tannaḥ	go protect, go towards, accesible or intelligible, that + us

Table 3. Dissimilar word in a word group

Relation	Words	Doesn't match
Celestial bodies	sūryaḥ, candraḥ, bālaḥ, pṛthvīḥ sun, moon, boy, earth	bālaḥ boy
Masculine names	arjunaḥ, rāmaḥ, rāvaṇaḥ, durgā Arjuna, Rama, Ravana, Durga	durgā Durga
Places in India	ayodhyā, kurukṣetra, mathurā, bhārata Ayodhya, Kurukshetra, Mathura, India	bhārata India

5.2 Compositionality

The idea of semantic compositionality can be traced to the early works of Indian philosophers of grammar such as Yäska [26]. Word Embeddings exhibit *additive compositionality* [22] capturing linear structures between words by elementwise addition of word vectors. Semantic relations between word pairs can be decomposed into an Element-wise addition of vector offsets sharing meaningful relationships. Algebraic operations on vector representations can express a word pairs analogy- "a is to b as c is to d" (d is unknown) is defined in Eq. 2, where w_i = word embedding for word i and $i \in a, b, c, d$

$$w_d = w_b + w_a - w_c \tag{2}$$

Table 4 shows interesting examples describing word pair analogies derived from the embeddings applying the operation in Eq. 2.

5.3 Visualizing Distributed Projections

Philosophical texts and scriptures of Sanskrit played an essential role in moulding world history. Many languages, traditions, and religions of the modern era

Table 4. Word pair analogy exhibiting compositionality (Eq. 2)

Operation	Result
vec(pārthaḥ) + vec(bālā) - vec(bālaḥ)	vec(rājaputrī)
vec(prince) + vec(young girl) - vec(young boy)	vec(princess)
vec(mātā) + vec(pituḥ) - vec(duhitā)	vec(putravatsalaḥ)
vec(mother) + vec(father's) - vec(daughter)	vec(devoted to son, son loving)
vec(puruṣaḥ) - vec(dharma)	vec(pāpāt)
vec(person) - vec(religion, virtue)	vec(evil or wicked)
vec(īśvaraḥ) - vec(manuṣyaḥ)	vec(prabhavaḥ)
vec(god) - vec(human)	vec(source, cause of existence)
vec(puruṣaḥ) - vec(prītiḥ)	vec(viśvātmā)
vec(person) - vec(sensation, pleasure)	vec(universal soul, super soul of all living entities)

Table 5. Visualised words and their meanings.

Word	आत्मा	योग	मोक्षः	मन्त्रः	बुद्धः	तपः	मृत्युः	विद्या
IAST	ātmā	yoga	mokṣaḥ	mantraḥ	buddhaḥ	tapaḥ	mṛtyuḥ	vidyā
Meaning	soul	yoga	liberation	mantra	buddha	meditation	death	knowledge
Word	शिवः	विष्णुः	वेदः	ब्रह्म	ॐ	पुनर्जन्म	निर्गुणः	वैराग्यं
IAST	śivaḥ	viṣṇuḥ	vedaḥ	brahma	oṃ	punarjanma	nirguṇaḥ	vairāgyam
Meaning	Shiva	Vishnu	Veda	Brahma	OM	rebirth	vicious	asceticism
Word	चेतना	जगत्	कर्म	वेदात्मा	महानात्मा	सर्वगः	जन्म	वेला
IAST	cetanā	jagat	karma	vedātmā	mahānātmā	sarvagaḥ	janma	velā
Meaning	senses	universe	karma	soul of veda	high-souled	heaven	birth	time

are born and established upon the spiritual and theological notions of Sanskrit literary compositions. To investigate these concepts, we visualise 2D projections of embeddings in a continuous space using bhtsne. Table 5 lists words and their respective meanings in English.

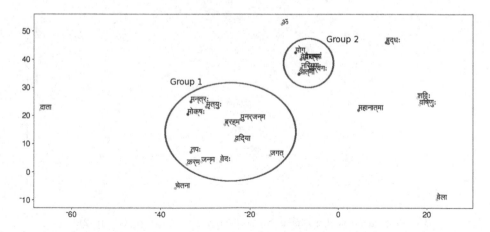

Fig. 3. Word vector groups distribution in embedding space.

Figure 3 illustrates the distribution of words in a continuous vector space. Embeddings having semantic similarities are nearby forming word groups, for example *Group* 1 and *Group* 2.

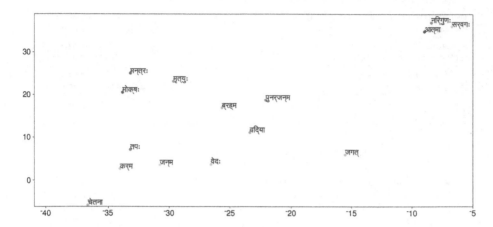

Fig. 4. Shared vector offsets for life-death relationship (Group 1).

Expressions like death, rebirth, liberation, karma, Veda, knowledge, sharing common concepts are in close vicinity. Figure 4 illustrates *Group* 1 corresponding to this word group.

Furthermore, notions such as vicious, asceticism, soul, heaven, yoga, are dispersed adjacent to each other (*Group* 2) represented in Fig. 5.

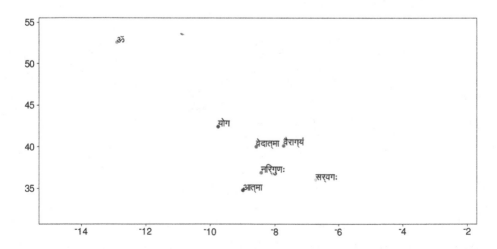

Fig. 5. Shared vector offsets for soul-yoga relationship (Group 2).

Feminine words and their corresponding masculine words from composition-ality task (Sect. 5.2) have embeddings vectors located at opposite directions. Additionally, feminine words tend to group together (Fig. 6).

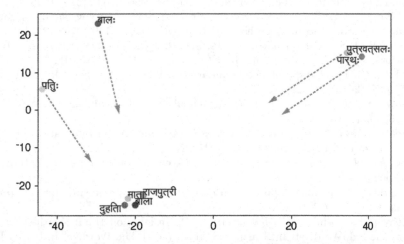

Fig. 6. Feminine and Masculine words. Direction represented from masculine to feminine words.

6 Conclusion

In our work, we present research on learning and representing word embeddings from a Sanskrit text corpora. We apply a recent development in word vector frameworks and also attempt to interpret possible semantic and syntactic relationships between vector offsets. Since the corpus used in this study preserve the contextual and lexical structures within sentences, we successfully assessed the quality of learned representation using analogical and similitude tasks. Given the complexity of Sanskrit morphology and the richness of its vocabulary, our framework performs reasonably well. In our future work, we will further extend our approach to generate *Sense embeddings* for words sense induction and disambiguation systems.

References

1. Begum, R., Husain, S., Dhwaj, A., Sharma, D.M., Bai, L., Sangal, R.: Dependency annotation scheme for Indian languages. In: IJCNLP, pp. 721–726 (2008)
2. Bengio, Y., Ducharme, R., Vincent, P., Jauvin, C.: A neural probabilistic language model. J. Mach. Learn. Res. **3**(Feb), 1137–1155 (2003)
3. Bharati, A., Chaitanya, V., Sangal, R., Ramakrishnamacharyulu, K.: Natural Language Processing: A Paninian Perspective. Prentice Hall of India Pvt. Ltd., New Delhi (1995)

4. Chowdhury, G.G.: Natural language processing. Ann. Rev. Inf. Sci. Technol. **37**(1), 51–89 (2003)
5. Donoho, D.L., et al.: High-dimensional data analysis: the curses and blessings of dimensionality. AMS Math Challenges Lect. **1**, 32 (2000)
6. Elman, J.L.: Finding structure in time. Cogn. Sci. **14**(2), 179–211 (1990)
7. Goldberg, Y., Levy, O.: word2vec explained: deriving Mikolov et al.'s negative-sampling word-embedding method. arXiv preprint arXiv:1402.3722 (2014)
8. Goyal, P., Huet, G.P., Kulkarni, A.P., Scharf, P.M., Bunker, R.: A distributed platform for Sanskrit processing. In: COLING, pp. 1011–1028 (2012)
9. Hellwig, O.: Detecting sentence boundaries in Sanskrit texts. In: Proceedings of COLING (2016)
10. Hellwig, O.: Improving the morphological analysis of classical Sanskrit. WSSANLP **2016**, 142 (2016)
11. Huet, G.: Towards computational processing of sanskrit. In: International Conference on Natural Language Processing (ICON). Citeseer (2003)
12. Jain, A.K.: Data clustering: 50 years beyond k-means. Pattern Recogn. Lett. **31**(8), 651–666 (2010)
13. Kak, S.C.: The paninian approach to natural language processing. Int. J. Approx. Reason. **1**(1), 117–130 (1987)
14. Kashyap, L., Joshi, S.R., Bhattacharyya, P.: Insights on Hindi Wordnet coming from the IndoWordNet. In: Dash, N.S. et al. (eds.) The WordNet in Indian Languages, pp. 19–44. Springer, Heidelberg (2017)
15. Kerschen, G., Golinval, J.C.: Feature extraction using auto-associative neural networks. Smart Mater. Struct. **13**(1), 211 (2003)
16. Krishna, A., Santra, B., Satuluri, P., Bandaru, S.P., Faldu, B., Singh, Y., Goyal, P.: Word segmentation in Sanskrit using path constrained random walks. In: Proceedings of COLING (2016)
17. Krishna, A., Satuluri, P., Sharma, S., Kumar, A., Goyal, P.: Compound type identification in sanskrit: what roles do the corpus and grammar play? WSSANLP **2016**, 1 (2016)
18. van der Maaten, L., Hinton, G.: Visualizing data using t-SNE. J. Mach. Learn. Res. **9**(Nov), 2579–2605 (2008)
19. Manning, C.D., Schütze, H., et al.: Foundations of Statistical Natural Language Processing, vol. 999. MIT Press, Cambridge (1999)
20. Mikolov, T., Kopecky, J., Burget, L., Glembek, O., et al.: Neural network based language models for highly inflective languages. In: IEEE International Conference on Acoustics, Speech and Signal Processing, ICASSP 2009, pp. 4725–4728. IEEE (2009)
21. Mikolov, T., Sutskever, I., Chen, K., Corrado, G.S., Dean, J.: Distributed representations of words and phrases and their compositionality. In: Advances in Neural Information Processing Systems, pp. 3111–3119 (2013)
22. Mikolov, T., Yih, W.T., Zweig, G.: Linguistic regularities in continuous space word representations. In: Hlt-naacl, vol. 13, pp. 746–751 (2013)
23. Mishra, A.: Modelling the grammatical circle of the pāṇinian system of Sanskrit grammar. In: Kulkarni, A., Huet, G. (eds.) ISCLS 2009. LNCS, vol. 5406, pp. 40–55. Springer, Heidelberg (2008). doi:10.1007/978-3-540-93885-9_4
24. Nandi, D., Pati, D., Rao, K.S.: Implicit processing of LP residual for language identification. Comput. Speech Lang. **41**, 68–87 (2017)
25. Pandey, R.K., Jha, G.N.: Error analysis of sahit-a statistical Sanskrit-Hindi translator. Procedia Comput. Sci. **96**, 495–501 (2016)

26. Staal, J.: Sanskrit philosophy of language. In: History of Linguistic Thought and Contemporary Linguistics, pp. 102–136 (1976)
27. Van Der Maaten, L.: Accelerating t-SNE using tree-based algorithms. J. Mach. Learn. Res. **15**(1), 3221–3245 (2014)
28. Žalik, K.R.: An efficient k-means clustering algorithm. Pattern Recogn. Lett. **29**(9), 1385–1391 (2008)
29. Zass, R., Shashua, A.: Nonnegative sparse PCA. Adv. Neural Inf. Process. Syst. **19**, 1561 (2007)

A Mobile-Sensor Fire Prevention System Based on the Internet of Things

Julio César Rosas[1], José Alfonso Aguilar[1]([✉]), Carolina Tripp-Barba[1], Roberto Espinosa[3], and Pedro Aguilar[2]

[1] Señales y Sistemas, Facultad de Informática Mazatlán, Universidad Autónoma de Sinaloa, México, 82120 Mazatlán, Mexico
{julior,ja.aguilar,c.tripp}@uas.edu.mx
[2] Centro de Valoración y Estudios Urbanos Escuela de Ingeniería Mazatlán, Universidad Autónoma de Sinaloa, México, 82120 Mazatlán, Mexico
pedro_a4@uas.edu.mx
[3] Departamento de Computación e Informática, Universidad La Frontera, Temuco, Chile
roberto.espinosa@ufrontera.cl

Abstract. The Internet of Things (IoT) is changing the industrial sectors, business models and processes. In this context, a special area positively affected by IoT is the prevention of risk factors. A permanently connected network of new products, machines, people and organizations, is a technological advance that can be oriented to combat the risks of fires in places of greater vulnerability in the business areas. This paper presents an application project regarding to a fire prevention system in places where temperature can be measured using the new IoT tools. This project and its novel advantages can be helpful in the development of preventive to increase safety in industry.

1 Introduction

The human being has been evolving through the years, a great part of this evolution brought with it important advances for the benefit of society, a transcendental one was the discovering of the fire. With the emergence of fire, important advances were made in the development tools, food processing and preservation. Nowadays, production models in factories bring with them new risk situations that can ocasionate damage in the factory and its workers. This can paralyse the factory activities, affecting life safety and originates environmental damage, which can become problems for the factory stability. Among these dangers we have the fire, defined as an uncontrolled fire occurrence that can affect or burn something that is not intended to burn. The usual damage that affects human life is the inhalation of smoke or by fading caused by intoxication and subsequent severe burns. With regard to fire protection and prevention, the recommendation are preventing the ignition of fire and managing the impact of fire. Fire needs three basic elements to begin to occur: fuel, combustion and oxygen, fuel sources are materials such as; paper, wood and flammable liquids such as gasoline. The

O. Gervasi et al. (Eds.): ICCSA 2017, Part VI, LNCS 10409, pp. 274–283, 2017.
DOI: 10.1007/978-3-319-62407-5_19

combustion can be produced by a spark or other source of heat and finally the oxygen. Just eliminating just one of the elements, a fire can be prevented.

In the industry, specifically in factoires, it is necessary the instalation of fire prevention systems, specially if the factory has places where there is no a constant presence of workers in order to avoid the spread of the fire in case on a fire siniester. In this sense, modern fire-fighting technology has produced several methods for detecting the presence of a fire and alerting humans to its existence, this metods includes fire detection and an alarm in case of fire. The detection and alarm systems aims to fast discover the fire and transmit the alarm-notice in order to start the fire extinction and human evacuation. At present, there are machines, cameras, microphones, clothing, accessories and medical equipment connected to the Internet linked by applications running on mobile devices by using the so-called Internet of Things. IoT is a system of interrelated computing devices, mechanical and digital machines, objects, animals or people that are provided with unique identifiers and the ability to transfer data over a network without requiring human-to-human or human-to-computer interaction. However, the IoT is much more than watches monitoring the heart rate and calories consumed by an athlete to subsequently generate performance statistics. The IoT can have positive implications by allowing the organization and monitoring of tools, machines and people. A thing, in the Internet of Things, can be a person with a heart monitor implant, a farm animal with a biochip transponder, an automobile that has built-in sensors to alert the driver when tire pressure is low or any other natural or man-made object that can be assigned an IP address and provided with the ability to transfer data over a network. Today, a commonly accepted definition for IoT is a dynamic global network infrastructure with self-configuring capabilities based on standard and interoperable communication protocols where physical and virtual "Things" have identities, physical attributes, and virtual personalities and use intelligent interfaces, and are seamlessly integrated into the information network. In this sense, in recent years the evolution of mobile communication has caused a considerable increase in the development an use of mobile applications in scenarios such as social safety, entertainment, social networks, educational fields and tools for prevention in accidents such as fires, earthquakes, hurricanes and tornadoes.

In this work we stated that not having a mobile technology applied in disasters or fire emergencies can affects the time-response in order to ameliorate the danger of a disaster. In this respect, if smoke sensors and mobile applications are combined, the material damaged and human fatalities can be minimized avoiding catastrophes such the one ocurred in the ABC kindergardten in Sonora, Mexico [5,6]. Where the fire claimed the lives of 49 children under 5 years old and another 106 were serious injured by burns, despite the fact that the infrastructure of the place strategically implemented smoke sensors that did not make any difference because these devices work only while there is fire.

Considering this situation, in this work we introduce our IoT Mobile-Sensor Fire Prevention System, combining the development of an mobile application for android systems and using sensors for heat detection, humidity measure and smoke detection. The system presented in this fire prevention research will make a difference when working with temperature sensors that will be connected to a server in real time to monitor the most vulnerable place and manage to eradicate a fire incident. The internet of things has come to evolve communication systems and will be an essential part of this work.

The paper is structured as follows: In Sect. 2 we briefly review the most important milestones to reach to the current state of the art regarding to fire prevention systems. In Sect. 3 we detail the design of the IoT Mobile-Sensor Fire Prevention System. Section 4 an application example. We present our conclusions and future work in Sect. 4.

2   Related Work

In this section latest fire accident detection technologies and intelligent prevention system are discussed. The progress on fire detection technologies has been substantial over the last decade due to advancement in sensors and microelectronics [1]. In [2] a review of progress in various emerging sensor technologies for fire detection and monitoring is elaborated. Mostly fire detection technologies are categorized into two groups, one is vision based [3,4] technique that analyzes video frames and process images to detect fire and another one is sensor based fire detection.

There are several preventive proposals for accidents that have a great social impact and are focused to safeguard human integrity. Among the recent projects, we have an application called "Mobile Guardian" [14], created in Mexico and is available for free download in Android Market, the application contains a panic button, when the user push it, the GPS is activated and send a message to Contacts that previously configured by the user. The service requires the Internet in order to know the location and the user must have at least one contact in the list to use it, the application configuration is easy, just enter the name, email and phone number. The user will receive a text message with the location and also an email with the same information.

With different implementations we find the mobile application "My Police" [18], the electronic app for cell phone, Ipad, etc. Known as Mi Polic'ia, it is an investigation of the Public Security Secretariat of Mexico City, which serves to assist and be a direct link with the emergency forces, since its launch in March 2013, has registered 104 thousand 983 downloads And answered 41, 332 calls to support citizens. The downloads are linked to complaints for the commission of crimes, which reports 1 495 complaints, while the preventive ones for the presence of suspicious persons registers 4 thousand 216. "My Police" details the SSPDF (Secretary of Public Security of the Mexico City) complies The goal of bringing citizens closer to the police authority. However, there is a national and international interest in these issues, which is why work of great importance has been structured.

The NFC mobile [12] life saving application (Madrid, Spain 2012) is based on Near Field Communication (NFC) technology, NFC enables the direct exchange of data between devices at close range and can invoke other wireless data transmission technologies WLAN-WiFi, Bluetooth, RFID, GPRS, HSDPA / HSPA + / HSUPA, etc. This technology was implemented in first time by Prof. Dr. Javier Areitio Bertoln, which is a research-professor of the Faculty of Engineering and Director of the Networks and Systems Research Group of the University of Deusto in 2011. Which in Sooft (NFC) are developing together with the business development company Kinguard Group for Tecfire Doth, a pioneer company and leader in the development, manufacture and approval of glazed protection systems. The solution is to incorporate NFC chips into the systems and products that Tecfire Doth installs in its construction projects. The chips are coded with practical product information (product data sheet, classification technician report, date of production, etc.) as well as information that will improve the safety conditions of a building that needs to be evacuated in case of fire. We can say that the IoT fits perfectly with NFC and radiofrequency services, leaving aside access management through software on a PC.

The Prosegur tool [12], project "Promobile", is a new application that allows users to manage their alarm from a mobile device, as well as to obtain images of their home or business at any time that requires it. The user can also customize the application, in addition to receiving and storing all the responses or warnings that are generated from the system. "Promobile", available in the mobile application markets, is developed for Prosegur Proview +, a security solution for the residential and SME sectors.

Tyco Integrated Fire and Security (Spain and Portugal): is a worldwide company in security and fire protection solutions, made the project VideoEdge Go, a mobile application of the American Dynamics brand for the remote management of video systems VideoEdge NVR surveillance. VideoEdge Go is compatible with iOS devices, including iPhone, iPad, iPod Touch and iPad Mini. Designed to ensure a higher level of security for VideoEdge users, the application provides remote monitoring and analysis capabilities that enable security personnel to receive real-time information about events and act quickly. In the "Alertcops" [17] research, which covers a mobile application, it is part of the SIMASC (Citizen Security Alerts Mobility System) technology platform. Its objective is to universalize access to public security services throughout the country. Society, For any user, irrespective of their language, origin or their hearing or vocal disabilities, can communicate with the Civil Guard and the National Police to alert them about a situation in which they are victims or witnesses. The projection of the mobile applications presented in recent years do not contain a great qualitative contribution in terms of fire accidents, since most only contain information about accidents or emergencies that may arise, as is the case of the first aid app. The red cross, the pre-defined content would not help in the event of a fire and would not serve as a preventive method.

The technological applications presented are very useful today, but it is important to mention that they work when the fire incident is already initiated and only minimizes the impact on the damages without presenting itself as a preventive tool to make any catastrophe void. It is necessary to emphasize that the most effective preventive method for the detection of a fire is the measurement of the temperature of the place. Having a real-time temperature de-rating of a place that needs to be monitored or which presents greater fire vulnerability, marks the difference between an accident, economic damages and loss of human life, or present a preventive tool to rule out any fire.

Summarizing, the Prosegur project is the work that approaches the present research, however, it only monitors a particular place without having a measurement base such as air quality, temperature, proximity or mobility sensors, and measurement Smoke or fire. It is possible to carry out an extensive and in depth investigation on the IoT tools that are being applied and to make an analysis to focus especially on the subject of interest. There is an immense need of implementation of automatic fire detection system to protect lives and assets from fire hazards, in our proposal, the Internet of Things will be integrated to structure an Android mobile application with tools that help to mitigate the tragedies caused by fires in areas where the temperature of the place can be measured and effective as a prevention method. This is described next.

3 Fire Prevention System

Since it is shown in the previous section several research papers have been published about fire detection and prevention system before. In our work is presented a mobile application developed for Android operating system named Ice Cream Sandwich version 4.0 and higher systems [15], the app is used to capture temperature data through a server which will collect information data from a temperature sensor that will be calibrated to the ambient temperature of the defined place (see Fig. 1). The Android app will be programmed to receive Push Notifications, these will be sent each time the sensor calibration is exceeded, each temperature change will be notified in the application and when its change in temperature exceeds the programmed threshold will activate an alert tone so that Preventive measures are taken before a possible fire begins. The remote communication will allow the user to know the temperature of the place without having to be present physically, if the user is outside the city, this could notify a third party to verify the risk zone in case it is Present an alert.

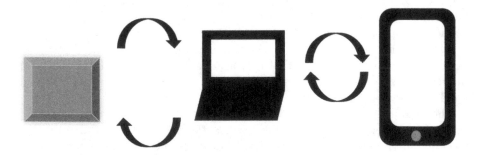

Fig. 1. Structure of remote communication

This work will integrate existing preventive systems and obtain a temperature reading as a preventive tool. This will interpret temperature information data, followed by smoke and after fire. So you will have a way to prevent fires before you can read the quality of the air by means of the smoke sensor that is currently installed. The sensor will be installed strategically at the highest risk location and where the temperature of the sector can be measured. With this dynamic system between user and the Internet of things, it is intended to take a positive turn in the current way of making effective fire prevention.

3.1   Deployment and Installation: Temperature Sensor

It will be taken as an example of application of the remote system for fire prevention, a standard business office, with the aim of implementing the temperature sensor. The sensor will be placed strategically in the most vulnerable areas of an office, which are, the centers of charge of electricity, installation of air conditioners and connectors to charge electronic devices. In the previous image, it is visualized and it is proposed to place the temperature sensor to capture the information. The placement in that space, proposes the agile and real time reading of the sector that is being remotely inspected (see Fig. 2).

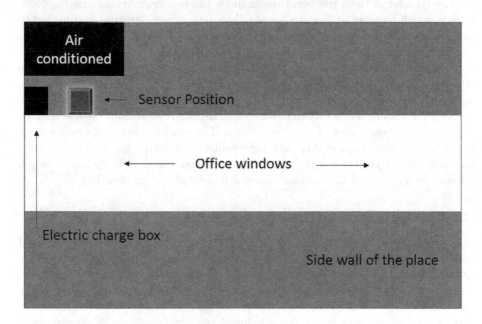

Fig. 2. Location of the temperature sensor in a standard office.

3.2   Transfer of Temperature Data

By specifying the position of the sensor in the most vulnerable place, possible errors in the data will be minimized when remote communication is established.

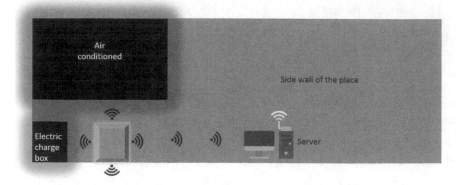

Fig. 3. Collection of temperature data.

The data transfer will be done through the Internet and the destination medium in this case will be a server (Cloud server or local server for testing), when collecting the data the server, it will replicate them to the mobile device that has the Android app (see Fig. 3).

The transfer of data temperature includes the functionality of existing technology, which consists of a smart sensor that detects temperature and humidity changes and the presence of smoke and activates an alarm in the event of a fire.

The emergency text message includes the buildings address, contact information of the buildings legal owner, and the indoor temperature detected, which is expected to make it possible for help to arrive more quickly. The owners of the eight buildings equipped with the IoT service will receive text messages twice a day (at 9 a.m. and 7 p.m.) that include real-time information on the in-door environmental conditions of their buildings. If any of the detected environmental variables (temperature, dust, etc.) exceed the standard level, building owners will immediately receive a warning via text message. Environmental information can also be monitored in real-time using a mobile phone application.

3.3   Data Reception and System Life Cycle

The server will get the temperature information in real time and will be updated within 8 s, the Android mobile application will be programmed to receive updated data. After that, the app will be programmed with a limit of temperature values, if the limit of temperature values is exceeded, an alert (alert push notification) will be sent notifying the user to take preventive measures against any possible fire event, if the value of temperature is stable, the application will update the data in the application in a passive way so that the user can visualize the information when it requires it. Is important to mention that the sensor must be already calibrated to the ambient temperature of the monitored place by means of a set operating tests.

The life cycle of the system would be completed by monitoring the temperature through the sensor, sending the information that the sensor will send to the server, and finally, the data collection of the application by means of data update through the server. On the subject of network security in communication, there will be a complete analysis of the different vulnerability scenarios that may be presented (see Fig. 4).

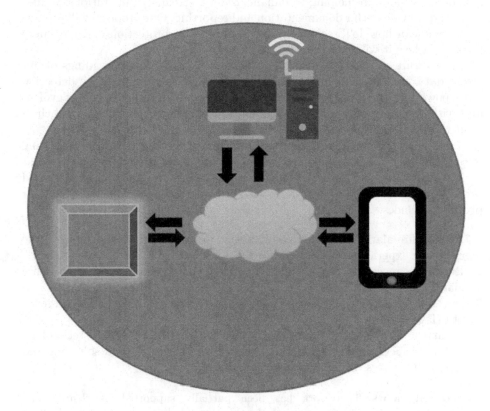

Fig. 4. Remote firewall life cycle.

4 Conclusion and Future Work

The Internet of Things (IoT) begins to transform and offer new technological projections to combat risks of all kinds. IoT technology provides added value to emergency response operations in terms of obtaining efficient cooperation, accurate situational awareness, and complete visibility of resources. In this project, we launch an a proposal that complements the preventive tools against existing fire risks. With the help of the Android mobile application, a collaborative work of remote monitoring is presented, which reduces the need for the user to be physically in the place that is being analyzed in terms of temperature information. The installation options are maximized, because tests can be performed in educational, private, industrial, hotel and personal services sectors.

Today seven billion devices are connected to the Internet. In 10 years that will grow to 30 billion connected devices. These technological advancements are having a transformative impact in virtually every industry. That includes fire protection and life safety, where connected fire alarm panels, smoke detectors, and other devices are helping to enhance safety, streamline operations, reduce costs and provide better documentation and reporting. Our proposal will provide an overview of how IoT technology is improving fire protection and life safety today and how it will evolve in the future.

As a future work is intended to integrate sensors to collect readings of different data: presence, air quality, motion detection and temperature detection more powerful. The future is today, not tomorrow, the Internet of things evolves and with evolution affected and benefited are human beings. The approach of modern technologies depends on our humanity.

Finally, in our work we show how Internet of Things technology is being applied in other industries and increasingly in fire, security and life safety. What technology, policies and protocols are in place to help provide data security and protection for cloud-based solutions with Android app development since our proposal includes:

1. Remote fire alarm system diagnostics that increase uptime and improve the customer experience.
2. Remote connectivity and a mobile app that enable faster, more precise fire alarm system inspections.
3. How fire protection and life-safety codes are changing to reflect the emergence of IoT technology.
4. What the future may hold for expanding the application of IoT technology in fire protection and life safety to deliver smart services and solutions and enable customers to improve their operations.

Acknowledgments. This work has been partially supported by: Universidad Autónoma de Sinaloa (México) through the Programa de Fomento y Apoyo a Proyectos de Investigación (PROFAPI2015/002). The work is derived from a research project in order to obtain B.S. Degree in computer systems by means of a thesis.

References

1. San-Miguel-Ayanz, J., Ravail, N.: Active fire detection for fire emergency management: potential and limitations for the operational use of remote sensing. Nat. Hazards **35**(3), 361–376 (2005)
2. Liu, Z., Kim, A.K.: Review of recent developments in fire detection technologies. J. Fire Protect. Eng. **13**(2), 129–151 (2003)
3. Celik, T., Demirel, H., Ozkaramanli, H., Uyguroglu, M.: Fire detection using statistical color model in video sequences. J. Vis. Commun. Image Represent. **18**(2), 176–185 (2007)
4. Dedeoglu, Y., Töreyin, B.U., Güdükbay, U., Cetin, A.E.: Real-time fire and flame detection in video. ICASSP **2**, 669–672 (2005)

5. Turati, M.: Cayeron al más hondo de los pozos(Víctimas del incendio de la guardería ABC). Proceso (México, DF) (2009)
6. Castillo, M.: The Case of ABC Day Care, Continuation of Tragedy (2010)
7. Soriano, J.E.A.: Android: Programación de dispositivos móviles a travs de ejemplos. Marcombo (2011)
8. Aponte Gómez, S., Dávila Ramírez, C.A.: Mobile operating systems: functionalities, effectiveness and useful applications in Colombia (2012)
9. Cataldi, Z., Lage, F.J.: TICs en Educación: Nuevas herramientas y nuevos paradigmas. In: VII Congreso de Tecnología en Educación y Educación en Tecnología (2012)
10. Cervantes, J.M., Luna, J.S., Sánchez, E.M., Díaz, R.V.: Aplicaciones Android para el control de sistemas mecatrónicos. Revista Iberoamericana de Producción Académica y Gestión Educativa (2012)
11. Espinosa Granados, A.: El trauma cívico: análisis sobre el impacto de la tragedia en la guardería ABC (2012)
12. Rodríguez, A.S.: Los cibermedios y los móviles: una relación de desconfianza (2013)
13. Barón Rubio, A.: PerSe: Person as Sensor (2014)
14. Fernández Medina, A., and Prim Sabriá, M.: Gestión de mecanismos Arduino controlados inalámbricamente por dispositivos Android (2014)
15. Srairi, M.A.: Utilización de servicios Web en dispositivos móviles Android (Doctoral dissertation) (2014)
16. Prosegur Smart (2014). http://www.prosegur.com.uy
17. Batuecas, F.A., Morales, C.M.: Alertcops: the new channel of citizen security alerts with the state security forces and bodies. In: SIC Magazine: Cybersecurity, Information Security and Privacy, pp. 64–65 (2015)
18. Secretaría de Seguridad Pública de la Ciudad de México (2015). http://www.miguelhidalgo.gob.mx/prensa.php?id=12

Geographical Information System for Patients, Neoplasms and Associated Environmental Contamination

Jesús Leonardo Soto-Sumuano[1,2(✉)],
Francisco Javier Olivera-Guerrero[3], José Alberto Tlacuilo-Parra[2],
Roberto Garibaldi Covarrubias[2,3,4], Hugo Romo-Rubio[2,3,4],
and Emmanuel Abundis-Gutierrez[5]

[1] Information Systems Department,
University of Guadalajara, Periférico Norte 799, Zapopan, Mexico
leonardo.lsoto@gmail.com
[2] Medical Investigation Division, High Specialty Medical Unit (UMAE),
Pediatric Hospital, IMSS National Medical Center, Guadalajara, Mexico
[3] Systems Developer and Professor, Zapopan, Jalisco, Mexico
[4] Division of Oncology, Civil Hospital OPD UdeG, Guadalajara, Mexico
[5] Mexican Society for Non-Ionizing Radiation Protection, CEO,
Guadalajara, Mexico

Abstract. This application offers a computerized solution to the registration and geo-referenced visualization of patients, neoplasms and associated environmental contamination. The statistical information generated allows physicians and hospitals to evaluate in a timely manner the behavior of the disease and its association with the pollutants, also by means of algorithms to obtain clusters that allow the delimitation of high risk areas. In general, the application allows the study of pediatric patients with different neoplasms and different contaminants. The particular case presented here is the study and analysis of the electromagnetic contamination associated with leukemia in the infant population of the metropolitan area of Guadalajara, the application allows the registration of the general information of the patient as well as the measurements of electric and magnetic field in high and low frequency. The results shows the geographical distribution of patients, radiation values received, hospitals, schools, high voltage lines, telecommunication antennas and high risk areas for contamination using maps, graphs and data tables. As a conclusion, we now have a real-time tool that facilitates the creation of prevention policies to combat the impact of radiation as one of the possible causes in the development of leukemia in the child population considered as an additional element of health risk public.

Keywords: Database · Environmental burden of disease · Cluster · Electromagnetic pollution · Geo-referencing · Non-ionizing radiation · Leukemia

© Springer International Publishing AG 2017
O. Gervasi et al. (Eds.): ICCSA 2017, Part VI, LNCS 10409, pp. 284–298, 2017.
DOI: 10.1007/978-3-319-62407-5_20

1 Introduction

In a report by the World Health Organization (WHO) on 16 June 2006, it is estimated that more than 33% of diseases of children under five are due to exposure to environmental hazards. Preventing these risks could save the lives of many people every year, including four million children, especially in developing countries. The report also shows that much of these environmental risks can be avoided through targeted interventions [1].

One tool for the quantification and measurement of environmental impact is the assessment of the environmental burden of disease. This tool assesses the burden of this disease attributable to environmental risk factors. It is a fact that the increase of evidence and links between health and the environment becomes new opportunities to quantify the impact on health due to the environment at the population level.

The burden of disease due to environmental risk factors was included from the outset in the Global Burden of Disease Studies for 1990 and was first published in 1996. Reference was made for environmental risks such as: Water and Sanitation, Indoor Air Pollution and Outdoor Air Pollution [2].

This type of studies represented an important step forward with respect to what had been done previously, consisting essentially in monitoring the levels and trends of certain environmental pollutants, in order not to exceed internationally established and accepted levels of safety.

However, electromagnetic pollution is not considered as a risk in this study, despite the fact that the burden of disease and damage has been assessed at global levels [3], shown during the event held at the National Academy of Medicine in Mexico The National Institute of Public Health (INSP) and the Institute of Health Metrics and Assessment of the University of Washington (IHME) presented the results of the study. *The burden of disease, injuries and risk factors in Mexico 1990–2013: results at national and state level* [4].

The INSP and the IHME also explained the magnitude of the studies that allow to make of the information generated by the researchers a public good of use for the decision making. Even so, "the need to have geo-referenced fine data to design health policies focused on local problems and to investigate possible reductions in the burden of disease at different socioeconomic levels" was stated.

Taking into account this need, it was proposed to design an application to record geo-referenced data of different neoplasms and contaminants as well as pediatric patients to help health sector institutions to assess the magnitude of local problems and to be able to reduce the taking into account the different socioeconomic levels of the population. In addition to allowing the design of more effective policies for problem solving, with more equity and justice [3].

This work is limited to quantifying the burden associated with environmental risk factors and that in general due to the lack or insufficiency of specific and relevant data it has been considered difficult to calculate. However, this quantification constitutes important evidence for environmental health decision-making.

In this context the general application developed in this project offers a computerized solution to the registration and geo-referenced visualization of patients, neoplasias and their associated environmental contamination. It aims to generate statistical information to determine a relationship between the pediatric patient's illness and the environmental contamination, this information contributes to make decisions in the treatment that each doctor can indicate to his patient, as well as to help prevent more cases with Risks of generating this disease.

With this application you can store the patient's general data, his most relevant medical history that the doctor obtains in the consultations for his medical follow-up and above all allows the recording and storage of contaminating factors. The neoplasms addressed in this first application are acute lymphoblastic leukemia and acute myeloid leukemia, and contamination by non-ionizing electromagnetic radiation.

2   Justification

SirGeoS, the application developed, and was carried out with the support and advice of specialist from the IMSS Pediatric Hospital of the National Medical Center of the West in Guadalajara (Hospital de Pediatría del IMSS del Centro Médico Nacional de Occidente en Guadalajara) and the Civil Hospital of Guadalajara (Hospital Civil de Guadalajara). The arguments that justify this development are:

(1) The increase in electromagnetic pollution, which is recognized and classified by the WHO International Agency for Research on Cancer (IARC) as 2B "possibly carcinogenic to humans". Electromagnetic pollution, also known as "electrosmog", is understood as the presence of various forms of electromagnetic energy in the environment, which, because of their magnitude and time of exposure, can cause risk, damage or annoyance to people, ecosystems or property in certain circumstances [4].
(2) The high vulnerability of children to all types of contamination is a concern for health institutions due to the growth in the annual incidence rate (TIA). Data from studies in Mexico on acute leukemia show rates ranging from 5.76 to 10.12 cases/100,000 inhabitants which is well above average [5].
(3) The absence of this type of application in IMSS's pediatric hospitals that allows the keeping of a registry control of data of neoplasias, patients and risk factors which must be related to the patient as possible enhancers that trigger or generate diseases in children [6–9].
(4) The increase in the percentage of children's diseases associated with exposure to environmental hazards [10, 11].

3   Materials and Methods

3.1   Methodology

The application development is established with the combined methodologies of prototype, waterfall and spiral (Fig. 1). This combination is justified because optimal results have been obtained in other processes based on research systems of similar structure. Thanks to this combination of methodologies, the project has a development cycle that allows reanalyzing and redesigning parts of the system based on the requirements and as data is obtained.

All this begins with a survey analysis of requirements of the project. The use of the waterfall methodology was established in this section of the project to make the first design of the database.

In turn, the use of prototyping methodology is implemented as it is a project that includes a research process and data analysis, while generating

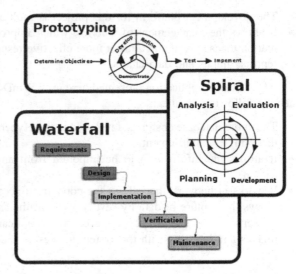

Fig. 1. Methodologies

prototypes of project modules which are worked on in conjunction with the medical staff that establishes the data capture and recording conditions, as well as the researcher in charge who guides the pollutants that need to be set within the system, with their measurements, names, values and geo-referential position if required.

The same rotation of information and the structure that is established in the system makes it necessary to use the spiral methodology since it is common to make some changes in the design of the system as it is part of what is being discovered within the research itself.

Statistical methods in medical research are highly classified. It is particularly useful for medical researchers who deal with data and provides a key resource for medical and statistical libraries.

3.2 Used Tools

The system is a desktop application developed with the programming language WinDev version 20. In turn WinDev is an integrated development environment that allows the creation of applications based on a runtime (framework) and has the following characteristics:

- Works with 32 and 64 bits in Windows and Linux, with a native Framework and no execution costs (free distribution).
- It allows access to databases through connection to various database engines, natively and under a unified environment and structure.
- It is a 5th generation language.
- Uses HyperFileSQL which is the database engine included within the development environment.

- The development platform for this project is multiuser and works with Windows.
- It allows the management of Queries, which improves the performance for information management and a faster more effective result retrieval, either within forms, reports, graphs and data charts.

The SirGeoS application presented here uses WinDev as a development tool which has the following features:

- The entire database design is developed in the HyperFileSQL HFSQL engine within the WinDev environment.
- It allows the information to be imported from an Excel file from a preexisting format.
- It works in network that allows the concurrent access to the information and also allows the capture of data by several users at the same time.
- Implements Google Maps as a web map application server that provides map images and interacts with the system to present the elements within a map in real time.

3.3   Location of the Project

This project was established as source to obtain the data of the metropolitan area of Guadalajara and 6 of its municipalities, Guadalajara, Zapopan, Tlaquepaque, Tonalá, Tlajomulco, and El Salto.

In the initial phase, a geo-referenced census of the cases of lymphoblastic leukemia and acute myeloid leukemia of the Metropolitan Area of Guadalajara (ZMG) was carried out for the first time in Mexico. The census represents 95% of all the cases of the ZMG, for which processes were established to obtain intra-domiciliary measurement values following the protocols of measurement of electromagnetic pollution levels in high and low frequency. Low-frequency magnetic field measurements and high-frequency electric field measurements at the home of recognized and treated patients with leukemia are done using HI-3604 ETS-LINDGREN equipment [12].

3.4   Development of the System

The development of the system consisted of the following development processes:

1. First, the system requirements were obtained.
2. The analysis and design of the project was developed starting with a process diagram of the elements of the application (Fig. 2).
3. Once the requirements were obtained and with the help of the process diagram, the design of the system database was carried out.
4. Capture modules were designed and created within the system to record the information of each patient and the contamination factors as well as schools and hospitals.
5. Reports, tables, graphs and maps were designed and created that show the results of the information registered in the system.

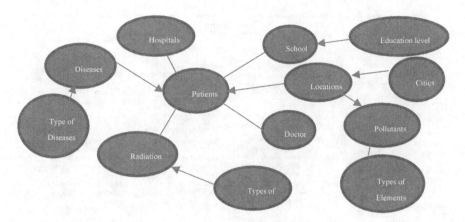

Fig. 2. Detailed process diagram of the application

The system is designed to allow the capture by the doctor in the office when he is doing the consultation, thus obtaining the necessary information of the patient's medical history, the user can capture the environmental and electromagnetic pollution measurements, which are recorded by location of the patient or the pollutants and create the geographical distribution.

3.5 Detailed Application Block Diagram

To make the structural part of the system clearer, a block diagram has been created which allows us to describe in a simple and quick way the structure of the most important modules and processes that are developed in the application (Fig. 3).

First we have the "inputs" block which shows us all the input requirements that the system needs to obtain and record the information in the database:

- Doctors, diseases, patients, hospitals, schools, cities, municipalities, pollutants, measurements.

Of these elements it is necessary to capture its location by means of the address and/or its geographical location with the territorial coordinates of latitude and longitude. In addition, additional data such as name, registry keys, registration dates, classification, units of measure, types and status must be registered.

Where patients are classified as cases and controls, where a "control" is a healthy child who lives in the area of the domicile of the sick child that we refer to as a "case".

In the block of the "Processes" it integrates all the information that the application uses for processing of algorithms, classification of information, geographical distribution of elements by means of maps.

In the section of "outputs" we find the product that generates as output the system.

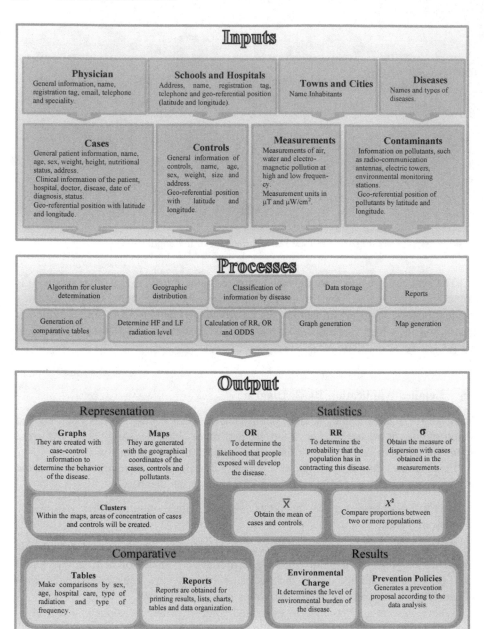

Fig. 3. System architecture

3.6   Operation

The application is focused for use in the area of pediatrics. For now the application is used in the Department of Hemato-Oncology, UMAE Pediatric Hospital CMNO IMSS, Guadalajara, we show the system logo (Fig. 4). The user can access the application using a user and password.

Fig. 4. System logo

The system establishes as the main part the capture of the catalog of patients to process information and generate results and it is imperative that previously the different catalogs ac company the patient, such as the capture of cities, municipalities, doctors, schools, hospitals, Type of disease, diseases, contaminating factors and units of measurement, which are required for filling each patient's form.

In the system there is a general menu where you can access elements of the catalog, in the submenu of catalogs are shown options to capture the following values:

- Educational levels.
- Schools.
- Locations.
- Municipalities.
- Hospitals.

- Networks / radiation elements.
- Types of networks.
- Types of measurement.
- Diseases.
- Measurement units.

In addition the system is able to import data that is in an Excel table following an established format of the information. For new cases the application has an information capture screen of all data catalogs (Fig. 5).

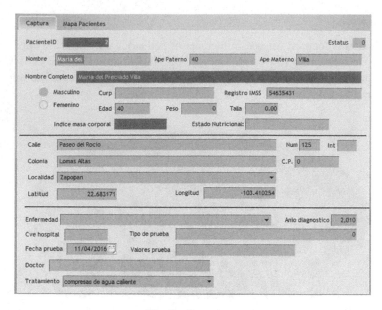

Fig. 5. Cap screen

The following are the icons used in the map according to the status or type of element:

	Child with HIGH status		Radio Antennas
	Child with low and high frequency radiation status		School
	Child with high radiation in high and low frequency		Hospital
	Child with high radiation in high frequency		Lines of transmission
	Child with high radiation at low frequency		High tension cables
	Child with excluded status		Electrical substations
	Child with deceased status		Environmental measurement points
	Telephony Antennas		Wifi antennas

4   Results

The following are examples of results that the system emits after the processing of the information by means of charts and graphs (Table 1).

Table 1. Radiation values

	Cases	% Total cases
Radiation: electric field		
$R \geq 1$ $\mu W/cm^2$	28	33.33%
$0.5 \leq R < 1$ $\mu W/cm^2$	15	17.86%
$R < 0.5$ $\mu W/cm^2$	41	48.81%
Radiation: magnetic field		
$R \geq 0.3$ μT	27	32.14%
$0.1 \leq R < 0.3$ μT	12	14.29%
$R < 0.1$ μT	45	53.57%

The statistical analysis of values of two cases were obtained in the measurement process and are classified in high and low frequency, obtained the result of standard deviation, the mean, sample variance, minimum and maximum how data to use to modeling the knowledge for predictive with purposes of prevention [7, 8] (Table 2).

Table 2. Data analysis of radiation HF and LF [13–16]

	Radiation HF	Radiation LF
Mean	135.7063929	0.25195238
Standard error	110.0487111	0.03814278
Median	0.534	0.09
Mode	0.294	0.05
Standard deviation	1008.613098	0.3495844
Sample variance	1017300.381	0.12220925
Kurtosis	79.38118599	7.67766415
Skewness	8.824306928	2.48204528
Range	9144.998	1.99
Minimum	0.002	0.01
Maximum	9145	2
Sum	11399.337	21.164
Count	84	84

The charts show the cases with radiation values classified in 3 options, both for electric field and for magnetic field. Values in high (≥ 1 $\mu W/cm^2$) and low (≥ 0.3 μT) frequency (Fig. 6).

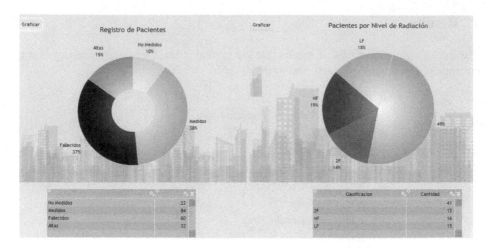

Fig. 6. Graphs

Maps generated with information from patients, schools, hospitals and pollutants.

The map option filters the data to be displayed according to the selected option of the disease in the access section of the system (Figs. 7 and 8); in addition, this window shows filters for the classification of the elements that will be displayed on the map.

Fig. 7. Map with all elements

Fig. 8. Map option filters

The window shows the map which is a "Google Maps" API inserted inside the application for the geo-referential management of the information.

In the comparative results charts the values obtained in the measurement process are classified in high and low frequency, and in turn in high and low radiation for boys and girls (Table 3); in addition to the classification of the disease in ALL Acute Lymphoblastic Disease) and AML (Acute Myeloid Leukemia) [17–19].

Table 3. Statistical charts of measurements in high and low frequency of electromagnetic radiation.

High radiation				Low radiation			
Boys		Girls		Boys		Girls	
LLA	LMA	LLA	LMA	LLA	LMA	LLA	LMA
High frequency							
13	2	11	2	22	3	30	1
15.48%	2.38%	13.10%	2.38%	26.19%	3.57%	35.71%	1.19%
Low frequency							
15	2	7	3	20	3	34	0
17.86%	2.38%	8.33%	3.57%	23.81%	3.57%	40.48%	0.00%

According to the frequency, another classification was made that is the main factor the status of the measurement (Table 4) according to the ranges allowed by the WHO in Mexico in high and low radiation.

Table 4. Clasification of measurement by status

High frequency		Low frequency		Mixed		High frequency	Low frequency
High radiation	Low radiation	High radiation	Low radiation	High radiation mixed	Low radiation mixed	HF high radiation	LF high radiation
28	56	27	57	12	41	16	15
33.33%	66.67%	32.14%	67.86%	14.29%	48.81%	19.05%	17.86%

As it can be seen in the chart, almost 15% of the cases of the children population measured have a very high combined exposure in high and low frequency.

5   Conclusions

This version creates the bases of registration of the information necessary to evaluate the environmental burden of disease in the Guadalajara metropolitan area. In the current version are implemented maps geo-referenced information of patients, diseases and environmental pollution associated with electromagnetic fields.

We have a tool that allows you to identify and analyze in real time, the existence of areas of high risk for the development of leukemia among children, resulting from exposure to contaminants environmental (air or electromagnetic radiation), considered to be a risk for public health to the healthy subjects, allowing to create policies of prevention or mitigation of the effect [20–23].

The collaboration between engineers and doctors allowed the development of a specialized application, which makes evident, measure and analyze the existing interaction between environmental pollution and the development of diseases, using information geo-referenced, useful for decision-making in health [24–26].

References

1. Osseiran, N.: Organización Mundial de la Salud (2006). http://www.who.int/mediacentre/news/releases/2006/pr32/es/. Accessed 10 Jan 2017
2. Placeres DMR: Biblioteca Virtual en Salud de Cuba (2014). http://www.bvs.sld.cu/revistas/hie/vol52_2_14/hie01214.htm. Accessed 10 Jan 2017
3. Bermejo DPM: Carga de la enfermedad ambiental, Biblioteca Virtual en Salud de Cuba (2014). http://www.bvs.sld.cu/revistas/hie/vol42_3_04/hig01304.htm. Accessed 10 Jan 2017
4. Wilkinson, J.D., Gonzalez, A., Wohler-Torres, B., Fleming, L.E., MacKinnon, J., Trapido, E., Button, J., Peace, S.: Cancer incidence among hispanic children in the United States. Rev. Panam. Salud. Publica **18**, 5–13 (2005)
5. Sumuano, L.S., et al.: Geographical distribution of childhood acute leukaemia in the metropolitan area of Guadalajara, Mexico and its correlation with the wireless and high voltage networks. In: Progress in Electromagnetics Research Symposium, Praga, Checoslovaquia, pp. 1245–1249 (2015)

6. Perez-Saldivar, M.L., Fajardo-Gutierrez, A.F., Bernaldez-Rios, R., Martinez-Avalos, A., Medina-Sanson, A., Espinosa-Hernandez, L., et al.: Childhood acute leukemias are frequent in Mexico city: descriptive epidemiology. BMC Cancer **11**, 355 (2011)
7. Ferlay, J., Soerjomataram, I., Ervik, M., Dikshit, R., Eser, S., Mathers, C., Rebelo, M., Parkin, D.M., Forman, D., Bray, F.: GLOBOCAN 2012 v1.0, Cancer Incidence and Mortality Worldwide: IARC Cancer Base No. 11 [Internet]. International Agency for Research on Cancer, Lyon (2013)
8. Perez-Cuevas, R., Doubova, S.V., Zapata-Tarres, M., Flores-Hernandez, S., Frazier, L., Rodriguez-Galindo, C., et al.: Scaling up cancer care for children without medical insurance in developing countries: the case of Mexico. Pediatr. Blood Cancer **60**, 196–203 (2013)
9. Wilkinson, J.D., Gonzalez, A., Wohler-Torres, B., Fleming, L.E., MacKinnon, J., Trapido, E., Button, J., Peace, S.: Cancer incidence among hispanic children in the United States. Rev. Panam. Salud Publica **18**, 5–13 (2005)
10. Fajardo-Gutierrez, A., Juarez-Ocaña, S., Gonzalez-Miranda, G., Palma-Padilla, V., Carreon-Cruz, R., Ortega-Alvarez, M.C., Mejia-Arangure, J.M.: Incidence of cancer in children residing in ten jurisdictions of the Mexican Republic: importance of the cancer registry (a population based study). BMC Cancer **7**, 68 (2007)
11. Belson, M., Kingsley, B., Holmes, A.: Risk factors for acute leukemia in children: a review. Environ. Health Perspect. **115**, 138–145 (2007)
12. ETS-Lindgren. ETS-Lindgren (2017). http://www.ets-lindgren.com/HI-3604. Accessed 10 Jan 2017
13. Martínez, P.C.C., Guzmán, L.G.G.: El Valor de la Estadística para la Salud Pública, Marzo (2003)
14. Martinez De Jordan, M., Ramos, R., Paola, S.: Análisis estadístico de la población infantil con cáncer en los principales centros urbanos del país (2009)
15. Balmón, M.A.: Guía práctica de análisis de datos: IFAPA (2006)
16. Martín, Z.H.: Métodos de análisis de datos Logroño: Universidad de la Rioja, servicio de publicaciones (2012)
17. Perez-Saldivar, M.L., Fajardo-Gutierrez, A.F., Bernaldez-Rios, R., Martinez-Avalos, A., Medina-Sanson, A., Espinosa-Hernandez, L., et al.: Childhood acute leukemias are frequent in Mexico city: descriptive epidemiology. BMC Cancer **11**, 355 (2011)
18. Ferlay, J., Soerjomataram, I., Ervik, M., Dikshit, R., Eser, S., Mathers, C., Rebelo, M., Parkin, D.M., Forman, D., Bray, F.: GLOBOCAN 2012 v1.0, Cancer Incidence and Mortality Worldwide: IARC Cancer Base No. 11 [Internet]. International Agency for Research on Cancer, Lyon (2013)
19. Perez-Cuevas, R., Doubova, S.V., Zapata-Tarres, M., Flores-Hernandez, S., Frazier, L., Rodriguez-Galindo, C., et al.: Scaling up cancer care for children without medical insurance in developing countries: the case of Mexico. Pediatr. Blood Cancer **60**, 196–203 (2013)
20. Fajardo-Gutierrez, A., Juarez-Ocaña, S., Gonzalez-Miranda, G., Palma-Padilla, V., Carreon-Cruz, R., Ortega-Alvarez, M.C., Mejia-Arangure, J.M.: Incidence of cancer in children residing in ten jurisdictions of the Mexican Republic: importance of the cancer registry (a population based study). BMC Cancer **7**, 68 (2007)
21. Belson, M., Kingsley, B., Holmes, A.: Risk factors for acute leukemia in children: a review. Environ. Health Perspect. **115**, 138–145 (2007)
22. Perez-Saldivar, M.L., Ortega-Alvarez, M.C., Fajardo-Gutierrez, A., Bernaldez-Rios, R., del Campo-Martinez, M.D.L.A., Medina-Sanson, A., Palomo-Colli, M.A.: Father's occupational exposure to carcinogenic agents and childhood acute leukemia: a new method to assess exposure (a case control study). BMC Cancer **8**, 7 (2008)

23. Dockerty, J.D., Sharple, K.J., Borman, B.: An assessment of spatial clustering of leukemias and lymphomas among young people in New Zealand. J. Epidemiol. Community Health **53**, 154–158 (1999)
24. Jarup, L.: Health and environment information systems for exposure and disease mapping, and risk assessment. Environ. Health Perspect. **112**(3), 995–997 (2004)
25. Instituto Nacional de Estadistica y Geografia (INEGI): Conteo de poblacion y vivienda (2010). http://inegi.org.mx/sistemas/mexicocifras
26. Geraci, M., Eden, T.O., Alston, R.D., Moran, A., Arora, R.S., Birch, J.M.: Geographical and temporal distribution of cancer survival in teenagers and young adults in England. Br. J. Cancer **101**, 1939–1945 (2009)

Implementation and Performance Comparison of a Four-Bit Ripple-Carry Adder Using Different MOS Current Mode Logic Topologies

Naman Saxena$^{(\boxtimes)}$ ⓘ, Shruti Dutta ⓘ, Neeta Pandey ⓘ,
and Kirti Gupta ⓘ

Electrical Department, Delhi Technological University, Delhi 110042, India
{namansaxena, shrutidutta}@dtu.ac.in

Abstract. In this paper, we have implemented a four-bit ripple carry adder using three different MOS Current Mode logic (MCML) topologies, namely conventional MCML, triple-tail cell based MCML, and quad-cell based MCML. The ripple-carry adder has been designed using four full adder circuits that essentially comprise of Sum and Carry circuit. The design of Sum and Carry circuits based on XOR gates and multiplexers has been proposed and implemented using the three specified topologies and a performance comparison is also presented. As the circuit has multiple inputs, quad cell based MCML implementation has shown the most promising performance in terms of power consumption and output voltage v/s temperature stability. However, the output voltage is most susceptible to noise in this topology. A deeper analysis of the circuits revealed that the number of transistors used is least in the conventional MCML based implementation while the triple tail based topology has the minimum delay.

Keywords: MOS current mode logic (MCML) · Triple-tail based MCML · Quad cell based MCML · Pull-down network · Differential pair · Ripple carry adder

1 Introduction

With the recent advancements in VLSI design and computer architecture, there is a perpetual demand for suitable solutions to design faster and low power circuits. Conventional CMOS logic family is unable to deliver satisfactorily when used for the implementation of complex circuits due to their large switching noises and high-power consumption pertaining to high operating voltages [1]. Source Coupled Logic families are a suitable alternative due to their faster performance and ability to be operated at lower voltages by manipulating the sizes of the transistors employed [2, 3]. Also, SCL based circuits provide greater immunity towards switching noises along with high-current resolution. MOS current mode logic (MCML) family is a differential-input based Source coupled logic family [4–6]. The inputs constitute of both the input and its complement. MCML based circuits find applications in signal processors, optical fibers, ADC/DACs and communication systems [7].

© Springer International Publishing AG 2017
O. Gervasi et al. (Eds.): ICCSA 2017, Part VI, LNCS 10409, pp. 299–313, 2017.
DOI: 10.1007/978-3-319-62407-5_21

Typically, a conventional MCML topology consists of a differential transistor pair, a current source and a load (usually capacitor) which basically converts the current into voltage [8]. The circuit produces both the output and its complement. To reduce the power consumption, either the biasing current or voltage can be minimized. As diminishing the current would cause instability, the operating voltage is reduced to 1.1 V which leads to significant reduction in power consumption. The supply voltage can be reduced by using multi-threshold levels in conventional MCML design.

Triple-tail cell and quad-cell based MCML topologies are extensions of the conventional MCML design which use suitably sized, one and two tail transistors respectively [9]. Quad-cell based MCML is motivated by the need of reducing the number of stacked levels, as number of inputs are increased [10].

A four-bit ripple carry adder is a commonly used combinational circuit and uses four full adders [11, 12]. The sum function can be implemented using a three-input XOR gate while the Carry function can be implemented using a combination of AND and OR gates or by manipulating the inputs of a 4×1 multiplexer [13–15].

In this paper, first we delve into the details of conventional, triple-tail cell and quad cell based MCML circuit design and implementation. In the next section, we discuss the implementation of a four-bit ripple-carry adder using full adder circuits. The full adder circuit has been implemented through XOR, AND, OR gates and the design and execution of these elements using the above-mentioned topologies has been discussed in the subsequent sections. Finally, a performance comparison has been presented and simulation results are concluded. The simulations have been carried out using 0.18 μm CMOS technology parameters in ORCAD PSPICE [16]. The performance of the different topologies was evaluated based on power consumption, propagation delay, area overhead, variation of output voltage with respect to temperature, and the susceptibility of the circuit to noise.

2   MOS Current Mode Logic Topologies

2.1   Conventional MCML

The traditional MCML implementation consists of a differential pull-down network (PDN), a current source and pull-up load transistors [2, 3]. The input and its complement are given to the PDN. This class of SCL family generates both the output and it's complement which makes it particularly suitable for circuits which make use of inverted outputs and inputs [2]. It is a current driven circuit. If the input is high, the transistor M1 is switched ON while M2 is switched OFF. Therefore, the output voltage, $V_O = V_{DD} - I_{SS}.R_D$ is equal to zero. The pull-up load transistors determine the value of drain resistance based on the values of voltage swing and current steered [4] (Fig. 1).

2.2   Triple-Tail Cell Concept

The triple tail topology is faster than the conventional MCML and reduces the latency of the circuit significantly [9]. The triple tail fundamental cell consists of three transistors (M3, M4, M5) as shown in Fig. 2. The size of the middle transistor (M5) is

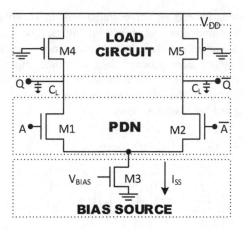

Fig. 1. Basic conventional MCML implementation [2]

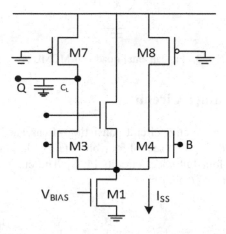

Fig. 2. Fundamental triple-tail MCML cell [9]

selected suitably, usually N times larger than the remaining transistors to allow a greater flow of current through it [13, 15].

In the case of triple tail implementation, if the tail transistor is switched ON the entire current flows through it as its sizing is very large as compared to the other transistors and hence the fundamental cell is activated only when the tail is deactivated.

2.3 Quad-Cell Based MCML Implementation

This topology can be considered an extension of the triple tail based concept. In MCML circuits, as the number of inputs increase, the number of stacked levels increase owing to the series gating approach [10]. Hence, this topology allows the number of stacks to be reduced by two and subsequently reduces the number of voltage sources to be employed. Instead of three transistors in the fundamental cell, this concept makes uses four transistors as shown in Fig. 3. The middle two transistors (M3–M4) are sized in a manner that they draw N times more current than the other two transistors. This topology is generally beneficial for multiple input designs. If Y and/or Z is high, the current flows through transistors M3 and/or M4 and the transistors (M1 and M2) remains deactivated. The only scenario where transistors M1–M2 are activated is when M3 and M4 are switched low.

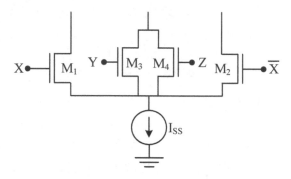

Fig. 3. Fundamental quad-cell MCML [10]

3   Ripple Carry Adder Circuit

A ripple-carry adder is an iterative circuit consisting of cascaded full adder circuits as shown in Fig. 4. Considering two sets of four-bit data, A [0:3] and B [0:3], which are bitwise inputs into the four full adder circuits [4, 8]. The circuit gives a four-bit sum output and an output carry (C_{OUT}) bit.

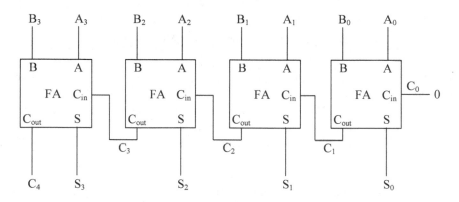

Fig. 4. Four-bit ripple carry adder [12]

The least significant bits are summed first and then their C_{OUT} is connected to the C_{IN} of the second adder corresponding to the next most significant bit. This process carried out till the final C_{OUT} is generated. The propagation delay is cumulative as the second adder cannot work till C_{OUT} is generated by the first adder. Therefore, the total delay can be calculated by adding individual delays produced by each of the full adders [12, 14].

The gate level schematic is as shown in Fig. 5. The Sum circuit is implemented through two, 2-input XOR gates [5, 10] while the Carry function is implemented through a combination of AND and OR gates. The Sum and Carry expressions are as follows:

$$S = \bar{X}.\bar{Y}.C_{in} + \bar{X}.Y.\overline{C_{in}} + X.\bar{Y}.\overline{C_{in}} + X.Y.C_{in} \tag{1}$$

$$S = \bar{X}.(\bar{Y}.C_{in} + Y.\overline{C_{in}}) + X.(\bar{Y}.\overline{C_{in}} + Y.C_{in}) \tag{2}$$

$$S = \bar{X}.(Y \oplus C_{in}) + X.\overline{(Y \oplus C_{in})} \tag{3}$$

$$S = X \oplus Y \oplus C_{in} \tag{4}$$

$$C_{out} = \bar{X}.Y.C_{in} + X.\bar{Y}.C_{in} + X.Y.\overline{C_{in}} + X.Y.C_{in} \tag{5}$$

$$C_{out} = (\bar{X}.Y.C_{in} + X.Y.C_{in}) + (X.\bar{Y}.C_{in} + X.Y.C_{in}) + (X.Y.\overline{C_{in}} + X.Y.C_{in}) \tag{6}$$

$$C_{out} = Y.C_{in} + X.C_{in} + X.Y \tag{7}$$

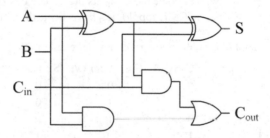

Fig. 5. Gate-level schematic of full adder [12]

There are numerous ways of implementing a full adder such as using two half adders and an OR gate, using two 4 × 1 multiplexers or through the typical two-input gate implementation [9, 13]. In this paper, we have implemented the full adder using two 4 × 1 multiplexers, manipulating their inputs to meet the circuit requirements. The inputs of the 4 × 1 multiplexers corresponding to the Sum and the Carry circuits has been changed as shown in Fig. 6. In the case of Sum circuit the Boolean expression corresponds to that of a three input XOR gate however, in the case of Carry circuit, the Boolean expression cannot be represented using a single gate.

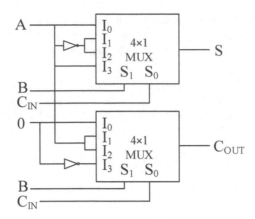

Fig. 6. Full adder circuit implementation using two 4 × 1 multiplexer

4   Full Adder Circuit Design and Implementation

To execute a full adder, we require a three input XOR gate to carry out the Sum functionality and a circuit for Carry, both of which are implemented using 4 × 1 multiplexers. The sum and carry circuit work as per the truth table values shown in Table 1. In the sum circuit, it can be concluded that the output is 1, when the input has odd number of 1's and, is zero when the input has even number of 1's. In case of the Carry circuit, the output is 1 when two or more inputs are 1. The design of the Sum and carry circuits using the various MCML topologies have been illustrated below.

Table 1.   Truth table for a full adder circuit [12]

Input bit (A)	Input bit (B)	Carry I/P (C_{IN})	Sum O/P (S)	Carry O/P (C_{OUT})
0	0	0	0	0
0	0	1	1	0
0	1	0	1	0
0	1	1	0	1
1	0	0	1	0
1	0	1	0	1
1	1	0	0	1
1	1	1	1	1

4.1   Conventional MCML

A three input XOR gate has been implemented using conventional MCML to perform Sum function as shown in Fig. 7. The structure can be understood as a binary tree in which the number of transistors on each level is determined using 2^n, where n is the level and is limited by the number of inputs [4]. As we have implemented a three-input XOR gate, the top-most level has $2^3 = 8$ transistors, the following row has $2^2 = 4$ transistors and so on. It can be observed from the figure, that the output is given by the Boolean expression,

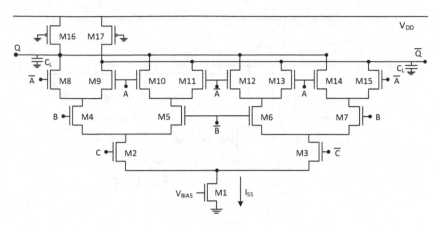

Fig. 7. Sum circuit implementation using conventional MCML topology

$$Q = \bar{A}.B.C + A.\bar{B}.C + \bar{A}.\bar{B}.\bar{C} + A.B.\bar{C} \tag{8}$$

This clearly testifies that the function of the Sum circuit is successfully implemented using the circuit as shown in Fig. 7. Let us consider the following cases to understand the implementation in a better manner.

Case 1: A, B and C are Zero

Starting from the lowest level, as C is zero, M3 is switched ON. As M3 is connected to M6 and M7, and B is zero, therefore, M6 is switched ON. Now as M6 is connected to M12 and M13, amongst which M12 is switched ON as A is zero, therefore the output voltage is given by

$$V_O = V_{DD} - I_{SS}.R_D \tag{9}$$

The output voltage becomes zero as expected. The current I_{SS} flows through these transistors and hence makes the output voltage zero as $V_{DD} = I_{SS}.R_D$.

Case 2: A is High While B and C are Zero

Following the above convention, transistors M3 and M6 are switched ON. As A is high, therefore M12 is switched OFF while M13 is switched ON. The output branch is connected to M12, and since no current flows through it, the output voltage $V_O = V_{DD}$ or high output (as expected). In a similar fashion, other cases can also be explained using similar analysis.

For the carry circuit implementation, the inputs of the 4×1 multiplexer have been changed as shown in Fig. 8. The select lines are B and C while $I_0 = 0$, $I_1 = I_2 = A$ and $I_3 = 1$. The output equation of the given circuit is given by:

$$Q = \bar{A}.B.C + A.C + A.B.\bar{C} \tag{10}$$

On further evaluation, it can be found that Eq. (10) is same as the C_{OUT} equation listed in Eq. (7). Hence, C_{OUT} functionality is achieved as shown above.

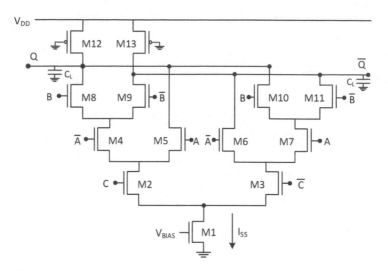

Fig. 8. Carry circuit implementation using conventional MCML topology

4.2   Triple-Tail Based MCML

Triple-tail MCML consists of three transistors in the fundamental cell where the tail transistor is sized N times more than the remaining two transistors [9]. In case of triple-tail cell MCML, we can implement a 2 × 1 multiplexer using two fundamental cells. However, in order to achieve the functionality of the sum or carry circuit, we have to used two 2 × 1 multiplexers whose inputs have been adjusted accordingly [13]. In the case of Sum circuit, the three-input XOR gate is obtained by designing two, 2-input XOR gate using 2 × 1 multiplexer and then cascading them as shown in Fig. 9.

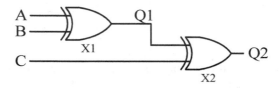

Fig. 9. Gate-level schematic of a three input XOR gate using two 2-input XOR gates

The transistor level implementation consists of four fundamental cells where the tail transistor of the first two cells is driven through input B while the other two tails are driven by the output of the first XOR gate as shown in Fig. 10. Two fundamental cells are given input A while the other two are given input C.

Let us consider the following cases to understand the working of the circuit.

Case 1: A, B and C are Zero
As B is zero, transistor M13 is switched OFF and hence that fundamental cell is activated. As A is zero, transistor M5 is switched ON and hence the output voltage

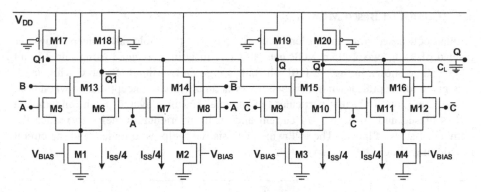

Fig. 10. XOR implementation using triple-tail cell based MCML

obtained is zero ($V_O = V_{DD} - I_{SS}.R_D$ as $V_{DD} = I_{SS}.R_D$). This output Q1 is given to the tails of the other two fundamental cells. As Q1 is zero, therefore transistor M15 is switched OFF and that fundamental cell is activated. The value of C input is also zero, hence among M9 and M10, M9 is switched ON. As this transistor is connected to Q, the current flows through this transistor and the output voltage becomes zero.

Case 2: A is High, B and C are Zero

As B is zero, transistor M13 is switched OFF and hence that fundamental cell is activated. As A is high, transistor M6 is switched ON and hence the output voltage obtained is high ($V_O = V_{DD}$ as $I_{SS} = 0$). This output Q1 is given to the tails of the other two fundamental cell. As Q1 is high, therefore transistor M16 is switched OFF and that fundamental cell is activated. The value of C input is also zero, hence among M11 and M12, M12 is switched ON. As M11 transistor is connected to Q, the current does not flow through this transistor and the output voltage becomes high. The other cases can be explained in a similar manner.

In case of the carry circuit, similar approach has been adopted. The desired circuit is obtained using two 2×1 multiplexers where the inputs are adjusted as shown in the transistor level design in Fig. 11.

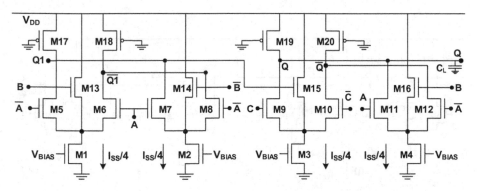

Fig. 11. Carry circuit implementation using triple-tail cell MCML topology

4.3   Quad Cell Based MCML

A quad cell based implementation is an extension of the triple tail concept, however instead of using two 2-input XOR gate, this circuit is implemented using one 3-input XOR gate as shown in Fig. 12 [10]. The tails are given two inputs B and C while input A is given to the pull-down transistors. It works on a similar concept that is, when both B and C are low, only that particular fundamental cell is activated. Each fundamental cell is biased and sized to $I_{SS}/4$ current and such an orientation helps to reduce the number of stacked layers. The following analysis will help us to understand the circuit functioning better.

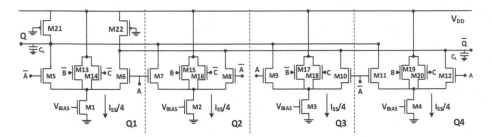

Fig. 12. XOR implementation using quad cell based MCML

Case 1: When A is Low, B and C are High
As B and C are high, fundamental cell Q1 is activated. As A is low, therefore M5 is turned ON and as the output branch is connected to this transistor, current flows through this branch and the output voltage is equal to zero.

Case 2: When A is High, B and C are Zero
As B and C both are zero, the fundamental cell Q4 is activated. As A is high, transistor M12 is in saturation mode and hence the output branch Q consisting of transistor M11 is turned OFF. Therefore, as the current through this transistor is zero, the output voltage is high (VDD) as expected. Using this analysis, all the input cases can be explained.

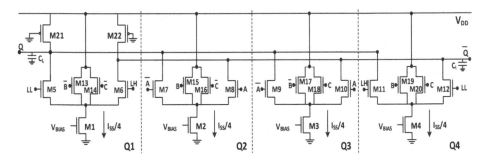

Fig. 13. Carry circuit implementation using quad cell based MCML

The carry circuit can be designed on similar grounds using a 4×1 multiplexer implemented using four fundamental quad cells as shown in Fig. 13.

5   Simulation Results

A four-bit ripple-carry adder is implemented using 0.18 μm CMOS technology on PSPICE A/D with a voltage swing of 400 mV. The simulation environment parameters are as shown in Table 2

Table 2. Simulation Parameter

Process corner	Typical
Supply voltage	1.7 V (Conventional MCML)
	1.1 V (Triple tail and quad cell MCML)
Data frequency	100 MHz
Output load	50 fF
Bias current	30 μA
Temperature	27 °C

The waveforms of input data (A, B, C_{IN}) along with the Sum and Carry outputs for a single full adder circuit are shown in Fig. 14. The waveforms are observed to be following the truth tables the Sum and Carry expressions.

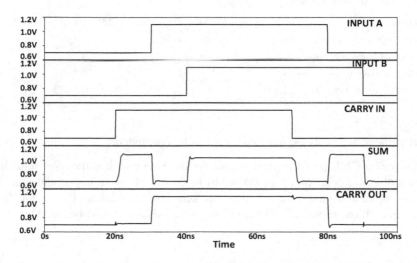

Fig. 14. Waveforms obtained from the implementation of a full adder circuit using quad-cell based MCML

6 Performance Comparison

The three MCML based topologies, namely conventional MCML, Triple-tail cell and quad cell based MCML are used to implement a four-bit ripple carry adder and the different performance parameters such as propagation delay, power consumption, and DC sweep at different temperatures are mapped for each of the circuit (Tables 3, 4 and 5).

Table 3. Comparison of different parameter for the Sum circuit, S [4]

Parameters	Conventional MCML	Triple tail MCML	Quad cell MCML
Power consumption (μW)	204	198	132
Propagation delay (ps)	27	22	24
Power delay product (pJ)	5.508	4.356	3.168

Table 4. Comparison of different parameter for the Carry circuit

Parameters	Conventional MCML	Triple-tail MCML	Quad cell MCML
Power consumption (μW)	816	1188	528
Propagation delay (ps)	14	11	12
Power delay product (pJ)	11.424	13.068	6336

Table 5. Comparison of various performance parameters of the complete four-bit ripple carry adder circuit using different MCML topologies

Parameters	Conventional MCML	Triple tail MCML	Quad cell MCML
Power consumption (μW)	1632	1980	1056
Number of transistors used	136	200	176
Propagation delay (ps)	14	11	12
Power delay product (pJ)	22.848	21.780	12.672

From the above analysis, the following can be concluded:

- Quad cell MCML based implementation for the Sum circuit is the most suitable in terms of Power delay product followed by triple-tail cell based implementation. The delay is least in the case of triple-tail cell implementation. Hence for faster implementation, triple-tail cell based Sum circuit design would be suitable.

Table 6. Comparison of noise output at different temperatures for a four-bit ripple carry adder

Temperature	Conventional MCML	Triple tail cell MCML	Quad cell MCML
−20 °C	14.4 nV	18 nV	25.5 nV
27 °C	16.6 nV	21 nV	26.2 nV
100 °C	19.8 nV	25 nV	28.6 nV

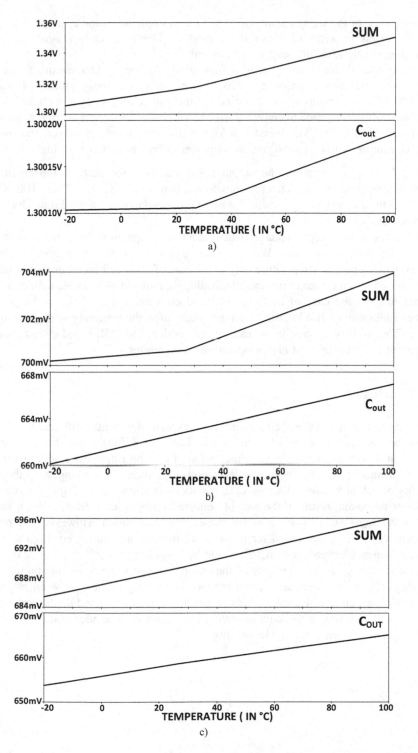

Fig. 15. Variation in output voltage with change in temperature, (a) conventional MCML, (b) triple-tail cell based MCML, (c) quad cell based MCML

- In the case of the Carry circuit, Quad cell based implementation again produces the best results in terms of power delay product. However, conventional quad cell implementation has the least area overhead.
- In the overall implementation of a four-bit ripple carry adder circuit, Quad cell produces the most promising results in terms of power delay product. It shows 71.875% better results when compared to triple-tail cell topology and 80.30% better results when compared to conventional MCML. Also, the area reduction in the case of conventional MCML based topology is the maximum. However, the overall performance is not so satisfying as its power delay product is very high.

We have also compared the output noise variation for each of the circuits at different temperatures. Three temperature points, that is −20 °C, 27 °C, and 100 °C are taken and the corresponding variation in noise produced in each of the circuits has been analyzed (Table 6).

Changes in the output voltage with respect to temperature have been shown in Fig. 15 for the given circuits. We can observe that the output voltage shows linear change with respect to temperature only in the case of quad cell based implementation. Also, the slope is the least in this case, indicating that the circuit is stable with changing temperature. In the case of triple-tail cell and conventional MCML, the slopes are steeper and a sudden rise in output voltage occurs after the operating voltage becomes 27 °C. The slope is steepest in the case of conventional MCML based circuit proving the circuit to be unstable at higher operating temperatures.

7   Conclusion

The design of Sum and Carry circuits has been explained using the different topologies to implement a full adder which forms the building block for the four-bit ripple carry adder. Quad cell based MCML implementation of four-bit ripple carry adder gives the best performance results. It has the least power product delay along with the best stability of output voltage with respect to variations in temperature. Triple-tail cell also produces promising results if the area of concern is propagation delay. Also, it shows almost the same stability as quad cell based implementation. However, the power consumption and the area used for this type of implementation is very high as compared to others. Conventional MCML would be a suitable option if chip size is the area of concern. Also, the susceptibility of the output voltage to noise is minimum in this topology. However, its remaining performance is not very suitable to be employed in complex digital circuits. Hence, the new proposal for quad cell based implementation is a suitable alternative to make circuits having low power consumption and delay, and at the same time showing admirable stability.

References

1. Saxena, N., Dutta, S., Pandey, N., Gupta, K.: Implementation of asynchronous pipeline using transmission gate logic. In: 2016 International Conference on Computational Techniques in Information and Communication Technologies (ICCTICT), pp. 101–106 (2016)
2. Gupta, K., Sridhar, R., Chaudhary, J.: Performance comparison of MCML and PFSCL gates in 0.18 μm CMOS technology. In: International Conference on Computer and Communication Technology (ICCCT), vol. 1, no. 1, pp. 230–233 (2011)
3. Saxena, N., Dutta, S., Pandey, N.: An efficient hybrid PFSCL based implementation of asynchronous pipeline. i-Manag. J. Circuits Syst. **4**(3), 6–14 (2016)
4. Gupta, K., Pandey, N.: A novel active shunt-peaked MCML-based high speed four-bit ripple-carry adder. In: 2011 International Conference on Multimedia, Signal Processing and Communication Technologies, vol. 1, pp. 285–289 (2011)
5. Gupta, K., Pandey, N., Gupta, M.: Analysis and design of MOS current mode logic exclusive-OR gate using triple-tail cells. Microelectron. J. **44**(6), 561–567 (2013)
6. Wu, X.: Low power DCVSL circuits employing AC power supply. Sci. Chin. Ser. F: Inf. Sci. **45**(3), 232–240 (2002)
7. Alioto, M., Pancioni, L., Rocchi, S., Vignoli, V.: Exploiting hysteresys in MCML circuits. IEEE Trans. Circuits Syst. II Express Briefs **53**(11), 1170–1174 (2006)
8. Musa, O., Shams, M.: An efficient delay model for MOS current-mode logic automated design and optimization. IEEE Trans. Circuits Syst. I Regul. Pap. **57**(8), 2041–2052 (2010)
9. Gupta, K., Pandey, N., Gupta, M.: MCML D-latch using triple-tail cells: analysis and design. Hindawi J. Act. Passiv. Electron. Compon. **2013** (2013)
10. Pandey, N., Gupta, K., Choudhary, B.: New proposal for MCML based three-input logic implementation. Hindawi J. VLSI Des. **2016** (2016)
11. Alioto, M., Pancioni, L., Rocchi, S., Vignoli, V.: Power-delay-area-noise margin tradeoffs in positive-feedback MOS current-mode logic. IEEE Trans. Circuits Syst. I Regul. Pap. **54**(9), 1916–1928 (2007)
12. Rabaey, J.M., Chandrakasan, A., Nikolic, B.: Digital Integrated Circuits. Pearson Education, Upper Saddle River (2003)
13. Gupta, K., Pandey, N., Gupta, M.: Low-voltage MOS current mode logic multiplexer. Radioeng. J. **22**(1), 259–268 (2013)
14. Kang, S.-M., Leblebici, Y.: CMOS Digital Integrated Circuits: Analysis and Design. Tata McGraw-Hill, New York (2003)
15. Gupta, K., Pandey, N., Gupta, M.: Multithreshold MOS current mode logic based asynchronous pipeline circuits. ISRN Electron. **2012** (2012)
16. Pandey, N., Dutta, S., Saxena, N.: A comparative study on electronic design automation tools. In: 2016 International Conference in Mechanical and Automation Engineering, Delhi Technological University, Delhi, India, pp. 1–7 (2016)

Evaluation of Facial Expressions for Automatic Interpretation and Classification of Motivation by Means of Computer Techniques

Catalina Alejandra Vázquez Rodríguez[(⊠)] and Raúl Pinto Elías

Centro Nacional de Investigación y Desarrollo Tecnológico, Cuernavaca, Mexico
{cvazquez, rpinto}@cenidet.edu.mx

Abstract. Facial expressions provide non-verbal information about people's emotional and mental states without the need for verbal communication, so the extraction and automatic tracking of facial components are the main tasks that artificial vision systems must solve in the realm of Human behavior, detection of facial expressions, man-machine interfaces, among other areas. In this work are proposed to use Kinect and the facetracking library to create a system that automatically locates the face and perform an interpretation of its elements to detect the motivation in an activity. The evaluation of the tests of this system obtained for cases where the system determined that the subjects were paying attention was 89.79%. For the cases where the system evaluated that the test subjects did not pay attention was 90.72%.

Keywords: Engagement · Kinect · Automatic assessment · Facial expressions · 3D · FACS · Artificial vision

1 Introduction

Motivation is the technique that ensures that students demonstrate interest in new teacher imparting knowledge to contribute to the academic development [1]; the lack of interest is identified as one of the main causes of failure in the learning process, numerous studies have shown that there is no learning without interest or engagement [2]; both, theoretical learning and teachers, agree that students who have learning stakeholders more quickly and more effectively than those who are not [3]. Interest is a feeling or emotion that makes the attention is focused on an object, an event or a process [4]. Therefore it is important to know if a student is interested in the class since the higher interest better performance. However, interest will not always be expressed verbally can also be expressed through facial expressions and posture may represent various characteristics if a person is interested or disinterested during the learning process [5]. FACS (Facial Action Coding System) was designed by Paul Ekman as a guide for classifying facial expressions; instead of naming the muscles individually, "Units of Action" relating to specified areas of the face whose movement can be seen relatively easy; for example the head movement, eye movement [6].

In this paper are proposed to solve the automatic assessment of engagement and attention by classifying in motivated, demotivated or neutral, through facial

© Springer International Publishing AG 2017
O. Gervasi et al. (Eds.): ICCSA 2017, Part VI, LNCS 10409, pp. 314–324, 2017.
DOI: 10.1007/978-3-319-62407-5_22

expressions. The resolution of this problem entails a work in several stages; First it is necessary to detect the face, later to detect each of the facial elements with which it will work to extract its information, finally the engagement can be clasificated and inter-pretated. Based on some facial and behavioral expressions, twenty students while watch a one of four experimental videos for automatic interpretation of engagement.

2   Features of Engagement and Disengagement

People are able to express their mood gestures. There are features that provide infor-mation on whether a person is interested in the activity being performed, for example the eyes hardly find a person interested in if eyes closed and even if that person is yawning and slouching. In contrast a person who keeps his head straight attending activities with the focused look and a stance that does not show fatigue is certainly an interested person. However also it is possible for a person present expressions that seem to belong to more than one category and therefore are used rules of decision to analyze the possible combinations of nonverbal. Figure 1 shows the rules of combi-nations of facial elements.

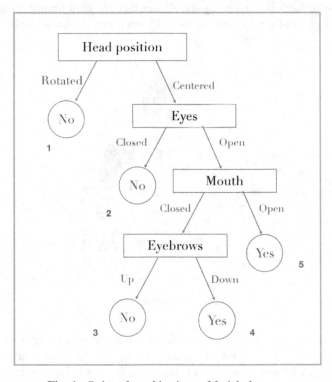

Fig. 1. Rules of combinations of facial elements.

It is impossible with a facial element isolated decide to whether attention is being put, for that reason they were analyzed together. The interpretation of facial elements is a stage in which each element, mouth, eyes, face and eyebrow are evaluated based on their numerical value, each of these attributes has different values to evaluate and the set of all of them gives an interpretation: engagement, disengagement or neutral.

For each element are made the corresponsive interpretation. For the mouth is evaluated whether it is open or closed, this to analyze if the mouth is in a neutral position or yawning since a person who is yawning constantly might not be interested. For the face were analyzed two movements pitch and yaw, the pitch movement evaluates if the head is turning up or down, while the yaw movement evaluates if the head is turned sideways. This is for the system to decide if a person is paying attention. For eyes it was evaluated whether they are open or closed, since a person with closed eyes is definitely not paying attention even could be sleeping. For the eyebrows, it has been found in the literature that the eyebrows in a furrowed state of the inner part usually denote concentration, so it was decided to analyze if the eyebrows are wrinkled.

3   Extraction, Classification and Interpretation

The extraction of faces is done using the Kinect face tracking library. In the Fig. 2 is shown the detection of the face through the Kinect facial detection mesh.

Fig. 2. Facial detection.

Once the face has been detected, the facial elements are extracted to know their numerical values. There is a coding system developed by Ekman and Friesen, published in 1978 [7] FACS (Facial Action Coding System), with this system are identified the muscular movements and lines of expression that produce momentary changes in the appearance of the face. These movements are called Action Units (AU). In this work the facetracking action units are used, which are 4 units comprising the eyebrows, the mouth and the lower jaw. Table 1 shows that all AU's generate values from −1 to 1. In this work some facetracking action units are used which include the eyebrows, mouth, eyes and head. Table 1 are shown the facial elements, the corresponding AU, the numerical value and descriptive images.

Table 1. Facial element and values taken.

Facial element	AU Kinect	Negative values	Neutral value	Positive values	Image
Eyebrows	BrowLower	-1 a 0 Eyebrows up	0	1 a 0 Eyebrows down	
Mouth	MouthCornerLift	-1 a 0 Closed mouth	0	1 a 0 Open mouth	
Eyes	UpperEyelid	-1 a 0 Closedeyes	0	1 a 0 Openeyes	
Eyes	LowerEyelid	-1 a 0 Closedeyes	0	1 a 0 Openeyes	
Head	Pitch	-45° a 0° Head down	0	0° a 45° Head up	
Head	Yaw	-45° a 0° Head turned left	0	0° a 45° Head turned right	

The behavior of the facial elements is evaluated by the historical, *i.e.* if the mouth is open for a period of time or the eyes are closed. The calculation of the historical is divided into 6 stages:

Stage 1. Instance capture. It captures 31 instances where the data is stored, which correspond to each of the elements with which it is working and in each record is contained the numerical value of each element. The recording of these 31 data in the block takes approximately two seconds.

Stage 2. The data enter a discriminant function, which is responsible for assigning a class (interested, disinterested and neutral) to each of the records of the 31 data block. The predominant class of records is assigned to the whole block, *i.e.* if the 31 records 25 were assigned to the interested class, the whole block will belong to the interested class. The evaluation of the block will be sent to the decision tree.

Stage 3. The decision tree receives the predominant class of each block, then they are evaluated according to the established rules.

Stage 4. It is decided to which class belongs the block.

Stage 5. The blocks are being received to create the history. Once the video ends, a percentage of the attention is calculated with all the blocks in the course of the test time.

Stage 6. Finally with all the evaluations, a percentage of attention is calculated as follows (See Eq. 1).

$$x = \frac{b}{a} * 100 \tag{1}$$

where a is the total number of evaluations performed by the system during the video, b is the number of times the system has given an interested result and x is the percentage of attention paid to the activity during the total time of the video. Figure 3 shows the process with which the historical is carried out and based on it, a conclusion is reached that is motivated, demotivated or neutral.

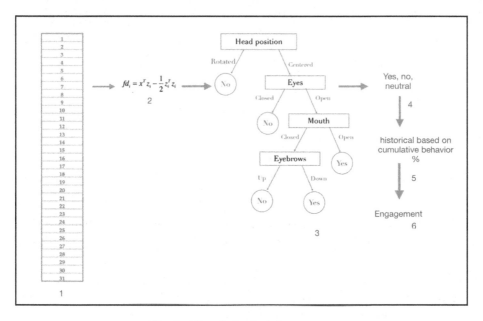

Fig. 3. Historical calculation process.

In this work we used a discriminant function by distance [8]. In Eq. 2 the average of the class is calculated using all the data of the training set, this to obtain a reference of the average of all the data pertaining to each one of the classes.

$$Z_i = \frac{1}{p}\sum_{j=1}^{p} x_{ij} \tag{2}$$

Equation 3 calculates the data to be evaluated, minus the means of class, this generates a result for each class that has.

$$d(x, z_i) = \sqrt{(x - z_i)^T (x - z_i)} \tag{3}$$

From Eqs. 4 to 8 is moved from the equation of the Euclidean distance to factors obtaining, by maximizing the discriminating factor between the data for a more precise classification.

$$(d(x, z_i))^2 = \left(\sqrt{(x - z_i)^T (x - z_i)} \right)^2 \tag{4}$$

$$(d(x, z_i))^2 = (x - z_i)^T (x - z_i) = ||x - z_i||^2 \tag{5}$$

$$(d(x, z_i))^2 = x^T x - 2x^T z_i + z_i^T z_i \tag{6}$$

$$d(x, z_i) = x^T z_i - \frac{1}{2} z_i^T z_i \tag{7}$$

$$fd_i = x^T z_i - \frac{1}{2} z_i^T z_i \tag{8}$$

Finally, with Eq. 9, the class membership of the new data is determined. The highest value given by the discriminant function will indicate membership.

$$x \in a_i \quad sii \quad fd_i(x) > fd_j(x) \quad \forall i, j, \quad i \neq j \tag{9}$$

Where:

   Z middle of class
   p amount of data by class
   x data to be evaluated.

4 Results

This section shows the tests performed and the results obtained with the artificial vision system for the automatic recognition of motivation where twenty subject were evaluated. To evaluate the subjects' attention, four test videos were used, which were reproduced for participants. Table 2 shows the cover of the videos used in these tests.

Table 2. Cover of the videos and description used in these tests.

Number video	Video	Description
1		The length of this video is 1 minute with 40 seconds is about airplanes
2		The duration of this video is 3 minutes with 19 seconds is about the evolution of the earth
3		The duration of this video is 3 minutes with 3 seconds is about the transformation from caterpillar to butterfly
4		The length of this video is 11 minutes with 44 seconds is about blackboards

Figure 4 shows test subject one, which was recorded while viewing video number two. As can be appreciated, the test subject was attentive attention while performing the task. This process was performed for each of the 20 subjects.

Figure 5 shows the graph of the attention given and how it was modified in blocks of approximately two seconds. Through the system it is concluded that he was interested during the 42.55% of the activity. The historical is obtained by calculating Eq. 1, always taking into account the previous value obtained to perform the calculation.

Figure 6 shows the graph where each bar corresponds to the block containing 31 data, there are three colors on the graph that correspond to the attention. The section covering the red color corresponds to the set of blocks where no attention was paid during the activity. The section covering the yellow color corresponds to a neutral state and finally the green color for the blocks where interest was shown.

Figure 7 shows how each of the bars in Fig. 6 is formed. Each bar contains the information of a block, at the top of the image a vector is shown which corresponds to the first block of results Of test subject one; Said bar has 78% of values with result "yes", therefore of the vector of 31 data, 24 correspond to yes and 7 to no.

Table 3 shows the confusion matrix where 89.79% corresponds to the correct classification in motivated, 91.66% corresponds to the correct classification in unmotivated. The 10.21% corresponds to the incorrect classification for the class motivate. Finally 8.34% corresponds to the incorrect classification for class unmotivated.

Fig. 4. Test subject one.

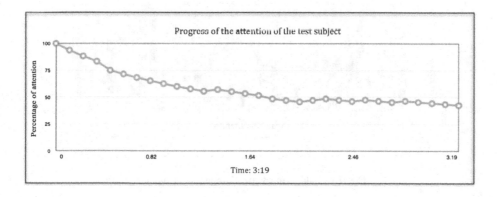

Fig. 5. Historical attention the subject one.

The data for the confusion matrix were taken from the results of the test subjects and the videos they viewed as well as from the identification of interest. Table 4 shows the breakdown of the data used in the confusion matrix.

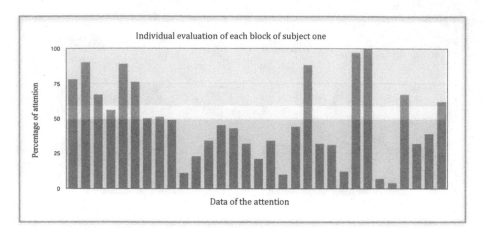

Fig. 6. Evaluation of attention without historical for subject one. (Color figure online)

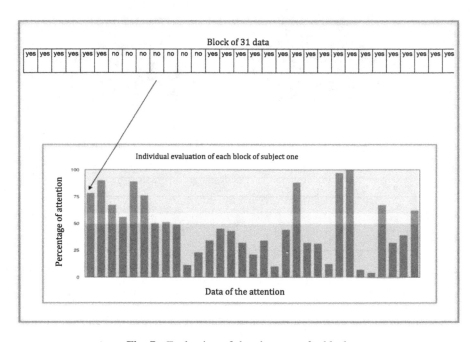

Fig. 7. Evaluation of the elements of a block.

Table 3. Matrix of confusion for attention.

Engagement	Yes	No
Yes	89.79%	10.21%
No	8.34%	91.66%

Table 4. Breakdown of the data used in the confusion matrix.

Subject	Number of videos viewed	Qualification matches	Observations
1	4	3	The error in the interpretation was because the individual is drinking water from a bottle
2	3	3	Correct identification on all videos
3	4	2	Too much direct illumination which caused errors in the detection of the elements
4	4	3	The eyebrow of the individual is very high in nature
5	2	1	The individual was eating a candy so the mouth element was not correctly interpreted
6	4	4	Correct identification on all videos
7	2	1	Too much direct illumination which caused errors in the detection of the elements
8	2	1	The subject wore thick frame lenses, which caused confusion in the system
9	3	3	Correct identification on all videos
10	4	3	The subject constantly touched the face so there was no correct interpretation of the elements
11	4	3	The individual wore glasses, which caused confusion in the system
12	4	4	Too much direct illumination which caused errors in the detection of the elements
13	3	3	Correct identification on all videos
14	3	1	The individual had very small eyes, so much that they seemed to be closed
15	2	2	Correct identification on all videos
16	4	3	Too much direct illumination which caused errors in the detection of the elements
17	2	2	Correct identification on all videos
18	3	2	Too much direct illumination which caused errors in the detection of the elements
19	4	4	Correct identification on all videos
20	4	4	Correct identification on all videos

5   Conclusions

The implemented system can analyze the behavior while an individual performs a test, in this case four videos were used to determine the attention of the individuals and in this way to discover the motivation that they had by the activity. The data bank was created for this work and the subjects of test were twenty, of which its behavior during the experimentation has been recorded to corroborate the results obtained by the system. In the experimentation it was obtained that for cases where the system determined that subjects were paying attention and were actually paying attention was 89.79%. For the cases where the system evaluated that the subjects of test did not pay attention and in fact did not do it was of 90.72%. It is important to mention that in using accessories such as lenses can influence the detection of the elements.

References

1. Tapia, J.A.: School guidance in schools, Madrid, pp. 209–242 (2005)
2. Pereira, N., Luisa, M.: Motivation: theoretical perspectives and some considerations of their importance in the educational field. Universidad de Costa Rica, San Pedro, Montes de Oca, Costa Rica, Educación, vol. 33, no. 2, pp. 153–170 (2009)
3. Freitas-Magalhães, A.: Microexpression y macroexpression. In: Ramachandran, V.S. (ed.) Enciclopedia del comportamiento Humano, vol. 2, pp. 173–183. Elsevier/Academic Press, Oxford (2012). ISBN 978-0-12-375000-6
4. Eckman, P.: FACS (Facial Action Coding System), Annual Review of Psychology (1979)
5. Boyás, V.S.: Interview for final project of the subject introduction to artificial intelligence. CENIDET, Mexico (2014)
6. Cho, J., Mirzaei, S., Oberg, J., Kastner, R.: FPGA-based face detection system using Haar classifiers. In: Proceedings of the ACM/SIGDA International Symposium on Field Programmable Gate Arrays, FPGA 2009, pp. 103–112. ACM (2009)
7. Stocchi, L.: 3D Facial Expressions Recognition Using the Microsoft Kinect. In: 18th International Conference on Image Processing (ICIP), School of Computing, Dublin City University, Dublin, Ireland, pp. 773–776. IEEE (2014)
8. Tejera, F.M.H., Navarro, J.J.L.: Recognition of classification and learning forms, chap. 3. Universidad de Las Palmas Gran Canaria (2008)

Nearness and Influence Based Link Prediction (NILP) in Distributed Platform

Ranjan Kumar Behera[1]($\boxtimes$), Lov Kumar[1], Monalisa Jena[2], Sambit Mahapatra[1], Abhishek Sai Shukla[1], and Santanu Kumar Rath[1]

[1] Department of Computer Science and Engineering,
National Institute of Technology Rourkela, Rourkela 769008, Odisha, India
jranjanb.19@gmail.com, lovkumar505@gmail.com, sambit9238@gmail.com,
deva.abhi96@gmail.com, srath@nitrkl.ac.in
[2] Department of Information and Communication Technology,
Fakir Mohan University, Balasore 756001, Odisha, India
bmonalisa.26@gmail.com

Abstract. Link prediction is a trending research direction in the field of social network analysis due to its vast application especially in the field of network evolution analysis like discovering missing links, identifying positive and negative links etc. It is also used for recommending commodities in e-commerce sites and suggesting friends in online social network. The objective of this paper is to predict hidden or missing links in a directed or undirected, unweighted social network using distributed platform like Spark, which is found to be both accurate and reasonably robust due to its scalable nature. This approach is found to be effective as compared to conventional methods as it considers number of parameters together such as the influence of a node, community structure of the network and the shortest paths between nodes. This approach is found to be efficient as compared to other path-based approaches as it has comparatively less time complexity and preferable over other node-based approaches owing to its accuracy with reasonable computational time.

Keywords: Online social network · Spark · Link prediction · Similarity · Community detection

1 Introduction

In recent years, Social Network Analysis (SNA) have become an emerging research topic due to the far insights that can be inferred about existing social structures. A major issue in SNA is its exponentially increasing size. Thus, even a high performance system fails to do computation due to its limited computational resources in comparison to the huge size of the data. Distributed computing platforms are quite necessary to handle such enormous data in SNA. Basically, social networks are dynamic in nature, since new edges and vertices

© Springer International Publishing AG 2017
O. Gervasi et al. (Eds.): ICCSA 2017, Part VI, LNCS 10409, pp. 325–334, 2017.
DOI: 10.1007/978-3-319-62407-5_23

are added to the network over time. So, it is very important to study the association between two specific nodes. Link prediction is a process by which future possible links between nodes are being predicted. It is the process of analyzing network evolution which basically depends on the existing structural information in the network. This paper presents a method for link prediction in large-scale social networks, which can be helpful in recommender systems for e-commerce sites, predicting missing links in terrorist networks, suggesting friends in social networking sites etc. [1].

The presented method basically utilizes two main features of the network i.e., local features, based on network structure and global features, based on community structure [2] of the network and influence of each node in the network. To start with, the method first divides the given network into different communities. Then, for each pair of node in a single community a score is being assigned which indicates the probability of that link being realized in recent future. This score is assigned based on certain parameters like mutual neighbors, shortest distance, eigen-vector centrality of nodes. As nodes belonging to same community are more probable of getting associated, proximity score is give to node pairs of same community only. Features like small-world model [3] and scale-free networks [4] have been helpful in reducing the computational complexity of the proposed method.

In subsequent sections, the paper has been organized as follows: In Sect. 2, the related works in link prediction has been discussed. Section 3 presents the motivation behind the proposed research in the field of link prediction. The methodology used for link prediction is presented in Sect. 4. The concept of RDD used in spark has been elaborated in Sect. 5. Sections 6 explains the proposed algorithm. Experimental analysis and comparative study has been presented in Sect. 7. Conclusion and future scope of the work is presented in Sect. 8.

2   Related Work

Liben Nowell and Kleinberg, are among the first few to carry out an extensive research on the uses of topological features of a graph for link prediction [5]. They had concluded that the topological information is highly productive in comparision with random predictor. Many other literatures like the former ones have assigned proximity scores between two nodes, specifying their probability of getting associated [6, 7].

In other recent works on similarity measures, path-based proximity evaluations are either discarded or left for future works due to its high computational cost for large complex networks. Based on small-world hypothesis Papadimitrou, has attempted to solved link prediction problem by traversing all paths of a bounded length in the given network [10]. An effective method known as PropFlow, introduced by Birtan et al. is a variation of PageRank [8]. The basic principle of PropFlow is to include more localized measure of propagation and to minimize the topological noise far from source node. In another approach, it is identified that the most strongly connected components are more desirable due to their high clustering coefficient [9]. Kossinets and Watts analyzed a large-scale social network empirically for finding solution for link prediction [10]. The working principle, behind their proposed method is that two persons having more number of mutual friends are more likely to be associated in the future. Facebook is the one real world example of this approach, where users tend to send or accept friend request on the basis of number of mutual friends thus forming a community. Extensive analysis on disparate networks implies that neighbor similarity plays a measure role in predicting the missing links [11]. In another approach, it is argued that different common neighbors play different roles leading to different contributions to similarity between two nodes [1]. On the basis of this argument, a local Naive Bayes model is being proposed for link prediction. In their literature, Xu Feng proposed that performance of the prediction model can be improved by analyzing its relationships which can be obtained from the network structure [12]. Their proposed method is based on way the structure and clustering in the network affects the prediction of future probable links. They observed that, precision of the method increases significantly with the growth of clustering and structure of the network.

It may be observed that most of the works for link prediction have a limitation of scalability, which is to be overcome in the present scenario due to the exponential increase in size of social networks. The previously proposed methods have trade off with accuracy and execution time to a greater extent. This is due to the fact that they have not been considered number of major influencing factor like importance of nodes, community structure of the graph, path structure between two nodes etc.

3   Motivation

As mentioned earlier, this paper introduces a novel approach to link prediction by considering community structure and influence of nodes in the network, along with mutual neighbours and geodesics between nodes. In a social network, inter-community nodes tend to be associated within themselves more rapidly than intra-community nodes. A node with high influence is prone to make more links than the one with lower influence as more nodes tend to be associated with the former one. The traditional algorithms like preferential attachment [13] and rooted page rank [14] etc. don't consider all these factors together.

To measure the influence of nodes in a network the proposed method uses eigen value centrality measures. Eigen value centrality indicates the extent to which a node is associated with other important nodes [15]. Basically it implies how many popular friends one have rather than number of friends as in a social structure one wants to get associated with an influential person. While calculating geodesics path lengths more than 6 units have been ignored owing to six degree separation of nodes in social networks [16].

4   Methodology

Various existing similarity index used for link prediction problem are presented in Table 1. Following Notations are used to describe the equation of similarity index presented.

- $LP(a, b)$: Link Prediction score between node 'a' and 'b'
- $N(a)$: Neighbouring nodes of 'a'
- β^i: Small constant value raised to path length 'i'
- γ: Real number between 0 and 1
- $path_{a,b}^i$: Path of length i between 'a' and 'b'
- d: Average of $|N(a)| * |N(b)|$ over all node pairs 'a' and 'b'.

Table 1. Different approaches for link prediction

Similarity index	Equation	Description	Complexity
Common neighbour	$LP(a,b) = N(a) \cap N(b)$	More the number of neighbors shared by 2 nodes more is the chance of them getting associated in near future	$O(Nk^2)$
Jaccard's coefficient	$LP(a,b) = \frac{N(a) \cap N(b)}{N(a) \cup N(b)}$	Computes the similarity based on common neighbours	$O(N^2)$
Preferential attachment	$LP(a,b) = N(a) * N(b)$	Nodes with high degree are more likely to form new links	$O(N^2k^2)$
Adamic/adar	$LP(a,b) = \sum_{c \in N(a) \cap N(b)} \frac{1}{log(N(c))}$	A person having more friends is less likely to help in forming relationships between his/her friends than someone with fewer friends	$O(Nk^2)$
Katz measure	$LP(a,b) = \sum_{i=1}^{\infty} \beta^i \left|path_{a,b}^i\right|$	Two nodes having more number of intermediate paths are more prone to be connected in future	$O(N^3)$
Sim rank	$LP_{SR}(x,y) = \gamma \frac{\sum_{a \in N(x)} \sum_{b \in N(y)} LP(a,b)}{|N(x)||N(y)|}$	Nodes with similar neighbours are more prone to get linked in future	$O(N^2d)$

5   SPARK Programming Model

SPARK is an open source cluster computing framework with in-memory processing to speed up analytic applications, upto 100 times faster compared to other

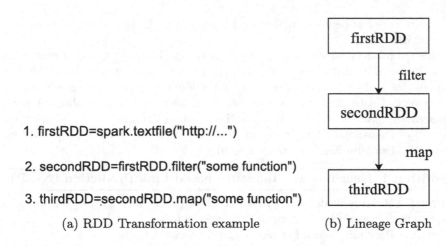

1. firstRDD=spark.textfile("http://...")

2. secondRDD=firstRDD.filter("some function")

3. thirdRDD=secondRDD.map("some function")

(a) RDD Transformation example   (b) Lineage Graph

Fig. 1. An Example for illustrating spark programming model

technologies in the market today. It doesn't have its own distributed file system but, can use Hadoop Distributed File System (HDFS). It has built-in modules for streaming, SQL, machine learning and graph processing.

Resilient Distributed Dataset(RDD) is the basic abstraction on which Spark programming model works. RDDs are immutable and partitioned collection of records, which can only be created by coarse grained operations (map, filter etc.) i.e. the operations that are applied on all elements in a datasets. RDDs generally include two types of operations i.e., transformation and action. Transformations create a new dataset from an existing one and Actions return a value to the driver program after running a computation on the dataset. RDDs can only be created by reading data from a stable storage such as HDFS or by transformations on existing RDDs. map, flatMap, filter are some examples of transformation and collect, saveAsTextFile, count are some examples of Action in spark. Since, RDDs are created over a set of transformations, it logs those transformations, rather than actual data. Graph of transformations to produce one RDD is called as Lineage Graph. In case of loss of some partitions of RDD, we can replay the transformation on that partition in lineage to achieve the same computation, rather than replicating data across multiple nodes. RDD Transformations are lazily evaluated. They do not compute their results right away. Instead, they just remember the transformations applied to some base dataset (e.g. a file). The transformations are only computed based on lineage graph, when an action requires a result to be returned to the driver program (Fig. 1).

6   Proposed Method

The proposed method for link prediction is a mixed approach of local and global features of the network. It has been implemented in Spark in order to be useful in analyze real world large scale data, as presented in Algorithm 1. Here, a proximity score has been calculated for the currently disconnected node-pairs of the network.

This approach considers edge-list of the network G and eigen-vector centrality of each node as its initial input. Firstly, the network gets divided into different communities and then the shortest paths of length upto 4 are calculated, in the network. Customized ring search has been used to calculate shortest path betweeen any pair of nodes as it has a time complexity of $O(n)$ in comparison to $O(n^3)$ of traditional BFS algorithm. The execution flow for the proposed link prediction algorithm has been presented in Fig. 2.

Algorithm 1. Nearness and Influence based Link Prediction (NILP)

Input: The large scale social network $G = (V, E)$ in edge list format and eigen-vector centrality for each node in vector form.
Output: Proximity score between i and j where A[i,j]=0
i.e. node pair between which no edge exists in the given graph.

Step 1: Edge list is processed in flatMap transformation which makes an undirected graph bidirectional. This step can be ignored in case of a directed graph.
Step 2: The processed edge list is passed on to a detect community using Girvan Newmann algorithm.
Step 2: reduceByKey transformation creates an adjacency list from the edge list generated in Step 1.
Step 3: flatMap transformation converts the adjacency list(from Step 2) to a modified adjacency list i.e. key = [Destination Node, Source Node] Value = [Colour Code, Distance, Destination Neighbour List, Source Neighbour List]
Step 4: flatMap and reduceByKey transformations are repeated to calculate geodesics upto length 6 in the given network.
Step 5: The output is then passed to another flatMap transformation where, the link prediction score is evaluated using the proposed method(Equation 6)

$$LS = \frac{1}{d}\left(EC(a) + EC(b)\right)\frac{\sum_{z\in\{N(a)\cap N(b)\}} EC(z)}{\sum_{k\in\{N(a)\cup N(b)\}} EC(k)} \tag{1}$$

where LS is the predicted proximity score between nodes a and b. EC(a) and EC(b) are the eigen-vector centrality scores of nodes a and b respectively. d is the shortest distance between a and b(d is 6 for nodes with the shortest path length more than 6 based on small-world hypothesis [16]).

---

7   Experimental Setup and Comparative Analysis

7.1   Experimental Setup

Proposed link prediction method has been implemented on five real world social network datasets as listed in Table 2. All the experiments have been carried out

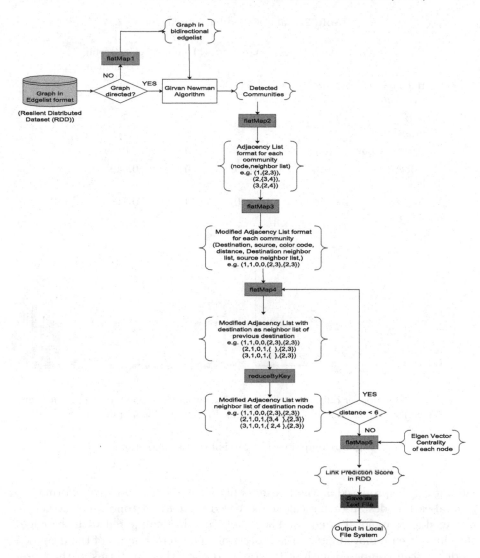

Fig. 2. Execution flow for nearness and influence based link prediction (NILP)

on a cluster of 5 computing nodes each with i7 processor 3.4 GHz clock speed. Master node has 1 TB hard disk and 10 GB RAM while, 4 worker nodes each with 1 TB hard disk and 20 GB of RAM.

7.2 Comparative Analysis

In order to check the performance of the proposed algorithm, at the beginning of the processing 10 percent of randomly chosen edges are removed from the network. Proximity score between currently unconnected nodes are then calcu-

Table 2. Dataset used for experiment

Datasets	No. of nodes	No. of edges	Diameter	Clustering coefficient
Hamsterster friendship	1858	12534	14	0.901
Jazz musician	1133	2742	6	0.52
Ego facebook	2888	2981	9	0.0359
Adolescent health	2539	12969	10	0.142
Arxiv astro-ph	18771	198050	14	0.318

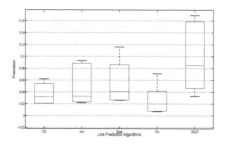

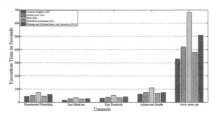

(a) Precision boxplot for different Link Prediction Algorithms

(b) Execution time comparison for different Link Prediction Algorithms

Fig. 3. Comparative analysis for link prediction

lated using the proposed method in spark platform. Global and local information around each nodes are being calculated based the current topological structure obtained after removal of edges. The results are then compared with the edges that have been removed before the processing. The performance of the proposed model is then compared with other standard algorithms in terms of their precision and execution times. Graphical comparison has been presented in Fig. 3. The execution times of different methods are given in Table 4.

Figure 3a presents the boxplot of precision values of different link prediction methods. Here it can be clearly observed that the proposed method performs well as compared to both local-based and global-based approaches in terms of precision. In the given datasets, NILP is approximately twice better than the Adamic Adar. Thus, its precision is effectively higher than all node based approaches. The precision of NILP is also quite higher than katz, thus proving itself to be better than the path based approaches. Figure 3b presents the comparison of execution times for different link prediction approaches. Here, proposed method is quite faster than simple path based approach like katz measure. However, the computational speed of NILP is comparable with adamic adar and little higher than other node based approaches.

From the detailed observation and analysis, it can be concluded that NILP outperforms all the path based approaches in terms of both precision and computational time and it has better precesion value than all the node based link prediction algorithms. From Fig. 3a, it can be observed that NILP performs faster than Katz with higher precision value. Link Prediction using Preferential Attachment algorithm is found to have lowest precision value. When execution time is the major parameter for evaluation, time-accuracy trade off comes into picture. It may be observed that NILP has higher computational complexity for small scale datasets. However, It is the most preferable algorithm for processing large scale real world complex network. The scalable nature and spark implementation of NILP, might be the suitable combination, which can perform better on large-scale networks than rest of the techniques.

Table 3. Precision for different link prection approach

Network	CN	AA	Katz	PA	NILP
Hamsterster friendship	0.021	0.022	0.026	0.020	0.085
Jazz musician	0.021	0.024	0.041	0.007	0.051
Ego facebook	0.032	0.033	0.027	0.008	0.033
Adolescent health	0.051	0.087	0.076	0.032	0.169
Arxiv astro-ph	0.062	0.093	0.116	0.071	0.155

Table 4. Execution time in sec.

Network	CN	AA	Katz	PA	NILP
Hamsterster friendship	44	53	74	46	59
Jazz musician	16	27	36	25	28
Ego facebook	32	39	53	36	41
Adolescent health	62	76	110	69	75
Arxiv astro-ph	329	420	680	379	509

8   Conclusion and Future Work

In this paper, an effort has been made to enhance the accuracy and speed of the link prediction algorithm, which is evident from the comparative study with large data sets. The method aims to enhance accuracy of link predictions by considering community structure, influence of nodes and neighbor similarity of nodes in the given network. In order to increase the robustness of the method, the method is implemented in spark platform to distribute the computation task in multiple nodes.

This work can be further extended to find similar nodes in the given graph considering contents of the nodes along with the positioning of the nodes. A measure of influence of one community on the other can also be measured by considering the few links between these communities and their edge betweenness value.

References

1. Liu, Z., Zhang, Q.-M., Lü, L., Zhou, T.: Link prediction in complex networks: a local naïve bayes model. EPL (Europhys. Lett.) **96**(4), 48007 (2011)
2. Newman, M.E.J.: The structure and function of complex networks. SIAM Rev. **45**(2), 167–256 (2003)
3. Hiromoto, R.E.: Parallelism and complexity of a small-world network model. Int. J. Comput. **15**(2), 72–83 (2016)
4. Warren, C.P., Sander, L.M., Sokolov, I.M.: Geography in a scale-free network model. Phys. Rev. E **66**(5), 056105 (2002)
5. Guy, I.: Social recommender systems. In: Ricci, F., Rokach, L., Shapira, B. (eds.) Recommender Systems Handbook, pp. 511–543. Springer, Heidelberg (2015)
6. Koren, Y., North, S.C., Volinsky, C.: Measuring and extracting proximity in networks. In: Proceedings of the 12th ACM SIGKDD International Conference on Knowledge Discovery and Data Mining, pp. 245–255. ACM (2006)
7. Sarkar, P., Moore, A.W., Prakash, A.: Fast incremental proximity search in large graphs. In: Proceedings of the 25th International Conference on Machine Learning, pp. 896–903. ACM (2008)
8. Bütün, E., Kaya, M., Alhajj, R.: A new topological metric for link prediction in directed, weighted and temporal networks. In: 2016 IEEE/ACM International Conference on Advances in Social Networks Analysis and Mining (ASONAM), pp. 954–959. IEEE (2016)
9. Yang, J., Leskovec, J.: Defining and evaluating network communities based on ground-truth. Knowl. Inf. Syst. **42**(1), 181–213 (2015)
10. Kossinets, G., Watts, D.J.: Empirical analysis of an evolving social network. Science **311**(5757), 88–90 (2006)
11. Lü, L., Jin, C.-H., Zhou, T.: Similarity index based on local paths for link prediction of complex networks. Phys. Rev. E **80**(4), 046122 (2009)
12. Feng, X., Zhao, J.C., Xu, K.: Link prediction in complex networks: a clustering perspective. Eur. Phys. J. B **85**(1), 3 (2012)
13. Jeong, H., Néda, Z., Barabási, A.-L.: Measuring preferential attachment in evolving networks. EPL (Eur. Lett.) **61**(4), 567 (2003)
14. Liben-Nowell, D., Kleinberg, J.: The link-prediction problem for social networks. J. Assoc. Inf. Sci. Technol. **58**(7), 1019–1031 (2007)
15. Ruhnau, B.: Eigenvector-centralitya node-centrality? Soc. Netw. **22**(4), 357–365 (2000)
16. Wang, X.F., Chen, G.: Complex networks: small-world, scale-free and beyond. IEEE Circuits Syst. Mag. **3**(1), 6–20 (2003)

Requirements Engineering for Cloud Systems: A Mapping Study Design

Fernando Wanderley[1,2](✉), Eric Souza[2], Miguel Goulão[2], Joao Araujo[2],
Gilberto Cysneiros[3], and Ananya Misra[4]

[1] CCT, Universidade Católica de Pernambuco, UNICAP, Recife, Brazil
fernando@unicap.br
[2] NOVA LINCS, Universidade Nova de Lisboa, UNL, Lisbon, Portugal
[3] Universidade Federal Rural de Pernambuco, UFRPE, Recife, Brazil
[4] Middle East Technical University, NCC, Ankara, Turkey

Abstract. Cloud Computing gets increasingly established in industrial practice as an option for modelling cost-efficient and demand-oriented information systems. Despite the increasing acceptance of cloud computing within the industry, many important questions remain unanswered. Issues related to the best software architectures decisions for cloud-based systems are faced with the question of appropriate techniques applying at early phase like requirements engineering. The goal of this paper defines a design of a mapping study to verify and identify the existence of relevant research gaps, which refers mainly to requirements models, tools or methods for cloud systems (SaaS). The conclusion of this mapping study design reinforces and actively encourages the necessity of the complete execution (and replication) of a systematic mapping study regarding the synergy of requirements engineering (e.g.: with model-driven issues) applied for a cloud computing.

Keywords: Requirements engineering · Cloud systems · SaaS · Systematic review · Mapping study

1 Introduction

Cloud computing is considered a benefit for the small businesses because through it they will have access to technologies that before weren't accessible for them in terms of money spending; and these is an advantage for them because they can start competing with other small businesses or even with big ones. The cost implied for someone to come and fix/install an application will be cut down and the company will save money, it is cheaper to use applications that are on cloud then to buy other ones, there is the possibility to use one multi-application cloud service for all the needs of the company, the applications that exist on the cloud will integrate perfectly within the company because of the API that is helping to find the application that is compatible with the companies goals. Because cloud computing is updated regularly the company doesn't need to spend money for this. Cloud computing is a way for companies to cut the expenses of the company.

© Springer International Publishing AG 2017
O. Gervasi et al. (Eds.): ICCSA 2017, Part VI, LNCS 10409, pp. 335–349, 2017.
DOI: 10.1007/978-3-319-62407-5_24

Therefore, cloud computing has the potential to meet both enterprises' and individual end-users' needs, as observed by Marston *et al.* [4] and Kim [1]. While cloud computing has already found its way into industrial practice, there continue to be significant deficits in the scientific basis [1]. One such shortfall is requirements engineering (RE) for cloud computing. While some initial research initiatives have been carried out under the sub-domain of Software as a Service (SaaS) [1,3], none has yet been made for cloud computing overall.

Cloud computing brings numerous challenges in this area since the traditional methods need to be adapted and new RE methods [5], including the requirements modelling, has to be investigated. More specifically, the success of adopting the new paradigm highly depends on the degree to which requirements are correctly understood by both service providers and consumers (in cloud context) [5,6]. Thus, as per context described above, the primary goal of this paper is to produce a consistent design of a systematic mapping study [8]. The protocol proposed here has an intentionality to find out a brief overview of the current practice, in an industry and academic contexts, for requirements engineering (eliciting and modelling) approaches for cloud computing services (especially for SaaS platform).

Thus, a protocol was developed to define the main guidelines for conducting this study. According to Brereton et al. [9], a systematic mapping study is used to describe the kinds of research activity that have been undertaken and describes the studies rather than extracting specific details. That is, it does provide a context for the later synthesis. According to Kitchenham [7], a systematic map is a method that can be conducted to get an overview of a particular research area. After this, the state of evidence on specific topics can be investigated using a systematic review, if necessary. According to Budgen and others [10], in systematic mapping study, the research question itself is likely to be much broader than in a systematic review. It is necessary, to address the wider scope of the study. For constraints questions (this paper reports a just pilot) for a systematic mapping study in the available literature, including academic and industrial publications. It's important to detach that our first goal was to verify the research gaps in the requirements engineering deeply (especially requirements elicitation and modelling activities for cloud systems - SaaS platform). The mapping study [7–9] objective was to verify if exists relevant gaps, challenges and opportunities for requirements engineering researchers, as well as to guide practitioners regarding describing what is involved in the adoption of current cloud technologies.

Our research method and findings are described in the remainder of this paper. First, we discuss a brief background (Sect. 2) on cloud computing and related concepts; Next (Sect. 3), we describe the systematic mapping study process, including the search protocol (search string and index research databases), the adopted criteria and some interesting data related to the study. In Sect. 4 we present the pilot execution. Section 5 presents the result of the pilot execution. The next section (Sect. 6) we discussion main threats to validity identified in the review. Section 7 describe a brief discussion of results and, finally, (Sect. 8) the conclusions of this study.

2   Background

Cloud computing is a way of computing that has as main base sharing computing resources instead of having local servers or personal devices to give access to applications. The word cloud from cloud computing is used as a another name for the internet, so cloud computing means a type of internet based computing, where services are delivered to organizations with the help of the internet. The concept of cloud did not arise as a new technology model but as the integration of technologies from the past [3], which resulted in a new way to use and provide computing power as a service through the Internet.

The impact of cloud computing brings changes not only in terms of the global performances of a company, but also in terms of internal organization, especially in the IT department. This opportunity is modifying the usual methods of back-up for data, cloud computing is bringing new tools and new perspectives of evolution for the company that is using it. *"Clouds are a large pool of easily usable and accessible virtualized resources (such as hardware, development platforms and/or services). These resources can be dynamically reconfigured to adjust to a variable load (scale), allowing also for optimum resource utilization. This pool of resources is typically exploited by a pay-per-use model in which guarantees are offered by the Infrastructure Provider by means of customized Service Level Agreements SLAs"* [3,4].

One technical commonly cited and largely accepted definition is provided by the United States Government's National Institute of Standards and Technologies (NIST) [2], which in its 16th and final report related to this area in 2011:

"Cloud computing is a model for enabling ubiquitous, convenient, on-demand network access to a shared pool of configurable computing resources (e.g., networks, servers, storage, applications, and services) that can be rapidly provisioned and released with minimal management effort or service provider interaction". The cloud model, mentioned in the NIST's definition, is composed of five essential characteristics: on-demand self-service, broad network access, resource pooling, rapid elasticity and measured service.

3   A Systematic Mapping Study Process

In this study, as a mapping study pilot to verify the feasibility of the future replication, we followed a formal systematic literature review process [7–9]. A systematic mapping study (as well a Systematic Literature process) proposes a fair assessment of the research topic as it uses a rigorous and reliable review methodology, together with auditing tasks to reduce the researcher bias [9].

There are several reasons to perform a systematic re view, and the usual ones are [8,9]:

- To review existing evidences about a reminder or a technology;
- To identify research gaps in current research;

– To provide a framework/background for new research activities and;
– To support the generation of new hypotheses.

Based on the motivation described above, the second reason fits the purpose of this review. Our goal in this definition of a pilot of the mapping study is very if exist relevant research gaps, forwarding to a replication (more rigorous) in a systematic study of this topic. Furthermore, the initial results achieved with the review can also provide a background for new researchers interested in requirements engineering (eliciting methods, models or tools) for cloud computing environments.

The systematic review described here was based on Kitchenham and Charters' guidelines [7], which is divided in three main phases: Planning, Conduction and Reporting. Each of these phases contains a sequence of stages, but the execution of the overall process involves iteration, feedback, and refinement of the defined process, according to Fig. 1.

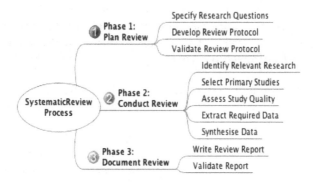

Fig. 1. Mind map of systematic review.

3.1   Specify Research Question

As described before, the objective of this review is to find out answers about the *current state of the art and current challenges of the requirements engineering approaches for cloud computing.*

According to the systematic review process [7,8], we frame our research goal according to the **PICOC** (**P**opulation, **I**ntervention, **C**omparison, **O**utcome, and **C**ontext) structure (Table 1).

Thus, the research goal was defined as:

"What requirements engineering approaches have been proposed for cloud computing?"

Table 1. PIPOC applying

Population	The most recent works related with the treatment of the requirements engineering approaches (elicitation, analysis or modelling tasks) with the new issues and challenges introduced by cloud computing environment. Thus, Requirements Engineers and Requirements Engineering researchers compose the population, seeking to provide a set of new challenges through of the current state of the art, as well as the others stakeholders like Cloud Service Providers, Cloud Service Consumers and Cloud Server Creators
Intervention	Requirements engineering approaches for development of the cloud-based systems
Comparison	Not applicable: our intention is to classify the existing requirements engineering approaches for cloud computing to identify challenges and the current state of the art, not to compare the approaches with other approaches
Outcomes	The objective of this study is revealing existing gaps between requirements engineering approaches and the new dynamic environment of the cloud computing
Context	Research papers. We are working in a research context with experts in the domain as well as other practitioners, academics, consultants and students

3.2   Search String and Research Sources

Based on the structure and the research question, keywords were extracted and used to search the primary studies. Furthermore, sophisticated search strings could then be constructed using Boolean *AND* and *OR* operators. Thus, we final search string was defined as: *"requirements engineering" AND "cloud computing"*. In the next sections (Discussion) we'll detail the reason and a primary rationale to simplify the research string only in these two keywords.

The search for primary studies was based on the following digital libraries: ACM Digital Library, IEEE Computer Society Digital Library, SpringerLink, and Science Direct. These searches had as target some journals and conferences, which are detailed in *Appendix A*, but if relevant results from different journals or conferences were found, they were not discarded. These libraries were chosen because they are some of the most relevant sources in software engineering [8, 9].

3.3   Primary Study Selection

We defined inclusion and exclusion criteria to help selecting the relevant studies for analysis and data extraction through reading of following studies' sections: title, abstract, and conclusion. We included peer-reviewed papers from journals and conferences, that presented requirements engineering approaches for cloud computing (I1). Additionally, we plan to use snowball search by including relevant studies cited by authors of the papers we read during the conduction process

(I2). On the other hand, we excluded informal literature (slide shows, conference reviews, informal reports), secondary and tertiary studies (reviews, surveys) and studies from conferences, workshops and journals without peer-review (E1), duplicated studies (E2), studies that did not answer the research questions(E3), studies that were not written in English (E4), and papers that were not available for download from the source bases (E5).

Aiming at improving the understanding of the area and facilitating the data extraction, we decomposed our Research Question according to two perspectives: context and validation. In relation to context perspective, we wish to analyse: *What is the requirements model used by approach?*, *What is the requirements engineering area that the approach has focused?*, and *What is the coverage of approach to cloud infrastructure with respect to SaaS, Paas or IaaS?*

Regarding the validation perspective, we wish to analyze: *How is the approach evaluated?*

3.4 Quality Assessment

Although there is no agreed definition of what a high quality study is, there is a common agreement that the quality of the chosen primary studies is critical for obtaining trustable results in systematic literature reviews and mapping studies. Thus, we will select studies published in the best conferences and journals of area. In *Appendix A* we described some of these journals and conferences.

4   Pilot Execution

This work aims to perform a pilot of research protocol to evaluate its complete applicability in the future. In addition, this paper is a deliverable to Empirical Software Engineering discipline of the doctoral course in Science Computer of New University of Lisbon.

For execution of the pilot, we ran the search string in IEEE Xplore (returning 48 studies), ACM Digital Library (returning 4 studies), Science Direct (returning 97 studies) and SpringerLink (returning 270 studies). Furthermore, we selected a subset of studies (first four papers) returned from each source library to evaluate the research protocol. Table 2 depicts the set of evaluated studies.

4.1 Filtering Process

After to define the studies that should be evaluated, we applied the inclusion and exclusion criteria (with exception of I2 - snowballing approach) and read the title, abstract and conclusion of selected studies; we discovered that some topics of the papers were different to that of our study. These filtering process decreased the number of selected studies from 12 to 9. The Table 3 lists the result of the filtering process.

Once the filtering process was completed, we verified the journals and conferences where the studies were published to assess the quality of our research. This process eliminate more 2 studies (ID8 and ID9).

Table 2. Set of evaluated studies

ID	Paper title	Source
1	Table of Contents (CLEI)	IEEE
2	Modelling secure cloud systems based on system requirements	IEEE
3	Requirements Engineering process for Software-as-a-Service (SaaS) cloud environment	IEEE
4	Pattern-Based Support for Context Establishment and Asset Identification of the ISO 27000 in the Field of Cloud Computing	IEEE
5	Crowd-centric Requirements Engineering	ACM
6	Cloud adoption: prioritizing obstacles and obstacles resolution tactics using AHP	ACM
7	Towards bridging the communication gap between consumers and providers in the cloud	ACM
8	Cloud adoption: a goal-oriented requirements engineering approach	ACM
9	Global Collaboration Requirement Analysis System in Cloud Computing	ScienceDirect
10	A goal-oriented simulation approach for obtaining good private cloud-based system architectures	ScienceDirect
11	Energy-Aware Profiling for Cloud Computing Environments	ScienceDirect
12	Cost-aware challenges for workflow scheduling approaches in cloud computing environments: Taxonomy and opportunities	ScienceDirect
13	Requirements Engineering for Cloud Computing in University Using i*(iStar) Hierarchy Method	SpringerLink
14	Requirements Engineering for Security, Privacy and Services in Cloud Environments	SpringerLink
15	A Methodology for the Development and Verification of Access Control Systems in Cloud Computing	SpringerLink
16	Requirements Engineering for Cloud Computing: A Comparison Framework	SpringerLink

Table 3. Filtering process

Criteria	IEEEXplore	ACM	ScienceDirect	Springerlink	Total
I1	ID1, ID2, ID3, ID4	ID5, ID6, ID7, ID8,	ID9, ID10, ID11, ID12	ID13, ID14, ID15, ID16	16
E1	ID1	-	-	-	15
E2	-	-	-	-	15
E3	-	-	ID11, ID12	ID15	12
E4	-	-	-	-	13
E5	-	-	-	ID13, ID14, ID16	9

5   Pilot Result

This section presents the results of the mapping study (piloting) for the resulting research questions. Initially, we present a demographic data and, soon after, the research result.

5.1   Demographic Data

The goal of our demographic analysis was verify where the requirements engineering for cloud approaches have been published. The Fig. 2 shows the distribution in relation to the research databases, using as main databases of our protocol. The SpringerLink has the major percentage (64%), the second one was the Science Direct with 23%, after that the IEEE with 12% and with 4 studies (1%) of total the ACM portal.

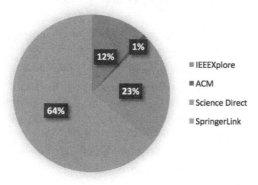

Library Distribution

Fig. 2. Library distribution

The result of this pilot shows that most publications have been published in conferences (86%), follows to journals (14%). Despite the count of selected studies is not statistically significant, we suspected that most part of publication is in conferences because the approaches are not mature enough, maybe because the research area is still recent. The Fig. 3 presents the venue distributions.

5.2   What is the Requirements Model Used by Approach?

The authors did not specify a requirement model used by their approaches in 3 studies (ID3, ID5, and ID7). The Fig. 4 presents the distribution of studies according the requirements model. In ID3 the authors evolve by considering a CMMI modification by adding a new element in such a process, specifically devoted to SaaS and proposed a modification in the traditional Requirements

Venue Distribution

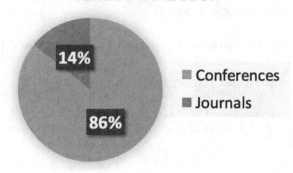

Fig. 3. Venue distribution

RE model

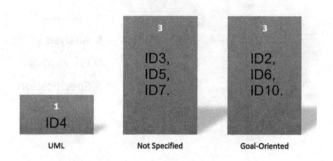

Fig. 4. Requirement model distribution

Engineering process. The ID5 presents a method for requirements engineering where users become primary contributors, resulting in higher-quality requirements and increased user satisfaction. Finally, the study ID7 presents a platform which will act as a cloud resources marketplace, allowing consumers to input their needs and providing them with matching cloud services. The most requirement model used in the approaches was goal-oriented (ID2, ID6, and ID10). In ID2, the authors demonstrate how components of the cloud infrastructure can be identified from existing security requirements models using goal-oriented model. The ID6 proposes a novel systematic method for prioritising obstacles and their resolution tactics using Analytical Hierarchy Process (AHP). Finally, the study ID10 proposes a goal-oriented simulation approach for cloud-based system design whereby stakeholder goals are captured, together with such domain characteristics as workflows, and used in creating a simulation model as a proxy for the cloud-based system architecture. The ID4 is the unique study using UML. It presents a way to support the asset identification described in ISO 27005 focusing on the scope of cloud computing systems. The ISO 27005 is a well-established series of information security standards.

5.3 What is the Requirements Engineering Area the Approach Has Focused?

The two most Requirements Engineering area which the studies have focused is Requirements Analysis (37%) and Requirements Elicitation (27%), followed by Requirements Prioritization (18%) and Requirements Specification (18%). The Fig. 5 presents the distribution of studies according the Requirements Engineering Area.

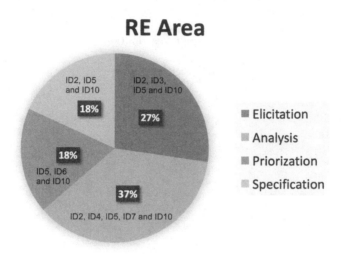

Fig. 5. Requirements engineering area distribution

5.4 What is the Coverage of Approach with Respect to Cloud Service Model (SaaS, Paas or IaaS)?

Regarding the coverage of approaches with respect to Cloud Service Model, we verified that SaaS was the most researched cloud layer with 5 studies (ID2, ID3, ID4, ID7, and ID10), followed by PaaS with two studies (ID4 and ID10). We also note that two studies does not specify the cloud service layers (ID4 and ID10) and no study focuses on IaaS. It is important to notice that all studies focusing on PaaS, also focus on SaaS. The Fig. 5 depicts the Cloud Coverage Distribution (Fig. 6).

5.5 How is the Approach Evaluated?

The most part of approaches found have been evaluated through examples (57%) and, subsequently, case study (29%). This suggests that the area of research is still immature. The Fig. 7 depicts the study evaluation Distribution.

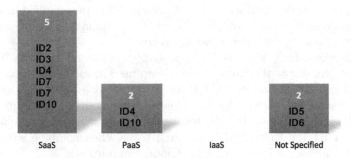

Fig. 6. Cloud coverage distribution

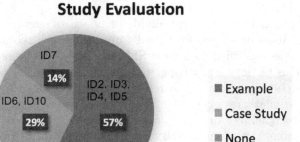

Fig. 7. Evaluation distribution

6 Threats to Validity

The threat to validity of this study are related to potential problems in the completeness of our search queries, the primary studies selection process, and potential inaccuracies in data extraction, classification and interpretation.

Because of this, all steps were independently validated by Ph.D. students who are not authors of this paper. The goal was to mitigate possible bias in selecting or interpreting the studies, hence minimizing the risk of not including the relevant papers or not interpreting the author's goal correctly.

7 Discussion

This work had an initial motivation: to elaborate a protocol of a mapping study with focus on identify relevant research gaps in requirements modelling strategies (models or tools) for cloud systems, specially to SaaS platform. So, the first search string that we had defined cited keywords as: "model-driven" and their variations with "Requirements Analysis" OR "Requirements Elicitation" for examples, but the first calibration test of this string showed us a minimal set of works (models or tools) using model-driven techniques applied in Requirements Engineering for Cloud Systems. This is the rationale to define our search string more embracing with these two keywords.

One first reason for this after-clap is probably correlated with the "Cloud" - is new hype technology - this term has also been used in various contexts such as describing large ATM networks in the 1990s. However, it was after Google's CEO Eric Schmidt used the word to describe the business model of providing services across the Internet in 2006, that the term really started to gain popularity. So, it's a good indication that is a hot research topic when combine with Model-Driven in Requirements Engineering for Cloud Computing.

What is point out here for us, which a good challenge for new Requirements Engineers Researcher, of course, with a clear care of execute a systematic study to prove this hypothesis.

Regarding to increase our suspicion about no exists relevant primary studies point out RE tools of models for Cloud Computing, the recent tertiary study [21] reported has identified 53 unique SLR from 64 publications in the period 2006–2014. This tertiary study paper represents the first ever tertiary study in the RE research literature and just reinforce the necessity of the execution of a formal and systematic review process in our topic here: requirements engineering (eliciting and modelling) for cloud Computing.

Also according this recent tertiary study, two central SLR(s) was performed with the goal to identify gaps in Security Requirements (Non-Functional Requirements) for cloud systems and cloud infrastructure.

Thus, there are many primary studies referencing the methods (or process) and tools applied in requirements analysis, see Fig. 7, (notedly non-functional requirements - security requirements - and important studies with goal-oriented approaches to verify the essential conflicts in choice of cloud providers to establish the cost involved.

Another important point of this protocol is the exclusion criteria definition, the quality assessment. Because, the requirements engineering with model-driven for cloud is very recent topic (all these areas together). For this reason, some papers probably are founded in work in progress, and thus, indexed in Workshops or in a not very relevant conferences. In others words, this protocol requires a reflection about the quality assessment of the "work in progress" intending improve the set of selected papers and consequently their discussion about our research questions.

And a final and not more important question was an limited access of the important papers (e.g.: ID13, ID14 e ID16) letting us with a just choice of the send an email to authors. So, we need more time, waiting their considerations in send us his papers.

8 Conclusion

The goal of this work was identify the feasibility of the protocol in a future replication on the systematic literature review process. This paper aims to perform a pilot execution of a research protocol to evaluate the complete applicability of a systematic mapping study. Our research question, *"What requirements engineering approaches have been proposed for cloud computing?"*, was analysed from the point of view of its context and evaluation perspectives. From the initial 419 papers obtained by querying the most relevant four research libraries, we selected a subset these studies only for evaluation. The results obtained in this pilot encourages to the full execution of research as future work.

Acknowledgment. This research is supported by the UNICAP, NOVA LINCS Research Laboratory (Ref. UID/CEC/04516/2013), and the programa Ciência sem Fronteiras from CAPES.

Appendix A: Journals and Conferences

The selected target journals were:

- ACM Computing Survey;
- Requirements Engineering Journal;
- Annals of Software Engineering;
- IEEE Software;
- IEEE Transactions on Software Engineering;
- International Journal of Systems and Service oriented Engineering;
- Journal of Network and Computer Applications;
- Information and Software Technology;
  Journal of Systems and Software;
- Software and System Modelling;
- Computers and Electrical Engineering;
- Software Practice and Experience; and
- Future Generation Computer Systems;

And the target conferences were:

- Computer Software and Applications Conference (COMPSAC);
- International Requirements Engineering (RE);
- International Conference on Software Engineering (ICSE);
- International Conference on Cloud Computing (CLOUD);
- International Conference on Web Services (ICWS);
- International Conference on Service Computing (SCC).

References

1. Kim, W.: Cloud computing: today and tomorrow. J. Object Technol. **8**(1), 65–72 (2009)
2. NIST Definition of cloud computing v15, NIST, (ed.) National Institute of Standards and Technology, Gaithersburg, MD (2009)
3. Buyya, R., Yeo, C.S., Venugopal, S., Broberg, J., Brandic, I.: Cloud computing and emerging IT platforms: vision, hype, and reality for delivering computing as the 5th utility. Future Gener. Comput. Syst. **25**(6), 599–616 (2009)
4. Marston, S., Li, Z., Bandyopadhyay, S., Zhang, J., Ghalsasi, A.: Cloud computing - the business perspective. Elsevier Decis. Support Syst. **51**, 176–189 (2011)
5. Verlaine, B., Jureta, I.J., Faulkner, S.: Towards conceptual foundations of requirements engineering for services. In: Proceedings of RCIS 2011: 5th International Conference on Research Challenges in Information Science, Gosier, pp. 1–11 (2011)
6. Liu, L., Yu, E., Mei, H.: Guest editorial: special section on requirements engineering for services - challenges and practices. IEEE Trans. Serv. Comput. **2**(4), 318–319 (2009)
7. Kitchenham, B.: Guidelines for performing Systematic Literature Reviews in Software Engineering. EBSE Technical report, Version 2.3 (2007)
8. Petersen, K., Feldt R., Mujtaba, S., Mattsson, M.: Systematic mapping studies in software engineering. In: 12th International Conference on Evaluation and Assessment in Software Engineering (2008)
9. Brereton, P., Kitchenham, B., Budgen D., Turner M., Khalil, M.: Lessons from applying the systematic literature review process within the software engineering domain. J. Syst. Softw. pp. 571–583 (2007)
10. Budgen, D., Turner, M., Brereton, P., Kitchenham, B.: Using mapping studies in software engineering. Proc. PPIG **2008**, 195–204 (2008)
11. Vaquero, L.M., Rodero-Merino, L., Caceres, J., Lindner, M.: A break in the clouds: towards a cloud definition. ACM SIGCOMM Comput. Commun. Rev. **39**, 1 January 2009
12. Alford, T., Morton, G.: The Economics of Cloud Computing. Booz Allen Hamilton (2009)
13. Wang, L., von Laszewski, G., Younge, A., He, X., Kunze, M., Tao, J., Fu, C.: Cloud computing: a perspective study. New Gener. Comput. **28**(2), 137–146 (2010)
14. Chen, Y., Li, X., Chen, F.: Overview and analysis of cloud computing research and application. In: 2011 International Conference on E-Business and E-Government (ICEE), pp. 1–4 (2011)
15. Candea, G., Bucur, S., Zamfir, C.: Automated software testing as a service. In: Proceedings of 1st ACM Symposium on Cloud Computing. SoCC 2010, pp. 155–160. ACM, New York (2010)
16. Chan, W.K., Mei, L., Zhang, Z.: Modeling and testing of cloud applications. In: Services Computing Conference, APSCC 2009, pp. 111–118. IEEE Asia-Pacific (2009)
17. Armbrust, M., Fox, A., Griffith, R., Joseph, A.D., Katz, R., Konwinski, A., Lee, G., Patterson, D., Rabkin, A., Stoica, I., Zaharia, M.: A view of cloud computing. Commun. ACM **53**(4), 50–58 (2010)
18. Rimal, B.P., Choi, E., Lumb, I.: A taxonomy and survey of cloud computing systems. In: Fifth International Joint Conference on INC, IMS and IDC, NCM 2009, pp. 44–51 (2009)

19. Chung, L., et al.: A goal-oriented simulation approach for obtaining good private cloud-based system architectures. J. Syst. Softw. (2012)
20. Zardari, S., Bahsoon, R.: a goal-oriented requirements engineering approach. In: Proceedings of 2nd International Workshop on Software Engineering for Cloud Computing, pp. 29–35
21. Bano, M., Zowghi, D., Ikram, N.: Systematic reviews in requirements engineering: a tertiary study. In: EmpiRE, pp. 9–16 (2014)

Improved Energy-Efficient Target Coverage in Wireless Sensor Networks

Bhawani S. Panda[1], Bijaya K. Bhatta[1(✉)], and Sambit Kumar Mishra[2]

[1] Indian Institute of Technology Delhi, New Delhi, India
{bspanda,bijaya}@maths.iitd.ac.in
[2] National Institute of Technology, Rourkela, India
skmishra.nitrkl@gmail.com

Abstract. Achieving optimal field coverage is a significant challenge in various sensor network applications. In some specific situations, the sensor field (target) may have coverage gaps due to the random deployment of sensors; hence, the optimized level of target coverage cannot be obtained. Given a set of sensors in the plane, the target coverage problem is to separate the sensor into different groups and provide them specific time intervals, so that the coverage lifetime can be maximized. Here, the constraint is that the network should be connected. Presently, target coverage problem is widely studied due to its lot of practical application in Wireless Sensor Network (WSN). This paper focuses on target coverage problem along with the minimum energy usage of the network so that the lifetime of the whole network can be increased. Since constructing a minimum connected target coverage problem is known to be NP-Complete, so several heuristics, as well as approximation algorithms, have been proposed. Here, we propose a heuristic for connected target coverage problem in WSN. We compare the performance of our heuristic with the existing heuristic, which states that our algorithm performs better than the existing algorithm for connected target coverage problem. Again, we have implemented the 2-connected target coverage properties for the network which provide fault tolerance as well as robustness to the network. So, we propose one algorithm which gives the target coverage along with 2-connectivity.

Keywords: Wireless Sensor Networks · Topology control · Minimum spanning tree · Target coverage

1 Introduction

The Wireless Sensor Network constitute a larger set of sensor nodes with different capabilities. These nodes are supplied with sensing, computing and transmitting environmental information. Some existing facts are to be monitored for the deployment of sensors in WSN. WSN also works in a verity of domains such as water quality monitoring, health monitoring, and environmental monitoring. Each sensor has a sensing area and communicating area. Reduction of consumption of power [5,6] and extending the network lifetime are the main challenging

© Springer International Publishing AG 2017
O. Gervasi et al. (Eds.): ICCSA 2017, Part VI, LNCS 10409, pp. 350–362, 2017.
DOI: 10.1007/978-3-319-62407-5_25

activity in wireless networks. Here, the primary intention is to expand the network lifetime of the whole system. If sensor S1 can sense a target T1, then the target T1 is covered by the sensor S1, and if the sensor S2 is within the communicating range of a sensor S1, then that sensor S1 can communicate with sensor S2.

Our main approach for lifetime maximization of WSN is to maximize the number of target coverage sets. Then, at a time sensors of only one target coverage set are in active state and remaining all sensors are in sleeping state. Because of sensors, those are not in the active state not losing their energy; more energy remains saved. One of the most important limitations of WSN is that each node is restricted by a certain energy level. That is why scheduling of sensor nodes for the target coverage set is most important. Sensing is the major responsibility of the sensor network [8]. So, sensors those are in the active state must be connected to communicate with each other in the network. Single connectivity among sensors is not adequate for the network system in light of the fact that a single defect could break down the entire system. Therefore a 2-connected target coverage problem is discussed in this paper to achieve fault tolerance. Similarly, single target coverage is additionally not adequate. We can find several literature surveys related to target coverage from [3,4,10,11,18].

Connectivity suggests there is a path among any active sensors pair and 2-connected means connectivity still accomplish after a single failure. 2-target coverage implies each target is covered by not less than two sensors. The outline of this paper is arranged as follows. Section 1.1 describes the preliminaries and modeling of the problem. Section 2 describes the related work and also explain the variants of a coverage problem. In Sect. 3, we provide a heuristic algorithm for energy-efficient target coverage problem. Section 4 shows simulation results between existing algorithm and our purposed heuristics. In Sect. 5, we give a procedure for establishment of 2-connectivity with example, and Sect. 6 tells the conclusion of the research.

1.1 Preliminaries and Modeling the Coverage Problem

Let $S = \{s_1, s_2...s_n\}$ be a set of n-radio stations placed in an Euclidean space and $T = \{t_1, t_2, ...t_m\}$ be a set of targets covered by the sensors in set S. Here $d(x, y)$ is the Euclidean distance from the node x to node y. The range assigned for each node $r \in R$ is given by range of r, which is a positive real no. A signal transmitted from one node s to another node t with power p_s is attenuated by a factor $p_s = p_s/dist(s, t)^{\alpha}$ where $dist(s, t)$ is the euclidean distance between the node s to t and α is the distance power gradient, which may vary between 1 to 6 depending upon various environmental factors. In ideal scenario α is 2.

Here, every sensor deployed over the network are homogeneous from the point of view of their ranges (sensing and communication range), and also their energy. Two unique ranges are allotted to each and every sensor, and those are sensing and communication range. Every target situated at the previously noted position are expected to be continuously covered and periodically monitored by a set of the arbitrary number of sensors. A pair of sensors is connected if they are within their communication range. Here, the network lifetime is described as the time duration till all target nodes are covered by a set of sensor nodes. For the energy conservation of every sensor and to increase the network lifetime, one of the methods is to create the sensor nodes into different subsets or sensor covers, such that each subset sensor cover covers all the targets, and among the sensors in each set cover, there is a path to other sensor in that set cover, such that every sensors in each set cover are connected among them.

Definition 1. Target Coverage Set Given M different targets and N different sensors in WSN. A subset of sensor node is said to be a target coverage set if those sensor nodes in the subset covers all the targets.

Definition 2. Connected Target Coverage problem. Let M be a set of targets and N be a set of sensors in the WSN. It is obliged to schedule the sensor activity in such a manner that

- each target is covered by not less than one active sensor node at any point of time
- from each active sensor node, existence of a path to all active sensor nodes is must and
- the lifetime of the entire network is amplified.

The illustration of this problem is in Fig. 1 where 13 sensors and 11 targets are deployed in the field of WSN. If the targets are inside the circle of a sensor, then those targets are covered by the specified sensor, dark solid circles are active sensor nodes, and rest are for sleep sensor nodes. But, to make all active nodes connected, some sleeping sensing nodes may change their state to active state.

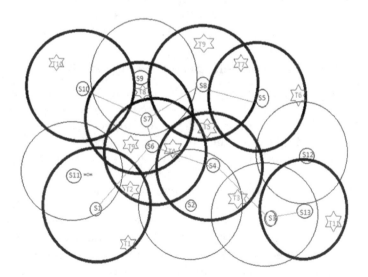

Fig. 1. A sensor network with targets

In Fig. 1, S_1; S_4; S_5; S_6; S_7; S_8; S_{10}; S_{13} are active nodes, but to make all active sensors to be connected, sensor S_3 will became in active state. The lines between sensors are used to denote the path.

Definition 3. 2-Connected Target Coverage problem.

Given M different of targets and N different of sensors in an energy limited WSN. This definition is same as Definition 2, here only additional thing is that after single sensor failure, there must exist a path to all active sensors from each active sensor node and the lifetime of the entire network is amplified.

Modeling Coverage Using Graph Theory. Let $G(S \cup T, E)$ be an undirected bipartite graph, where $S = \{s_1, s_2, ...s_n\}$ is set of sensors and $T = \{t_1, t_2, ...t_m\}$ is set of targets. For each sensor s and each target $t \epsilon T$, if s can cover t then $(s, t) \epsilon E$.

2   Related Work

Different researchers proposed different formulations as well as different algorithms based on coverage problem. The construction of connected target coverage algorithms found from various aspects. Some of these aspects are discussed below. Cardei and Wu [4] have suggested the generalization of target coverage problem which is proven to be NP-hard, where they presents an proficient technique to increase the network lifetime by arranging the sensors into a maximum number of distinct set covers so that each and every cover set fully covered all targets. The extended version of this work is present in [7], where the sensor network life time can be further improved by organizing the sensors into non-disjoint set covers. Wang et al. [16] has gave the idea of coverage and connectivity in 3-dimensional networks. Authors in [15] provide the placement mechanism of sensors in WSN for the conservation of energy. To improve the network lifetime of the system, they have proposed an algorithm along with a sink mobility model to spread the entire network system range.

The researcher proposed in [10] is the first article where for a point coverage problem an approximation algorithm is evaluated. Here, the objective is to cover a set of target points. However, in this work they only concentrate on target coverage but not connectivity. Li et al. further extend in [12] an already addressed maximum set covers problem [10], where they imposed an extra requirement, each sensor cover set is connected to the Base-Station. They additionally presented the Connected Set Covers issue and demonstrate that the problem is NP-complete in nature due to its large solution space, and then they proposed three different solutions for the effective solution of the problem. Among these three solutions, first two are centralized, and the third one is distributed and localized. The first solution is based on integer programming and the second solution is based on a Greedy technique. Abrams et al. [1] have introduced an algorithm $1 - 1/e$, in which they have also proved that the ratio can not be improved more that $15/16$ unless P = NP. Cheng and Gong [7] have suggested a combination of linear programming based and approximation algorithm with a ratio l. Deshpande et al. [9] have studied a particular case where each target is surrounded by at most d-sensors. They have performed a practical approximation algorithm with ratio $1/d + 2(\alpha/(d+2)) \times (1 - 1/d)$ where α signifies the approximation ratio of the semi-definite programming based algorithm for Max k-Cut.

Luo et al. [14] have formulate the two issues as dominating problems on bipartite graphs and they proved the dominating problem is NP-hard in the situation where each sensor can cover not more than d1 targets. They showed that both the problems are remains NP-hard even when $d_1 = 2$ or $d_2 = 2$. On the other side, they also explain how these two problems can be solved in polynomial time for some specific cases. For a number of groups $K_1 = 3$ this problem is NP-hard.

Variants of Target Coverage Problem. In literature, lots of variants of the target coverage problem have been studied depending upon the various propertied of the network like, connectedness, coverage, robustness etc. Here, some variant of target coverage problems are explained below.

1. k-connected target coverage problem: This is an important problem in the deployment activity of WSN where it is difficult to maintain target coverage with proper utilization of energy and to achieve fault-tolerant. This problem which is NP-hard in nature deals with the building of energy efficient and fault-tolerant target coverage set [13]. In literature we can find various heuristic algorithms for this problem.
2. m-coverage problem: This query is very useful for fault-tolerance point of view. This problem is to detect the least number of active sensor nodes making sure that any target is surrounded by not less than m-distinct sensor nodes. Because of the NP-hard nature of this problem several heuristics and approximation algorithms has been purposed.
3. k-Connected m-coverage target coverage problem: This issue deals with fault-tolerant and energy-efficient target coverage where the minimum number of active sensors are there to form m-coverage for each target and there are at least k-disjoint paths in between any pair of active sensor nodes. This mentioned problem is NP-hard in nature [12].
4. r-hop k-connected target coverage problem: The principal goal of this issue is to provide the fault tolerance along with robust, this r-hop k-connected is usually used to enhance the robustness of the backbone [17].

3   Proposed Algorithm

This target coverage problem is a well known NP-Complete problem. Because of this reason researchers gave various heuristics and approximation algorithms and here, we also propose a new heuristic. We propose Algorithm 1 to find out the maximum number of target cover set one by one. Here, each sensor s has some energy initially which is represented as E_s, CE_s is the current energy or battery life of sensor s, and N_T is the number of targets covered by the sensor s. At a particular time, the target which is covered by least number of sensors is referred to as critical target.

An additional terminology called granularity which is most effective for this heuristic. We propose a greedy algorithm for the granularity calculation in Algorithm 2. This algorithm first finds the cardinality of minimal set of sensors those covers all the targets. We explain the importance of granularity with an example below. Suppose, a WSN has a set sensors $S = [s_1; s_2; s_3]$ and targets $T = [t_1; t_2; t_3]$. Here, s_1 covers $[t_1; t_2]$, s_2 covers $[t_2; t_3]$ and s_3 covers $[t_1; t_3]$. Here, each sensor has an energy level of 1 time unit and each sensor can be placed in at most one target coverage set. So, the target coverage set is $s_1; s_2$ and the network lifetime will be 1 time unit. But, if we assign a value 0.5 to granularity, then the network lifetime increases as follows. Granularity $= 0.5$ means a sensor can be placed in two target coverage sets or in other words, a sensor can be active in a target coverage set for 0.5 time unit. So, the target coverage set is $s_1; s_2; s_2; s_3; s_1; s_3$ and the network lifetime becomes 1.5 time unit.

In this algorithm, we start with a empty target coverage set. We keep on adding the sensors to it until it covers all the targets with minimum energy. We first find the critical target which is not covered till now and then find a sensor such that the energy of the sensor is greater than the $\frac{1}{C}$ of initial energy along with $CE_s \times N_T$ is maximum, if such sensor doesn't exist then take the sensor with maximum current energy and covers the critical target. Continue this process until all targets are being covered. After finding the complete target coverage set optimise it (find sensors from target coverage set such that after removing then the target coverage set covers all the targets). Finally this algorithm returns the optimised target coverage set.

4 Experimental Results

Here, we have evaluate the performance of the introduced scheme with the existing scheme described in [2]. We have performed all the simulation using Matlab tool. We have considered a maximum of 100 number of sensor nodes located in an area of 200×200 m. The sensors are homogeneous in nature and initially have the same battery life. The default simulation parameters used for our scheme is tabulated in Table 1. In this section, we have examined the following parameters, (a) The impact of network lifetime with varying sensing range from (5–25) m. (b) Comparison of the lifetime with 100 targets by varying the number of targets starting from 50. (c) Comparison of lifetime with 100 sensors by changing the number of sensors starting from 50.

The simulation is based on the same network with a fixed number of sensors and varies the number of targets and vice-versa deployed. The sensing range capacity of all sensors is considered to be equal. The main purpose is to study the effect when we increase in the number of cover sets for certain granularity (w) that will maximize the network lifetime.

Algorithm 1. Maximum Target Coverage

Input: A Graph $G(S \cup T, E)$ where $S = \{s_1, s_2, \ldots, s_n\}$ is the set of sensors and
$T = \{t_1, t_2, \ldots, t_m\}$ is the set of targets
Output: A feasible target coverage with minimum energy
begin
    Initialize $T = (V, E')$, where $E' = \phi$,
    SENSORS = ALL SENSORS, TCOVERAGES = $\emptyset$
    CALCULATE THE GRANULARITY VALUE w using Algorithm 2;
    while $\exists$ *a target node* **do**
        ONETARCOVER = $\emptyset$;
        TARGETS = ALL TARGETS;
        while $TARGETS \mathrel{!=} \emptyset$ **do**
            Find a critical target T_k, among uncovered target;
            Select all sensor $l_i \subset S$ from SENSORS such that $l_i \geq \frac{E_i}{C} i$
            if $|l_i| \geq 1$ **then**
                Find a sensor s from l_i where $CE_s * N_T$ is maximum
            end
            else
                Otherwise, Select a sensor s with maximum $CE_s * N_T$
            end
            ONETARCOVER = ONETARCOVER $\cup$ s;
            SENSOR =SENSOR $\setminus$ s
            TARGET = TARGET $\setminus t$, $\forall$ targets t covered by s
        end
    end
    MINIMIZE $ONETARCOVER$ // check if removal of any sensor can still produce the
    target coverage
    TCOVERAGES = TCOVERAGES $\cup$ S
    for *all* $s_i \in ONETARCOVER$ **do**
        $LIFE_i = LIFE_i$ - w
    end
    return TCOVERAGES
end

Algorithm 2. Granularity Calculation

Input: A Graph G(S $\cup$ T, E) where S = $\{s_1, s_2 \ldots s_n\}$ is set of sensors and T
    = $\{t_1, t_2 \ldots t_m\}$ is set of targets
Output: Returns the granularity w
begin
    Initialize Targets = $\emptyset$, $Minimalset = \emptyset$
    for *all sensors s in Sensor sorted in non-increasing coverage quality* **do**
        for *all targets Tk in Target* **do**
            if *sensor s cover target Tk* **then**
                $Target = Target \cup Tk$
            end
        end
        if *all targets have been covered in Target* **then**
            break
        end
        else
            $Minimalset = Minimalset \cup s$
        end
    end
end
w = $1/|Minimalset|$
return w

Table 1. Simulation parameters

Parameter	Value
Initial energy	10 J
Network area	200×200
Maximum number of sensors	100
Maximum number of targets	100
Channel rate	250 kbps

Simulation 1: In the first experiment, we have studied the influence of increase on sensors (each sensor might have different sensing range) with lifetime. That is if it shows the network lifetime changes when we change the number of sensing nodes. We have taken 100 number of sensors distributed randomly in a network of size 200×200 m. We varied the number of sensors from 10 to 100 to know the sensing range of the network.

From Fig. 2, it is seen that sensing ranges have a great influence on network lifetime. With the increase in the sensing ranges, the lifetime of a network gets increased so as the network complexity. In the simulation, we have taken the parameters for sensing range (5 m to 25 m) with varying the sensors from 10 to 100 with an interval of 10.

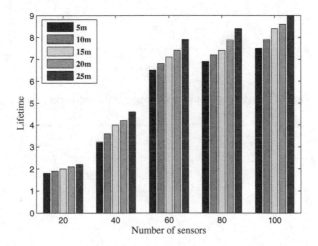

Fig. 2. Calculating lifetime with varying the number of sensors and their sensing range

Simulation 2: Here, we have assumed that the number of sensors is fixed, and the number of targets varies. In the second simulation, the number of sensors is set to 80 with an increasing number of targets from 50 to 100 by an increment of 5.

Table 2 shows the comparison table between the Maximum Coverage Heuristic (MCH) [2] and our proposed heuristic Energy Efficient Connected Target

Coverage (EECTC) where the lifetime of a network is calculated in terms of a time unit. Figure 3 shows the comparison graph of this simulation. From Table 2, it is inferred that with an increase in the number of targets, the network lifetime gets decreased.

Table 2.

Targets	50	55	60	65	70	75	80	85	90	95	100
MCH	7.8	7.6	7.6	7.4	7.4	7.4	7.2	7.2	7.2	7	7
EECTC	8.2	8.2	8	8	7.8	7.6	7.4	7.4	7.2	7.2	7.2

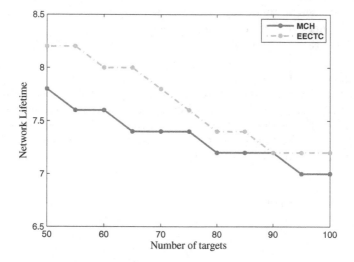

Fig. 3. Increasing number of target making target fixed

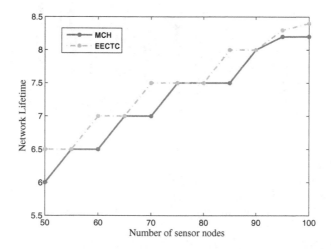

Fig. 4. Increasing number of sensor making sensors fixed

Simulation 3: Number of targets are fixed, and the number of sensors varies. In the third simulation, the number of target is 50 and increasing the number of sensors from 50 to 100 by five each time are deployed. Table 3 is the comparison table between the MCH [2] and EECTC with fixed targets and different sensors where the lifetime of the network is calculated in terms of a time unit. Figure 4 shows the comparison graph of this simulation.

Table 3.

Sensors	50	55	60	65	70	75	80	85	90	95	100
MCH	6	6.5	6.5	7	7	7.5	7.5	7.5	8	8.15	8.15
EECTC	6.5	6.5	7	7	7.5	7.5	7.5	8	8	8.25	8.4

Algorithm 3. Establishment of 2 Connectivity

Input: A Graph $G(S_A, E)$ where $S = \{s_1, s_2, ..., s_a\}$ is set of active
    sensors, $S_I = \{s_1, s_2, ..., s_b\}$ is the set of inactive sensors and E = { } is an empty set
    of edges or links
Output: Returns a 2-connected network generated by sensor nodes
begin
    for *i = 1 to a-1* do
        for *j = i+1 to a* do
            if *(distance(s_i, s_j)* $\leq r$) then
            | join s_i with s_j
            end
        end
    end
    if(All sensors in set S forms a 2-connected network)
    return S;
    for *i = 1 to a* do
        if *(degree(s_i) == 0)* then
            s_i broadcast a message to all reachable inactive nodes in S_I and get the node n
            $\in S_I$ has more number of active nodes(if at least 1 then $flag = 1$) excluding s_i
            if *flag == 1* then
            | join s_i with n
            end
        else
            | Randomly choose a node n belongs to S_I with maximum degree
        end
        join all reachable active nodes to n
        $S = S \cup n$
        $S_I = S_I - n$
        $a = a + 1$
        $b = b - 1$
        flag = 0;
        if(All sensors in set S forms a 2-connected network)
        return S;
        end
    end
    return False: 2-connectivity is not possible with the given sensors;
end

5   Establishment of 2-Connectivity

High connectivity is essential for fault tolerance purpose and to make the network reliable. A network is said to be 2-connected network if there are at least two different paths among any pair of active sensors. After finding the Maximum target coverage using Algorithm 1 we are establishing 2-connectivity among the active sensors by adding some of the inactive sensors if needed, which maintains fault-tolerance. Once Algorithm 1 generated the target coverage set, then Algorithm 3 finds the 2-connectivity among the sensors using active and inactive sensors set.

Greedy Algorithm for Establishment of 2 Connectivity: This algorithm established 2-connectivity among the active sensors (means, among the sensors of a target coverage set) by changing some inactive sensors to active sensors.

Suppose, we have 'N' sensors in the network, 'r' is the sensing range of a sensor, 'S_A' is the set of active sensors and 'a' is its cardinality, 'S_I' is the set of inactive sensors and 'b' is its cardinality. The algorithm first find all the active nodes who are within their communication range, and connect them. If after the connection the connected sensors maintains 2-connectivity then we are done else start choosing some of the sensors from inactive sensors which are in communication range and move them to the active sensor from inactive sensor set. Continue the above process till the induced network by the active sensor is 2-connected. Initially, flag = 0 and the algorithm for the establishment of 2-connectivity is in Algorithm 3.

We have an example as follows in Fig. 4, where a target coverage set having sensors $s_1, s_2, ..., s_{11}$ which are active sensors and the sensors $n_1, n_2, ..., n_{19}$ which are inactive sensors. So, we have total 30 sensors node here (Fig. 5).

Fig. 5. A sensor network establishing 2-connectivity

6   Conclusion

This paper tries to overcome the difficulty of maximizing the wireless sensor network lifetime while maintaining connectivity.

Here, we have also studied the non-disjoint target coverage problem in the sensor network by introducing the term granularity and then we have proposed a heuristic algorithm for maximizing WSN lifetime, which is then compared with the result in [2]. As shown in simulation results, the network lifetime calculated by the heuristic algorithm that we proposed is more than as compared to lifetime found in [2] in most cases. So, our heuristic algorithm improved the performance. Study of various problems listed in our variants of problem section, and we can propose various heuristics as well as the approximation algorithm.

References

1. Abrams, Z., Goel, A., Plotkin, S.: Set k-cover algorithms for energy efficient monitoring in wireless sensor networks. In: Proceedings of the 3rd International Symposium on Information Processing in Sensor Networks, pp. 424–432. ACM (2004)
2. Bajaj, D., et al.: Maximum coverage heuristics (MCH) for target coverage problem in wireless sensor network. In: IEEE International Advance Computing Conference (IACC), pp. 300–305 (2014)
3. Cardei, M., Du, D.Z.: Improving wireless sensor network lifetime through power aware organization. Wirel. Netw. **11**(3), 333–340 (2005)
4. Cardei, M., Wu, J.: Energy-efficient coverage problems in wireless ad-hoc sensor networks. Comput. Commun. **29**(4), 413–420 (2006)
5. Mishra, S., Swain, R.R., Samal, T.K., Kabat, M.R.: CS-ATMA: a hybrid single channel MAC layer protocol for wireless sensor networks. In: Jain, L.C., Behera, H.S., Mandal, J.K., Mohapatra, D.P. (eds.) Computational Intelligence in Data Mining - Volume 3. SIST, vol. 33, pp. 271–279. Springer, New Delhi (2015). doi:10. 1007/978-81-322-2202-6_24
6. Panda, B.S., Shetty, D.P.: An incremental power greedy heuristic for strong minimum energy topology in wireless sensor networks. In: Natarajan, R., Ojo, A. (eds.) ICDCIT 2011. LNCS, vol. 6536, pp. 187–196. Springer, Heidelberg (2011). doi:10. 1007/978-3-642-19056-8_13
7. Cheng, M.X., Gong, X.: Maximum lifetime coverage preserving scheduling algorithms in sensor networks. J. Glob. Optim. **51**(3), 447–462 (2011)
8. Swain, R.R., Mishra, S., Samal, T.K., Kabat, M.R.: An energy efficient advertisement based multichannel distributed MAC protocol for wireless sensor networks (Adv-MMAC). In: Wireless Personal Communications, pp. 1–28 (2016)
9. Deshpande, A., Khuller, S., Malekian, A., Toossi, M.: Energy efficient monitoring in sensor networks. In: Laber, E.S., Bornstein, C., Nogueira, L.T., Faria, L. (eds.) LATIN 2008. LNCS, vol. 4957, pp. 436–448. Springer, Heidelberg (2008). doi:10. 1007/978-3-540-78773-0_38
10. Gil, J.M., Han, Y.H.: A target coverage scheduling scheme based on genetic algorithms in directional sensor networks. Sensors **11**(2), 1888–1906 (2011)
11. Keskin, M.E., Altinel, I.K., Aras, N.: Wireless sensor network design by lifetime maximisation: an empirical evaluation of integrating major design issues and sink mobility. Int. J. Sensor Netw. **20**(3), 131–146 (2016)

12. Li, D., Cao, J., Liu, D., Yu, Y., Sun, H.: Algorithms for the m-coverage problem and k-connected m-coverage problem in wireless sensor networks. In: Li, K., Jesshope, C., Jin, H., Gaudiot, J.-L. (eds.) NPC 2007. LNCS, vol. 4672, pp. 250–259. Springer, Heidelberg (2007). doi:10.1007/978-3-540-74784-0_26

13. Li, D., Cao, J., Liu, M., Zheng, Y.: K-connected target coverage problem in wireless sensor networks. In: Dress, A., Xu, Y., Zhu, B. (eds.) COCOA 2007. LNCS, vol. 4616, pp. 20–31. Springer, Heidelberg (2007). doi:10.1007/978-3-540-73556-4_5

14. Luo, W., Wang, J., Guo, J., Chen, J.: Parameterized complexity of max-lifetime target coverage in wireless sensor networks. Theor. Comput. Sci. **518**, 32–41 (2014)

15. Singhal, S., Bharti, S.K., Kaushik, G.: Localization in wireless sensor network: an optimistic approach. Int. J. Comput. Sci. Inf. Technol. **4**(2), 224–226 (2013)

16. Wang, Y., Li, F., Dahlberg, T.A.: Energy-efficient topology control for three-dimensional sensor networks. Int. J. Sensor Netw. **4**(1–2), 68–78 (2008)

17. Zheng, C., Yin, L., Zhang, Y.: Constructing r-hop 2-connected dominating sets for fault-tolerant backbone in wireless sensor networks. In: 8th International Conference on Wireless Communications, Networking and Mobile Computing (WiCOM), pp. 1–4. IEEE (2012)

18. Zhou, Z., Das, S., Gupta, H.: Connected k-coverage problem in sensor networks. In: 13th International Conference on Computer Communications and Networks (ICCCN), pp. 373–378. IEEE (2004)

Standard Propagation Model Tuning for Path Loss Predictions in Built-Up Environments

Segun I. Popoola[1(✉)], Aderemi A. Atayero[1], Nasir Faruk[2],
Carlos T. Calafate[3], Lukman A. Olawoyin[2], and Victor O. Matthews[1]

[1] Electrical and Information Engineering, Covenant University, Ota, Nigeria
segun.popoola@stu.cu.edu.ng, {atayero,
victor.matthews}@covenantuniversity.edu.ng
[2] Department of Telecommunication Science,
University of Ilorin, Ilorin, Nigeria
{faruk.n,olawoyin.la}@unilorin.edu.ng
[3] Department of Computer Engineering,
Technical University of Valencia, Valencia, Spain
calafate@disca.upv.es

Abstract. This paper provides a simple optimization procedure using ATOLL planning tool for Standard Propagation Model (SPM). Measurement campaigns were conducted to collect Received Signal Strength (RSS) data over commercial base stations operating at 1800 MHz. The prediction accuracy of widely used models were assessed. The models provided high prediction errors. The optimization procedure involves the use of Digital Terrain Model (DTM), clutter classes, clutter heights, vector maps, scanned images, and Web Map Service (WMS). A Logarithmic weighting function was used to calculate the weight of the clutter loss on each pixel from the pixel with the receiver in the direction of the transmitter, up to the defined maximum distance. The approach has proven promising by achieving high accuracy and minimizing the prediction errors by 47.4%.

Keywords: Model tuning · Path loss · RF network planning · Radio propagation

1 Introduction

For an efficient and effective radio access network planning, a knowledge of the characteristics of radio wave propagation in a built-up environment is required [1, 2]. Due to the medium of propagating Electromagnetic (EM) waves and the antecedent effect such as Reflections, diffractions, and scatterings which are dominant during propagation in urban areas as a results of physical obstructions (clutter) in the propagation environment. The interaction of transmitted EM waves with building walls, surfaces of bill boards and other artificial structures, and the bodies of moving objects usually results in reception of multiple copies of the transmitted signals at the receivers which can be termed as multipath [3]. The multiple copies reach the mobile station from different directions with different time delays, leading to signal fading.

© Springer International Publishing AG 2017
O. Gervasi et al. (Eds.): ICCSA 2017, Part VI, LNCS 10409, pp. 363–375, 2017.
DOI: 10.1007/978-3-319-62407-5_26

Consequently, the complex nature of urban environments is responsible for the random and linear time-varying characteristics of the radio channel [4].

Signal fading in wireless communication may be at small scale or large scale [3]. Small-scale signal fading occurs as a result of multipath with an effect of rapid variation in the received signal strength with time [5]. On the other hand, large-scale signal fading is the attenuation of the mean signal power by virtue of the position of the receiver relative to the transmitter. This is also known as path loss [6]. Since radio engineers do not have control over the man-made environment, it is important to properly quantify the resulting power loss for an efficient radio network.

Path loss prediction models are mathematical formulation of the propagation channel as a function of the separation distance between the transmitter and the receiver, the frequency of transmission, the antenna heights of both transmitter and receiver, and other environmental factors. Radio network planners rely on path loss prediction models to ensure an acceptable Quality of Service (QoS) and satisfactory customer experience. In order to avoid incessant call drops and other network connectivity issues, the received signal level at the receiver must be greater than the minimum sensitivity of the receiver. The reception sensitivity varies with different types of receiver. Estimation of signal path loss is one of the main components of the link budget of a wireless communication system [3].

Radio propagation models have been broadly categorized into two: deterministic and empirical models. Deterministic models [7–10] have proven to be more accurate but they require a detailed information about the propagation environment while empirical models [11–13] are simple and requires less computation effort. They are formulated based on extensive measurement campaigns which makes them to be highly environment-dependent. Although several empirical path loss models have been proposed in the literature [13–22], it is still very difficult to synthesize a global model which is suitable for every built-up area. The use of an empirical path loss model for wireless network design in an environment other than the one it was intended for will produce significant prediction errors. Large prediction errors will consequently lead to poor network coverage as RF engineers are much likely to situate the base stations at inappropriate locations.

In this paper, we seek to investigate the prediction accuracy of widely available empirical path loss prediction models in built-up environment of Lagos, Nigeria. Measurement campaigns were conducted to collect received signal strength data over commercial base stations operating at 1800 MHz. In addition, we proposed an adaptable empirical model for path loss predictions by tuning the parameters of SPM to give a better representation of the local environment.

The remaining part of the paper is organized as follows: Sect. 2 reviews the available empirical path loss models suitable for 1800 MHz frequency band; Sect. 3 describes the materials and the methodology used in this study; Sect. 4 gives the results and discusses the findings while the paper is concluded in Sect. 5.

2 Empirical Path Loss Models

2.1 Okumura-Hata Model

Okumura-Hata model is an empirical formulation of the graphical path loss data for 150–1500 MHz band [11]. The separation distance between the transmitter and the receiver ranges from 1 to 20 km. The standard formula for median path loss in urban areas is given by Eq. (1):

$$PL_{urban}(dB) = 69.55 + 26.16 \times \log(f_c) - 13.82 \times \log(h_t) \\ - a(h_r) + [44.9 - 6.55 \log(h_t)] \times \log(d) \qquad (1)$$

The correction factors for mobile antenna height in a built-up environment is given by Eqs. (2) and (3)

$$a(h_r) = 8.29[\log(1.54h_r)]^2 - 1.1 \quad \text{for} \ \ f_c \leq 300 \, \text{MHz} \qquad (2)$$

$$a(h_r) = 3.2[\log(11.75h_r)]^2 - 4.97 \quad \text{for} \ \ f_c \geq 300 \, \text{MHz} \qquad (3)$$

where

f_c = Frequency (in MHz) from 150 MHz to 1500 MHz
h_t = Effective transmitter antenna height in meters
h_r = Effective receiver antenna height in meters
$d = T_x - R_x$ separation distance in km

The path loss for a suburban area is given by Eq. (4)

$$PL_{suburban} = PL_{urban}(dB) - 2[\log(\frac{f_c}{28})]^2 - 5.4 \qquad (4)$$

For an open rural area, Eq. (5) is used

$$PL_{rural} = PL_{urban}(dB) - 4.78[\log(f_c)]^2 - 18.33 \times \log(f_c) - 40.98 \qquad (5)$$

2.2 COST 231-Hata Model

COST 231 extends Hata model to cover the frequency range of 1500 to 2000 MHz [12]. The transmitter height varies from 30 to 200 m. The height of the receiver can be between 1 and 10 m. The path loss model is given by Eqs. (6) and (7):

$$PL(dB) = 46.3 + 33.9 \times \log(f_c) - 13.82 \times \log(h_t) - a(h_r) + [44.9 - 6.55 \log(h_t)] \\ \times \log(d) + C_m \qquad (6)$$

$$a(h_r) = 8.29[\log(1.54h_r)]^2 - 1.1 \quad \text{for} \quad f_c \leq 300\,\text{MHz} \tag{7}$$

$$a(h_r) = 3.2[\log(11.75h_r)]^2 - 4.97 \quad \text{for} \quad f_c \geq 300\,\text{MHz} \tag{8}$$

$$C_m = \begin{cases} 0\,\text{dB} & \text{for medium-sized city and suburban areas} \\ 3\,\text{dB} & \text{for metropolitan areas} \end{cases} \tag{9}$$

Range of parameters are as follows:

f: 1500–2000 MHz
h_t: 30–200 m
h_r: 1–10 m
d: 1–20 km

2.3   ECC-33 Model

This model is suited for the Ultra-High Frequency (UHF) band. Here, path loss is given by Eq. (10) [23]:

$$PL(dB) = A_{fs} + A_{bm} - G_t - G_r \tag{10}$$

where

A_{fs} = free space attenuation
A_{bm} = basic median path loss
G_t = transmitter antenna height gain factor
G_r = receiver antenna height gain factor

$$A_{fs} = 92.4 + 20\,\log(d) + 20\,\log(d) \tag{11}$$

$$A_{bm} = 20.41 + 9.83 \times \log(d) + 7.894 \times \log(f) + 9.56[\log(f)]^2 \tag{12}$$

$$G_t = \{log\left(\frac{h_t}{200}\right)\}\{13.958 + 5.8 \times (\log d)^2\} \tag{13}$$

For medium-sized cities,

$$G_r = \{42.57 + 13.7 \times \log(f)\}\{\log(h_r) - 0.585\} \tag{14}$$

For large cities,

$$G_r = 0.759h_r - 1.862 \tag{15}$$

2.4   Egli Path Loss Model

This model is applicable to frequency range of 40 to 900 MHz at a maximum separation distance of 60 km [4]. It was derived from measurement data on UHF and VHF television transmissions in several large cities. The formula of Egli path loss model is given in Eq. (16):

$$PL(dB) = 20 \times \log(f) + 40 \times \log(d) - 20 \times \log(h_t) + 73.6 - 10 \times \log(h_r) \quad (16)$$

2.5   Standard Propagation Model

SPM was developed based on the Hata path loss formulas [11, 12]. This empirical model is suitable for path loss predictions in the 150–1500 MHz frequency band. It determines the large-scale fading of received signal strength over a distance range of 1–20 km. Therefore, it is appropriate for mobile channel characterization of popular cellular technologies such as GSM [13].

The received signal strength is given by Eq. (17):

$$
\begin{aligned}
P_r = P_t - \{ &K_1 + K_2 \log(d) + K_3 \log(h_t) + K_4.DiffractionLoss \\
&+ K_5 log(d).log(h_t) + K_6.h_r + K_7 \log(h_r) + K_{clutter}.f_{clutter} + K_{hill} \}
\end{aligned}
\quad (17)
$$

The model parameters are defined as follows:

P_r = Received power in dBm
P_t = Transmitted power (EIRP) in dBm
K_1 = Constant offset in dB
K_2 = Multiplying factor for log (d)
d = Separation distance (in meters)
K_3 = Multiplying factor for log (h_t)
h_t = Effective transmitter antenna height (in meters)
K_4 = Multiplying factor for diffraction calculation
K_5 = Multiplying factor for $log(d).log(h_t)$
K_6 = Multiplying factor for h_r
K_7 = Multiplying factor for log (h_r)
h_r = Effective mobile receiver antenna height (in meters)
$K_{clutter}$ = Multiplying factor for $f_{clutter}$
$f_{clutter}$ = Average of the weighted losses due to clutter
K_{hill} = Corrective factor for hilly regions

The Hata path loss model is represented by Eq. (18) [8]:

$$
\begin{aligned}
PL(dB) = &A_1 + A_2 log(f) + A_3 log(h_t) + [B_1 + B_2 log(h_t) + B_3.h_t][log(d)] - a(h_r) \\
&- C_{clutter}
\end{aligned}
$$

$$(18)$$

The definition of parameters is as follows:

$A_1 \ldots B_3$: Hata parameters
f: Frequency in MHz
h_t: Effective transmitter antenna height in metres
d: Separation distance in km
h_r: Mobile receiver height in meters
$a(h_r)$: Mobile receiver antenna height correction factor in dB
$C_{clutter}$: Clutter correction function

Although distance is usually expressed in km in Hata formulas, SPM accepts distance values in meters. The generic values of the Hata model parameters are the ones stated below:

$$A_1 = \begin{cases} 69.55 & for \ 900\,MHz \\ 46.30 & for \ 1800\,MHz \end{cases}$$

$$A_2 = \begin{cases} 26.16 & for \ 900\,MHz \\ 33.90 & for \ 1800\,MHz \end{cases}$$

$A_3 = -13.82$
$B_1 = 44.90$
$B_2 = -6.55$
$B_3 = 0$

Therefore, the path loss model for GSM technologies that operate in the 900 MHz band becomes Eq. (19)

$$PL(dB) = 69.55 + 26.16\,log(f) - 13.82\,log(h_t) + [44.9 - 6.55\,log(h_t)][log(d)] - a(h_r) - C_{clutter} \tag{19}$$

On the other hand, the path loss predictions for the DCS 1800 counterpart is given by Eq. (20) [9]:

$$PL(dB) = 46.3 + 33.9\,log(f) - 13.82\,log(h_t) + [44.9 - 6.55\,log(h_t)][log(d)] - a(h_r) - C_{clutter} \tag{20}$$

SPM ignored the effects of diffraction, clutter, and terrain to produce Eq. (21). It assumed that appropriate settings of A_1 and K_1, which account for only one clutter class, will cater for the influence of these external factors on signal propagation. The correction function for the mobile receiver antenna height was also ignored for $h_r \leq 1.5\,m$ since it has negligible values for an average mobile antenna height. The resulting path loss model is given in Eq. (5):

$$PL(dB) = A_1 + A_2 log(f) + A_3 log(h_t) + [B_1 + B_2 log(h_t)][log(d)] \qquad (21)$$

The SPM formula can be further reduced to Eq. (22):

$$PL(dB) = K_1 + K_2 \log(d) + K_3 \log(h_t) + K_5 log(d).log(h_t) + K_6.h_r + K_7 \log(h_r) \quad (22)$$

Presenting the reduced Hata equation as a model which accepts distance input in m as in SPM, we have Eq. (7):

$$PL(dB) = A_1 + A_2 log(f) - 3B_1 + [A_3 - 3B_2][log(h_t)] + B_1 log(d) + B_2 log(h_t).log(d) \qquad (23)$$

Equating the coefficients of Eqs. (22) and (23):

$K_1 = A_1 + A_2 log(f) - 3B_1$
$K_2 = B_1$
$K_1 = A_3 - 3B_2$
$K_5 = B_2$
$K_6 = K_7 = 0$

Hence, the SPM mathematical repr esentation of the DCS 1800 MHz mobile channel is given by Eq. (24):

$$PL(dB) = 22 + 44.9 \, log(d) + 5.83 \, log(h_t) - [6.55 \, log(d).log(h_t)] \qquad (24)$$

3 Materials and Methods

3.1 Field Measurements and Data Collection

Measurement campaigns were conducted in the built-up areas of Lagos, Nigeria at 1800 MHz. Lagos is a popular urban center in Nigeria, located on coordinates 6°31′ 28.22″ N, 3°22′45.17″ E. The metropolis has a land area of 385.9 square miles, 90.75% of which is categorized as urban. It is composed of high-rise residential apartments built close to each other with building heights of between 2–4 floors.

A drive test was performed along pre-planned survey routes to collect received signal strength data in the built-up environment. A TEMS W995 phone was interfaced with TEMS investigation software [24], Garmin Global Positioning System (GPS) and MapInfo Pro [25] on a laptop. The laptop has the following specifications: Intel Core i5-3210MCPU@2.50 GHz speed with 4 GB RAM and 64-bit Windows 7 operating system.

The mobile station was locked on a single Broadcast Control Channel (BCCH) to avoid the measurement of interfering signals. Since the commercial base stations use directional antennas, measurements were carefully planned to cover only areas in the direction of the main beam. In order to overcome the effects of fast fading, a minimum of 36 samples were collected over distance of 40λ satisfy the Lee criterion.

A maximum vehicle speed of 40 km/h was ensured during the survey to eliminate Doppler effects. The measurements were taken under normal climatic conditions to ensure high-quality data collection. Also, a good radio frequency clearance was maintained.

3.2 Standard Propagation Model Tuning

This section describes the SPM tuning process using field measurements obtained from the built-up environment. This process involves the use of geographic data which include DTM, clutter classes, clutter heights, vector maps, scanned images, and WMS raster-format geo data files. The DTM provides information about the elevation of the ground over sea level. The information about the land cover or land use is obtained from the clutter classes. Clutter height maps offers more precise information about the altitude of clutter over DTM with one altitude defined per pixel. Vector maps are maps with possible routes which are defined as vectors. Scanned images such as road maps and satellite images provides a precise background representing the actual physical environment. WMS raster-format geo data files are raster images from a WMS server. These geographic data helps in verifying the correspondence between the area under investigation and the field measurements. The whole procedure was carried out in ATOLL [26].

The measurement data were filtered to eliminate the points that are the least representative of the survey area while retaining a number of points that is both representative and large enough to provide statistically valid results. Such outliers include the measurement points in clutter classes that are of no significance in terms of the propagation model to be tuned. Also, we disregarded extreme signal levels, those that were too close to the transmitter, and those that suffered from too much diffraction. Furthermore, data points that came from behind directional antennas were not included.

The tuning process considered both the 'near transmitter' and 'far from transmitter' conditions. The effective transmitter antenna height was calculated using the ground slope at the receiver as represented by Eq. (25)

$$H_{Txeff} = (H_{Tx} + H_{OTx}) - H_{ORx} + (Kxd) \tag{25}$$

where

H_{ORx} = the ground elevation above sea level at the receiver (m)
K = the ground slope calculated over a minimum distance from the receiver

The diffraction loss over the transmitter-receiver profile was calculated using the Millington method. The average clutter height specified for each clutter class was used for the transmitter-receiver profile in the calculation of the diffraction edge. The losses due to clutter was estimated over a maximum distance from the receiver using Eq. (26):

$$f(clutter) = \sum_{i=1}^{n} L_i w_i \tag{26}$$

where,

    L = loss due to clutter (in dB).
    w = weight determined through the weighting function.
    n = number of points taken into account over the profile.

Weighting function is the mathematical formula used to calculate the weight of the clutter loss on each pixel from the pixel with the receiver in the direction of the transmitter, up to the defined maximum distance. Logarithmic weighting function employed is given by Eq. (27):

$$w_i = \frac{\log(\frac{d_i}{D} + 1)}{\sum_{j=1}^{n} \log(\frac{d_i}{D} + 1)} \tag{27}$$

where, d_i is the distance between the receiver and the ith point and D is the maximum distance defined. The maximum distance indicates the distance from the receiver for which clutter losses will be considered via a weighting function, with an effect on the influence of clutter on total losses which diminishes with distance from the receiver. This value was set to 150 m for the urban environment.

    The model tuning was carried out using ATOLL planning tool with the aim of minimizing the mean error and standard deviation of measured values versus calculated values. A mathematical solution was obtained for both Line of Sight (LOS) and Non-Line of Sight (NLOS) conditions.

3.3 Statistical Analysis of Prediction Accuracy of Models

Two different approaches were employed to evaluate the performance of the model tuning. First, the prediction results were appraised based on the path loss data used for tuning. Second, the prediction accuracy of the tuned SPM was adjudged relative to other models and the measurement data. The following statistical performance metrics were used: Mean Absolute Error (MAE), Root Mean Square Error (RMSE), and Standard Deviation (SDE). The mathematical expressions of these metrics are given by Eq. (28)–(30).

$$MAE = \frac{1}{n} \sum_{i=1}^{n} (PL_i^m - PL_i^p) \tag{28}$$

$$RMSE = \sqrt{\frac{1}{n} \sum_{i=1}^{n} (PL_i^m - PL_i^p)^2} \tag{29}$$

$$SD = \sqrt{\frac{1}{n} \sum_{i=1}^{n} (|PL_i^m - PL_i^p| - \mu)^2} \tag{30}$$

where, μ = the mean prediction error in decibel.

4 Results and Discussions

Figures 1 and 2 provide the comparison of path loss prediction at Locations A and B respectively. In both Figures, Egli model under-estimates the path loss and ECC-33 over-estimates the path loss. The Okumura-Hata and COST 231 models are the base line SPM, these models provides good fitness at distances above 400 m, below which they underestimate the path loss. However, the tuned SPM provides better predictions along all the routes, across all distances. Table 1 provides the tuned parameters for the tuned model. The SPM ignored the effects of diffraction, clutter, and terrain. This

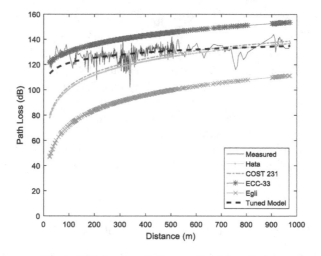

Fig. 1. Path loss prediction results at Location A.

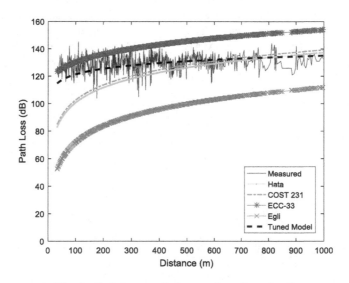

Fig. 2. Path loss prediction results at Location B

Table 1. SPM tuning result for built-up environments

Parameter	Description	Before tuning	After tuning
H_t method	-	Abs spot H_t	Slope at receiver
Diffraction method	-	Deygout	Millington
K_1 (LOS)	-	22	63
K_1 (NLOS)	-	22	76
K_2 (LOS)	log (d)	44.9	18
K_2 (NLOS)	log (d)	44.9	20
K_3	log (h_t)	5.83	18.2
K_4	Diffraction	1	0
K_5	log (d).log (h_t)	−6.55	−10
Additional losses	Clutter losses	0	50

assumes single clutter class, while, the turned model, considers the diffraction losses. This increases the clutter losses from 0 dB in the case of conventional SPM to 50 dB as in the turned model. The tuned SPM is expressed mathematically in Eq. (31) and (32). Equation (31) is applicable to LOS scenarios while Eq. (32) can be used for NLOS propagation scenarios in built-up areas.

$$PL(dB) = 113 + 18 \times log(d) + 18.2 \times log(h_t) - [10\,log(d).log(h_t)] \qquad (31)$$

$$PL(dB) = 126 + 20 \times log(d) + 18.2 \times log(h_t) - [10\,log(d).log(h_t)] \qquad (32)$$

Tables 2 and 3 provide the statistical evaluation of path loss at Locations A and B. For Location A, RMSE values of 13.75 dB, 12.48 dB, 14.12 dB and 38.21 dB respectively for Hata, COST 231, ECC-33 and Egli models. These are higher than the acceptable 6 dB threshold values defined in [7]. While, for Location B, Hata, COST 231, ECC-33

Table 2. Statistical evaluation of path loss predictions at Ikeja, Lagos (Location A)

	ME (dB)	RMSE (dB)	SD (dB)
Hata	10.71	13.75	8.63
COST 231	9.33	12.48	8.28
ECC-33	12.75	14.12	6.06
Egli	36.69	38.21	10.67
Tuned SPM	4.10	5.72	3.99

Table 3. Statistical evaluation of path loss predictions at Ikoyi, Lagos (Location B)

	ME (dB)	RMSE (dB)	SD (dB)
Hata	11.87	15.60	10.13
COST 231	10.78	14.42	9.58
ECC-33	12.57	14.45	7.12
Egli	37.18	39.32	12.79
Tuned SPM	4.52	5.92	3.83

and Egli modes obtained RMSE values of 15.60 dB, 14.42 dB, 14.45 dB and 37.18 dB respectively, wit COST 231 model providing the last error amongst the four contending models, across the locations. It worth noting that despite the high prediction error of the ECC-3 model across the locations, the model has the least standard deviation error. However, the tuned SPM achieves RMSE of 5.92 dB and 5.72 dB for the two locations. These values are 47.4% and 39.6% decrease in prediction error when compared with COST 231 model along the locations.

5   Conclusion

This paper presents measurement campaigns of EM waves in the GSM 1800 MHz band within built-up areas. A simple optimization procedure using ATOLL planning tool for standard propagation model is provided. The optimization process involves the use of DTM, clutter classes, clutter heights, vector maps, scanned images, and WMS. A Logarithmic weighting function was used to calculate the weight of the clutter loss on each pixel from the pixel with the receiver in the direction of the transmitter, up to the defined maximum distance. The approach has proven promising by achieving high accuracy and minimizing the prediction errors by 47.4%.

Acknowledgment. The authors wish to appreciate the Center for Research, Innovation, and Discovery (CU-CRID) of Covenant University, Ota, Nigeria for partly funding of this research.

References

1. Lempiäinen, J., Manninen, M.: Radio Interface System Planning for GSM/GPRS/UMTS. Kluwer Academic Publishers, Dordrecht (2001)
2. Laiho, J., Wacker, A., Novosad, T.: Radio Network Planning and Optimisation for UMTS. Wiley, New York (2002)
3. Rappaport, T.S.: Wireless Communications: Principles and Practice. Prentice Hall PTR, Upper Saddle River (2002)
4. Parsons, J.D.: The Mobile Radio Propagation Channel, 2nd edn. Wiley, New York (2000)
5. Clark, R.H.: A Statistical Description of Mobile Radio Reception. BSTJ **47**, 957–1000 (1968)
6. Mishra, A.J.: Advanced Cellular Networks Planning and Optimization 2G/2.5G/3G... Evolution to 4G, pp. 1–12. Wiley, New York (2007). ISBN 13 978-0-470-01471-4
7. Luebbers, R.J.: Propagation prediction for hilly terrain using GTD wedge diffraction. IEEE Trans. Antennas Propag. **32**(9), 951–955 (1984)
8. Mohtashami, V., Shishegar, A.A.: Modified wavefront decomposition method for fast and accurate ray-tracing simulation. Microw. Antennas Propag. IET **6**(3), 295–304 (2012)
9. Hufford, G.A.: An integral equation approach to the problem of wave propagation over an irregular surface. Q. J. Mech. Appl. Math. **9**(4), 391–404 (1952)
10. Zelley, C.A., Constantinou, C.C.: A three-dimensional parabolic equation applied to VHF/UHF propagation over irregular terrain. IEEE Trans. Antennas Propag. **47**(10), 1586–1596 (1999)

11. Masahara, H.: Empirical formula for propagation loss in land-mobile radio services. IEEE Trans. Veh. Technol. **29**(3), 317–325 (1980)
12. COST 231 Project: Urban Transmission Loss Models for Mobile Radio in the 900 and 1800 MHz band, COST 231 TD (90) 119 Rev. 2, The Hague, Netherlands (1991)
13. Popoola, S.I., Oseni, O.F.: Performance evaluation of radio propagation models on GSM network in urban area of Lagos, Nigeria. Int. J. Sci. Eng. Res. **5**(6), 1212–1217 (2014)
14. Popoola, S.I., Oseni, O.F.: Empirical path loss models for GSM network deployment in Makurdi, Nigeria. Int. Refereed J. Eng. Sci. **3**(6), 85–94 (2014)
15. Oseni, O.F., Popoola, S.I., Abolade, R.O., Adegbola, O.A.: Comparative analysis of received signal strength prediction models for radio network planning of GSM 900 MHz in Ilorin, Nigeria. Int. J. Innov. Technol. Explor. Eng. (IJITEE) **4**(3), 45–50 (2014)
16. Faruk, N., Ayeni, A.A., Adediran, Y.A.: Error bounds of empirical path loss models at VHF/UHF bands in Kwara State, Nigeria. In: Proceedings of IEEE EUROCON Conference, Croatia, pp. 602–607 (2013)
17. Faruk, N., Adediran, Y.A., Ayeni, A.A.: On the study of empirical path loss models for accurate prediction of TV signal for secondary users. Prog. Electromagn. Res. (PIER) B USA **49**, 155–176 (2013)
18. Rath, H.K., Verma, S., Simha, A., Karandikar, A.: Path loss model for Indian terrain - empirical approach. In: 2016 Twenty Second National Conference on Communication (NCC), Guwahati, pp. 1–6 (2016)
19. Al Salameh, M.S.H., Al-Zu'bi, M.M.: Prediction of radio wave propagation for wireless cellular networks in Jordan. In: 2015 7th International Conference on Knowledge and Smart Technology (KST), Chonburi, pp. 149–154 (2015)
20. Nimavat, V.D., Kulkarni, G.R.: Simulation and performance evaluation of GSM propagation channel under the urban, suburban and rural environments. In: 2012 International Conference on Communication, Information and Computing Technology (ICCICT), Mumbai, pp. 1–5 (2012)
21. Ibhaze, A.E., Ajose, S.O., Atayero, A.A.A., Idachaba, F.E.: Developing smart cities through optimal wireless mobile network. In: 2016 IEEE International Conference on Emerging Technologies and Innovative Business Practices for the Transformation of Societies (EmergiTech), Balaclava, pp. 118–123 (2016)
22. Ibhaze, A.E., Imoize, A.L., Ajose, S.O., John, S.N., Ndujiuba, C.U., Idachaba, F.E.: An empirical propagation model for path loss prediction at 2100 MHz in a dense urban environment. Indian J. Sci. Technol. **8**(1) (2017)
23. Abhayawardhana, V.S., Wassell, I.J., Crosbsy, D., Sellars, M.P., Brown, M.G.: Comparison of empirical propagation path loss models for fixed wireless access systems. In: IEEE Vehicular Technology Conference, Spring, vol. 1, pp. 73–77 (2005)
24. TEMS: Testing, Monitoring, and Analytics Software, Ashburn, VA, USA. www.tems.com
25. Pitney Bowes: MapInfo Pro. http://www.pitneybowes.com/us/location-intelligence/geographic-information-systems/mapinfo-pro.html
26. Forsk: ATOLL 3.2.0 Radio Planning and Optimization Software, France. www.forsk.com

Workshop on Sustainability Performance Assessment: Models, Approaches and Applications Toward Interdisciplinary and Integrated Solutions (SPA 2017)

Ecosystem Services and the Natura 2000 Network: A Study Concerning a Green Infrastructure Based on Ecological Corridors in the Metropolitan City of Cagliari

Ignazio Cannas[✉] and Corrado Zoppi

Department of Civil Engineering and Architecture, University of Cagliari,
Cagliari, Italy
{ignazio.cannas,zoppi}@unica.it

Abstract. An important set of ecosystem services (ESs) delivered by green infrastructure (GI) is based on habitats and species protection and enhancement, that is on maintaining and improving biodiversity. Indeed, the second objective of the European Biodiversity Strategy recommends that ecosystems and their services are maintained and enhanced by establishing GI and restoring at least a 15% of the ecosystems which show up significant decay. From this perspective, habitat fragmentation can be considered one the most outstanding causes of a decreasing attitude of GI towards the delivery of habitat-based ESs, since it weakens the capacity to deliver such services by undermining the networking potential of habitats. In this study, we propose a study concerning the Metropolitan City of Cagliari which includes seventeen municipalities into a unique system of metropolitan government. Sixteen Natura 2000 sites (N2Ss) are located in the City, which amount to about 30% of the metropolitan area. We propose a methodological approach to identify ecological corridors (ECs) between N2Ss, based on the prioritization of functional land patches related to their suitability to deliver ESs concerning biodiversity maintenance and enhancement. The methodology consists of two steps: i. identifying the most suitable patches to be included in ECs on the basis of their accessibility, that is, on their negative attitude towards contributing to landscape fragmentation; ii. assessing, through a discrete-choice-model, the ECs identified through point i in terms of their suitability to be included in a metropolitan GI, starting from the territorial taxonomy based on biodiversity characteristics connected to N2Ss, habitat suitability, and recreational and landscape potentials.

Keywords: Ecological corridors · Green infrastructure · Ecosystem services · Dichotomous-choice models

1 Introduction

ECs are important spatial elements that can improve the structural consistency of ecological networks. ECs identify and implement connectivity concerning migration of wild species, their geographical distribution, and related genetic exchange. As connective elements, ECs should minimize high environmental pressure-related threats

© Springer International Publishing AG 2017
O. Gervasi et al. (Eds.): ICCSA 2017, Part VI, LNCS 10409, pp. 379–400, 2017.
DOI: 10.1007/978-3-319-62407-5_27

generated by humans, such as agricultural and forestry practices, pollution, spread of exotic species, improperly-built infrastructure and urbanization. These threats are likely to cause significant negative impacts on the environment and fragmentation of the ecosystems matrix [15].

Baudry and Merriam [5] maintain that the movement of the species in an ecological network framework depends on factors related to the wild fauna and flora, and it occurs through connections that they define as ECs. Moreover, they distinguish between two conceptual reference categories related to ECs, i.e. "connectedness" and "connectivity". The first concept regards the physical contiguity between ecosystems and/or species populations. The second is based on two components: the first is a structural element which depends on the spatial position of ecosystems, on their physical continuity, and on the presence, type and size of natural and anthropic elements; the second is a functional element which regards the perception scale of species, and their ecological and behavioral needs, including their degree of specialization. There is a substantial difference between physical and territorial aspects, and ecological and functional aspects, concerning the concept of spatial "restitching" in terms of wild species mobility [4, 15, 16].

A working definition of GI is the following: "[A] strategically planned network of natural and semi-natural areas with other environmental features designed and managed to deliver a wide range of ESs. It incorporates green spaces (or blue if aquatic ecosystems are concerned) and other physical features in terrestrial (including coastal) and marine areas. On land, GI is present in rural and urban settings".[1] Moreover, "The work done over the last 25 years to establish and consolidate the network means that the backbone of the EU's GI is already in place. It is a reservoir of biodiversity that can be drawn upon to repopulate and revitalize degraded environments and catalyze the development of GI. This will also help reduce the fragmentation of the ecosystems, improving the connectivity between sites in the Natura 2000 network and thus achieving the objectives of Article 10 of the Habitats Directive".[2] These citations entail that the GI concept is strictly related to ESs. Furthermore, a GI implies an operating spatial network of areas that are worth protecting [28]. Urban and regional spatial planning can effectively address the question of setting-up, developing and monitoring GI and ESs networks. This also implies that GI is very important to restore biodiversity and to reduce ecosystems fragmentation, and their capacity to deliver ESs [17]. As a consequence, a comprehensive objective of management policies of GI can be identified by the issue of the role played by GI to promote and improve the supply of habitats restoration and the delivery of ESs [19, 28].

In this study, we propose a methodological approach to identify ECs connecting N2Ss, based on the prioritization of functional land patches related to their suitability to deliver ESs concerning biodiversity maintenance and enhancement. The methodology consists of two steps: i. identifying the most suitable patches to be included in ECs on the basis of their accessibility, that is, on their negative attitude towards contributing to

---

[1] Communication from the Commission to the European Parliament, the Council, the European Economic and Social Committee and the Committee of the Regions (GI) (COM (2013) 249 final).

[2] Ibid.

landscape fragmentation; ii. assessing, through a discrete-choice-model, the ECs identified through point i in terms of their suitability to be included in a metropolitan GI, starting from the territorial taxonomy based on biodiversity characteristics connected to N2Ss, conservation value, and recreational and landscape potentials.

We implement the methodology with reference to the Metropolitan City of Cagliari, an administrative area located in Southern Sardinia, Italy (Fig. 1).

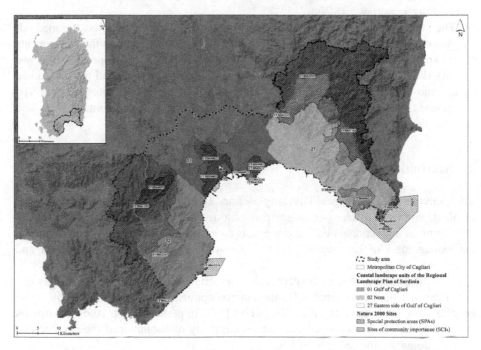

Fig. 1. The study area: the Metropolitan City of Cagliari and its extensions related to N2Ss and the Regional Landscape Plan units.

The Metropolitan City of Cagliari was established by the Sardinian Regional Law n. 2016/2. It is located in Southern Sardinia, and includes seventeen municipalities. The total area amounts to 1,247 km^2. The most important regional strategic transportation nodes are located in the Metropolitan City of Cagliari, such as the Port of Cagliari and the International Airport of Elmas. Land cover classes in the metropolitan area, based on level 1 of the CORINE[3] Land Cover (CLC) [18], show the following proportions: 10%, artificial; 32%, agricultural; 52%, forest and semi-natural areas; 3%, wetlands; and, 3%, water bodies.

Artificial areas, wetlands and water bodies, are located in the central position of the Metropolitan City and surrounded by rural areas. Forests and semi-natural areas are

---

[3] CORINE is the acronym of COoRdination de l'INformation sur l'Environnement [Coordination of the information concerning the environment].

mainly placed on the Western and Eastern sides. The resident population amounts to about 430,000 inhabitants.

The study area includes the Metropolitan City of Cagliari with minor extensions concerning three coastal landscape units of the Regional Landscape Plan of Sardinia, and the boundaries of the N2Ss located in the Metropolitan City (Fig. 1). The size of the study area is about 1,786 km^2. Moreover, thirty municipalities and nineteen N2Ss are included in the study area: six Special protection areas (SPAs) and thirteen Sites of Community importance (SCIs).

The study is organized as follows. In the second section, we discuss the methodology we use to address the two steps mentioned above. We describe the components of the spatial database that we use to apply our methodological approach, how we identify the ECs and how we detect the relations between the ECs and the GI studied by Lai and Leone [26]. Outcomes and findings are discussed in the third section. The conclusions present planning policy implications, put in evidence some caveats and propose suggestions for future research.

2 Methodology

The Convention on Biological Diversity[4] defines an ecosystem as "a dynamic complex of plant, animal and micro-organism communities and their non-living environment interacting as a functional unit", and suggests an ecosystem-based approach integrating the management of land, water and living resources, to promote conservation and sustainable uses.

Scientific methodologies involve processes, functions and interactions between organisms and their environment. Humans are components of ecosystems and they gain benefits from them, in terms of available ESs [37]. In particular, our study focuses on ECs as spatial elements that support the connectivity of habitats and their protection, and, by doing so, the delivery of ESs.

Prioritising spatial elements as ECs implies the definition of their spatial configuration and connectivity. This entails the identification of the spatial structure which protects biodiversity functions and their long-term-persistence [42].

Since an EC is inherently a spatial connection, its identification involves spatial data modelling and planning methods, which aim to identify and prioritize spatial elements to maximize the supply of ESs by granting the most effective flow of species. We address this question by identifying a set of potential ECs based on different input

---

[4] The Nairobi Conference for the Adoption of the Agreed Text of the Convention on Biological Diversity was held on 22 May 1992. As it is stated in the official Internet site of the Convention (https://www.cbd.int/convention/), "The Convention was opened for signature on 5 June 1992 at the United Nations Conference on Environment and Development (the Rio "Earth Summit"). It remained open for signature until 4 June 1993, by which time it had received 168 signatures. The Convention entered into force on 29 December 1993, which was 90 days after the 30th ratification. The first session of the Conference of the Parties was scheduled for 28 November–9 December 1994 in the Bahamas" [accessed 14 June 2017].

layers (a potential habitat suitability and an ecological integrity combined in a resistance map).

The second methodological point is related to the definition of an operative relation to identify ECs as parts of a GI. Lai and Leone [26] propose a taxonomy of the entire Cagliari Metropolitan City based on three factors as follows. The first, conservation value, is defined in terms of habitats of community interest (under the provisions of the "Habitats" Directive, n. 92/43/EEC); the second, natural value, is related to the biodiversity capacity of supplying ESs; the third, recreation value, accounts for the feelings and perceptions of the users of the open spaces of the study area in terms of recreation potential (Table 1).

ECs are selected by connecting the N2Ss located in the Cagliari Metropolitan City by prioritizing functional land patches identifying the most suitable parcels to be included in ECs on the basis of their accessibility, that is, on their negative attitude towards contributing to landscape fragmentation.

Through an overlay mapping of Lai and Leone's taxonomy and the ECs selected in Sect. 2.1, we identify a dichotomous categorization of patches belonging to the ECs. The value associated to each patch is 1 if the patch is included in the GI and 0 otherwise.

This section is divided into two subsections which address the two objectives of the study, concerning ECs and the GI-related characteristics of ECs.

In the first subsection, we describe, step by step, how we identify the ECs of the Cagliari Metropolitan area. We use the land cover map of Sardinia and the Sardinian regional monitoring system. We define the meaning of ecological integrity and describe the GIS tool we use to implement spatial analysis. Moreover, we discuss the issue of habitat suitability mapping versus resistance mapping, and how we integrate resistance and ecological connectivity mapping to identify the ECs.

The second subsection describes the discrete-choice methodological approach we use in this study to assess the ECs identified through the prioritization of the most accessible patches in the study area. Our assessment is based on a discrete-choice-model (DCM), namely a Logit DCM model (LM). The LM is implemented to identify ECs in terms of their suitability to be included in a metropolitan GI, starting from the territorial taxonomy proposed by Lai and Leone [26].

We use a discrete-choice approach since the dependent variable, which indicates whether or not a land parcel belongs to the metropolitan GI, takes values that can be easily divided into two groups, which are mutually exclusive. In this case, dichotomous DCMs are the most adequate, whereas regression models would be preferred if it was not possible to identify data groupings.

On the side of the explanatory variables, there is no loss of information since all the values of the variables used in the model are not subject to any transformation.

In this study, we use an LM model to implement spatial analysis, in order to assess if, and to what extent, ECs connecting N2Ss can be identified as part of a metropolitan GI, whose characteristics are tentatively defined through the taxonomy proposed by Lai and Leone [26]. We think that this is a step further in the use of DCMs, since we use this methodological approach, implemented in several scientific and technical contexts, as we show in the second subsection, in the field of spatial analysis and regional and city planning.

Table 1. Definition of variables and descriptive statistics.

Variable	Definition	Mean	St. dev.
EC-GI	Discrete variable – A land parcel belonging to a EC: • 0 if it is not included in the metropolitan GI; • 1 if it belongs to the metropolitan GI as well	0.4008	0.4901
CONS_VAL	Continuous variable in the interval [0,1]. Presence of natural habitat types of Community interest (as listed in Annex I of the Habitats Directive) and conservation importance thereof CONS_VAL = 0 for areas where no habitats of Community interest have been identified; else CONS_VAL = P*(R + T + K) [normalized in the interval [0,1] where: • priority habitats P = 1.5 in case of priority habitat, P = 1 in case of non priority habitat; • rarity R = [1, 5] depending on the number of Natura 2000 standard data forms in which the habitat is listed within the regional Natura 2000 network; the higher the number of occurrences, the lower the value of R; • threats T = [1, 5] depending on the number of threats recorded in the standard data forms for the Natura 2000 sites in our study area; the higher the number of threats, the higher the value of T; • knowledge K = [1, 4] depending on the level of current knowledge (e.g. number of onsite surveys, existence of up-to-date and reliable monitoring data) of a given habitat within the regional Natura 2000 network; the lower the knowledge, the higher the value of K	0.06373	0.1464
NAT_VAL	Continuous variable in the interval [0,1]. Potential capability of biodiversity to supply final ecosystem services in face of threats and pressures it is subject to The value was calculated using the software "InVEST"[a], tool "Habitat quality". Data inputs for the model were: • land cover types as per the 2008 Regional land cover map (rasterized); • raster maps of ten threats selected on the basis of their spatial character among those listed in the standard data forms for the Natura 2000 sites in our study area. The ten selected threats are as follows: cultivation; grazing; removal of forest undergrowth; salt works; paths, tracks, and cycling tracks; roads and motorways; airports; urbanized areas; discharges; fire and fire suppression;	0.6239	0.3140

(continued)

Table 1. (*continued*)

Variable	Definition	Mean	St. dev.
	• weights and decay distance for each threat from expert judgments; • sensitivity of each land cover type to each threat from expert judgments; • accessibility to sources of degradation, in terms of relative protection to habitats provided by legal institutions. The three categories we used are as follows: natural parks, areas protected and managed by the regional Forestry Agency, Natura 2000 sites		
RECR_VAL	Continuous variable in the interval [0,1]. Recreational attractiveness of landscapes and natural habitats. The average photo-user-days per year between 2010 and 2014 was calculated using the software "InVEST" (tool "Recreation") and a 3-km grid, and subsequently normalized in the interval [0,1]	0.3613	0.3000

[a]InVEST (Integrated Valuation of Ecosystem Services and Tradeoffs) is a free of cost software product licensed under the BSD open source license. As indicated in the InVEST-related documentation available online at http://data.naturalcapitalproject.org/nightly-build/invest-users-guide/html/index.html [accessed 14 June 2017], "InVEST is a tool for exploring how changes in ecosystems are likely to lead to changes in benefits that flow to people". InVEST is developed by the Natural Capital Project, whose partners are: Woods Institute for the Environment and Department of Biology of Stanford University; Institute on the Environment of Minnesota University; The Nature Conservancy; and, The World Wildlife Fund (WWF).

2.1 The Ecological Corridors

The Land Cover Map of Sardinia. The Autonomous Region of Sardinia (ARS) published, in 2008, a land cover map at the 1:25,000 scale. This map is a geographical database of the land covers of Sardinia, whose classification was adapted to the local situation considering the standard code of the CLC. Areal and linear elements are included in the database. The areal elements are identified by CLC classes, up to level 5. The linear elements represent a potential hydrographic network (e.g. canals and waterways, rivers, streams and ditches), and the transportation network (railway and road networks).

The Monitoring System of ARS. Starting from 2008, the ARS [3] has implemented a monitoring system related to the conservation status of habitats and species of Community interest [3]. For each Sardinian N2S, species-specific values of habitat suitability are mapped by associating them to land cover classes. These maps regard only the areas included in the N2Ss. Species-specific values of habitat suitability have been derived directly from the "Rete Ecologica Nazionale" (REN) [11], that is the National Ecological Network. In the REN, the list of analyzed fauna species does not include all species listed in the Habitats Directive and contained in the standard data forms of the

N2Ss. For this reason, the habitat suitability values are not identified for all species, and the CLC evaluation does not include all species. The suitability values are defined as follows: 0 (unsuitable), i.e., spatial elements that do not meet the ecological requirements of species; 1 (low suitability), i.e., spatial elements that support the presence of species discontinuously (through time); 2 (average suitability), i.e., spatial elements that support the presence of species, even though they are not their optimal locations; 3 (high suitability), i.e., spatial elements that are the best locations for permanent presence of species.

The Ecological Integrity Values. Burkhard et al. [12, 13] proposed a matrix where qualitative values are assigned to 44 land cover types as regards their capacity to provide ESs on a scale from 0 (no relevant capacity) to 5 (very highly relevant capacity). In the matrix, they also assess seven indicators related to the ecological integrity, associated to land cover types. The indicators represent the main components of the ecosystem functionality, by describing structures and processes relevant for the long-term functionality and the self-organizing capacity of ecosystems. The values of ecological integrity describe issues related to structures, as numbers and characteristics of species (biotic diversity) and physical habitat components (abiotic heterogeneity), and processes concerning ecosystem energy budgets (exergy capture), matter budgets (nutrient storage and loss) and water budgets (biotic water flows and metabolic efficiency).

High values of ecological integrity are found as regards different land cover types, whereas very low or no relevant capacities correspond to land cover types characterized by significant anthropic impacts (e.g., urbanized fabric, industrial or commercial areas, mining sites and landfills).

The Identification of the ECs. Several studies approach the identification of ECs through a class of algorithms named "least-cost path" (LCP) algorithms [1, 6, 19, 27, 28, 41, 47]. This approach allows to prioritize patches to define ECs between N2Ss. Adriaensen et al. [1] proposes a useful explanation related to this approach as follows. Two raster layers (a source layer and a friction/resistance layer) are the model inputs. The source layer represents patches needed to be analyzed in terms of connectivity. The resistance layer represents, in each cell of a grid, a resistance value, that is a cost variable identified by the attributes of the land cover type of the cell. An LCP path represent the less expensive way that is likely to be chosen by species to move across a surface in order to shift from a patch to another. In the LCP approach, species are assumed to perceive the movement across the surface as a burden, that is a cost, represented by the resistance opposed to the movement, which is the compound value of the energy spent to move, the mortality risk, the negative impact on future reproductive potential.

Sawyer [41] claims that the LCP approach is particularly attractive to analyze and design habitat corridors based on the identification of the impacts of habitats on species movement.

The effectiveness of LCP analysis can be affected by the quality of input data. Expert opinions can help building a resistance values map as well as other techniques.

LCP analyses are based on subjective interpretations, but the approach is very useful for land use managers to identify where potential corridors fit environmental priorities defined as regards a particular landscape.

Our connectivity analyses are based on the following steps: firstly, we map habitat suitability outside the N2Ss using data available in the literature [3] and related to ecological integrity [12, 13]; secondly, we use the inverse of the habitat suitability map integrated by the ecological integrity map to set a resistance map; thirdly, we use the Linkage Mapper[5] tool in order to identify the least-cost corridors between core areas as the paths of minimum cost, and, by doing so, to identify potential ECs.

Linkage Mapper is a GIS tool which analyzes the habitat connectivity of the regional wildlife scenarios. Linkage Mapper uses vector maps of core habitat areas and raster maps of movement resistance. In the resistance map, cells represent values reflecting the energetic cost, difficulty, or mortality risk of moving across the landscape. Using cost-weighted distance (CWD) analyses, Linkage Mapper produces maps of total movement resistance accumulated designing animals' movement between specific core areas [36].

In the following paragraphs, we explain how we generate the final resistance map. Each raster map is created with a resolution of 625 m^2 per pixel (a pixel = 25 25 m^2), which is the optimal resolution for the Land Cover Map of Sardinia. Moreover, we always use the level 3 of the CLC.

Habitat Suitability Mapping. Habitat suitability can be defined as the probability that species use habitats. Models of habitat suitability can help improving networks and evaluating connectivity. These models show the distribution of suitable habitats or resource patches in a landscape. A suitability map represents the probability of patches to be used by a given species [11, 45]. Habitat suitability indexes are generated from expert opinions, and related to core environmental variables [22, 47]. In our study, we derived a global habitat suitability value related to a CLC class from data available in the monitoring system of ARS [3], by calculating its weighted mean value concerning the species associated in a CLC class. By doing so, we identify globally potential suitable areas to host species of Community interest with reference to the global suitability values, not only within the N2Ss, but also within the entire study area.

Ecological Integrity Mapping. The ecological index of Burkhard et al. [12, 13] describes the functionality concerning relevant structures and processes as regards the long-term functionality and the self-organizing capacity of ecosystems. We assume that land patches showing high values of ecological integrity, which implies that they provide suitable habitats for different species, are very effective in supporting transitions. Under this perspective, the ecological index can be understood as a value of suitability.

Resistance Mapping. Resistance reflects the effects of morphological characteristics of the landscape on species movement and mortality, in terms of hampering flows of species, energy, and material [19, 21, 22, 27]. Resistance parameters are retrieved in the

---

[5] Linkage Mapper can be downloaded from http://www.circuitscape.org/linkagemapper [accessed 14 June 2017].

relevant literature or estimated through data concerning the uses of habitats (such as habitat suitability) [6, 22]. Resistance values can be determined by the cell characteristics, such as land cover or housing density, combined with species-specific landscape resistance models. As a consequence, a study area could be divided into more-suitable and less-suitable elements. The less-suitable elements could be considered the most resistant [22]. Under this perspective, LaRue and Nielsen (2008, cited in [47), suggest setting resistance values as the inverse of habitat suitability values [19].

In this study, we build a resistance raster map representing negative attitudes as regards contribution to landscape fragmentation, through the following steps. Firstly, we generate the inverse raster map of habitat suitability and ecological integrity; secondly, we rescale these raster maps from 1 to 100 [19]; thirdly, we create a new raster map by summing up the generated raster maps; and, finally, we rescale the last raster map from 1 to 100.[6] This combined map, processed for the whole study area, shows, on the one side, a global perception scale of species, and, on the other side, a morphological perception coming from land uses and the ecosystem services that they supply.

Connectivity Analysis. In Linkage Mapper, we load a shapefile of core areas related to the N2Ss and the resistance raster map generated according to the procedure described above. Linkage Mapper generates maps of total resistance related to the modelled species movement across specific core areas. Linkage Mapper modelling is based on the following steps: firstly, adjacent core areas are identified; secondly, a network of core areas is defined by using adjacency and distance data; thirdly, CWD analyses and LCP algorithms are implemented; and, finally, least-cost corridors are identified and represented into a single map, by a single CWD-based raster map for all core areas [36].

Linkage Mapper computes least-cost corridors as the sum of CWD based on the raster maps related to pairs of core areas. For example, if A and B are two core areas, the least-cost corridors are normalized as follows:

$$NLCC_{AB} = CWD_A + CWD_B - LCD_{AB} \qquad (1)$$

where $NLCC_{AB}$ is the normalized least-cost corridor connecting core areas A and B, CWD_A is the cost-weighted distance from core area A, CWD_B is the cost-weighted distance from core area B, and LCD_{AB} is the cost-weighted distance accumulated by moving along the potential least-cost path connecting the two core areas [36].

2.2   The DCM Methodology

DCMs concern the interpretation of phenomena represented by a finite number of possible occurrences, that is by a set of mutually exclusive alternative outcomes. McFadden [31, 33] generalized the approach of Williams [46] by implementing into Williams' framework the classic microeconomic theory related to agent's choice.

---

[6] We set 1 for the lowest value of resistance and 100 for the highest as 100 as suggested on the user guide of Linkage Mapper to create a resistance map [36].

The utility maximizing agent's behavior of classic microeconomic theory is implemented by McFadden into the Williams' framework by including several characteristics related to agents, which are heterogeneous and which may or may not be known to the modeller; in case they are unknown, they are considered as random features [31, 33, 35]. Textbook DCMs are usually derived from these theoretical references [7, 39, 43], since they assume that the agent's behavior implies the awareness, on behalf of the modeller, that information on the agents' choices is not complete and that agents' behaviors are not consistent with perfect rationality [44].

A further theoretical assumption we take as granted in the model we implement in this study is that the random component of an agent's utility function is unrelated to the random component of the other agents. We propose this restrictive assumption since data related to dependent and explanatory variables of the model are deterministically and independently defined by the modeller, and, as a consequence, the distribution of the random terms should be characterized by the following (where: x is the vector of explanatory variables; X is the matrix of observations related to explanatory variables): i. $E(\varepsilon \mid x) = 0$ (i.e., the error terms have a 0 conditional mean), ii. $Var(\varepsilon) = \sigma^2$ (i.e.: the error term has the same variance at each observation), and iii. $E[\varepsilon_i \varepsilon_j \mid X] = 0$ (i.e.: the error terms are uncorrelated between observations).[7]

DCMs are used in several fields, because they are very effective, from the theoretical and empirical points of view, to assess a large number of issues related to several scientific fields.[8]

DCMs are commonly implemented to analyze multiple outcomes related to a set of factors which need to be tested as possible determinants. For example, Bockstael et al. [9] use an LM to assess the choice concerning seawater and freshwater beaches, whereas Bockstael et al. [10] problematize the McFadden's approach [29, 30] by putting in evidence that an issue, that needs to be addressed in a proper way, is represented by the case of nested steps, which may occur when dichotomous or multiple outcomes are conditioned by previous dichotomous or multiple outcomes "if…then…else…": in terms of recreational activity, if fishing is chosen in the first place as a dichotomous alternative instead of gaming, then underwater fishing and fishing with a rod or a line is the next nested alternative, etc.

McFadden, who won the Nobel Prize in 2000 (with Heckman) addresses the question of discrete choice from different standpoints. Among many issues, he studies the demand for services related to urban transport [29], the multiple choice between different urban transportation modes [32], the multiple choice concerning phone services [34], dichotomous and multiple-choice problems that involve public administration at different levels [30].

Moreover, Pavlopoulos et al. [40] use DCMs to analyze wage mobility across Europe, and Ambrogi et al. [2] assess, through DCMs, the impact of sets of exogenous determinants on the effectiveness of clinical therapies.

In this study, we implement an LM, which is generally used to study phenomena characterized by dichotomous observations, that is, observations which can take just

---

[7] What is proposed in this theoretical session is largely based on Cherchi [14].

[8] For the applications of the DCMs approach in several fields see Bhat [8].

two possible outcomes, represented by the 0 and 1 numerical values. The assumption of the LMs is that these phenomena are related to a set of explanatory variables through a logistic probability function.

We operationalize an LM in order to assess the relationships between regional green infrastructure and ecological corridors as follows.

1. A dichotomous variable is defined, which is related to land parcels included into ecological corridors (ECs) which connect Natura 2000 sites located in the Metropolitan City of Cagliari, whose dichotomous nature concerns the fact that a parcel does belong (dichotomous variable equal to 1), or does not belong (dichotomous variable equal to 0), to a regional green infrastructure (RGI).
2. The explanatory variables of the LM are related to the factors assumed by Lai and Leone [26] as determinants of the inclusion of a land parcel into an RGI.
3. The discussion of the estimates of the LM entails an assessment of how, and to what extent, land parcels included into ECs are part of an RGI as well, or, in other words, if, and to what extent, the regional Natura 2000 network can be identified as a part of a RGI.

The LM described below is based on Greene's [23, pp. 666–672], Nerlove and Press' [38], and Zoppi and Lai's [48] studies.

We consider a set of two events, Y_0 and Y_1, with probability of event $Y_i = j$ given by:[9]

$$\text{Prob}(Y_i = 1) = \frac{e^{\beta'x}}{1 + e^{\beta'x}} \tag{2}$$

$$\text{Prob}(Y_i = 0) = \frac{1}{1 + e^{\beta'x}} \tag{3}$$

where $j \in \{0, 1\}$, β is a vector of coefficients β_h, $h \in \{1, ..., N\}$, and x is a vector of characteristics x_h, $h \in \{1, ..., N\}$, of the land parcel k where the event Y_i occurs, $k \in \{1, ..., M\}$. The coefficients of vector β are estimated by solving the maximization problem of the following log-likelihood function, ln L:

$$\ln L = \sum_{k=1}^{M} \sum_{j=0}^{1} d_{kj} \ln \text{Prob}(Y_i = j) \tag{4}$$

(where $d_{kj} = 1$ if the event j occurs, and $d_{kj} = 0$ otherwise), in the coefficients β's. These coefficients will appear in (4) through the expressions (2) and (3) of Prob $(Y_i = j)$.

---

[9] If we define $\beta^* = \beta + q$ for any nonzero vector q, the identical set of probabilities result, as the terms involving q all drop out. A convenient normalization that solves the problem is to assume that vector $\beta_0 = 0$. The probability for $Y_i = 0$ is therefore given by (2) [24: 666].

The derivatives of (4) with respect to the coefficients β's have the following form:

$$\frac{\partial \ln L}{\partial \beta_h} = \sum_{k=1}^{M} [d_{kj} - \text{Prob}(Y_i = j)] x_h. \tag{5}$$

The values of the vectors of coefficients of vector $\boldsymbol{\beta}$ which maximize (4) are the solution of the system which comes from equalizing to zero the derivatives expressed by (5). The values of the vector of coefficients of vector $\boldsymbol{\beta}$ make it possible to calculate the marginal effects of a change of the characteristics x on the probability that the event j occurs, $\frac{\partial \text{Prob}(Y_i)}{\partial x_h}$., as follows:

$$\frac{\partial \text{Prob}(Y_i)}{\partial x_h} = [\text{Prob}(Y_i)] \left\{ \beta_h - \sum_{l=1}^{N} [\text{Prob}(Y_i)] \beta_l \right\}. \tag{6}$$

The estimate of the model makes it possible to calculate the marginal effects of (6), e.g. with reference to the average values of the coefficient of vector **x**, and the probabilities of the events j's. Moreover, the model makes it possible to estimate the standard errors of the estimates of the coefficients of vector $\boldsymbol{\beta}$ and of the marginal effects of (6).

3   Results and Discussion

In this section, we present and discuss the results of the implementation of the methodological approaches implemented in order to identify the ECs of the Metropolitan City of Cagliari and to characterize them in terms of their relationship with the metropolitan GI defined by Lai and Leone [26]. The first subsection concerns the ECs, the second subsection shows and discusses the results as regards the GI-related characterization.

3.1   The ECs of the Metropolitan City of Cagliari

The results generated by the implementation of the methodology to identify the ECs are the following: i. a composite raster map of linkage maps, where each cell represents the minimum value of all individual normalized corridor layers; ii. a shapefile representing the normalized least-cost corridors.

By extending the analysis beyond the limits of our study area, in order to avoid biases coming from the edge effect, we obtain: i. a map where CWD values are included in the range between 0 and 260,746 km; ii. a shapefile where twenty-four linear corridors are identified.

A field containing the ratio of CWD to Euclidean distance is reported in the attribute table of the shapefile of the linear corridors. The resulting corridors can be valued by this quality metric [36], since the linear corridors are identified starting from a resistance map. For example, the corridor which connects the core areas 7 and 9

(Fig. 2) which has a length of 14 km, shows a cost-weighted ratio of 32.3, due to the presence of several urbanized areas and a high value of obstacles in the initial resistance map. On the other hand, the corridor which connects the core areas 1 and 9, which has a length of 19 km, presents a cost-weighted ratio of 5.4, due to the presence of low resistance values in the initial resistance map.

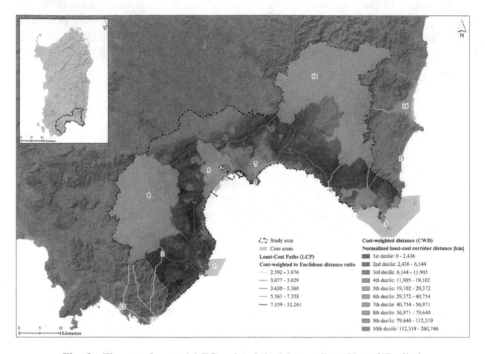

Fig. 2. The set of potential ECs related the Metropolitan City of Cagliari.

We define the spatial boundaries of the ECs by classifying the raster map of the normalized cumulated CWD values in ten deciles and we choose the first decile as a threshold value for a land parcel to be included in an EC. The first decile is equal to 2.4 km (Fig. 2), that identify an area of 245 km^2, which is roughly 14% of the total study area. Regarding the first CLC class, the ECs show the following proportions: 0.71%, artificial (1.75 km^2); 36.24%, agricultural (88.78 km^2); 61.07%, forest and semi-natural areas (149.62 km^2); 0.75%, wetlands (1.83 km^2; 1.22%, water bodies (3.00 km^2). We overlap the ECs, defined by the first decile, with the spatial taxonomy, implemented by Lai and Leone [26] (Fig. 3), as discuss as follows.

3.2   Characterization of the ECs with Reference to the Metropolitan GI

By means of the dichotomous categorization and the values of factors of the spatial taxonomy implemented by Lai and Leone [26] we operationalize the model described by (2) thru (6) as follows:

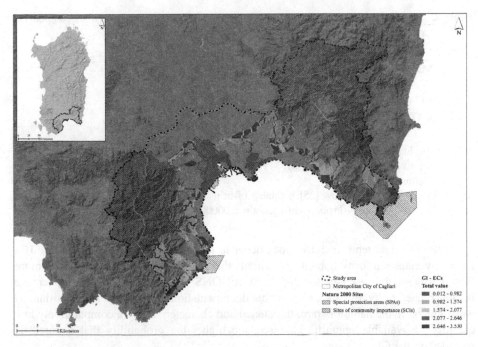

Legend:
- Study area
- Metropolitan City of Cagliari
- **Natura 2000 Sites**
- Special protection areas (SPAs)
- Sites of community importance (SCIs)

GI - ECs
Total value
- 0.012 - 0.982
- 0.982 - 1.574
- 1.574 - 2.077
- 2.077 - 2.646
- 2.646 - 3.530

Fig. 3. The overlapping map of the selected ECs and the metropolitan GI identified in the study by Lai and Leone [26].

1. the dichotomous variable Y_i takes the values 1 and 0, as indicated above;
2. the components of vector of characteristics x_h, $h \in \{1, ..., N\}$, are the three factors of the spatial taxonomy implemented by Lai and Leone [26], which entails that $N = 3$;
3. the overlay mapping identifies 6233 land parcels k where the event Y_i occurs, $k \in \{1, ..., 6233\}$.

Descriptive statistics concerning Y and x's are reported in Table 1.

Estimates of the marginal effects of the x's variables on the probability that a land parcel belonging to an EC belongs to the metropolitan GI as well (the $Y_i = 1$ event), based on (5), are shown in Table 2.

The LM estimate performs very well in terms of goodness of fit, since both, the log-likelihood ratio and Hosmer and Lemeshow [24] chi square-based tests indicate no significant difference between the observed and estimated probability distributions of the event Y_i (Table 2).

In the following paragraphs, the outcomes related to the impacts of variables x's on the $Y_i = 1$ event are discussed. The outcomes are quantitatively consistent with each other and in terms of sign (a positive sign is expected in all three cases) as well.

Conservation Value. The presence of habitats of Community interest, as listed in Annex I of the Habitats Directive, is very important in terms of inclusion in the metropolitan GI of a land parcel located in an EC. The probability of inclusion grows

Table 2. Marginal effects on the probabilities of the $Y_i = 1$ event of the variables components of vector **x**.

Variable	Marginal effect	z-statistic	Hypothesis test: marginal effect = 0
Marginal effect on probability of $Y_i = 1$, $\partial Prob (Y_i = 1)/\partial x_h$, Prob $(Y_i = 1) = 0.401$			
CONS_VAL	1.14774	15.462	0.0000
NAT_VAL	1.62997	39.921	0.0000
RECR_VAL	1.33653	33.693	0.0000

Log-likelihood goodness-of-fit test.
Log-likelihood ratio = 4332.987 – Prob. > chi-square = 0.00000 (3 degrees of freedom).
Hosmer and Lemeshow [25] goodness-of-fit test.
HL = 444.91669 – Prob. > chi-square = 0.00000 (8 degrees of freedom).

by 11.4% if a one tenth increase does occur at the mean value of CONS_VAL. If a parcel overlaps a priority habitat, this entails that the probability grows by 50% more than if it is not the case (see the definition of CONS_VAL in Table 1). Moreover, the habitat-related rarity, threats and available documentation and studies, are important as well. The more the habitat is rare, threatened and characterized by a comparatively low quantity of available scientific analyses, the higher the probability that a parcel is included in the GI.

Natural Value. The potential of biodiversity to supply final ecosystem services by opposing threats and environmental pressures shows a positive impact on the suitability to be included in the metropolitan GI of a land parcel located in an EC. This is consistent with expectations, since ECs are identified by parcels comparatively more capable to host species protected under the provisions of the Habitats Directive, and to supply ESs. The estimate of the LM indicates that the probability of a parcel to be included in the metropolitan GI increases by 16.3% in case it shows a 10% increase in its natural value, which is identified by its habitat quality as defined in Table 1 and by Lai and Leone [26], and spatially implemented by the software InVEST. The natural value depends on: i. the capacity of supplying ESs by different land cover types; ii. the related environmental threats, their decay distance and the sensitivity of land cover types to each threat; and, iii. the degree of protection to habitats granted under the provisions of established legal acts (natural parks, areas protected and managed by the Sardinian Forestry Agency and N2Ss).

The normalized natural value, variable NAT_VAL in Tables 1 and 2, shows a comparatively higher impact than conservation value, and should be carefully considered as a point of reference for environmental planning policies to identify ECs in the context of a regional or metropolitan GI.

Recreation Value. The recreation value (variable RECR_VAL in Tables 1 and 2) is identified by the interest for natural amenities, landscapes and habitats as expressed by the users through the pictures posted on social networks, collected by InVEST and reported on a 3-km spatial grid. Variable RECR_VAL is normalized in the [0,1] interval and shows an impact comparatively higher than conservation value and lower than natural value.

Even though this measure of natural and environmental attractiveness is not a direct source of suggestions concerning planning policies, however its positive impact on the probability of a land parcel located in an EC to be included in the metropolitan GI recommends planners and policy-makers take account very carefully of the attractiveness issue in the definition and implementation of environmental protection-related plan actions.

4 Conclusions

This study presents and discusses a methodology to model ECs which connect the nodes of the Natura 2000 network, that is SCIs and special areas of conservation (SACs) identified under the provisions of the Habitats Directive itself, as well as SPAs, identified under the provisions of the "Birds" Directive (Directive 2009/147/EC). We implement the methodology to identify ECs in the Metropolitan City of Cagliari through connectivity and ecological integrity measures related to species, habitats, land cover types and capacity of providing ecosystem services.

Restoring and maintaining functional connections in ecological networks is a crucial challenge as regards spatial planning and landscape ecology. Habitat patches located outside of protected areas play a connective role in building a spatial network of sites endowed with environmental values. Therefore, we implement a methodological approach to prioritize key patches around protected areas with reference to their connectivity potential. Our results show that the use of Linkage Mapper can be very effective to identify sets of ECs. Moreover, we find that several species, with different behaviors, can benefit from improved connectivity entailed by the definition of ECs and their identification as worth protecting natural areas. The metropolitan government should be aware of the presence of ECs and should take into account their strategic role in terms of environmental protection and conservation of natural heritage. The importance of ecological priorities such as areas belonging to ECs should be recognized and supported in decision-making processes concerning metropolitan planning policies. Whereas species-specific and site-specific conservation strategies are implemented into spatial plans of protected areas, these measures should be extended outside their boundaries in order to address the issue of biodiversity conservation, by integrating into land use management practices the objective of preserving and improving landscape connectivity.

Secondly, we assess the ECs in terms of their suitability to be included in the metropolitan GI identified through the methodology proposed and implemented by Lai and Leone [26]. Our appraisal is based on a LM which detect if, and to what extent, the factors used to identify the metropolitan GI influence the inclusion in the GI of land parcels belonging to ECs. Our findings are quite consistent with expectations in terms of significance of the estimated LM, which entails that metropolitan planning policies aiming at identifying ECs as part of a GI should strengthen GI-related characteristics of ECs, since, as Table 2 shows, at present only 40% of the ECs are part of the metropolitan GI as identified by Lai and Leone [26].

This outcome implies a set of policy recommendations for the metropolitan government in order to improve and strengthen the GI characteristics of the ECs of the N2Ss. These recommendations aim to enhance the features of the metropolitan Natura 2000 network in terms of its suitability to be part of the metropolitan GI.

The first recommendation concerns the presence of habitats of Community interest, and, among these, of priority habitats as well, which are located outside N2Ss. These habitats are not protected under the provisions of the Habitats and Birds Directives and endangered by pressures coming from planning policies whose goals are related to new developments, e.g. new residential or productive settlements. National, regional and metropolitan plan actions should aim at increasing scientific knowledge on these areas and at identifying environmental protection measures implemented through adequate planning rules. Moreover, the metropolitan administration should lobby with the European Union and the national government to extend the environmental protection regime under the provisions of the Habitats and Birds Directives, in order to grant a higher level of protection to the ECs connecting the N2Ss. This is a general open issue related to the Natura 2000 network, which is only characterized by the identification of the nodes (SCIs, SPAs and SACs) and related conservation measures, whereas almost nothing has yet been done with reference to the branches (i.e. ECs).

The second recommendation is related to the capability of biodiversity to supply ESs. This feature should be protected and enhanced to make ECs suitable to be included in the metropolitan GI. This entails a number of measures such as: improving and extending scientific knowledge related to the complex relationships between land cover types and ESs productivity; increasing the level of protection of areas characterized by the presence of highly productive ecosystems; mitigating or even preventing future land-taking processes; and, implementing planning policies based on the normative concept of maintaining and possibly increasing the production of goods and services supplied by ecosystems. The ESs-based planning approach would entail the use of natural protection measures and the implementation of controls to detect and prevent damages to ecosystems and their productive potential.

Finally, we find that recreation value matters. The attractiveness of areas within the metropolitan GI needs to spread out and to involve ECs. The impact of the degree of attractiveness on ECs is more important than conservation value, since its marginal effect on the probability of a land parcel belonging to an EC to be included in the metropolitan GI is 40% higher. Since attractiveness is much more volatile than the other factors and somehow subjective, we would recommend to consider this issue for future research, since addressing this question would entail studies which go beyond the content and the goal of this study.

There is an important issue that needs to be discussed with reference to the implementation of the ESs-based approach to spatial planning concerning the definition and implementation of a GI inclusive of ECs that we propose in this study. Protection measures related to certain types of ESs may possibly generate negative impacts on the supply of other ESs. For example, conservation measures concerning habitats and species within N2Ss may conflict with agricultural production (fodder and crops) and pastures, which entails to address the trade-off between supporting and provisioning ESs, as they are classified in the relevant literature on ESs [see, among many: 20, 37]. This is an important issue for future research, which is particularly important in the context of the discussion of this study, since recreation value, which is very important for the definition and implementation of the metropolitan GI and related ECs, can suffer dramatically from conservation measures related to habitats and species which prevent accessibility to attractive sites. In the literature, the trade-off question is brilliantly discussed by Kovács et al. [25] as regards three N2Ss located in the Great Hungarian Plain.

Our methodological approach can be easily exported to other Italian and European Union (EU) regional contexts, since the Natura 2000 network is established in all the EU countries under the provisions of the Habitats Directive and the N2Ss are always identified through the Standard Data Form defined by the Decision of 11 July 2011 (2011/484/EU), even if national contexts are characterized by different institutional frameworks. Our position is that the still-not-defined question of establishing and implementing ECs as branches connecting the nodes of the Natura 2000 network should be addressed through the identification of a local GI.

Acknowledgements. This paper is written within the Research Program "Natura 2000: Assessment of management plans and definition of ecological corridors as a complex network", funded by the Autonomous Region of Sardinia for the period 2015–2018, under the provisions of the Call for the presentation of "Projects related to fundamental or basic research" of the year 2013, implemented at the Department of Civil and Environmental Engineering and Architecture (DICAAR) of the University of Cagliari, Italy.

Ignazio Cannas and Corrado Zoppi have made substantial contributions to the paper's conception and design, introduction, discussion, and conclusions. Ignazio Cannas has taken care of Sects. 2.1 and 3.1; Corrado Zoppi has taken care of Sects. 2.2 and 3.2.

References

1. Adriaensen, F., Chardon, J.P., De Blust, G., Swinnen, E., Villalba, S., Gulinck, H., Matthysen, E.: The application of 'least-cost' modelling as a functional landscape model. Landsc. Urban Plan. **64**, 233–247 (2003). doi:10.1016/S0169-2046(02)00242-6
2. Ambrogi, F., Biganzoli, E., Boracchi, P.: Estimating crude cumulative incidences through multinomial logit regression on discrete cause-specific hazards. Comput. Stat. Data Anal. **53**, 2767–2779 (2009). doi:10.1016/j.csda.2009.01.001
3. ARS (Autonomous Region of Sardinia) [RAS (Regione Autonoma della Sardegna)]. Realizzazione del sistema di monitoraggio dello stato di conservazione degli habitat e delle specie di interesse comunitario della Regione Autonoma della Sardegna [Implementation of the monitoring system of the conservation state of habitats and species of Community importance of the Autonomous Region of Sardinia]. Autonomous Region of Sardinia, Cagliari (2010)
4. Battisti, C.: Frammentazione ambientale, connettività, reti ecologiche. Un contributo teorico e metodologico con particolare riferimento alla fauna selvatica. [Environmental fragmentation, connectivity, ecological networks. A technical and methodological contribution with particolar reference to wildlife species]. Provincia di Roma, Assessorato alle politiche ambientali, Agricoltura e Protezione civile [Province of Rome, Board of environmental policy, agriculture and civil defence]. Province of Rome, Rome (2004)

5. Baudry, J., Merriam, H.G.: Connectivity and connectedness: functional versus structural patterns in landscapes. In: Proceedings of the 2nd IALE Seminar "Connectivity in Landscape Ecology", vol. 29, pp. 23–28. Münsterche Geographische Arbeiten (1988)

6. Beier, P., Majka, D.R., Newell, S.L.: Uncertainty analysis of least-cost modeling for designing wildlife linkages. Ecol. Appl. **19**, 2067–2077 (2009). doi:10.1890/08-1898.1

7. Ben-Akiva, M., Lerman, S.: Discrete Choice Analysis: Theory and Application to Travel Demand (Transportation Studies). The MIT Press, Cambridge (1985)

8. Bhat, C.: Econometric choice formulations: alternative model structures, estimation techniques, and emerging directions. In: Axhausen, K.W. (ed.) Moving Through Nets: The Physical and Social Dimensions of Travel – Selected papers from the 10th International Conference on Travel Behavior Research, pp. 45–80. Elsevier, Oxford, United Kingdom (2007)

9. Bockstael, N.E., Strand, I.E., Hanemann, W.M.: Time and the recreational demand model. Am. J. Agric. Econ. **69**, 293–302 (1987)

10. Bockstael, N.E., McConnell, K.E., Strand, I.E.: Recreation. In: Braden, J.B., Kolstad, C.D. (eds.) Measuring the Demand for Environmental Quality, pp. 227–270. North Holland, Amsterdam (1991)

11. Boitani, L., Corsi, F., Falcucci, A., Maiorano, L., Marzetti, I., Masi, M., Montemaggiori, A., Ottaviani, D., Reggiani, G., Rondinini, C.: Rete Ecologica Nazionale. Un approccio alla conservazione dei vertebrati italiani. Relazione finale [The National ecological network. An approach to conservation of Italian vertebrate species. Final report]. Ministry of Environment and Protection of Natural Resources, Rome (2002)

12. Burkhard, B., Kroll, F., Müller, F., Windhorst, W.: Landscapes' capacities to provide ecosystem services – a concept for land-cover based assessments. Landsc. Online **15**, 1–22 (2009). doi:10.3097/LO.200915

13. Burkhard, B., Kroll, F., Nedkov, S., Müller, F.: Mapping ecosystem service supply, demand and budgets. Ecol. Indic. **2**, 7–29 (2012). doi:10.1016/j.ecolind.2011.06.019

14. Cherchi, E.: Modelling individual preferences, State of the art, recent advances and future directions. Resource paper prepared for the Workshop on "Methodological developments in activity-travel behavior analysis" at the 12th International Conference on Travel Behavior Research held in Jaipur, India, 13–18 December 2009. https://iatbr2009.asu.edu/ocs/custom/resource/W5_R1_Modelling%20individual%-20preferences,%20State%20of%20the%20art.pdf. Accessed 14 June 2017

15. D'Ambrogi, S., Nazzini, L.: Monitoraggio ISPRA 2012: La rete ecologica nella pianificazione territoriale [The ecological network in spatial planning]. In: Reticula, n. 3, ISPRA (2013). http://www.isprambiente.gov.it/files/pubblicazioni/-periodicitecnici/reticula/Reticula_n3.pdf. Accessed 14 June 2017

16. D'Ambrogi, S., Gori, M., Guccione, M., Nazzini, L.: Implementazione della connettività ecologica sul territorio: il monitoraggio [Implementation of ecological connectivity in the territory: the monitoring process]. ISPRA 2014. In: Reticula, n. 9. ISPRA (2015). http://www.isprambiente.gov.it/it/pubblicazioni/periodici-tecnici/reticula/Reti-cula_n9.pdf. Accessed 14 June 2017

17. European Commission's Directorate-General Environment. The Multifunctionality of Green Infrastructure. Science for Environment Policy, In-depth Reports (2012). http://ec.europa.eu/environment/nature/ecosystems/docs/Green_Infrastruc-ture.pdf. Accessed 14 June 2017

18. EEA (European Environment Agency) CORINE Land Cover (2013). http://www.eea.europa.eu/publications/COR0-landcover. Accessed 14 June 2017

19. EEA (European Environment Agency) Spatial analysis of green infrastructure in Europe. In: EEA Technical report 2/2014. Publications Office of the European Union, Luxembourg (2014). doi:10.2800/11170

20. Fisher, B., Turner, R.K., Morling, P.: Defining and classifying ecosystem services for decision-making. Ecol. Econ. **68**, 643–653 (2009). doi:10.1016/j.ecolecon.2008.09.014
21. Forman, R.T.T.: Land Mosaics. The Ecology of Landscapes and Regions. Cambridge University Press, Cambridge (1995)
22. Graves, T., Chandler, R.B., Royle, J.A., Beier, P., Kendall, K.C.: Estimating landscape resistance to dispersal. Landsc. Ecol. **29**, 1201–1211 (2014). doi:10.1007/s10980-014-0056-5
23. Greene, W.H.: Econometric Analysis. Macmillan, New York (1993)
24. Hosmer, D.W., Lemeshow, S.: Applied Logistic Regression. Wiley, New York (1989)
25. Kovács, E., Kelemen, K., Kalóczkai, A., Margóczi, K., Pataki, G., Gébert, J., Málovics, G., Balázs, B., Roboz, A., Krasznai Kovács, E., Mihók, B.: Understanding the links between ecosystem service trade-offs and conflicts in protected areas. Ecosyst. Serv. **12**, 117–127 (2015). doi:10.1016/j.ecoser.2014.09.012
26. Lai, S., Leone, F.: Bridging biodiversity conservation objectives with landscape planning through green infrastructures: a case study from Sardinia, Italy. Paper Presented at 17th International Conference on Computational Science and Applications (ICCSA 2017) Held in Trieste, Italy, 3–6 July 2017, Forthcoming (2017)
27. Lechner, A.M., Sprod, D., Carter, O., Lefroy, E.C.: Characterising landscape connectivity for conservation planning using a dispersal guild approach. Landsc. Ecol. **32**, 99 (2017). doi:10.1007/s10980-016-0431-5
28. Liquete, C., Kleeschulte, S., Dige, G., Maes, J., Grizzetti, B., Olah, B., Zulian, G.: Mapping green infrastructure based on ecosystem services, and ecological networks: a pan-european case study. Environ. Sci. Policy **54**, 268–280 (2015). doi:10.1016/j.envsci.2015.07.009
29. McFadden, D.: The measurement of urban travel demand. J. Public Econ. **3**, 303–328 (1974). doi:10.1016/0047-2727(74)90003-6
30. McFadden, D.: The revealed preferences of a government bureaucracy: empirical evidence. Bell J. Econ. Manag. Sci. **7**, 55–72 (1976). doi:10.2307/3003190
31. McFadden, D.: Modelling the choice of residential location. In: Karlqvist, A., Lundqvist, L., Snickars, F., Weibull, J.W. (eds.) Spatial Interaction Theory and Planning Models, pp. 75–96. North Holland, Amsterdam (1978)
32. McFadden, D., Train, K.: The goods/leisure tradeoff and disaggregate work trip mode choice models. Transp. Res. **12**, 349–353 (1978). doi:10.1016/0041-1647(78)90011-4
33. McFadden, D.: Econometric models for probabilistic choice among products. J. Bus. **53**, 13–29 (1980)
34. McFadden, D., Train, K., Ben-Akiva, M.: The demand for local telephone service: a fully discrete model of residential calling patterns and service choices. RAND J. Econ. **18**, 109–123 (1987)
35. McFadden, D.: Disaggregate behavioral travel demand's RUM side a 30-year retrospective. In: Hensher, D. (ed.) Travel Behavior Research: The Leading Edge, pp. 17–63. Pergamon Press, Oxford (2000)
36. McRae, B.H., Kavanagh, D.M.: Linkage Mapper Connectivity Analysis Software. The Nature Conservancy, Seattle, WA (2011). http://www.circuit-scape.org/linkagemapper. Accessed 14 June 2017
37. Millennium Ecosystem Assessment: Ecosystems and Human Well-being: A Framework for Assessment. Island Press, Washington, DC (2003)
38. Nerlove, M., Press, S.: Univariate and multivariate log-linear and logistic models. In: Report No. R1306-EDA/NIH. RAND Corporation, Santa Monica, CA, United States (1973)
39. de Ortúzar, D.J., Willumsen, L.G.: Modelling Transport, 3rd edn. Wiley, Chichester (2001)
40. Pavlopoulos, D., Muffels, R., Vermunt, J.K.: Wage mobility in Europe: a comparative analysis using restricted multinomial logit regression. Qual. Quant. **44**, 115–129 (2010)

41. Sawyer, S.C., Epps, C.W., Brashares, J.S.: Placing linkages among fragmented habitats: do least-cost models reflect how animals use landscapes? J. Appl. Ecol. **48**, 668–678 (2011). doi:10.1111/j.1365-2664.2011.01970.x
42. Snäll, T., Lehtomäki, J., Arponen, A., Elith, J., Moilanen, A.: Green infrastructure design based on spatial conservation prioritization and modelling of biodiversity features and ecosystem services. Environ. Manag. **57**, 251–256 (2016). doi:10.1007/s00267-015-0613-y
43. Train, K.: Discrete Choice Methods with Simulation, 2nd edn. Cambridge University Press, Cambridge (2009)
44. Tversky, A.: Elimination by aspects: a theory of choice. Psychol. Rev. **79**, 281–299 (1972). doi:10.1037/h0032955
45. Wang, Y.H., Yang, K.C., Bridgman, C.L., Lin, L.K.: Habitat suitability modelling to correlate gene flow with landscape connectivity. Landsc. Ecol. **23**, 989–1000 (2008). doi:10.1007/s10980-008-9262-3
46. Williams, H.C.W.L.: On the formation of travel demand models and economic evaluation measures of user benefit. Environ. Plan. A **9**, 285–344 (1977)
47. Zeller, K.A., McGarigal, K., Whiteley, A.R.: Estimating landscape resistance to movement: a review. Landsc. Ecol. **27**, 777–797 (2012). doi:10.1007/s10980-012-9737-0
48. Zoppi, C., Lai, S.: Differentials in the regional operational program expenditure for public services and infrastructure in the coastal cities of Sardinia (Italy) analyzed in the ruling context of the Regional Landscape Plan. Land Use Policy **30**, 286–304 (2013). doi:10.1016/j.landusepol.2012.03.017

Planning with Ecosystem Services in the Natura 2000 Network of the Metropolitan City of Cagliari

Maddalena Floris$^{(\boxtimes)}$ and Daniela Ruggeri

Department of Civil Engineering and Architecture, University of Cagliari,
Cagliari, Italy
{maddalena.floris,daniela.ruggeri}@unica.it

Abstract. Ecosystem Services (ESs) contribute to the human well-being, and, according to the recent classification of the Common International Classification of Ecosystem Services (CICES), they can be categorised into provisioning, regulating and cultural services.

This paper investigates functions and benefits of ESs, especially regulating services such as Carbon sequestration and Water purification, which are essential to life. In Carbon sequestration, the atmospheric CO_2, produced by human activities, is removed from the atmosphere through natural processes and stored by terrestrial ecosystems in the soil. In Water purification, nutrients and other pollutants (i.e. metals, oils and sediments), introduced into inland waters and coastal and marine ecosystems, are processed and filtered out through water flow along forests, riparian zones and wetland.

Land-use changes can have important effects on both Carbon sequestration and Water purification services.

The aim of this paper is to identify benefits of the Natura 2000 Network to ESs in urban context; for this purpose we focus on a case study in the context of the Metropolitan City of Cagliari (Sardinia, Italy). The Natura 2000 Sites involve a thirty percent of the metropolitan land area and the results of the assessments of Carbon sequestration and Water purification show that areas including Natura 2000 Sites are highly potentially suitable to contain both ESs.

The knowledge of the potential distribution and the quantity of ESs is crucial and can lead the decision-makers to a new planning approach, which is the key for sustainable urban development.

Keywords: Ecosystem Services · Urban planning · Natura 2000 Network

1 Introduction

Ecosystems, through chemical, physical and biological processes, provide a unique support to the quality of life. These processes are recognised as Ecosystem Services (ESs), i.e. material and immaterial benefits provided by ecosystems to humans. The

© Springer International Publishing AG 2017
O. Gervasi et al. (Eds.): ICCSA 2017, Part VI, LNCS 10409, pp. 401–415, 2017.
DOI: 10.1007/978-3-319-62407-5_28

knowledge referred to ESs has considerably improved, starting from their conceptual introduction. The major global initiatives to analyse ESs provided by terrestrial ecosystems are the Millennium Ecosystem Assessment (MEA) [15], in 2005, and The Economics of Ecosystems and Biodiversity (TEEB) [21], in 2010. Both have influenced European environmental policies such as the EU biodiversity strategy until 2020 [10]. Through Action no. 5 of this strategy, the European Union invites Member States to map and assess the state of the ESs on their territories. For this reason, the Group Mapping and Assessment of Ecosystems and their Services (MAES) in 2013 set up to support Member States assessing and mapping ecosystems and their services in their national territory, adopting the Common International Classification of Ecosystem Services (CICES) [11] as a common base [9]. This is an ongoing project developed by the European Environment Agency (EEA). The first release, on December 2009, was followed by other versions, the latest being CICES version 4.3, on January 2013. Both TEEB and CICES assume the MEA initiative as a starting point. The hierarchical structure of CICES, organized into five levels, shows a specific vision related to all local peculiarities.

The ESs, according to the CICES, are divided into three categories: provisioning services, which include goods and raw materials, such as food, water, timber, fiber, fuel, but also genetic material and ornamental species; regulating services, which include services, such as carbon sequestration, air quality, water purification, soil formation, pollination, waste assimilation; and, cultural services, which include non-material benefits, such as the heritage and cultural identity, spiritual enrichment and intellectual, aesthetic and recreational values.

This paper focuses on two regulating ESs, mainly provided by soil: Carbon sequestration and Water purification, which represent an important key for improving health and well-being. Scientific researchers have focused on the productive capacity of the soil, without considering the wide range of ESs which the soil can provide [6]. The soil, because of physical, chemical and biological fertility, operates unique functions and ESs in the production of food and biomass, in the storage and transformation of mineral, organic material, water, energy and chemical substances, in order to filter water and pollutants.

The soil represents the base of human activities, besides being the habitat of most of the biosphere organisms. Moreover, the soil is an essential component of the critical area of the Earth, that is the layer which extends from the outer limit of vegetation up to the area where groundwater circulates. The processes of soil formation are so extremely slow that it can be considered as a non-renewable resource [5]. Consequently, informing decision-makers about the consequences on the supply of ESs due to the land-use changes is decisive. ESs mapping is attracting a growing interest into urban planning, but its effective importance in decision-making is still limited [3].

The case study involves the context of the Metropolitan City of Cagliari (MCC), in Southern Sardinia (Italy), that is characterised by the presence of several Natura 2000 Sites, although it is the highest regional urbanised area.

In order to analyse these ESs, we refer our approach to the InVEST[1] software, specifically for the possibility of a preliminary observation of input data. InVEST software allows to quantify and map the provision of ESs, and explore how changes in ecosystems are likely to lead changes for benefits involving human well-being [17].

This paper is organised as follows. In Sect. 2, firstly we describe the case study; secondly, we describe our approaches to assess the Carbon sequestration and the Water purification ESs, each consisting of four main parts: description of the ES, methodology, materials and results obtained. In Sect. 3, we discuss the results obtained in Sect. 2 and referred to the assessment of Carbon sequestration and Water purification. Finally, in Sect. 4, we conclude discussing the potential integration of ESs into the planning practise.

2   An Approach to Assess Regulating Ecosystem Services

2.1   The Case Study

The case study analysed in this paper involves the metropolitan context of Cagliari, in Southern Sardinia (Italy). The total area is the union of the metropolitan boundaries, three coastal landscape units of the Regional Landscape Plan of Sardinia, and the Natura 2000 Sites (see Fig. 1). There are 18 Natura 2000 Sites, whereof 16 are totally or partially included into the MCC, and 31 municipalities are partially or totally involved, whereof 17 belong to the MCC, where about 430000 inhabitants live. Notwithstanding the human presence, the Natura 2000 Sites involve about the 30% of the context.

We analyse the context with reference of two ESs: Carbon sequestration and Water purification. Both are here assessed on the basis of the Land cover map[2], developed by the Autonomous Region of Sardinia [Regione Autonoma della Sardegna (RAS)], and shown in Fig. 2 using level 1 of the CORINE Land Cover (CLC) nomenclature. In particular, this figure shows that there is a symmetrical distribution of land uses, referring to the core area. Artificial surfaces are mainly in the central position and amount to 7.2%. Agricultural areas, which surround the Artificial surfaces, amount to 24.3%. Forest and semi-natural areas are predominant in the western and eastern sides; they amount to 64.7%, which constitute the biggest part of the context. Wetlands are in the central position and are located close to the highest urbanised area, in a coastal position; they amount to 2.2%. Finally, Water bodies are mainly in the central side and amount to 1.6%.

---

[1] InVEST (Integrated Valuation of Ecosystem Services and Tradeoffs) is a free of cost software product, licensed under the BSD open source licence. InVEST is developed by the Natural Capital Project (NCP), whose partners are: the Woods Institute for the Environment and Department of Biology of Stanford University; the Institute on the Environment of Minnesota University; the Nature Conservancy; and, the World Wildlife Fund (WWF). InVEST-related documentation is available online at http://data.naturalcapitalproject.org/nightly-build/invest-users-guide/html/index. html [accessed April 10, 2017].

[2] http://www.sardegnageoportale.it/index.php?xsl=1594&s=40&v=9&c=8936&na=1&n=100 [accessed June 19, 2017].

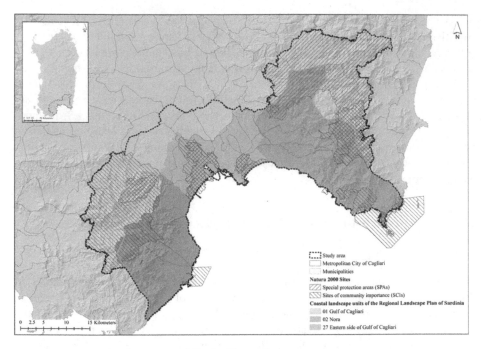

Fig. 1. The study area

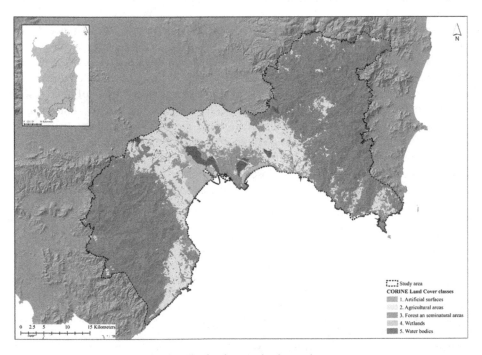

Fig. 2. The land cover in the study area

2.2 The Methodology to Assess Carbon Sequestration

The growing concern on the climate change, caused by increasing concentrations of greenhouse gases in the atmosphere, has recently encouraged research into the assessment of the soil organic carbon.

According to the National Oceanic and Atmospheric Administration of the USA, carbon dioxide (CO_2) concentration in atmosphere has increased from a pre-industrial value of about 280 parts per million (ppm) to 396 ppm in 2014, while in the last decade the annual average increase equalled 2.04 ppm per year. Human activities are the main cause of the increased concentration of CO_2 and other gases in the atmosphere [16].

In the last few years different European instrument and national programs have been adopted to achieve of the Kyoto targets for the reduction of greenhouse gas emissions.

The European Climate Change Program (ECCP) in 2000 identifies the necessary measures to achieve the Kyoto targets, particularly the second ECCP Report assigns to the CO_2 sequestration a prominent role against climate change [8].

Indeed, the Kyoto Protocol identifies Carbon sequestration among factors reducing CO_2 in the atmosphere, together with containment measures and reduction of emissions.

During their regular growth cycle, plants, withdraw CO_2 from of the atmosphere and store it in plant biomass and soil in form of organic carbon. This is a natural phenomenon where the atmospheric CO_2, produced from human or natural activities, is captured and stored in wood and soil through plants releasing water and oxygen in the atmosphere. Therefore, sustainable use of soil resources plays an important role in policies concerning mitigation of climate change.

CO_2 sequestration varies both with variations in environmental conditions (temperature, light availability, etc.) and with characteristics of the species (foliar surface, growth rates, etc.) as well as of the individual (age, state of health, etc.) [16]. The soil ecosystem is the largest carbon pool, this is 3.3 times higher than the atmospheric quantity [14]. It is therefore the most important pool but it is also the most influenced by human action.

Assessing the organic carbon stored in the soil is of strategic importance for land use policies with a view to reducing land take and considering it as common and non-renewable resource that exerts functions and produces ESs. Following the Kyoto Protocol, the Good Practice Guidance for Land Use, Land-Use Change and Forestry [18] identifies five carbon pools in the forest ecosystems, in order to account carbon released into the atmosphere and carbon adsorbed by terrestrial ecosystems:

- above-ground biomass, consisting of tissues that constitute the aerial parts of the plants (stems, branches and stumps, bark, leaves, seeds and fruits);
- below-ground biomass, constituted by root system of plant;
- deadwood, represented by the coarsest plant residues;
- litter, consisting from the thinner plant residues (leaves, flowers and inflorescences, fruits and seed heads, twigs, etc.), not yet decomposed;
- soil, composed of the organic carbon present in the mineral and organic layers, including the thinner roots.

Methodology. Similarly to the InVEST model of "Carbon storage and sequestration: Climate regulation", the methodology here proposed is based on the concept that each land cover has a specific value of carbon stored. The input data analysis needs to be improved in the future in order to allow and analysis with the InVEST model. With the purpouse to estimate the potential capacity of each land cover class to store carbon, we have re-elaborated the data provided by AGRIS[3], obtained from direct organic carbon analyses in soil in some specific area in Sardinia. The potential capacity of each land cover class to store the carbon is obtained from the aggregation of the carbon stored in the soil and of the carbon stored in the dead mass. According to Dyson [7] we assume that the potential capacity to store carbon in the artificial land cover classes is equal to zero.

Materials. The inputs used in this methodology are: a land use (or land cover) map and the carbon value stored in the soil for each land use or land cover class.

Land Cover of RAS. The RAS updated its land cover map on 2008. This map is a geographical database of land covers of Sardinia, classified in accordance with the CLC taxonomy up to the level 5. We use until level 5 to generate a carbon stock map.

Carbon Pools[4]. A table of land cover classes (Table 1), containing data on carbon stored in the soil pools for each class:

- c_soil: organic carbon in soil [g/kg];
- c_dead: organic carbon in dead mass [g/kg].

The organic carbon table is structured as follows: i. land cover classes; ii. the organic carbon (g/kg) for each land cover class, this value is obtained with the sum of organic carbon in soil and organic carbon in dead mass.

Results. The carbon stock map (Fig. 3) shows the potential carbon stock distribution in the metropolitan context (g/kg).

The high values (green) imply high carbon stock in soil, which represents a positive aspect, while low values (red) imply low carbon stock in soil, which represents a negative aspect.

The capacity potential of each soil to capture and store (CO_2) varies from 0 g/kg for Artificial surfaces and 88 g/kg for the Forest and semi-natural areas, hence confirming that the ecosystem value of soils is influenced by their use.

---

[3] Data were extracted by the Database Survey of Sardinia (DBSS) Agris, made by the Agency in collaboration with the Laore Agency and the Universities of Cagliari and Sassari.

[4] These are derived: from the surveys conducted for the project "Charter of the land units and land use capability – First Lot" (2011–2013) by: Agris Sardinia for Muravera-Castiadas area; Laore Sardinia for the Arzana and Nurra south area; University of Cagliari for the Pula-Capoterra area and University of Sassari for Nurra area north and south, funded by the Department EE.LL. Finance and Planning of the Autonomous Region of Sardinia, and from the historical archives of aforementioned institutions on occasions of other studies and surveys.

Table 1. Sinthesys of the organic carbon related to the land cover map

Land cover code	Organic carbon [g/kg]
1111, 1211, 1212, 1221, 1222, 1223, 1224, 123, 124, 131, 1321, 1322, 141, 1421, 143, 2124, 3222, 3311, 3315, 332, 5111, 5112, 5122, 5211, 5212, 5231	0.0
133, 3313	5.6
2111, 2112, 2121, 2123	11.8
3121	11.9
333, 1422	12.4
241	13.8
3241	14.0
223, 2411	15.3
221, 2412	20.1
244	21.7
31122	22.6
231	23.5
222	24.1
3242	25.7
242	30.0
31121	30.5
243	31.1
3231	35.4
1112, 1121, 1122	37.5
3111	37.7
422	37.9
3232	39.2
421, 423	39.7
321	47.9
411	54.5
3221	87.4
313, 3122	88.0

2.3 The Methodology to Assess Water Purification

We assess the Water purification ES with reference to the retention of nitrogen. Nitrogen is a key element of ecosystems and a major regulator for ecological conditions and functioning in the biosphere, being the major constituent of air of the Earth's atmosphere. Nitrogen is defined "nutrient" because it is essential to the growth of many organisms and vegetables. In the last century, due to human activities, the natural nitrogen cycle has been significantly altered, with several consequences: N_2O trace in gas emissions and changes in the carbon balance, public health alarms, water bodies eutrophication, groundwater quality and consequent effects on biodiversity [4]. When it rains, water flows through the territory carrying pollutants from these surfaces into

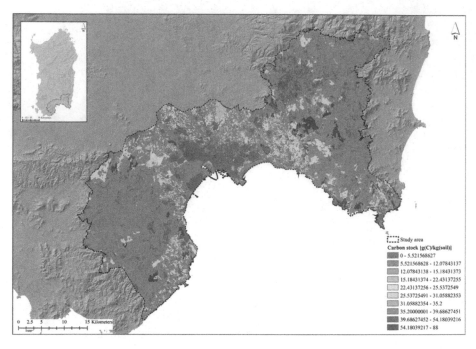

Fig. 3. The map of potential carbon stock (Color figure online)

streams, rivers, lakes, and, finally, into the sea. Water quality affects people through numerous pathways, from drinking water to recreation to commercial fisheries, directly affecting their health or well-being, and for aquatic ecosystems that have a low capacity to adapt to these nutrient loads.

The Directive 91/676/CEE, concerning the protection of waters against pollution caused by nitrates from agricultural sources, defines "pollution" as "the discharge, directly or indirectly, of nitrogen compounds from agricultural sources into the aquatic environment, the results of which are such as to cause hazards to human health, harm to living resources and to aquatic ecosystems, damage to amenities or interference with other legitimate uses of water". Moreover, the Directive 2000/60/EC, establishing a framework for Community action in the field of water policy, underlines that Community waters are under increasing pressure from the continuous growth in demand for sufficient quantities of good quality water for all purposes.

Consequently, water purification is a strategic regulating service. Ecosystems contribute to water purification by removing sediments, nutrients and pesticides from surface water runoff through deposition, filtration, infiltration and adsorption [23]; in this way, pollutants can be retained or degraded before they reach the stream. These services can be provided in several ways: vegetation can remove pollutants by transferring them into their tissues or by liberating them back in the environment in an alternative form; soils can trap and store soluble pollutants; wetlands can flow slowly long enough for pollutants to be taken up by the vegetation; also, riparian vegetation has a specific importance, acting as the last barrier before pollutants reach the stream.

Nitrogen loads can be caused by both punctual and diffuse sources. Punctual sources could be factories, sewage treatment plants, underground mines, oil tanks; diffuse sources could be urban areas, agricultural areas, industrial zones.

Methodology. This methodology is grounded on the "Nutrient Delivery Ratio" (NDR) model of the InVEST software, which refers to the retention of nutrients such as nitrogen and phosphorus in order to assess the water purification ecosystem services. The NDR model describes the movement of masses of nutrient through space, and the steady-state flow of nutrients through empirical relationships. The NDR model maps nutrient sources from watersheds and their flows to the stream. It is based on the concept that each element of the watershed is characterised by its nutrient load and its nutrient delivery ratio, which is a function of its upslope area and its downslope flow path. Downslope flow is related to the retention efficiency of land cover [17].

Materials. The inputs required by the NDR model are: a digital elevation model (DEM); a land use (or land cover) map; a shapefile of watersheds; a raster related to the potential runoff; a biophysical table, containing fields where land uses (or land covers) are associated to the nutrient loads. In this study, we only analyse the nutrients transported by the surface flow.

The DEM. A DEM is the representation of the distribution of units of the territory in digital format, with an elevation value for each cell. The digital elevation model is typically produced in raster format by associating each pixel to the absolute height. We use the DEM available on the Internet at Sardegna Geoportale[5]. This raster file has the resolution of 10 m.

The Land Cover of RAS. The RAS updated its land cover map on 2008. This map is a geographical database of land covers of Sardinia classified in accordance with the CLC taxonomy up to the level 5. We use the level 3 of CLC classification to generate a raster map of land cover.

Watersheds. The watersheds are defined in the Management plan of the river basin district of Sardinia[6] [20]. In this study, we consider a shapefile containing sixty-four watersheds, so that the total analysed area amounts to 6,422 km^2. The biggest watersheds are Bacino del Rio Flumendosa (1,842 km^2) and Bacino del Flumini Mannu (1,276 km^2); the smallest are localised mainly along the coastline.

Nutrient Runoff Proxy. In relation to the watersheds, we select a hundred-nine rain gauge stations which measure punctually runoff values. We consider a three-year-period to calculate a mean annual precipitation. We interpolate this punctual value of mean annual precipitation with the geostatistical kriging method [19] in order to represent the spatial distribution of the runoff and generate a raster dataset representing the variability in runoff potential defined as the mean of the three-annual-period of

---

[5] Sardegna geoportale is available at http://www.sardegnageoportale.it/index.php?xsl=1594&s=40&v=9&c=8936&na=1&n=100 [accessed June 19, 2017].

[6] *Piano di Gestione del Distretto Idrografico della Sardegna*, available at http://www.regione.sardegna.it/index.php?xsl=510&s=304398&v=2&c=6703&t=1&tb=6695&st=7 [accessed June 19, 2017].

precipitation. Finally, we generate a raster map of the runoff potential index by normalising the interpolated raster dataset, dividing by the average value.

The Biophysical Table. The NDR model requires a table containing data on water quality coefficients (Table 2). Coefficients of hydro-chemical parameters can be found in the literature in order to associate them to each land cover class.

Table 2. Sinthesys of the biophysical table

Land cover code	Load_N [kg*ha^{-1}*yr^{-1}]	Eff_N [-]	Crit_len_N [m]
111, 112, 121, 122, 123, 124, 131, 132, 133, 141, 142, 143	6.0	0.05	10
211, 212, 221, 222, 223, 224, 231, 241, 242, 243, 244	5.0	0.2	25
311	1.8	0.9	300
312	2.0	0.5	300
313	1.8	0.8	300
321	5.0	0.4	150
322	1.5	0.7	150
323	1.5	0.7	300
324	1.8	0.8	150
331, 332	0	0	10
333, 411, 421, 422, 423	0.8	0.6	10
511, 512, 521, 523	1.0	0	10

In this study, we compile the biophysical table referring to the study of Bachmann Vargas [2]. We adapt some values to the study context (i.e. the loads corresponding to 321 code are modified in accordance with those concerning agriculture areas, since in this context there are few activities related to intensive animal farming). The biophysical table is structured as follows: i. land cover classes in our study area; ii. nitrogen loads for each land cover class, as sources of nutrients (Load_N); iii. maximum retention efficiency for each land cover class, expressed as a proportion of the amount of nutrient from upstream (Eff_N); iv. maximum capacity distance in which a land cover class can retain nutrients.

Results. The NDR model generates a final raster map (Fig. 4) where each pixel shows how much nutrient, measured in kg/pixel, eventually reaches the stream, and a shapefile where the total nutrient value is aggregated per watershed as the sum of the contribution from all pixels within the watershed. Two fields, related to total nutrient loads and total nutrient exports, measured in kg/yr, are contained in the shapefile.

The map shown in Fig. 4 indicates the contribution of soil and vegetation to purifying water through the removal of nutrient pollutants from runoff, showing at the pixel level how much load from each pixel can reach the stream (measured in kg/pixel). In particular, the map shows the susceptibility of specific parts of the territory to purify

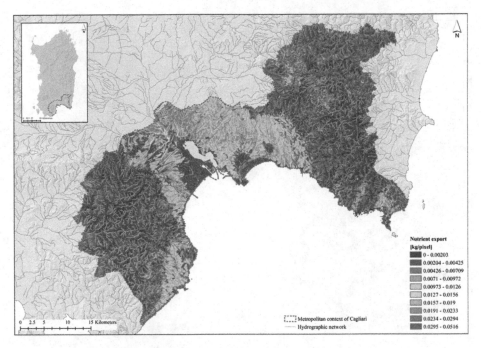

Fig. 4. The Nitrogen export raster map generated by the NDR model

the water: there are high values of nitrogen export where the purification is less present, or absent at all, because pollutants can reach the streams; on the contrary, there are low values of nitrogen export where the purification is higher, because pollutants are retained.

Because the model has a small number of parameters, outputs are generally highly sensitive to inputs. Thus, errors in the empirical load parameter values have a large effect on predictions. Similarly, the retention efficiency values are based on empirical studies, and factors affecting these values (like slope or intra-annual variability) are averaged [17]. Consequently, it should be taken into account that the biophysical table contains values stemming from other studies and not directly referred to this context.

3 Results and Discussion

Two ESs, Carbon sequestration and Water purification, have been evaluated in the context of the Metropolitan city of Cagliari, in order to identifying areas which contain the highest capacity to provide both ESs and areas with high vulnerability to provide these ESs.

After these evaluations, we elaborated the maps in order to combine them into a final map (Fig. 5) showing a comprehensive analysis of these regulating services, with a qualitative meaning. The procedure has been implemented as follows. Firstly, the Carbon stock map (Fig. 3) has been converted from shapefile to raster (cell size

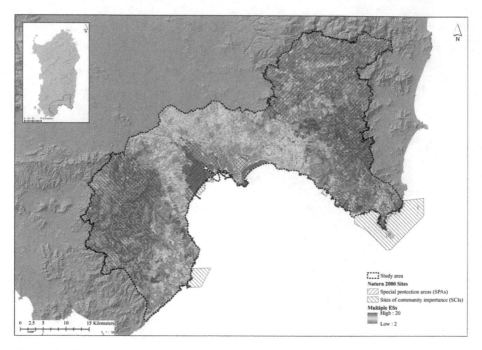

Fig. 5. Overlay of the Carbon sequestration and Water purification maps with reference to the Natura 2000 Sites (Color figure online)

10×10 m), where high values imply high quantities of carbon stock in soil, which represents a positive aspect, while low values imply low quantities of carbon stock in soil, which represents a negative aspect. By contrast, Nitrogen export map (Fig. 4) is already a raster (cell size 10×10 m), where high values imply that high nutrient export can reach the stream, which represents a negative aspect, while low values imply that low nutrient export can reach the stream, which represents a positive aspect. Both raster maps have been reclassified in 10 classes by using the Jenks' natural breaks classification method. The map of Carbon sequestration shows value from 1 (negative aspects) to 10 (positive aspects); while, the Nitrogen export map shows value from 1 (positive aspects) to 10 (negative aspect).

In order to compare the range of values, the values of the Nitrogen export map have been inverted. Finally, we create a combined map (Fig. 5), which consists in the overlay of the Carbon sequestration and Water purification maps with reference to the Natura 2000 Sites. This map shows values classified from 2 up to 20, with a colorimetric ramp from red to green. In this map, pixels mapped with red colour represent land parcels where the minimum value for each ESs is equal to 1, and that is due to the low presence of both ESs; while, pixels mapped with green colour represent land parcels where the maximum value for each ESs is equal to 10, and that is due to the high presence of both ESs.

This kind of map clearly shows, in a qualitative way, that the role of the Natura 2000 Network to provide these regulating ESs is very relevant. Indeed, in this case study, especially in the west and in the east side, the highest percentage of green values is contained into the Natura 2000 Sites, or is contained into areas very close to the Sites.

The awareness of the importance of ecosystem dynamics in changing environments and their consequences for the sustainable provision of ecosystem services is clearly underpinned in recent European conservation management strategies and policy [12].

Notwithstanding this kind of approach can seem simple, or trivial, since it makes several simplifications and assumptions, it can be very useful in order to be aware of threats associated when the anthropic pressure increases, and to understand where ESs are substantially provided and where land parcels need more protection and where address strict rules. However, this approach can contribute to taking into account the conservation of natural phenomena in planning processes, and it confirms the strategic role of the Natura 2000 Network providing benefits for human well-being.

4 Conclusions

From Rio 1992 up to the European Biodiversity Strategy to 2020, the natural capital is recognised as an insurance for human well-being. The assessment of benefits provided by the natural capital, by means of the identification and assessment of ecosystems and their services, is one of the more recent challenges for scientific research and for institutional practice in next few years. As reflected on the report *The European Environment State and Outlook 2010*, the territories with most ESs are usually more resilient and less vulnerable to extreme natural events, and, consequently, they can better bear impacts [1].

The world's protected areas are a unique source of biodiversity and involve the most productive areas of ESs [13]. In rural and urban planning the interest for ESs mapping is growing, mostly for their contribution to decision-making process, to secure the sustainable use of natural resources.

In this paper, we have analysed two regulating ESs, Carbon sequestration and Water purification, with the purpose to identify areas where their presence is relevant and areas where their presence is lacking. We have focused on a case study which involves the context of the Metropolitan City of Cagliari, Southern Sardinia (Italy), characterised by the presence of several sites of the Natura 2000 Network, although it is the highest regional urbanised area. The results have shown, in a qualitative way, that the Sites, and their surrounded context, involve the highest percentage of areas potentially relevant for these ESs and this is a very important issue to take into account in planning processes.

Land-use changes affect physical and chemical properties of soils, and consequently carbon stock and water quality in the mid- to long-term time scale. Predicting carbon stock and nitrogen retention quantitatively for environmental planning is a complex task. Models to quantify these services are either rather complex and require several data as input, which are often not available during the planning stage, or which simplify the involved processes with reference to the data available [22]. This is an important issue to take into account in planning processes. With this information, we can not only identify the beneficiaries and providers of services, but also enhance the local vocations of the territory considering the specific conditions.

The aim of this paper is to propose an ecosystem approach to support government decisions and land management, within a metropolitan context. This approach, which considers the soil as a finite and non-renewable resource, is able to integrate the evaluation of environmental benefits provided by the soil. The ESs concept holds the potential to improve communication about natural environment matters and their integration into political decision-making.

Acknowledgements. This essay is written within the Research Program "Natura 2000: Assessment of management plans and definition of ecological corridors as a complex network", funded by the Autonomous Region of Sardinia for the period 2015–2018, under the provisions of the Call for the presentation of "Projects related to fundamental or basic research" of the year 2013, implemented at the Department of Civil and Environmental Engineering and Architecture (DICAAR) of the University of Cagliari, Italy.

Maddalena Floris and Daniela Ruggeri have made substantial contributions to the paper's conception and design, introduction, discussion and conclusions.

Maddalena Floris has taken care of Sect. 2.2; Daniela Ruggeri has taken care of Sect. 2.3.

References

1. Assennato, F., De Toni, A., Di Leginio, M., Fumanti, F., Munafò, M., Sallustio, L., Strollo, A.: Azione B1 – I servizi ecosistemici del suolo – Review (2015). http://www.sam4cp.eu/
2. Bachmann Vargas, P.: Ecosystem services modeling as a tool for ecosystem assessment and support for decision making process in Aysén region, Chile (Northern Patagonia). Master thesis (2013). http://www.academia.edu/5148764/Ecosystem_services_modeling_as_a_tool_for_ecosystem_assessment_and_support_for_decision_making_process_in_Aysén_region_Chile_Northern_Patagonia_
3. Barò, F., Palomo, I., Zulian, G., Vizcaino, P., Haase, D., Gòmez-Baggethun, E.: Mapping ecosystem service capacity, flow and demand for landscape and urban planning: a case study in the Barcelona metropolitan region. Land Use Policy **57**, 405–417 (2016). doi:10.1016/j.landusepol.2016.06.006. Elsevier
4. Breuer, L., Vaché, K.B., Julich, S., Frede, H.G.: Current concepts in nitrogen dynamics for mesoscale catchments. Hydrol. Sci. J. **53**(5), 1059–1074 (2008). doi:10.1623/hysj.53.5.1059
5. Daily, G.C.: Nature's Services: Societal Dependence on Natural Ecosystems. Island Press, Washington (1997)
6. Dominati, E., Mackay, A., Green, S., Patterson, M.: The value of soil services for nutrient management. In Adding to the Knowledge base for the Nutrient Manager. In: 24th Annual Fertiliser and Lime Research Centre Workshop, Occasional report, vol. 24, pp. 1–8. Massey University, Palmerston North, New Zealand (2011)
7. Dyson, K.E. (ed.): Inventory and projections of UK emissions by sources and removals by sinks due to land use, land use change and forestry, Annual report (2009). http://nora.nerc.ac.uk/9396/1/Defra_Report_2009.pdf
8. European Climate Change Program (ECCP). Second ECCP Progress Report. Can we meet our Kyoto targets? (2003). https://ec.europa.eu/clima/sites/clima/files/eccp/docs/second_eccp_report_en.pdf

9. European Union. Mapping and Assessment of Ecosystems and their Services. An analytical framework for ecosystem assessments under Action 5 of the EU Biodiversity Strategy to 2020 (2013). doi:10.2779/12398
10. European Union. The UE Biodiversity Strategy to 2020, Luxembourg (2011). doi:10.2779/ 39229, http://ec.europa.eu/environment/nature/info/pubs/docs/brochures/2020%20Biod% 20brochure%20final%20lowres.pdf
11. Haines-Yong, R., Potschin, M.: CICES V4.3 – Revised report prepared following consultation on CICES Version 4, August–December 2012. EEA Framework Contract No EEA/IEA/09/003 (2013). https://unstats.un.org/unsd/envaccounting/seearev/GCComments/ CICES_Report.pdf
12. Haslett, J.R., Berry, P.M., Bela, G., Jongman, R.H., Pataki, G., Samways, M.J., Zobel, M.: Changing conservation strategies in Europe: a framework integrating ecosystem services and dynamics. Biodivers. Conserv. **19**, 2963–2977 (2010). doi:10.1007/s10531-009-9743-y
13. Istituto Superiore per la Protezione e la Ricerca Ambientale (ISPRA) Biodiversità e attività sugli ecosistemi, pp. 144–184 (2011). http://www.isprambiente.gov.it/files/pubblicazioni/ statoambiente/tematiche2011/02_BiodiversitA_e_attivita_sugli_ecosistemi_2011.pdf/view
14. Lal, R.: Soil carbon sequestration impacts on global climate change and food security. Science **304**(5677), 1623–1627 (2004). doi:10.1126/science.1097396
15. Leemans, R., De Groot, R.S.: Millennium Ecosystem Assessment: Ecosystems and Human Well-Being: A Framework for Assessment. Island Press, Washington, Covelo, London (2003). (Millenium assessment contribution), https://islandpress.org/book/ecosystems-and-human-well-being-0?prod_id=474
16. Mirabile, M., Bianco, P.M., Silli, V., Brini, S., Chiesura, A., Vitullo, M., Ciccarese, L., De Lauretis, R., Gaudioso, D., Istituto Superiore per la Protezione e la Ricerca Ambientale: Linee guida di forestazione urbana sostenibile per Roma, ISPRA, Manuali e Linee Guida, 129/2015 (2015)
17. Natural Capital Project (NCP) InVEST User Guide (2015). http://data.naturalcapitalproject. org/nightly-build/invest-users-guide/html/#
18. Penman, J., Gytarsky, M., Hiraishi, T., Krug, T., Kruger, D., Pipatti, R., Buendia, L., Miwa, K., Ngara, T., Tanabe, K., Wagner, F.: Good Practice Guidance for Land Use, Land-Use Change and Forestry, The Intergovernmental Panel on Climate Change, Global Environmental Strategies (2003)
19. Phillips, D.L., Dolph, J., Marks, D.: A comparison of geostatistical procedures for spatial analysis of precipitation in mountainous terrain. Agric. For. Meteorol. **58**, 119–141 (1992). Elsevier Science Publishers, Amsterdam
20. Regione Autonoma della Sardegna (RAS), Direzione generale, Agenzia regionale del distretto idrografico della Sardegna [Autonomous Region of Sardinia, General Direction, Regional Agency of the Hydrographic District of Sardinia]. Riesame e aggiornamento del Piano di gestione del distretto idrografico della Sardegna, 2° ciclo di pianificazione 2016–2020 [Review and update of the Management Plan of the Hydrographic District of Sardinia, 2nd cicle of planning 2016–2020] (2016). http://www.regione.sardegna.it/index.php?xsl= 510&s=304398&v=2&c=6703&t=1&tb=6695&st=7
21. The Economics of Ecosystems and Biodiversity (TEEB). In: Pushpam, K. (ed.) The Economics of Ecosystems and Biodiversity Ecological and Economic Foundations. Earthscan, London and Washington (2010). http://www.teebweb.org/our-publications/teeb-study-reports/ecological-and-economic-foundations/
22. Trepel, M., Palmeri, L.: Quantifying nitrogen retention in surface flow wetlands for environmental planning at the landscape-scale. Ecol. Eng. **19**(2), 127–140 (2002). Elsevier
23. Zhang, X., Liu, X., Zhang, M., Dahlgren, R.A., Eitzel, M.: A review of vegetated buffers and a meta-analysis of their mitigation efficacy in reducing nonpoint source pollution. J. Environ. Qual. **39**, 76–84 (2010). doi:10.2134/jeq2008.0496

Citizen Engagement for Sustainable Development of Port Cities: The Public Debate About Development Projects of Livorno Port

Claudia Casini[⊠]

University of Pisa, Largo Lucio Lazzarino, 56122 Pisa, Italy
claudia.casini@ing.unipi.it

Abstract. The need to apply the paradigm of sustainable development appears clearly relevant in the port cities to heal the potential and real conflict between the port area and urban area and enhance the innovative potential of these margin areas.

The paper analyzes the process of public debate as a policy tool to raise awareness of the port development projects to the portual, local and territorial communities and to improve the sustainability performance of the urban system. The paper describes the first case of regional public debate organized to the meaning of the Tuscan regional law on participation, held in Livorno in 2016.

The aim of the paper is to describe multiple lessons that can be learned by the first experience of regional public debate in Tuscany, both in relation with French experiences of *debat public*, both to renforce the future use of public debate tool at a national scale in Italy.

Although not all the frictions are solved, the public debate tool allows improving the environmental sustainability of the projects, increasing and disseminating information on future projects, making monitoring of design and construction more transparent. In essence, this first experience described confirmed that the public debate tool could bring citizens nearest to institutions by increasing social cohesion of the communities involved.

Keywords: Portual areas · Public debate · Social sustainability

1 Sustainability Policies for the Port Cities

The challenge of sustainability has now invested strongly all urban areas of the planet, including the coastal and portual ones [5]; the paradigm of sustainability has slowly asserted and has allowed to innovatively interpret urban systems [2, 6] proposing innovative policies and actions to regenerate both physical city and the community that animates it [9].

Port cities have intensely experienced changes and contradictions of globalization both in its expansion phase that in time of crisis, because, in them, ancient urban structures and commercial, maritime, logistics and industrial activities are set together in small areas; these functions simultaneously generate economic, environmental and social problems, and a great potential in terms of socio-economic and sustainable innovation of the port [8].

© Springer International Publishing AG 2017
O. Gervasi et al. (Eds.): ICCSA 2017, Part VI, LNCS 10409, pp. 416–429, 2017.
DOI: 10.1007/978-3-319-62407-5_29

A harbor strongly characterizes the local production system: it creates wealth and employment, attracts entrepreneurial resources, and promotes territorial development at the scale of the port city but also to the regional and national scale.

In addition to the direct impacts, indirect and induced impacts of the ports are the most interesting from the point of economically and socially because they over time transform society gravitating around the harbor causing not only an increase in employment, but also an improvement and growth of types of work request.

The social dimension of sustainable development of port areas can therefore be defined through direct and indirect employment in the port companies and those connected, to the wellness of port workers, to the living conditions of neighboring areas and to virtuous or conflictual interaction between the harbor and city [3].

Port areas have traditionally been placed close to urban areas and have played an important role in their economic and at large cultural development, thanks to the movement of people and ideas that gave vitality to existing communities; today the mutation of port activities, however, has also generated challenges problems to solve, such as those of greenhouse gas emissions from ships and traffic generated by maritime traffic, of noise pollution, of the placement of high activity industrial risk in close contact with the town [11].

In addition, the oldest part of the port and its links to the city are precious public spaces because they are accessible and enjoyable by citizens with existing functions; the strictly commercial and industrial part of the port is instead often inaccessible, separated from the city by a wall.

Moreover, the port area planning has to answers to questions that basically belong to different scales: globalization and competition in the global market orient activities and the size of the port facilities in accordance with international trends (gigantism, containerization), and it is increasingly important to comply with EU legislation especially in environmental matters; at the same time the economic role of the port requires to look for regional integration and the proximity of the port to the city center requires to pay attention to local equilibrium, especially from an environmental point of view: expectations of users and local residents demand more effective interventions [3].

Finally, port authorities consider the environmental sustainability of port activities as an element of economic competitiveness, and the environmental awareness is slowly spreading among port operators [4].

Green nature (and image) of shipping is useful to ports in order to maintain its competitiveness by reducing friction with communities, working together to maintain a good livability of the area surrounding the port. In order to have a peaceful coexistence between ports and cities, the concept of Social License to Operate (SLO) is of great importance. The SLO is the social acceptance of port activities by local communities [10].

In case the citizens do not approve the port activities, it could later transform into a lack of political support to the port and finally condition the expansion of the harbor or the increase in the port activity. In this situation seems important that the port is able to have a positive relation with the local society and overcame the isolation in which remained in the past.

To reach this goal ports count with soft values, that are the non-socioeconomic values, which include historical, sociological, artistic and cultural sub-functions, giving to the ports an additional advantage. They are not just industrial areas on the waterfront but, in the majority of cases, the main identity element of the city [11].

For this reason, the physical structures that perform these functions, like the port centers, are becoming increasingly widespread and participatory practices developed alongside the traditional planning are becoming more frequent: to inform, consult and involve port and local community becomes a competitive factor for the realization of development projects [1].

2   The Public Debate

The Public Debate is not a participatory process in strict sense because it usually does not include a co-design stage, but it still aims to build consensus and to the growth of community awareness.

The Public Debate is a process of information, discussion and public discussion that develops regarding a large infrastructure project; its purpose is to evaluate possible alternatives to the project, to address fears and conflicts in an open and transparent way, to provide answers to the concerns of the community [9].

The project submitted to Public Debate should not be completely defined, so that it is still possible to choose between alternative design assumptions, including the so-called "Option 0": participants discuss both on whether to build the work (if) both on project's details (how).

The information to citizens about the project should be complete, adequate, capillary and simultaneously understood by a general audience. The debate managers must create a real debate between the government, the public and private proponents, associations and citizens. Citizens must therefore be able to express their opinions and be able to verify if and how these affect the subsequent technical and political decisions.

A public debate is developed through moments and spaces to meet, study and discuss about all relevant aspects of the project, using pre-established rules and times and should conclude with a clear decision.

The proposer can gather information and useful elements to better develop the project in relation to the characteristics of the territory and to the reference development model. The proposer of the work is not bound to respect the outcome of the Public Debate, but is committed publicly to give it the utmost consideration and to argue why will own the results or rather deviate from them.

Typically the Public Debate is managed by a neutral and competent responsible, sufficiently authoritative to give credibility to the process.

2.1   Practices and Regulations

The Public Debate is used in Quebec and in Australia and, as regards Europe, in France, where the Débat Public been set up and regulated in several stages from 1995 onwards [9].

In Italy, in recent years, Public Debates have been experienced about the touristic project of regeneration of the Tuscan village of Castelfalfi (2007), for two highways in Genoa (2009) and Bologna (2016) and for the redevelopment of the city center of Termoli (2016). These processes have been carried out without a mandatory regulatory reference; this is something positive because strong motivation were moving proposers and institutions involved; on the other hand, the absence of a reference standard may have caused the lack of definition of significant characteristic elements and the lack of comparability of experiences.

In the same years, some regulatory devices concerning Public Debate have been developed: Tuscany Region law introduced the Public Debate process since 2007; increasing conflict for the construction of major projects and reflection on the events of the high-speed rail in the Val di Susa, have meant that the Public Debate was in fact introduced as mandatory in the new national Code of Contracts, legislative Decree 18 April 2016, n. 50 (art. 22).

The Tuscany Region has a regional law on the promotion of public participation since 2007; It was changed in 2013 after five years of implementation and was replaced by LR46/2013. The first version of the law provided for the possibility of activating the Regional Public Debate on large public works or matters of great environmental and social impact on the life of the whole regional community; Unfortunately, in the years of application of the Regional Law 69/2007, the DPR has never been activated on any of the works discussed at that time.

For this reason, in the second version of the Regional Law, the LR46/2013 currently in force, the Regional Public Debate was made compulsory for specific types of works: works of public initiative involving total investments of more than 50 million euros, localization predictions of national works contained in regional plans which involve a total investment of more than EUR 50 million euros, works of private initiative involving total investments exceeding 50 million euros (In this case, the regional authority involves the promoter because they cooperate and co-finance the construction of the Public Debate).

The Regional Public Debate is voluntary for works involving total investments of between 10 and 50 million euro, and that in any case have relevant profiles of regional interest.

The conclusion of the Public Debate is a condition for the start of the environmental impact assessment procedure.

3 The Public Debate on Livorno Port Development Projects

3.1 Context Conditions

The city of Livorno was born around its port, and its urban and socio-economic configuration has always been affected by it; the relationship between the port and the city is constantly changing and, in the last century, was often confrontational [7].

The territorial government plans that determine the physical layout of the port areas are the Port Development Plan, developed by Port Authority and approved by Tuscany

Region, and territorial plans of competent local authorities (Municipality, Province, Region).

In 2015, after about 60 years of discussions, Tuscany Region has approved the new Port Development Plan, a strategic plan that still contains the projects of some specific works.

Following the approval of the port plan, the first regional public debate was launched, concerning two major interventions of transformation planned for the port of Livorno (see Fig. 1):

Fig. 1. The areas affected by the Public Debate: Platform Europe to the left (north) and Maritime terminal to the right (south). Source: Our elaboration on photos Livorno Port Authority-Scovavento

- the first phase of construction of Platform Europe, a large expansion to the sea that wills double the extension of the port area;
- the project for the Maritime Terminal area, a focal point between the port areas destined to cruise traffic and ferries and historical center of the city of Livorno.

With regard to the European Platform, a Feasibility Study is the basis of a contract for the project financing; with regard to the Maritime Terminal, there is a masterplan and an implementation plan will be drawn up and be approved by the City Council.

The cost of the works of the first phase of the Platform Europe is about 870 million euro, of which 540 will come from public funds and 330 from private investment; the time required for the execution is about four years after the approval of the project. The cost of the entire platform is estimated at around 1.3 billion euro.

As for the Maritime Station, the costs of implementation should be around the 130 million euros (these will be added the cost of the redevelopment of the Old Fortress, the most important monument in this area).

Platform Europe. The public consultation process concerned the construction of the new terminal for containers, the works of defense by the waves and the roads and rails that link Platform Europe with existing infrastructures (see Fig. 2).

Fig. 2. The configuration of the first phase of the Platform Europe (to the left) and the final configuration (to the right). Source: Port Authority of Livorno-Scovavento

The Strategic Environmental Assessment procedure of Port Plan required numerous specific studies on hydrological and coastal dynamics, consumption of raw materials, interference with neighboring natural ecosystems, quality of port and coastal waters, dredging, atmospheric emissions, increase in noise and waste, energy consumption, changes in coastal landscape, public health and accident risk. For each of these topics the most critical aspects have been identified and mitigation or compensation measures have been suggested [3].

Studies have also deepened the socio-economic effects of the project: the construction of the container terminal of the European Platform will have a major economic impact on the economy of the city of Livorno and will create new business opportunities in different sectors from transport and logistics.

The project will be submitted again to the environmental assessment procedures in the future, in a more advanced state of designing.

Maritime Terminal. Livorno is the entry point from the sea in Tuscany: tourists can easily reach Florence, Pisa, Lucca and Siena; ferries offer daily connections with Sardinia, Corsica and Capraia and a weekly connection with Spain and Morocco.

Urban space is currently arranged in a functional way to handling and parking of motor vehicles and historical buildings are not valorized properly.

The masterplan designs a substantial reorganization of the area with the aim of creating a deep link between the port areas and the urban areas (see Fig. 3) through the construction of a new building for the maritime terminal, the development of typically urban functions (commercial, service, office, tourism), the planning of a new system of accessibility and the regeneration of historical and cultural heritages.

Fig. 3. The area of the Maritime Terminal (in the middle), the ferry harbor (on the quay) and the city's historic center (almost a pentagon). Source: our elaboration on Tuscany Region orthophotos.

The actions planned do not produce specific critical environmental issues; socio-economic impacts will be significant because the planned investments will generate a substantial increase in employment both during the construction phase and in the operating phase [7].

3.2   The Process of Public Debate

The Public Debate on development projects of the port of Livorno was held in three stages, as requested by the regional law:

- the preparatory phase (September 2015–March 2016) has provided for the definition of the roles (manager, staff...) and the definition of basic documents (dossier, synthesis...);
- the public phase (March 2016–July 2016) has provided for the communication activities, the organization of meetings and workshops and the preparation of the final report;
- the phase of the response (August–November 2016) has permitted to the Port Authority to respond to the requests made by the Public Debate (abandoning the project, proposing changes or confirming the project as it is).

The public phase of the public consultation process has provided for the organization of a launch meeting, a workshop with local stakeholders and one with the territorial ones, three thematic laboratory (on Platform Europe, on maritime terminal and on environmental impacts), a closing meeting. In conjunction with laboratory, interesting visits to the port areas and port center were organized.

Workshops and laboratories were managed by alternating plenary moments, for projects explanations and specific technical interventions, and moments of discussion groups, followed by restitution in plenary. Each participant was able to express his personal opinions by filling out special forms, and could then them discuss the worktables.

It was also possible to send written contributions outside of meetings through the "notes of the stakeholders".

The duration of the public phase was a critical element of the process: Just over three months are not enough to really mobilize the whole community and develop a complex and articulated debate on all the issues at stake. Probably the communication activities should have been anticipated and the time available for the public stage should have been devoted only to visits and public meetings.

The public debate has cost around € 130,000, of which 50,000 are funded by the Regional Authority for public participation, and 80,000 funded by the Port Authority of Livorno, the proposing administration of the works.

The cost of the process is far less than the cost of a French public debate and is a very small percentage of the value of works in debate; Despite this, some criticisms have been made to the process by some local politicians who thought the cost too high, without taking into account the good results that the public debate could obtain.

Great communication activities have been used, both traditional (press releases, advertisements on radio and TV, newspaper advertising) and innovative (social media management).

The discussion was thorough and sincere and brought out some hot topics perceived as problematic. The participants were able to express their opinions, even the most dubious or critical. Although during the public debate also very critical voices have been heard, the zero option was substantially excluded because the two works had already been approved in the regional and municipal plans.

Participation was numerically quite satisfactory but still limited to active and informed enough citizens; the subject was perceived as very complex.

Technical Languages. Information about the two projects were given to participants through a debate dossier, drawn up by the proposer according to the instructions of the Regional Authority for Public Participation, a summary of the dossier, an interactive table with information dedicated to the public debate at the port center.

A long and complex work of information translation in a non-technical language was needed, both for the preparation of the dossier, that for the preparation of experts' interventions during workshops and laboratories.

During the public consultation process, the contributions and the questions of the participants were collected in all their forms, have been transformed into questions

	Comments		Questions (?)	Suggestions (!)	TOTAL
	Positive (+)	Negative (-)			
Platform Europe	37	57	175	28	297
Maritime Terminal	87	104	144	35	370

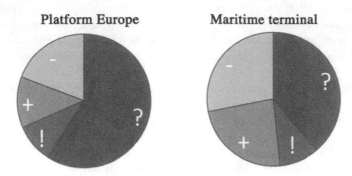

Fig. 4. Kind of comments, questions and suggestions for the two projects. Source: elaboration from www.dibattitoinporto.it

through an elaboration by categories and have generated the FAQ -Frequently Asked Questions (see Fig. 4), made available on the website www.dibattitoinporto.it.

Outcomes. At the conclusion of the public phase, the principal facilitator and manager of the Debate has drafted a final report giving account of the outcomes of the discussions, and has delivered it to the Regional Authority for Public Participation and to the Livorno Port Authority.

The Port Authority has therefore produced a written response in which partially upheld the suggestions emerged from the public debate.

With regard to the Platform Europe, the Livorno Port Authority has decided to confirm the project, however, by accepting the majority of the requests relating to the governance of the successive phases of planning, implementation and monitoring of the work.

As for the implementation plan of the Maritime Terminal and the following projects related to it, the Livorno Port Authority has agreed to change some significant elements of the master plan proposed. Also in this case, most of the requests relating to the governance of the successive phases of design and transformation of the areas were accepted.

The conclusion of the public consultation process has coincided with the approval of the reform of Port Authorities (in this case the merger of the AP of Livorno and Piombino and the changing the authority's governance) and the extension of the deadline concerning both Platform Europe that Maritime terminal public notice.

The ongoing reorganization of the new Port Authority system will make it possible to implement the requirements set out in the Public Debate.

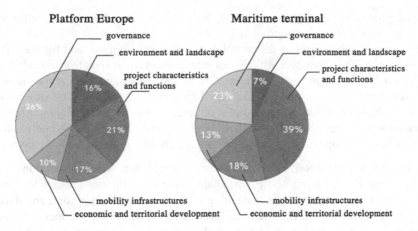

Fig. 5. Percentage distribution of macro-themes for the two projects. Source: elaboration from www.dbattitoinporto.it

The observations and questions of citizens who participated in the workshops were grouped into macro-themes (see Fig. 5):

- project characteristics and functions;
- mobility infrastructures;
- environment and landscape;
- economic and territorial development;
- governance.

Environmental Outcomes. The participants have discussed environmental issues intensively. Two types of prevailing attitudes have occurred: on the one hand many people have expressed concern about the environmental impacts of development projects; on the other hand, some participants have also criticized the lack of ambition towards environmental sustainability of the projects, which involve the use of standard technologies and the achievement of general objectives, in a "business as usual" scenario, for example with regard to the port energy system.

Preliminary technical studies seemed to some people too detailed to be understood by anyone, in another sense too general because of the poor level of detailed development project (a masterplan and a feasibility study).

During the public debate there has been a conflict, not violent but still deep and often unconscious by the same actors, largely due to two different approaches to sustainability: on one hand the *weak sustainability* of the institutions (municipality and Port Authority) who plan the development of the port according to the schemes of neoclassical economics, favoring globalization and competition on the naval gigantism, managing environmental externalities through mitigations and compensations, with a *techno-centric approach*; on the other hand, *the strong sustainability* of the organized groups of citizens, not willing to sacrifice the (already compromised) integrity of the natural capital of the territory in exchange for a future socio-economic development, according to an *eco-centric approach*.

However, at the request of the participants, the Port Authority and the Region of Tuscany, are going to create an institutional environmental and social Observatory, that aims to verify the design and implementation of the works and to monitor the environmental components ex-ante, ex-post and during the construction. The Observatory will be a technical structure composed of technicians from the different administrations (local authorities, Region and Port Authority), environmental agencies and universities; also interested environmental associations and stakeholder groups should be involved.

Socio-economic Outcomes. The economic and social issues were discussed in a very polarized way. On one hand, some participants questioned the real reasons for justify the projects, assuming an over-dimensioning of the development forecasts and thus of the planning strategies. On the other hand, the works have been put under accusation for the low ratio between new jobs potentially created and the necessary investments.

It is interesting to note that those who were concerned about the environmental impact of the projects were not willing to consider the positive socio-economic effects as compensation, while those who focused on the socio-economic aspects did not seem concerned about the environmental effects.

Territorial Cooperation: From Process to Project Governance. At the beginning of the Public Debate the relations between territorial institutions were quite conflicting, because of concerns identified during the approval stage of planning instruments.

For this reason, from the beginning of the DP, a coordination table, which met in the most significant moments of the process, was activated. It included the Livorno Port Authority, the Regional Authority for public participation, the Tuscany Region, the Municipality of Livorno and the Province of Livorno.

The theme of institutional relations prominently emerged from the demands of citizens, who consider the cooperation between public bodies one of the keys to the success of development projects and, on the contrary, fear the lack of communication as a source of delays or failures. For this reason most of the requests for the future concern the establishment of an effective multi-level governance context in the design and implementation of projects.

As a result of the participants requests, a coordination (political and technical) board will be activated; the coordination board will check the consistency of the projects with other local projects and municipal planning, will frame the work of private operators involved organizing events for information and public discussions and specific participatory processes, will report the progress of the projects to the port community, the local and regional communities, will monitor the territorial impact and compensation of environmental impacts, informing stakeholders on the progress of studies, designs and worksite phases.

An element of great interest was the request of the participants to think and formulate policies according to regional scale, expanding the reference framework to the area behind the port and the whole Tuscan coast until Piombino, just in tune with the

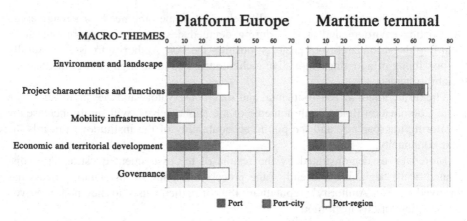

Fig. 6. Percentage distribution of FAQ for the two projects according to the macro-themes and spatial scales to which they refer. Source: elaboration from www.dbattitoinporto.it

new structure created by the law of the national port system reform, which provides for the unification between the port of Livorno and the port of Piombino (Fig. 6).

4 Conclusions

The public debate has been a two-way process: the Port Authority has tried to convey as much information as possible to the participants, and simultaneously also received a very important feedback on questions, proposals and issues raised by citizens.

Concrete problems related to the implementation of the Port Master Plan have been addressed.

The organization of the meetings has enabled a very significant direct relationship between institutions and citizens, noting indications of projects improvement in order to make them responsive to the business and community objectives.

Public debate has really impacted on the various dimensions of sustainability in the relationship between city and port of Livorno and has increased the sustainability performance of the entire system composed of territory, institutions and communities.

First, it has increased the awareness and dissemination of information on port development projects, giving the opportunity to the port community and the local and regional communities to increase their knowledge, also through direct attendance of the port areas usually landlocked. This element has been demonstrated by the results of the evaluation questionnaires distributed at the beginning and end of the process: 73% of participants said that the public debate process has helped him to better understand the projects, 14% said that the process has helped him to form an opinion, 11% said that it did not increase his knowledge and only 2% said that his ideas are more confused.

About a third of the participants has improved its opinion on the proposal for Platform Europe, about half did not change his opinion. 44% of participants improved their opinion about the plan for the Maritime Terminal.

Moreover, the outcome of the public debate reaffirmed the need for strengthening the existing environmental studies during detailed design, and set ambitious goals regarding the sustainability of projects, pushing the Port Authority to assess carefully the possibility of adopting innovative technologies and arise high environmental sustainability targets.

The institution of an environmental and social observatory and the stabilization of a Local Coordination Board, both outcome of the public debate, will surely increase the monitoring transparency and the public accountability of the institutions towards the local community.

Moreover, as demonstrated by the results of the assessment questionnaires distributed at the beginning and end of the process, the public consultation process has improved the Port Authority's reputation: 48% of participants said they had improved their opinion on this institution.

However, some distances were not redressed by the carrying out of the public consultation process: in particular the distance between the weak sustainability of development that characterizes the institutions and the strong sustainability which for some groups of citizens should have the local territorial strategic vision.

The analysis and reflection about the described experience, the first public debate carried out under the Tuscan regional law, can draw some lessons for the next public debates, both at a regional and at a national scale.

First of all, it is reasonable to ask whether the economic threshold is the only valid criterion for deciding whether a work is subject to mandatory public debate or not. In the Livorno studio case, the public debate was mandatory for the platform Europe project, the most complex topic to be discussed for an unprepared citizen, to be realized in an area spatially far from the city; the public debate was not mandatory for the area of the maritime terminal, where the discussion was rather animated and rich because the project was less defined and the area more in contact with the city center.

Secondly, the problem of the design definition level of the work to be submitted to public debate has arisen: the platform Europe project, at the feasibility study stage, was well defined by the technical point of view, so it was more difficult to discuss it for a non expert participant, and the zero option was not viable; the design of the maritime terminal area was less definite and therefore more easily questionable by untrained citizens, although even in this case the zero option was in fact excluded.

Lastly, the results of the public debate will need to be monitored in the long run, in order not to create disempowerment in the participants and to respond to the skeptical criticisms of this approach.

References

1. AIVP. The Port Center: Step-by-step operational guide (2016). www.aivp.org Accessed 13 May 2017
2. Bologna, G.: Manuale della sostenibilità. Edizioni Ambiente srl, Milan (2005)
3. Casini, C.: La valutazione ambientale strategica dei piani regolatori portuali. Il caso di Livorno, in XXXIV Conferenza Italiana di Scienze Regionali (2013)

 4. ESPO. Trends in EU Ports Governance (2016). www.espo.be. Accessed 13 May 2017
 5. Girard, L.F.: Toward a smart sustainable development of port cities/areas: the role of the "historic urban landscape" approach. Sustainability **5**, 4329–4348 (2013)
 6. Lombardini, G.: Visioni della sostenibilità. FrancoAngeli, Milan (2016)
 7. Massa, M.: A cura di, Livorno: un porto e la sua città - Progetti e Studi. Debatte, Livorno (2015)
 8. Ravetz, J.: New futures for older ports: synergistic development in a global urban system. Sustainability **5**, 5100–5118 (2013)
 9. Romano, I.: Cosa fare, come fare Decidere insieme per praticare davvero la democrazia. Chiarelettere, Milano (2012)
10. Thomson, I., Boutilier, R.G.: Modelling and measuring the social license to operate: fruits of a dialogue between theory and practice (2011)
11. Van Hooydonk, E.: Soft Values of Seaports: A Strategy for the Restoration of Public Support for Seaports, Coronet Books Incorporated (2007)

Market Prices and Institutional Values

Comparison for Tax Purposes Through GIS Instrument

Gianluigi De Mare[✉], Antonio Nesticò, Maria Macchiaroli,
and Luigi Dolores

University of Salerno, Via Giovanni Paolo II, 132, Fisciano, SA, Italy
{gdemare,anestico,mmacchiaroli,ldolores}@unisa.it

Abstract. In Italy, the institutional analysis of the real estate market values is carried out by the Inland Revenue (government agency), through the Observatory of the Real Estate Market (OMI). In the last ten years, the average values (for types of real estates: houses, warehouses, garages, etc.) reported by the OMI are much closer to the market prices really recorded through the housing market sale contracts. Therefore, the reform of taxes on the real estate, recently strongly required by the European Commission, intends to take as reference the OMI values to increase the level of equalization in the taxation. This measure wants to correlate taxes to the real market value of the property and not to the land register value, which is completely distant from the real prices: this is true both for historical reasons (the latest update of land register values dates to several years ago) and for the evolution that the market has suffered especially in the big cities because of metropolitan and transport infrastructure development.

This paper intends to verify the reliability of the OMI values compared to actual market prices and, at the same time, intends to control the possibility to equalize the fiscal mechanism considering the same tax revenue, as the Government claims to be able to do.

The intent is to avoid the sacrifice of the less affluent segments of the population benefiting the lobbies of high-quality property owners using these modern mechanisms of the tax system.

In this model, has been implemented an informative dataset in GIS mode. The use of GIS instrument makes it easier to verify the differential between government data and market prices.

Keywords: Estimation of the real estate · Real estate taxation · Dataset of values · GIS for the housing market

1 Introduction and Objectives

The proposed study is divided into three parts.

The first part concerns the classification of more than 500 deeds of sale, collected from 2008 to date in the Campania Region (Italy), in a computerized database divided

---

The work was developed equally among the authors.

© Springer International Publishing AG 2017
O. Gervasi et al. (Eds.): ICCSA 2017, Part VI, LNCS 10409, pp. 430–440, 2017.
DOI: 10.1007/978-3-319-62407-5_30

into the following sections: features of the deeds of sale, features of the properties and features of the contractors.

The features of the deeds of sale are used to identify it: this includes information about the notary who has notarized the deed; the archive number; the number of collection; the registration date and the purchase price. There is also a hyperlink to the PDF version of each deed, to let read the whole sales document. This section is important to univocally identify the transaction and to recorder the real price paid for the property and the date of the sale (think to the importance of the temporal horizon in the Appraisal applications).

The characteristics of the properties are related to the type of sold property, the address, the land registry data as well as a short typological description. These are only partially contained in the sale contract (for example the commercial area is calculated through the measure of the plan of property according to DPR 138/1998) and they supply some information about the property to create an appraisal profile of the good.

The features of the contractors concern the entity (legal or natural) of seller and buyer (in Italy sale contracts between legal contractors are more reliable for fiscal reasons, so it becomes important to collect this kind of information).

The second part of the work concerns the implementation in the GIS instrument of data contained in the computerized database. In fact, all the information collected were geo-referenced. Subsequently, the part of database related to the town of Salerno has been illustrated. In this municipality, they were collected 66 purchase agreements.

Finally, the third part of the research concerns the comparison between the prices implemented within GIS interface and the values of Observatory of the Real Estate Market, thus giving rise to a reliability test model of the OMI's values based on prices directly recorded through the sale contracts[1].

2 The Computerized Database

The deed of sale is a contract in writing, with which the contractors (seller and buyer) transfer the ownership of the good and undertake to pay the price, to deliver the good and to give the guarantees provided by the law. The sales contract is governed by the articles of the Civil Code, from 1470 to 1509. Article 1470 states that "the sale is the contract concerning the transfer of ownership of a thing or the transfer of another right in exchange for the payment of a price".

In Appraisal, the deed of sale has an important role; in fact, two Appraisal's postulates state that: "the estimate method is unique and comparative" and that "the price is the foundation of every estimate" [1]; this means that there cannot exist estimates if these are not based on a comparison with other goods that have already been subject of a sale in the real market; in this case, in the deed of sale is reported the merchant information useful to proceed with the comparison.

In Italy, the registration documents of the trades have never been reliable for issues related to the tax authorities. Because property taxes are paid in proportion to the

[1] It would like to thank Eng. Paolo Risi for his collaboration to the project.

amount of the transaction, historically it was not convenient to declare the real price; so, it was registering a price lower than the real in order to pay less tax; the only constraint was to declare a price higher than the land register value (automatically calculated by the government), because, if the amount was at or below the land register value, the Inland Revenue (like government agency) intervened imposing a penalty on the transaction. This was the habit: the deeds of sale always contained prices next to the land register values and distant from the real market value; thus, these contracts were useless for value estimates. This has resulted in limited development at national level of the studies and statistical applications [2, 3], unlike what happened in the Anglo-Saxon world [4, 5], that is characterized by greater transparency in this kind of transactions and by the possibility to implement a more representative database of the real market.

However, the Italian scenario is changed between 2006 and 2008 because some laws were introduced to increase the transparency of transactions [6]. For the purposes of this study they were selected only deeds of sale registered after January 1st, 2008.

As I said, the database is organized in three main sections, as shown in the diagram of Fig. 1.

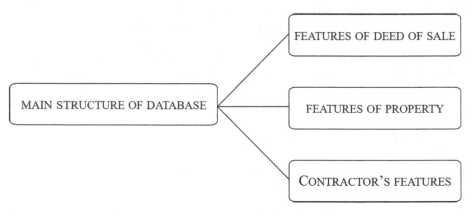

Fig. 1. Structure of the computerized database

The excel interface of the information system created is presented like in Tables 1, 2 and 3 shown below.

Table 1. Excel extract from the computerized system – Section 1

Features of deed of sale						
ID	Notary	Archive number	Number of collection	Registration date	Date of deed of sale	Price (€)
1	A	26118	7093		22/10/2008	184.000
2	B	50			14/10/2010	26.354
3	C	38384	17444	24/09/2009	22/09/2009	175.000
4	D	61965	12937	19/02/2010	17/02/2010	118.800
5	E	66152	17574	16/09/2008	09/09/2008	150.000
6	F	35800	6445	15/07/2010	15/07/2010	165.000
...	...	...	...	...	...	...

Table 2. Excel extract from the computerized system - Section 2

Features of property										Area			
ID	Municipality	Prov	Sheet	Parcel	Sub.	Address	Street number	Description	Floor	(vani)	(m^2)	Plan	Commercial area (m^2)
1	Baronissi	SA	14	422	14	Via Indipendenza	9	Flat	3	5		Hyperlink	
					25			Basement	S1		33		
2	Pellezzano	SA	7	1886	20	Via Gentile	9	Flat	4	3		Hyperlink	
3	Fisciano	SA	19	166	13	Via Roma	3	Flat	PT	6,5		Hyperlink	117,5
					25			Garage	S1		30		
4	Fisciano	SA	20	2200	9	Via Fratelli Napoli		Flat	2	3,5		Hyperlink	
5	Solofra	AV	5	929		Via San Gaetano		House	S1-PT-1-2			Hyperlink	
6	Salerno	SA	66	35	4	Via Calata San Vito	36	Flat	2	4,5		Hyperlink	110
...	...	...	...	...	...	...	...	...	...	...	...	...	...

Table 3. Excel extract from the computerized system - Section 3

Contractor's features						
ID	Seller	Legal entity	Natural person	Buyer	Legal entity	Natural person
1			1			1
2	G	1		B		1
3	G		1	B		1
4	G	1		B		1
5		1				1
6	G		1	B		1
...	...	...	...	...	...	...

3   GIS for the Representation of Market Prices

The use of GIS for analysis and the estimate studies is known in the bibliography [7].

In the case that interests us, the geographical representation instrument is used for constituting a user interface easy to manage in evaluation questions even for non-professional persons, like municipal employees for the verification of taxation levels of real estate assets.

As mentioned, 66 properties were collected on the territory of Salerno, the city chosen as example for the proposed application.

The result of the geo-referenced information is summarized in Fig. 2 (next page).

It has proceeded, therefore, to export the vector data in GIS. The graphic interface has been enhanced by downloading a plug-in supported by the software QGIS called OpenLayers. This plug-in allows us to attach the maps directly from the Web from different sources: OpenStreetMap, Google Maps, Bing Maps, Apple Maps, MapQuest, OSM/Stamen. Among these, it is chosen the Google map, especially the satellite view (Google Satellite), geo-referencing it compared to the same reference system (Fig. 3).

Completed the GIS export is therefore almost completed the connection between the computerized databases and the geographical data; this means that any geographic

Fig. 2. Properties surveyed in the database and geo-referenced in the GIS instrument

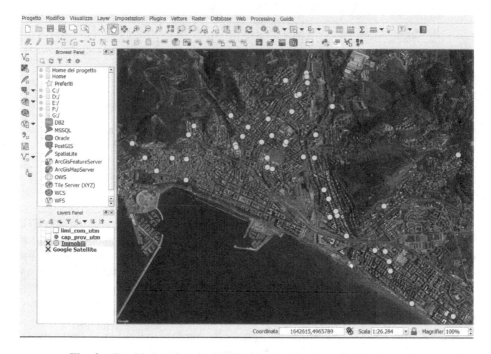

Fig. 3. Graphic interface in QGIS with identification of some properties

datum represented on the map is associated with the respective data string of the computerized database.

The main feature of a GIS instrument, in fact, it is precisely the ability to associate attributes to the components of a map; therefore, in this case the data present within the computerized databases were associated with each point that representing a building (see Fig. 4).

Immobili :: Features total: 66, filtered: 66, selected: 0

	STRINGA	testo_id	ID	Pertinenza	Atto	Indirizzo	n° civico	Tipologia	Piano	Superficie
1	6	6	6	No	C:\Users\P...	Via Calata ...	36	Appartame...	2	110
2	9	9	9	No	C:\Users\P...	Via C. Sor...	27	Appartame...	2	87
3	23	23	23	No	C:\Users\P...	Via Madon...	80	Appartame...	3	-
4	80	80	80	No	C:\Users\P...	Vicolo Giu...	14	Appartame...	3	-
5	98	98	98	Si	C:\Users\P...	Via Fabrizi...	79	Appartame...	5	132
6	124	124	124	No	C:\Users\P...	Via Paolo ...	22	Appartame...	6	67
7	154	154	154	No	C:\Users\P...	Via Sichel...	22	Appartame...	1	115
8	172	172	172	No	C:\Users\P...	Via Irno	211	Appartame...	1	61
9	181	181	181	No	C:\Users\P...	Via Giovan...	64	Appartame...	2	91
10	191	191	191	No	C:\Users\P...	Via Vincen...	17	Appartame...	3	62
11	200	200	200	No	C:\Users\P...	Via Luigi C...	57	Appartame...	3	63
12	206	206	206	No	C:\Users\P...	Via Ernest...	4	Appartame...	3	113
13	225	225	225	No	C:\Users\P...	Via Casa D...	24	Appartame...	PT	97
14	244	244	244	No	C:\Users\P...	Via Madon...	21	Appartame...	6	148

Fig. 4. Excerpt of the attribute table

The component of the computerized database is descriptive, alphanumeric, non-spatial, it expresses the value of a quantity, and it is manifested in several attributes that describe the characteristics of the property. The GIS instrument automatically associates a new ID for each geographic feature, so it was appropriate to link the ID automatically generated by the GIS with that built in database.

4 The Institutional Market Values and the Real Market Prices

One of the major innovations introduced by the reform of the land register was the creation of the Property Market Observatory (OMI) within the Land Agency (D. Lgs. 30th August 1999, n. 300 art. 64, paragraph 3) that, in addition to the management of cadastral data, also contains the office where public deeds are stored ("Conservatoria" of property registers).

The OMI is responsible for the production of updated data related to the sale and rent values of real estate on a local scale but also national, thereby providing a support for analysis and studies in the appraisal sector.

This instrument is run by the Central Directorate of the Property Market Observatory which is composed of the Observatory Area of Market Values, of the Methodologies Observatory Office, of the Bank Data Management Office, of the Analysis and Studies Office.

Regardless of the draft revision of the urban Land Registry promoted by the Government, the OMI has divided each municipality in representative areas of the urban structure (Central Area, Near Centre Area, Suburban, Rural). In each area, they have been identified homogeneous zones with respect to property values.

For each OMI zone, they have been identified:

- the maximum and the minimum real estate value; their relative deviation should never exceed the multiplier of 1,5 excluding the particular situations that go beyond ordinariness;
- the type or types prevalent;
- the prevailing conservation status (good, normal, poor).

Fig. 5. The OMI zones for the town of Salerno

For the city of Salerno, the subdivision zones for the market analyzes are illustrated in Fig. 5.

It is chosen the B12 area for the development of the test model for its significant concentration of data about prices recorded directly from the market.

The OMI for the area of interest provides the following technical and commercial indicators relating to the last update (first half 2016), as shown in Table 4.

To proceed to the comparison between the OMI values and the market prices, it is chosen the most common kind of construction, the civil one, getting the following average datum of € 2.500 for the B12 interest area.

Table 4. OMI market values

Database of real estate quotation							
Answer to your interrogation: year 2016 - semester 1							
Province:	SA						
Municipality:	Salerno						
Zone:	Central/Railway Station - DALMAZIA street - E. CATERINA street - PRUDENTE street- NADDEO square- LANZALONE street -P.SSA SICHELGAITA street- C. SORGENTE street						
Zone code:	B12						
Most common typology:	Civil buildings						
Function:	Residential						
Typology	Conservation state	Market value (€/m^2)		Area (Gross/Net)	Rent Value (€/ m^2 a month)		Area (Gross/Net)
		Min	Max		Min	Max	
Civil buildings	Normal	2.300	2.700	G	5,5	8,0	G
Cheap buildings	Normal	2.200	2.600	G	5,1	7,6	G
Parking garages	Normal	1.250	1.700	G	4,0	5,8	G
Private garages	Normal	1.500	2.100	G	5,1	7,1	G

The surveyed market prices, opportunely updated to the first half of 2016 through the time series extracted from appropriate bibliography [8], are arranged around the mean value as shown in Fig. 6 (2D model) and Fig. 7 (3D model, next page).

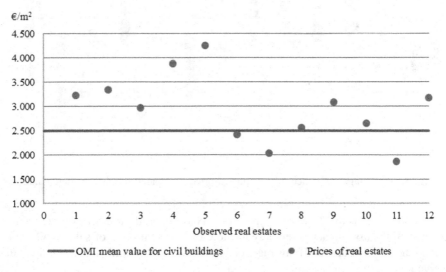

Fig. 6. OMI main value for the B12 zone and updated prices

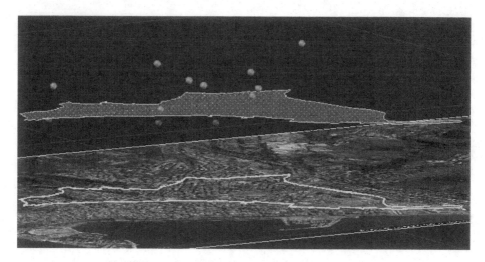

Fig. 7. OMI main value (plane in yellow) and the updated prices (higher prices in green, lower prices in red) (Color figure online)

It is immediately evident that the average datum is basically less than the recorded market prices, according to a percentage deviation that is shown in Fig. 8.

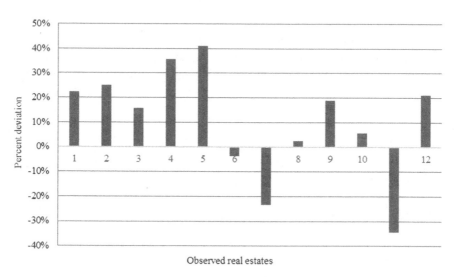

Fig. 8. Percentage difference of the updated prices compared to OMI average datum

This shows that, for 9 owners, a possible taxation based on OMI average value is reductive of the tax impact potentially anchored to the real price of sale. While, for 3 owners the use of the OMI average datum entail a tax burden greater than that deserved.

It outlines therefore a tax scenario that should guarantee a mild taxation for most owners. However, some citizens would be in a position disadvantaged, forced to disburse a surplus than the actual value of their assets; this is even more strange if we consider that they own the properties with the worse characteristics in the survey sample. So, the reform would lead to an increase in taxes for the poorer segments of the population and therefore holders of less valuable properties.

5 Conclusions

The computerized database built with over 500 contracts of sale signed after January 1st, 2008 is an important database for monitoring the real estate market in the region of Campania (Italy). With the model for data classification, through hyperlinks, we can directly view the documents from which the classified information is taken. The GIS interface of the database allows us to bring on the map all the data, greatly facilitating the management of the same also for operators that do not have specific skills in appraisal.

The application of the two instruments, carried out by way of example for the town of Salerno, has made possible to compare the level of residential real estate market recorded from supervisory government bodies (Tax Agency) with the real market prices. This is a fundamental prerequisite to check the consistency of government data that will be used as a reference for the modification of the tax base in the tax asset. The comparison between institutional values and market prices shows that the government guidelines are on average lower than real market. However, for a minority of properties, the average governmental datum exceeds the actual value and for this reason the adoption of this would lead to an unjust taxation. A compensatory model is being processed [9–11]; it allows you to review the distortions, to maintain constant, the overall revenue that the government gets from property taxes (like the Government claims to be able to do), ensuring all owners to pay in relation to the actual value of the property.

References

1. Medici, G.: Principi di estimo, Calderini (1972)
2. Mare, G., Manganelli, B., Nesticò, A.: Dynamic analysis of the property market in the city of Avellino (Italy). In: Murgante, B., Misra, S., Carlini, M., Torre, C.M., Nguyen, H.-Q., Taniar, D., Apduhan, B.O., Gervasi, O. (eds.) ICCSA 2013. LNCS, vol. 7973, pp. 509–523. Springer, Heidelberg (2013). doi:10.1007/978-3-642-39646-5_37
3. Marella, G., Antoniucci, G.: Small town resilience: housing market crisis and urban density in Italy. Land Use Policy 59, 580–588 (2016). doi:10.1016/j.landusepol.2016.10.004
4. Monfils, S., Hauglustaine, J.: Introduction of behavioral parameterization in the EPC calculation method and assessment of five typical urban houses in Wallonia, Belgium. Sustainability 8(11), 1205 (2016). doi:10.3390/su8111205
5. Liow, K.H., Newell, G.: Real estate global beta and spillovers: an international study. Econ. Model. 59, 297–313 (2016). doi:10.1016/j.econmod.2016.08.001

6. Bencardino, M., Nesticò, A.: Demographic changes and real estate values. A quantitative model for analyzing the urban-rural linkages. Sustainability 9(4), 536 (2017). doi:10.3390/su9040536. MDPI AG, Basel, Switzerland

7. Bencardino, M., Granata, M.F., Nesticò, A., Salvati, L.: Urban growth and real estate income. A comparison of analytical models. In: Gervasi, O., et al. (eds.) ICCSA 2016. LNCS, vol. 9788, pp. 151–166. Springer, Cham (2016). doi:10.1007/978-3-319-42111-7_13

8. Tamborrino, M.: Come si stima il valore degli immobili, il Sole 24 ore (2016)

9. Granata, M.F., De Mare, G., Nesticò, A.: Weak and strong compensation for the prioritization of public investments: multidimensional analysis for pools. Sustainability (Switzerland) 7(12), 16022–16038 (2015). doi:10.3390/su71215798

10. Nesticò, A., Macchiaroli, M., Pipolo, O.: Costs and benefits in the recovery of historic buildings: the application of an economic model. Sustainability (Switzerland) 7(11), 14661–14676 (2015). doi:10.3390/su71114661

11. Guarini, M.R., Buccarini, C., Battisti, F.: Technical and economic evaluation of a building recovery by public-private partnership in Rome (Italy). In: Stanghellini, S., Morano, P., Bottero, M., Oppio, A. (eds.) Appraisal: From Theory to Practice. GET, pp. 101–115. Springer, Cham (2017). doi:10.1007/978-3-319-49676-4_8

Daily Temperature and Precipitation Prediction Using Neuro-Fuzzy Networks and Weather Generators

Vito Telesca[1,2(✉)], Donatella Caniani[1], Stefania Calace[1],
Lucia Marotta[1], and Ignazio M. Mancini[1]

[1] School of Engineering, University of Basilicata,
Viale dell'Ateneo Lucano n.10, 85100 Potenza, Italy
{vito.telesca,donatella.caniani,stefania.calace,
ignazio.mancini}@unibas.it,
lucia.marotta.87@gmail.com
[2] CMCC - Euro-Mediterranean Center on Climate Change, Lecce, Italy

Abstract. In this study, a stochastic weather generator and a neuro-fuzzy network were developed to generate precipitation and air temperature (max-mean-min) on a daily basis for meteorological stations of the coastal area of the Basilicata region (Southern Italy). Several simulations were carried out to build an optimal model, whose efficiency was evaluated with the calculation of the root mean square error (RMSE) and the mean absolute error (MAE), which were both obtained comparing simulated and observed values. Subsequently, the developed neuro-fuzzy model was applied to generate other weather variables, such as relative humidity, solar radiation and wind velocity. The simulations showed the good performance of the neuro-fuzzy network in the data filling of the available time series. The evaluation of the performance was made by comparing the values of RMSE and MAE obtained with the neuro-fuzzy method developed in this study and literature methods such as multiple linear regressions and trend line.

Keywords: Climate change · Weather generators · Neuro-fuzzy network

1 Introduction

Weather forecasting, particularly for the case of daily precipitation occurrence and amount and daily air temperature, plays a leading role in several fields, such as agriculture and industry, because climate change impact on human life and its projection are relevant for the definition of correct prediction strategies. In this context, weather generators (WGs) and soft computing techniques (i.e., neuro-fuzzy network) have found widespread application in meteorology and hydrology, in studies of climate variability or climate change and weather forecasting [1–4]. Indeed, the application of WGs may help to take decisions for planning tasks when available weather data are limited; according to them, precipitation is a primary climate pattern that affects the rest of the variables. Thus, the first step consists of modelling daily precipitation; its generation requires a range of models whose combination and configuration depend on

© Springer International Publishing AG 2017
O. Gervasi et al. (Eds.): ICCSA 2017, Part VI, LNCS 10409, pp. 441–455, 2017.
DOI: 10.1007/978-3-319-62407-5_31

the process and temporal and spatial scales involved: 1. empirical statistical models, based on stochastic models that are calibrated from actual data; these models reproduce annual, monthly and daily precipitation data, resembling actual values. 2. dynamic meteorological models, which incorporate complex non-linear partial differential equations representing different physical processes and are used for weather forecasting. 3. intermediate stochastic models, which incorporate a limited number of parameters, determined from actual data collected at short time intervals (for example hourly data) and are used to represent complicated physical phenomena associated with storm precipitation.

In literature, many soft computing techniques have been developed to study weather forecast. However, the concept of "climate prediction" is very general; in fact, numerous studies analyze different aspects, such as:

1. Weather forecasting system using concept of soft computing with general indication about the topic, complete forecasting for near future not for singular variable considering different parameters that influence the process.

    Sharma et al. [2] focuses on the comparison between ANFIS (adaptive neuro-fuzzy inference system) and MLR (multiple linear regression) for monthly temperature prediction (mean - minimum - maximum), wind speed and pressure. The results show that the ANFIS accuracy is higher, with a smaller RMSE. Amanullah and Khanaa [5] compared the ANFIS and ARIMA (autoregressive integrated moving average) techniques for the study of historical series. The analyzed data are the following: average daily temperature, pressure and wind speed for the first, and temperature, dew point, relative humidity, wind direction and pressure for the latter. The evaluation of RMSE and MAE confirmed the performance of ANFIS that gives a better fitting of the studied trend of climatic variables.

2. Many researchers have applied the soft computing techniques for the prediction of single climatic variables, focusing on parameters, such as precipitation [6], solar radiation [7], wind speed [8], and surface temperature [9]. All these studies have shown that these techniques offer better results than those traditionally used.

3. Weather forecasting systems using concept of soft computing have been applied for singular specific variables, particularly for air temperature.

    Al-Matarneh et al. [10] studied the prediction of the air temperature with neural networks and fuzzy logic, using the historical data of the day t−1 for predicting the temperature of the day t. Using different input data, such as temperature, precipitation and relative humidity for the first and only one temperature for the second, both models generate a minimum MAE, for which they can be considered suitable for temperature forecasting. Kisi and Shiri [11] built an ANN and an ANFIS for the prediction of long-term monthly air temperature using geographical inputs. The data of 20 and 10 weather stations have been respectively used for the training and the testing in a temporal range going from 1989 to 2000. The parameters used as input are: number of the months, station latitude, longitude and altitude. The results exhibit a better determination coefficient for the ANN rather than for the ANFIS model, because the first predicts values closer to the expected temperatures.

The present study focuses on empirical statistical models [4], which usually represent daily weather sequences, particularly on a type of two-part model for generating daily precipitation at a specific site (step 1. a model for generating wet and dry events; step 2. a model for assigning a precipitation amount to a wet day). Here, the performance of a new Daily Precipitation Stochastic Generator (DPSG) has been presented [3]. The calibration is made on 30 years (1961–1990) and 100 years are simulated.

Furthermore, a neuro-fuzzy network is developed to generate air temperature (max-mean-min) on a daily basis. Particularly, this study aims at developing a specific model for the generation of the temperature value of the day i + 1 starting from the parameters relating to the previous day. This choice is based on the extensive use of this time relationship in literature studies [2]. The chosen weather variables are the following: relative humidity, solar radiation and wind velocity. Usually, the training period is 30 years, but in this work not all the four required variables for the model are available for the training years (1961–1990). Therefore, due to this problem, the years 2000–2002 and 2005–2006 (5 years) are used for the calibration and 2008–2009 (2 years) are used for the validation of the model. The dataset is divided into two groups because it is important that the training phase encloses a time series higher than that used for the validation. Moreover, to evaluate the potentiality of the developed neuro-fuzzy network for data filling procedure, the year 2010, for which different missing spells are simulated, is used. This section is important since the data filling is considered a decisive step for the use of time series data characterized by missing data [12].

The process has exploited also in this case the temporal relationship starting from the data at time t to generate those at time t + 1. At the end, the performance of the built model is defined computing the RMSE (Root Mean Square Error), the correlation between expected values and simulated ones and the MAE (Mean Absolute Error).

2 Dataset

Weather variables dataset available for agro-meteorological stations throughout the Basilicata Region are used thanks to ALSIA (Basilicata agency for agricultural development and innovation). Particularly, the models are validated for a time series (1959–2014) recorded at the meteorological station of Policoro (MT) - Basilicata – Italy, chosen because of its availability dataset for Climatological analyses for the required minimum series length as suggested by WMO. It is possible apply this model to others meteorological stations calibrating the dataset with specific conditions of the site.

3 Methods

In this work, a new Daily Precipitation Stochastic Generator (DPSG) based on a multivariate quasi-stationary and weakly depending stochastic process, is presented. DPSG, based on different hypotheses [3], is made of two-part model for rainfall simulation which consists of the first-order Markov chain to simulate precipitation occurrence using parameters of two transitional probabilities from a wet day to a wet day p(w|w) and from a dry day to a wet day p(w|d), and a two-parameter Weibull probability function for non-zero amounts [3, 4]. Then, through the application of artificial intelligence techniques, neuro-fuzzy network is developed to daily air temperature to solve complex problems, that have no solution using traditional techniques. The neural network consists of a set of single units, called "neurons", interconnected between them through the "synapses" or connections. Figure 1 shows the working model of a neuron in the network.

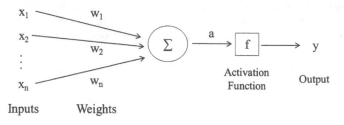

Fig. 1. Neuron model

The input variables $[x_1,...,x_n]$ are connected to the specific weights $[w_1,...,w_n]$, and then they come to the independent unit for generate an output through its activation function.

The neural network is seen as a "black box" where input data are introduced and from which the outputs are generated without request for additional information. In fact, the neural network is applied in many situations in which the relationship is unknown or it is not easily obtainable in terms of time or processing. The identification of the relationship occurs in the training step, the main stage of the process. During this phase a large set of inputs, and corresponding outputs, are provided to the network in order to ensure the right quantity of useful examples. Once the neural network is created through the training, it is subjected to the testing phase, in order to assess the acquired skills. The process is based on the availability of a series of known input and output data, different from those used for training. The new inputs are supplied to the new network, which processes the data and provides new outputs. Control is based on the evaluation of the error between the data generated and expected. If the error is acceptable, the model is considered reliable. Otherwise, there are problems in the training phase. The training phase has a key role in the construction of a neural network, so it is necessary provide more pieces of information to ensure the right number of examples to derive the relationship.

The fuzzy logic is a logic that overcomes the limits of classic logic of true or false [13]. It attributes to an element a degree of membership between 0 and 1 to a fuzzy set.

The fuzzy logic is based on the presence of fuzzy rules that link the various inputs present to provide the final output. A general rule consists of an antecedent, containing the basic conditions (IF), and a consequent (THEN). The combination of this two elements gives the ANFIS (Adaptive Neuro-Fuzzy Inference System) [6, 8, 14], which behaves as a neural network and express the relationship between inputs and outputs by fuzzy rules. Figure 2 shows the ANFIS structure that is characterized by 5 different levels [15].

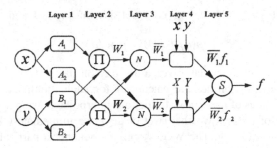

Fig. 2. ANFIS structure

In the study, during the training a large set of inputs and corresponding outputs are provided to the network, so that the network can have enough examples to know the relationship between input and output. Validation phase consists in evaluating whether the error between known and generated outputs is acceptable.

4   Application of DPSG to Daily Precipitation

This study focuses on empirical statistical models, which usually represent daily weather sequences, particularly on a type of Two-part model for generating daily precipitation at a specific site. The model [3] reproduces annual, monthly and daily precipitation data resembling actual data values, but, an outstanding problem associated with their use is that they are single-location, or point-process, models. Therefore, using these methods for simultaneous simulation of weather sequences at multiple points, for example to evaluate regional hydrological or agricultural behavior, the quite strong spatial correlations in daily weather data must be considered [4].

Two part model [3] for daily precipitation consists of two basic steps: first, a model for generating wet and dry events (rainy and non-rainy days); and second, a model for assigning a precipitation amount to a wet day.

Step I: precipitation occurrence.

A first-order two state Markov chain is used to stochastically generate dry and wet days; it usually captures the distribution of wet spells as well as higher order models [16].

The generation of precipitation is based on two assumptions. One is that the rain condition on day i is related to the rain condition on day i−1, and the other is that the amount of rain on rainy days is described by a suitable distribution function.

The first assumption describes a type of model called Markov chain. Defining p(w/w) as the probability of a wet day on day i given a wet day on day i−1, and p(w/d) as the probability of a wet day on day i given a dry day on day i−1, then the two complementary transition probabilities are shown in the Eq. 1:

$$p\left(\frac{d}{w}\right) = 1 - p\left(\frac{w}{w}\right) \tag{1}$$

that is the probability of a dry day given a wet day on day i−1 and

$$p(d/d) = 1 - p(w/d) \tag{2}$$

that is the probability of a dry day given a dry day on day i−1.

These transition probabilities are calculated for each month at each location of interest [3]. Daily values of these probabilities are interpolated using spline functions. If we know the state of today's weather (wet or dry), we immediately know the probability of a wet day tomorrow. The WG determines whether a particular day is wet or dry by subtracting p(w/w) or p(w/d) from a random number between 0 and 1. If the result is greater than zero, the generator assumes no rain on that day. If it is less than or equal to zero, rain is assumed to have occurred, and the amount of rain is determined using a distribution function for rain amounts on wet days. Quadratic spline functions are used for daily interpolation of monthly probabilities of a wet day given a previous wet day and a wet day given a previous dry day.

As transitional probabilities are conditional, the following expression holds:

$$fwet = p(w/d) \cdot (1 - fwet) + p(w/w) \cdot fwet \tag{3}$$

The two transitional probabilities are estimated for each available data source as follows:

since the monthly occurrence of precipitation is available the monthly frequency of wet days fwet, can be determined and the transitional probability of a wet day after a dry one p(w/d) for each month is estimated according to the following empirical expression:

$$p(w/d) = 0 \quad \text{if } fwet = 0 \tag{4}$$

$$p(w/d) = a_2 \cdot p(w) + a_1 \quad \text{if } fwet > 0 \tag{5}$$

where a_1 and a_2 are two-specific coefficients, given by linear regression.

fwet, p(w/d) and p(w/w) on monthly basis are evaluated for a lot of simulated years, but, it's more important the simulation of spells on a daily basis using the following relations:

$$A = RND(0; 1) - p(w/d) \tag{6}$$

$$A = RND(0; 1) - p(w/w) \tag{7}$$

in which RND(0;1) is a random number between 0 and 1.

The first relation is used if the previous day is dry, instead the second is used if the previous day is wet. Then, if A <= 0 the day is wet, instead if A > 0 the day is dry.

Step II: precipitation amount assigned to a wet day.

Nonzero precipitation amounts are simulated here using the exponential distribution. The Gamma and Weibull precipitation distribution functions are selected because their site-specific shape can be estimated from the expected amount of wet day precipitation per month as, respectively, shown in Selker and Haith [17]. Therefore, the model's distribution function varies from month to month. In this study, the amount of precipitation is assumed to follow a Weibull distribution. According to Rodriguez [18] the two-parameter Weibull precipitation distribution may be converted to a single parameter distribution using the following expression:

$$W(x, \zeta) = 1 - \exp\left[-\left(\Gamma(1+1/\zeta)\frac{x}{\mu}\right)^{\zeta}\right]$$
(8)

where x is the daily amount of precipitation, $\mu = P/k\zeta$ is the expected monthly amount of wet day precipitation, k is the number of wet days in a month, and is adimensionless parameter related with the coefficient of variation (CV) in the following way [4]:

$$1 + CV^2 = \frac{\Gamma(1+2/\zeta)}{\Gamma(1+1/\zeta)}$$
(9)

5 Results and Discussion

5.1 Application of DPSG to Daily Precipitation

Initially, the quality control of precipitation series is made by homogeneity tests (Run Test); then, to evaluate the goodness of generated weather data, a validation procedure has been developed. This procedure consists of different graphical analyses evaluating the correspondence of historical and generated data (RMSE_MEAN = 0.1292; RMSE_SD = 0.5719), comparing monthly means and monthly standard deviations of precipitation occurrence. The obtained results, shown in Fig. 3, suggest the capability of the first-order Markov chain in reproducing monthly statistical patterns, particularly the occurrence of wet and dry days, which are required for operational purposes in engineering.

5.2 Application of Neuro-Fuzzy to Air Temperature

In literature, there are several studies showing the relationship between parameters of day i and the same parameters of the next day $i + 1$ [10]. Starting from this, a lot of calculations are carried out to compare different relationship. The basic models have shown the best results in the ratio $i - i + 1$, which was used permanently in the model. The model is built with the Matlab 7.6.0 software using normalized dataset. Using this form of dataset with values between 0 and 1, the system performance is better with an exiguous error.

The choice of input variables, evaluated in respect of the influence on the forecast temperature is very important. At first, the analysis of climatic variables that influence the temperature trend is carried out, and for this reason, the following four input variables are considered:

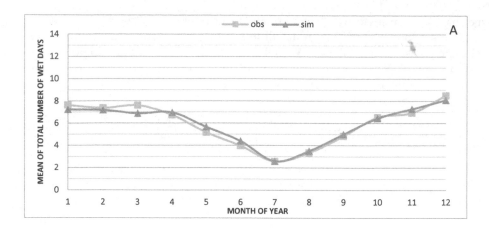

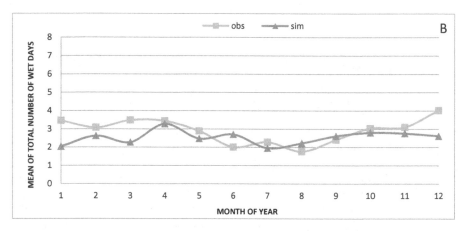

Fig. 3. Comparison between monthly total number of wet days observed and simulated mean (A) and standard deviation (B)

INPUT: Temperature (i), Relative Humidity (i), Solar Radiation (i), Wind speed (i) OUTPUT: Temperature (i + 1).

For this scheme of the neuro-fuzzy network, considering the need to have 4 available data series for 30 years (1961–1990), the possibility of missing data is very high. This has deflected the choice of the training period (2000–2002 and 2005–2006), and then validation (2008–2009). The *genfis2* Matlab function is run through a structure FIS Sugeno type using subtractive clustering demanding set of input data and output in a separate way. In addition to the series of input and output, the genfis2 function also contains a vector that represents the range of influence of the cluster center, called *radii*. The main objective in the use of this specific function has been to find the optimum combination of the values of radii in order to obtain the lowest error, thus providing the correct value of the temperature of the day i + 1. The combination has been found through a parfor cycle that identified the minimum error for each combination of 5 values between 0 and 1. The optimal combination for the training of the network that includes mean temperature is the following:

[mean temperature (i) relative humidity (i) solar radiation (i) wind velocity (i) mean
temperature (i + 1)] = [0.4; 1; 0.7; 0.7; 0.1].

This particular combination associates a high influence to the values of relative humidity in the first place, followed by the parameters of the solar radiation and wind speed. In Fig. 4 it is possible to assess how the real mean temperature trend is reproduced by the neuro-fuzzy network with the simulated temperature.

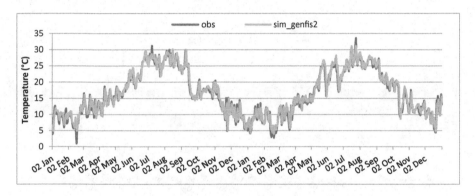

Fig. 4. Real and simulated mean temperature trend for the 2008–2009

The same calculations are made for maximum and minimum temperature, and the optimal combinations are the following:

[minimum temperature (i) relative humidity (i) solar radiation (i) wind velocity (i)
minimum tempetature (i + 1)] = [1; 1; 0.64; 0.28; 0.1].

The principal weight is related to the minimum temperature and relative humidity of the day i.

In the case of the maximum temperature, the greatest influence is associated to relative humidity and wind velocity of the day i.

[maximum temperature (i) relative humidity (i) solar radiation (i) wind velocity (i)
maximum temperature (i + 1)] = [0.4; 1; 0.7; 1; 0.1].

Figures 5 and 6 show the measured trends of minimum and maximum temperature and the ones simulated by genfis2.

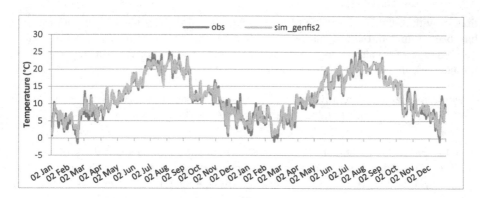

Fig. 5. Trend of observed and simulated (by genfis2) minimum temperature for years 2008–2010

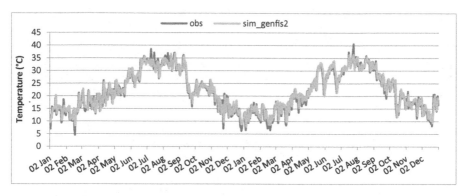

Fig. 6. Trend of observed and simulated (by genfis2) maximum temperature for years 2008–2010

Different parameters have been used to the performance of the designed network:

- RMSE (Root Mean Square Error), used to compute the performance of the simulation, is valued between real and simulated data. The equation used is the following:

$$\text{RMSE} = \sqrt{\frac{1}{n} \sum\nolimits_{i=1}^{n} (O_i - P_i)^2} \qquad (10)$$

where:

n = number of total observations
O_i = observed data
P_i = simulated data.

- CORRELATION identifies the link between simulated and real data.
- MAE (Mean Absolute Error) is the difference between real and simulated data. It is expressed by the following equation:

$$MAE = \frac{1}{n}\sum_{i=1}^{n} |O_i - P_i| \qquad (11)$$

where:
n = number of total observations
O_i = observed data
P_i = simulated data.

The parameters are evaluated for the mean - minimum - maximum temperature and are summarized in Table 1.

Table 1. Comparison among simulation of mean, minimum and maximum temperature

Method	Mean	Minimum	L. Regression
RMSE	1.6047	1.7534	2.3433
Correlation	0.9731	0.9606	0.9549
MAE	1.2365	1.3394	1.8239

The correlation values are high in all cases and the MAE are low and therefore acceptable.

The neuro-fuzzy network can provide the value of the temperature of the day *i*, but can also provide the future values of the three other variables that constitute the inputs of the network. The values associated with each network are shown below, in Figs. 7, 8 and 9, respectively.

Relative Humidity

[mean temperature (i) rel. humidity (i) solar radiation (i) wind velocity (i) relative humidity (i + 1)] = [1; 0.85; 0.85; 0.4; 0.1]

Solar Radiation

[mean temperature (i) rel. humidity (i) solar radiation (i) wind velocity (i) solar radiation (i + 1)] = [1; 0.7; 1; 0.1; 0.25]

Wind Velocity

[mean temperature (i) rel. humidity (i) solar radiation (i) wind velocity (i) wind velocity (i + 1)] = [0.55; 0.7; 0.4; 0.4; 0.55]

The results are good; in fact, the errors are acceptable. However, for the wind velocity, the maximum error is equal to 2.47 km/h and it has a low correlation, equal to 0.5052. In this particular case, the simulation generates a trend that does not include the use of the wind velocity. Indeed, the results show that the wind velocity is an aleatory variable, and it is not easily predictable, because it is influenced by other variables at different scales.

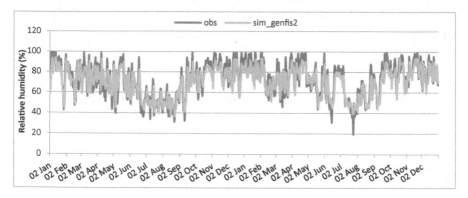

Fig. 7. Trend of observed and simulated (by genfis2) relative humidity for years 2008–2010

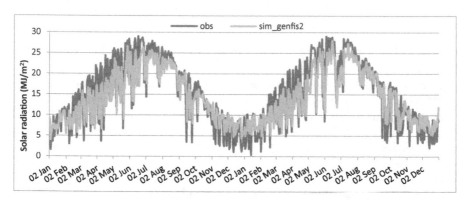

Fig. 8. Trend of observed and simulated (by genfis2) solar radiation for years 2008–2010

After the testing of the best network for predicting each variable, several simulations are run to apply the network for the data filling. In this case, the validation of the data filling method is carried out through a comparison among different data filling methods (trend line, linear regression). The use of the all variables (DOY – RAIN – RAD – RHMEAN – WINDS) gives the best results, as can be seen in Fig. 10.

The values of optimal configuration (multiple linear regression) are:

– MAE: 3.24662
– RMSE: 4.09961
– R^2: 0.6364

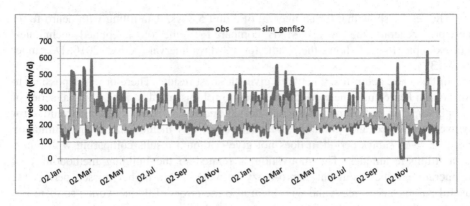

Fig. 9. Trend of observed and simulated (by genfis2) wind velocity for years 2008–2010

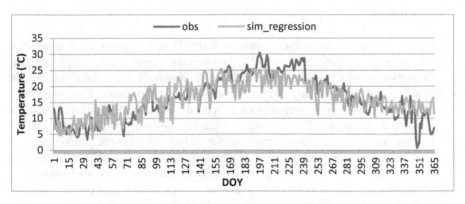

Fig. 10. Trend of observed and simulated (by regression) mean temperature using 5 independent variables: DOY, RAIN, RAD, RHMEAN, WINDS for the year 2010

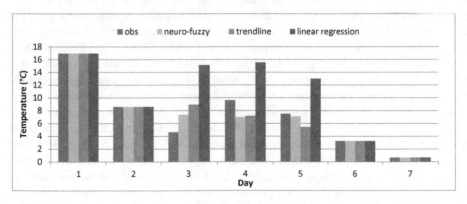

Fig. 11. Comparison among different data filling systems for the mean temperature using a 3 days interval for year 2010

The results show that for an interval of 3 and 5 days, data filling with neuro-fuzzy network generate the lowest MAE compared to the other applied techniques. For example, Fig. 11 shows the results for a 3 days interval (period: 2010, December 11th–13th).

In this case, neuro-fuzzy network gives the best results. Therefore, it is possible to affirm that neuro-fuzzy network can be used to fill missing data inside annual series of temperature; in fact, among different tested models it returns the minimum error between observed and simulated data. However, this method has a time limitation because over 7 days period, it does not give optimum results comparing with others data filling techniques. The motivation is related to the building methodology of temperature values using the neuro-fuzzy network. For a higher missed period, the useful parameters increase and with them also the errors of the networks, so there is a temporal limit for this procedure.

6 Conclusion

In this paper, a stochastic weather generator (DPSG) and a neuro-fuzzy network is developed to generate precipitation and air temperature (max-mean-min) on a daily basis in particular fora meteorological station (Policoro) of the coastal area of the Basilicata Region (Southern Italy). In this developed weather model, all the parameters in DPSG demonstrate a good performance in reproducing monthly statistical patterns. At this moment, the obtained results suggest the capability of the first-order Markov chain in reproducing precipitation occurrence for daily Mediterranean rainfall series. In fact, good validation results obtained show that DPSG can be considered an accurate tool for the generation of meteorological data (precipitation at first) in Mediterranean climate. Moreover, DPSG can be built to provide synthetic series of primary weather data, such as minimum and maximum air temperature.

Furthermore, a neuro-fuzzy network is designed for predicting the daily air temperature (mean – min – max) starting from known meteorological data. The network allows the prediction of the temperature of the day i starting from the climatic parameters (air temperature, relative humidity, solar radiation and the wind speed) of the day i−1. The best network is found by computing different performance error values between observed and simulated data (MAE, CORRELATION, RMSE). The MAE of final networks are equal to 1.2–1.3–1.8 °C for mean – minimum – maximum temperature, respectively. In particular, it is possible to associate a weight to the different element of the structure, as an influence factor; this result allows identifying the weight of different parameters on the result. Another important issue concerning climate forecasting is the problem of data filling. As known, it is very important the use of complete and available series for a climatological study. A comparison with two traditional techniques (trendline and linear regression) is made, in order to show the potentiality of the developed method for data filling. The results show that there is a maximum temporal limit of seven consecutive when neuro-fuzzy is the best method, instead above which the error increases and the linear regression become the optimal technique for data filling.

Acknowledgements. The study was carried out in the framework of the project Smart Basilicata in Smart Cities and Communities and Social Innovation" (MIUR n.84/Ric 2012, PON 2007–2013).

References

1. Rahul, G.K., Khurana, M.: A comparative study review of soft computing approach in weather forecasting. Int. J. Soft Comput. Eng. (IJSCE), **2**(5) (2012). ISSN: 2231–2307
2. Sharma, M., Mathew, L., Chatterji, S.: Weather forecasting using soft computing and statistical techniques. Int. J. Adv. Res. Electr. Electron. Instrum. Eng. **3**(7) (2014)
3. Marotta, L., Telesca, V.: A stochastic weather generator for daily climate variable analysis. IAHR Congress 2013 Tsinghua University Press, Beijing (2013)
4. Castellvì, F., Mormeneo, I., Perez, P.J.: Generation of daily amounts of precipitation from standard climatic data: a case study for Argentina. J. Hydrol. **289**(2004), 286–302 (2003)
5. Amanullah, M., Khanaa, V.K.: Application of soft computing techniques in weather forecasting: ANN approach. Int. J. Adv. Res. **2**(1), 212–219 (2014)
6. Mekanik, F., Imteaz, M.A., Talei, A.: Seasonal rainfall forecasting by adaptive network–based fuzzy inference system (ANFIS) using large scale climate signals. Clim. Dyn. **46**(9–10), 3097–3111 (2015)
7. Wang, L., Kisi, O., Zounemat-Kermani, M., Zhu, Z., Gong, W., Niu, Z., Liu, H., Liu, Z.: Prediction of solar radiation in China using different adaptive neuro-fuzzy methods and M5 model tree. Int. J. Climatol. (2016)
8. Liu, J., Wang, X., Lu, Y.: A novel hybrid methodology for short-term wind power forecasting based on adaptive neuro-fuzzy inference system. Renew. Energy 1–10 (2016)
9. Bilgili, M.: Prediction of soil temperature using regression and artificial neural network models. Meteorol. Atmos. Phys. **110**, 59–70 (2010)
10. Al-Matarneh, L., Sheta, A., Bani-Ahmad, S., Alshaer, J., Al-Oqily, I.: Development of temperature-based weather forecasting models using neural networks and fuzzy logic. Int. J. Multimedia Ubiquit. Eng. **9**(12), 343–366 (2014)
11. Kisi, O., Shiri, J.: Prediction of long-term monthly air temperature using geographical inputs. Int. J. Climatol. **34**(1), 179–186 (2014)
12. Kotsiantis, S., Kostoulas, A., Lykoudis, S., Argiriou, A., Menagias, K.: Filling missing temperature values in weather data banks. In: 2nd IET International Conference on Intelligent Environments (2006)
13. Caniani, D., Lioi, D.S., Mancini, I.M., Masi, S.: Application of fuzzy logic and sensitivity analysis for soil contamination hazard classification. Waste Manag. **31**(3), 583–594 (2011)
14. Sdao, F., Lioi, D.S., Pascale, S., Caniani, D., Mancini, I.M.: Landslide susceptibility assessment by using a neuro-fuzzy model: a case study in the Rupestrian heritage rich area of Matera. Nat. Hazards Earth Syst. Sci. **13**(2), 395–407 (2013)
15. Roger, J.: ANFIS: adaptive – network – based fuzzy inference system. Trans. Syst. Man Cybern. **23**(3) (1993)
16. Wilks, D.S.: Interannual variability and extreme-value characteristics of several stochastic daily precipitation models. Agric. For. Meteorol. **93**(3), 153–169 (1999)
17. Selker, J.S., Haith, D.A.: Development and testing of simple parameter precipitation distributions. Water Resour. Res. **26**(11), 2733–2740 (1990)
18. Rodriguez, R.N.: A guide to the Burr type XII distributions. Biometrika **64**, 129–134 (1977)

Bridging Biodiversity Conservation Objectives with Landscape Planning Through Green Infrastructures: A Case Study from Sardinia, Italy

Sabrina Lai[(⊠)] and Federica Leone

Dipartimento di Ingegneria Civile, Ambientale e Architettura (DICAAR),
University of Cagliari, Cagliari, Italy
{sabrinalai,federicaleone}@unica.it

Abstract. The definition of Green Infrastructure (GI) provided by the European Commission in its 2013 Communication "Green Infrastructure: Enhancing Europe's Natural Capital" regards GI as a network having the Natura 2000 sites at its core, able of delivering numerous ecosystem services, and "strategically planned", stressing the importance of GI in integrating ecological connectivity, biodiversity conservation, and multi-functionality of ecosystems. Consequently, the spatial identification and management of GI is an important issue in planning, and especially in landscape planning as understood in the European Landscape Convention.

Building on a previous work by Arcidiacono et al. (2016), this paper tests a methodology whereby the spatial configuration of a GI is identified in relation to four aspects (conservation value, natural value, recreation value, anthropic heritage) which summarize the multifaceted character of landscape. The methodology is tested in the Italian region of Sardinia, by applying it in the coastal landscape units defined in the Regional Landscape Plan currently in force which overlap the metropolitan area of Cagliari.

We argue that this methodology can effectively help integrate biodiversity conservation objectives into spatial planning by implementing article 10 of the Habitats Directive, stating that relevant features of the landscape should be managed so as to improve the ecological coherence of the Natura 2000 network.

Keywords: Green infrastructure · Landscape planning · Ecosystem services

1 Introduction

A fundamental definition of Green Infrastructure (GI) was provided by the European Commission in its Communication "Green Infrastructure: Enhancing Europe's Natural Capital" [1], where GI is regarded as a network having Natura 2000 sites at its core, able of delivering numerous ecosystem services, and "strategically planned", hence stressing the importance of GI's in integrating ecological connectivity, biodiversity conservation, and multi-functionality of ecosystems. This definition particularly highlights three important aspects: first, the idea of a network; second, planning and

© Springer International Publishing AG 2017
O. Gervasi et al. (Eds.): ICCSA 2017, Part VI, LNCS 10409, pp. 456–472, 2017.
DOI: 10.1007/978-3-319-62407-5_32

management issues; third, the ecosystem services concept [2]. Moreover, in the European context the importance of GI is remarked by the European Union (EU) in its Biodiversity Strategy, whose target no. 2 regards GI as a key element to maintain and enhance ecosystems and their services. In addition, action no. 6 promotes the use of GI through the development of a European "Green Infrastructure Strategy" and of a strategic framework in order to identify priorities for ecosystem restoration in each Member State [3].

From this conceptual perspective, the concept of GI is closely connected with the broader issue of biodiversity conservation for three main reasons. First, the identification of a GI focuses on the designation and maintenance of natural and semi-natural areas in developed, and sometimes even built-up, landscapes. Second, it entails the development of ecological connections among different habitats so as to allow species movements. Third, it uses a language that can be understood by people that play a key role on urbanization processes, such as planners and private businesses [4]. Consequently, the spatial identification and management of GI is an important issue in planning, and especially in landscape planning as understood in the European Landscape Convention. As Snäll et al. [5] put it, a spatially explicit approach is needed in designing a GI, because only such an approach "can support land managers' decisions in real-world situations at the operational level". Hence, the spatial identification and management of GI represent a significant issue for planning at various scale levels. As a matter of fact, the integration of GI within planning policies can support decisions having implications for conservation and protection of landscapes and environment, in terms of knowledge about territory [6] and in relation to their capacity of combining ecological, social and cultural functions [7].

Within this framework, in this study we analyze multifunctional GI and discuss the issue of its spatial configuration in the case study of the metropolitan city of Cagliari, Italy. In addition, we argue that the identification of GI can ease the integration of biodiversity conservation within planning policies in order to promote the implementation of Council Directive 92/43/EEC "on the conservation of natural habitats and of wild fauna and flora" (the so-called "Habitats" Directive), whose article no. 10 states that Member States must promote the management of key elements of landscape which are significant for natural biodiversity.

This article is structured into five sections. The first reviews the relevant literature to identify the open issues to which the research is contributing. The second provides information on the case study and defines the methodological approach. The results are presented in the fourth section, while the last section discusses the results, identifies strengths and caveats of the methodological approach and provides directions for future research.

2 Literature Review: Key Constituents of Multifunctional Green Infrastructures

According to the European Commission [8, p. 1], "GI [...] promotes integrated spatial planning by identifying multifunctional zones and by incorporating habitat restoration measures into various land-use plans and policies." From this perspective, multi-functionality represents a key element in the spatial definition of a GI due to the multi-functional use of natural capital that allows to address multiple purposes, among which prominent are biodiversity conservation and ecosystem services production [7].

Various research has advocated GI as a means to ensure ecological connectivity, in connection to ecological corridors [9, 10] either to ensure connections among protected areas [11], or, and in opposition to grey infrastructures such as transportation networks, to ensure ecological functions between and within cities and towns, hence with a view to benefitting human populations and economies first [12]. Such views risk emphasizing the ecological function of GI while leaving other functions in the background. Incorporating GI within spatial planning therefore serves the purpose of "accounting for trade-offs and synergies among multiple ecosystem services in a spatially explicit context" [5].

Building on a previous work by Arcidiacono et al. [13], we assume that the spatial configuration of a multifunctional GI, able to maintain and enhance both natural resources and elements upon which the relations between people and places are grounded, can be identified in relation to four aspects which summarize the multi-faceted character of landscape.

The first, conservation value, accounts for the presence of natural habitat types of Community interest, listed in Annex I of the Habitats Directive. According to art. 1 of the Habitats Directive, a natural habitat is deemed of Community interest when it is endangered or threatened with extinction in its natural range, or has a small natural range, or exhibits typical characteristics of one or more of the nine biogeographical regions to which Member States of the EU belong. Moreover, since 1999 the European Commission has produced an interpretation manual of such habitats [14] and some Member States have tailored the EU manual to their national and local specificities and produced national or regional maps of natural habitats of Community interest.

The second, natural value, accounts for biodiversity in a broader sense, beyond the intrinsic conservation value implicit in the definition of the Habitats Directive. The concept of ecosystem services has recently taken hold worldwide to refer to those goods and services provided by nature that sustain life and human well-being, and hence has met with some criticisms from ecologists and environmental scientists because it focuses on the capability of biodiversity to satisfy human needs, therefore neglecting its intrinsic value and leading to commodification of nature [15, 16]. Various categorizations of ecosystem services have been proposed so far; among the most widely used, the "Common International Classification of Ecosystem Services" (CICES) only considers final goods and services, that is those for which a human demand exists (in accordance with Boyd and Banzhaf [17]), and groups them into three main categories: provisioning, regulating, and cultural services. In accordance with Müller [18], and with Fisher and Turner [19], other categorizations, such as the "Millennium Ecosystem Assessment" (MA) [20] and "The Economics of Ecosystems and Biodiversity" (TEEB) [21], also include a fourth group (labeled supporting services, or habitat services) that accounts for ecological functions and integrity, not directly "consumed" by people but necessary for ecosystems to produce final goods and services. Within this framework, natural value in this study accounts for biodiversity's quality, which implies its ecological integrity, current levels of ecosystem functions, its capability to supply human-demanded ecosystem services notwith-standing pressures and threats to habitats.

The third, recreation value, is a final ecosystem service part of the cultural services group. In the MA taxonomy [20, p. 58], this group includes different kinds of non-material benefits derived from ecosystems such as "spiritual enrichment, cognitive development, reflection, recreation, and aesthetic experiences". Recreation, in partic-ular, accounts for the fact that landscapes and natural habitats are among the factors that people take into account when deciding where they want to spend their holidays or just some leisure time, which not only positively affects on residents' and tourists' quality of life and wellbeing, but also impacts directly and indirectly on local economies. In contrast to the other cultural ecosystem services, recreational services can be measured through economic indicators [22]. The TEEB [21], for instance, suggests that the recreational value related to biodiversity could be evaluated in monetary terms through travel cost methods, which have indeed been used by several scholars (among many, [23–25]), sometimes in combination with willingness to pay exercises [26] or with contingent valuation methods [27]. A different approach is that of non-monetary evaluation, regarded as able to capture in a broader and multidimensional way people's understanding and valuing of non-tangible ecosystem services [28]. Various methods can be applied in non-monetary evaluations, and two frameworks, one [29] that cat-egorizes them into three broad groups: quantitative, qualitative and deliberative, and one that looks at deliberative methods only [28], have been proposed. Among such methods, quantitative methods (either non-consultative or consultative), in principle, could easily be used to assess recreation values, as they merely imply collecting data on tourists and visitors from official statistics (as in [30]) or carrying out ad-hoc surveys. However, because of the costs and time needed to gather them, such data are often unavailable; as a consequence, social-media based approaches have recently been proposed that estimate visitor preferences and numbers using the number of pictures uploaded to social media such as Flickr [31, 32], also coupled with Instagram [33], as a proxy.

The fourth, anthropic heritage, accounts for the interactions between natural and human factors as understood in the European Landscape Convention, whereby land-scape is a complex system that includes not only individual historic monuments and landmarks, but also minor spots of land [34] that have contributed to shaping the identity of European cultures. Anthropic heritage, in this sense, is not restricted to archaeological remains, historic monuments, listed buildings, or areas of outstanding beauty, but is grounded on people's perception of their territories, and on the recog-nition that different places show different characters. In compliance with the Con-vention, whose implementation varies across countries [35], landscapes are to be protected, managed and planned. Consequently, landscape plans are the tools whereby landscapes are interpreted, anthropic heritage is identified and protection devices taking various forms that span from strict regulations to soft guidelines to orient future sce-narios are devised.

3   Materials and Methods

3.1   Case Study

The scale of a whole city has been advocated [36] as the minimum ideal scale for spatial planning to set up GI aiming at preserving biodiversity, maintaining ecosystem services. We chose as a case study the metropolitan city of Cagliari, in the Italian region of Sardinia, where a Regional Landscape Plan (RLP), set up in compliance with the Italian Landscape Code (Decree Enacted by Law no. 42 of 2004, which implements the European Landscape Convention in Italy), has been in force since 2006 and where an extensive Natura 2000 network, covering almost 19% of the regional land area, has been designated.

The metropolitan city of Cagliari is partially included within three out of the 27 Coastal Landscape Units identified in, and ruled by, the RLP. Such plan does not give explicit provisions for setting up a regional GI; however, it does provide building and planning restrictions in sensitive contexts so as to preserve ecological functions (art. 23 and 26 of the plan implementation code), and it gives directions on the necessity to integrate Natura 2000 sites within a single coherent network (art. 34 of the plan implementation code). The study area overlaps 20 Natura 2000 sites, out of which 13 sites of community importance (SCIs) and 7 special areas of conservation (SACs).

Our study area, mapped in Fig. 1, therefore encloses (i) the metropolitan city area of Cagliari, as well as (ii) the entirety of the three coastal landscape units that overlap the metropolitan city area, and (iii) all of the 20 Natura 2000 sites that either are completely included therein, or partially overlap it.

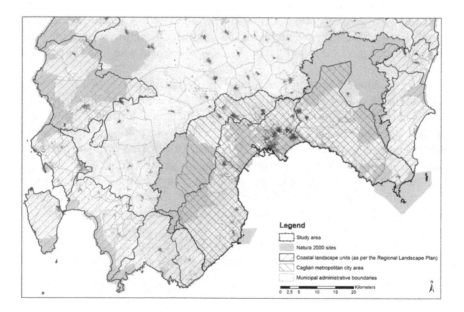

Fig. 1. Study area

3.2   Methodology

The four constituent values that, in our framework, can be used to identify the spatial configuration of a GI were calculated and mapped in a GIS-based environment.

Conservation Value. Conservation value (CONS_VAL) accounts for the presence of natural habitat types of Community interest. As mentioned, such habitats have been enlisted in Annex I of the Habitats Directive on the basis of their being rare, or threatened, or typical within a given biogeographical region; moreover, the Directive gives prominent importance to a small number of habitats classed as "priority" habitats. Hence, and building on a recent report [37, pp. 27–28] that ranks the importance of habitats of Community interest in Sardinia with a view to defining a regional monitoring plan, we calculate this value as follows:

- for areas where no habitats of Community interest have been identified:

$$CONS\_VAL = 0 \tag{1}$$

- for areas hosting habitats of Community interest:

$$CONS\_VAL = P * (R + T + K) \tag{2}$$

where:

- P indicates whether a given habitat is enlisted as priority habitat (P = 1.5 in case of priority habitat, P = 1 in case of non priority habitat);
- R denotes rarity, which, for each habitat of Community interest, can be evaluated on the basis of the number of Natura 2000 sites in which the presence of the habitat has been recorded in the standard data forms [38]. The figures can be retrieved from the official website of the European Environment Agency[1] and normalized in the interval $(1 \div 5)$; the lower the number of occurrences, the higher the value of R;
- T stands for threats, which are recorded in each Natura 2000 standard data form. For each Natura 2000 site, the number of threats recorded in the standard data forms is counted and normalized it in the interval $(1 \div 5)$; the higher the number of threats, the higher the value of T;
- K stands for knowledge: since reliable and up-to-date information gathered through on-site surveys is not available for every habitat of Community interest and for every Natura 2000 site, we deem the level of knowledge important from a conservationist's standpoint. The level of knowledge was assessed by experts within a recent regional monitoring project titled "Monitoring the conservation status of habitats and species of Community interest within Natura 2000 sites in Sardinia" [39, pp. 42–44]. For each habitat, the level of knowledge was therefore classed as "good", "acceptable", "insufficient", "poor". We converted these judgments into values in the interval $(1 \div 4)$, where the worse the level of knowledge, the higher the value of K. We chose to assign a

---

[1] http://natura2000.eea.europa.eu.

maximum score (4) lower than those of both R and T (5) due to the fact that K (contrary to R and T) depends on subjective assessments.

As a results, where habitats of Community interest are present the value can range from 1 (minimum conservation value) to 21 (max conservation value). Such values were subsequently normalized, hence CONS_VAL takes values in the $(0 \div 1)$ interval.

The following two spatial datasets were used: the first, the so-called "carta della natura" ("Nature map", [40]), has a scale of 1:50,000, and makes use of the CORINE biotopes nomenclature, while the second, the so-called "carta degli habitat" ("Habitat map", [41]), has a scale of 1:10,000 and maps habitats of Community interest by using the Habitats Directive taxonomy within Natura 2000 sites only. The interoperability between the two taxonomies was handled through a conversion tool made available by the Italian Superior Institute for Environmental Protection and Research [42], which allowed us to map habitats of Community interests also outside Natura 2000 sites using the "Nature map".

Natural Value. Natural value (NAT_VAL) accounts for the potential capability of biodiversity to supply final ecosystem services in face of threats and pressures. For this purpose, we used the software "InVEST"[2], tool "Habitat quality", which produces habitat quality maps by combining information on land covers and threats to biodiversity, on the basis of the assumption that areas having high values of habitat quality can better support biodiversity. The input data required by the tool and used to feed the model are as follows:

1. A raster land cover map. We used the 2008 Land Cover Map produced by the Regional administration of Sardinia that we first reclassified at the third level of the CORINE[3] taxonomy and next converted into a raster map having cellsize 25 * 25 m.
2. A list of current threats to biodiversity, and for each threat a weight (which denotes the threat's relative importance), a decay distance and function. After examining the standard data forms of the 20 Natura 2000 sites included in our study area, we selected those threats that generate negative impacts on the land zone of these sites (marine areas are out of the scope of this study) and that can be mapped. As a result, we obtained a list of ten pressures and threats; for weights and decay distances, we delivered a questionnaire to local experts in the field of biodiversity and environmental impact assessment. In the questionnaire, the weight was to be expressed using a "Likert" scale 1–5, hence grading the relative importance of a given threat, while the decay distance was to be provided in kilometers. For each threat, we next averaged both the weights and the decay distances provided by the surveyed experts; moreover, the weights were normalized in the $(0 \div 1)$ interval as required by InVEST. Table 1 provides a list of the ten selected threats, as well as their

---

[2] InVEST is a free software program developed within the Natural Capital Project and available at http://www.naturalcapitalproject.org/invest/.

[3] CORINE is acronym of "COoRdination de l'INformation sur l'Environnement", French for "Coordination of information concerning the environment".

Table 1. Threats to biodiversity in the study area, parameters for the InVEST model (weight, decay distance and function), and spatial data sources

Code	Threat name	Weight	Decay distance (km)	Decay function	Data source[a]
T01	Cultivation	0.58	1.63	linear	2008 Land Cover Map
T02	Grazing	0.68	0.58	linear	2008 Land Cover Map
T03	Removal of forest undergrowth	0.79	0.65	linear	2008 Land Cover Map
T04	Salt works	0.63	0.83	linear	2008 Land Cover Map
T05	Paths, tracks	0.53	0.55	linear	Regional multi-precision database
T06	Roads, motorways	0.95	3.00	linear	Regional multi-precision database
T07	Airports	0.95	4.75	linear	2008 Land Cover Map
T08	Urbanized areas	0.95	3.25	linear	2008 Land Cover Map
T09	Discharges	1.00	3.50	linear	2008 Land Cover Map
T10	Fire	0.95	2.05	linear	2011–2015 Fire maps

[a]All of the spatial datasets can be freely downloaded from the regional geoportal http://www.sardegnageoportale.it.

averaged weights and decay distances, and, as for weights only, also normalized. The decay function was always set as "linear".

3. A raster map for each current threat source. Table 1 lists, in detail, the data sources used.

4. A vector map representing accessibility to sources of degradation, in terms of relative protection to habitats provided by legal institutions. We considered regional and national parks, as well as areas protected and managed by the public regional forestry agency, as having the highest protection and hence lowest accessibility level (score 0.2); a second level we used was that of Natura 2000 sites (score 0.5); all the rest of the study area was considered as completely accessible (score 1). All the maps were available from the regional geoportal.

5. A matrix listing habitat types (where habitats represent resources and conditions present in an area that can support the life of given organisms, and therefore are not restricted to those of Community interest accounted for by CONS_VAL) and their sensitivity to each threat. The sensitivity of each habitat to each threat was developed using a two step expert-based approach: first, each land cover code (at the third level of the CORINE taxonomy) was given a trichotomous value (1 if the land cover could be intrinsically regarded as habitat; 0.5 if it could be considered habitat contingent on external factors; else 0); second, for each land cover code that could be considered as habitat, a score representing that land cover's sensitivity to each threat was assigned. An excerpt of this matrix is provided in Table 2.

6. A so-called "half-saturation constant", having default value 0.5 in the InVEST tool.

Table 2. An excerpt of the sensitivity matrix concerning agricultural land covers

Land cover code	Habitat score	T01	T02	T03	T04	T05	T06	T07	T08	T09	T10
211	0.5	0	0.5	0	0	0	0.5	0	0.5	0.5	0.5
212	0.5	0	0.5	0	0	0	0.5	0	0.5	0.5	0.5
221	0.5	0	0.5	0	0	0	0.5	0	0.5	0.5	0.5
222	0.5	0	0.5	0	0	0	0.5	0	0.5	0.5	0.5
223	0.5	0	0.5	0	0	0	0.5	0	0.5	0.5	0.5
231	1	1	0.5	0	0	0.5	1	0.2	0.5	1	1
241	0.5	0	0.5	0	0	0	0.5	0	0.5	0.5	0.5
242	0.5	0	0.5	0	0	0	0.5	0	0.5	0.5	0.5
243	1	0.5	1	0.5	0	1	1	0.2	1	1	1
244	1	0.5	0.5	1	0	1	1	0.2	1	1	0.5

Recreation Value. Recreation value (REC_VAL) is connected to people's (both locals' and tourists') appreciation of nature and biodiversity. In the absence of official data on visitors' numbers, we used the software "InVEST", tool "Recreation model". This tool gathers data from the social media Flickr, whose users upload geotagged pictures on the platform, and counts the total photo-user-days in specific locations (either cells or polygons) by using location, username and date in which the images were shot so as to avoid double counting. The unit of measure is therefore "photo-user-day" (PUD); one PUD means that, in a given spatial unit and on a specific day, one unique photographer took at least one photo. The study area was gridded using a square 100-meter grid; for each cell, the average PUD per year between 2010 and 2014 was calculated by the model, and subsequently normalized in the interval (0 ÷ 1).

Anthropic Heritage. Anthropic heritage (ANTH_HER) takes account of the landscape assets protected under the RLP, in force in Sardinia since 2006. For each protection level defined in the plan, a score was assigned in the (0 ÷ 1) interval depending on the level of restriction stemming from the plan implementation code, having also regarded to other restrictions originating from national and regional legislation. The full list of protection levels is provided in Table 3, together with reference to the articles of the implementation code that provide rules and directions for each protection level, and finally the score we assigned, expressing the anthropic heritage value. As for the spatial layout, the full spatial dataset of the protection levels established by the RLP is retrievable from the regional geoportal. It is worth noting that parcels subject to multiple protection levels were assigned the score corresponding to the strictest protection level in force in that parcel.

Total Value. In each specific location, the total value corresponds to the sum of the four above-listed values (CONS_VAL, NAT_VAL, REC_VAL, ANTH_HER), each ranging in the (0 ÷ 1) interval; as a result, total value can range in the interval (0 ÷ 4). The total value was calculated through a GIS geoprocessing tool after converting the

Table 3. Anthropic heritage; types of landscape protection levels established in Sardinia by the RLP, and value assigned on the basis of the restrictions in force

	Type	Plan implementation code: ruling articles	Value
Environmental assets	Coastal strip	8, 17, 18, 19, 20	1
	Coves, cliffs and small islands	8, 17, 18	0.8
	Sand dunes and beaches	8, 17, 18	0.8
	Coastal wetlands	8, 17, 18	0.8
	Areas above 900 m	8, 17, 18	0.8
	Lakes, reservoirs, wetlands and their 300-m buffers	8, 17, 18	1
	Rivers, creeks and their 150-m buffers	8, 17, 18	1
	Areas of significant importance for wild animals	17, 18, 38, 39, 40	0.2
	Areas of significant importance for plant species	17, 18, 38, 39, 40	0.2
	Grottos and caves	8, 17, 18	0.8
	Monumental trees	8, 17, 18	0.2
	Natural monuments (as per regional law 1989/31)	8, 17, 18	0.5
	National parks and marine protected areas	8, 17, 18	0.5
	Volcanoes	8, 17, 18	0.5
Historic and cultural assets	Listed buildings and areas (art.146 Decree 42/2004)	8	0.8
	Listed archaeological heritage	8, 47	1
	Archaeological areas subject to building restrictions	8, 47	0.5
	Areas with prehistoric, historic, cultural remnants	8, 47, 48, 49, 50	1
	Historic districts	8, 47, 51, 52, 53	0.8
	Traditional Sardinian farmer's building complexes	8, 47, 51, 52, 54	0.8

two raster maps (NAT_VAL and REC_VAL) into vector maps (CONS_VAL and ANTH_HER were already vector).

4   Results

The results of our model are represented in Figs. 2 and 3, where the former displays the spatial distribution of each of the four single values, while the latter maps the spatial distribution of the total value, and therefore can be used to delineate the spatial configuration of the GI within our study area.

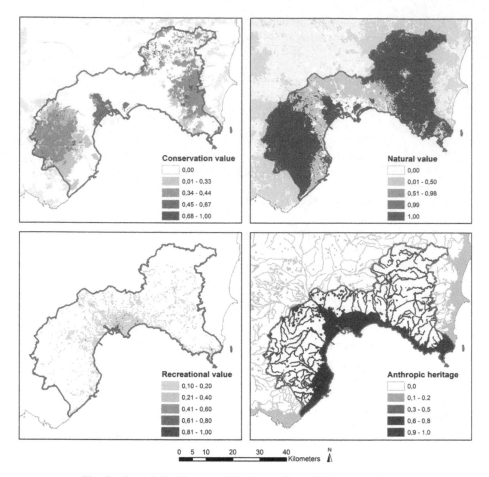

Fig. 2. Spatial distribution of the four values within the study area

Areas taking non-zero values in conservation value are mostly, but not exclusively, located within Natura 2000 sites and highly spatially clustered in their immediate surroundings.

As for natural values, two large clusters taking extremely high values are notable in the eastern and western parts of the study area, as well as two smaller clusters in the middle (corresponding to two wetlands surrounded by the built-up core of the metropolitan city); contrary to what happens with conservation values, the rest of the area does not have a zero value, meaning that, although not of Community interest, some middle-quality habitats act as "bridges" between areas having highest values.

With regard to recreation value, approximately 96% of the cells display zero value, meaning that for the vast majority of the study area the average PUD value is zero. Cells taking non-zero values, and especially cells taking the highest values, are highly clustered within the city of Cagliari; moreover, a linear constellation of non-zero values is clearly visible along the coastline, while the rest of the study area only shows scattered non-zero values.

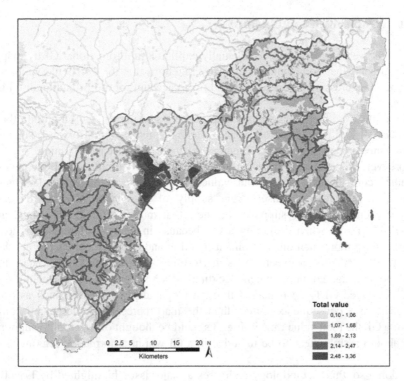

Fig. 3. Spatial distribution of the total value within the study area (classification: Jenks algorithm)

Finally, the anthropic heritage map shows that the most part of the study area takes either the minimum (0) or the maximum (1) value; only some small scattered areas take intermediate values. Moreover, areas taking the maximum value are largely dominated by two main environmental assets, characterized by strict restrictions on new development and land use changes. The first type, "Coastal strip", is protected under the RLP (article 20 of the plan implementation code), which forbids any kind of new development while allowing restoration or renewal of existent buildings. The second type, comprising "[listed] Rivers, creeks and their 150-m buffers", is protected under national law, whereas the RLP merely makes it clear which areas are to be preserved and protected as belonging to this type. A third type of asset that brings about the maximum value, but less significant in terms of size, comprises both "Listed archaeological heritage" and "Areas with prehistoric, historic, cultural remnants".

The total value map (Fig. 3) shows that no land parcel achieves the maximum total score (i.e., 4), which also implies that no land parcel simultaneously achieves the maximum score in each of the four values above presented. Areas taking the highest values consist mainly of rivers and creeks together with their 150-m buffers, wetlands, as well as large forest areas and bits of coastal areas.

5   Discussion and Conclusions

In this study we attempted to make spatially explicit a multifunctional GI by mapping four constituents that correspond to as many functions that the GI should support and ensure, hence addressing an open issue on how to account for multifunctionality [43] in designing a GI.

The results from the methodological approach tested in this study highlight that the four values vary differently across space, which is pretty straightforward, since each value captures a specific aspect or function relevant to landscape planning.

Moreover, it is also not surprising that no land parcel simultaneously achieves the maximum score in each of the four values that express as many single functions (high-quality biodiversity conservation, supply of ecosystem services, recreation, landscape protection). This supports the view that multifunctionality is an ideal (or "elusive" [44]) goal when designing a GI, because in reality different areas tend to perform one (or more than one) dominant function and complement each other, hence some spatial tradeoffs between areas performing different functions need to be understood and agreed upon when deciding which areas are to be included in a spatially-designed GI to be managed through a spatial planning tool such as a landscape plan. As a consequence, rather than the map representing the spatial configuration of a GI, the total value map in Fig. 3 should be thought of as a tool to support the choice about possible areas to be included in a GI within a normative regional spatial plan.

By doing so, this methodology addresses a major issue highlighted by Lovell and Tayor [36], that of a "limited success" in institutionalizing GI, because in Italy landscape plans are to be prepared and adopted by an institution to fulfill an obligation by law through a process in which participatory processes are mandatory in the strategic environmental assessment framework as per European Directive 2001/42/EC. Such participatory processes would engage ecosystem services beneficiaries [45], which include but are not limited to local communities. Ecosystem services beneficiaries' knowledge, expectations, and priorities were not included in the methodology here implemented, which is solely grounded on official, scientific datasets (e.g. as for the selection of threats when assessing NAT_VAL) or on expert judgment (e.g. on prioritizing threats or assessing sensitivity weights, again in connection with NAT_VAL). Therefore, future research could address the issue of taking ecosystem services beneficiaries into account, for instance as regards possible weights to assign to each constituent value to reflect their priorities.

Finally, Habitats Directive's objectives have been incorporated in the methodology primarily via CONS_VAL and secondarily via NAT_VAL, with the aim to help integrate biodiversity conservation objectives into spatial planning, so as to ease implementation of article 10 of the Habitats Directive, which states that relevant features of the landscape should be managed to improve the ecological coherence of the Natura 2000 network. Future researches could therefore explore how the methodology would work at a larger scale, compatible with Natura 2000 network's spatial layout (for instance, biogeographical regions or disconnected parts thereof) and what the relationship is between the multifunctional GI as here identified and the ecological network connecting Natura 2000 sites as implied in article 10 of the Habitats Directive, comprising linear and continuous structures as well as stepping stones.

Acknowledgments. This study was supported by the Research Program "Natura 2000: Assessment of management plans and definition of ecological corridors as a complex network," funded by the Autonomous Region of Sardinia for the period 2015–2018, under the Call for "Projects related to fundamental or basic research" of the year 2013, implemented at the Department of Civil and Environmental Engineering and Architecture (DICAAR) of the University of Cagliari, Italy.

This paper is to be equally attributed to Sabrina Lai and Federica Leone, who collaboratively designed the paper and jointly wrote Sects. 1 and 5. Federica Leone has written Sect. 2, while Sabrina Lai has written Sects. 3 and 4.

References

1. European Commission: Communication from the Commission to the European Parliament, the Council, the Economic and Social Committee and the Committee of the Regions. Green infrastructure (GI). Enhancing Europe's natural capital (2013). http://ec.europa.eu/environment/nature/ecosystems/docs/green_infrastructures/1_EN_ACT_part1_v5.pdf. Accessed 22 Mar 2017

2. Liquete, C., Kleeschulte, S., Dige, G., Maes, J., Grizzetti, B., Olah, B., Zulian, G.: Mapping green infrastructure based on ecosystem services and ecological networks: a Pan-European case study. Environ. Sci. Policy **54**, 268–280 (2015). doi:10.1016/j.envsci.2015.07.009

3. European Commission: Communication from the Commission to the European Parliament, the Council, the Economic and Social Committee and the Committee of the Regions. Our life insurance, our natural capital: an EU biodiversity strategy to 2020 (2011). http://eur-lex.europa.eu/legal-content/EN/TXT/PDF/?uri=CELEX:52011DC0244. Accessed 22 Mar 2017

4. Garmendia, E., Apostolopoulou, E., Adams, W.M., Bormpoudakis, D.: Biodiversity and green infrastructure in Europe: boundary object or ecological trap? Land Use Policy **56**, 315–319 (2016). doi:10.1016/j.landusepol.2016.04.003

5. Snäll, T., Lehtomäki, J., Arponen, A., Elith, J., Moilanen, A.: Green infrastructure design based on spatial conservation prioritization and modeling of biodiversity features and ecosystem services. Environ Manag. **57**, 251–256 (2016). doi:10.1007/s00267-015-0613-y

6. Wickham, J.D., Riitters, K.H., Wade, T.G., Vogt, P.: A national assessment of green infrastructure and change for the conterminous United States using morphological image processing. Landsc. Urban Plan. **94**, 186–195 (2010). doi:10.1016/j.landurbplan.2009.10.003

7. Spanò, M., Gentile, F., Davies, C., Lafortezza, R.: The DPSIR framework in support of green infrastructure planning: a case study in Southern Italy. Land Use Policy **61**, 242–250 (2017). doi:10.1016/j.landusepol.2016.10.051

8. European Commission: The multifunctionality of green infrastructure. Science for Environment Policy. In-depth Report (2012). http://ec.europa.eu/environment/nature/ecosystems/docs/Green_Infrastructure.pdf. Accessed 22 Mar 2017

9. Hansen, R., Pauleit, S.: From multifunctionality to multiple ecosystem services? A conceptual framework for multifunctionality in green infrastructure planning for urban areas. Ambio **43**, 516–529 (2014). doi:10.1007/s13280-014-0510-2

10. Chang, Q., Li, X., Huang, X., Wu, J.: A GIS-based green infrastructure planning for sustainable urban land use and spatial development. Procedia Environ. Sci. **12**, 491–498 (2012). doi:10.1016/j.proenv.2012.01.308

11. Kilbane, S.: Green infrastructure: planning a national green network for Australia. J. Landsc. Architect. **8**, 64–73 (2013). doi:10.1080/18626033.2013.798930

12. Mell, I.C.: Green infrastructure: concepts and planning. FORUM Ejournal, **8**, 69–80 (2008). http://research.ncl.ac.uk/forum/v8i1/green%20infrastructure.pdf. Accessed 22 Mar 2017

13. Arcidiacono, A., Ronchi, S., Salata, S.: Managing multiple ecosystem services for landscape conservation: a green infrastructure in Lombardy Region. Procedia Eng. **161**, 2297–2303 (2016). doi:10.1016/j.proeng.2016.08.831

14. European Commission, DG Environment: Interpretation manual of European Union habitats - EUR28 (2013). http://ec.europa.eu/environment/nature/legislation/habitatsdirective/docs/Int_Manual_EU28.pdf. Accessed 22 Mar 2017

15. Gómez-Baggethun, E., de Groot, R., Lomas, P.L., Montes, C.: The history of ecosystem services in economic theory and practice: from early notions to markets and payment schemes. Ecol. Econ. **69**, 1209–1218 (2010). doi:10.1016/j.ecolecon.2009.11.007

16. Peterson, M.J., Hall, D.M., Feldpausch-Parker, A.M., Peterson, T.R.: Obscuring ecosystem function with application of the ecosystem services concept. Conserv. Biol. **24**, 113–119 (2010). doi:10.1111/j.1523-1739.2009.01305.x

17. Boyd, J., Banzhaf, S.: What are ecosystem services? Ecol. Econ. **63**, 616–626 (2007). doi:10.1016/j.ecolecon.2007.01.002

18. Müller, F.: Indicating ecosystem and landscape organisation. Ecol. Indic. **5**, 280–294 (2005). doi:10.1016/j.ecolind.2005.03.017

19. Fisher, B., Turner, R.K.: Ecosystem services: classification for valuation. Biol. Conserv. **141**, 1167–1169 (2008). doi:10.1016/j.biocon.2008.02.019

20. Millennium Ecosystem Assessment: Ecosystems And Human Well-Being: A Framework for Assessment. Island Press, Washington, DC (2003)

21. Kumar, P. (ed.): The Economics of Ecosystems and Biodiversity: Ecological and economic Foundations. Routledge, New York (2011)

22. Milcu, A.I., Hanspach, J., Abson, D., Fischer, J.: Cultural ecosystem services: a literature review and prospects for future research. Ecol. Soc. **18**, 44 (2013). doi:10.5751/ES-05790-180344

23. Gürlük, S., Rehber, E.: A travel cost study to estimate recreational value for a bird refuge at Lake Manyas, Turkey. J. Environ. Manag. **88**, 1350–1360 (2008). doi:10.1016/j.jenvman.2007.07.017

24. Martín-López, B., Gómez-Baggethun, E., Lomas, P.L., Montes, C.: Effects of spatial and temporal scales on cultural services valuation. J. Environ. Manag. **90**, 1050–1059 (2009). doi:10.1016/j.jenvman.2008.03.013

25. Lankia, T., Kopperoinen, L., Pouta, E., Neuvonen, M.: Valuing recreational ecosystem service flow in Finland. J. Outdoor Recreat. Tourism **10**, 14–28 (2015). doi:10.1016/j.jort.2015.04.006

26. van Berkel, D.B., Verburg, P.H.: Spatial quantification and valuation of cultural ecosystem services in an agricultural landscape. Ecol. Indic. **37**, 163–174 (2014). doi:10.1016/j.ecolind.2012.06.025

27. Jobstvogt, N., Watson, V., Kenter, J.O.: Looking below the surface: the cultural ecosystem service values of UK marine protected areas (MPAs). Ecosyst. Serv. **10**, 97–110 (2014). doi:10.1016/j.ecoser.2014.09.006

28. Kenter, J.O.: Deliberative and non-monetary valuation: a review of methods. In: Laurence Mee Centre for People and the Sea, Working Papers 2014-02 (2014). http://www.sams.ac.uk/lmc/working-papers/kenter-valuation-review. Accessed 22 Mar 2017

29. Kelemen, E., García-Llorente, M., Pataki, G., Martín-López, B., Gómez-Baggethun, E.: Non-monetary techniques for the valuation of ecosystem service. In: Potschin, M., Jax, K. (eds.) OpenNESS Ref. Book. EC FP7 Grant Agreement no. 308428 (2014). http://www.openness-project.eu/sites/default/files/SP-Non-monetary-valuation.pdf. Accessed 22 Mar 2017

30. Eagles, P., McLean, D., Stabler, M.: Estimating the tourism volume and value in protected areas in Canada and the USA. George Wright Forum **17**, 62–76 (2000)

31. Wood, S.A., Guerry, A.D., Silver, J.M., Lacayo, M.: Using social media to quantify nature-based tourism and recreation. Sci. Rep. **3**, 2976 (2013). doi:10.1038/srep02976

32. Sonter, L.J., Watson, K.B., Wood, S.A., Ricketts, T.H.: Spatial and temporal dynamics and value of nature-based recreation, estimated via social media. PlosOne **11**, 1–16 (2016). doi:10.1371/journal.pone.0162372

33. Hausmann, A., Toivonen, T., Slotow, R., Tenkanen, H., Moilanen, A., Heikinheimo, V., Di Minin, E.: Social media data can be used to understand tourists' preferences for nature-based experiences in protected areas. Conserv. Lett. (2017). doi:10.1111/conl.12343

34. De Montis, A.: Measuring the performance of planning: the conformance of Italian landscape planning practices with the European Landscape Convention. Eur. Plan. Stud. **24**, 1727–1745 (2016). doi:10.1080/09654313.2016.1178215

35. De Montis, A.: Impacts of the European Landscape Convention on national planning systems: a comparative investigation of six case studies. Landsc. Urban Plan. **124**, 53–65 (2014). doi:10.1016/j.landurbplan.2014.01.005

36. Lovell, S.T., Taylor, J.R.: Supplying Urban Ecosystem Services through Multifunctional Green Infrastructure in the United States. Landsc. Ecol. **28**, 1447–1493 (2013). doi:10.1007/s10980-013-9912-y

37. CRITERIA Ltd. Consultants, TEMI ingegneria per la sostenibilità Ltd. Consultants: Monitoraggio dello stato di conservazione degli habitat e delle specie di importanza comunitaria presenti nei siti della Rete Natura 2000 in Sardegna. Definizione della rete di monitoraggio. Volume 2: Piano di monitoragio degli habitat e delle specie vegetali [Monitoring the conservation status of habitats and species of Community interest within Natura 2000 sites in Sardinia. Defining a monitoring system. Volume 2: Monitoring plan for habitats and plant species] (2014). Unpublished report

38. European Commission: Commission Implementing Decision of 11 July 2011 concerning a site information format for Natura 2000 sites (2011). http://eur-lex.europa.eu/legal-content/EN/TXT/PDF/?uri=CELEX:32011D0484. Accessed 22 Mar 2017

39. CRITERIA Ltd. Consultants, TEMI ingegneria per la sostenibilità Ltd. Consultants: Monitoraggio dello stato di conservazione degli habitat e delle specie di importanza comunitaria presenti nei siti della Rete Natura 2000 in Sardegna. Elaborazione rapporto di sintesi sullo stato di conservazione di habitat e specie (Linea 4, 4.c.1) [Monitoring the conservation status of habitats and species of Community interest within Natura 2000 sites in Sardinia. Synthesis report on conservation status of habitats and species (Deliverable 4.c.1)] (2014). Unpublished report

40. Camarda, I., Laureti, L., Angelini, P., Capogrossi, R., Carta, L., Brunu, A.: Il Sistema Carta della Natura della Sardegna. ISPRA, Serie Rapporti, 222/2015 [The Nature map of Sardinia. ISPRA, Report series no. 225/2015] (2015). http://www.isprambiente.gov.it/files/pubblicazioni/rapporti/R_222_15.pdf. Accessed 22 Mar 2017

41. CRITERIA Ltd. Consultants, TEMI ingegneria per la sostenibilità Ltd. consultants: Servizio di monitoraggio dello stato di conservazione degli habitat e delle specie di importanza comunitaria presenti nei siti della Rete Natura 2000 in Sardegna. Shapefile (Linea 6) [Tender: monitoring the conservation status of habitats and species of Community interest within Natura 2000 sites in Sardinia. Shapefiles (Deliverable 6)] (2014). Unpublished data

42. ISPRA: Tabelle delle corrispondenze in uso nel Sistema Carta della Natura [Correspondence table for codes used within the Nature map system] (2013). http://www.isprambiente.gov.it/files/biodiversita/Tabella_Corrispondenze_181213.xls. Accessed 22 Mar 2017
43. Newell, J.P., Seymour, M., Yee, T., Renteria, J., Longcore, T., Wolch, J.R., Shishkovsky, A.: Green Alley Programs: planning for a sustainable urban infrastructure? Cities **31**, 144–155 (2013). doi:10.1016/j.cities.2012.07.004
44. Meerow, S., Newell, J.P.: Spatial Planning for Multifunctional Green Infrastructure: Growing Resilience in Detroit. Landsc. Urban Plan. **159**, 62–75 (2017). doi:10.1016/j.landurbplan.2016.10.005
45. Landsberg, F., Ozment, S., Stickler, M., Henninger, N., Treweek, J., Venn, O., Mock, G.: Ecosystem services review for impact assessment: introduction and guide to scoping. WRI working paper, World Resources Institute, Washington DC (USA) (2011)

Assessing the Investments Sustainability After the New Code on Public Contracts

Manuela Rebaudengo[✉]🆔 and Francesco Prizzon🆔

Interuniversity Department of Regional and Urban Studies and Planning,
Politecnico di Torino, Viale Mattioli 39, 10125 Turin, Italy
{manuela.rebaudengo,francesco.prizzon}@polito.it

Abstract. The new Code on Contracts regulations reaffirmed the central role of assessment in procedures for public works construction. However, it has significantly distorted the sense of the previous rules where the preliminary investment assessment was essential and was implemented through the feasibility study. What happened, then, to the feasibility studies? Maybe the intentions of Legislative Decree no. 50/2016 were mixed, but the preliminary project and the feasibility study was erased, now at least formally. Why has the legislature felt the need to eliminate such an important study? So far was the feasibility study to identify the most proper management aspects, not the preliminary project, not the final draft or the executive project. The Italian tradition has unfortunately too often led to the creation of works with no real requirement to be satisfied without worrying about the sustainability of the operating costs. The only thought is on allocated resources available for construction. This has generated works partially unused, slightly used, and/or often obsolescent because the contracting authorities try to confirm that the asset portion is actually used while decreasing resources. This paper focuses on multi-annual programming, which has seen the most relevant changes while investigating the economic and procedural implications consequential to the new Code.

Keywords: Feasibility studies · Three-year programme · Sustainability assessment · New code on public contracts

1 Introduction

The new Code on Public Works, Services, and Supply Contracts regulations (Legislative Decree no. 50/2016) reaffirmed the central role of assessment [1] in the multi-annual programming, in the design of buildings, in the selection of tenders, and in the correct allocation of risks in public-private partnership transactions. However, it has significantly distorted the sense of the previous rules where the preliminary investment assessment was essential and was implemented through the feasibility study. What represented the feasibility study? As detailed into the following pages, the feasibility study was an ex-ante assessment showing the investment sustainability (i.e.

This paper is to be attributed in equal parts to the authors.

O. Gervasi et al. (Eds.): ICCSA 2017, Part VI, LNCS 10409, pp. 473–484, 2017.
DOI: 10.1007/978-3-319-62407-5_33

technical, economic and financial sustainability), drawn up before the preliminary project. It was a fundamental document in the whole PPP's process, especially in Project Financing. As you know, Project Finance is a long-term contract used to finance, by the private sector, the construction and the management of public works. Feasibility study was erased, now at least formally.

2 Assessment for the Multi-annual Programming Investment: The Signs Legislation

The old rule for the multi-annual programming of works (art. 128 of Legislative Decree no. 163/2006 as amended, hereinafter the Code on contracts [2]) regulated the use of the feasibility studies for the inclusion of interventions in the three-year programme and in the annual list. Paragraph 2 stated, in fact, that *the three-year programme is currently implementing feasibility studies*. The concept was better explained in the MD October 24, 2014 in which in subparagraph 2 describes the inclusion in the three-year programme of each intervention amounting equal to or less than 10 million Euros. *These interventions were prepared synthetic studies* within the meaning of Article 14, subparagraph 1 of Presidential Decree no. 207/2010 as amended, hereinafter Implementing Regulation [3].

For operations involving an amount exceeding 10 million Euros, *there are feasibility studies* according to Art. 14, sub paragraph 2. of Implementing Regulation. The subparagraph 6, then contains rules for the public work inclusion in annual list: for those less than 1 million euro, it requires the prior approval of at least a feasibility study. For those equal to or more than 1 million euro at least the preliminary project […] except for maintenance work for which it is enough to mention the interventions accompanied by the summary estimate of costs. For the work referred to in Article 153 (Project Finance) the feasibility study is enough (requirement introduced in 2008 by the 3rd corrective to the Code).

The Legislative Decree no. 50/2016 [4] (hereafter the New Code of Contracts) governs the initial phase. This is called now *the program of acquisitions of contracting*, in art. 21 subparagraph 3: *for work amounting equal to or more than one million euro for inclusion in the annual list contracting authorities previously approved the technical and economic feasibility project*. Nothing is specified for inclusion in the three-year programme.

What happened, then, to the feasibility studies? Maybe the intentions of Legislative Decree no. 50 were mixed, but the preliminary project and the feasibility study was erased. Now at least formally. Let's see what the long-awaited Infrastructure, Environmental Protection and Cultural Heritage MD, has on the contents of the three design levels referred to the art. 23 subparagraph 3.

The historical breakdown into the preliminary project, final design and executive project was present from the Merloni Law [5] (no. 109/1994 and subsequent amendments). This was well-established both in operational terms and in term of content.

The two words *feasibility study* appear in the Legislative Decree no. 50 only three times to art. 46 (*Economic Operators for architectural and engineering services* - about how to show requirements to submit a tender). The two words *preliminary project* have disappeared altogether. The words *technical and economic feasibility project* and *feasibility project* remain. Will there be a difference? Maybe not, although it could indicate a different level of detail. Perhaps the document consists of two studies drawn up in two successive phases.

Why has the legislature felt the need to eliminate such an important study? So far was the feasibility study to identify the most proper management aspects, not the preliminary project, not the final draft or the executive project. This was a feasibility study to determine private palatability for the construction of works and its management. This identifies the attractiveness of a concession contract and not a procurement contract.

3 The Role of Feasibility Studies Before the Approval of Legislative Decree No. 50/2016

National legislation has been about feasibility studies since 1994 when they appear in the text of the Merloni Law. Art. 14 stipulates that the *three-year programme is currently implementing feasibility studies and identification and quantification of needs* [...]. A strong player was the introduction (by Law 144/1999) of *Assessment and Verification of Public Investment Units* as well as the provisions of Article. 4 according to which the *feasibility study for works amounting more than 20 billion lire is the normal preliminary tool for the recruitment of investment decisions by public authorities.*

The subsequent Decree of the Ministry of Public Works no. 5374 of 21 June 2000 [6] establishes different levels of detail for the first time depending on the amount of the work: [...] *To interventions amounting less than 20 billion lire, subjects [...] shall prepare synthetic studies [...]; to interventions amounting more than 20 billion lire subjects [...] shall prepare feasibility studies.*

The *"Guide for certified studies"* written in 2001 by the Assessment and Verification of Public Investment Regional Units [7], updated in 2003, clearly defines the minimum contents and the structure of the studies for their certification. The Guide identified macro areas of analysis for the preliminary verification of feasibility: (a) *technique*; (b) *economic and social*; (c) *financial*; (d) *environmental*; and (e) *administrative – procedural* feasibility.

Overall, we can say that since 2003 the regulations on the feasibility studies has remained essentially unchanged until December 2008 when Legislative Decree no. 152 [8] modified the discipline of the works—especially for Project Financing procedures involved the role of the feasibility studies.

In 2009 the Regulatory Authority of Public Contracts published the Determination no. 1 *"Guidelines for the preparation of a feasibility study"* [9] enhancing its function in particular for public-private partnership operations.

Finally, in 2010, the Implementing Rules of the Code of Contracts established the contents into binding normative terms (art. 14). Subparagraph 1 presents contents of a study to be used for programming. Subparagraph 2 shows the contents of a study to be used for tender base.

The following figure identifies the minimum contents related to the feasibility study at different time points (1999; 2001/2003; 2010) (Fig. 1).

CIPE delibera no. 106/1999 CIPE delibera no. 135/1999	Guide for certified studies 2001 as amended in 2003	Implementing Rules 207/2010 article 14 subparagraphs 1		Implementing Rules 207/2010 article 14 subparagraphs 2
territorial and socio-economic framework	preliminary analysis and possible alternatives analysis	functional, technical, operational, economic and financial characteristics	EXPLANATORY REPORT	territorial and socio-economic framework
target groups current and expected demand analysis	technical sustainability	possible alternatives analysis, compared to the identified solution	EXPLANATORY REPORT	current and expected supply-demande analysis
current and planned supply analysis	environmental performance	detailed investigation to check the realization through public private partnership contracts	EXPLANATORY REPORT	design alternatives analysis
possible alternatives analysis	financial sustainability	present status quo analysis	EXPLANATORY REPORT	environmental impact preliminary analysis
management costs during operation analysis	social and economic sustainability	work requirements for environmental sustainability and landscape compatibility	TECHNICAL REPORT	functional and technical characteristics
discounted cash flow analysis	procedural check		TECHNICAL REPORT	work requirements for environmental sustainability and landscape compatibility
cost benefit analysis	sensitivity and risk analysis		TECHNICAL REPORT	summary analysis of construction techniques
environmental impacts analysis			TECHNICAL REPORT	timeline
			TECHNICAL REPORT	preliminary estimate of costs
			TECHNICAL AND ECONOMICAL REPORT	detailed investigation to check the realization through public private partnership contracts
			TECHNICAL AND ECONOMICAL REPORT	business plan
			TECHNICAL AND ECONOMICAL REPORT	cost-benefit analysis
			TECHNICAL AND ECONOMICAL REPORT	pricing scheme
			TECHNICAL AND ECONOMICAL REPORT	preliminary agreeement schemes
			PLANS	

Fig. 1. Regulatory requirements and feasibility studies contents

As mentioned, the Presidential Decree no. 207/2010 distinguished between minimum content required in cases where the feasibility study is placed based on tender (art. 14, sub 2) and general content (art. 14, sub 1) for three-year programme.

Thus, there was a Piedmont Region adopting in 2012 the *Regional Guidelines on Feasibility Studies* [10]; the will to regulate locally the Feasibility Studies derived from the belief that the increasing scarcity of resources available to public administrations always requires the use of prior verification tools especially in defining the implications (technical, economic and procedural) that characterize the management of public works.

The document was drawn up in line with the national regulatory framework and must be understood as a reference tool to be followed for the preparation of all feasibility studies of public works or public interest in cases where both affected the Piedmont Region. It was also adopted by the Italian Institute for innovation and transparency of procurement and environmental compatibility (ITACA) [11] and it shall be consulted by national Authority on public contracts in drawing up some guidelines and deliberations.

In the past, the Feasibility Study has evaluated and selected the government assistance or public-private to finance. For example, in the case of Piedmont Region, different calls from 2004 to 2007 [12–14] involving different Regional Directorates were evaluated for intervention projects and their contribution [5–7]. For example, the 2006 assessment, identifies each requested insight (1.3, 2.1, 2.2, 4.1, 4.2, 4.3, 4.4, 5.1, 6.1, 6.2, 7.1, 7.2) related to the (at the time informal) Regional Guidelines on Feasibility Studies and some scores (Fig. 2) were obtained from feasibility studies prepared for admission to the contributions (Fig. 3).

CONTENS	1.3	2.1	2.2	4.1	4.2	4.3	4.4	5.1	6.1	6.2	7.1	7.2
max	4,00	2,00	2,00	4,00	4,00	8,00	8,00	4,00	2,00	2,00		
min	0,00	0,00	0,00	0,00	0,00	0,00	0,00	0,00	0,00	0,00		
average	2,60	1,48	1,61	1,85	1,72	3,78	3,79	2,07	1,33	1,27		
median	3,00	1,50	2,00	2,00	2,00	4,00	4,00	2,00	1,50	1,00		
std. dev.	1,08	0,54	0,52	1,37	1,33	2,14	2,34	1,34	0,61	0,62		
score [FS_C]	65%	74%	81%	46%	43%	47%	47%	52%	66%	64%		

CONTENS	1.3	2.1	2.2	4.1	4.2	4.3	4.4	5.1	6.1	6.2	7.1	7.2
max	4,00	2,00	2,00	4,00	4,00	6,00	8,00	4,00	2,00	2,00	1,00	1,00
min	0,00	0,00	0,00	0,00	0,00	0,00	0,00	0,00	0,00	0,00	0,00	0,00
average	2,62	1,52	1,59	1,97	1,72	2,97	3,84	2,17	1,27	1,18	0,26	0,21
median	3,00	1,50	1,75	2,00	2,00	3,00	4,00	2,00	1,00	1,00	0,00	0,00
std. dev.	1,2	0,52	0,5	1,35	1,4	1,79	2,29	1,29	0,55	0,67	0,38	0,35
score [FS_B]	66%	76%	80%	49%	43%	49%	48%	54%	63%	59%	26%	21%

CONTENS	1.3	2.1	2.2	4.1	4.2	4.3	4.4	5.1	6.1	6.2	7.1	7.2
max	4,00	2,00	2,00	4,00	4,00	5,00	7,00	2,00	2,00	1,50	2,00	2,00
min	0,00	0,00	0,00	0,00	0,00	0,00	0,00	0,00	0,00	0,00	0,00	0,00
average	2,29	1,60	1,40	1,79	1,46	2,21	2,58	0,77	1,10	0,83	0,63	0,33
median	2,00	1,75	1,50	1,50	1,00	2,50	3,00	0,75	1,00	1,00	0,75	0,00
std. dev.	1,48	0,51	0,62	1,70	1,56	1,56	2,07	0,70	0,52	0,44	0,64	0,62
score [FS_A]	57%	80%	70%	45%	36%	44%	37%	39%	55%	55%	32%	17%

Fig. 2. Evaluation of feasibility studies (2006): scores analysis

	basic Tab	basic FS	simplified FS	complete FS
	size range less than 800.000€	size range of 800.001€ to 2.000.000€	size range of 2.000.001€ to 10.000.000€	size range exceeding 10.000.001€
studies no.	778	150	180	12
max	40,00	37,00	40,00	32,25
min	2,00	1,00	1,00	6,00
average	23,91	21,51	21,28	17,00
median	24,00	21,00	21,75	16,88
std. dev.	7,50	8,71	8,63	8,33

Fig. 3. Assessed feasibility studies during the 2006 regional call: synthetic scores analysis

The most interesting information in Fig. 2 is the evaluations (very low) of work management contents: the grey table cells show scores insufficient on sustainability assessment. Contents must be always more than 50%; in all three size the lowest scores are from 4.1 to 5.1 contents, all referred to management of work during operation [14, 15].

Perhaps the professionals are very good at designing the work, but not at foreseeing the management implications—both in terms of costs and in operational terms. The feasibility study has always had this function: to investigate management costs in advance for a greater awareness of the costs to be incurred over the life of the asset. Thus far, it has always suggested that a substantial investment is the initial one, while the reverse is true. The operating costs (such as utilities, maintenance, staff, etc.) correspond to about 4 times [16, 17] the initial investment in the work lifetime!

To those who say that the FS legitimizes the interventions, it can be answered that more often the actions are dictated by political needs than the services needed to satisfy while still implemented regardless of the feasibility study.

Waiting on the MD on the project levels, the article no. 14 of Presidential Decree 207/10 on the feasibility studies is still in force… but studies do not exist anymore!

At the moment, one can only think that the issue of assessing the sustainability of investments is delegated to the technical and economic feasibility project yet one does not yet know in detail what it is. Perhaps an "old" preliminary project on which the feasibility study content will be added? Or a feasibility study with more technical content of the past? This second possibility was already recognized both by the Authority and the Regulations since they are the responsible official and could request that the feasibility study were attached to design documents normally drafted in the preliminary project.

What is certain is that the feasibility study was never a design level—the contents have always been different but essential for the good development of the three successive level of project.

However, as if to confirm this second chance, please note that in the draft law for the transposition of European Directives on public procurement and concession was included among the delegation criteria. The feasibility study, while continuing to represent document place based on tender, should have a higher level of detail than provided by law. On the one hand, if this arrangement aims at reducing the frequency of overly simplified preliminary projects with little content—especially for its listing in the three-year programme—then the other requires more content, and it will cost more and will cause problems to the contracting authorities that, due to scarcity of resources, is difficult to proceed to external commitment.

Article no. 23, subparagraph 5 includes discussions about design levels: the technical and economic feasibility project identifies, among several solutions, which one offers the best balance between costs and benefits for the community. A project, at least as we used to think in technical terms, can only be referred to a timely intervention; it says here that it identifies the solution that has the best balance between costs and benefits.

This suggests that the feasibility study has just changed its name and unfortunately incorporated in the first design level. In terms of content, it is clear that the reference to the issues of multi-criteria and of cost-benefit analysis both provided insights to art. 14 of the old Implementing Regulations.

Provocatively, should the feasibility study have to select an intervention based on a single criterion that is assessable through benefit-costs ratio (BCR) indicator maximization?

4 Unfinished Works in Italy

The Italian tradition has unfortunately too often led to the creation of works with no real requirement to be satisfied without worrying about the sustainability of the operating costs [1, 18]. The only thought is on allocated resources available for construction. This has generated works partially unused, slightly used, and/or often obsolescent because the contracting authorities try to confirm that the asset portion is actually used while decreasing resources.

What are the *unfinished works*? According to art. 44-bis of the Legislative Decree 201/2011, these are works whose construction had been started but had not been completed or is not usable by the community for at least one of the following causes: (a) lack of funds; (b) technical reasons; (c) supervening new technical regulations or laws; (d) the bankruptcy of a contractor; (e) lack of interest at the completion by the

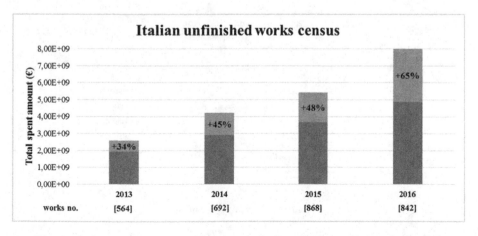

Fig. 4. The unfinished works: national analysis from 2013 (numbers and amounts)

Contracting Authority [18]. Subsequently, in March 2013, the Ministry of Infrastructure and Transport published the MD no. 42/2013 to regulate the action of a census-register of unfinished public works. It provides that (art. 3, c.1) by March 31 of each year should be drawn up by each contracting authority where they list the works that are subsequently published by June 30 of that year.

The national situation is that one synthetically represented in Fig. 4 (national breakdown by year), in Fig. 5 (national breakdown by year) and Fig. 6 (last census).

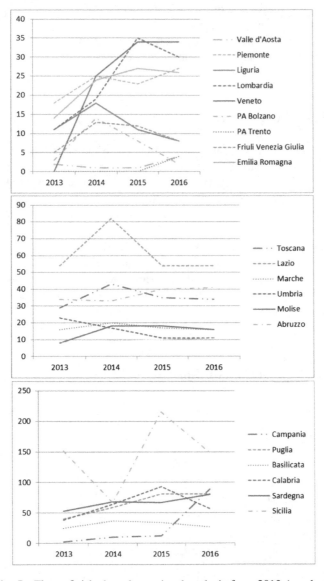

Fig. 5. The unfinished works: regional analysis from 2013 (numbers)

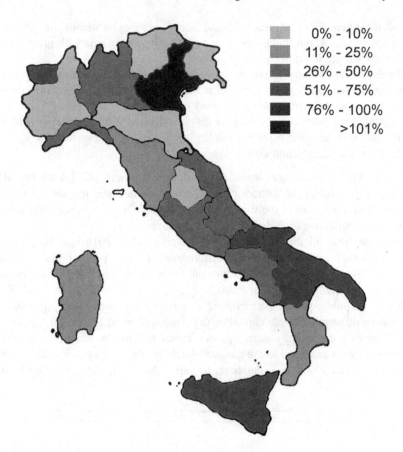

	0% - 10%
	11% - 25%
	26% - 50%
	51% - 75%
	76% - 100%
	>101%

Fig. 6. The unfinished works last census by region (% of spent amounts to completion)

5 Unfinished Works and Three-Year Programme: What Kind of Sustainability?

Further complicating the issue of evaluation in the process of programming of the works, the legislator has decided to facilitate the completion of unfinished works, introducing the theme to art. 21 subparagraph 2: the unfinished public works are included in the three-year programme [...] for the purpose of their completion or for the identification of alternative solutions such as reuse, even resized, supply as remuneration for the construction of another public work, the sale or the demolition.

These contents are a real assessment concentration:

(A) The cost estimate for completion of the work (are sufficient and reliable the indications contained in published lists? There were more than less 800 works surveyed at the end of 2015 (census 2016) without considering those under State jurisdiction. These correspond to about over 3 billion euro spent but they need at least 1.5 billion euro for completion;

(B) The identification of alternative reuse solutions (maybe means choice between several solutions, but this is certainly a more complex theme with greater interest than the simple identification;

(C) The definition of a possible downsizing (here, to use the old terminology, it would be useful a feasibility study to consider the technical and management aspects of a change of use or partial use of the work);

(D) Estimating the asset value, assuming that it is used as remuneration or sold, but at the same time also the asset palatability according to its transformability;

(E) The estimated demolition costs.

Considering that the current legislation is not the previous one (in the old MD on programming you read: the authorities *shall take into account* the same (unfinished works) for the preparation of the three-year programme. Now, however, the new code states that the works *are included in the three-year programme.*

This means that all the unfinished works surveyed in 2016 will be part of the next three-year programme, and this will require a large part of the assessments mentioned above. It seems very difficult to put aside the causes that led to their failure to complete—primarily the lack of funds.

What of the public-private partnership? This requires a careful analysis looking for works that could attract private capital in the construction and/or in the management. Figure 7 shows a preliminary scenario, first result of sample analysis conducted on some Regions, then extended to the national database. Under way, some more detailed analysis on last census national database, to verify the assumption of figure below.

	[A]	[B]	[C]	[D]	[E]	
	COMPLETION	ALTERNATIVE REUSE	DOWNSIZING	SALE OR SUPPLY AS REMUNERATION	DEMOLITION	TOTAL
census 2016 (no. of work)	50,00%	10,00%	20,00%	15,00%	5,00%	100,00%
which in PPP	5,00%	50,00%	20,00%	1,50%	2,50%	11,85%
census 2016 (needed amount)	€ 757.343.209	€ 189.335.802	€ 242.349.827	€ 113.601.481	€ 75.734.321	€ 1.378.364.641
which in PPP	€ 37.867.160	€ 94.667.901	€ 48.469.965	€ 1.704.022	€ 1.893.358	€ 184.602.407
						13,39%

Fig. 7. Potential reuse and completion (nationwide)

Even the public-private partnership has important assessments. Until now, all depth is through the feasibility studies—primarily those concerning the cost-effectiveness, financial sustainability and risk assessment. If the first two issues are verified and verifiable through the business plan, then the determination and the assessment of risks and also the verification of appropriately allocation are still partially new themes.

The Authority Determination no. 10 of 2015 was already diffusely occupied, and it treated the theme of the value for money definition that was calculated through the public sector comparator, for example. In technical terms, the psc can be defined as a hypothetical realization cost adjusted with a risk component. According to this technique, the true cost of realization of the public body is given by the net present value of costs (VANc) plus the net present value risk (VANr). If the private partner can minimize the risks transferred to him by the contracting authority—for example through the respect of the construction budget, construction time, maintenance costs, etc.—then the administration will choose a concession contract and not a procurement contract. Until six months ago, a proper risk assessment can be accomplished through the development of a risk matrix to build during the preparation of the feasibility study. This now offers good chances in the technical and economic feasibility project.

6 Final Thoughts

Published data from the first corrective code is expected by the end of April 2017. The decree on the design level will hopefully change the terminology to correspond to a change in content, but the insights related to the economic and financial sustainability continue to be central to the planning of public investment and in the case of the use of public-private partnership.

In monetary terms, the preparation of a feasibility study costs about 1/3 of an "old" preliminary draft, and it takes about half the time required for the first phase of the project. If all these activities would in fact be postponed to the first planning level (technical and economic feasibility of the project), then there would be a concentrated increase in the single initial solution costs.

Hopeful the first corrective code (or the expected MD on level contents?) will explain what kind of insights are needed to be effective the programming phase. Certainly we're expecting a formal amendment, not a substantial change, because make disappear a such relevant follow-up, appears to be unwisely!

References

1. Prizzon, F., Rebaudengo, M.: Public investments evaluation: what lies ahead? click day vs. selection. LaborEst n. 10 (2015). ISSN 1973-7688. ISBN 978-88-72-21-380-3. ISSN 2421-3187 (online)
2. Ordinary Supplement to GURI no. 100 of 2 May 2006
3. Ordinary Supplement to GURI no. 288 of 10 December 2010
4. Ordinary Supplement to GURI no. 10 of 19 April 2016
5. Ordinary Supplement to GURI no. 41 of 19 February 1994
6. Ordinary Supplement to GURI no. 148 of 27 June 2000
7. http://www.regioni.it/fascicoli_conferen/Presidenti/2001/20010308/08012001_studi_fat.htm
8. Ordinary Supplement to GURI no. 227 of 2 October 2008
9. http://www.anticorruzione.it/portal/public/classic/AttivitaAutorita/AttiDellAutorita/_Atto?ca=3665
10. Regional Executive Order (DGR) 28 March 2012
11. http://www.itaca.org/documenti/news/LG%20ITACA%20SDF_Completo_240113.pdf

12. Prizzon, F., Rebaudengo, M., Taccone, G., Talarico, A.: La programmazione di opere pubbliche in piemonte: il caso dei programmi integrati di sviluppo locale (P.I.S.L.). Paper in Atti della XXVII Conferenza Italiana Di Scienze Regionali (AISRe), Pisa (2006). ISBN 88-87788-07-3

13. Rebaudengo, M.: I P.I.S.L. della Regione Piemonte: Proposta di classificazione tematica attraverso l'analisi del discriminante. Paper in atti della XXVIII Conferenza Italiana Di Scienze Regionali (AISRe), Bolzano (2007). ISBN 88-87788-08-1

14. Prizzon, F., Rebaudengo, M., Taccone, G., Talarico, A.: Il ruolo degli studi di fattibilità per la selezione dei programmi integrati: il caso dei P.I.S.L. della regione piemonte. Paper in atti della XXVIII Conferenza Italiana Di Scienze Regionali (AISRe), Bolzano (2007). ISBN 88-87788-08-1

15. Rebaudengo, M., Taccone, G.: Esperienze di programmazione regionale a confronto: PISL e PTI della regione piemonte. Paper in atti della XXX Conferenza Italiana Di Scienze Regionali (AISRe), Firenze (2009)

16. Haroun, A.E.: Maintenance cost estimation: application of activity-based costing as a fair estimate method. J. Q. Maint. Eng. **21**(3), 258–270 (2015). doi:10.1108/JQME-04-2015-0015

17. Higgins, L.R., Mobley, R.K., Wikoff, D.: Estimating repair and maintenance costs. In: Maintenance Engineering Handbook, 7 edn. McGraw-Hill Professional (2008, 2002, 1995, 1988, 1977, 1966, 1957). AccessEngineering

18. Prizzon F., Rebaudengo M.: Unfinished public works: a national heritage to develop? LaborEst n.11 (2015). ISSN 1973-7688. ISBN 978-88-72-21-380-3. ISSN 2421-3187 (online)

Quality of Life and Social Inclusion of Inland Areas: A Multidimensional Approach to Performance Policies and Planning Assessment

Carmelina Prete[(✉)], Mario Cozzi, Mauro Viccaro,
and Severino Romano

SAFE - School of Agricultural Food Forestry and Environmental Sciences,
University of Basilicata, 85100 Potenza, Italy
carmelina.prete@unibas.it

Abstract. Improving quality of life and social inclusion is one of the priorities of national and community policies. Adopting the approach based on capabilities-functioning, the aim of this paper is to measure a *Quality of Life* (*QoL*) index of communities, selecting the specific variables that may influence quality of life of inland areas, such as: economic opportunities, health care, education, cultural and leisure activities, work-life balance, health and environmental protection. A *QoL* index for three dimensions (economic, social and environmental) and a global *QoL* were calculated using a non-compensatory method. The values obtained are included in the 70–130 range. The model, applied to the Basilicata region (131 municipalities), takes values in a range between 93 and 105, with 61% of municipalities with a global *QoL* below the regional average (=100). It tends to assume lower values in inland areas: 62% of inland areas are characterized by a global *QoL* below the regional average, due in part to fewer economic opportunities and social services, but also to the presence of major landslides and seismicity risks, against a greater health and environmental protection. The opportunity to assess the quality of life through an index, over time, may help *policy makers* addressing policies and evaluating their effects. Furthermore, an analysis of spatial autocorrelation helps identify different clusters and spatial outliers, useful for the identification of areas requiring priority interventions and future actions, which should take into account a balanced growth of the economic, social and environmental dimensions related to the quality of life.

Keywords: Quality of life · Social inclusion · Multidimensional approach · Non-compensatory method · Spatial autocorrelation

1 Introduction

Inland areas, which are predominantly rural, have suffered, especially in the last 50 years, a process of marginalization, which has led to a gradual decline in employment and productivity, reduced share capital, land abandonment, and consequently, the loss of soil protection and a landscape modification. This contributed to public negative

© Springer International Publishing AG 2017
O. Gervasi et al. (Eds.): ICCSA 2017, Part VI, LNCS 10409, pp. 485–500, 2017.
DOI: 10.1007/978-3-319-62407-5_34

stereotyping considering inland areas as "peripheral" zones subject to a negative relationship between centre and periphery, which concerns access to services and other economic opportunities, social interaction, culture [1].

To contrast this process of marginalisation, the National Strategy for the Internal Areas (SNAI) was launched in 2012, with focus towards:

- an *intensive* development, with the increase in well-being and social inclusion of those living in these areas;
- An *extensive* development, with the increase in labour demand and utilization of territorial capital.

Therefore, SNAI is set up as a long-term strategy, with a territorial value, to improve social inclusion in a number of multidimensional results, through the provision of public goods and services, by first ensuring socially shared essential standards for everyone and then improving the welfare of less advantaged groups [2].

This means identifying an approach for the evaluation of the performance of territorial type, according to a multidimensional approach. Many features related to people's lives actually depend on the territory in which they live that influences the "socially acceptable level" and implies inequality polarization at a regional scale, thus making it necessary and desirable to carry out a more detailed analysis, with a spatial value.

On the basis of the issues and strategies discussed, the objective of this work is to measure the *Quality of Life (QoL)* of people at a local level, based on the existing opportunities in the territory. This is feasible through an approach that provides for the selection and manipulation of variables that can influence the quality of life of inland areas [3, 4], such as: economic opportunities, health care services, education, cultural and leisure activities, work-life balance, health and environmental protection.

This approach can be referred to the theory proposed by Sen [5], based on Capabilities-Functionings. This theory leads us from the space of goods, income, utilities towards the space of the building blocks of living, functionings, which are the set of actions and conditions that affect lives (health, education, nutrition, etc.), and capabilities representing the set of functionings that can be reached on the basis of the type of life, given by opportunities and freedom of choice [6]. Many authors [7, 8] argue, in fact, that conventional measures based on income, wealth and consumption, are insufficient to assess human well-being, since they exclude a large category of key factors, such as environment, health, social inclusion, etc. In fact, the Stiglitz Report [9] laid the foundations for a multidimensional approach to the welfare estimates. This approach is generally regarded as a significant enrichment for policy analysis; on the other hand, there is no consensus on how to define the most appropriate multidimensional space. Following different paths, several studies attempt to calculate an index of quality of life starting from the existing opportunities in the territory [10, 11]. In recent years, there has been growing interest in the compilation of composite indicators of well-being at the local level [12–14].

Based on the above, the innovative element of the research lies in the implementation of a methodological framework able to obtain real-world summary data (in terms of quality of life), by using a non-compensatory method to aggregate variables, and to enhance the territorial units in which indicators have balanced levels of performance in the three dimensions (economic, social and environmental) of sustainable development. This aggregation function indicates in some detail the imbalances (highlighted by means of a spatial autocorrelation analysis), so as to to verify, in the specific case, the existence of functional relations between what happens in a specific location of the space and what happens in other positions. This enables the identification of the outliers and "homogeneous" areas, characterized by a favourable condition, (in a positive sense) or by marginality (in a negative sense), compared to the general condition (in this case the regional one).

The results represent a tool to assist *policy makers*, within a general concept of equalization of quality of life, to have a redistribution of resources aimed at levelling out imbalances among territories and encouraging social inclusiveness.

2 Materials and Methods

2.1 Model Implementation

The assumed model is based on the relationship between the level of quality of life of the individuals living in the *i-th* municipality (QoL_i) and the level of existing opportunities in a given area (t_r), including the services s_r provided in the *i-th* area.

The basic assumption is that the individual well-being may be expressed as:

$$QoL_i = f(\bar{y}, t_r) \tag{1}$$

Where $t_r = f(s_r)$, $\bar{y}$ is the vector of individual conditions (employment, gender, etc.) that is considered exogenous to the model.

The indicators that most contribute to define levels of *QoL,* are important to emphasise the territorial disparities in well-being [3, 4], depending on the availability of data at the required level of detail, which is quite high in the present analysis. The dataset applied to develop the model includes a set of **basic indicators** derived from different sources (National Institute of Statistics - ISTAT, property market Observatory, regional technical map - CTR, Higher Institute for Environmental Protection and Research - ISPRA, river basin authority, etc.) that have been grouped in **thematic areas** and further categorised based on the relevant **dimensions** (economic, social and environmental) (see Fig. 1) (Appendix – List of indicators included in the model).

The economic dimension is meant as the level of wealth owned by a single individual or a population; you can express it in both income and equity terms. The

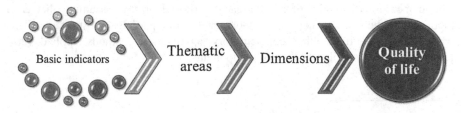

Fig. 1. Grouping process

indicators - relating the economic dimension - concern the number of bank branches and the average estate prices as proxy of the economic well-being and of the economic opportunity of an area. Indeed, the assumption is that the number of bank branches in a municipality is proportional to the population and to the amount of operating volumes (loans and deposits). The average estate prices of the last five years reflect the economic dynamism of an area and depend, for instance, on population trends and on the level of the "services and quality" provided [15].

The concept of social sustainability, defined as the ability to guarantee human welfare conditions (safety, health, education) equally distributed, was considered in terms of spread and proximity to services/facilities/activities that exercise a decisive influence not only on the everyday life organisation of a community, but also on its mobility and degree of external dependence. The presence of healthcare settings is an essential condition influencing citizens' security, or their possibility to receive preventive care services and appropriate treatment. These services are widespread, although access to them may vary for the citizens of different municipalities. Other factors were included, such as the spread and proximity of education services, recreational facilities (camping sites, sports structures, playgrounds) and cultural activities (libraries, cinema, museums, theatres, etc.), non-decentralised departments (courts, chambers of commerce, etc.).

To take into account proximity, the travel time to reach different services was calculated by the isochrones method, using GIS, via the Network Analysis [16]. Among daily trips that influence the organisation of everyday life, those related to work or study were shown to be prevailing, so they were used to derive the homework mobility rate and the mean journey time.

The environmental dimension is expressed in the form of natural capital and/or natural heritage[1], to be understood in the ability to provide essential goods and services for human well-being. Consequently, in order to outline this dimension, the following indicators were considered: population equivalents [18] that reflect the estimated pollutant load produced by domestic and economic activities; the proximity to waste dumps and industrial areas that may affect the environmental health; the availability and extent of areas characterized by high ecological-natural value; and the presence of factors of environmental risk (hydro-geological and seismic risks).

2.2   Aggregation of Indicators by a Non-compensatory Method

A non-compensatory approach, i.e. the method of the coefficient of variation penalty [19], was applied in order to develop the composite indicator. This method enables a synthetic measure of quality of life for each territorial unit x_i, assuming that each component of the QoL is not substitutable or is only partially substitutable. This approach, different from other compensatory aggregation methods applied [20], requires a balanced supply of all basic components.

---

[1] In economic terms, natural capital can be seen as a resource to manage and increase, while the inheritance is the resource within the transmission concept [17].

The method involves standardising indicators by using a transformation criterion to release them from their units of measurement and variability [21]. Therefore, basic indicators have been corrected so as to be ranged within the same scale, by transforming each indicator in a standardised variable with an average of 100 and a mean square deviation of 10; the values obtained will be approximately comprised within the range 70–130.

Thus, once the matrix $X = \{x_{ij}\}$ of n rows (territorial units) and m columns (basic indicators) was constructed, the next step was the matrix $Z = \{z_{ij}\}$:

$$z_{ij} = 100 \pm \frac{(x_{ij} - M_{x_j})}{S_{x_j}} 10 \tag{2}$$

Where $M_{x_j} = \frac{\sum_{i=1}^{n} x_{ij}}{n}$ is the average and $S = \sqrt{\frac{\sum_{i=1}^{n}(x_{ij} - M_{x_j})^2}{n}}$ is the mean square deviation.

Then, the aggregation function (Mazziotta-Pareto index - MPI) was "corrected" by a penalty coefficient that depends, for each territorial unit, on the degree of variability of indicators from the mean value ("horizontal variability").

$$MPI_i^{+/-} = M_{z_i} \pm S_{z_i} cv_i \tag{3}$$

The arithmetic mean (M_{z_i}) of standardised indicators is corrected by subtracting an amount (the product $S_{z_i} cv_i$) proportional to the mean square deviation, and is a direct function of the coefficient of variation.

This variability, measured by the coefficient of variation (cv_i), penalises the scoring of the units with the highest imbalance between the values of indicators and, hence, an imbalanced supply. The use of standardised deviations (S_{z_i}) enables a robust measure that is not influenced by the elimination of a single basic indicator [19]. The main disadvantage lies in the possibility of making only 'relative' comparisons of the values of units over time, with respect to the average. The method has been applied to calculate the QoL for each dimension: economic (EcQoL), social (SocQoL), and environmental (EnvQoL) - and then to calculate a global QoL (TotQoL) that takes into account all basic indicators.

2.3   Spatial Autocorrelation Analysis

To give to research a specific imprint referred to the geographical context and territorial dynamics, the analysis was accompanied by spatial processing in order to further contribute to the knowledge and research in the area. Thus, although the examination of maps may lead to the rational conclusions on the presence (or absence) of spatial dependence, this analysis enables some conclusions to be drawn, independently of the technique used for the representation (for example in the identification of classes). So in this paper we explore a procedure for creating statistically robust hot/cold spot maps. It is possible apply several methods, that include the application of point pattern analysis techniques to identify for spatial clustering, spatial dispersion, spatial autocorrelation, and Local Indicators of Spatial Association (LISA).

A primary step that is often avoided but is fundamental to the detection of clusters of points/features are global tests, that indicate if clustering and dipersion exist in the original point/feature distribution. For example, whether the features show evidence of clustering or are randomly distributed, and how dispersed the distribution is within the three dimensions of QoL. There are several approaches for analysing a point or feature distribution for spatial randomness. Most of them incorporate the basic principles of hypothesis testing and classical statistics, where the initial assumption is that the point/feature distribution is one of complete spatial randomness (CSR). By setting the CSR assumption as the null hypothesis the point distribution can be compared against a set significance level to accept or reject the null hypothesis. So several techniques and algorithms have been developed and are used in practise for the generation of hotspot maps, used mainly for creating hotspot maps of crime [22, 23], all of which have different merits. These mainly relate to their ease of use, application to different types of events, visual results and interpretation.

In particular, in the case of aggregate counts within a certain geographic area, the use of the spatial autocorrelation technique, Moran's I, is suggested to test for clustering [23]. The spatial autocorrelation analysis expressed the spatial concentration of similar values (in the case of positive interdependence) or different values (in the case of negative interdependence). Therefore, the geo-referenced dataset was tested for the presence of global spatial autocorrelation, Moran's I [24], considering the entire set of observations, indicates the trend of analysed data to focus (or less) in space. Moran's works by comparing the value at any one location with the value at all other locations. The significance of the results can then be tested against a theoretical distribution (one that is normally distributed) by dividing by its theoretical standard deviation [23].

Furtheremore LISA statistics have been described as being particurarly suited to identyfing crime hotspots, but they are applied also to different contexts [25–29]. LISA statistics assess the local association between data by comparing local averages to global averages. For this reason they are useful in adding definition to hot/cold spots, and placing a special limit on these areas of highest/lower concentration [23]. Therefore the presence of local spatial association was also tested within the analyzed data, for characterizing geographically the area with different types of correlation. Five distinct situations were detected:

- Hot spots: observations with a high value of the variable under study with similar neighbours (if high-high);
- Cold spots: observations characterized by a low value of the variable under study with similar neighbours (if low-low);
- Spatial outliers: observations with a high value of the variable under study but with neighbours characterized by low values for the same variable (if high-low);
- Spatial outliers: observations characterized by a low value of the variable under study but with neighbours characterized by high values for the same variable (case low-high);
- Observations that do not have local autocorrelation situations significantly different from zero (no significance).

More specifically, by applying the Getis-Ord Gi* [30] and the Local Moran I [31] to the three selected dimensions, it is possible to calculate the degree of similarity with respect to other nearby observations, counting - at the same time - the statistical significance.

The expression that characterizes the Getis-Ord Gi * I is the following:

$$G_i^* = \frac{\sum_{j=1}^{n} w_{i,j} x_j - \bar{X} \sum_{j=1}^{n} w_{i,j}}{S \sqrt{\frac{\left[n \sum_{j=1}^{n} w_{i,j}^2 - \left(\sum_{j=1}^{n} w_{i,j} \right)^2 \right]}{n-1}}} \tag{4}$$

Where x_j is the attribute value for feature j, $w_{i,j}$ is the spatial weight between feature i and j, n is equal to the total number of feature and $\bar{X} = \frac{\sum_{j=1}^{n} x_j}{n}$ $S = \sqrt{\frac{\sum_{j=1}^{n} x_j^2}{n} - (\bar{X})^2}$.

The expression that characterizes the Anselin Local Moran I is the following:

$$I_i = \frac{x_i - \bar{X}}{S_i^2} \sum_{j=1, j \neq i}^{n} w_{i,j} (x_j - \bar{X}) \tag{5}$$

Where x_i is the attribute value for feature i, $\bar{X}$ is the average of the corresponding value, $w_{i,j}$ is the spatial weight between feature i and j, and $S_i^2 = \frac{\sum_{j=1, j \neq i}^{n} (x_j - \bar{X})^2}{n-1} - \bar{X}^2$, n is equal to the total number of features.

3   Results

The model, applied to the Basilicata region, assumes a *TotQoL* variable in a range of values comprised between 93 and 105 (see Table 1), with 61% of municipalities characterized by a *TotQoL* below the average (=100) (Fig. 2a).

Table 1. Descriptive statistics of *EcQoL, SocQoL, EnvQoL* and *TotQoL*

	EcQoL	SocQoL	EnQoL	TotQoL
Min	91	88	73	93
Max	130	113	109	105
Mean	100	100	100	100
St Dev	5.9	4.1	4.5	2.5

By analyzing the different dimensions, the results were the following:

- The EcQoL (91–130) is characterized by a wide variation range (St. Dev. = 5.9) with a max value that is considerably spaced from the average (Table 1), but with 53% of municipalities characterized by a value of EcQoL below the average (Fig. 2a). This means that these values, although high, affect very few municipalities in relation to the general condition that appears to be below the regional average or otherwise around the mean.

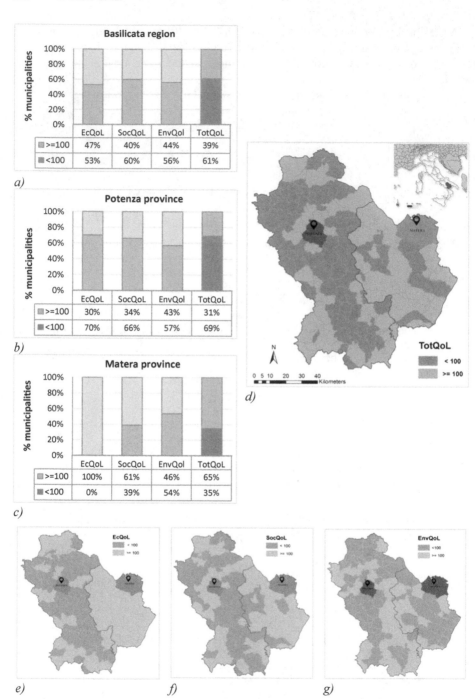

Fig. 2. *QoL* at regional and provincial level

- The *SocQoL* (88–113) is characterized by a less wide variation range (St. Dev. = 4.1) with min and max that are almost equally distanced (Table 1), with 60% of municipalities characterized by a *SocQoL* below the average (Fig. 2a).
- The *EnvQoL* (73–109) is characterized by a slightly wider variation range than the *SocQoL* (St. Dev. = 4.5), but with a min that is considerably spaced from the average (Table 1); 56% of municipalities are characterized by an *EnvQoL* below the average (Fig. 2a).

Table 2. Moran's Index for *EcQoL, SocQoL, EnvQoL* e *TotQoL*

	Moran's I	z-score	p-value
EcQoL	0.184203	3.929312	0.000085
SocQoL	0.174758	3.646187	0.000266
EnvQoL	0.267113	5.674260	0.000000

Looking at the territories of the two provinces (Potenza and Matera), there is a considerable difference, with a percentage of municipalities with *TotQol* respectively of 69% and 35% below the regional average (Fig. 2b and c); this difference is maintained even if you examine the three dimensions individually (Fig. 2b and c). In fact, while the province of Potenza is characterized by a prevalence of municipalities with a low quality of life, the province of Matera has a prevalence of municipalities with a quality of life above the regional average (100% of municipalities for the economic dimension) with the exception of the environmental dimension (Fig. 2b and c).

Maps show the geographical distribution of the index and its components (Fig. 2d, e, f and g).

Applying Moran's I that assumes values between −1 and +1, (and the relevant z-score and p-values– a measure of statistical reliability), in the three considered dimensions, higher values than zero were found (z-score > 2.58 and p-value < 0.01 with a significance level of 99%) (Table 2); the dimension with the highest spatial autocorrelation is *EnvQoL*, followed by *EcQoL* and *SocQoL*.

The Anselin Local Moran and the Getis-Ord Gi* enabled to obtain negative and positive clusters (Fig. 3), by setting a precise significance level. The Getis-Ord Gi* compared to Anselin Local Moran leads to the identification of clusters located in the same positions but affecting more municipalities (cold spots and hot spots).

In general, for the *EcQoL* and *SocQoL* you may experience a partial overlap of the clusters. In particular, for the *SocQoL* it is possible to identify the hot spots in the regional centres (Potenza and Matera); for the *EcQoL* the hot spots are identified in Matera and along the Ionian coast. The environmental dimension affects some areas situated in the southern part of the region. For *EnvQoL*, cold spots are delineated in the most industrialized areas (Potenza, Vulture, Matera and Metapontum), and hot spots in the southernmost part of the region (Fig. 3b, d, and f).

The Anselin Local Moran I can be detected when the spatial outliers for economic dimension in the municipalities of Potenza (high-low) and Lauria (high-low) and the *SocQoL* in the municipalities of Venosa (high-low) and Cancellara (low-high) (Fig. 3a, c and e).

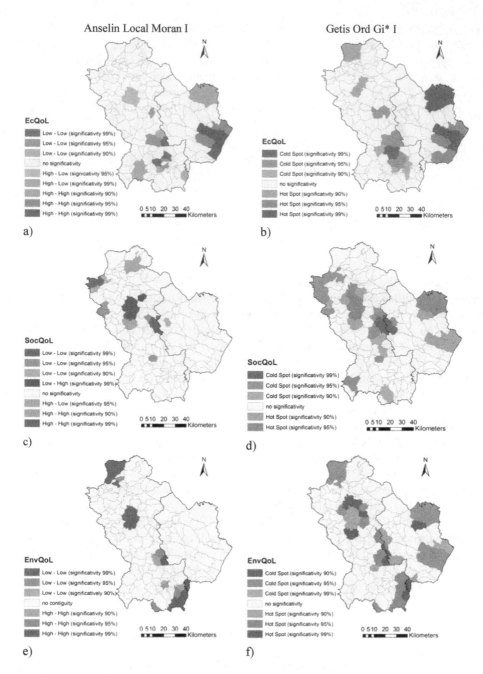

Fig. 3. Anselin Local Moran I and Getis-Ord Gi* I of *EcQoL*, *SocQoL* and *EnvQoL*

4   Discussion

First of all, the analysis of the data reveals a significant difference between the two provinces, partly related to the morphological diversity of the territory: the province of Potenza is characterized by a mainly mountainous (Apennines) and hilly territory (clay soils in 45.13% of the region, subject to erosion resulting in landslides), while the flat part (8% of the region) is concentrated in the province of Matera along the Ionian coast. Considering that the regional population is mostly concentrated in large centres, the distribution in percentage is the following: 56% live in the 12 largest towns in the region, 27% live in medium-sized centres, namely those between 5,000 and 9,999 inhabitants, and the remaining 17% live in small towns, which are mostly concentrated in the province of Potenza (82 municipalities out of 100 are below 5,000 inhabitants, of which 52 below 2,000 inhabitants).

Furthermore the analysis allows to delineate areas that offer ecosystem, natural, landscape and cultural resources with a direct impact on the welfare of people: they are highest in the suburbs and minimum in the central agglomerations. On the other hand, in some areas vulnerability conditions still persist as a result of risk factors (related to landslides and seismicity of the area), which affect, for example, the quality of road infrastructure, thus aggravating the socio-economic situation (there is a series of cold spots in the central portion, in north-western and southern areas of the region – Fig. 3a, b, c, and d).

In particular, the municipality of Potenza (hub) outlines a traditional monocentric and polarized structure of the province, meaning a relatively large dominant centre surrounded by a general economically stagnant condition. In fact, at the hub you can record an increase in population, which moves in order to have access to commercial and public interest services that have a higher concentration in the urban centre, where wider demographic dimensions guarantee market demand and adequate catchment. Matera's hub, instead, shows a much more uniform situation in the province but a more balanced offer of services, thus indicating a functional relationship between different municipalities.

By comparing the national classification of Inland Areas based on their peripherality from essential services [32], the variables identified for calculating the QoL allow a more complete and accurate reading of the sub-regional territory. Different areas can actually have a positive or negative connotation in relation to the general context, depending on the dimension concerned. The factors considered, in fact, allow to discriminate in a more precise manner the imbalances on the territory, highlighting, for example, the areas that have developed autonomously, in terms of many important services, even though - or maybe simply because – they are distant from the hubs. An example is the hot spot of municipalities along the Ionian coast, whose *QoL* conditions are above the regional average, although they are distant from the two regional centres and their areas are classified as peripheral in the national classification (Fig. 3a and b).

Moreover, Inland Areas include not only weaknesses, as the hot spots related to the environmental dimension characterize territories that may be less "attractive" in relation to the level of services offered (Fig. 3e and f); they also involve strengths, related to their still unexploited potentials (this is the case of the areas of great natural value that could offer important opportunities for tourism, recreation and gastronomy). This reinforces the strategy targeted towards interventions aimed at improving services, at least with the aim of ensuring sufficient life opportunities to maintain and attract population for territory stewardship.

In fact, results confirm that there is a difference in the population rate between 18 and 35 years[2] equal to +1.47% ($EcQoL$) and +0.62 ($SocQoL$), respectively, of hot spots compared to cold spots; it follows that socio-economic characteristics are the main factor influencing the residential choices (compared for example to issues, such as environmental health) of that population segment which must "ensure" a reversal of demographic trends.

5 Conclusions

The methodological framework developed is an important tool to support the actions already undertaken and those that still need to be undertaken, with a view to smoothing out regional imbalances and promoting social inclusion, with special reference to EU funds managed by the regions (Rural Development Programme 2014–2020) and resources targeted specifically by the Stability laws of 2014 and 2015 [33].

The results facilitate decision-making: global QoL provides an overall idea, a measure of the gap in quality of life of each territory with respect to the regional average. In addition, the observation of the index components enables the definition of more specific guidelines on which to focus the attention and the available resources to foster a balanced growth of the three dimensions of sustainable development. The non-compensatory nature of index components ensures the possibility of identifying peculiarities (both positive and negative), and rewarding the areas characterized by a balance of all indicators, assuming that these indicators/dimensions are not substitutable.

The results reveal a level of quality of life that tends to decrease from the centres to the internal areas, with the possibility to distinguish clusters under marginal conditions from clusters under favourable conditions. In addition to morphological and demographic characteristics of the territory, the presence of basic and leisure services turns out to be differentiated. On average, the populations of municipalities in inland areas take more time and resources than urban municipalities to access different services offered (hospitals and health care facilities, cultural, sports and recreational opportunities), where there is a greater presence of those services. It is particularly interesting to point out that the two provinces are not homogeneous: Potenza's hub has a territorial structure that tends to be more monocentric with respect to that of Matera, which is instead more aligned towards a polycentric territorial development model. At regional level, polycentrism implies the promotion of complementary and interdependent municipalities' networks that can ensure rural environment integration.

---

[2] It has been empirically demonstrated that there is a positive relationship between the proportion of population and the factors related to quality of life, in particular the 18–35 age group, which is particularly sensitive to the level of quality of life [4, 34, 35]; therefore it represents a proxy for assessing the effectiveness of the proposed model.

In order to encourage development opportunities, the authors consider it necessary to pay particular attention to the problems related to the presence and accessibility of services. The possibility of ensuring the coverage or capillarity of services would lead to reduce the migratory balance, troubling especially in more marginal areas. In a perspective of local development policy, the abandonment of these areas could undermine the maintenance of the territory reducing "non-market" services (ecosystem services).

In fact, as shown by the analysis, municipalities of inland areas (mostly rural) are also those with a greater ecological and natural value but also with major landslide risks. The smallest municipalities (mainly concentrated in the province of Potenza) turn out to be the most sensitive and, therefore, would require more attention, for example, by promoting forms of association between municipalities (Dgls. 267/2000), as intended into the SNAI.

Although the methodology makes it possible to obtain significant results and a solid foundation of knowledge useful to determine specific and targeted policy instruments, it is believed that there is room for fine-tuning the indicators to be considered (e.g. qualitative indicators). It would be useful, furthermore, to carry out more detailed assessments both in space and time, the latter being useful for determining the growth/decline trends. In this sense, a limit may be represented by the availability of data.

Another important issue to be addressed in future studies is the possibility of linking the information acquired through the methodology applied in this study with the interventions implemented in previous years (with particular reference to the old rural development programs). At present, there is indeed no information relating to the actions already taken. Such a step would lead to even outline the possible paths to be pursued for future programs.

In conclusion, the proposed framework can present a useful tool in the current political context in the implementation of actions aimed at gradually reducing regional disparities in terms of quality of life, that follow these goals:

- address of interventions, which should take into account balanced growth of the (economic, social and environmental) dimensions of quality of life;
- ex-ante and ex-post effects evaluation of the carried out interventions, as a synthetic "measure" of achievements in terms of improving the quality of life;
- identification and, if necessary, redistribution of the areas that need priority interventions and resources.

Appendix

Dimensions	Thematic areas	Polarity	Indicators	Calculation method	Unit of measure	Data source	Reporting year
Economic dimension	Economic opportunities of the territory	+	Average purchase prices of real estate (PPR)	-	€	Italian real estate market monitor	2010–2014
		+	Bank branches number (BBN)	-	Number	Bank of italy	2015
Social dimension	Proximity to places of work/study	-	Mobility rate domicile-work/study (MDWS)	(Number of individuals who move to another town/residents' number) * 100	%	Elaboration on ISTAT data – census population 2011	2011
		+	Proximity rate domicile-Work/study (PDWS)	(Number of individuals who employ <15 min. to reach work/study/residents number) * 100	%	Elaboration on ISTAT data – census population 2011	2011
	Spread and proximity to health and educational facilities	-	Mileage time to reach hospital structures (MTH)	-	Minutes	Our GIS processing - CTR	2015
		-	Mileage time to reach secondary schools (MTS)	-	Minutes	Our GIS processing - CTR	2015
		+	Percentage of education services (PEd)	(Number of schools/population density) * 100	%	Our GIS processing - CTR	2015
	Proximity to non-decentralized services	-	Mileage time to reach administrative offices (MTA)	-	Minutes	Our GIS processing - CTR	2015
	Cultural and recreational facilities	-	Mileage time to reach cultural activities (MTC)	-	Minutes	Our GIS processing - CTR	2015
		-	Mileage time to reach green spaces (MTG)	-	Minutes	Our GIS processing - CTR	2015
		+	Percentage of sport facilities (PSp)	-	%	Our GIS processing - CTR	2015
		+	Percentage of free time facilities (PFT)	-	%	Our GIS processing - CTR	2015
	Broadband access	+	Percentage of population coverage with access to internet between 2 Mbps e 20 Mbps (PAI)	Number of individuals with access to the internet between 2 Mbps and 20 Mbps/resident number) * 100	%	www.Infratelitalia.it	2015
Environmental dimension	Environmental health	-	Inhabitant Equivalent Total (IET)	1 inhabitant equivalent = 60 g of Bod5	Inhabitants	ISTAT	2009
		+	Distance from Industrial areas (DI)	-	Meters	Our GIS processing - CTR	2015
		+	Distance from Landfills (DL)	-	Meters	Our GIS processing - CTR	2015
	Nature conservancy	+	Percentage of areas percentage with High Ecological-natural value (AHE)	(High ecological-natural areas/municipal areas) * 100	%	ISPRA	2010
	Risk factors	-	Landslide risk (LR)	Landslide risk areas (R1, R2, R3, R4/municipal areas) * 100	%	Basin Authority	2015
		-	Seismic risk (SR)	-	Classes	www.utsbasilicata.it	2012

References

1. Dematteis, G.: Montagna e aree interne nelle politiche di coesione territoriale italiane ed europee. Territorio, Washington, D.C. (2013)
2. Barca, F.: Un'agenda per la riforma della politica di coesione. Una politica di sviluppo rivolta ai luoghi per rispondere alle sfide e alle aspettative dell'Unione Europea (2009)
3. Cagliero, R., Cristiano, S., Pierangeli, F., Tarangioli, S.: Evaluating the improvement of quality of life in rural areas. In: 122nd EAAE Seminar Evidencebased Agricultural and Rural Policy Making: Methodological and Empirical Challenges of Policy Evaluation, pp. 17–18 (2011)
4. Boncinelli, F., Pagnotta, G., Riccioli, F., Casini, L.: The determinants of quality of life in rural areas from a geographic perspective: the case of tuscany. Rev. Urban Reg. Dev. Stud. 27(2), 104–117 (2015). doi:10.1111/rurd.12035
5. Sen, A.: Equality of what? In: McMurrin, S. (ed.) Tanner Lectures on Human Values, vol. 1. Cambridge University Press, Cambridge (1980)
6. Sen, A.: Capability and well-Being73. Qual. Life, 30 (1993)
7. Kuznets, S.S.: Economic Growth of Nations (1971)
8. Giovannini, E., Hall, J., Mira d'Ercole, M.: Measuring well-being and societal progress. In: Conference Beyond GDP-Measuring Progress, True Wealth, and the Well-Being of Nations. European Parliament, Brussels (2007)
9. Stiglitz, J.E., Sen, A.K., Fitoussi, J.: Report by the commission on the measurement of economic performance and social Progress (2009). http://www.stiglitz-sen-fitoussi.fr/documents/rapport_anglais.pdf. Accessed 26 Apr 2016
10. Nuvolati, G.: Resident and non-resident populations: quality of life, mobility and time policies. J. Reg. Anal. Policy 33(2), 67–84 (2003)
11. Brereton, F., Bullock, C., Clinch, J.P., Scott, M.: Rural change and individual well-being the case of Ireland and rural quality of life. Eur. Urban Reg. Stud. 18(2), 203–227 (2011). doi:10.1177/0969776411399346
12. Costanza, R., Erickson, J., Fligger, K., Adams, A., et al.: Estimates of the genuine progress indicator (GPI) for Vermont, Chittenden County and burlington, from 1950 to 2000. Ecol. Econ. 51(1), 139–155 (2004). doi:10.1016/j.ecolecon.2004.04.009
13. Pulselli, F.M., Ciampalini, F., Tiezzi, E., Zappia, C.: The index of sustainable economic welfare (ISEW) for a local authority: a case study in Italy. Ecol. Econ. 60(1), 271–281 (2006). doi:10.1016/j.ecolecon.2005.12.004
14. Chelli, F.M., Ciommi, M., Emili, A., Gigliarano, C., Taralli, S.: Comparing equitable and sustainable well-being (BES) across the italian provinces. a factor analysis-based approach. Riv. Ital. economia demogr. Stat. 69(3), 61–72 (2015)
15. Rosen, S.: Hedonic prices and implicit markets: product differentiation in pure competition. J. Polit. Econ. 82(1), 34–55 (1974)
16. Wang, Z.J., Shi, P.J., Li, W.: Study of central cities service scope based on time accessibility in gansu province. In: Advanced Materials Research, vol. 524, pp. 2854–2860. Trans Tech Publications (2012)
17. Godard, O.: Environnement, modes de coordination et systèmes de légitimité: analyse de la catégorie de patrimoine naturel. Rev. économique 41(2), 215–241 (1990). doi:10.2307/3501802
18. ISTAT: Stime del carico inquinante delle acque reflue urbane (2016). http://www.istat.it/it/archivio/41920

19. Mazziotta, M., Pareto, A.: Methods for constructing non-compensatory composite indices: a comparative study. Forum Soc. Econ. **45**(2–3), 213–229 (2015). doi:10.1080/07360932. 2014.996912. Routledge

20. Cozzi, M., Romano, S., Viccaro, M., Prete, C., Persiani, G.: Wildlife agriculture interactions, spatial analysis and trade-off between environmental sustainability and risk of economic damage. In: Vastola, A. (ed.) The Sustainability of Agro-Food and Natural Resource Systems in the Mediterranean Basin, pp. 209–224. Springer, Cham (2015). doi:10.1007/978-3-319-16357-4_14

21. Delvecchio, F.: Scale di misura e indicatori sociali. Cacucci editore, Bari (1995)

22. Anselin, L.: Local indicators of spatial association-LISA. Geograph. Anal. **27**(2), 93–115 (1995). doi:10.1111/j.1538-4632.1995.tb00338.x

23. Chainey, S., Reid, S., Stuart, N.: When is a hotspot a hotspot? A procedure for creating statistically robust hotspot maps of crime. In: Kidner, D., Higgs, G., White, S. (eds.) Innovations in GIS 9: Socio-Economic Applications of Geographic Information Science, pp. 21–36. Taylor & Francis, England (2002)

24. Moran, P.: The interpretation of statistical map. J. R. Stat. Soc. B **10**, 243–251 (1948)

25. Brachert, M., Titze, M., Kubis, A.: Identifying industrial clusters from a multidimensional perspective: methodical aspects with an application to Germany. Pap. Reg. Sci. **90**(2), 419–439 (2011). doi:10.1111/j.1435-5957.2011.00356.x

26. Carroll, M.C., Reid, N., Smith, B.W.: Location quotients versus spatial autocorrelation in identifying potential cluster regions. Ann. Reg. Sci. **42**(2), 449–463 (2008). doi:10.1007/s00168-007-0163-1

27. Mitra, R., Buliung, R.N., Faulkner, G.E.: Spatial clustering and the temporal mobility of walking school trips in the Greater Toronto Area. Can. Health Place **16**(4), 646–655 (2010). doi:10.1016/j.healthplace.2010.01.009

28. Prasannakumar, V., Vijith, H., Charutha, R., Geetha, N.: Spatio-temporal clustering of road accidents: GIS based analysis and assessment. Procedia-Soc. Behav. Sci. **21**, 317–325 (2011). doi:10.1016/j.sbspro.2011.07.020

29. Romano, S., Cozzi, M., Viccaro, M., Persiani, G.: A geostatistical multicriteria approach to rural area classification: from the European perspective to the local implementation. Agric. Agric. Sci. Procedia **8**, 499–508 (2016)

30. Anselin, L., Griffiths, E., Tita, G.: Crime mapping and hot spot analysis. Environ. Criminol. Crime Anal. 97–116 (2008)

31. Getis, A., Ord, J.: The analysis of spatial association by use of distance statistics. Geogr. Anal. **17**, 81–88 (1992). doi:10.1111/j.1538-4632.1992.tb00261.x

32. Agenzia per la coesione territoriale Classificazione dei Comuni italiani secondo la metodologia per la definizione delle Aree Interne (2014). http://www.agenziacoesione.gov. it/it/arint/OpenAreeInterne

33. IFEL: I Comuni della Strategia Nazionale Aree Interne. Studi e Ricerche, prima edizione – 2015 (2015)

34. Blanchflower, D.G., Oswald, A.J.: The Rising Well-Being of the Young. Working Paper No 6102, NBER Working Paper Series. National Bureau of Economic Research, Cambridge (1997)

35. Barber, T.: Participation, citizenship, and well-being: engaging with young people, making a difference. Young-Nord. J. Youth Res. **17**(1), 25–40 (2009)

An Operative Framework to Support Implementation Plan Design Applied in Transnational Cooperation Project

Francesco Scorza[✉] and Beatrice Giuzio

Laboratory of Urban and Regional Systems Engineering, School of Engineering, University of Basilicata, 10, Viale dell'Ateneo Lucano, 85100 Potenza, Italy
francesco.scorza@unibas.it, beatricegiuzio@hotmail.it

Abstract. Territorial cooperation across national borders has the task to bring a real European added value [1, 2] in pursuing the goal of territorial cohesion and in balancing spatial development of the EU territory [3] through common development strategies. Although INTERREG programs do not constitute a big share of the Structural Fund budget (only 2.5% of the total Structural Funds Budget for 2007–2013 and the 2.8% of the total of the European Cohesion Policy budget for 2014–2020 [4], they play a key role in the main stream policies development. One of the innovations promoted in the 2014–2020 programming period is to reinforce the operational dimension of INTERREG measures and projects. In fact, INTERREG Projects not only required to demonstrate a positive impact on the development of transnational approaches and solutions to targeted issues but their concrete actions design and, possibly, consequent implementation must be carried out on a common cross-border basis. Such procedural schema is based on the design of an Implementation Plan (IP) at project partners level and its implementation within project timeframe. Hence the need for a detailed and organized methodological schema which could support and enable the effectiveness application of IPs delivered by transnational cooperation projects. This paper provides contributions to support and improve the IPs design practices according to several techniques and recommendations also retrieved in previous relevant transnational cooperation experiences.

Keywords: Transnational cooperation · Action Plan · INTERREG programs

1 Introduction

Since EU included the Transnational Cooperation among its priority of territorial cohesion (2007–2013 programming period) the focus on interregional cooperation project increased relevance for Public Administrations and the wide community of eligible actors. Numbers of cooperation experiences was already developed with a variable impact in generating operative contributions in regional development, while, probably, a highest result in exchanging good practices was performed.

One of the innovations promoted in the 2014–2020 programming period is to reinforce the operational dimension of INTERREG measures and projects. In fact,

© Springer International Publishing AG 2017
O. Gervasi et al. (Eds.): ICCSA 2017, Part VI, LNCS 10409, pp. 501–516, 2017.
DOI: 10.1007/978-3-319-62407-5_35

INTERREG Projects not only required to demonstrate a positive impact on the development of transnational approaches and solutions to targeted issues but their concrete actions design. The design phase is strictly linked with the implementation phase in the last generation of Interreg projects where implementation of foreseen actions has to be carried out on a common cross-border basis as a compulsory task for project partners.

A relevant project output for every INTERREG Projects becomes the Implementation Plan (IP)[1]. At program level it represents a relevant innovation shifting the transnational cooperation approach from the focus of learning experiences and producing common (often 'methodological') solutions to the implementation of concrete actions defined and shared in a transnational debate.

Designed at project partners level, the IP has to be implemented within project timeframe (second phase for standard project). Hence the need for a detailed and organized methodological schema which could support and enable the effectiveness application of IPs delivered by transnational cooperation projects. This paper provides contributions to support and improve the IPs design practices according to several techniques and recommendations also retrieved in previous relevant transnational cooperation experiences.

Most of the European Programs provide an IP model, which is a kind of "roadmap" that can offer a number of useful solutions to address the topics identified at the beginning of the single project. The purpose is to provide operational output whose application will be feasible in the future, i.e. after project closure. It means that the information and monitoring about implementation is out of the project level and is solely entrusted to the single partner. A very weak point for the structure of EU transnational cooperation strategy.

Very often the IPs proposed are skinny and synthetic, as they are information gatherers that lend themselves to monitoring and evaluation operations, in order to understand if and how the system work. Even the INTERREG Program requires a project scheme, that we will define an IP with minimum requirements, set to provide outputs useful to the second stage of the program dedicated to the monitoring and the implementation [4].

It appears to be lacking, in the various European programs, an exhaustive, comprehensive and common approach to follow in order to pursue the main goal.

We wondered, what can be an Effectiveness-Based Approach? To answer this question we asked ourselves, what do we expect from an IP? What actions have planned and implemented the successful projects? What kind of methodologies did the successful projects use?

Of course, the paper have not the claim to propose an exhaustive methodology that takes into account all possible implementation contexts or the issues developed by the projects. Rather it represents a reflection applied within a concrete experience[2] that

---

[1] In this work we refer to Implementation Plan and or Action Plan considering the two concepts as synonymous.

[2] We refer to the work delivered by the research group of the LISUT laboratory in the School of Engineering of the University of Basilicata within the scientific support provided to the Province of Potenza during the implementation of LOCARBO project [5].

looks at standardized and known approaches and their integration into the proposal of a synthetic grid to be used as a basis for the IPs design within international partnerships.

2 What We Expect from an Implementation Plan: A Preliminary Analysis

Let's try to describe the general schema for the IP where each element represents a synthesis of the complex work of operative design and should be considered strongly integrated with all the other sections.

This schema could be improved according with specific needs or thematic requirements but it is also a basis for the development of the IP design. The objective of this proposal is to share, at project level, a common and effective grid in order to define a procedure based on well-known methodologies and practices and to allow an easy comparison of different elaborations delivered by individual Project Partner.

The main structure of the IP:

  i. Strategic approach
 ii. Vision, goals and results
iii. Implementation procedure
 iv. INPUTS analysis
  v. Indicators and sources of verification
 vi. External conditions assessment.

2.1 Strategic Approach

Starting from the "ambitions" to be addressed in the IP, coming as a result of detailed analysis of the context, a clear identification of strategic framework for the operation should be defined through the statement of "overall objectives" and main achievement strategies to be implemented. A strategy should identify an homogeneous field of application (i.e. training people; CO2 reduction investments; etc.) aggregating specific objectives, outcomes and effects to which we look at in order to face main troubles.

We refer to the wide bibliography on strategic planning (amoung others some relevant authors: [6–11]) and especially to the vision that an affective planning/project process should start from the elicitation of problems defined according to the lack of the three main planning principles of: equity; effectiveness; conservation of not reproducible resources (this is a specific feature of the contributions by Las Casas [12, 13]). It means that it is possible to detect a problem if a measurement the non-satisfaction of one of these principles is provided.

2.2 Vision, Goals and Results

For each strategy, the expected intervention scenario (in other words the vision) has to include a set of operative objectives (selected according with the strategic priorities), activities and related outcomes which has to be defined according to general selection criteria. We consider 'selection criteria' those horizontal principles of "Feasibility" and "Effectiveness" to be demonstrated through adequate clarification of:

- role and responsibilities of the AP promoters;
- role and function of involved Stakeholders;
- clear identification of (financial, human, natural etc.) resources necessary to carry on the operations;
- clear determination of time necessary to carry on the activities;
- comparison of alternatives compared in a quantitative grid of indicators;
- elicitation of external conditions influencing the process (positively or negatively).

2.3 Implementation Procedure

Implementation is an essential part of the strategic planning process.

The implementation procedure is based on series of activities to be realized in order to produce expected results [14]. However its main focus should remain on the future actions planned to ensure that the benefits of the exchange of experience are not lost after the project's end. These actions could, for instance, include local meetings to mobilize relevant stakeholders and preparation of bids to be submitted in local/regional/national programs, launch of new initiatives in the region, measures to modify a specific policy document/instrument.

It is interesting distinguish in top-down to policy development, conceptualizing a policy objective to be applied at the lower tiers of territory, to a bottom-up approach development focusing on how the notion is locally and regionally understood, interpreted and put into practice [15].

2.4 INPUTS Analysis

The IP has to include a detailed analysis of resources to be provided for the realization of the implementation procedure.

INPUTS could be described in categories:

- financial resources,
- human resources,
- environmental resources.

Other categories can be included according to the specific fields of action.

2.5 Indicators and Sources of Verification

It is important define "performance indicators", a huge literature could support the selection of an indicator matrix according to implementation field [16–18].

Considering IP indicators, as a general rule, each result should be measured by at least one indicator and each objectives should be accompanied by a set of indicators allowing to report the impacts generated at IP scale.

Each indicator will be accompanied by the definition of the Source of Verification: it means the source providing data for measurements (which database is available or which documents have been draw up or can be obtained elsewhere to prove that the output has been achieved? What records voucher for the cost entailed, consumption of materials, use of the equipment, inputs of personnel, etc., have been used to finalized the activities? And so on.)

As a general example, we refer to the case of LOCARBO project identifying the condition in which the strategy is "*to improve energy performance of residential buildings*". In this strategy we have different specific objectives with their indicators and source of verification. In the following table is reported a synthetic example referred to the declared hypothesis.

Concerning results we can define the following situation (Tables 1 and 2).

Table 1. LOCARBO example shows strategy between objectives – indicator – sources of verification.

Objectives	Indicator	Sources of verification
To improve heating systems	Reduction of energy bill for eating (€) m3 of natural gas saved	Regular Survey to households Data from regular survey to households compared with pre-investments statistics
To promote responsible use of heating systems in households	Effective hours of use per day No of households respecting the technical level of comfort temperature/Total No of households	Regular survey to households Regular survey to households

Table 2. LOCARBO example shows strategy between results – indicator – sources of verification.

Results	Indicator	Sources of verification
New eating plants in each residential apartment	No of eating system installed € invested m2 of residential surface equipped with new eating systems	Project implementation data
Renovation of buildings insulation	m2 of vertical external surfaces renewed with high performance insulation materials No of apartments insulated	Project implementation data
"Improving users' behavior" campaign	No of training events realized No of participants No of households participating	Attendance sheet Project data
.....	.....	.....

2.6   External Conditions Assessment

According to the operative condition of a short time frame for the IP' implementation, an effective design of the IP should include the assessment of internal and external constraints/conditions.

Starting from a comprehensive view of responsibilities and INPUTs the designers should pay attention to the constraints (especially the external ones) that could influence the achievement of results and – consequently – of declared objectives. It means that each actor involved in the process has to clearly identify the external factors influencing the implementation process.

We can define "external factors" the whole of actions, tools, decisions, approvals etc. depending on the responsibility of actors not directly included in the implementation process, by whom depends the possibility to implement an activity or the success to achieve an objective through a set of results.

For instance, among such external conditions the "normative and regulatory authorization process" generally represents a relevant potential constraint for infrastructural activities; "ROPs financing procedures" could be external conditions for financing operations and activities; "national and European rules for public tenders" could influence the process of purchasing equipment or technologies necessary for specific activities included in the IP.

Another example of external conditions affecting the goals' achievement could be the level of involvement of local groups (i.e. the community of users benefitting of a specific services), or the capacity to communicate and disseminate achieved results letting people aware of a new territorial conditions or services etc.; or the confidence of local entrepreneurship in investing on targeted sectors delivering clusters or PPP agreements.

The IP, after declaring such external condition, should be sensitive in monitoring such components during implementation phase in order to adjust on-going the previsions ensuring the global achievement of desired objectives stated in the AP's strategy.

3   A Toolkit of Techniques and Approaches from Previous Experiences

In order to stimulate a positive integration between standardized approach and case studies in developing IP we present a short analysis of a set of former Cooperation Projects selected because oriented to deliver an Action Plan among the main project outputs.

Such small list of experienced allowed to investigates the frequency of standardized approach in concrete applications. We refer to well-known methodologies and approaches such as SWOT, Risk Assessment, Policy recommendation etc. widely used in transnational cooperation experiences.

3.1 Some Relevant Cases from Transnational Cooperation Practices

In this chapter, we analyze a set of Projects developed under the framework of EU Cooperation programs (in particular INTERRG IVC) oriented to deliver an Implementation Plans (i.e. Action Plans) considered as the operative dimension through which the project results will be applied at territorial level.

The analysis was not only addressed to the contents and the characterizing sectorial analysis included in the IPs but mainly to the procedural aspects describing the IPs design and development process.

The whole of reference projects come from the Good Practices Database of INTERREG IVC Program [19] and other relevant databases [20] (Table 3).

3.2 Applied Techniques and Approaches

The following list represents the operative steps, techniques and methodological tools applied in the process of IP development, identified through the analysis of the selected transnational cooperation project experiences.

Mental Mapping (brainstorming): A good way to identify ideas and also relevant risks can be an open brainstorming session at one of the partner meetings. All partners should be involved in this process to suggest ideas or raise their awareness about possible risks, and to identify as many relevant risks as possible, the impact in order to identify possible solution or understand if is the case to change activities, roles, times or budget into the action plan.

SWOT: Identify and analyze strengths, weaknesses, opportunities, and threats of single action is a strategy helpful to detect the internal and external factors that are favorable and unfavorable to achieve the fixed objective.

Interview and Questionnaire: In order to receive more in-depth information concerning the current state of the art of the regions, semi structured interviews are hold with stakeholders in each region. The target of the people interviewed are representatives of the various stakeholder groups and come from all sectors of society. From the results of the interviews it is possible to identify the main trends.

Networking and Stakeholders Engagement: There are different action to involve the stakeholder in the project and there are also different levels of stakeholders engagement [28] that can be resumed in a matrix where can catalog every stakeholder in a category, such as Unaware: Unaware of project and potential impacts, Resistant: Aware of project and potential impacts and resistant to change, Neutral: Aware of project yet neither supportive nor resistant, Supportive: Aware of project and potential impacts and supportive to change, Leading: Aware of project and potential impacts and actively engaged in ensuring the project is a success; and define the current level of engagement and the desired one.

Workshop and Seminars: A workshop session is important in the EU project to link the "expert vision" for each thematic to every participant region. Furthermore, a workshop is a way to engage and networking with stakeholders.

Table 3. Some relevant cases from transnational cooperation practices.

Project	Characterizing features
SUSTAIN: Working across land and sea boundaries Environmental and risk protection - Water management (www.sustain-eu.net) [21]	The topic of the project is the use of a wood shredder to convert organic, agricultural waste into material suitable for surfacing local footpaths. Key success factors were discussions with the agriculturalists as to the (financial) benefits accruing, specific trainings and on-the-job experience of working with the machinery
MiSRaR: A study on the environmental, economic and social impacts of climate change in Greece Environmental and risk protection - Natural and Technological risk (www.misrar.nl) [22]	The MiSRaR project is about protecting people, environment and property against the destructive impact of natural and technological hazards. MiSRaR lays down the principles of risk mitigation, discusses how mitigation processes should be launched, how risk and capability assessment should be undertaken, and provides ideas for drafting mitigation plans, for financing of and lobbying for implementation actions, and for monitoring, enforcement and evaluation
PRESERVE: Peer Reviews for green and sustainable regions through EUROPE | Environment and risk prevention - Cultural heritage and landscape (www.preserve.aer.eu) [23]	The PRESERVE project aims to make their tourism strategies more sustainable and to set new standards for regional actors. The partners intends to increase the effectiveness of the regional development policies starting with cultural heritage and the countryside as central factors of the development. This project is into the best practice because, after the project, the local economy has been revived and previously deserted areas have seen an uplift
PERIURBANparks: Improving Environmental Conditions in Suburban Areas | Environment and risk prevention - Biodiversity and preservation of natural heritage (including air quality) (www. periurbanparks.eu) [24]	PERIURBANparks is a regional initiative project, which uses interregional exchange of experiences to improve policies on management of natural suburban areas. PERIURBAN focuses specifically on policy and management solutions to mitigate pressures on biodiversity. Focus on the creation and management of parks in natural suburban areas, in line with European environment policy and redevelopment in suburban areas, can impact positively on the environment and on halting biodiversity loss
ENERCITEE: European networks, experience and recommendations helping cities and citizens to become Energy	EnercitEE contributed to the improvement of local and regional policies and provided

(*continued*)

Table 3. (*continued*)

Efficient | Environment and risk prevention - Energy and sustainable transport (www. enercitee.eu) [25]	assistance in the transfer of knowledge on energy efficiency and sustainable transport
TOURage: (no in best practice) **Developing Senior Tourism in Remote Regions** | Innovation and the knowledge economy - Entrepreneurship and SMEs (www.tourage. eu) [26]	TOURAGE project has the overall purpose to improve sustainable regional economy by developing senior tourism. In 2014 the PESTO project which is coordinated by the E.N.T.E.R. network selected TOURAGE as one of the 20 best European tourism projects enhancing sustainable tourism development. The selection was made from a project pool containing around 200 projects
BOO - Games: Boosting European Games Industry | Innovation and the knowledge economy - Entrepreneurship and SMEs (www.boogames.eu) [27]	Games could represent a new source of growth for Europe economy but many regions are still missing adequate policies + funding schemes which could sustain this market. Further, funding + support mechanisms often do not meet the special needs of the small + innovative game developers. Aim of the BOO - Games project is to support the pub.reg.development authorities in understanding the importance of the games industry for the Europ.economy

Transferring Good Practices and Instruments: Transferring good practices and instrument from one region to another it means contribute to enrich and renew the way they implement their policy. The interaction of different actors and the exchanges of knowledge and know-how can generate, in the best way, the learning process, he main expected outcomes of the INTERREG projects. The most important potential for learning was identified within improving and creating new channels of information, the exchange of experience and learning, as the cross-border initiatives can be used as 'laboratories' through which trans-national ideas can be channeled and tested [29, 30].

Exchanges Experiences or Stages: The transfer of practice and simple instrument is generally not sufficient to ensure a policy effect so it will be crucial exchange also experience (or staff for a period) on the different policy framework of their region. INTERREG projects provide the possibility for its participants to experience active collective learning and a beneficial exchange of experiences. In this regard, the experience of working in practice with the concept of polycentric development seems to have enabled a large majority of participants to improve their understanding of the concept [15]. It is through this strategic approach that the cooperation can achieve more structural changes in each participant region [31].

Sub Project Call for Proposal: Sometimes the project structure include call for proposal in order to involve sub-themes that explore different sides of the same coin. The EU program must explicitly state the need for a sub project call and must provide specific economic resources for this action. Sub-projects have to be more focused on a specific regional policy issue. It is evident that these sub-themes are interlinked in many ways, within and even between the main priorities. Therefore projects can propose a cross-sectoral and integrated approach where appropriate. Sub-projects should be jointly and interregionally developed and should affect determinedly on the policies of the involved regions and the EU.

Setting Out a Toolkit Instruments: The project action will develop a methodology or produce toolkits applicable in other framework. To create a toolkit it is essential to identify first the necessity requirement and then the solution strategy, based on clear and precise steps. The creation of a joint understanding of the whole proposed process is therefore a crucial prerequisite for the success of the toolkit.

Peer Review: A Peer Review method helps regional authorities to understand how well their policies are working and support them in making improvements. During the project, the review team holds meetings with key stakeholders in order to ascertain the strength, weaknesses and potential of the policies. After the visit the review team prepares a report with a series of recommendations and set of benchmarks on how to improve the policies.

Policy Recommendation (transferable approach): One of the main outputs/actions of the EU projects is declaring to address (influence) local policy instrument as a real step for the challenge. It means transferring the successful results of an initiative to appropriate decision-makers in order to promote changes and show institutions a better way. Design a new approach and test it to assess whether it can actually be replicated in the future is the basis of a policy recommendation.

The following matrix represents the frequency of methodological steps in the sample set of projects (Table 4):

Table 4. Project matrix: frequency of methodological steps.

Project	Mental Mapping	SWOT	Interview and questionnaire	Networking and stakeholders engagement	Workshop and seminars	Transferring good practices and instrument	Exchanges experiences or stages	Sub project call	Setting out toolkits	Peer-review	Policy recommendation
SUSTAIN					x	x	x		x		x
MiSRaR					x	x			x		
PRESERVE		x		x	x	x	x		x	x	x
PERIURBAN parks				x	x	x			x		x
EnercitEE					x	x	x	x			x
TOURage	x	x		x		x	x				x
BOO-Games		x	x	x							x

3.3   Some Comprehensive Consideration Concerning the Assessment Process

All selected projects aim to address local policy instrument: the goal is transferring the successful outcomes of the project activities to appropriate decision-makers in order to promote the change. This is possible with operational proposals that modify the implementation procedures of spatial development policy in the region.

The success of the SWOT method, output of many of the projects examined, is mainly owed to its simplicity and its flexibility. Its implementation does not require technical knowledge and skills: a correlation is made between the internal factors, strengths and weaknesses of the organization, and the external factors, opportunities and threats. An effort can be made to exploit opportunities and overcome weaknesses and at the same time for the organization to protect itself from the threats of the external environment through the development of contingency plans. The SWOT have also drawbacks, as the possible length of the lists of factors that have to be taken into account in the analysis, the lack of prioritization of factors, the statement not obligatory based on data or analysis.

The aim of seminars and practical workshops is to provide the necessary guidance on the issues to be addressed through the expert intervention. In the session teamwork can create synergies and networking among the participants. The benefit described along with the simplicity and the few financial resources required for the organization, are the reasons why the seminars and the workshops are successful actions.

The networking, the involvement and the engagement of stakeholders, the identification of good practices and instruments to exchange and the willingness to exchange experience are the recurrent and relevant methodological steps in every EU project.

Interviews and questionnaires are a not very used in transnational cooperation project practices.

The practice of survey is used for snatch information concerning the current state of the art of the subject in question and recognize the prevalent needs in the current climate. Build interview form is outlined from desk researches addressed to understand how structured the questions and to whom to submit it. This survey instrument is just a step and can't be a unique and exhaustive methodology to accomplish the whole process.

The mental mapping session is the group decision-making technique designed to generate a large number of creative ideas through an interactive process and identify during the meeting potential risk and possible solution. Brainstorming brings team members' diverse experience into play. It increases the richness of ideas explored, which means that you can often find better solutions to the problems that you face. This step require the fully understand by all team members of the objective of the brainstorming session and their active participation. Involve participants, develop an enthusiastic climate, encourage the creative thinking, build ideas, identify the risk, focus on the solution is not so simple and if is not well planned will produce confused output.

Peer Review is not so much used as a methodological tool because it requires the involvement of local actors not only in the final phase, but during the whole duration of the project. The results is the product of joint action and compromises between the parties, a lot of work and meetings. However, surely the fact that it is a joint work makes it more adherent to what can be done, in other words it will not find any obstacles because the actions are agreed. Moreover, the constructive reviews between project promoters and actuators will bring decisive action emerging from the real needs.

Also the sub-project call for proposal doesn't often use: it is a mode explicitly required from the programs and must provide specific economic resources for the action. The diffusion depends strongly by the program and call requirements.

Setting out a toolkit, on the one hand, require hard work to develop a methodology that fits well with the project objective and it is difficult set up this toolkit so as to adapt to several contexts, on the other hand, we want to try to propose an operating scheme that can be applied in AP cases of transnational projects, which is consistent with what INTERREG requires, which integrates the techniques according to the objective to be pursued and is oriented to monitoring.

The single action can be combined with others according with the type of project, the objective fixed and the results to be achieved. For instance, there may be some interesting combinations:

We can say that SWOT method can support and improve (in terms of rationality contribution) the brainstorming and the survey actions:

- The brainstorming technique can be used within SWOT method, because it can channel the information.
- SWOT analysis can be used prior to scenario building to highlight some important factors and help focus the questionnaire and the interviews.

The organization of workshops and seminars favor the networking and the involvement of stakeholders, and are also a good way to exchange practices and knowledge, instrument and experiences. These actions are closely interrelated with each other.

Setting out a toolkit, in terms of a set of recommendations or instruments, or follow a peer review methodological tools depends from project level governance.

In general we can affirm that the combinations of such approaches must aim at two main key criterion features of "Feasibility" and "Effectiveness":

- the short and the long term impact;
- relevance of the action, understood as what makes the difference to other projects.

4 Benchmarking the INTERREG IVc AP Template with the Proposed Model

The INTERREG Programme *(InterregProgrammeManual)* [4], in order to give a clear information and a guide for the application, provide a sample template for the Action Plan.

This is structured in three different parts:

- General information: to decode the project, its name and its type, the partners involved and the identification of the contact person;

- Policy context: what does the project intend to do? About which policy instrument intend to impact?
- Details of the actions envisaged: for the single "Action" there is a checklist that start from the individuation of the background, as the findings from interregional policy learning that are transformed into actions, the description of the "Action" to be implemented, the players involved, the timeframe and the costs (with the indication of the funding source).

Comparing those required information, that we intend as a minimum structure to consider in operative projects, several differences emerge.

The INTERREG Model leads to a not integrated operation: the actions, considered as single element of the plan has to be considered as a stand alone component folly identified with budget and implementation frame. No information are required on the impacts generated by each action and the way we intend to monitor such externalities (positive or negative). The main attention is on demonstrating the connections of the AP with the Policy Instrument and with the policy framework where the AP should be implemented. This is strongly coherent with the strategic view of the INTERREG Program but doesn't represent a supporting stage to improve the IP under a methodological or procedural pint of view.

Finally, the operation dimension is strongly reinforced pushing the Action components over a comprehensive and strategic view of the overall operation. Also stake holders depend by the Action and are defined players.

Our model looks more oriented to ensure the quality of the whole design process and is strongly based on the integrated approach codified by the Logical Framework Approach. In facts the LFA [32] can be very useful for guiding project design and implementation and it is in full agreement with the will of propose an operating scheme for transnational projects, which is consistent with INTERREG requirements. It integrates the techniques and analytical stages according to the objective structure to be pursued and is oriented to monitoring.

The basic ideas behind the LFA are simple and common for any design process: clarity about what and how achieved; degree of depth on the knowledge of the pursue objective choosing monitoring system; and clarity about assumptions and risk condition so succeed to arise or change.

The LFA limitation is that such methodology could lead to a rigid and bureaucratically controlled project design that becomes disconnected from field realities and changing situations. However, the LFA is easy to use more adaptively, as needing future finalization and probably revision, and project management prioritizes annual reviews and logframe updating. So, in practice, a summarized logframe will be useful to provide an overview of the project and for those making decisions about project funding.

Considering the above remarks and integrating the formal recommendation coming from INTERREG Manual for project development we define the final structure of the IP:

- Section A:
  General information: to decode the project, its name and its type, the partners involved and the identification of the contact person;

- Section B:
  Policy context: what does the project intend to do? About which policy instrument intend to impact?
- Section C:

  i. The Strategic approach
  ii. Vision, goals and results
  iii. Implementation procedure
  iv. INPUTS analysis
  v. Indicators and sources of verification
  vi. External conditions assessment
  vii. The IP GANNT.

5   Final Considerations

The IP has the purpose of providing details on how the cooperation can influence and improve the policy instrument tackled. It specifies the nature of the actions to be implemented, their timeframe, the players involved, the costs (if any), procedures and funding sources (if any).

A well-structured Action Plan could define the quality of the whole transnational cooperation project. The first step for objective oriented project planning and management is to define the overall goal. The starting point is to identify the general situation that will be improved.

The general situation is defined through documentary information and data, the identification of the likely primary stakeholders and other interest group (representatives of locally affected communities, national or local government authorities, politicians, civil society organizations and businesses) and what they want to stake in the involvement, the scope of the project, the range of issues that will be addressed and the necessary time. This initial information runs for defining and guiding the exhaustive situation analysis and design steps. After indicating the main objectives, it is required think about the expected impacts of AP implementation considered in a short, medium and long term perspectives with reference to the local context (on going process, policies, experiences).

In this way, all the elements provide a comprehensive framework to determine the feasibility and it is possible define clearly what the design process will be, in terms of who will be involved, how and a what stage, what information needs to be collected and how the final design will be checked with key stakeholders. Indeed, stakeholders help build a common understanding of the current status and the desired changes in practices and outcomes.

From the target level of definition of the overall and strategic goal we can go down to the specific objective, results and scheduled activities.

The list of activities should be drawn up following, not only the objectives but also a financial plan and human resource plan to realize those activities implement the same in a participatory manner. Draw up a finance plan means ensure adequate cash flow. Draw up the operational budget means monitor the recurring expenditure and income effectively. Draw up the capital budget if required is ensure efficient utilization of the resources for the objectives. The strategic human resources planning should serve as a link between human resources management and the overall strategic plan of an organization.

References

1. Mairate, A.: The 'added value' of European union cohesion policy. Reg. Policy **40**, 167–177 (2006)
2. Colomb, C.: The added value of transnational cooperation: towards a new framework for evaluating learning and policy change. Planning Practice & Research **22**, 347–372 (2007). Routledge, T&F Group
3. da European Commission: Available budget 2014–2020, 15 September 2015. http://ec. europa.eu/regional_policy/en/funding/available-budget/. Accessed 30 April 2017
4. Interreg Programme Manual (s.d.). www.interregeurope.eu - Programme Manual. www. interregeurope.eu. https://www.interregeurope.eu/fileadmin/user_upload/documents/Call_ related_documents/Interreg_Europe_Programme_manual.pdf
5. Attolico, A., Scorza, F.: A transnational cooperation perspective for "low carbon economy. In: Gervasi, O., Murgante, B., Misra, S., Rocha, A.M.A.C., Torre, C., Taniar, D., Apduhan, B.O., Stankova, E., Wang, S. (eds.) ICCSA 2016. LNCS, vol. 9786, pp. 636–641. Springer, Cham (2016). doi:10.1007/978-3-319-42085-1_54
6. Archibugi, F.: Planning theory: reconstruction or requiem for planning? Eur. Plan. Stud. **12** (3), 425–444 (2004). doi:10.1080/0965431042000194994
7. Chadwick, G.F.: A Systems View of Planning. Pergamon, Oxford (1971)
8. Alexander, E.R.: Approaches to planning: introducing current planning theories Concepts and Issues. Gordon and Breach, New York (1992)
9. Faludi, A.: A Decision-Centered View of Environmental Planning. Pergamon, Oxford (1987)
10. Mintzberg, H.: The Fall and Rise of Strategic Planning, pp. 107–114. Harvard Business Review, Brighton (2007)
11. Schoeffler, S., Buzzell, R.D., Heany, D.F.: Impact of Strategic Planning on Profit Performance, pp. 137–145. Harvard Business Review, Brighton (1974)
12. Casas, G.L., Scorza, F.: Sustainable planning: a methodological toolkit. In: Gervasi, O., et al. (eds.) ICCSA 2016. LNCS, vol. 9786, pp. 627–635. Springer, Cham (2016). doi:10.1007/ 978-3-319-42085-1_53
13. Las Casas, G., Murgante, B., Scorza, F.: Regional local development strategies benefiting from open data and open tools and an outlook on the renewable energy sources contribution. In: Papa, R., Fistola, R. (eds.) Smart Energy in the Smart City. GET, pp. 275–290. Springer, Cham (2016). doi:10.1007/978-3-319-31157-9_14
14. Blom-Hansen, J.: Principals, agents, and the implementation of EU cohesion policy. J. Eur. Public Policy **12**, 624–648 (2005)
15. Lähteenmäki-Smith, D.: Collective learning through transnational co-operation–the case of Interreg IIIB. Nordregio Working Paper, Stockholm (2006)
16. Organization for Economic Cooperation and Development. In: Handbook on Constructing Composite Indicators: Methodology and User Guide. OECD Publishing, Paris (2008)
17. Organization for Economic Cooperation and Development: How's Life?: Measuring Well-Being. OECD Publishing, Paris (2011)

18. Perotto, E., Canziani, R., Marchesi, R., Butelli, P.: Environmental performance, indicators and measurement uncertainty in EMS context: a case study. J. Clean. Prod. 16(4), 517–530 (2008). doi:10.1016/j.jclepro.2007.01.004
19. Good Practices (s.d). www.interreg4c.eu. http://www.interreg4c.eu/good-practices/index.html. Accessed 10 May 2017
20. Approved project Database (s.d). www.interreg4c.eu. http://www.interreg4c.eu/projects/index.html. Accessed 10 May 2017
21. SUSTAIN Project (s.d.). www.sustain-eu.net Sustain: (http://www.sustain-eu.net/what_are_we_doing/index.htm). Accessed 30 April 2017
22. MISRAL Project (s.d.). www.misrar.nl MiSRaR: http://www.misrar.nl/publications/brochure_1_on_risk_assessment. Accessed 30 April 2017
23. PRESERVE Project (s.d.). www.preserve.aer.eu Preserve: http://preserve.aer.eu/it.html. Accessed 30 April 2017
24. PERIURBAN Project (s.d.). www.periurbanparks.eu PeriurbanParks: http://www.periurbanparks.eu/live/index.php?a=open&id=4c99fc98837e2&ids=4c8ff07964a15&l=en. Accessed 30 April 2017
25. ENERCITEE Project (s.d.). www.enercitee.eu EnercitEE: http://enercitee.eu/#&panel1-4. Accessed 30 April 2017
26. TOURAGE Project (s.d.). www.tourage.eu TourAge: http://www.tourage.eu/. Accessed 30 April 2017
27. BOOGAMES Project (s.d.). www.boogames.eu BooGames: http://www.boogames.eu/. Accessed 30 April 2017
28. Bryson, J.M.: What to do when stakeholders matter. Public Manag. Rev. 6, 21–53 (2007)
29. Dasì, F.: Territorial Governance. ESPON 232 (2006)
30. Nadin, D., Stead, V.: European spatial planning systems. Soc. Models Learn. DISP 44, 35–47 (2012)
31. Faludi, A.: European territorial cooperation and learning. Disp – Plan. Rev. 44, 3–10 (2012)
32. Prasad, D.P.: LFA. Rural Project Management, Chap. 3 (2008)

A Renewed Rational Approach from Liquid Society Towards Anti-fragile Planning

Giuseppe Las Casas and Francesco Scorza[✉]

Laboratory of Urban and Regional Systems Engineering, School of Engineering,
University of Basilicata, 10 Viale dell'Ateneo Lucano, 85100 Potenza, Italy
{giuseppe.lascasas, francesco.scorza}@unibas.it

Abstract. Starting from the Blečić and Cecchini book *"Verso una pianificazione antifragile"* [1], this paper will identify main arguments that: (i) help to deal with the conflicts of a complex society that weakens the connections between pieces of society on all level; (ii) recognize in Z. Bauman thought the elements of concern that characterize the liquidity of our society and its negative connection with urban and regional planning; (iii) highlight in "anti-fragile planning" an innovation instance for the discipline promoting new approaches that starting from the reduction of territorial vulnerability (resistent), are able to promote the regeneration of utility functions (resilient) by involving local communities in a collective form of creativity strategic development form. In one word: anti-fragile.

Keywords: Urbanism · Strategic planning · Resilience · Anti-fragility

1 Introduction

Through a careful consideration on *"Verso una pianificazione antifragile"*, the recent book by Blečić and Cecchini [1], our proposal for a renewed approach in planning rationality [2–4] has identified enrichment with the comparison with two terms that point to two opposing situations:

- the instance of anti-fragile urbanism;
- the society that becomes liquid.

The first refers to an aspiration: overcoming the concept of resilience to engage in creativity forms that make durable the anthropic territorial systems. In the second, through the writings of Bauman [5] is described a society that becomes liquid where the link between individuals and groups become weaker. Differently, the anti-fragile planning by Blečić and Cecchini [1, 6] proposes the research for shared values and scenarios, towards which the system develops its regenerative and creative capacities.

The relevant topics we identified:

- to help in dealing with the conflicts of a complex society that weakens the connections among pieces of society on all levels;
- to recognize in Z. Bauman thought the elements of concern that characterize the liquidity of our society and its negative connection with urban and regional planning;

© Springer International Publishing AG 2017
O. Gervasi et al. (Eds.): ICCSA 2017, Part VI, LNCS 10409, pp. 517–526, 2017.
DOI: 10.1007/978-3-319-62407-5_36

– to highlight in "anti-fragile planning" an innovation instance for the discipline promoting new approaches that, starting from the reduction of territorial vulnerability (resistent), are able to promote the regeneration of utility functions (resilient) by involving local communities in a collective form of creativity in strategic development proposals. In one word: anti-fragile.

– to identify the three principles of planning: *i. equity; ii. efficiency; iii. resource conservation*, as a right for citizens and as a mean to support the identification of a-priori rationality in planning.

As a case study, we cite an experience where conflicts between economic development and natural resources conservation prepare a dramatic consumption of natural resources and socio/cultural values whose intensity allow us to classify them as "disaster". Of course, this is a disaster in a broad sense. It represents a rapid and deep transformation [7, 8] which includes the loss of a complex economic and social tissue in a weak inland area of Basilicata Region (the Agri Valley) where those peculiar characters are consistent with a system of natural resources and landscapes, actually under a not conscious exploitation pressure.

Our thinking is based on three stages of analysis tending to highlight:

• the emphasis on the conflicts and following complexities (the complexity as a daughter of the conflict at different levels: individual's conflicts, conflicts inside the social groups and between groups);

• the diffusion and the intensity of this conflict that we look at to explain, even partially, the breakdown of the social connections that generates liquid society and, consequently, generate complexity and the unpredictability of the evolutionary processes at local and regional level (Connections break, Individualism and liquid assets);

• in trying to answer the question: how to think about the future without foreseeing it? Here resides, in our opinion, the research for an anti-fragile planning that investigates the essential structure of shared development scenario for cities identifying in a renewed planning rationality the minimum set of investments allowing local communities to regenerate, innovate and reconstruct links (anti-fragile and tactical strategies).

In facts, when to the physical dimension of the territory forms is added the necessity to take into account the behaviors of the individuals living there and the socio-economic values associated with them, the level of complexity increases dramatically.

In conclusions, we will try to demonstrate how this consideration can lead to innovations characterized by a strong applicability and a strong disciplinary approach; and how can support a renewed rational approach in the practice of managing the public decision process on territorial transformation.

2 From the Resilient City to Anti-fragile Planning

In the research for renewing tools and models of urban planning, the concept of "resilience" has come to light according to different approaches: from the ecosystemic and behavioral vision by Holling et al. [9], to the general vision of adaptation capacity of a system to external perturbations proposed by many authors ([10–12]). Others conceive resilience as a descriptor of the evolutionary propensity of complex systems [13] and as an opportunity for a development "more" sustainable, ecological and landscape-sensitive [14].

In this perspective, we linked the assessment of urban resilience to the prevention, reduction/mitigation and management of natural hazards (DRRs, DRMs): a major field of action for local government instruments and a global challenge.

The view that authors consider useful for urban and regional planning could be defined "performative", or linked to the ability of a system to quickly reestablish its service level at time T_0, lost as a result of a perturbative event T_1 [15].

There is an effort in the international community to produce "resilience complying" governance tools (at least to adapt existing ones) by coagulating technical and political working groups applying to Disaster Risk Reduction (DRR) and Disaster Risk Management (DRM).

A renewed approach to DRM comes from the "Resilience Approach". We refer in particular to the "Sendai Framework for Disaster Risk Reduction 2015-2030" (SFDRR) [16]. The SFDRR was adopted at the United Nations Third World Conference held in Senday, Japan in March 2015. It is the outcome of UNISDR stakeholders' consultations, started in 2012 and the resulting intergovernmental negotiation phase, launched in July 2014. The SFDRR replaces the previous "Hyogo Framework for Action" (HFA) 2005–2015: Building the Resilience of Nations and Communities to Disasters [17] which promoted disaster risk mitigation and management at global level by integrating policies defined through the International Decade for Natural Disaster Reduction (1989), the Yokohama Strategy for a Safer World, adopted in 1994 and the International Strategy for Disaster Reduction in 1999.

Analyzing the SFDRR, a global strategy emerges to guide and promote an inclusive approach as a key tool for achieving the four proposed operational priorities:

- Priority 1: Understanding disaster risk
- Priority 2: Strengthening disaster risk governance to manage disaster risk
- Priority 3: Investing in disaster risk reduction for resilience
- Priority 4: Enhancing disaster preparedness for effective response and to "Build Back Better" in recovery, rehabilitation and reconstruction.

What is clear is that the DRM approach focuses on the definition of preparatory actions based on an inclusive and participatory multi-actor structure in which the involvement of local communities is one of the major future challenges. The main community oriented actions are:

- empowerment;
- engagement;
- investments (from the scale of soft actions to large infrastructure investments) to reducing territorial vulnerability in all sectors.

These references to international agreements, in fact, lead the actions of governments and territorial transformations towards a planning system defining medium-term "road maps" in which the role of local communities contribute to ensure the feasibility and effectiveness of the results. As a consequence of this reinforced role of local communities is the strengthen of the links between individuals and groups and the conservation of social, economic and environmental security against liquidity paradigm.

It results a governance system that coagulates the actions of individuals and interest groups towards the shared goals of reducing vulnerability of human systems. This complex process generates knowledge, skills and development opportunities for such engaged communities. In other words, it is a "creative" process of programming and transformations governance in terms of new fields made available by the action of the local authorities in favor of citizens' initiative.

This vision makes more coherent our proposal with the concept of antifragility proposed by Cecchini and Blečić [1] which goes beyond a performance view of to the concept of resilience suggesting a perspective of creative multi-actor engagement.

The United Nations New Urban Agenda [17–19] places among the key elements of its vision the focus on the reduction and management of natural and man-made risks, the protection of ecosystem resources and the promotion of "civic engagement" as a process of participation and inclusion in a territorial governance system that raises the role of urban and regional planning. It explicitly refers to the plan as a rational tool for adopting sustainable, people-centered, age- and gender-responsive and integrated approaches to urban and territorial development by implementing policies, strategies, capacity development and actions at all levels, based on fundamental drivers of change, including:

(i)   Developing and implementing urban policies at the appropriate level, including in local–national and multi-stakeholder partnerships, building integrated systems of cities and human settlements, and promoting cooperation among all levels of government to enable them to achieve sustainable integrated urban development;

(ii) Strengthening urban governance, with sound institutions and mechanisms that empower and include urban stakeholders, as well as appropriate checks and balances, providing predictability and coherence in urban development plans to enable social inclusion, sustained, inclusive and sustainable economic growth, and environmental protection;

(iii) Reinvigorating long-term and integrated urban and territorial planning and design in order to optimize the spatial dimension of the urban form and deliver the positive outcomes of urbanization;

(iv) The support of effective, innovative and sustainable financing frameworks and instruments enabling strengthened municipal finance and local fiscal systems in order to create, sustain and share the value generated by sustainable urban development in an inclusive manner.

We completely agree with these UN recommendations highlighting the profound contrasts that exist with a "in-progress" vision of the resilient city concept. What is most worrying is the transposition of the concept of liquid society explained by

Baumann[1] in an evolution of the contemporary urbanism. One could think that a liquid city is the maximum level of a resilient city because of its capacity of maximum adaptability to any form of stress. In our view liquid city id based on the casual braking down of functional, affective, traditional links connecting parts of the city without any opportunity of a good regeneration.

On the contrary we see this improper association among resilience, liquidity and transformation concepts as leading to a degenerative discussion where prospects for urban development and urban regeneration should facilitate urban transformation looking for new conveniences. It's the case where public-private partnership promote transformations out of predefined accords (the plans) and consequently, often, generate small group conveniences rather than collective interest.

As Blecic and Cecchini express [1], as the tension in predicting future events remains a feature of planning, we are facing with increasing levels of uncertainty depending on the complexity of assessing the dynamic processes of the city as well as the impossibility to entirely grasp individual behaviors and choices. The approach of bounded rationality (see [20, 21]) remains valid within these last considerations especially in the research for comparing decisions and alternative scenarios with the three principles discussed in the following paragraph.

3 Conflicts, Liquid Society and Anti-fragile Strategies

Eco says [22]: *Is there a way to survive to liquidity? Yes there is. We have to realize we live in a liquid society and in order to understand it and also – hopefully- to overcame it we need new tools. The problem remains in the politic and in the most part of the intelligentsia that still not understood the extend of the phenomenon. Baumann remains a "vox clamantis in deserto".*

We will focus on conflicts considered as a liquefaction factor that considers the instability of agreements among the components of society, from which depend territorial planning choices.

The experience of the PSI (Intermunicipal Structural Plan) of the Val d'Agri [4, 23] has offered us the opportunity to verify how the conflicts' overlap are related to the most dramatic problem: the conflict between nature conservation and industrial activities characterized by significant emissions and dramatic socio-economic effects. In the case of Agri Valley, it is possible to describe a high level of complexity in conflicts overlapping among the groups of actors, among individuals inside each group and, finally, any individual incoherence.

Probably in this complex structure of conflicts lies the complexity that, according to Simon [20], limits the possibility of knowledge about the planning context and imposes limitations to plan rationality (Fig. 1).

---

[1] It would be a worse hypothesis within this reason to consider Baumann's idea as a model for contemporary society by committing the error of those who confuse the negative position with which the author interprets with an inevitable aim to which the covenant must be adapted Social with critical consequences for urban planning and its principles.

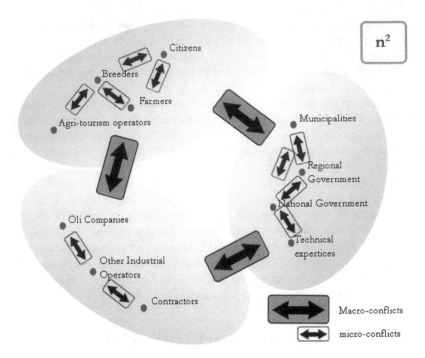

Fig. 1. Conflicts between groups and actors - the Val d'Agri case

The conflicts examined in the case of the Val d'Agri can be identified on two levels: those emerging between "codifiable" groups - that is to say those categories of decision makers interacting within plan decision making process; and internal conflicts to those macro groups marking of disaggregated social tissue (i.e. "liquidity"). This happens due to the absence of appropriate means of knowledge on sensitive issues: pollution, health risks, loss of identity resources, a persistent disadvantage of opportunities for residents (in terms of work, services, infrastructures) in front of a territory facing an "announced disaster".

Our disciplinary position is based on the statement that the new tools referred by Eco [22] are those of a renewed approach to the plan rationality in order to reduce the impact of the lack of knowledge that determines the limits of rationality according to Simon [20, 21].

To answer Eco's questions, in our experience of PSI development we experimented two operational levels:

- Anti-fragile strategies to be confronted with communities;
- Procedures that clarify an implementation accompanied by monitoring of transformations (tactics).

This approach [24] is based on the application of the following three principles, as a solid anchor point for the identification process of the hierarchical structure of problems based on a cause-effect relation and the consequent formulation of a hierarchical structure of the objectives (we call it "Program Structure").

Following Faludi [25], in fact, the question arises whether the problem is identified before of the objective according with the assumption that "the problem is what is meant to achieve an objective and the objective exists in order to remove the problem" and the answer was given by the three planning principles [26]:

- equity;
- efficacy;
- unsupported resource conservation.

Their acceptance is given as universal as they correspond to the essence of the social contract that would ensure the cohabitation of members of a community. The constitutions of many countries rest on them. So in applying these basic principles we affirm that in all that cases in which one or more of these principles are unrespected "problems" emerge, and to remove those problems is aimed objectives elicitation and planning strategies.

It seems that the need for planning secure territory represents a statement of aggregative intentions that arise with the natural disasters increase, international agreements, widespread literature.

In our view it clearly appears that Bauman's liquefaction is precisely about tampering with these principles in favor of searching for the prevalence of an individual over another. Consequently, any strategy could be usefully be proposed in the field of public decision, without defending the application of those fundamental principles.

In addition, the acceptance of the three principles determines in every case another level of complexity: the need to manage competing principles such as equity vs effectiveness and ending with creating further conflict between ethics and individual or group interests.

It imposes a reflection on the disciplinary renewal of planning in order to face an operative and working role that oscillates between two opposite dimension: idealism and compromise bending to the interests of emerging powers leading to the negation of the plan[2].

4 Reasearch Perspectives

Considering the periodic focus on safety and security, from a disciplinary point of view, our position looks to highlight that instance of *preparedness*, usually invoked in the context of Civil Protection Plans and systematically extended to a comprehensive planning scenario. The research for Minimal Urban Structures (MUS) [27] and the definition of limitations in transformation levels related to guarantee territorial and urban security is a good example of current practices which represents an argument for legitimation of planning discipline. Indeed the most suitable intervention are those that realize safety with maximum efficacy.

---

[2] The cases of decisions that in the name of acquired rights, or the need to cover public debts, or to accomplish instances of flexibility are available to introduce variants of the plan which, in fact, deny its robustness.

The implementation of this renewed approach to plan rationality can take place on two levels:

- *"a tavolino" (on the desk)* it means as a theoretical exercise of planning developed in a technical dimension avoiding the debate with decision makers and relevant stakeholder
- on the field developing appropriate participation.

In the first case uncertainty remains, while working on field, accepted the ethical principle of sharing uncertainty and in accordance with the most recent international guides [18], comes out the usefulness of promoting participation. Participation represents a necessary condition for urban and territorial governance in a creative direction but should be extended beyond the only stakeholders that traditionally have the possibility to decide. It has to include the other groups that traditionally remain the "object" of the decisions. Of course, it should include an extensive use of suitable knowledge-structuring tools [4].

We mention Leone and Zoppi [28]: *"Participation and participatory process remain a central element of modern society, representing a prerequisite and a democratic right in Western nations. However, ..., the participatory practices has resulted in some criticism and problematic aspects due to the ambivalent nature and concept of participation. .... Governments, sometimes, implement inclusive procedure in order to reinforce the existing power relations..."*. They affirm the right to inclusive participation and, at the same time, the great risk of unfair exploitation of the participative processes. Anti-fragile strategies then become the ones that go in the direction of consolidating positive right and, with reference to the three principles of rationality in plan, ensure rules and appropriate tools for the implementation of participatory forms within a planning process that will eventually finalize investment, protection and transformation choices.

Acknowledgement. The research work was developed in the framework of the contribution of the Engineering Laboratory of Urban and Territorial Systems of the School of Engineering of the University of Basilicata to the Basilicata Region for the formation of the Inter-Municipal Structural Plan of the Val d'Agri. The paper represents a joint reflection of the authors (Sects. 1, 4). Section 2 is mainly edited by Francesco Scorza while Sect. 3 by Giuseppe Las Casas.

References

1. Blečić, I., Cecchini, A.: Verso una pianificazione antifragile. FrancoAngeli, Milano (2016). ISBN 978-88-917-2775-6
2. Las Casas, G.B., Sansone, A.: Un approccio rinnovato alla razionalità nel piano. In: Depilano, G. (ed.) Politiche e strumenti per il recupero urbano. EdicomEdizioni, Monfalcone (2004)
3. Las Casas, G.B., Scorza, F.: Un approccio "context-based" e "valutazione integrata" per il futuro della programmazione operativa regionale in Europa". In: Bramanti, A., Salone, C. (eds.) Lo Sviluppo Territoriale Nell'economia Della Conoscenza: Teorie, Attori Strategie, Collana Scienze Regionali, vol. 41. FrancoAngeli, Milano (2009)

4. Los Casas, G., Scorza, F.: Sustainable planning: a methodological toolkit. In: Gervasi, O., et al. (eds.) ICCSA 2016. LNCS, vol. 9786, pp. 627–635. Springer, Cham (2016). doi:10. 1007/978-3-319-42085-1_53
5. Bauman, Z.: Liquid Modernity. Wiley, Hoboken (2013)
6. Blečić, I., Cecchini, A.: On the antifragility of cities and of their buildings. City, Territ. Archit. 4(1), 3 (2017). doi:10.1186/s40410-016-0059-4
7. Petrillo, A.S., Prosperi, D.C.: Metaphors from the resilience literature: guidance for planners. In: Proceedings REAL CORP 2011, Tagungsband 18–20 May 2011, Essen (2011). ISBN: 978-3-9503110-0-6. http://www.corp.at
8. Prosperi, D.C., Morgado, S.: Resilience and transformation: can we have both? In: Proceedings REAL CORP 2011, Tagungsband 18–20 May 2011, Essen. (2011). ISBN: 978-3-9503110-0-6. http://www.corp.at
9. Holling, C.S., Gunderson, L., Lance, H.: Resilience and adaptive cycles. In: Gunderson, L., Holling, C.S. (eds.) Panarchy: Understanding Transformations in Human and Natural Systems. Island Press, Washington, D.C. (2002)
10. Pickett, S.T.A., Cadenasso, M.L., Grove, J.M.: Resilient cities, meaning, models, and metaphor for integrating the ecological, socio-economic, and planning realms. Landscape Urban Plan. 69, 369–384 (2004)
11. Carpenter, S.R., Westley, F., Turner, G.: Surrogates for resilience of social–ecological systems. Ecosystems 8, 941–944 (2005)
12. Wilkinson, C.: Social-ecological resilience: insights and issues for planning theory. Plan. Theory 11(2), 148–169 (2012). doi:10.1177/1473095211426274
13. Davoudi, S.: Resilience: a bridging concept or a dead end? Plan. Theory Pract. 13(2), 299–307 (2012)
14. Ricci, M.: Nuovi Paradigmi. LISt Lab, Trento (2012)
15. Las Casas, G., Scardaccione, G.: Contributi per una Geografia del Rischio sismico: analisi della vulnerabilità e danno differito. In: Di Gangi, M. (ed.) Modelli e metodi per l'analisi delle reti di trasporto in condizioni di emergenza: contributi metodologici ed applicativi. Potenza, Ermes (2005). ISBN 9788887687705
16. UNISDR: Sendai framework for disaster risk reduction 2015–2030, Geneva, Switzerland (2015)
17. UN HABITAT: International Guidelines on Urban and Territorial Planning, UN-Habitat (2015)
18. UN HABITAT: New Urban Agenda, UN-Habitat (2016)
19. UN HABITAT: Action Framework for Implementation of the New Urban Agenda UN UN-Habitat (2017)
20. Simon, H.A.: Models of bounded rationality: empirically grounded economic reason. MIT press, Cambridge (1982)
21. Simon, H.A.: Theories of bounded rationality. Decis. Organ. 1(1), 161–176 (1972)
22. Eco, U.: La società liquida. Con questa idea Bauman illustra l'assenza di qualunque riferimento "solido" per l'uomo di oggi. Con conseguenze ancora tutte da capire. Repubblica L'Espresso (2015). http://espresso.repubblica.it/opinioni/la-bustina-di-minerva/2015/05/27/news/la-societa-liquida-1.214625?refresh_ce. Accessed 12 May 2017
23. Las Casas, G.B., Scorza, F.: I conflitti fra lo sviluppo economico e l'ambiente: strumenti di controllo. In: Atti della XIX Conferenza nazionale SIU, Cambiamenti, Responsabilità e strumenti per l'urbanistica a servizio del paese, Catania 16–18 Giugno 2016. Planum Publisher, Roma-Milano (2017)
24. Las Casas, G.B.: L'etica della Razionalità" In: Urbanistica e Informazioni, vol. 144 (1995)
25. Faludi, A.: A decision-centred view of environmental planning. Elsevier, Amsterdam (1985)

26. Las Casas, G.: Governo del territorio innovare la ricerca per innovare l'esercizio professionale. In: Francini, M. (ed.) Modelli di sviluppo di paesaggi rurali di pregio ambientale. Franco Angeli, Milano (2011). ISBN 9788856840308
27. Fabietti, W. (ed.): Vulnerabilità e trasformazione dello spazio urbano, vol. 8. Alinea Editrice, Florence (1999)
28. Leone, F., Zoppi, C.: Participatory Processes and Spatial Planning. The Regional Landscape Plan of Sardinia, The Regional Landscape Plan of Sardinia. Franco Angeli, Milano (2016). ISBN 9788891740984

Conflicts Between Environmental Protection and Energy Regeneration of the Historic Heritage in the Case of the City of Matera: Tools for Assessing and Dimensioning of Sustainable Energy Action Plans (SEAP)

Francesco Scorza[✉], Luigi Santopietro, Beatrice Giuzio,
Federico Amato, Beniamino Murgante, and Giuseppe Las Casas

Laboratory of Urban and Regional Systems Engineering, School of Engineering,
University of Basilicata, 10, Viale dell'Ateneo Lucano, 85100 Potenza, Italy
{francesco.scorza, luigi.santopietro, beatrice.giuzio,
federico.amato, beniamino.murgante,
giuseppe.lascasas}@unibas.it

Abstract. The effort towards the reduction of energy consumption, reduction of emissions and the adoption of Renewable Energy production technologies produced significant spatial and urban transformations. In terms of environmental impact assessment, a structural contradiction between a system of governance that promotes renewable plants, an economic system ready to invest huge resources and high profitability, a weak system of territorial planning rules and instruments of landscape protection not yet adequate to govern such transformations.

This paper proposes a local case study (the city of Matera) where the ex-ante evaluation of investment programs for the energy regeneration of the public housing stock under the Covenant of Mayors has to be compared with the preservation objectives of an unique historical settlements ("*i sassi*"). In fact, the city, elected European Capital of Culture 2019, has characteristics of unique historical and architectural value of historical value. On it they act the signs of a PRG dated and the management rules of the UNESCO site most recently adopted (2014).

The Municipalities adopted the Sustainable Energy Action Plan (SEAP) - a new category of instrument of urban government which includes strategies and methods of urban transformation - but the intervention scenario not considered the integration of RES plants and technologies with historical settlements.

This paper, starting from remote sensing assessment of local radiation index, proposes a methodology to improve the integration between the issue of implementing RES at urban scale and to preserve traditional settlements in a sustainable perspective.

Keywords: Sustainable energy planning · RES · Urban renewal

© Springer International Publishing AG 2017
O. Gervasi et al. (Eds.): ICCSA 2017, Part VI, LNCS 10409, pp. 527–539, 2017.
DOI: 10.1007/978-3-319-62407-5_37

1 Introduction

This work proposes an in-depth analysis of the potential use of energy-saving technologies in urban areas based on the Global Horizontal Irradiance Index (GHI) an assessment [1] (it is a measure of solar radiation incident in a territorial scale through remote data) and consequent scenarios of Renewable Energy Sources (RES) technologies use within the European Union policy framework: the Covenant of Mayors [2, 3].

The adoption of RES technologies is a topical issue about several scenarios: development and adoption of innovative technologies, economic dynamics and RES resource development, local contributions to climate change adaptation, regeneration processes and sustainable urban development.

With reference to the paper's goals, the relationship between potential energy production related with the installation of photovoltaic and solar thermal technologies is compared with restrictions and criteria for the buildings changes and urban spaces transformability.

The case study of the Municipality of Matera highlights, on the one hand, the high attitude to reduce energy consumption and to regenerate public buildings, in the other hand, this situation is contrasted with a strict system of rules for the historic settlement protection. Then, it is necessary to balance these instances in order to define an intervention scenario well-matched with sustainability concept that includes environmental, social and local identity issues.

The 2008 marked a significant milestone in the climate change topic [4, 5] in the EU policies framework: the establishment by the European Commission of the Covenant of Mayors [2].

The first citizens together, voluntarily, with their communities join in to act concretely and locally to reduce energy consumption and emissions following the EU 202020 target [6]. This target means overcome 20% reduction of CO_2 emissions [7]. When the agreement is signed, the PAES is formed: Action Plan for Sustainable Energy.

In January 2015, Matera adopted its own PAES, aiming to reduce CO_2 emissions by 20.5% through a planned series of actions to reduce consumption and to improve the use of renewable resources, both in the public and in the private sector.

These actions are inside a wider program and management process related to the European Capital of Culture Matera's nomination [8, 9].

In this paper, according with Matera's PAES and its rules in terms of public investment policies for energy saving, has been applied a procedure to understand the scenario related to photovoltaic technologies. Starting from a punctual estimation of energy productivity on a set of public buildings selected (without architectural conservation constraints), an intervention scenario has been designed which allows an increase in emission targets of CO_2 declared in the PAES [10]. The result is schematically shown in an update proposal intervention form about the PAES photovoltaic technology.

The first section of this work introduces a general overview of the EU measures taken to counter climate change, from the early 1990s to the latest ones. Starting from the EU general framework, we have gone through an analysis of the state of implementation of the European policies of the Covenant of Mayors in Basilicata, highlighting the level of accession to the COM and the elaboration of the PAES of the Lucan municipalities. It is a feature characterizing the presence in the Basilicata Region of a public entity: Società Energetica Lucana S.P.A. (SEL), which provides technical support in the drafting of the PAES. In the second part, the PAES of the Municipality of Matera has been analyzed, with particular attention to: electrical energy consumption of public and municipal buildings (municipal and provincial), the expected consumption reduction targets for 2020, the actions taken during 2009–2012 (which we call "A") and those scheduled during the period 2013–2020 (which we call "B") and the verification of what has been done without the official monitoring report. The third part of this work describes the used data and the related analysis and evaluation processes. In addition to the remarkable reproducibility of the procedures adopted in this research, the conclusions show that the proposal developed in this paper contributes to raising the target of energy efficiency and environmental sustainability set for 2020 by the Municipality of Matera: a further "topic" for the role Of European Capital of Culture 2019.

2 From European Policy on Climate Change to Sustainable Energy Action Plans

The European Union in the early 1990s has acted globally to counteract climate change on the Earth considering that climate in the early 1980s was defined, according to ONU, "shared resource of mankind".

European climate policy could be expressed with following dates: in 1992 European Union with all of its members joined the United Nations Framework Convention on Climate Change (UNFCCC) [11], the main international treaty on climate change. Successively in 1997 there was the Kyoto Protocol subscription [12]: the first step on reducing greenhouse gas emissions. In 1998 there were set the reduction targets of emissions for 15 European States of that period, choosing as common target the reduction of 8% according to the 1990s levels. The period 2003–2013 was characterized by the accession of several regulatory instruments to encourage an improvement of Kyoto goals. In period 2013–2020 European Union joined Climate Change Package (Integrated Energy and Climate Change Package, IECCP) [6]. IECCP reserves Member States of EU to be achieved by 2020 following goals:

- Energy production from renewable sources equal to 20% of energy consumption and use of biofuels equal to 10% in transports;
- Reduction of greenhouse gas emissions equal to 20% according to 1990;
- Reduction of energy consumption equal to 20% according to baseline scenario improving energy efficiency.

In December 2015, it was joined Paris agreement [13], adopted by all parts of UNFCCC: it is the first-ever universal, legally binding global climate deal enter into force by 2020.

European Union established key objectives every ten years to 2020 [6]: reducing equal to 20% greenhouse gas emissions according to 1990, improving equal to 20% the percentage of renewable energies and improving at least 27% energy efficiency. Instead for 2030 the European Union established following objectives: reducing at least 40% greenhouse gas emissions according to 1990, improving at least 27% the percentage of renewable energies and improving at least 27% energy efficiency [2]. No later than 2050, European Union aims to reduce its emissions substantially about 80–95% according to 1990s levels in the endeavor required from advanced countries [2].

2.1 Covenant of Mayors

The Covenant of Mayors is a singular movement "bottom up" that is successful with wide margin of success mobilizing a lot of local and regional authority, encouraging to develop action plans and directing their efforts to climate change mitigation. In 2008, after the adoption of the Integrated Energy and Climate Change Package EU 2020, European Commission promoted Covenant of Mayors to endorse and supporting efforts made by local authorities in implementation of policies in the area of renewable energy.

The new Covenant of Mayors for Climate and Energy was launched on 15 October 2015. In that occasion, fundamental pillars reinforced were shown: climate change mitigation, adaptation and access to secure, sustainable and affordable energy for all. Signatories have a shared vision to 2050: accelerating the decarburization of our territories, strengthening our capacities to adapt to unavoidable climate change impacts, thus making our territories more resilient, increasing energy efficiency and the use of renewable energy sources on our territories, thus ensuring universal access to secure, sustainable and affordable energy services for all.

By 2030, signatories commit to reducing carbon emissions across their territory by at least 40%, to increasing their resilience to the impacts of climate change. Following the successful of Covenant of Mayors, in 2014 started Mayors Adapt [14], that is based on the same model of governance, encouraging policy commitments and adopting prevention actions preparing cities to unavoidable climate change. At the end of 2015 the initiatives merged into the Covenant of Mayors for Climate and Energy, that adopted EU 2030 objectives and an integrated approach to mitigation and adaptation to climate change. Adaptation to climate change is necessary to contrast their negative effects and save resources. Turn own political commitments into real activities, the signatories of the Covenant of Mayors have to edit Baseline Emission Inventory and a Risk and Vulnerability Assessment. The signatories, within two years from their accession, undertake to edit a SECAP that outlines the main action that authorities have to plan to do.

3   The Actualization of Covenant of Mayors in Basilicata (IT): The SEAP of Matera

Regionally, reduction of energy consumption is one of main goal of PIEAR – Energy Regional Environmental Plan [15]. The Basilicata Region wants to achieve, targets set by UE and Italian government, an increase of energy efficiency that allows in 2020 a reduction of energy demand in the amount of 20% according to the same expected from this period[1]. The PIEAR provides that increase of electricity production from RES, expected to 2020, will be achieved with following planning:

- Wind farm for 981 MWe;
- PV systems for 359 MWe;
- Biomass installations for 50 MWe;
- Hydroelectric installations for 40 MWe.

Starting from January 2017, 63[2] Municipalities joined the Covenant of Mayors (at 2016), where 33 are in province of Potenza and 33 in province of Matera.

In the following feature we have overall view of Municipalities joined to Covenant of Mayors (Fig. 1).

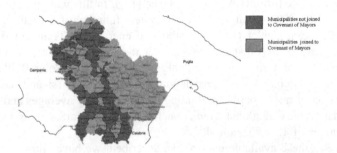

Fig. 1. Basilicata municipalities and signature to Covenant of Mayors.

3.1 The 'Società Energetica Lucana' (SEL) and the Support for Editing SEAP to Local Authorities

Regional Council in Basilicata with regional law (see [16]), n. 13 of 31-07-2006 promoted the foundation of SEL, a completely public company, that manages and improves the energy demand encouraging RES installation. SEL provides support to local authorities for editing SEAP from signature to Covenant of Mayors to following steps as monitoring and evaluation of achieved goals. Municipalities turned to SEL are exclusively from province of Potenza.

The province of Potenza, on 25 September 2012, was identified as support structure to Covenant of Mayors. Main goal of the Province is encouraging municipalities to join Covenant of Mayors, helping them to processing SEAP.

The Province of Potenza reinforced an important international promotion of its energy policy and adaptation to climate change considering local resilience [17, 18].

Since 2016, it is leader of an international cooperation project: LOCARBO (Novel roles of regional and LOcal authorities in supporting energy consumers' behaviour change towards a low CARBOn economy" - Program INTERREG EUROPE

---

[1] The energy demand expected is given by ENEA Trisaia, September 2011.

[2] Information collected from website http://www.tuttitalia.it/associazioni/patto-dei-sindaci/basilicata/ (Accessed on January 2017).

2014–2020) [17]. The project aim is improving instrument of policy-making to increase energy efficiency joined to build environment through participation of stakeholder and users.

4 From GHI to Pre-sizing of the Solar Power Production Scenario

In this work, we have considered some inclined and non-flat photovoltaic (PV) systems, so it was necessary to search for an inclination of the PV systems that was optimal to obtain the best efficiency in terms of solar radiation, considering that the optimal angle of inclination β of PV systems depends on the latitude φ and solar declination δ with the formula $\beta_{ott} = \varphi - \delta$ (see [19]). In this way, we have chosen to adopt an optimal inclination angle of the PV system function of the latitude according to the Joint Research Center (JRT) – Institute of energy and Transport (IET) of EU Commission (see [20, 21]) that achieved a solar radiation database from climatologic data homogenized for Europe and available in the European Solar Radiation Atlas, using the r.sun model and the interpolation techniques s.vol.rst and s.surf.rst. The database consists of raster maps representing twelve monthly averages and one annual average of daily sums of global irradiation for horizontal surfaces, as well as those inclined at angles of 15°, 25°, and 40°.

According to these available data, above described, we have chosen to adopt the ENEA guide "Progettare e installare un impianto fotovoltaico" (2008) (see [22]). The produced energy of the PV system is given by the expression:

$$Ep = H \times S \times Eff.pv \times Eff.inv = H \times Pnom \times (1 - Ppv) \times (1 - Pinv)$$

Where:

- Ppv are the losses (thermal, optical, resistive, falling on diodes, mismatch) of PV system, measured to a first approximation about 15%;
- Pinv are the losses (resistive, switching, magnetic, power control circuits) of the inverter that are provisionally assumed of about 10%;
- Pnom is the power rating of the PV system necessary to produce the energy Ep;
- H is solar irradiation on the surfaces (S).

5 An Application of a Case Study: The City of Matera

The short description showed below about the local SEAP of Matera (see [23]), is the reference where we have taken into consideration the data about the energy local production.

There is an analysis of the urban local planning useful to identify the historical heritage. It is important to remember that Matera is a UNESCO site with a management planning, that is an instrument to preserve the historical buildings. For this reason, we want to build PV systems in accordance with safeguard of the historical town. So we have considered only the public buildings that had electric and thermic consumptions, ruling out those on which there have just been carried out operations or it was possible to make a prediction of them. The final part deals with the sizing and an estimate of costs of the operations even in relation with criteria adopted in accordance with best condition of PV systems.

5.1 The SEAP Prediction and the Proposal for Intervention

From 2009 to 2012 the energy local production refereed to local PV systems under 100 kWp, is increased by 429,45%, passing from 1.495,40 MWh/year in 2009 to 7.917,34 MWh/year in 2012 (Table 1).

Table 1. PV local energy production in period A

Local PV energy production - period A		
Production MWh	Reduction CO2 t/a	Percentage contribution to the CO2% reduction
6.421,94	2.039,80	0,85%

In period A it is expected an increase of the local electric energy production using PV systems with promotion and the their promotion. The aim of this action is to promote the installation of PV systems on public buildings (Table 2).

Table 2. PV local energy production in period B

Local PV Energy production - period B		
Risparmio MWh/a	Reduction CO2 t/a	Percentage contribution to the CO2% reduction
1.495,40	474,98	0,16%

As a critical feature of Matera SEAP, no many efforts are promoted to increase users' capacity to 'self energy management' practices [24].

As regards the proposal of operations, using RSDI Geoportal of the Basilicata Region (see [25]) it was possible to derive with the conceptual map of the DBGT Geotopographic Database in shape file format, especially the classes "edifici" and "unità volumetriche." In the following figures are represented the logical criteria of the buildings selection and the proposal of operations are shown (Figs. 2 and 3).

5.2 Proposal of Operation and Estimate of Costs

To each building previously selected it was associated GHI value [1], that will be adopted for the sizing of the PV system necessary to copy entirely the energy needs have shown in the table of consumption inside the SEAP.

According to the SEAP predictions, there will be a RES production by 100% in 2020 compared to the last consumptions of 2012. Therefore, we started to the last

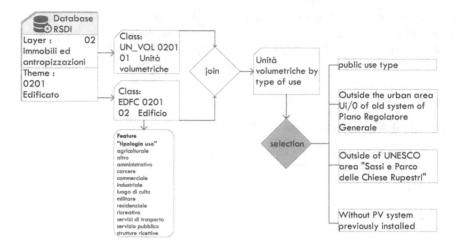

Fig. 2. Logic scheme of the selection of the proposal operations

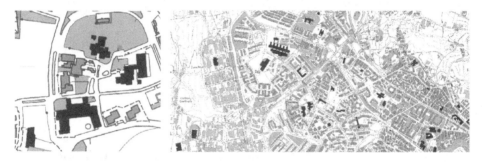

Fig. 3. Some representations of the urban areas where the public buildings have been selected for the PV system energetic estimate

available data of electric energy consumption and we carried out a general sizing of the PV system using the ENEA guide (cfr. [22]). In Table 3 data and results for each building are shown.

The hypothesis suggested in this work and shown in Table 4, consists of the integration of the action N27-PE.1B provided by SEAP of Matera town and referred to the buildings that are potentially changeable, that is those of public use selected according to qualifying criteria for installation of PV systems with aim of the energetic need of each building using the GHI application estimated according to the remote sensing.

After the pre-sizing it was possible to develop a general economic estimate of the operations necessary to the implementation of these systems. According to the SEAP of Matera town the investment cost planned for middle market is 3500 €/kWp.

Table 3. Total covered surface of the chosen public buildings (source SEAP MATERA [23])

Building	Energetic consumption SEAP 2012 kWh/year	PV system surface m^2	% of covered surface
I.T.C.G Stella via E. Mattei	88504	168,95	4,63
I.T.I.S. via Virgilio	163698	312,49	2,4
Istituto Alberghiero via Castello	60411	113,5	9,18
Istituto Professionale maschile-Da Rondinelle	29117	55,43	0,87
Istituto Ragioneria	75579	143,82	6,43
Istituto Tecnico Agrario	98637	187,78	3,72
Liceo Artistico via Cappuccini	33958	63,73	4,24
Liceo Classico via Nazioni Unite	39532	75,22	3,2
Liceo Scientifico viale Europa	79242	150,79	5,1
Patrimonio - via Aldo Moro, snc	550643	1047,81	58,38
Patrimonio Piazza Monte Grappa 11	86	0,16	0,01
Patrimonio piazza S. Agnese 7	16752	31,44	3,4
Patrimonio via G. Saragat 44A	11508	21,9	2,63
Polizia Locale - via Trabaci	64027	121,73	11,33
Scula Materna viale Quercia 3	7555	14,38	2,15
Scuola - via Frangione 4	2924	5,49	0,93
Scuola Elementare via Bramante 8	16854	32,07	1,59
Scuola Elementare via Guglielmo Marconi2	38867	73,96	4,89
Scuola Elementare via Lucrezio snc	9610	18,34	1,36
Scuola Elementare via Nitti	25326	48,19	2,95
Scuola Elementare via S. Pardo	24376	46,38	12,05
Scuola Materna via Cererie snc	5510	10,48	1,17
Scuola Materna via Emilia snc	7390	14,06	1,08
Scuola Materna via Meucci 2	4319	8,22	3,27
Scuola Materna via Morelli Marcello snc	3832	7,2	0,66
Scuola Materna via S.Giovanni Mate 3	2821	5,37	1,16
Scuola Materna via Vulture snc	6261	11,79	2,07
Scuola Media ed Elementare (Francesco T.)	35652	67,84	3,98
Scuola Media Francesco Torrata	40914	77,85	3,3
Scuola Media Giovanni Pascoli viale Parini snc	64079	121,93	17,57
Scuola Media Nicola Festa via Lanera, 59	34732	66,09	3,58
Scuola Media via Fermi 8	12006	0,86	0,03
Scuola Superiore Professionale Femminile Isabella Morra	69535	132,32	5,33
Uffici Giudiziari via Aldo Moro snc	676466	1287,24	29,93

Table 4. Action N27-PE.1B for the estimated buildings

ACTION CODE N27-PE.1B	SECTOR	LOCAL ENERGY PRODUCTION
	AREA OF INTEREST	Renewable energies
	KEYWORDS	**Promotion and encouragement to installation of PV systems on public and private buildings and other renewable energies systems (RES)**
	CONNECTED	N31-AL.4B; N11-EP.1B
Description	The action considers installation of PV systems on the covered surface of each municipal/province public building, to satisfay the energy needs according to the last consuption inside the SEAP. PV system installed will be with following characteristics: losses of PV system measured to a first approximation about 15%; losses of the inverter that are provvisionally assumed of about 10% and an optimally-inclined surface of 35°. This action will agree with: decrising consuption of elecrticity, decrising emissions of CO_2 and reducing the cost of the electricity consuption. Planned use of covered surfaces, will make it possible to exploit only the covered surface necessary to achieve totally energy needs of each building, so the leftover covered surface will be available to other RES.	
Area of Interest	Production from renewable energies	
Policy instruments	National and Regional Incentives	
Responsible for implementation	Municipatility of Matera, Province of Matera	
Activation period and	2013-2020	
Costs	€ 5.333.517,891	
Percentage of CO_2 provided (t/year)	846,01 tCO_2/year	
Renewable energy production provided (MWh)	100% of the building energy production	
Monitoring	Marker	Power rating of PV installed
	Frequency of monitoring	Annual
	Instruments and systems for monitoring	Analisys of Atlasole web data
	Supervaisor of monitoring activity	Energy Manager/ Energy Bureau

If we wanted to know and update the costs analysis, we have to consider the information inside contained in "Prezzario Regionale delle OOPP della Basilicata"[3] 2015 edition. It is considered appropriate to compare the estimate the SEAP prices of Matera town and "Prezzario Regionale delle OOPP of Basilicata", according to the national scale indicated in the article of Sole 24 Ore on 30 September 2016 written by Dario Aquaro. In the following scheme the estimate costs of operations are shown (Table 5).

---

[3] Price list of public buildings of Basilicata edition 2015 where in R section: "Sistemi per lo sfruttamento delle energie rinnovabili e l'uso razionale delle fonti energetiche", are described single items wich are part of RES technologies.

Table 5. Estimate investments costs

Estimate of collective investments			
kW installed	Annual MWh produced	Total emission reduction of tCO 2/year	Estimate cost €
1.828,00	2.400,00	846,01	5.333.517,891

6 Conclusions

Our proposal discussed in this work, refers to the GHI application, that was obtained through remote sensing data PVGIS-CMSAF[4] in order to define a pre-sizing procedure of PV systems inside the operation strategy provided by the SEAP.

The case study of Matera town, allowed to consider the necessity to compare the issues to contribute and promote energy renewal of existing buildings with the preservation issues connected to the historical and architectonical values of traditional built patrimony.

The town of Matera is a UNESCO site [26] is a relevant case study as it is characterized by a wide presence of public buildings inside the historical urban area. A work hypothesis is that of the target achievement of the energetic needs of the public heritage. Thanks to this work, it was possible to verify the possibility of this target even in presence of a wide preservation concerning the urban area in Matera town.

Through this methodological procedure it is possible build RES technologies according to the preservation of the historical heritage. This problem can be overcame if we take in consideration a dimension and an energetic balance on a larger scale. In fact, we suggest inside the SEAP, to achieve the objective of energetic efficiency and CO2 reduction with operations that are concentrated only on a portion of local public patrimony, respecting the preservation and the landscape values non-negotiable landscape value. It will be useful to compare such procedural schema with methods and models for real estate estimation (as recent concrete applications [27–29]) considering that energy renewal increase the market values of building contributing to generate urban renewal as additional value to take into account in such planning activities (cfr. [30]).

References

1. GHI: Solar Radiation Data. http://re.jrc.ec.europa.eu/pvgis/solres/solrespvgis.htm. Accessed 19 Jan 2017
2. Covenant of Mayors for Climate and Energy: http://www.covenantofmayors.eu/index_en. html. Accessed 19 Jan 2017
3. The Action Plans Catalogue of the Covenant of Mayors gathers all SEAPs (Sustainable Energy Action Plans submitted under the 2020 Covenant) and SECAPs (Sustainable Energy and Climate Action Plans to be submitted under the 2030 Covenant). http://www. covenantofmayors.eu/actions/sustainable-energy-action-plans_en.html. Accessed 18 Jan 2017

---

[4] Optimally inclinate surface according with Joint Research Center EU for the Photovoltaic Geographical Information System.

4. Negoziati sul clima. http://www.isprambiente.gov.it/it/temi/cambiamenti-climatici/politiche-sul-clima-e-scenari-emissivi. Accessed 22 Jan 2017
5. Azione dell'UE .per il clima. http://ec.europa.eu/clima/citizens/eu_it. Accessed 21 Jan 2017
6. Pacchetto per il clima e l'energia 2020. https://ec.europa.eu/clima/policies/strategies/2020_it. Accessed 20 Jan 2017
7. CO2 in ambito internazionale. http://www.gse.it/it/Gas%20e%20servizi%20energetici/Aste%20CO2/CO2%20in%20ambito%20internazionale/Pagine/default.aspx. Accessed 21 Jan 2017
8. Matera to be 2019 European Capital of Culture in Italy. https://ec.europa.eu/culture/news/2014/matera-be-2019-european-capital-culture-italy_en. Accessed 20 Jan 2017
9. Matera 2019 capitale della cultura europea. http://www.matera-basilicata2019.it/it/. Accessed 20 Jan 2017
10. Amato, F., Martellozzo, F., Murgante, B., Nolè, G.: Urban solar energy potential in europe. In: Gervasi, O., et al. (eds.) ICCSA 2016. LNCS, vol. 9788, pp. 443–453. Springer, Cham (2016). doi:10.1007/978-3-319-42111-7_34
11. United Nations Framework Convention on Climate Change - FCCC/INFORMAL/84 GE.05-62220 (E) 200705. https://unfccc.int/resource/docs/convkp/conveng.pdf. Accessed 23 Jan 2017
12. Kyoto Protocol to the United Nations Framework Convention on Climate Change, 11 December 1997. http://unfccc.int/resource/docs/convkp/kpeng.pdf. Accessed 21 Jan 2017
13. Paris Agreement. https://ec.europa.eu/clima/policies/international/negotiations/paris_en. Accessed 24 Jan 2017
14. Mayors Adapt (iniziativa del Patto dei Sindaci sull'adattamento al cambiamento climatico). http://seap-alps.eu/hp3114/Mayors-Adapt-l-iniziativa-del-Patto-dei-Sindaci-sull-adattamento-al-cambiamento-climatico.htm. Accessed 19 Jan 2017
15. PIEAR - Piano Energetico Ambientale Regionale, published on BUR n. 2 of 16 January 2010
16. BUR Basilicata, 4 August 2006. http://www.societaenergeticalucana.it/
17. Attolico, A., Scorza, F.: A transnational cooperation perspective for "low carbon economy". In: Gervasi, O., et al. (eds.) ICCSA 2016. LNCS, vol. 9786, pp. 636–641. Springer, Cham (2016). doi:10.1007/978-3-319-42085-1_54
18. Scorza, F., Attolico, A., Moretti, V., Smaldone, R., Donofrio, D., Laguardia, G.: Growing sustainable behaviors in local communities through smart monitoring systems for energy efficiency: RENERGY outcomes. In: Murgante, B., et al. (eds.) ICCSA 2014. LNCS, vol. 8580, pp. 787–793. Springer, Cham (2014). doi:10.1007/978-3-319-09129-7_57
19. Stanciu, C., Stanciu, D.: Optimum tilt angle for flat plate collectors all over the world – a declination dependence formula and comparisons of three solar radiation models. Energy Convers. Manag. **81**, 133–143 (2014)
20. Šúri, M., Huld, T.A., Dunlop, E.D., Ossenbrink, H.A.: Potential of solar electricity generation in the European Union member states and candidate countries. Sol. Energy **81**, 1295–1305 (2007)
21. Huld, T., Müller, R., Gambardella, A.: A new solar radiation database for estimating PV performance in Europe and Africa. Sol. Energy **86**, 1803–1815 (2012)
22. Vivoli, P.F.: Progettare e installare un impianto fotovoltaico. ENEA (2008)
23. SEAP of Matera (2015). http://www.pattodeisindaci.eu/about/signatories_it.html?city_id=5572&seap. Accessed 24 Jan 2017
24. Scorza, F.: Towards self energy-management and sustainable citizens' engagement in local energy efficiency agenda. Int. J. Agricult. Environ. Inf. Syst. (IJAEIS) **7**(1), 44–53 (2016)
25. Cartographic Database by Basilicata Region. http://rsdi.regione.basilicata.it/webGis2/SpecificaDBT.html. Accessed 26 Jan 2017

26. The Sassi and the Park of the Rupestrian Churches of Matera. http://whc.unesco.org/en/list/670/. Accessed 22 Jan 2017
27. Morano, P., Tajani, F.: The break-even analysis applied to urban renewal investments: a model to evaluate the share of social housing financially sustainable for private investors. Habitat Int. **59**, 10–20 (2017)
28. Tajani, F., Morano, P.: Evaluation of vacant and redundant public properties and risk control. A model for the definition of the optimal mix of eligible functions. J. Prop. Invest. Financ. **35**(1), 75–100 (2017)
29. Tajani, F., Morano, P., Locurcio, M., Torre, C.: Data-driven techniques for mass appraisals. Applications to the residential market of the city of Bari (Italy). Int. J. Bus. Intell. Data Min. **11**(2), 109–129 (2016)
30. Zoppi, C., Argiolas, M., Lai, S.: Factors influencing the value of houses: estimates for the city of Cagliari, Italy. Land Use Policy **42**, 367–380 (2015). https://doi.org/10.1016/j.landusepol.2014.08.012

Measuring Territorial Specialization in Tourism Sector: The Basilicata Region Case Study

Francesco Scorza[✉], Ylenia Fortino, Beatrice Giuzio,
Beniamino Murgante, and Giuseppe Las Casas

Laboratory of Urban and Regional Systems Engineering, School of Engineering,
University of Basilicata, 10, Viale dell'Ateneo Lucano, 85100 Potenza, Italy
{francesco.scorza,ylenia.fortino,beatrice.giuzio,
beniamino.murgante,giuseppe.lascasas}@unibas.it

Abstract. From the beginning of the 21st century, following major European and global initiatives such as the Millennium Ecosystem Assessment (2005) [1, 2] and The Economics of Ecosystem and Biodiversity [3], the idea that Ecosystem services could be used as a decision support tool, it gained considerable importance in several fields: from economy to public policy, from territorial planning to environmental assessment. This research is part of the methodological framework of an important strategic reference: the Millennium Ecosystem Assessment, and international research project which rank the ecosystem services to identify the state of the latter, accessing the consequences of ecosystem changes on human well-being. According to the MA [1], they are grouped into four categories: supplying services (food and fiber production, water production, biological and cosmetics production etc.), regulation services (maintenance of the air quality, climate regulation, flood regulation, erosion or drought, pollination, water purification, etc.), cultural services (cultural diversity, recreational or spiritual services, aesthetic values, ecotourism etc.), and support services (carbon sequestration, soil formation, etc.). This classification in the most widespread and used in ecosystem studies because is easy to understand its logic and it can define fixed categories. Starting from this classification, the work done contributes to build interpretative models for the evaluation of a relevant part of the fourth class of ecosystem services: the territorial tourism attractiveness.

Keywords: Ecosystem services · Millennium Ecosystem Assessment · Tourist attractiveness

1 Introduction

This research aims to assess the territorial specialization level of the Basilicata region in terms of tourism attractiveness through the construction of a synthetic territorial index. The results can be used to support resource planning processes for the upgrading of the tourism sector. This improvement is in terms of "local specializations" identified with the presence of natural, cultural and landscape resources (tourism attractors) and a

© Springer International Publishing AG 2017
O. Gervasi et al. (Eds.): ICCSA 2017, Part VI, LNCS 10409, pp. 540–553, 2017.
DOI: 10.1007/978-3-319-62407-5_38

supply system of proper specific services. The attention to territorial specialization belongs to the strategic setting of the 2014–2020 European Union Cohesion Policy [4–7], the Smart Specialization Strategy (cf. [8]) and the Regional Operational Programs according to the indications already formalized by Barca [9]. This work wants to be an attempt to build decision-making tools (DSS) (i.e. ToolKit [10]) that allow to apply the place-based approach [9] (or context-based approach according to Las Casas and Scorza definition [11, 12]) in an interpretative system of the territory elaborated at regional scale. The model of territorial interpretation built with this research is based on open data and open source tools, so it guarantees a high replicability of the evaluation procedures and the results extension to other contexts and case studies. The main result of this research is a synthetic territorial indicator of the tourism specialization level in the Basilicata region built using the spatial analysis tools and techniques. The analytical model used is included in the suite proposed by the Invest (Integrated Valuation of Ecosystem Services and Tradeoffs) software [13]. After a brief description of the territorial context studied, highlighting the Basilicata's uniqueness in terms of tourism attractions, a suitable work was done to collect data and delivering elaborations in order to outline the comprehensive territorial index which, in this specific case, takes the significance of regional tourism attractiveness.

2 The Implementation Context of the Project

The case study focuses on Basilicata region, in particular on tourism promotion strategies and local resources. Basilicata territory is rich for natural habitats, cultural values [14] and traditions that make possible to indicate tourism development as the main key element for socio-economic progress of the entire region. Tourism has gained a consistent weight in the economic and productive system of Basilicata thanks to significant public and private investment. This situation has led to a significant increase in the number of beds and new accommodation facilities, with positive effects on the entire hospitality chains and a substantial increase of tourism demand, with the consequent strengthening of the tourist flows.

From 2016 emerge, with the role of major regional tourist destinations, the Metapontino area, focusing in seaside tourism, and Matera city, which is confirmed as the main cultural-historical attractive pole. Positive signals also come from other areas of the region where new tourist attractors are promoted, such as the Vulture area, the National Park of the Appennino Lucano, the city of Potenza - the main cultural services center - and, in general, all the coastal destinations.

The attraction system declines thematic areas (history, culture, landscape, etc.), each capable to generate significant added value depending on the communication modes, presentation and usage that will be realized in the short term: MATERA 2019 (see [15–17]).[1]

---

[1] We considered as main data source the datasets and the thematic reports distributed by the Regional Agency for Territorial Promotion of Basilicata Region APT (http://www.aptbasilicata.it/Dati-statistici-2016-2013.2094.html accessed Maj 2017.

In terms of receptive offer at 2016, the 80.9% of the tourism accommodation are not hotels (B&B, vacation houses) and the 56% of beds are in hotels. Should be highlighted that, mostly in recent years, a lot of this new accommodations are small-scale exercises (2–3 rooms and about 5–6 beds), and most of them are non-business type (B&B, houses and holiday apartments) (Fig. 1).

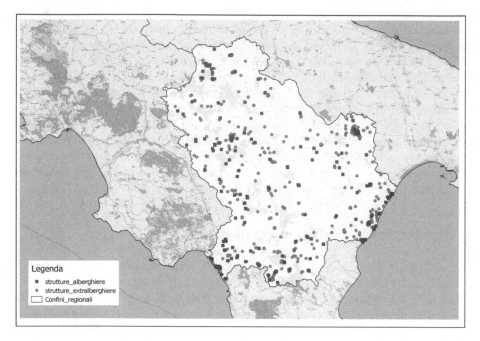

Fig. 1. Accommodation distribution in Basilicata, 2016.

3   Tourism as Ecosystem Service: INVEST, a Spatial Evaluation Tool

According to the most widespread understanding of "ecosystem" (see [1, 13, 18]), we refer to chemical, physical and biological processes that provide support (in terms of raw materials, food, environment and cultural) to the human quality of life. These processes are recognized as "ecosystem services".

The main global initiatives for the analysis of ecosystems are the Millennium Ecosystem Assessment (MA 2005) [1] and The Economics of Ecosystems and Biodiversity (TEEB 2010) [3, 19]; Both have influenced European environmental policies, namely Action Five of the EU Biodiversity Strategy until 2020. [20] Through this action, the European Union invites Member States to assess the state of ecosystems in their territory. For this reason, the Group of Mapping and Evaluation of Ecosystems and Their Services (MAES) is active for Member States to carry out this activity within their territory by adopting the Common International Classification of Ecosystem Services (CISES) [21]. This work helps to build interpretative models for the evaluation of a relevant part of the fourth class of ecosystem services: the territorial tourism attractiveness.

The INVEST[2] model (Integrated Valuation of Ecosystem Services and Tradeoffs) [13] was used as an instrument for assessing the tourist attraction of the Basilicata region. It is an open source software developed by the Stanford University Natural Capital (NCP) project and The University of Minnesota. This tool allows quantifying and mapping the consistency of ecosystem services and interpreting changes in ecosystems and relationships on human well-being. Invest includes eighteen distinct models and, in particular, for the purpose of the analysis carried out, the "Recreation and Tourism" package was applied. The software uses a graphical interface, elaborates a multiple linear regression model that includes geospatial functions on a complex system of input variables, which makes possible the interpretation of regional tourism specialization level.

4   The Methodological Approach: Multiple Linear Regression

For the purposes of this research work, input variables have been grouped into four domains of interest: natural heritage, cultural heritage, accommodation, tourism and social media. Based on these variables, the software processes a linear regression model. The regression equation used is the following:

$$y = \beta_0 + \beta_1 x1 + \beta_2 x2 + \ldots + \beta_i xi + e_i \quad i = 1, \ldots, N$$

where: β_i are the linear regression coefficients; xi are the territorial components considered as predictive variables to input into the software; y matches with the expected value of the model, which in the specific case is the Basilicata region tourism specialization level.

Linear regression analysis is a technique that allows to analyse the linear relationship between a dependent variable (or response variable) and one or more independent variables (or predictors). It is an asymmetric methodology that is based on the hypothesis of the existence of a cause-effect relationship between several variables. The equation shown here contributes to the formation of a global index: the regional tourism attraction index (in this work). The template was applied on each domain and on proper combinations of domains to identify the most significant variable combination. The determination coefficient, better known as R2, is used as a measure of the good adaptation of the multiple linear regression model. This is a value between 0 and 1 and expresses the relationship between the variance explained by the model and the total variance. If the result is close to 1, it means that the predictors (input variables) are a good interpreter of the dependent variable value in the sample; if it is close to 0, they don't.

---

[2] INVEST- related documentation is available online at http://data.naturalcapitalproject.org/nightly-build/invest-users-guide/html/index.html.

5   Application of the Invest Model: Interpretative Process and Results Achieved

To outline a regional tourist attraction index, the predictive variables, have been grouped into domains of interest: natural heritage, cultural heritage, accommodation, social media. The template was applied on each domain of interest and on proper combinations of domains. Concerning the natural heritage, each geographic component it is associated with an appropriate geospatial function where the model calculates the linear regression value (Table 1).

Table 1. Iterative processes and spatial functions to apply the linear regression model.

Territorial components	Geospazial function
Protected zone	polygon_area_coverage
Coast line	line_intersect_length
Hiking routes	line_intersect_length
Hydrographic basin	polygon_area_coverage
Hydrographic lattice	polygon_area_coverage

The table below shows the result of the model and allows to recognize the specialization of the southwest of the region in terms of offer of naturalistic-environmental attractors [22]; it also highlights the role of other peculiar landscapes that characterize the area such as Monte Vulture, the graves area in Matera and part of the Parco Nazionale dell'Appennino Lucano. The contribution of coastal areas to many protected areas is also visible in terms of attractiveness (Fig. 2 and Table 2).

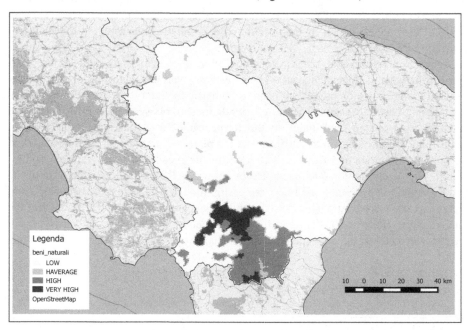

Fig. 2. Natural heritage and its impact on the regional tourist attraction index.

Table 2. Territorial components and associated linear regression coefficients

Territorial components	β_i
Intercept	+6.487 e-02
Oasis	+0.000 e+00
Areas_euap	+3.184 e-08
Coast_line	+2.074 e-03
Attractions	−3.536 e-17
Riverside	−9.983 e-06
Naturalistic trails	+5.741 e-05
Areas_protect	−2.724 e-09

Concerning the cultural heritage category and its territorial components we delivered a selection of data available for the analysis. The result shows the distribution of the historical/cultural sites and the contribution that, in particular, the historical centers and the archaeological sites have on the regional tourism specialization. Matera is the main historical/cultural attraction pole in Basilicata while Potenza is the center of cultural services (Fig. 3 and Table 3).

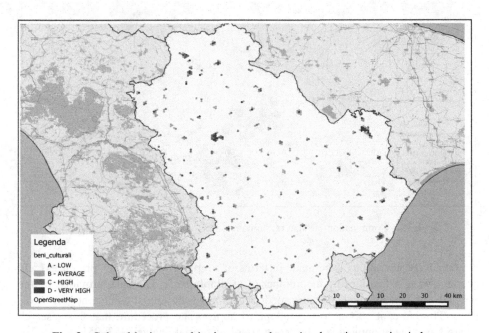

Fig. 3. Cultural heritage and its impact on the regional tourist attraction index.

It is interesting to consider the combination of "natural heritage" and "cultural heritage" domains. The model allows to specify areas whit the integration of several tourism specialization sector of the entire regional territory. (see Fig. 4 and Table 4)

Table 3. Territorial components and associated linear regression coefficients

Territorial components	β_i
Intercept	+2.959 e-02
Cultural heritage	+0.000 e+00
Archaeological sites	+2.384 e-06
Theatres	+3.031 e-04
Pro_Loco	+2.279 e-04
Museums	+1.875 e-04
Storic_centres	+2.529 e-06
Associations	+3.047 e-02

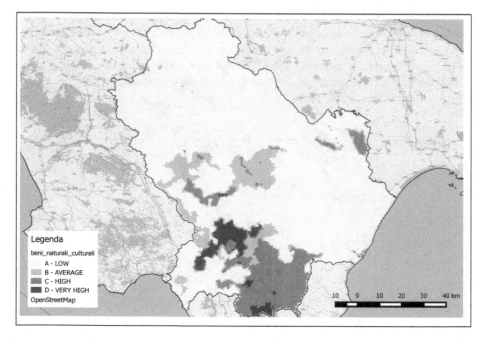

Fig. 4. Combination of "natural heritage" and "cultural heritage" domains

Other considerations were made in terms of tourist flows (arrival - presences) and accommodation facilities by the number of available beds. The data, provided by the APT, doesn't make available an immediate vision of the tourist consistency in Basilicata. Therefore, it was necessary adopt an appropriate procedure that this research group has already applied in other contexts of territorial analysis [23–25].

To simplify, in this section are reported only the details of the territorialization process of accommodation facilities from the APT list, which includes a total of 1024 tourism facilities (including Hotels, B&B, camping etc.). By a geocoding algorithm on each accommodation facility, it was possible to build a punctual dataset made of information about the number of beds and the services offered to the tourists (Fig. 5).

Table 4. Territorial components and associated linear regression coefficients

Territorial components	β_i
Intercept	+2.747 e-02
Oasis	+7.957 e-15
Areas_euap	+4.033 e-08
Areas_protect	+8.858 e-10
Coast_line	+2.012 e-03
Attractions	+2.220 e-16
Riverside	−1.790 e-06
Cultural heritage	+0.000 e+00
Archaeological sites	+2.373 e-06
Theatres	+3.036 e-04
Pro_Loco	+2.232 e-01
Museums	+1.872 e-04
Storic_centres	+2.532 e-06
Associations	+3.070 e-02

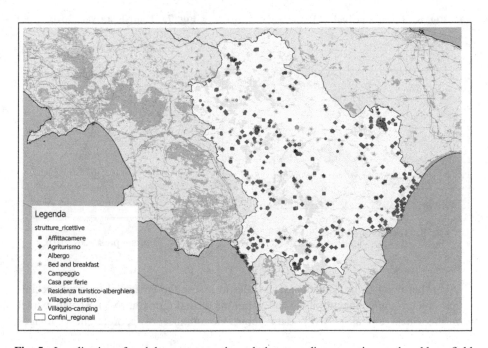

Fig. 5. Localization of each host structure through the geocoding operation on the address field.

Based on this database, information on tourist presences and tourist arrivals were distributed in proportion to the number of beds offered by each accommodation facility. This operation includes a significant approximation but it allowed to overcome the traditional aggregation of data per over-municipal tourist areas[3] in order to provide a result in a detailed-scale more useful for the research purposes.

Density measures are made in 2013–2016 period through the Kernel method (See footnote 3)[4].

The pictures (Figs. 6, 7, 8 and 9) shows how the tourist flow has moved from the main seaside tourism destinations (Basilicata coast) to the cultural centers, in particular, to Matera, designated Capital of European Culture 2019.

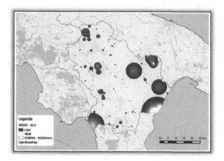

Fig. 6. Tourists arrivals 2013.

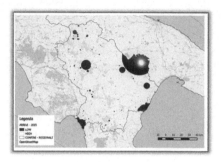

Fig. 7. Tourists arrivals 2016.

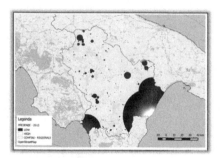

Fig. 8. Tourists presences 2013.

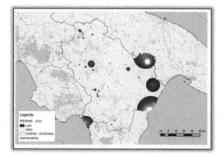

Fig. 9. Tourists presences 2016.

A further point to be considered, in order to assess the territorial tourism specialization level, is the connection between social media and tourism. INVEST can extract from the Flickr social network data about the number of georeferenced images

---

[3] The APT does not provide a municipal breakdown of data on arrivals at presences, due to privacy issues, but in terms of tourist areas (see Fig. 6) that aggregate more Municipality on the regional territory.

[4] The density was calculated on the arrivals and tourist presences data considering the intensity of the phenomenon the value of arrivals and attendances proportional to the number of beds per hotel and a radius of influence inversely proportional to the rarity of hotel types existent in Basilicata.

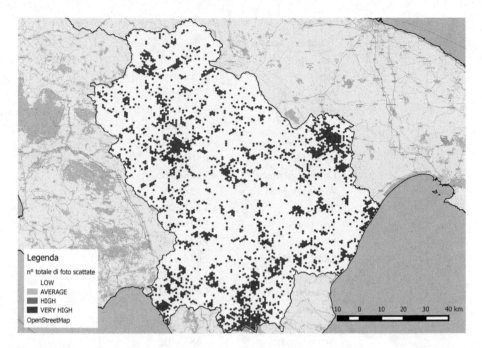

Fig. 10. Main tourists destination sites discovered by the Flickr social network on the basis of the number of photos taken per day.

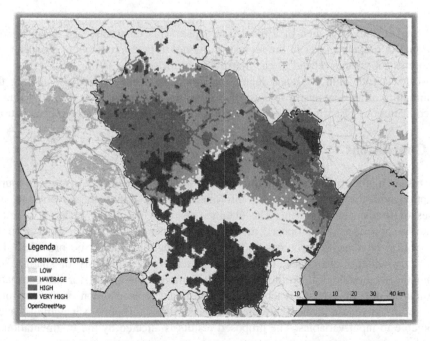

Fig. 11. The regional territory in terms of tourist attractiveness.

Table 5. Territorial components and associated linear regression coefficients.

Territorial components	β_i
Intercept	+2.833 e-02
Oasis	−2.018 e+01
Areas_euap	+3.935e-08
Areas_protect	+3.027 e-09
Coast_line	+1.827 e-03
Attractions	+2.220 e-16
Riverside	−1.500 e-06
Cultural heritage	+3.568 e+06
Archaeological sites	+2.228 e-06
Theatres	+3.267 e-04
Pro_Loco	+2.038 e-01
Museums	+7.674 e-05
Storic_centres	+2.048 e-06
Associations	+2.174 e-02
Restaurants	+4.672 e-02
Villages	+1.535 e-01
Seaside resorts	+3.803 e-01
Accommodation	+3.585 e-02
State roads	+2.589 e-05
High speed stations	−5.748 e-08
Airports	−3.737 e-08

published in the study area in a fixed time-lapse. This data was used as an unconventional information layer in the regression model to take into account a measure of the interest of visitors related to each territorial attractor (Fig. 10).

In the last iteration the linear regression model has been applied considering the combination of all domains of interest. It is possible understand (Fig. 11 and Table 5) the contributions of each individual domain and the different levels of specialization.

6   Conclusions

This research proposes a global interpretation of the regional territory in terms of tourist attraction with the ambition that this information could support the governance processes of tourism development. The obtained result is promising in terms of interpretation capacity of the proposed model but, in order to improve it, more accurate information on tourism resources and services categories should be provided. The research highlights incomplete information in terms of availability of spatial data. About the methodological profile, the INVEST model was useful for the geospatial features offered. In order to reach a more significant representation of the territorial tourism attractiveness levels, there are few possibilities to define a weighing system associable to input variables. The proposed results is affected by the fragmentation of the APT agency's information.

The final results don't coincide with the APT statistics but even more evident is the fact that the APT regional macro aggregations proposed for the touristic flows assessment and the regional tourism consistency are inappropriate to capture the levels of Basilicata tourism specialization, that recently has undergone many deeply transformations that, in the specific case of Basilicata region, should be compered with the spatial distribution of natural risks, including land take, and spatial planning system [26–28] together with real estate estimations [29–32].

The last thing to take into consideration is the use of data from the Flickr social network in the INVEST model. This is an innovative way of including unconventional information within spatial analysis.

Concerning with the relation between the obtained results and the actual tourism development governance system in Basilicata it is evident how the actual inter-municipal specialization areas defined by APT cannot reach the level of specialization we obtained through the analysis. It means that a renovation of the territorial policies concerning tourism management and governance should be delivered according to the attractive potential of territorial areas excluding municipal administrative borders as markers to represent territorial specialization (Fig. 12).

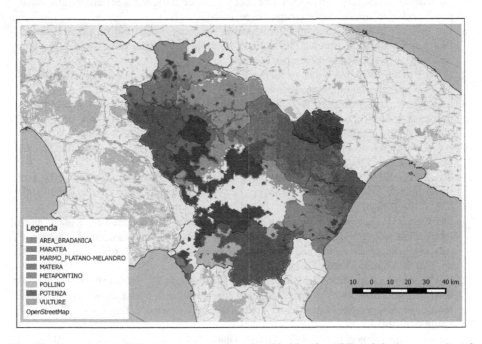

Fig. 12. Comparison of the tourist aggregations provided by the APT and the homogeneity of the result returned by Invest.

A potential extension of research is to rebuild the interpretative system in another scale in order to extend the spatial analysis area to an interregional or national context.

References

1. Millennium Ecosystem Assessment: Ecosystems and Human Well-Being: Synthesis. Island Press, Washington, DC (2005)
2. Leemans, R., De Groot, R.S.: Millennium Ecosystem Assessment: Ecosystems and Human Well-Being: A Framework for Assessment. Island Press, Washington (2003). (Millenium Assessment Contribution). https://islandpress.org/book/ecosystems-and-human-well-being-0?prod_id=474
3. Kumar, P.: The Economics of Ecosystems and Biodiversity (TEEB): The Economics of Ecosys-Tems and Biodiversity: Ecological and Economic Foundations. Earthscan, London (2010). http://www.teebweb.org/our-publications/teeb-study-reports/eco-logical-and-economic-foundations/
4. Regio, D.G.: The Programming Period 2014–2020—Monitoring and Evaluation of European Cohesion Policy (European Regional Development Fund and Cohesion Fund) - Concepts and Recommendations, Guidance Document, Directorate-General for Regional Policy, European Commission, Brussels (2011)
5. EC: Green Paper on Territorial Cohesion Turning Territorial Diversity into Strength Communication from the Commission to the Council, the European Parliament, the Committee of the Regions and the European Economic And Social Committee. European Commission COM (2008) 616 final. Brussels, 6 October 2008. http://ec.europa.eu/regional_policy/archive/consultation/terco/paper_terco_en.pdf
6. EC: Investing in Europe's Future: Fifth Report on Economic, Social and Territorial Cohesion. European Commission. Brussels (2010). http://ec.europa.eu/regional_policy/sources/docoffic/official/reports/cohesion5/pdf/5cr_part1_en.pdf
7. EC: EUROPE 2020 - A Strategy for Smart, Sustainable and Inclusive Growth. Communication from the Commission. Brussels (2010a)
8. McCann, P., Ortega-Argilés, R.: Smart specialization, regional growth and applications to European Union cohesion policy. Region. Stud. 49(8), 1291–1302 (2015). doi:10.1080/00343404.2013.799769
9. Barca, F.: An Agenda for a Reformed Cohesion Policy: A Place-Based Approach to Meeting European Union Challenges and Expectations (2009)
10. Casas, G.L., Scorza, F.: Sustainable planning: a methodological toolkit. In: Gervasi, O., et al. (eds.) ICCSA 2016. LNCS, vol. 9786, pp. 627–635. Springer, Cham (2016). doi:10.1007/978-3-319-42085-1_53
11. Las Casas, G., Scorza, F.: Un approccio "contex based" e "valutazione integrata" per il futuro della programmazione operativa regionale in Europa. In: Bramanti, A., Salone, C. (eds.) Lo sviluppo territoriale nell'economia della conoscenza: teorie, attori strategie, vol. 42, pp. 253–274. FrancoAngeli Editore (2009)
12. Las Casas, G., Lombardo, S., Murgante, B., Pontrandolfi, P., Scorza, F.: Open data for territorial specialization assessment territorial specialization in attracting local development funds: an assessment. procedure based on open data and open tools. Tema. J. Land Use Mob. Environ. (2014)
13. Natural Capital Project (NCP): InVEST User Guide (2015). http://data.naturalcapitalpro-ject.org/nightly-build/invest-users-guide/html/
14. Amato, F., Martellozzo, F., Nolè, G., Murgante, B.: Preserving cultural heritage by supporting landscape planning with quantitative predictions of soil consumption. J. Cult. Herit. 23, 44–54 (2015)
15. Lucio, A., Gennaro, I.: Capitale Europea della Cultura 2019. Un'analisi delle candidature italiane, pp. 141–158. http://doi.org/10.1446/78862

16. Mininni, M., Dicillo, C.: Urban policies and cultural policies for Matera en route to 2019— Politiche urbane e politiche culturali per Matera verso il 2019. Territorio (73) (2015)

17. Pontrandolfi, P., Scorza, F.: Sustainable urban regeneration policy making: inclusive participation practice. In: Gervasi, O., et al. (eds.) ICCSA 2016. LNCS, vol. 9788, pp. 552–560. Springer, Cham (2016). doi:10.1007/978-3-319-42111-7_44

18. Leone, F., Zoppi, C.: Conservation measures and loss of ecosystem services: a study concerning the Sardinian Natura 2000 network. Sustainability (Switzerland) 8(10) (2016). http://doi.org/10.3390/su8101061

19. De Groot, R., Brander, L., van der Ploeg, S., Costanza, R., Bernard, F., Braat, L., ... van Beukering, P.: Global estimates of the value of ecosystems and their services in monetary units. Ecosyst. Serv. 1(1), 50–61 (2012). http://doi.org/10.1016/j.ecoser.2012.07.005

20. European Union: The UE Biodiversity Strategy to 2020, Luxembourg (2011). doi:10.2779/39229. http://ec.europa.eu/environment/nature/info/pubs/docs/bro-chures/2020%20Biod%20brochure%20final%20lowres.pdf

21. Haines-Yong, R., Potschin, M.: CICES V4.3 – Revised report prepared following consultation on CICES Version 4, August–December 2012. EEA Framework Contract No. EEA/IEA/09/003 (2013). https://unstats.un.org/unsd/envaccounting/seearev/GCCom-ments/CICES_Report.pdf

22. PTR (Piano Turistico Regionale) Wikipedia: www.regione.basilicata.it/giunta/files/docs/DOCUMENT_FILE_523485.pdf

23. http://www.aptbasilicata.it/Dati-statistici-2016-2013.2094.0.html, https://support.google.com/fusiontables/answer/2571232?hl=en

24. Amato, F., Pontrandolfi, P., Murgante, B.: Using spatiotemporal analysis in urban sprawl assessment and prediction. In: Murgante, B., et al. (eds.) ICCSA 2014. LNCS, vol. 8580, pp. 758–773. Springer, Cham (2014). doi:10.1007/978-3-319-09129-7_55

25. Di Palma, F., Amato, F., Nolè, G., Martellozzo, F., Murgante, B.: A SMAP supervised classification of landsat images for urban sprawl evaluation. ISPRS Int. J. Geo-Inf. 5, 109 (2016)

26. Amato, F., Martellozzo, F., Nolè, G., Murgante, B.: Preserving cultural heritage by supporting landscape planning with quantitative predictions of soil consumption. J. Cult. Herit. (2015). doi:10.1016/j.culher.2015.12.009

27. Amato, F., Pontrandolfi, P., Murgante, B.: Supporting planning activities with the assessment and the prediction of urban sprawl using spatio-temporal analysis. Ecol. Inform. 30, 365–378 (2015)

28. Amato, F., Maimone, B.A., Martellozzo, F., Nolè, G., Murgante, B.: The effects of urban policies on the development of urban areas. Sustainability 8, 297 (2016)

29. Morano, P., Tajani, F.: The break-even analysis applied to urban renewal investments: a model to evaluate the share of social housing financially sustainable for private investors. Habitat Int. 59, 10–20 (2017)

30. Tajani, F., Morano, P.: Evaluation of vacant and redundant public properties and risk control. A model for the definition of the optimal mix of eligible functions. J. Prop. Invest. Financ. 35(1), 75–100 (2017)

31. Morano, P., Tajani, F., Locurcio, M.: GIS application and econometric analysis for the verification of the financial feasibility of roof-top wind turbines in the city of Bari (Italy). Renew. Sustain. Energy Rev. (2017). doi:10.1016/j.rser.2016.12.005

32. Tajani, F., Morano, P., Locurcio, M., Torre, C.: Data-driven techniques for mass appraisals. Applications to the residential market of the city of Bari (Italy). Int. J. Bus. Intell. Data Min. 11(2), 109–129 (2016)

Four Perspectives of Applied Sustainability: Research Implications and Possible Integrations

Jolanta Dvarioniene[1], Valentin Grecu[2], Sabrina Lai[3] (iD),
and Francesco Scorza[4(✉)] (iD)

[1] Kaunas University of Technology, Kaunas, Lithuania
jolanta.dvarioniene@ktu.lt
[2] Lucian Blaga University of Sibiu, Sibiu, Romania
valentin.grecu@ulbsibiu.ro
[3] Department of Civil and Environmental Engineering and Architecture,
University of Cagliari, Cagliari, Italy
sabrinalai@unica.it
[4] School of Engineering, University of Basilicata, Potenza, Italy
francesco.scorza@unibas.it

Abstract. How is applied sustainability being understood and implemented in academics' and practitioners' circles? Participants to the workshop "Sustainability Performance Assessment" within the ICCSA 2017 conference are confronted with this overarching question, which they address from their specific backgrounds and theoretical standpoints.

In this article, the organizers of the workshop first introduce the debate in which the workshop positions itself, and next offer a glimpse of how the main question can be addressed from four different academic perspectives and disciplines. The variety of perspectives here presented is only a small reflection of the broader diversity of themes and topics tackled by articles submitted to the workshop. Therefore, it provides the reader with an outlook over the wider academic debates that surround the sustainability concept, in a concerted effort to operationalize the concept itself.

Keywords: Applied sustainability · Energy efficiency · Eco-business · Rational-creative planning

1 Introduction

This paper originates from the debate generated by the scientific workshop "Sustainability Performance Assessment: models, approaches and applications toward inter-disciplinary and integrated solutions. SPA2017". The workshop aims to stimulate a scientific debate on methodologies, research reports, case study assessment concerning the multifaceted combinations between research and application domains in a multi- and interdisciplinary way concerning the following general questions:

© Springer International Publishing AG 2017
O. Gervasi et al. (Eds.): ICCSA 2017, Part VI, LNCS 10409, pp. 554–563, 2017.
DOI: 10.1007/978-3-319-62407-5_39

- How to enhance effectiveness in policy making, planning, development programs?
- How to assess sustainability through place-based approach? And what innovations in methods and practices are being implemented?
- Do assessment matrices help? What can we learn from a comparison between different quantitative and qualitative approaches in sustainability evaluation?
- What can we learn from failures and from discussing success examples, by drawing on a critical appraisal of ongoing concrete practices?

Participants to the workshop, fully committed with the above mentioned challenges, propose in the following pages a synthetic view of their current research domains including some operative research dimensions and considerations concerning further development.

The result is a fragmented work looking for integration. It means that authors convey on the multi- and interdisciplinary character of applied sustainability and they promote a wider scientific effort in integrating fields and implementation domain in order to contribute effectively to those up to date issues.

2 Sustainability and Energy in Buildings

Sustainability still remains the main paradigm of future development, and also plays an important role in the energy sector, as sustainable energy planning requires a cautionary approach to resource use. Energy is the life blood of our society. The well-being of our people, industry and economy depends on safe, secure, sustainable and affordable energy. At the same time, energy-related emissions account for almost 80% of the EU's total greenhouse gas emissions.

In fact, the main EU directives and national policies are designed to increase energy sustainability through an increased share of renewable energy sources (RES) and a more efficient use of energy resources. However, the approach to sustainable development has a more general conception, where the different aspects (social, economic, environmental and technological) should be taken into account comprehensively. A sustainable, safe and competitive energy supply properly addressing climate change mitigation are prominent EU energy policy's key priorities as well as for each Member State.

The strategy "Europe 2020", which outlines the EU's actions to address climate change and the energy requirements set three fundamental quantitative targets to be achieved by 2020: cut 20% greenhouse gas emissions, generate 20% energy from renewable sources and cut 20% energy consumption [1]. Still, in this framework energy consumption has a major role, since reducing consumptions is the main instrument to achieve a steady reduction in CO_2 emissions.

Many European local/regional actors struggle with developing targeted, implementation oriented policies addressing low carbon challenges. This holds particularly for energy wasting buildings irrespective of their ownership or use. Since buildings are responsible for 40% of energy consumption in the EU, this is a highly relevant issue in the European context.

Lithuanian energy-intensive economy has to change in order to increase energy efficiency and the use of RES as well as their share in the primary energy balance: this is the overriding objective of investments. Achieving this objective will help to address both climate change and environmental challenges and challenges related to the safety of energy provision. Investments into energy efficiency of public and residential buildings are expected to save up to one third of heat energy. Despite some recent positive changes in energy intensity (EI), the consumption of energy in Lithuania is much higher than the average in the EU or old EU Member States. EI amounted to 311.05 kg of oil equivalent per EUR 1,000 in Lithuania in 2010 (the EU average was 152.08 kg of oil equivalent per EUR 1,000) (source: Eurostat). Compared to other EU Member States, Lithuania is among the states with the most inefficient energy consumption. The above data show the magnitude of energy saving potential and call for enforced measures. 66% of the Lithuanian population live in multi-apartment buildings, of which around 60% were built during the last four decades of the last century. The average thermal energy consumption amounts to 160180 kWh/m^2 per year, whereas the standards in newly constructed buildings (i.e. built after 1993) is around 8090 kWh/m^2 per year.

The necessity to improve energy efficiency in buildings is illustrated also by the magnitude of public spending on compensations for heating of housing, which increases every year. According to a recent study published by the Government, the comparison of expenses on heating and income received by consumer's shows that the Lithuanian population spend probably the largest share of their income on heating of their housing. Successful renovation of residential buildings requires a wider use of technological innovation in this area, namely district-wide renovation, more inventive promotion and management of complex renovation. The targeted economic energy saving potential is 5.2 TWh in residential buildings and their engineering systems and 2.5 TWh in public buildings and their engineering systems [2]. However, all new energy efficiency implementations should be assessed using the life cycle approach. Life cycle assessment (LCA) is a quantitative method evaluating the environmental impacts of products or services during their whole life cycle, from cradle to grave [3]. The life cycle impact assessment conducted for the different building renovation scenarios shows that energy use is very closely linked to the environmental burden associated with each of the renovation scenarios [4].

3 Eco-Business Intelligence for Sustainable Management

During the past years, enterprises have been facing an increasing pressure to broaden the focus of sustainability and accountability in business performance beyond that of financial performance. Fear of loss of sales, regulations that incorporate societal mandates, and a potential decline in reputation if a firm does not have a tangible commitment to corporate sustainability management are just a few of the sources that raise the demands for sustainability management [5, 6].

Starting with the late 1980s, companies began to engage into multi-party collaborative initiatives to link their businesses with environmental concerns, due to the shift in corporate environmental management from traditional nonstrategic, reactive approaches to more strategic, proactive stances [7].

It is widely accepted that companies are facing the challenge to contribute to a better life today without compromising the chances of future generations to satisfy their own needs. Commitment and continuous improvement in the environmental, social and economic performance must be constantly demonstrated, given that the pressure exerted by customers, business partners, governmental and non-governmental organizations, internationalization and globalization of the economy and homogenization of standards is growing. These trends, combined with the benefits of implementing sustainable business processes have transformed the transition towards the sustainable organization into a generalized movement [8].

Under increasing competition and growing pressure of chaotic changes that occur in the market, companies are increasingly required to develop systems for collecting external data and process that data into relevant information for decision making, thus increasing their competitiveness and success in business. Therefore, business intelligence is a concept that spans strategy, applications and databases, rather than a system or a product. Data collection, recovery and analysis is used to highlight weaknesses and strengths of competitors and of the own company, it helps to understand trends and it gives a broader perspective of the business environment. Business intelligence can be addressed by everyone who needs information to better development, not only by big companies but also small and medium enterprises, non-governmental organizations, higher education institutions or public bodies.

Eco-business intelligence is the capacity of people, processes and applications/tools to organize business information, to facilitate consistent access to them and analyse them to improve management decisions, for better performance management of the organizations that are increasingly pressed to synchronize their processes and services with a sustainable development agenda, through the development, testing and implementation of decision-support software [9].

Companies can boost shareholder value, increase market share and gain added value by implementing eco-business intelligence systems for adopting sustainable practices. Moreover, new markets have emerged due to the growing demand for "green" products and the visionary entrepreneurs that adopted sustainable practices already reap the rewards of these efforts. Large and small companies are learning that sustainable business practices not only help the environment but also can improve profitability by pursuing higher efficiency, fewer harmful side-effects, and better relationships with the community and more.

As argued above, over the last few years the society has slightly changed its demands, requesting environmentally friendly, highly customizable products. Simultaneously, it takes less and less time to deliver the finished product after the clients' order, as the manufacturing time is constantly decreasing. Therefore, the manufacturing processes and the products are becoming more and more complex. This challenges companies to improve the efficiency of the product development, shorten the manufacturing and delivery times at a cost which is compatible with the demand. A sustainable company must be able to easily adapt its manufacturing technologies to the client's needs. This implies having an interdisciplinary know-how in process development and flexible manufacturing systems. The flexible manufacturing systems are a suitable link between the market requirements and the current state of development in micro production technologies [8].

To adjust to the market needs and implement an effective flexible manufacturing system, or to identify other sources for competitive advantage, companies need to take informed decisions. Having an eco-business intelligence system that identifies, extracts and analyses data both from the company, but also from the external environment, to provide real support for business decisions, is an essential ingredient of success.

Eco-intelligent tools help determining the organization's strategies, identifying the perceptions and capabilities of the competitors, analysing the effectiveness of current operations, deploying long-term prospects for environmental action and establishing indicators and key variables for organizational health, security and natural growth of its assets [10]. The eco-intelligent business tools can be used to obtain competitive advantages by the organizations that seek to contribute to a better quality of life in the present without compromising the development and life quality of future generations.

4 Ecosystem Services as a Novel Approach to Sustainability

Since the beginning of the XXI century, following global initiatives such as The Economics of Ecosystems and Biodiversity and European ones such as the Millennium Ecosystem Assessment, the "Ecosystem Services" (ES) concept has been gaining increasing popularity in several fields, including environmental economics, public policies, spatial planning, and environmental assessments.

The idea whereby life quality and the very existence of human kind depends on nature's support is certainly much older than the XXI century: several scholars have attempted to trace the origin of the ES concept, dating it back to the 1970s and 1980s, when Wilson and Mattews [11] and Westman [12] introduced the expression "environmental services" and "nature services", and Ehrlich and Mooney [13] introduced the ES phrase. Nevertheless, Daily [14] goes as far as the ancient Greek times and attributes to Plato, in his dialogue Critias, the conception that human activities can lead to soil erosion and loss of water springs; hence, Plato would be the first to have referred to services provided by nature in terms of loss caused by humans.

Notwithstanding the recent popularity of the ES concept, or possibly even as a reaction to it, in recent years some strong criticism has emerged from those environmental scholars who argue that putting a quantitative (and even more so monetary) value on goods and services provided by nature entails neglecting the intrinsic, immeasurable value of nature (among various, [15]). In other words, according to this view, ecosystems and their functions are not being regarded as something that is valuable per se, but only insofar as they cater for human needs, hence underpinning an anthropocentric vision of the world. As a consequence, detractors of the ES concept claim that it supports commodification of nature.

To the contrary, advocates of the ES concept argue that, by allowing natural and environmental notions to permeate social science disciplines (among which, prominently, economics and spatial planning), it has played a significant communicative and pedagogic role [16]. Under this perspective, the ES concept has raised awareness of the importance of nature and biodiversity for well-being and quality of life and, by doing so, it has contributed to building consensus on biodiversity conservation among the general public and policy makers.

As far as the academics' and practitioners' communities are concerned, another important consequence is that, following the popularity of the ES concept, several studies and cooperation projects have been carried out with the aim to identify ESs and arrange them in standardized taxonomies, to develop both qualitative and quantitative assessment models and tools, to map and assess them at different scales. Such studies and analyses, ultimately, aim at understanding the unwanted consequences of human activities on nature, either with a backward look (descriptive studies, e.g. on the consequences of past land-cover changes on the provision of ES) or with a forward look (e.g. *ex-ante* assessments aiming at supporting decision-making processes or environmental assessments). For some ES (typically, for provisioning services such as production of crops, food, timber, or regulatory services such as provision of clean water and clean air) quantitative methodologies allowing to put a monetary value on the trade-offs between ES have been implemented. For instance, converting a forest into agricultural land would result in increasing food production while decreasing carbon sequestration or protection against soil erosion; hence, the tradeoff would be not only, in abstract terms, between two or more ES, but also, in practical terms, between their beneficiaries: on the one hand, farmers who would benefit from the land-use change, on the other hand, local/regional population, who could experience floods and global population, as the loss in carbon sequestration would contribute to climate change.

Notwithstanding such theoretical and methodological progresses, still very limited is the integration of ES in planning practice and theory [17]. Bridging this gap between scientific advancements and policy and planning tools would entail that benefits and disadvantages (both environmental and economic) stemming from the implementation of either plans or policies would be brought to the fore and would help in reconciling societal development needs and expectations with the maintenance of an appropriate level of environmental resources on which social and economic development is grounded.

5 Sustainability in Urban and Regional Planning: A Way to Re-affirm Rationality and Creativity

In the last decades, urban and regional planning discipline was confronted with a crucial dualism was: on the one hand, the crisis of planning in term of progressive distrust in planning practices at different scales; on the other hand, the statement that planning practice is a tool to tackle global challenges concerning sustainable development, risks management, climate change adaptation and mitigation. If we analyze international organizations' statements concerning global challenges [18–20] which imply the commitment of Governments in promoting tools and implementation strategies at national and local level, we can observe an increasing role of planning practice as a mean to achieve expected goals. This is strongly remarked in the United Nations New Urban Agenda [19–22] where the role of urban and regional planning connected with the promotion of "civic engagement" is regarded as a rational tool for adopting sustainable, people-centered, age- and gender-responsive and integrated approaches to urban and territorial development by:

1. Developing and implementing urban policies at the appropriate level, including local-national and multi-stakeholder partnerships, building integrated systems of cities and human settlements, and promoting cooperation among all levels of government to enable them to achieve sustainable integrated urban development;
2. Strengthening urban governance, with sound institutions and mechanisms that empower and include urban stakeholders, as well as appropriate checks and balances, providing predictability and coherence in urban development plans to enable social inclusion, sustained, inclusive and sustainable economic growth, and environmental protection;
3. Reinvigorating long-term and integrated urban and territorial planning and design in order to optimize the spatial dimension of the urban form and deliver the positive outcomes of urbanization;
4. The support of effective, innovative and sustainable financing frameworks and instruments enabling strengthened municipal finance and local fiscal systems in order to create, sustain and share the value generated by sustainable urban development in an inclusive manner.

Such international legitimation of planning represents a lever to address sustainability issues by a renewed planning practice based on a rational approach. It is a way to reaffirm the three main planning principles: equity; effectiveness; preservation of non-renewable resources, as a basis for achieving sustainability in human systems [23].

An interesting methodological perspective comes from the vision of anti-fragile planning developed by Blečić and Cecchini [24, 25]. It represents a contribution for a sustainable planning framework as it integrates the rational dimension of planning practices [26, 27] with inclusive participatory practices in order not only to produce shared responsibilities and wider commitment on shared strategies/scenarios, but also to favor a planning system in which individual (or group) creativity in considered as a value to manage transformation and to achieve desired goals.

Individual contribution to global challenges represents a way to achieve ambitious goals; with reference to environmental sustainability, it becomes a cultural attitude towards a system of values that become common rules of living within a plan. In this vision, the plan represents a community agreement for future development, rather than a list of constraints and regulations. This process calls for tools and action to foster knowledge sharing and empowering competences of local communities and stakeholders.

This idea contrasts with the liquid vision of society [28] and liquid planning. In this dimension, in favor of the interests of small groups, the negotiation of transformations in derogation to the plan should be allowed with consequent disadvantages for an overall rational vision for the city and the territory linked to shared criteria of a priori sustainability.

"Applied Sustainability" becomes an experimentation field for planning discipline and also a way to remark the role of planning as a right of citizens and not a procedural fulfillment for decision making systems.

6 Final Remarks

The four above sections described, although in a very concise manner, different research domains where the application of "sustainability principles" has been renewing former approaches, methodologies and practices. From energy saving and renovation programs for multi-apartments blocks offering an application domain for EU policies on CO_2 emission reduction and renewable energy production to the eco-innovation promoted in enterprises management system, from the ecosystem services approach and its positive implications in for planning practices to the international issues reinforcing the role of rational-creative planning in contributing to global challenges, many promising issues were considered as a baseline for further development of researches.

Sustainability provides a wide field to experiment integrated research and to apply results in a wide variety of implementation sectors as stated in the SPA workshop aims [29].

The boosting power comes from the international commitment in facing environmental, economic, social global issues. Today "sustainability" represents also an opportunity for research results to permeate the barriers between science and politics decision making systems in order to generate positive implementation of current practices, regulations and procedures on the basis of methodological and operative scientific contributions.

Acknowledgement. This article refers to the scientific workshop proposal "Sustainability Performance Assessment: models, approaches and applications toward interdisciplinary and integrated solutions. SPA2017" promoted by the four authors in the framework of ICCSA 2017 Conference. The authors collaboratively designed this piece of work and jointly wrote Sects. 1 and 6, while individual contributions are provided in separate sections as follows: Jolanta Dvarioniene wrote Sect. 2, Valentin Grecu wrote Sect. 3, Sabrina Lai wrote Sect. 4, and Francesco Scorza wrote Sect. 5.

References

1. European Commission: Communication from the Commission to the European Parliament, The Council, The Economic and Social Committee and the Committee of the Regions. Energy 2020. A Strategy for Competitive, Sustainable and Secure Energy. COM, p. 639 (2010)
2. Republic of Lithuania: Operational Programme for the European Union Funds' Investments in 2014–2020 (2014)
3. Khasreen, M.M., Banfill, P.F.G., Menzies, G.F.: Life-cycle assessment and the environmental impact of buildings: a review. Sustainability 1, 674–701 (2009). doi:10.3390/su1030674
4. Ortiz, O., Castells, F., Sonnemann, G.: Sustainability in the construction industry: a review of recent developments based on LCA. Constr. Build. Mater. 23, 28–39 (2009). doi:10.1016/j.conbuildmat.2007.11.012

5. Lee, K.H.: Corporate Sustainability and the Value of Corporations. Pakyoungsa, Seoul (2005)
6. Siegel, S.: Green management matters only if it yields more green: an economic/strategic perspective. Acad. Manag. Perspect. **23**, 5–16 (2009)
7. Hunt, C., Auster, E.: Proactive environmental management: avoiding the toxic trap. Sloan Manag. Rev. **31**, 7–18 (1990)
8. Petruse, R.E., Grecu, V., Chiliban, B.M.: Augmented reality applications in the transition towards the sustainable organization. In: Gervasi, O., et al. (eds.) ICCSA 2016. LNCS, vol. 9788, pp. 428–442. Springer, Cham (2016). doi:10.1007/978-3-319-42111-7_33
9. Grecu, V., Nate, S.: Managing sustainability with eco-business intelligence instruments. Manag. Sustain. Dev. **6**, 25–30 (2014). doi:10.2478/msd-2014-0003
10. Petrini, M., Pozzebon, M.: Managing sustainability with the support of business intelligence: Integrating socio-environmental indicators and organisational context. J. Strateg. Inf. Syst. **18**, 178–191 (2009). doi:10.1016/j.jsis.2009.06.001
11. Wilson, C.M., Matthews, W.H. (eds.): Man's impact on the global environment: assessment and recommendation for action. Report of the Study of Critical Environmental Problems (SCEP). MIT Press, Cambridge (1970)
12. Westman, W.: How much are nature's services worth? Science **197**, 960–964 (1977). doi:10.1126/science.197.4307.960
13. Ehrlich, P.R., Mooney, H.A.: Extinction, substitution, and ecosystem services. Bioscience **33**, 248–254 (1983)
14. Daily, G.C.: Nature's Services. Societal Dependence on Natural Ecosystems. Island Press, Washington, DC (1997)
15. Peterson, M.J., Hall, D.M., Feldpausch-Parker, A.M., Peterson, T.R.: Obscuring ecosystem function with application of the ecosystem services concept. Conserv. Biol. **24**, 113–119 (2010). doi:10.1111/j.1523-1739.2009.01305.x
16. Gómez-Baggethun, E., de Groot, R., Lomas, P.L., Montes, C.: The history of ecosystem services in economic theory and practice: from early notions to markets and payment schemes. Ecol. Econ. **69**, 1209–1218 (2010). doi:10.1016/j.ecolecon.2009.11.007
17. Sitas, N., Prozesky, H.E., Karen, J., Esler, K.J., Reyers, B.: Opportunities and challenges for mainstreaming ecosystem services in development planning: perspectives from a landscape level. Landscape Ecol. **29**, 1315–1331 (2014). doi:10.1007/s10980-013-9952-3
18. The Intergovernmental Panel on Climate Change Report (www.ipcc.ch), The United Nations Millennium Forum Declaration reports (www.un.org/millennium/declaration), Millennium Ecosystem Assessment Reports (www.milleniumassessment.org) and UNEP's Fourth Global Environment Outlook: environment for development report (www.unep.org/geo/geo4/)
19. UN Habitat: New urban agenda (2016). http://habitat3.org/the-new-urban-agenda/
20. UNISDR: Sendai Framework for Disaster Risk Reduction 2015–2030. UNISDR, Geneva (2015)
21. UN Habitat: International Guidelines on Urban and Territorial Planning. United Nations Human Settlements Programme (UN-Habitat), Nairobi, Kenia (2015)
22. UN Habitat: Action Framework for Implementation of the New Urban Agenda (2017). http://nua.unhabitat.org/list1.htm
23. Las Casas, G., Scorza, F.: A renewed rational approach between liquid society and anti-fragile planning. In: Gervasi, O., et al. (eds.) Computational Science and Its Applications - ICCSA 2017. Springer, Heidelberg (2017)
24. Blečić, I., Cecchini, A.: Verso una pianificazione antifragile [Towards Antifragile Planning]. FrancoAngeli, Milan (2016). ISBN 978-88-917-2775-6
25. Blečić, I., Cecchini, A.: On the antifragility of cities and of their buildings. City, Territ. Archit. **4**, 3 (2017). doi:10.1186/s40410-016-0059-4

26. Faludi, A.: A Decision-Centred View of Environmental Planning. Pergamon Press, Oxford (1985)
27. Casas, G.L., Scorza, F.: Sustainable planning: a methodological toolkit. In: Gervasi, O., et al. (eds.) ICCSA 2016. LNCS, vol. 9786, pp. 627–635. Springer, Cham (2016). doi:10.1007/978-3-319-42085-1_53
28. Bauman, Z.: Liquid Modernity. Politi Press, Cambridge (2012)
29. Scorza, F., Grecu, V.: Assessing sustainability: research directions and relevant issues. In: Gervasi, O., et al. (eds.) ICCSA 2016. LNCS, vol. 9786, pp. 642–647. Springer, Cham (2016). doi:10.1007/978-3-319-42085-1_55

Making Urban Regeneration Feasible: Tools and Procedures to Integrate Urban Agenda and UE Cohesion Regional Programs

Piergiuseppe Pontrandolfi[1] and Francesco Scorza[2(✉)]

[1] Department of European and Mediterranean Cultures, University of Basilicata,
Via San Rocco, 3, 75100 Matera, Italy
piergiuseppe.pontrandolfi@unibas.it
[2] School of Engineering, University of Basilicata,
Viale dell'Ateneo lucano 10, 85100 Potenza, Italy
francesco.scorza@unibas.it

Abstract. Attention to "participation" as an inclusive form for redesign and urban regeneration represents the interesting news of the national and international urban planning debate. In the plurality of experiences and approaches produced, this work tends to highlight the originalities of the approach developed within the CAST project at Potenza.

It is an initiative promoted by the world of association in a program of the Basilicata Region, aimed at promoting youth creativity. "The city as an object on which to express the creativity of local communities in terms of urban development and regeneration" is the main theme of the experience developed in about three years of activity on the territory that has generated a number of significant contributions: a proposal defining possible addresses for a city's urban regeneration process, experimentation in the "Poggio Tre Galli" and "Zona G" districts in Potenza and finally the design of a "Virtual Urban Center" prototype to accompany the widespread participation processes in the government process of the city.

In this paper, attention will be given to the proposal, starting with the experimentation of the VUC prototype, of the establishment of a UC for Potenza, in that moment, when the Municipal Administration is launching discussion of proposals towards the large container of transformation projects/urban regeneration represented by the ITI (Integrated Territorial Investment) that the City of Potenza is drafting under the ERDF Operational Programming 2014–2020.

Keywords: Participation · Urban regeneration · Governance

1 Introduction

There is a need for citizens to participate in the process of urban regeneration and to share decisions and strategies, two aspects that the European Union has given particular attention to the programming of EU funds in recent years.

This is to encourage experiences of deliberative democracy (evolution of participatory democracy) as a participatory way of "making a decision on the merits of

© Springer International Publishing AG 2017
O. Gervasi et al. (Eds.): ICCSA 2017, Part VI, LNCS 10409, pp. 564–572, 2017.
DOI: 10.1007/978-3-319-62407-5_40

a matter, but only after having discussed and scrutinizing it thoroughly by diluting the pro and contra/pro and cons of the various possible courses of Action, including the consequences, constraints, opportunities, values and interests in competition, possible clashes and sacrifices in the game" [1]. A form of democracy, the deliberative one, which is not a replica or a contrast to representative democracy.

The goal is not the "consensus building", that is, the acquisition of a consensus on a few decisions as commonly thought, but the "empowerment" in favor of citizens, understood as the ability to increase their being political and their ability to influence.

The participatory and deliberative participatory pathways are not improvised, they have to be structured, assisted and non-episodic, and the key role of local government is that, far from being reduced, it is placed in a more open and dynamic dimension.

As part of the CAST project [2, 3], Potenza has developed activities that go in the direction of promoting forms of active citizenship in the processes of government of the city activated by the world of association.

2 Tools for Effective Participation in Urban Regeneration Processes

If it is true that "participation" has conquered a significant role in the city's government processes, it is also worth mentioning how this process took place according to a variety of approaches, methodologies and practices so heterogeneous that resulted in a vagueness of meaning which does not allow us to capture in a codified dimension the potentialities attributable to these formal and informal practices.

The formalization of the participatory process is generally contextual, that is, depends on the application. It may appear to be a procedural fulfillment (for example, as the less virtuous experiences of partnership consultation tables into the programming process of EU-wide regional resources), such as negotiating choices that would otherwise conflict (consultation of interest groups on instances concerning the irreversible modification of territorial values compared to the expected benefits and inevitable needs) as a tool for building consensus among decision-makers and citizens on transformation hypotheses (grouping groups around the possibility of influencing decision-making by choosing explicit alternatives). The level of Arnstein's participation [4], which could be understood as a methodological organization of the participatory process, actually measures the contextual conditions in which it occurs.

A participation idea in the service of territorial planning that we consider useful this proposal is suggested by Leone and Zoppi [5]: a useful and complementary tool for defining shared strategies and effective plan choices.

In an asymmetric practice, we attribute value (ex-ante) to the contribution expected from the participatory process, while ex-post measures the indeterminacy by entrusting to the rationality of the technical exercise the closure of the plan or project proposal.

A work hypothesis, experienced within the CAST project - which we will say in the next paragraph - positively addresses the fragility of participation as presented so far. We say that if participation is developed according to a rigorous process based on established tools in which stakeholders, citizens and administrations interact on a critical evaluation of real needs expressed at an appropriate scale with the assistance of experts, the usefulness of the comparison increases.

This position defines the need for rigorous design of participatory processes to ensure their relevance to the results, effectiveness in the process of enabling groups and participants, systematization within the territorial governance within the institutional instruments for managing urban transformations and huge area. The participatory process support tool is the Logical Framework Approach [6–8] extended to other decision support tools appropriately selected in relation to specific contextual requirements.

3   Participation as a Collective Creativity Exercise

The CAST Project (Attractive Citizenship for Sustainable Territorial Development) is a largely completed experimentation of a participatory process for the definition of an urban regeneration program specifically for the city of Potenza (the activities of the project were also developed in Matera with different goals and activities).

It should be noted that CAST has been selected and funded by the Basilicata Region within a public call not focused on participatory or urban planning actions. In fact, the call was open to creative projects (in the sense of artistic creativity) and the design proposal offered participatory exercise as a creative community exercise for the design of urban space. This has resulted in a singularity within the regional initiative which has been widely discussed at the technical and political level during the development of project activities and has allowed all players in the game to mature a higher level of awareness of the contribution that participatory processes structured in terms of support to the governance actions of urban transformations. At the same time, this singularity demonstrates the lack of public authorities' attention to promoting and supporting bottom up initiatives for building strategic urban development scenarios outside the codified administrative procedures linked, for example, to the adoption of urban planning instruments or training of regional operational programs on EU resources. Therefore, an indicator of the success of the CAST project will be that it has contributed to the dissemination of participatory practices as a support tool for the definition of urban governance in the context of the Basilicata Region.

Based on the experience it is possible to consider the CAST project as an evolving participatory experience with the aim of identifying spaces and forms of technical deepening aimed at building sustainable urban development strategies shared by the actors in the field [9] and with reference to the technological infrastructure developed to support the process [2]. A participatory process supported by evolved ICT tools [10] which has generated a wide interest of the local community and public administration, and has contributed to the implementation of forms of Inclusive Smart Planning (such as the extension of the experience described by Murgante et al. [11]) [12].

The CAST project has proposed the goal of developing participatory processes based on the widespread use of participatory technologies, organized according to innovative open source architectures, social interaction and social mapping that have been widely documented by the authors in previous work [2, 3].

In the case of the city of Potenza, these actions have been developed (in time) in a critical phase of the process of defining short-medium urban development related to the elaboration of the Agenda Urbana. This has amplified the interest of the Municipal Administration which has identified the procedures and contents of the CAST project proposed for the construction of intervention cards and design directions.

This "usefulness" attributed by the public decision-maker to the participatory process could be related to contingent situations in the context of implementation, but has enabled the long-term perspective to establish the need to structure "institutionalized" participatory tools and processes to support urban governance according to the methodological model of the CAST project that looks at relevant examples in literature [13–15] and the application of the Logical Framework Approach as a rigorous process engineering tool.

Among the outputs of the CAST project we highlight three contributions proposed to the municipal administration for a future implementation:

1. an address document on urban regeneration issues for Potenza city;
2. a regeneration proposal for "Poggio Tre Galli and Zona G";
3. the project for the establishment of an Urban Center to support the processes of participation to the city government.

We will focus on the first and third topic of the CAST project referring to the bibliography regarding the proposed urban regeneration [2, 3].

In particular, the Urban Center project is the core element of the proposal to transform CAST's experience into an institutionalized approach to participatory practice within urban governance. An Urban Center project was developed as an excellence structure for promoting support for participatory services and urban design laboratories linked to promotion and internationalization functions.

This service, compatible with the procedures and investments of the new Urban Agenda of the Municipality of Potenza, is eligible for short-term investments that the municipality will develop through European resources. It is based on a flexible and soft budget organization which in the medium term can help itself by generating not only community services, but also value in attracting funding and resources (from European projects to facilitating private support and intangible events and infrastructure investments on the city).

4 Contributions for a Comprehensive Urban Agenda

The keystones of the Urban Agenda Strategy, as specifically contained in the 2014–2020 Partnership Agreement, are based on the redesign and modernization of urban services for residents and users of the city, the dissemination of practices and planning for the urban environment, social inclusion, strengthening of the city's capacity to enhance the valuable segments of local production chains, environmental protection and enhancement of cultural resources, prevention of environmental hazards (including seismic risk reduction), accelerating the innovation processes in particular of the Public Administration.

The address document prepared in the CAST project intended to offer some reflections on the launch of a structured urban regeneration process for the city of Potenza, based on the principles, guidelines and procedures set out at European and international level in terms of urban policies and considering experiences developed in other territory.

The theme of urban regeneration for the city of Potenza has been extensively described in the address document that develops merit assessments on a number of thematic areas of intervention and related criticalities within the urban area to form a strategic reference framework for the city. The operational validity of this paper lies in supporting the city's governance actions such as identifying priorities for intervention in a long-term vision of urban quality expressed in terms of public spaces, services to citizens, environmental sustainability and economic processes of transformation.

The stagnation of the real estate economy [16–19] and the significant contraction of public resources for new investments place the urgent need to promote urban regeneration policies based on the upgrading and completion of the existing city rather than on improbable prospects for further growth.

A rethinking of the city, leveraging some of the existing strengths, is aimed at promoting new urban policies based on some key elements:

- the containment of soil consumption [20–23] and land-take [24];
- the recovery and reuse of disused public buildings;
- the enhancement of the regional and strategic functions already present (the University and the Regional Hospital in the first place);
- the adaptation and improvement of private building heritage;
- the upgrading and increase of urban green areas and ecosystem services;
- the safeguard of the areas that still free in urban fabric;
- the implementation of more up-to-date energy [25] and waste policies;
- a sustainable urban mobility that can be fully used by citizens;
- a renewed urban welfare that favors inclusive policies and opens up new employment opportunities in innovative sectors;
- the upgrading of the extra-urban landscape.

The issues covered in the document represent a first contribution to initiate a joint and shared reflection on the future of our city, with particular attention to the theme of urban regeneration and the widespread participation of the city government.

In this regard, the related proposal for the establishment of the URBAN CENTER is a democratic instrument of active participation and social inclusion of great potential and innovative strength, which can play an active role in the formulation of urban policies, becoming a meeting place for networks and actors that contribute to defining sustainable forms of land use and transformation as well as the empowerment of active citizenship.

The contents of this document for urban regeneration have fueled the experimentation of a participatory Urban Workshop on urban regeneration issues involving the involvement of citizens and representatives of the city administration, cultural associations and volunteering in the city. The initiative builds on the experience of the 2010 Laboratory, initiated by a group of UNIBAS researchers. The urban area concerned is the western part of the city, comprising the district of Poggio Tre Galli, Zona G and the "Centro Studi" area, where many institutions for higher education and schools for compulsory education are located district.

Again, the Laboratory's activities have been developed in two ways. A more traditional one that envisaged the work of a small group of subjects (experts, neighborhood representatives, associations) and a more innovative one that provided for an interlocutor with a broader spectrum of subjects through the use of new information technologies and the use of social networks.

5 Urban Center Model

The requirement of an organizational structure aimed at stimulating and promoting participative actions based on codified approaches and methodologies to support the governance of urban regeneration processes emerged within the public participation actions developed in the CAST project. If CAST has created a virtual participatory center (Virtual Urban Center), the formalization of a expertise center related to the Municipal Administration has been proposed as a project that the Municipal Administration can put into the new Urban Agenda in order to activate the necessary resources. To pursue this goal, a preliminary research activity has been developed to clarify organizational and content aspects of an Urban Center, selecting some good practices at national level, and deepening themes: what mission to the UC of Potenza?; The manner of management and organization; The sources of financing of the various activities.

From the discussion of the UCs that have been activated not only in our country but also in the international arena, a different picture emerges about their mission, which does not correspond to ranks of unity, though always oriented towards two general objectives:

- activate shared and transparent paths of comparison on the city's transformation policies and communicate them easily in order to reach a wider target of citizens than the more commonly employed people;
- encouraging and disseminating innovative participatory democracy practices through the use of IT and web technologies "which identifies in Urban Center 3.0 a stable and symbolic place to aggregate heterogeneous and diffused subjects with new impetus towards the exercise of the principles of participatory democracy".

The UCT's aims, strongly related to each other, are designed to stand-in the knowledge of the city and its development, to promote methodologies and tools to facilitate participation in city government processes, to be the site where are shown the topics of interest about the city; Promoting activities aimed at stimulating a new "urban culture", supporting Public Administration in defining shared urban policies.

It is believed that the establishment of the UC can take place in the legal form of the "Nonprofit Association" and it must be promoted by the Municipal Administration in concert with other Entities, also under existing agreements with the University of Basilicata (UNITOWN Project).

The financial resources for the activation of the UC are to be found either through an initial contribution from the Administration (also on the resources of the 2014–2020

EU program), and by the management of the real estate assets (rent for exhibition, etc.) made available by the Administration for the new service structure. A very important aspect regarding the feasibility of the intervention lies in the ability of the UC to self-sustain through resources raised by contributions and/or deliveries from citizens, patronage or sponsorships, offers and donations from third parties, activities of crowd funding, from the allocation of income taxes.

6   Conclusive Considerations and Future Actions

Potenza Urban Center's proposal must be applied as an "implementing tool" for the definition of urban policies (in the widest sense of the term) by providing services focusing on the following keywords: informing, involving, communicating, documenting, supporting (Local Authorities to develop inclusive projects and support them in the preparation of urban planning instruments), promote (partnerships with public and private actors), internationalize.

In these areas, the UC, a neutral arena where practicing forms of participatory/deliberative democracy, can play a strategic role in engaging, discussing, confronting, disseminating, communicating stakeholders (companies, associations, individual citizens, professional orders, etc.) that animate the urban scene.

The UC must be able to activate structured, virtuous and replicable pathways of "active citizenship" aimed also at the application of the subsidiarity principle, envisaged by our constitution to Art. 118, where the Municipal Administration must support and enhance the autonomous initiative of citizens, individuals or associates, in pursuit of objectives of general interest.

The experience and proposals advanced in the CAST project look to the conditions for launching a new development season for the city based on a joint and shared view of the future, but also on the collective ability to influence the conditions that will allow for a shared strategic vision of development.

If a creative project (CAST) promoted by the associations and supported by the university research system gives the city addresses and operational tools to move on to a dimension where participation becomes structural and doesn't have sporadic experience for the municipal administrations, it would be inappropriate losing the opportunity to integrate these voluntary contributions with the actions and funding that the city is called upon to manage in the 2014-2020 programming cycle, renouncing dialogue with citizens in the formulation of choices.

Citizens' contributions, individual or organized, are essential to ensure successful conditions for the city's revival strategy. The condition is that they are involved and made participants, that is to say, in the conditions of participating in the decision-making processes and in making choices in ways that look to the development of forms and experiences of deliberative democracy.

Acknowledgement. The research work is part of the contribution that the authors provided to the implementation of the CAST project - Active Citizenship for Sustainable Development of the Territory. Work is a joint reflection of the authors (Sects. 1 and 6). Sections 2 and 3 are edited by Francesco Scorza, while Sects. 4 and 5 by Piergiuseppe Pontrandolfi.

References

1. Carcasson, M.: Beginning with the end in mind. A call for goal-driven deliberative practice. Occasional paper no. 2, Center for Advances in Public Engagement (2009)
2. Pontrandolfi, P., Scorza, F.: Sustainable urban regeneration policy making: inclusive participation practice. In: Gervasi, O., et al. (eds.) ICCSA 2016. LNCS, vol. 9788, pp. 552–560. Springer, Cham (2016). doi:10.1007/978-3-319-42111-7_44
3. Scorza, F., Pontrandolfi, P.: Citizen participation and technologies: the C.A.S.T. architecture. In: Gervasi, O., Murgante, B., Misra, S., Gavrilova, Marina L., Rocha, A.M.A.C., Torre, C., Taniar, D., Apduhan, Bernady O. (eds.) ICCSA 2015. LNCS, vol. 9156, pp. 747–755. Springer, Cham (2015). doi:10.1007/978-3-319-21407-8_53
4. Arnstein, S.R.: A ladder of citizen participation. J. Am. Inst. Plan. **35**(4), 216–224 (1969)
5. Leone, F., Zoppi, C.: Participatory Processes and Spatial Planning. The Regional Landscape Plan of Sardinia, Italy. Territorio geovernance e sostenibilità, Franco Angeli, Milano (2016). ISBN 9788891740984
6. Casas, G.L., Scorza, F.: Sustainable planning: a methodological toolkit. In: Gervasi, O., et al. (eds.) ICCSA 2016. LNCS, vol. 9786, pp. 627–635. Springer, Cham (2016). doi:10.1007/978-3-319-42085-1_53
7. Gasper, D.: Evaluating the 'logical framework approach' towards learning-oriented development evaluation. Publ. Adm. Dev. **20**(1), 17 (2000)
8. Aune, J.B.: Logical framework approach. Development Methods and Approaches, p. 214 (2000)
9. Luisi, D.: Dinamiche inclusive e costruzione dell'agency nelle politiche pubbliche partecipate. Archivio di studi urbani e regionali (2016)
10. Lanza, V., Prosperi, D.: Collaborative E-Governance: Describing and Pre-Calibrating the Digital Milieu in Urban and Regional Planning. Taylor and Francis, London (2009)
11. Murgante, B., Tilio, L., Lanza, V., Scorza, F.: Using participative GIS and e-tools for involving citizens of Marmo Platano-Melandro area in European programming activities. J. Balk. Near East. Stud. **13**(1), 97–115 (2011)
12. Pontrandolfi, P. Scorza, F.: Una sperimentazione di strumenti web-based per la partecipazione dei cittadini ai processi di rigenerazione urbana: l'infrastruttura ICT CAST e l'Urban Center Virtuale, in Atti della XIX conferenza nazionale SIU, Cambiamenti, Responsabilità e Strumenti per l'urbanistica al servizio del Paese. Catania 16-18-Giugno 2016. Planum Publisher, Roma-Milano (2017)
13. Ave, G.: Play it again Turin. Analisi del piano strategico di Torino come strumento di pianificazione della rigenerazione urbana. In: Martinelli, F. (ed.) La pianificazione strategica in Italia e in Europa: Metodologie ed esiti a confronto, Franco Angeli, Milan, pp. 35–67 (2005)
14. Gibelli, M.C.: Vivibilità e nuova urbanità nelle politiche e nei progetti di rigenerazione urbana. In: Boniburini, L. (a cura di) Alla ricerca della città vivibile, Alinea, Firenze, 75–90 (2009)
15. Pontrandolfi, P., et al.: L'esperienza dei laboratori di Urbanistica Partecipata a Potenza: una iniziativa promossa dai docenti dell'Università per una più efficace attuazione degli strumenti urbanistici. Tafter J. (26) (2010)
16. Morano, P., Tajani, F.: The break-even analysis applied to urban renewal investments: a model to evaluate the share of social housing financially sustainable for private investors. Habit. Int. **59**, 10–20 (2017)

17. Tajani, F., Morano, P.: Evaluation of vacant and redundant public properties and risk control. A model for the definition of the optimal mix of eligible functions. J. Prop. Invest. Finance **35**(1), 75–100 (2017)
18. Morano, P., Tajani, F., Locurcio, M.: GIS application and econometric analysis for the verification of the financial feasibility of roof-top wind turbines in the city of Bari (Italy). Renew. Sustain. Energy Rev. (2017). http://dx.doi.org/10.1016/j.rser.2016.12.005
19. Tajani, F., Morano, P., Locurcio, M., Torre, C.: Data-driven techniques for mass appraisals. Applications to the residential market of the city of Bari (Italy). Int. J. Bus. Intell. Data Min. **11**(2), 109–129 (2016)
20. Amato, F., Martellozzo, F., Nolè, G., Murgante, B.: Preserving cultural heritage by supporting landscape planning with quantitative predictions of soil consumption. J. Cult. Herit. doi:10.1016/j.culher.2015.12.009
21. Amato, F., Pontrandolfi, P., Murgante, B.: Using spatiotemporal analysis in urban sprawl assessment and prediction. In: Murgante, B., et al. (eds.) ICCSA 2014. LNCS, vol. 8580, pp. 758–773. Springer, Cham (2014). doi:10.1007/978-3-319-09129-7_55
22. Di Palma, F., Amato, F., Nolè, G., Martellozzo, F., Murgante, B.: A SMAP supervised classification of landsat images for urban sprawl evaluation. ISPRS Int. J. Geo-Inf. **5**, 109 (2016)
23. Amato, F., Pontrandolfi, P., Murgante, B.: Supporting planning activities with the assessment and the prediction of urban sprawl using spatio-temporal analysis. Ecol. Inform. **30**, 365–378 (2015)
24. Lai, S., Zoppi, C.: The influence of natura 2000 sites on land-taking processes at the regional level: an empirical analysis concerning Sardinia (Italy). Sustainability (Switzerland) **9**(2), 259 (2017). http://doi.org/10.3390/su9020259
25. Scorza, F.: Towards self energy-management and sustainable citizens' engagement in local energy efficiency agenda. Int. J. Agricult. Environ. Inf. Syst. (IJAEIS) **7**(1), 44–53 (2016)

Engaged Communities for Low Carbon Development Process

Alessandro Attolico[1(✉)], Ian Bloomfield[2], Jolanta Dvarioniene[3],
Inga Gurauskiene[3], Daina Kliaugaite[3], Balazs Mezosi[4],
and Francesco Scorza[5]

[1] Province of Potenza, Department of Territorial Planning and Development,
Environment and Civil Protection, IT, Potenza, Italy
alessandro.attolico@provinciapotenza.it
[2] Durham County Council, Durham, UK
ian.bloomfield@durham.gov.uk
[3] Institute of Environmental Engineering, Kaunas University of Technology,
Gedimino str. 50, 44239 Kaunas, Lithuania
{jolanta.dvarioniene,inga.gurauskiene}@ktu.lt
[4] Hungarian Innovation and Efficiency Nonprofit Ltd., MI6 HU,
Székesfehérvár, Hungary
[5] School of Engineering, University of Basilicata, Potenza, Italy
francesco.scorza@unibas.it

Abstract. Starting from stakeholder's involvement methodology, the paper
will present example of community/stakeholders engagement process in concrete actions and Energy efficiency programs. The selected case study will be
assessed according to common criteria oriented to identify the citizens role and
the degree of partnership in delivering innovations and investments.

Keywords: Low carbon economy · Sustainability · Stakeholders involvement

1 Introduction

The Roadmap to a Resource Efficient Europe [1] aims inter alia at 20% increase in the
share of renewable energy sources in EU's final energy consumption and a 20%
increase in energy efficiency. However, according to the Energy Efficiency Communication from the Commission [2] the EU will not be able to achieve this goal. What's
more, one third of the forecasted energy savings will be due to the lower than expected
growth during the financial crisis.

In such a European scene, extra efforts are needed to reach the EU targets. It is
acknowledged that bottom up initiatives have key importance in formulating
regional/national energy policies and demand side actors play a crucial role in this
process since they are able to give place based responses to the challenges. This
approach is supported by the 2030 Framework for Climate and Energy adopted by the
European Council [3], in which the importance of demand side management and
ensuring security for households and communities is emphasized. Innovative stakeholder involvement measures are emerging, but they are not yet taken up at a

© Springer International Publishing AG 2017
O. Gervasi et al. (Eds.): ICCSA 2017, Part VI, LNCS 10409, pp. 573–584, 2017.
DOI: 10.1007/978-3-319-62407-5_41

significant scale. Specifically, local and regional authorities embark to find their role in these processes, such as coordination, planning, service provision, monitoring and feedback to policy making but in this respect, there is huge room for development. The issue tackled by LOCARBO project, fully fits Priority Axis 3 Low Carbon Economy of the Interreg Europe Programme, which seeks change in improved implementation of regional development policies that incorporate actions to increase levels of energy efficiency including public buildings and the housing sector. Reduction of energy consumption by businesses and households is presented as a key field of action just as the introduction of ICT based solutions (through increasing the energy performance of public buildings or public awareness strategies).

The stakeholder involvement process can be considered as an investment in the future as this structured process can support the trust in the institutions and the creation open innovation networks facilitating the implementation of policy strategies and further cooperation in future projects [4, 5].

The overall objective of LOCARBO is improving policy instruments targeting demand driven initiatives to increase energy efficiency related to the built environment. This is to be achieved by finding innovative ways for regional/local authorities to support energy consumers' behaviour change. The challenge to involve and motivate stakeholders (especially energy consumers) is perceived broadly as a major problem for public authorities. Motivation and awareness of consumers are of high significance to influence their behaviour and support more conscious energy decisions.

This paper, besides contributing to the dissemination of the main contents of the project LOCARBO, discusses the stakeholder's involvement process delivered through the implementation phase. Such approach, based on strong methodological background, contributes in structuring project development and to reinforce the operative approach in managing stakeholder's participation in project actions and it appears to be transferable in other implementation context.

In the next section of the paper we propose some information concerning the LOCARBO project structure and main contents. Then we propose methodological approach in stakeholder engagement as an operative roadmap supporting implementation phase.

Conclusions regards framework consideration about project expectation and a preliminary appraisal of results in local actors engagement process.

2 Communities Improving Local Governance Models

The challenge to involve and motivate stakeholders (regarding especially energy consumers) is perceived broadly as a major problem for public authorities. In fact, motivation and awareness of consumers are of high significance to influence their behaviour and support more conscious energy decisions.

LOCARBO is unique in focusing its activities on bottom up initiatives and mainly because of the approach to handle 3 thematic pillars (services, organizational structures and technological solutions) in a fully integrated way. The 7 partners from Hungary, Italy, Lithuania, Portugal, Romania and the UK are aware that regional policies on energy efficiency can only be successful if pieces of the puzzle are brought together.

Against this background project partners intend to exchange experience on how local and regional authorities can effectively undertake innovative roles in supporting demand side driven energy efficiency practices focusing on the built environment and to use the results to elaborate action plans on improving relevant policy instruments. The project approaches the 3 thematic pillars (TPs) in a fully integrated way. Partners are aware that only by bringing pieces of the puzzle together (services, organizational structures and technological solutions) regional policies on EE can be successful.

These three pillars structure is characterizing the project. Each TP represents a domain of interactions and operative application for the partners:

- TP1: Supplementary services and products offered by authorities
- TP2: Innovative cooperation models
- TP3: Innovative smart technologies

The first TP represents the issues coming from role and responsibilities by the local authorities. Local authorities can play an active role in providing services and products in addressing energy issues that affect their local communities and are often looked to by local businesses and local residents for guidance and support on energy related issues. By identifying the energy products and supplementary services currently being offered by the local authority, and where practicable, making them more widely available, simple interventions and technologies can be introduced that have a great influence in behavioural change that contributes effectively towards a constantly improving 'energy aware' community.

Dramatic savings can be made from simple internal or consultant lead energy awareness campaigns that can not only reduce energy consumption and cost, but prove significant in encouraging behavioural change in management and staff. The identification of these services can be transferred and implemented into local policies and grouped into cost and manageability sectors by partners in such a way that best fits their organization.

The second TP fits whit the identification of innovative cooperation models to be implemented in concrete projects based on a primary role of local communities and citizens. Penetration of smart technologies and requirements of energy efficiency could be the leading factors when implementing the policy instruments in particular region. The current state and future trends on smart technologies penetration in the everyday life requires for the new cooperation models in order to empower those technologies in managing energy flows in order to increase energy efficiency. The proactive role of energy consumers is based on the energy awareness of local communities' and the ability to influence the way of policy instruments' implementation to the demand oriented direction. Therefore, the role of local authorities and all the stakeholders of energy sector should be oriented to finding the most appropriate cooperation models, which would be based on sustainability principles: economic, social and environmental ones through the whole value chain of energy supply. Sustainable value chain management is driven by the values of the final consumers.

The third TP regards the effective application of smart technologies in energy saving. In a general view the concept of "innovative technologies" in the sector of Energy Efficiency concern with a huge variety of technological solutions, equipment, procedures delivered both at hard (i.e. plants and infrastructure) and soft (i.e. software, web, and "smart" application) level of application. Therefore, the technology should be considered as a driver to enhance citizens' and communities' commitment in EE shared strategy.

The LOCARBO focus will be on "perceived benefits", "availability of risk sharing", "trust building process among stakeholders and local actors", "users' empowerment" considered as expected impact deriving from the adoption of technological plants or solutions.

Among the numerous projects that try to increase energy efficiency in the built environment LOCARBO is unique in focusing its activities on bottom up initiatives but mainly because of the approach to handle the 3 thematic pillars in a fully integrated way.

Partners are aware that only by bringing pieces of the puzzle together, services, organizational structures and technological solutions, regional policies on energy efficiency can be successful. The core idea of how local/regional authorities can become involved in these initiatives is a novel approach in working on sustainability.

3   Stakeholders Involvement: A Methodological Scheme

Since the overall objective of LOCARBO is improving policy instruments targeting demand driven initiatives to increase energy efficiency related to the built environment via finding innovative ways for regional/local authorities to support energy consumers' behaviour change, the stakeholders (SH) involvement process represents a strategic option in terms of achieving desirable effects in public-private-people engagement.

The SH involvement strategy should act as the baseline of all interactions with relevant decision-making bodies and actors of different target groups.

The methodological scheme of the involvement of the SH was designed to comply with widely acknowledged international methodology [6] and is based on the following steps:

(1) Identify: Find all relevant stakeholders
(2) Sort: Classify them regarding their level of interest and power to affect the project
(3) Plan: Create action plans for each to handle them accordingly
(4) Do: Execute action plans for stakeholders involvement and engagement
(5) Follow up: Monitor the process throughout the entire project period.

3.1   Identify

The purpose of the first step is to identify and assess essential SH. Based on the nature of the project, it is necessary to take into account all potential stakeholders, assess their impact and level of involvement to be able to prepare plans engaging them. For the identifying of SH it was used brainstorming method and already established partner network during similar project. As the project touches many aspects of the economy and our environment, a number of stakeholder type were identified. Government organizations and their decision-making bodies, as they have the means to implement policy instruments, which are going to be in the project. Public sector partners can be of help in the local implementation, communicating the regional attributes as they have a deeper insight to local matters, and they can help identifying key points of interests, where Policy instruments should take actions for maximum effect on energy efficiency. NGOs and Non-Profit organizations are also an option for establishing a deep spread stakeholder network, and communicating project goals/action items towards the target segments, who will be affected by the project results. Other actors can also be identified, such as the business sector, utility suppliers, knowledge base owners (experts, academics, research institutions), even private citizen groups. The task involves a lot of players and a set of different tools which must be used to drive the project onward.

3.2 Sort

The second step is to classify SH based on their power and interest in the project and its outcomes. As well-established tool for assessing and sorting stakeholders is the mapping procedure, where potential stakeholders can be easily measured on a two dimensional scale (Fig. 1).

- Vertical: Stakeholder power of influencing the project (Power)
- Horizontal: Stakeholder interest in influencing the project (Interest)
- Bubble size (+often included dimension): Cost accessing stakeholder

A basic power map, which helps defining the preferred strategy, is presented in the Fig. 1.

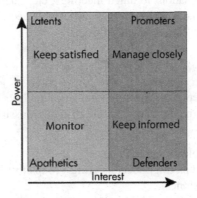

Fig. 1. Mapping procedure [7]

Preferred stakeholders types for mapping include: primary stakeholders (ultimately affected, either positively or negatively by the project actions); secondary stakeholders (are the 'intermediaries', i.e. persons or organizations who are indirectly, but still affected by the project); key stakeholders (who can also belong to the first two groups, and who have significant influence on or importance to the project).

There are a number of methods to be used in the actual mapping process. Classic power mapping, based on power to influence, the legitimacy of each stakeholder's relationship with the project, and the urgency of the stakeholder's claim on the project [8]. The mind mapping method, fitting as flexible way assessment to modern theories: ranking stakeholders based on needs and the relative importance of stakeholders to others in the network [9]. Assessing stakeholder expectations based on value hierarchies and Key Performance Areas (KPAs) [7].

Further dimensions can be included in the power mapping: power, support, influence and need.

3.3 Plan

After identifying and sorting stakeholders into different groups the next goal is to create action plans to handle them accordingly (i.e. communication methods).

As it is difficult and also not necessary to communicate with all stakeholders on an equal frequency and share information with them at the same level, the stakeholder involvement strategy must differentiate and choose preferred stakeholder groups to closely team up with. However, all "sub-preferred" stakeholders must also be involved at some point and some level to the project. Their engagement should be carried out by more conventional, one-way communication methods, such as newsletters, leaflets, etc. After identifying and sorting stakeholders into different groups, the next goal is to create action plans, communication methods to handle them accordingly.

It could be used formal communication methods as meetings, conference calls, newsletters/Email/Posters and informal communication. Informal communication can range from hallway conversations to lunch meetings, etc. Their use is only encouraged if the partner organization/stakeholder has a well-established connection with a project partner and the etiquette of the recipient allows it.

3.4 Do

When stakeholders are identified, and communication toolsets are established it is time to operate. The following methodology is one of many possible solutions to the same challenge.

Means for key stakeholders, promoters:
- Senior steering groups (high level steering groups to formulate the professional content of all milestones)
- Sponsor meetings and workshops (formal meetings to validate all milestones (and their contents))
- Thematic workgroups (based on research areas)
- Ad hoc thematic meetings for urgent decisions (such as finalization of questionnaires used to assess target segments)

Means for latents and defenders/other stakeholders:
- As they will be information providers, or "price takers" in the project, they must also be engaged in communication, however due to their possibly large numbers, one way communication form will be favoured, e.g. newsletters. etc.

All formal communication shall be supplemented by informal methods, which are to be used as "in between meetings" solutions for the key stakeholders.

3.5 Follow up

The last goal is to monitor the process through the entire project period. For that purpose the following three steps approach could be used. First step, the basic concept is to build on the natural interests of all stakeholders, and catalyse the fulfilment of these needs. The relevant stakeholders must be assessed their interests in the project, and communication must be built upon these fundamentals. Second step, build and maintain an approach that is ready to involve al PPs. The third step, projects distribution to the local level.

4 Case Study on Renovation of Multiapartment Buildings in Lithuania

The case study on renovation of multiapartment buildings in Lithuania has been chosen to disclose the advantages and disadvantages of existent situation in the renovation process in Lithuania.

The stakeholder analysis will help to evaluate additional efforts from the national perspective in order to increase the efficiency of renovation process.

The key challenge in Lithuania is to boost energy efficiency in the housing sector and to prove the quality of renovation process. The main regional problem regarding energy efficiency is the inefficient use of heat energy in the buildings. 66% of the Lithuanian population live in multiapartment buildings (around 60% of these buildings were constructed during the last four decades of the last century). The average thermal energy consumption 160.180 kWh/m2, whereas the standards in newly, constructed buildings built after 1993 is around 8090 kWh/m2 per year. The necessity to improve energy efficiency in buildings is illustrated also by the magnitude of public spending on compensations for heating of housing, which increases every year. In 2012, it amounted to LTL 169.5 million. According to a recent study published by the Government, the comparison of expenses on heating and income received by consumer's shows that the Lithuanian population spend probably the largest share of their income on heating of their housing [10].

The renovation process has been started since 2004. But the effect has not been very meaningful. Financing of the Energy Efficiency in 2007–2013 program JESSICA HF (EIB) -173 MEUR - Loans for renovation of multi-apartment buildings and student dormitories. Results of the period 2007–2013: 923 multi-apartment buildings and 16 student dormitories have been renovated [11].

The changes has been done since 2013 as the novel renovation model has been implemented: Since 2013 the novel model for the buildings refurbishment has been proposed. According to this model the Agency of Buildings Energy Efficiency, as the department of the Ministry of the Environment administrates all the national programs of buildings renovation and the funds intended for those activities. This Agency makes the selection of the buildings having the greatest potential and need for renovation in cooperation with the municipalities. The inhabitants, owners of the apartments are not obliged to take credits from the financial institutions (banks), the municipalities are responsible for the financing of the renovation activities (Fig. 2).

PROGRAM PROGRESS IN 2013-2016

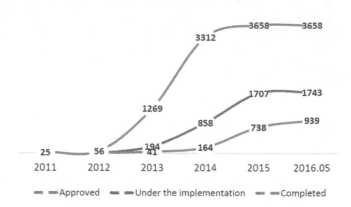

Fig. 2. Residential building renovation project pipeline, in units [12].

As in the period 2005–2012 – only 479 multiapartment buildings have been renovated, about 2000 buildings have been planned to be renovated until 2020 in order to reach the national goal. Therefore, the change in the financing measures has supported the reaching of this quantitative target. Since 2013, as the municipalities started to be initiators of the renovation process – 1507 multiapartment building have been renovated. There have been renovated 769 multi apartment building in 2016. Therefore the quantity of building with the need to be renovated has been reached.

However, the situation is positive concerning the amount of the buildings renovated and investment in the building sector, the problem of real energy efficiency remains. Evaluation of the reached energy efficiency is countable not measurable. As there is no real, monitoring system in Lithuania for the evaluation of the energy efficiency reached. The renovation process is oriented to the engagement of the building sector. It means that the primary goal is to renew the buildings based on the principal technologies in order to reach the efficiency described at the technical specification. The building companies are not obliged to guarantee particular level of energy efficiency in the coming years. There is identified the need to implement Life cycle thinking and sustainability approach in order to evaluate the current costs and the quality of technologies and materials implemented in the perspective of the longer period. The awareness of the final consumers, owners of multiaparatment buildings and other stakeholders is closely related to the proactive participation at this process.

According to the European Regional Development Fund (ERDF) [13] programme for Lithuania, supporting the shift towards a low carbon economy in all sectors Priority axis 4 is directed for promoting energy efficiency and production and use of renewable

energy. Investment priority 4.3. Supporting energy efficiency, smart energy management and RES use in public infrastructures, including in public buildings, and in the
housing sector. Based on this policy instrument the RAAIS and Working Plan, which
will be developed in the period of the project, will be based on heat energy efficiency
increase in public buildings and multiapartment houses (Table 1).

Table 1.   SHs mapping.

Roles	Potential players	PP stakeholders	Group of SHs
Creators of the policy instrument	*Local bodies/municipalities*	*Kaunas city municipality*	*Promoter*
	Regional bodies	*VIPA - Energy and Public Investment Development Agency;*	*Promoters*
	Government bodies	The Ministry of Energy of the Republic of Lithuania; Ministry of Environment of the Republic of Lithuania The ministry of Finance of the Republic of Lithuania Energy Agency	Promoter Promoter Promoter Defender
Adoption accelerators, instrument users of the Policy Instrument	Regional agencies	Housing energy efficiency agency (BETA);	Promoter
	NGOs	Association of Local Authorities in Lithuania	Defender
	Research institutes	Kaunas University of Technology Lithuanian Energy Institute	Promoter Promoter
	Civil partners	End users – private households	Apathetic/Defender
	Businesses/enterprise organizations	National association on buildings administration Lithuanian Energy Consultants Association Centre for the Education of Energetics	Apathetic Defender Defender
Players affected by the Policy Instrument	Individual businesses/enterprises	Ltd. „ESC" Ltd. ESO Ltd. CITUS Construction	Promoter Latent Promoter
	Households	Communities: Aleksotas district; Dainava district	Defender
	Civil and non-profit organizations	Association of Kaunas multiapartment owners communities Association of Kaunas community centers	Apathetic Apathetic

Low Carbon Economy [14] is oriented to the sustainability aspects, as not only environmental but also economic and social aspects should be included in evaluation and planning stages. Therefore, the stakeholder involvement could guarantee the integration of all relevant aspects and indicators in order to reach the sustainability in energy sector.

Based on the methodology described above, there have been identified the main stakeholders (SHs) and their roles in the increase of EE in Lithuania, Kaunas city [15] energy system.

The primary sorting of the SH had been done according the power mapping procedure, based a two dimensional scale: (a) power of influencing the multiapartment buildings renovation process; (b) their interest in doing so. The results shows what role of particular stakeholders in the renovation process: latents, promoters, apathetics, defenders.

Latent partners (as Ltd. ESO) have high power to influence the project, however low interest in the direct results. Their task is to keep the project on track, and monitor all activities. The main task is to cooperate with this company in implementing any innovative and smart measures ijn order to measure EE. This company is not directly concerned about EE, as it decrease the amount of the energy sold. However, the advanced view of this company declares the EE as the opportunity for new services. They are transferring to the selling of function instead of the amount of energy.

The promoters benefit directly from the project results and will be using the PIs (or will be building on them). They have a high interest in the project and they have power to influence it, this way they are the key stakeholders who have to be managed closely. This group of SHs integrates all the governmental and regional organizations; associations which are responsible for the administration of the renovation process; research institutions; and private organizations which are concerned about the profit from the increase of EE (Ltd. ESC – energy service company). Those SHs are very important in the implementation of the project; therefore, close collaboration will be done with them in order to improve the renovation process through innovative cooperation models, and integration of smart technologies.

The defenders will be affected by the project results, however, on an individual level they have low power to influence the project. This group of SHs is the most concerned about the EE increase results. All the final users and their associations, communities. That is the objective for the further research analysing bottom-up approach.

The apathetic are not interested (currently) in the project and have no desire or power to influence it. For instance, they be the local population who might get affected in the future by project results. In most cases it is enough to use passive, one-way communication methods to inform them about the potential positive (and/or negative) project outcomes. However it is very important that end users would not be in this group, therefore the system and Policy Instrument implementation needs for improvement in order the end-users could be proactive users in the multiapartment buildings renovation process.

The purpose of the next plan phase is to draw up high level action plans for each stakeholder identified and sorted previously. For each stakeholder a separate row is to be filled in with the PP's target related to the stakeholder and the high level definition of the approach to achieve it. That is the further steps in the project and the Local Living Labs are organised based on this analysis as the measure to develop Action Plan for the improvement of Policy Instrument.

5 Perspectives and Recommendations

The main idea of the LOCARBO project is to improve the Policy Instruments in such way, that it would reflect the real need of the energy consumers.

This is to be achieved by finding innovative ways for regional/local authorities to support energy consumers' behaviour change (see also [16]). The challenge to involve and motivate stakeholders (especially energy consumers) is perceived broadly as a major problem for public authorities. Motivation and awareness of consumers are of high significance to influence their behaviour and support more conscious energy decisions.

The initial research on stakeholder involvement is promising, as it leads to the development of the corresponding targets for each of the stakeholders and inclusion of the to the process of Action Plan.

The increase for renovation in Lithuania has been identified in previous years. However the proactive participation of the energy users is essential when ensuring the quality of building works and real EE increase.

It is recommended that the methodologies of stakeholder involvement would be used at the stage of Action Plan creation, or in the process of planning for it's implementation [17]. As it is addressed in LOCARBO project [18], stakeholders are essential for the project, as most of the time national and regional policy makers (as a main stakeholder group) have the means to implement project results, thus their involvement is essential for long term success and result sustainability. According to those preliminary results and reccomendations the LOCARBO experience will be traced as a good practics on applied sustainability in energy efficiency policy domain [19, 20].

Acknowledgement. This research was developed in the framework of the Project "Novel roles of regional and local authorities in supporting energy consumers' behaviour change towards a low carbon economy – LOCARBO" led by the Province of Potenza (Italy) and funded by the EU's European Regional Development fund through the INTERREG EUROPE, 2014–2020.

References

1. European Commission. Roadmap to a Resource Efficient Europe (COM(2011) 571) (2011)
2. European Commission. Energy Efficiency and its contribution to energy security and the 2030 Framework for climate and energy policy (COM (2014) 520) (2014)
3. European Commission. A policy framework for climate and energy in the period from 2020 to 2030, (COM(2014) 15) (2014)
4. Dvarioniene, J., et al.: Stakeholders involvement for energy conscious communities: the energy labs experience in 10 European communities. Renew. Energy **75**, 512–518 (2015)

5. Cosmi, C., et al.: A holistic approach to sustainable energy development at regional level: the RENERGY self-assessment methodology. Renew. Sustain. Energy Rev. **49**, 693–707 (2015)
6. Jeffrey, N.: Stakeholder engagement: a road map to meaningful engagement:# 2 in the doughty centre 'how to do corporate responsibility'series. Cranfield School of Management, Doughty Centre (2009)
7. Fletcher, A., et al.: Mapping stakeholder perceptions for a third sector organization. J. Intellect. Capital **4**(4), 505–527 (2003)
8. Mitchell, R.K., Agle, B.R., Wood, D.J.: Toward a theory of stakeholder identification and salience: defining the principle of who and what really counts. Acad. Manag. Rev. **22**(4), 853–888 (1997)
9. Cameron, B.G., Seher, T., Crawley, E.F.: Goals for space exploration based on stakeholder network value considerations. Acta Astronaut. **68**, 2088–2097 (2010)
10. Kveselis, V., Dzenajavičienė, E.F., Masaitis, S.: Analysis of energy development sustainability: the example of the Lithuanian district heating sector. Energy Policy **100**, 227–236 (2017). doi:10.1016/j.enpol.2016.10.019
11. Baležentis, T., Li, T., Streimikiene, D., Baležentis, A.: Is the Lithuanian economy approaching the goals of sustainable energy and climate change mitigation? Evidence from DEA-based environmental performance index. J. Cleaner Prod. **116**(10), 23–31 (2016). doi:10.1016/j.jclepro.2015.12.088
12. Public Investment Development Agency. http://www.vipa.lt/page/about_us
13. European Regional Development Fund. http://ec.europa.eu/regional_policy/en/funding/erdf/
14. Ürge-Vorsatz, D., et al.: Measuring multiple impacts of low-carbon energy options in a green economy context. Appl. Energy **179**(1), 1409–1426 (2016). doi:10.1016/j.apenergy.2016.07.027
15. Kaunas Municipality. Website. http://en.kaunas.lt/
16. Scorza, F.: Towards self energy-management and sustainable citizens' engagement in local energy efficiency agenda. Int. J. Agricu. Environ. Inf. Syst. (IJAEIS) **7**(1), 44–53 (2016)
17. Casas, G.L., Scorza, F.: Sustainable planning: a methodological toolkit. In: Gervasi, O., et al. (eds.) Computational Science and Its Applications – ICCSA 2016, pp. 627–635. Springer International Publishing, Cham (2016)
18. Attolico, A., Scorza, F.: A Transnational Cooperation Perspective for "Low Carbon Economy". In: Gervasi, O., et al. (eds.) Computational Science and Its Applications – ICCSA 2016 16th International Conference, Beijing, China, July 4-7, 2016, Proceedings, Part I, pp. 636–641. Springer International Publishing, Cham (2016)
19. Dvarioniene, J., Grecu, V., Lai, S., Scorza, F.: Four perspectives of appliedsustainability: research implications and possible integrations. In: Gervasi, O., et al. (eds.) ICCSA 2017, Part VI. LNCS 10409, pp. 554–563. Springer, Cham (2017)
20. Scorza, F., Grecu, V.: Assessing sustainability: research directions and relevant issues. In: Gervasi, O., et al. (eds.) ICCSA 2016. LNCS, vol. 9786, pp. 642–647. Springer, Cham (2016). doi:10.1007/978-3-319-42085-1_55

A Quantitative Measure of Habitat Quality to Support the Implementation of Sustainable Urban Planning Measures

Rosa Epifani[1], Federico Amato[1(✉)], Beniamino Murgante[1], and Gabriele Nolé[2]

[1] School of Engineering, University of Basilicata,
Viale dell'Ateneo Lucano 10, 85100 Potenza, Italy
rosaepifani@tiscali.it, {beniamino.murgante,federico.amato}@unibas.it
[2] IMAA, Italian National Research Council, C.da Santa Loja,
Tito Scalo, 85050 Potenza, Italy
gabriele.nole@imaa.cnr.it
https://www.unibas.it, https://www.imaa.cnr.it/

Abstract. The 2030 Agenda by United Nations highlights the necessity of undertake concrete actions to *"protect, restore and promote sustainable use of terrestrial ecosystems, sustainably manage forests, combat desertification, and halt and reverse land degradation and halt biodiversity loss"*. However, human activities on land use are strongly threatening habitat quality, causing their fragmentation and a dramatic loss of biodiversity all over the world. This paper proposes an application of the InVEST Habitat Quality model as a tool to support the definition of sustainable development policies able to favour the preservation of habitat structures while promoting their exploitation as cultural and landscape assets. The model is applied to the Basilicata Region (Southern Italy). Results show how modelling the impacts of human activities on biodiversity and ecosystem services can strongly help planning activities in distinguish those areas that should undergo to a conservation regime to preserve habitat integrity from those which are most prone to transformations, taking advantage by the social and economic benefit deriving from the human activities connected to their use.

Keywords: Biodiversity · Habitat quality · Land-use planning · Sustainable development · Spatial plannings

1 Introduction

Human activities on land-use - weather connected to agricultural production, natural resources extraction or urban expansion - are reshaping our planet, threatening natural habitats of fragmentation and biodiversity loss [1,2]. The negative impact of land-use practices in terms of changes in atmospheric composition, regional climates and on planet's ecosystem have been deeply investigated by scientists [3,4].

© Springer International Publishing AG 2017
O. Gervasi et al. (Eds.): ICCSA 2017, Part VI, LNCS 10409, pp. 585–600, 2017.
DOI: 10.1007/978-3-319-62407-5_42

For all these reasons cities, and therefore their correct planning, play an important role in reaching climate change adaptation and disaster risks reduction targets [5–7]. The Global Risk Report [8] highlights how three out of the top ten risks in terms of impact on the next decade are connected to/the environmental sphere, especially to water crisis, the failure of climate change adaptation policies and the loss of biological diversity. Several scenarios of biodiversity changes have been analysed [9] showing how a business as usual approach to its protection cannot stop its decline over the next decades [10,11]. Moreover, studies show that up to 88% of protected areas that will be threatened by urban expansion are located in low/moderate income countries [12].

At the same time, the rapid growth of urban areas has gone together with poorly planned urban patterns, generating a number of negative consequences on health, security and life quality [13]. Besides, urban growth generates air pollution, energy consumption and waste production [14–21]. Furthermore, urbanization constitutes one of the major threats to habitat integrity [22–24]. Urban growth is likely to happen in proximity to biologically diverse areas [25–27]. However, urban planning tools are usually not focused on the importance for human health of maintaining a biodiverse ecosystem. Protecting biodiversity means safeguarding the range of species and population in a specific habitat, and therefore underpinning site-specific ecosystem services. Its defence is even more important if the role of biodiversity in ensuring food availability and disaster risk management [28]. Indeed, alterations of natural habitats can led to a reduction of territorial resilience [29]. Thus, urban sprawl and land take phenomena have impacts on biogeochemical cycles such as the water and carbon ones, resulting in an increase of flood and landslide risks [30,31].

Recently, habitat quality maintenance and biodiversity protection become a crucial point in United Nations (UN) policies [32]. The Sustainable Development Goal 15 defined within the 2030 Agenda by UN considers as a top priority to protect, restore and promote sustainable use of terrestrial ecosystems, sustainably manage forests, combat desertification, and halt and reverse land degradation and halt biodiversity loss. Previous regulation on that issues were already provided in the Agenda 21 [33], which highlighted the necessity of taking actions for the maintenance of the biological diversity through the in situ conservation of habitats and ecosystems. Europe answered to the Agenda 21 propositions with the Habitats Directive CEE 94/43/EC and the Birds Directive 2009/147/CE, which together defined the Natura 2000 Network [34,35]. That is a network covering more than the 18% of the European Unions (EU) land area and the 6% of its marine territory, expanding across the 28 EUs countries.

Despite the institutional effort by EU and UN, habitat preservation is still not a priority in national and regional spatial policies, especially in Italy, where decision-makers seems unable to couple economic and social development with biodiversity and landscape protection [36]. Hence, the usage of tools able to measure the rate at which anthropogenic activities affects biodiversity conservation is highly recommendable [37,38]. In this way, a consistent framework would support planning and development choices, helping in distinguish those areas where a conservative regime could also favour economic development from those where socio-economic exploitation and biodiversity conservation are not companionable.

Within this context, this paper discusses the use of geographical modelling of habitat quality changes as a useful tool to assess the positive and negative interactions among natural environment and human-based activities or land-use practices. Among the different approaches to biodiversity conservation, several families of models are recognizable. Firstly, the coarse-filter models are based on habitat or vegetation-based representation. Further models are based on the maximisation of the number of species covered by a network of protected areas for a given budget, while other are founded on the analysis of the patterns of richness and endemism. The work presented in this paper proposes an application of a coarse-filter model as a robust methodology for the definition of knowledge framework on which develop sustainable conservation strategies for the protection of natural habitats and for the maintenance of biodiversity values.

2   Materials and Methods

A synthetic description of the study area is provided in order to underline relevant socio-economic and territorial criticality emerging in territorial decision-making processes we intended to support with additional information coming from the Habitat Quality Model interpretation. We present the model and the main variable considered with specific weights in the elaboration and we present the results obtained.

2.1   Study Area

The sample area selected for the analysis is the Basilicata Region (Fig. 1). Classified as a "lagging region" (among literature concerning EU cohesion classification and consequent investigation on the impact of development policies refer to [39–41] in the framework of EU Cohesion policy). Basilicata represents an interesting laboratory for the investigation of the conflicts between urban development - or, in other words, anthropic pressure on the environment and ecosystem - and preservation of non-reproducible resources in a sustainable view within a peculiar context characterized by a lack of economic development, a structural need for infrastructures, disperse and fragmented settlement, a wide range of natural and cultural values undermined by depopulation and ageing population trends. In a general view such characters appears to be common in several peripheral areas suffering the lack of a specific role in the wider urban structure of their country. To clarify this statement we point out the attention on the relationships between urban and rural areas, which represents a traditional research topic for planning disciplines. Anyway, we can affirm that the urban role is always well defined in the development models (according to list of clear economic values) while the rural areas are called upon to provide compensation contributions to

Fig. 1. The location, highlighted in red, of the Basilicata Region within the Italian regional boundaries. (Color figure online)

urban development resulting in the consumption of resources. Such exploitation - often not sustainable - is generally included in "compensation policies" or "rebalancing policies": one of the most representative example is European Union Cohesion Policy [42–44]. The operative ways through which such policies influence the territorial transformation process are mainly oriented to distribute benefits (in terms of economic resources, public services, infrastructures) in order to contribute to the essential role of such inland areas to the environmental balance between exploitation and preservation of natural resources. The case of Basilicata is relevant because it includes a wide range of contradictions: wide range of protected areas (including natural parks and reserves) and a deeper exploitation of oil fields; huge public investments in local development and a persistent socio-economic weakness [45]; availability of energetic sources (fossil and renewable) and a lack of a productive industrial system [46]. It emerges a territorial planning peculiar challenge to combine the principles of environmental protection to a sustainable development model based on the growth opportunities that a "weaker demand region" expresses according to its unique features (it means the combination of natural, cultural and social values). A way to detect and assess such territorial components is represented by the Habitat quality model and the effort to reinforce the territorial interpretation tools as Decision Support System is still a priority in regional and local planning system. As for the regulation framework concerning ecological networks and landscape protection, the regional urban and spatial planning Law 23/99 [47] considers as one of the object of urban and regional planning the analysis and the regulation of the natural and environmental systems. These are defined as the entire regional territory, except for the mobility and energy infrastructures and the built-up environment [48].

2.2 Habitat Quality Model

The quality of the habitats in the study area and their rarity rate have been evaluated through the use of the InVEST Habitat Quality model [49] available at [50]. The model framework couples analysis concerning the Land Use and Land Cover (LULC) with informations about the presence of threats to biodiversity conservation and protection in the study area. The application of the model allows the creation of raster maps identifying quality and rarity rates. Hence, it estimates the effect of each considered threat on the habitats analysed by considering also the indirect effects induced by a weighted combination of all the threats. It shall be stressed how threat is used to define every anthropogenic activity which could negatively affect the biodiversity preservation and, therefore, the habitat conservation. Thus, threats are spatial explicit variables which could cause the local extinction of one or more animal and plant species. The biologist Edward Owen Wilson quantified a loss of more than thirty thousand of species per year globally. He described the impact of human activity on the natural ecosystem with the acronym *HIPPO*, indicating as main threats for natural environment the Habitat loss, the introduction of Invasive species, Pollution, human Population growth and, finally, Overharvesting [51].

Quality and rarity are measured starting from LULC maps concerning the present land cover and an older reference coverage. To define which land cover class shall be considered as habitat, a binary approach could be adopted assigning value 1 for the land covers to be considered as habitat and value 0 in the other case. Nevertheless, to take into account the reduction of extinction risk of population due to the fragmentation of landscape, and therefore to the contribution in terms of biodiversity offered by those coverage which are non exhaustively and completely classifiable as natural habitat [52–54], the proposed model allows a non-binary classification of LULC through the assignment of a relative habitat suitability score defined in the range $[0, 1]$, where 1 corresponds to the LULC classes having the highest habitat suitability. Moreover, this approach considers the possibility that a species may prefer a given habitat. To evaluate the exposure of habitat to degradation, each threat is represented as a raster map which cells have a value representing the intensity of the threat or, if a binary approach is used, its presence/absence. In this paper, a binary approach for the evaluation of threats has been adopted.

Once habitat quality is evaluated, the model allows the measurement of their rarity. Rarity evaluation is done through two different phases. In the first one, the rarity of each single habitat identified through the LULC maps is carried out. Subsequently, an overall evaluation of the rarity of all the considered habitats is done. Hence, the final output is a raster map which grid cells are representing the overall rarity rate of the habitats in the considered pixel.

2.3   Materials and Data Preparation

As explained in the previous section, the Habitat Quality Model uses as input data raster grids concerning the LULC patterns in the study area at two different times and the threats to biodiversity conservation. The current LULC map and the historical coverage maps for the study area were derived from CORINE Land Cover data freely available at [55]. Two raster maps, respectively for 1990 and 2012, were derived for the Basilicata Region using a spatial resolution of 100 by 100 m (Fig. 2). A visual comparison among the two maps shows the absence of deep changes in the LULC, with the north-eastern part of the region, which is the hilly area, covered by forest in both the maps, while the coastal area in the south-eastern corner is mostly prone to crop production.

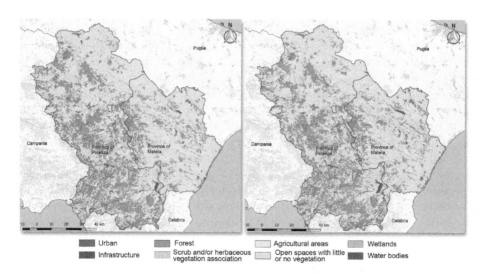

Fig. 2. Corine land cover maps for the years 1990 (left) and 2012 (right).

Table 1. Threats considered in the Habitat quality model and corresponding parameters.

Threat	Maximum distance	Weight	Decay
Urban areas	10.0	1.0	Exponential
Industrial areas	8.0	0.8	Exponential
Mining areas	6.0	0.7	Linear
Cultivated areas	8.0	0.7	Linear
Landfill	9.0	0.8	Exponential
Primary roads network	3.0	1.0	Linear
Secondary roads network	1.0	0.7	Linear
Railways	0.8	0.6	Linear
Electric network	0.5	0.5	Exponential

Subsequently, nine raster grids were prepared to analyse the different existing threats to the habitat conservation. Threats are used to identify the impact of anthropogenic activities on different habitats. To each threat correspond a binary raster map in which a value of 1 identifies the presence of the considered threat, while a value of 0 corresponds to the absence of the considered threat. Moreover, each threat is also defined by three parameters: the maximum distance, the weight and the decay. The first one indicates the maximum distance in kilometres at which the threat can have impacts on the habitat quality. The second one corresponds to a value in a range between 0 and 1 indicating the potential impact of the considered threat to each single habitat, where 1 indicates the maximum impact. Finally, the decay indicates whether a linear or an exponential function is used to characterise the impact of the threats on habitat. Table 1 shows a complete list of the considered threats and the corresponding parameters adopted.

The sensitivity of each habitat to each single threat is used to derive the overall threat level in each grid cell of the study area. The sensitivity is defined through a *.csv* file indicating a value in the range $[0, 1]$, where 1 corresponds to the higher sensitivity of the habitat to the corresponding threat. Finally, the accessibility of the habitats to the degradation source has been considered based on a *shapefile* derived considering physical, social and legal restrictions associated to each part of the study area. This accessibility map, as shown in Fig. 3, considers the presence of areas belonging to the Natura 2000 Network, flora and fauna, patch size of the LULC.

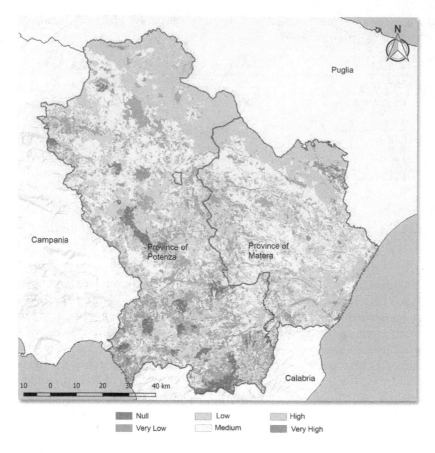

Fig. 3. The accessibility map shows the rate of protection, numerically evaluated in a range [0, 1], due to social or physical restriction or other specific regulations.

3   Results

Passing through an evaluation of the LULC, of the connected habitat and of the anthropogenic activities considered as threats, the Habitat Quality model provides as output a degradation index map and a quality index map. These two parameters can help in identifying the areas where planners should focus their attention to protect biodiversity. Moreover, the model provides a rarity index map, obtained considering the changes occurred in LULC by analysing the variation in the extension of the considered habitats in the period between the first LULC classification and the latter one.

Figure 4 shows the degradation index map and the quality index map obtained for Basilicata Region. The two maps are complementary, showing that the most degraded areas, and therefore those with a lower quality, are located in the North-Western and South-Eastern parts of the Region.

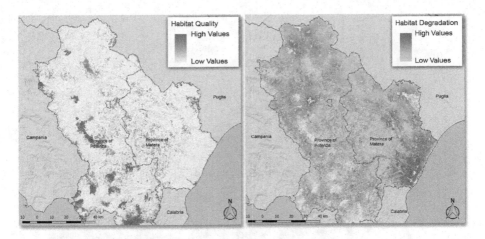

Fig. 4. The habitat quality index map (left) and the habitat degradation index map (right) for the Basilicata Region.

The central part of the region, belonging to the Apennine mountains, does not present criticalities as for these values. The figure highlights how the threats that mostly affected habitat quality and degradation are the presence of built-up areas, industrial sites and railways. Figure 5 shows the rarity index map, which shows the areas characterized by those LULC classes that downsized over time. A first overview of this map shows how the higher levels of rarity correspond to areas covered by coniferous and deciduous woods and sclerophyll vegetation. Similarly these areas are characterised by high quality values and a degradation value of almost 0.

To better evaluate the output of the model, zonal statistics for degradation index, quality index and rarity index maps were carried out both at the provincial and municipal level. Table 2 shows the average values of the indexes for the two provinces of Potenza and Matera.

Table 2. Averages values for degradation, quality and rarity indexes in the two provinces composing the Basilicata Region.

	Mean degradation value	Mean quality value	Mean rarity value
Province of Potenza	0.012	0.985	−0.128
Province of Matera	0.014	00.983	−0.234

The average degradation index is higher in the Province of Matera than in the Province of Potenza, while the average rarity index is higher in the Province of Potenza. Figure 4 illustrates the average indexes evaluated for all the Municipalities in the Region. The overall higher and lower values of degradation are

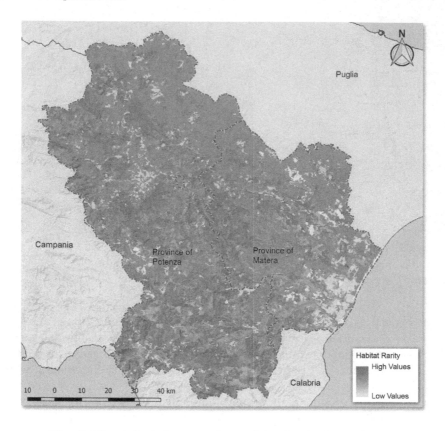

Fig. 5. The rarity index map obtained for the Basilicata Region.

measured respectively in the Municipality of Policoro (Province of Matera) and in the Municipality of Fardella (Province of Potenza). In the former, a degradation index of 0.094 is measurable, while in the latter a value of 0.003 is measured. Differently, considering the average values in the Municipalities the higher and lower mean values per Municipality are measured in Scanzano Jonico and Chiaromonte, with values of respectively 0.030 and 0.00004. The average degradation value of Scanzano Jonico is higher of both the Provincial mean (corresponding to 0.014) and the Regional one (corresponding to 0.013), while the value measured in Chiaromonte is lower of the Regional and Provincial mean (0.012). The mean values of the remaining Municipalities are all in a range between 0.005 and 0.010. Considering the quality index, it is possible to measure how almost the half of the Municipalities of the Region has mean values near to 1. The higher value is measured in the Municipality of Fardella, confirming what was already showed through the analysis of the degradation index. The lower mean value of the quality index is measured in the Municipality of Policoro. Its value, corresponding to 0.90, is lower than the Regional (0.984) and Provincial (0.980) one. Finally, it is important to consider how the rarity index can help in identifying the most threatened areas of the Region. Hence, considering the overall quality values, the most threatened Municipalities are Accettura, Irsina and Matera in the Province of Potenza and Campomaggiore, Pietrapertosa and Lagonegro in the Province of Potenza. In these Municipalities the rariti value is of about −4.40, showing how significant strategies for the protection of natural sites in those areas have to be considered as urgent and indispensable. When considering the mean values in the Municipalities, the lower values are measured in Policoro (−0.95) and Scanzano Jonico (−0.88). These values are higher of both the Provincial and regional means, respectively equal to −0.234 and −0.128.

4   Discussion and Conclusion

The aim of this study was to identify the most threatened areas of the Basilicata Region in order to allow decision-makers and urban planners to rapidly develop concrete and effective tools to support a better protection of the environmental, ecological and natural values of the regional landscapes by effective regulation of human actions affecting natural heritage and regional habitats. Hence, the output produced by the Habitat Quality model could be seen as a valuable tool for local authorities to support the development of planning policies able to follow a sustainable use of the territory, including the following general policy:

- Favouring reforestation and rejuvenation of degraded habitats avoiding human interventions which are not considering their induced impacts on natural resources;
- Coupling new urban development with green areas to mitigate the effects of the urban heat island, reducing the amount of greenhouse gases and air pollutants;
- Create ecological networks in urban areas through interventions of naturalistic engineering, aiming at the reduction of the fragmentation and degradation of urban ecosystems caused by overbuilding and excessive soil sealing;
- Controlling and limiting the construction of new plants considered as sources of pollution and adapt existing ones to the need of conservation of biodiversity and natural resources;
- Reducing the use of pesticides and eliminating unsustainable farming techniques by replacing them with the new organic farming methods. This approach is considered to be extremely effective as properly cultivated soils can become reservoirs of carbon dioxide (and thus contribute to reducing the greenhouse effect, acting positively on climate mitigation) and they may be affected in a lower extent by erosion and desertification, preserving the physical, chemical and microbiological properties that favour the maintenance of long-term fertility;
- Increasing the number of protected areas through the establishment of new parks and natural reserves to protect biodiversity.

These goals follow the targets of the Convention on Biological Diversity by United Nations [56].

Concerning the territorial dimensions included in the model, a set of controlling variables could be used in order to perform a scenario analysis on the following territorial patterns: Urban Areas, Industrial Areas, Mining Areas, Cultivated Areas, Landfill, Primary Roads Network, Secondary Roads Network, Railways, Electric Network; compared with natural protected areas and landscape networks in the Region. It means that through a detection of the territorial dynamics of the "threats" pattern it will be possible to highlight possible conflicts and main risks on the environmental system generated by on going transformation. In fact those factors are sensible to the current strategies incuded in the Regional Programs supporting the operative implementation of EU Cohesion policy at regional levels that in the short term several concrete actions and projects will be implemented in the region [57], probably without a comprehensive assessment tool-kit allowing the scenario analysis and an effective coordination of local responsibility in implementation processes.

It is possible to consider Habitat Quality Model as an additional layer to include in territorial interpretation of on-going transformation policies in order to achieve the goal to support the socio-economic development programs with effective territorial impact assessment tools detecting potential conflicts and reinforcing the governance at regional and local level. A relevant need in the low density settlement areas where the sensitivity level of natural ecosystems is higher and directly connected with the cultural and social structure of resident communities.

Other interesting feature to consider in order to reinforce the relevance of Habitat Quality Model results could be:

– Carbon stocking and climate change mitigation. Ecosystems regulate the Earth's climate through the greenhouse gas exchange with the atmosphere and absorption of about half of the CO_2 emissions attributable to the anthropogenic activities. It is crucial to protect them both from natural disasters such as fire and disease and from deterioration caused by human activities. Future evolution of this study should estimate the social value of carbon storage capacity of the terrestrial ecosystems to implement spatial plans to preserve the natural wells in the area (classified in aboveground biomass, belowground biomass, soil and deadwood) protecting them from human activities which do not take into account these values.
– Protection of the coastline from erosion and floods. To analyse that issue, further studies will have to quantify the benefits that natural habitats provide against coastal erosion and flooding to allow the definition of coastal development strategies that take into account their value.
– Protection and safeguard of the aesthetic and recreational values retained by the scenic areas. Recreation and tourism are important components of many national and local economies and contribute in countless ways to improve people's quality of life. A detailed analysis of the role of natural landscapes and protected areas as tourism attractors should be carried on to design integrated green networks connecting coastal areas with inland protected sites, reciprocally benefiting sharing their attractive potential.

The development a complete picture of the cited analysis will favour an extended comprehension of the territorial dynamics in the Basilicata Region in terms of interaction between its natural environment and the human activities/transformation. Future studies intend to pursue a more rational land use management supporting decision makers in planning processes [58], aiming at conservation and protection of the natural resources of the Region.

Author Contribution. Authors contributed equally to this work. Specifically, experiment design and writing of the manuscript was developed jointly by all authors. Rosa Epifani and Gabriele Nolé performed the experiments and analysed the data. Federico Amato and Beniamino Murgante interpreted the results and developed the discussion.

Conflict of Interests. The authors declare no conflict of interest.

References

1. Foley, J.A., Defries, R., Asner, G.P., Barford, C., Bonan, G., Carpenter, S.R., Chapin, F.S., Coe, M.T., Daily, G.C., Gibbs, H.K., Helkowski, J.H., Holloway, T., Howard, E.A., Kucharik, C.J., Monfreda, C., Patz, J.A., Prentice, I.C., Ramankutty, N., Snyder, P.K.: Global consequences of land use. Science **309**, 570–574 (2005). (New York, N.Y.)
2. Ramankutty, N., Coomes, O.T.: Land-use regime shifts: an analytical framework and agenda for future landuse research. Ecol. Soc. **21**(2) (2016)
3. Tilman, D., Fargione, J., Wolff, B., D'Antonio, C., Dobson, A., Howarth, R., Schindler, D., Schlesinger, W.H., Simberloff, D., Swackhamer, D.: Forecasting agriculturally driven global environmental change. Science **292**(5515), 281–284 (2001)
4. Vitousek, P.M., Mooney, H.A., Lubchenco, J., Melillo, J.M.: Human domination of earth's ecosystems. Science **277**(5325), 494–499 (1997)
5. Seto, K.C., Güneralp, B., Hutyra, L.R.: Global forecasts of urban expansion to 2030 and direct impacts on biodiversity and carbon pools. PNAS **109**(40), 16083–16088 (2012)
6. Lim, Y.K., Cai, M., Kalnay, E., Zhou, L.: Observational evidence of sensitivity of surface climate changes to land types and urbanization. Geophys. Res. Lett. **32** (2005)
7. Amato, L., Dello Buono, D., Izzi, F., La Scaleia, G., Maio, D.: HELP - an early warning dashboard System, built for the prevention, mitigation and assessment of disasters, with a flexible approach using open data and open source technologies (2016)
8. World Economic Forum: The global risk report 2016. Technical report, World Economic Forum, Geneva (2016)
9. Pereira, H.M., Leadley, P.W., Proença, V., Alkemade, R., Scharlemann, J.P.W., Fernandez-Manjarrés, J.F., Araújo, M.B., Balvanera, P., Biggs, R., Cheung, W.W.L., Chini, L., Cooper, H.D., Gilman, E.L., Guénette, S., Hurtt, G.C., Huntington, H.P., Mace, G.M., Oberdorff, T., Revenga, C., Rodrigues, P., Scholes, R.J., Sumaila, U.R., Walpole, M.: Scenarios for global biodiversity in the 21st century. Science **330**(6010), 1496–1501 (2010)
10. Wilcove, D.S., Rothstein, D., Dubow, J., Phillips, A., Losos, E.: Quantifying threats to imperiled species in the United States. BioScience **48**(8), 607–615 (1998)

11. Czech, B., Krausman, P.R., Devers, P.K.: Economic associations among causes of species endangerment in the United States. Bioscience **50**(7), 593 (2000)
12. Mcdonald, R.I., Kareiva, P., Forman, R.T.: The implications of current and future urbanization for global protected areas and biodiversity conservation. Biol. Conserv. **141**(6), 1695–1703 (2008)
13. Amato, F., Maimone, B.A., Martellozzo, F., Nolè, G., Murgante, B.: The effects of urban policies on the development of urban areas. Sustainability **8**, 297 (2016)
14. Scardaccione, G., Scorza, F., Casas, G.L., Murgante, B.: Spatial autocorrelation analysis for the evaluation of migration flows: the Italian case, pp. 62–76 (2010)
15. Scorza, F.: Towards self energy-management and sustainable citizens' engagement in local energy efficiency agenda. Int. J. Agricult. Environ. Inf. Syst. **7**(1), 44–53 (2016)
16. Casas, G.L., Scorza, F.: Sustainable planning: a methodological toolkit. In: Gervasi, O., Murgante, B., Misra, S., Rocha, A.M.A.C., Torre, C., Taniar, D., Apduhan, B.O., Stankova, E., Wang, S. (eds.) ICCSA 2016. LNCS, vol. 9786, pp. 627–635. Springer, Cham (2016). doi:10.1007/978-3-319-42085-1_53
17. Snyder, P., Delire, C., Foley, J.: Evaluating the influence of different vegetation biomes on the global climate. Clim. Dyn. **23**(3–4), 279–302 (2004)
18. Scorza, F., Casas, G.B.L., Murgante, B.: That's ReDO: ontologies and regional development planning. In: Murgante, B., Gervasi, O., Misra, S., Nedjah, N., Rocha, A.M.A.C., Taniar, D., Apduhan, B.O. (eds.) ICCSA 2012. LNCS, vol. 7334, pp. 640–652. Springer, Heidelberg (2012). doi:10.1007/978-3-642-31075-1_48
19. Murgante, B., Tilio, L., Scorza, F., Lanza, V.: Crowd-cloud tourism, new approaches to territorial marketing. In: Murgante, B., Gervasi, O., Iglesias, A., Taniar, D., Apduhan, B.O. (eds.) ICCSA 2011. LNCS, vol. 6783, pp. 265–276. Springer, Heidelberg (2011). doi:10.1007/978-3-642-21887-3_21
20. Murgante, B., Scorza, F.: Ontology and spatial planning. In: Murgante, B., Gervasi, O., Iglesias, A., Taniar, D., Apduhan, B.O. (eds.) ICCSA 2011. LNCS, vol. 6783, pp. 255–264. Springer, Heidelberg (2011). doi:10.1007/978-3-642-21887-3_20
21. Scorza, F., Casas, G.L., Carlucci, A.: Onto-planning: innovation for regional development planning within EU convergence framework. In: Murgante, B., Gervasi, O., Iglesias, A., Taniar, D., Apduhan, B.O. (eds.) ICCSA 2011. LNCS, vol. 6783, pp. 243–254. Springer, Heidelberg (2011). doi:10.1007/978-3-642-21887-3_19
22. Foley, J.A., Ramankutty, N., Brauman, K.A., Cassidy, E.S., Gerber, J.S., Johnston, M., Mueller, N.D., O'connell, C., Ray, D.K., West, P.C., Balzer, C., Bennett, E.M., Carpenter, S.R., Hill, J., Monfreda, C., Polasky, S., Rockström, J., Sheehan, J., Siebert, S., Tilman, D., Zaks, D.P.M.: Solutions for a cultivated planet. Nature **478**, 337–342 (2011)
23. Guo, X., Meng, M., Zhang, J., Chen, H.Y.H.: Vegetation change impacts on soil organic carbon chemical composition in subtropical forests. Sci. Rep. **6**(29607) (2016)
24. Li, H., Shen, H., Chen, L., Liu, T., Hu, H., Zhao, X., Zhou, L., Zhang, P., Fang, J.: Effects of shrub encroachment on soil organic carbon in global grasslands. Sci. Rep. **6**(28974) (2016)
25. Amato, F., Pontrandolfi, P., Murgante, B.: Supporting planning activities with the assessment and the prediction of urban sprawl using spatio-temporal analysis. Ecol. Inform. **30**, 365–378 (2015)
26. Martellozzo, F., Clarke, K.C.: Measuring urban sprawl, coalescence, and dispersal: a case study of Pordenone, Italy. Environ. Plan. **38**, 1085–1104 (2011)
27. Martellozzo, F.: Forecasting high correlation transition of agricultural landscapes into urban areas. Int. J. Agricult. Environ. Inf. Syst. **3**(2), 22–34 (2012)

28. FAO: The state of the world's land and water resources for food and agriculture (SOLAW) - managing systems at risk. In: Food and Agriculture Organization of the United Nations, Rome and Earthscan, London (2011)
29. Lombardini, G., Scorza, F.: Resilience and smartness of coastal regions. A tool for spatial evaluation. In: Gervasi, O., et al. (eds.) ICCSA 2016. LNCS, vol. 9788, pp. 530–541. Springer, Cham (2016). doi:10.1007/978-3-319-42111-7_42
30. Ramankutty, N., Evan, A.T., Monfreda, C., Foley, J.A.: Farming the planet: 1. Geographic distribution of global agricultural lands in the year 2000. Glob. Biogeochem. Cycles **22** (2008)
31. Romano, B., Zullo, F.: The urban transformation of Italy's adriatic coastal strip: fifty years of unsustainability. Land Use Policy **38**, 26–36 (2014)
32. United Nation - FCCC: adoption of the paris agreement. Proposal by the President. Technical report (2015)
33. United Nations Conference on Environment and Development: Agenda 21, Rio Declaration, Forest Principles. Technical report, United Nations, New York (1992)
34. European Parliament, Council of the European Union: Directive 2009/147/EC on the conservation of wild birds. Technical report (2009)
35. Council of the European Union: Council Directive 92/43/EEC on the conservation of natural habitats and of wild fauna and flora. Technical report (1992)
36. Zullo, F., Paolinelli, G., Fiordigigli, V., Romano, B.: Urban development in Tuscany land uptake and landscapes changes. TeMA - J. Land Use Mobil. Environ. **2**, 183–202 (2015)
37. Falcucci, A., Maiorano, L., Boitani, L.: Changes in land-use/land-cover patterns in Italy and their implications for biodiversity conservation. Landsc. Ecol. **22**(4), 617–631 (2007)
38. Torralba, M., Fagerholm, N., Burgess, P.J., Moreno, G., Plieninger, T.: Do European agroforestry systems enhance biodiversity and ecosystem services? A meta-analysis. Agricult. Ecosyst. Environ. **230**, 150–161 (2016)
39. Gil, C., Pascual, P., Rapún, M.: Regional allocation of structural funds in the European Union. Environ. Plan. C: Gov. Policy **20**(5), 655–677 (2002)
40. Gripaios, P., Bishop, P., Hart, T., McVittie, E.: Analysing the impact of objective 1 funding in Europe: a review. Environ. Plan. C: Gov. Policy **26**(3), 499–524 (2008)
41. Las Casas, G., Murgante, B., Scorza, F.: Regional local development strategies benefiting from open data and open tools and an outlook on the renewable energy sources contribution. In: Papa, R., Fistola, R. (eds.) Smart Energy in the Smart City, pp. 275–290. Springer International Publishing, Cham (2016). doi:10.1007/978-3-319-31157-9_14
42. Crescenzi, R., Giua, M.: The EU cohesion policy in context: does a bottom-up approach work in all regions? Environ. Plan. A **48**(11), 2340–2357 (2016)
43. Bachtler, J., McMaster, I.: EU cohesion policy and the role of the regions: investigating the influence of structural funds in the new member states. Environ. Plan. C: Gov. Policy **26**(2), 398–427 (2008)
44. Faludi, A.: EU territorial cohesion, a contradiction in terms. Plan. Theory Pract. **17**(2), 302–313 (2016)

45. Scorza, F., Casas, G.L.: Territorial specialization in attracting local development funds: an assessment procedure based on open data and open tools. In: Murgante, B., et al. (eds.) ICCSA 2014. LNCS, vol. 8580, pp. 750–757. Springer, Cham (2014). doi:10.1007/978-3-319-09129-7_54

46. Scorza, F., Attolico, A., Moretti, V., Smaldone, R., Donofrio, D., Laguardia, G.: Growing sustainable behaviors in local communities through smart monitoring systems for energy efficiency: RENERGY outcomes. In: Murgante, B., et al. (eds.) ICCSA 2014. LNCS, vol. 8580, pp. 787–793. Springer, Cham (2014). doi:10.1007/978-3-319-09129-7_57

47. Regione Basilicata: Legge Regionale, no. 23, del 11 August 1999

48. Scorza, F.: Improving EU cohesion policy: the spatial distribution analysis of regional development investments funded by EU structural funds 2007/2013 in Italy. In: Murgante, B., Misra, S., Carlini, M., Torre, C.M., Nguyen, H.-Q., Taniar, D., Apduhan, B.O., Gervasi, O. (eds.) ICCSA 2013. LNCS, vol. 7973, pp. 582–593. Springer, Heidelberg (2013). doi:10.1007/978-3-642-39646-5_42

49. Terrado, M., Sabater, S., Chaplin-Kramer, B., Mandle, L., Ziv, G., Acuña, V.: Model development for the assessment of terrestrial and aquatic habitat quality in conservation planning. Sci. Total Environ. 540, 63–70 (2016)

50. InVEST - Natural Capital Project. http://www.naturalcapitalproject.org/inv

51. Wilson, E.O.: The Future of Life, Reprint edn. Vintage, New York (2003)

52. Franklin, J.F., Lindenmayer, D.B.: Importance of matrix habitats in maintaining biological diversity. Proc. Natl. Acad. Sci. USA 106(2), 349–350 (2009)

53. Ricketts, T.H.: The matrix matters: effective isolation in fragmented landscapes. Am. Nat. 158(1), 87–99 (2001)

54. Ramankutty, N., Rhemtulla, J.: Can intensive farming save nature? Front. Ecol. Environ. 10, 455 (2012)

55. Copernicus Programme: CORINE Land Cover. http://land.copernicus.eu/pan-european/corine-land-cover

56. United Nations: Convention on biological diversity. Technical report (1992)

57. Casas, G.L., Murgante, B., Scorza, F.: Regional local development strategies benefiting from open data and open tools and an outlook on the renewable energy sources contribution. In: Papa, R., Fistola, R. (eds.) Smart Energy in the Smart City. Springer International Publishing, Cham (2016)

58. Las Casas, G., Sansone, A.: Una cultura della pianificazione in un approccio rinnovato alla razionalitá nel piano. Deplano G. Politiche e strumenti per il recupero urbano, Edicom Edizioni, Monfalcone (GO) (2004). ISBN 978-988

Incentives for Social Sustainability of the Third Sector in Lithuania

Dziugas Dvarionas(✉)

Faculty of Social Sciences, Vytautas Magnus University,
Jonavos Str. 66, 44191 Kaunas, Lithuania
dziugas.dvarionas@vdu.lt

Abstract. Built on European social cohesion policy mainstream, this paper reveals incentives of Third sector players' active involvement into economic growth in Lithuania. Emerging role of social enterprises needs a scientific focus on social economy categorisation in terms of opening a stage for scientific debates around sustainable development of non-governmental sector in Lithuania. The paper discusses themes, such as social economy recognition, inter-sectoral boundaries, social economy partnership and other factors of social sustainability.

Keywords: Social economy · Social capital · Third sector · Social enterprise · Satellite accounts · Social economy partnership

1 Introduction

Following reflection of European social cohesion policy mainstream in Lithuania, it can be noticed, that level of social economy recognition in Lithuanian legislative practice is rather weak. Taking transition period from 1988 up to now, social capital transformations in Lithuania did not serve as potential to create strong third sector organisations, capable to add to national economy. Despite the fact that there are almost twenty thousand active non-governmental organisations, their capacities for sustainable growth are limited because of big competition for existing recourses and lack of competences for social investment.

The role of small and medium-sized enterprises in the national economy is still not enough significant, what results in low economic power of Lithuanian middle class. Although, growing sector of small and medium enterprises in a year 2004–2007 in Lithuania gives a good basis to estimate economic growth of a sector, but still, the main problem remaining here is that small and medium enterprise (SME) and Third sector's organizations do not form a competitive social economic partnership, that finally would lead to strong and sustainable Third Sector (TS).

Financing and provision of public services is mainly centralized in Lithuania. There is only a little part of social services that is given to the TS through the local authorities using the mechanisms of public procurement and municipal calls for proposals. Major part of European Structural funds goes through state administrative institutions and remain in public sector.

© Springer International Publishing AG 2017
O. Gervasi et al. (Eds.): ICCSA 2017, Part VI, LNCS 10409, pp. 601–609, 2017.
DOI: 10.1007/978-3-319-62407-5_43

The aim of this article is to present and discuss socio-economic conditions for implementing widely recognized European social economy concepts and principles into Lithuanian economy for sustainable growth of community based social organisations.

2 Methodology

Last research of NGO sector in Lithuania shows, that currently no quality standards exists that would allow to distinguishing independent and sustainable social NGOs capable to start social enterprise activities in the country. Furthermore, there are no NGOs accountability and quality standards resulting in lack of public information about these organizations. The objective of this article is to contribute to a better understanding and positioning of the Third sector in Lithuania by showing distinguishing roles and functions of players in this part of economy, to better disclose the nature and types of their activities and interactions with players of other – governmental and business – sectors, through forming social economy partnerships for meeting societal demands.

As regards methodology, main analysis results of this research study were carried out from analysis of secondary source, reviews of policy papers and reports, NGO mapping exercise performed by the author and Third sector research data. The Framework Method of qualitative content analysis was used for large-scale social policy research. Analytical framework for this study was built around the category of social economy partnership, which serves as a drive for local development.

Innovativeness of this approach dwells in sub-themes derived from research study conducted - such as inter-sectoral boundaries and sector players. In terms of scientific categories, cultivating of this terminology seem to be a new approach for social economy research in Lithuania.

3 Level of Social Economy Recognition in Lithuania

The most recent definition of the term social economy is given in the Summary of the Report drawn up for the European Economic and Social Committee (EESC) by the International Centre of Research and Information on the Public, Social and Cooperative Economy.

"Social economy is the set of private, formally-organised enterprises, with autonomy of decision and freedom of membership, created to meet their members' needs through the market by producing goods and providing services, insurance and finance, where decision-making and any distribution of profits or surpluses among the members are not directly linked to the capital or fees contributed by each member, each of whom has one vote. The Social Economy also includes private, formally-organised organisations with autonomy of decision and freedom of membership that produce non-market services for households and whose surpluses, if any, cannot be appropriated by the economic agents that create, control or finance them" [1].

While most of old European Union member states are considering social economy players – cooperatives, mutual societies, associations and foundations – as players of national economy reflected in national accounts system, Lithuania still remains among countries were social economy recognition is scarcely presented. This can be explained so, that in a group of countries composed of Austria, the Czech Republic, Estonia, Germany, Hungary, Lithuania, the Netherlands and Slovenia the concept of the SE is little known or incipient, because relatively similar concepts of the Non-Profit Sector, Voluntary Sector and Non-Governmental Organizations (NGO) sector enjoy a greater level of practical use in official debates [8].

Since 2014 a public discussion is ongoing on the "Concept of Social Enterprise" and it's role in Lithuanian economy. The Concept determining social enterprises is adopted in 2015 by Ministry of Economy and further debates are going through annual Social Enterprise Summit organised by non-governmental organisations involving all social partners and stakeholders from government, business and the Third sectors.

4 Formation of Social Capital Resources

More than fifty years of soviet system and communist government with plenty of social restrictions formed a very strong underground sense of 'citizenship' among Lithuanians. This common sense could be watched during years 1989–1991 when changes in Lithuania were named a 'singing revolution'. Unfortunately, after fall of the regime, nothing really changed on political arena: the same professional politicians and corrupted bureaucracy remained on the top of administration of public sector and government. Also, former soviet leaders of trade and industries became the new-born capitalists of Lithuania. So, Lithuanian national economy became mostly shared between State (public) and Big business (industries, private companies) sectors. A kind of see-saw formed: private business started a huge financial support for parties in Parliament, after what politicians made their "best" to pass relevant laws in favour of their donors. By that time, Third Sector developed on its own, mainly on peoples voluntarily enthusiasm. This phenomenon, influenced by "Lithuanian introvert character resulted in a forming of total mistrust and criticism in society towards government and all forms of political parties" [6]. In fact, for people, who lived nearly half of their lives in soviet system, terms 'cooperation', 'cooperatives' is a reminiscence of 'kolkhozes' and socialism, what causes that the process of creating Cooperative society is still very slow in Lithuania.

Formation of social capital strongly depends on people's patterns of communication on grass roots level [7]. History wise, Lithuanian people always have been organised around some cultural centres, like palaces, estates, manors, culture centres, societies, groups and so forth. Thus, talking about development of social capital on local (community) level in Lithuania nowadays, the process of raising number of non-governmental organisations should be considered.

As a significant fact of social capital presence, the relatively big number of non-governmental organisations, approximately 15000, is registered in Lithuania, but the Third Sector representation on economic stage is very weak, it still remains invisible in terms of social economy activities and statistics on national satellite accounts.

Generally speaking, Lithuanian government is very rigid in keeping most of the functions for social services and intermediary labour market development, instead of passing this social sector to social organisations of the Third sector. On the other hand, the laws letting to establish a social enterprise under one of juridical statuses, requires from NGO a lot of efforts and financial costs in order to start an economic activity. This circumstance leads to situation, that trades, services and productions, which normally are performed by non-governmental organisations, like social care, social integration, training, environmental projects, tourism, culture events are produced and organised mostly by private business corporations.

5   Sector Boundaries and Positioning of Social Economy Actors

After regaining independency, Lithuania was continuously developing NGO legislation based on the historical heritage and experience of western democracies. It was the Law of Societal organisations, which was passed in the year 1995. But it was too complicated to legalize all variety of societal groups under this one law, and soon after this, another three laws for non-profit organizations were passed in 1996 – Law for Associations, Law for Foundations for Charity and Support and Law for Public Establishments [5].

From a legal perspective, in Lithuania recognition of TS is in place, because there is all necessary legislation in force to start any form of social economy activities. As regards financial perspective, a lot of improvements has to take place, because TS in Lithuania is financially deprived sector. National networks advocate the Third Sector organisations in Lithuania. Unfortunately, the main trigger at the present moment is external – financial motive. Third sector organisations depend on weak system of sharing scarce funds. Organisations are forced to compete in between. On one hand it is good, that only stronger NGOs will survive, but on the other hand, there are a lot of very good NGO (especially in social welfare sector), which are not able to manage project proposal for their organizations, so they work more voluntarily and ineffective.

Based on societal transformations since 1990, economy's sectoral boundaries mainly were set by two major sectors – government and private business sectors. Main strategic industries and objects were privatized by individual groups loyal to ruling parties. On the other hand, prosperous businesses were donating for elections and winning a necessary legislative support as a result. Beside this power-play of two major sectors, the Third sector developed on its voluntary initiatives and societal organizations [6]. Only after passing a Law for Public Establishments in 1996, the first wave of NGO started their economic activities.

The formation of social economy entities in Lithuania can be explained reviewing interactions between three main societal sectors (Fig. 1). The clearest understanding is about Private sector and its interactions with two other sectors – business is always business. The Governmental sector is even easier to understand – politicians will always try to remain on the thrones and will interact with other two sectors in a self-directed manner. Much more complicated picture can be observed in the Third sector, where roots of social economy are found. It is hard to distinguish where are the boundaries of "non-for-profit" activities in Lithuania. As the Third sector interacts with two other sectors, all possible variations of possible associations, mutual, cooperatives or foundations can be created (depends on the interests of the stakeholders or members). Though, the biggest challenge for Lithuanian democracy is acceptance of a principle of Social Economy Partnership (SEP) (Fig. 1), which is in the core of welfare State. On one hand, the SEP principle is based on social interest, and on the other hand, it is based on the principle of subsidiarity, which guarantees NGO and social cooperatives' involvement in public procurement.

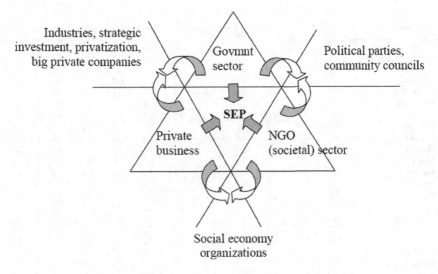

Fig. 1. Sectoral boundaries and social economy partnership [2]

While investigating NGO sector according existing laws in Lithuania we'll find different juridical forms that group of people can become in order to become economically active player in the market. These are:

- Cooperatives
- Credit unions
- Associations (under this status also other bodies can be founded: societal organisation, confederation, union, society)
- Foundations
- Public establishments

Overall, Lithuania has all necessary legislation in force for founding Third Sector organisations. The only problem here is that there are complicated procedures according post-law regulations how NGOs (voluntary and self-help organisations) can become self-sustainable (entrepreneurial) and economically reliable.

6   The Role of Social Enterprises in the National Economy

It is widely recognized and proven the fact, that "in democratic societies with strong and competitive third sector organizations, economic crisis is less likely to impact gross domestic product (GDP) rate significantly" [7]. This can be noticed (Fig. 2) when comparing countries of different economic development levels, but similar in size, e.g. France and Poland, or Denmark and Lithuania. In terms of economic stability there is a clear difference between "old" democracies with sustainable social economy sector and emerging civil societies and their developing social economy models in enlarged European Union. The bigger amount of active third sector organizations the stronger social economy sector in a country.

	2000	2001	2002	2003	2004	2005	2006	2007	2008	2009	2000-09
EU-27	3.9	2.0	1.2	1.3	2.5	2.0	3.2	3.0	0.5	-4.2	1.5
Euro area (EA-16)	3.9	1.9	0.9	0.8	2.2	1.7	3.0	2.8	0.5	-4.1	1.4
Belgium	3.7	0.8	1.4	0.8	3.2	1.7	2.7	2.9	1.0	-2.8	1.5
Bulgaria	5.7	4.2	4.7	5.5	6.7	6.4	6.5	6.4	6.2	-4.9	4.7
Czech Republic	3.6	2.5	1.9	3.6	4.5	6.3	6.8	6.1	2.5	-4.1	3.4
Denmark	3.5	0.7	0.5	0.4	2.3	2.4	3.4	1.7	-0.9	-4.7	0.9
Germany	3.2	1.2	0.0	-0.2	1.2	0.8	3.4	2.7	1.0	-4.7	0.9
Estonia	10.0	7.5	7.9	7.6	7.2	9.4	10.6	6.9	-5.1	-13.9	4.8
Ireland	9.7	5.7	6.5	4.4	4.6	6.0	5.3	5.6	-3.5	-7.6	3.7
Greece	4.5	4.2	3.4	5.9	4.4	2.3	4.5	4.3	1.3	-2.3	3.3
Spain	5.0	3.6	2.7	3.1	3.3	3.6	4.0	3.6	0.9	-3.7	2.6
France	3.9	1.9	1.0	1.1	2.5	1.9	2.2	2.4	0.2	-2.6	1.5
Italy	3.7	1.8	0.5	0.0	1.5	0.7	2.0	1.5	-1.3	-5.0	0.5
Cyprus	5.0	4.0	2.1	1.9	4.2	3.9	4.1	5.1	3.6	-1.7	3.2
Latvia	6.9	8.0	6.5	7.2	8.7	10.6	12.2	10.0	-4.2	-18.0	4.8
Lithuania	3.3	6.7	6.9	10.2	7.4	7.8	7.8	9.8	2.9	-14.7	4.8
Luxembourg	8.4	2.5	4.1	1.5	4.4	5.4	5.0	6.6	1.4	-3.7	3.6
Hungary	4.9	3.8	4.1	4.0	4.5	3.2	3.6	0.8	0.8	-6.7	2.3
Malta (1)	:	-1.6	2.6	-0.3	0.9	4.0	3.6	3.7	2.6	-2.1	1.5
Netherlands	3.9	1.9	0.1	0.3	2.2	2.0	3.4	3.9	1.9	-3.9	1.6
Austria	3.7	0.5	1.6	0.8	2.5	2.5	3.6	3.7	2.2	-3.9	1.7
Poland	4.3	1.2	1.4	3.9	5.3	3.6	6.2	6.8	5.1	1.7	4.0
Portugal	3.9	2.0	0.7	-0.9	1.6	0.8	1.4	2.4	0.0	-2.6	0.9
Romania	2.4	5.7	5.1	5.2	8.5	4.2	7.9	6.3	7.3	-7.1	4.6
Slovenia	4.4	2.8	4.0	2.8	4.3	4.5	5.9	6.9	3.7	-8.1	3.1
Slovakia	1.4	3.5	4.6	4.8	5.0	6.7	8.5	10.6	6.2	-4.7	4.7
Finland	5.3	2.3	1.8	2.0	4.1	2.9	4.4	5.3	0.9	-8.0	2.1
Sweden	4.5	1.3	2.5	2.3	4.2	3.2	4.3	3.3	-0.4	-5.1	2.0
United Kingdom	3.9	2.5	2.1	2.8	3.0	2.2	2.8	2.7	-0.1	-5.0	1.7
Iceland	4.3	3.9	0.1	2.4	7.7	7.5	4.6	6.0	1.0	-6.8	3.1
Norway	3.3	2.0	1.5	1.0	3.9	2.7	2.3	2.7	0.8	-1.4	1.9
Switzerland	3.6	1.2	0.4	-0.2	2.5	2.6	3.6	3.6	1.9	-1.9	1.7
Croatia	3.0	3.8	5.4	5.0	4.2	4.2	4.7	5.5	2.4	-5.8	3.2
FYR of Macedonia	4.5	-4.5	0.9	2.8	4.1	4.1	4.0	5.9	4.9	-0.7	2.6
Turkey	6.8	-5.7	6.6	4.9	9.4	8.4	6.9	4.7	0.4	-4.5	3.8
Japan	2.9	0.2	0.3	1.4	2.7	1.9	2.0	2.4	-1.2	-5.2	0.7
United States	4.1	1.1	1.8	2.5	3.6	3.1	2.7	1.9	0.0	-2.6	1.8

(1) Average growth 2001-2009.

Source: Eurostat (nama_gdp_k), Switzerland: Secrétariat de l'Etat à l'Economie,

Japan: Bureau of Economic Analysis, United States: Economic and Social Research Institute

Fig. 2. Comparison of GDP growth rates: Europe, Japan and United States. [4]

The bigger part of GDP is created by social economy players the lesser drop-down in GDP rate is noticed. The Lithuanian case can be taken as an example of such a misbalance. Following the last updates from Lithuanian statistics department (Fig. 3) the whole dynamic of GDP rate changes through period of global economic crisis can be observed.

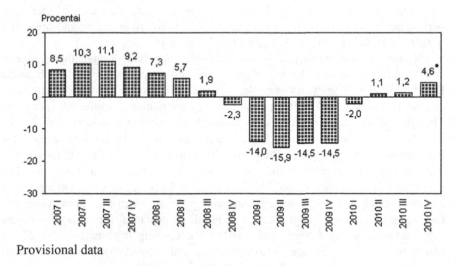

Provisional data

Fig. 3. GDP rate dynamic in Lithuania [3]

The period from the middle of the year 2008 until the III quarter of 2010 illustrates a deep bottom of the crisis. These illustrations rightly confirm the earlier discussed fact of economically weak third sector organizations in Lithuania.

7 Conclusions

As milestones for further research of sector (private, public and third sector) interactions in terms of social economy development in Lithuania, these conclusions can be taken for further discussions about social economy:

- The most challenging political task regarding social economy recognition in Lithuania raises in front of ruling parties, government and social partners in civil dialog – employers' and civil society organizations. The task is to come to common agreement around social economy sector development. Because, while most of old European Union member states are considering social economy players as a significant part of national economy, Lithuania still needs to recognise and settle social economy as a separate economic sector.
- Lithuanian national economy is mostly shared between State (public) and big private business (big industries, private companies, construction) sectors. This unequal share of national resources between two sectors resulted to a creation of a weak

third sector organizations and slow Cooperative society development in Lithuania. As most of the social services and intermediary labour market institutions are provided by Lithuanian government, the Third sector organizations, being administratively weak, are kept away from process public procurement. This is due to complex legislative and administrative procedures, which requires from NGO a lot of efforts and financial costs in order to start an economic activity. This circumstance leads to situation, that trades, services and productions, which normally are performed by social economy actors and organizations are produced and organised mostly by private business corporations or by state institutions.

- Small and medium enterprises in Lithuania are supported through the European structural funds, managed by one of public administrative institutions – Lithuanian Business support Agency. As examples of social economy partnerships in Lithuania can be mentioned LEADER + partnerships in rural municipalities of Lithuania. Though, having over fifty established partnerships all over the country, only few of them are really linked to local social needs and uses a local social capital for implementing strategic goals. This shows, that now it is up to Lithuanian government to suggest the next scheme social economy support in rural Lithuania. The suggestion for government could be forming The Third sector's support structures and institutions, e.g., Local strategy partnerships in Northern Ireland or Australian Government AusAID programme for Funding schemes for non-government organisations, or UK Government's policy on creating opportunities for SMEs and third sector organisations is to encourage and support these organisations to compete for public sector contracts where this is consistent with value for money policy and the UK regulations and EU Procurement Directives.

- Helping the Third sector organizations to obtain status of social enterprise the existing Lithuanian Law of Social Enterprises should be split into two separate laws – one for social firms (for disabled people) and another for the rest of social enterprises. Or by other means, there should be established a major state support for NGO's to go through all complex documentation and business plan development procedures.

- In order to maintain development of sustainable social economy, much attention should be given to the support of the third sector organisations – mutual funds, community development organizations, cooperatives and associations. Only when national government will see the civil society players as negotiation partners in fiscal policy, only then social economy will be noticed in annual balance of national accounts. Effectiveness of the social economy can be reached under condition of prioritising local development support on local community level.

References

1. EESC. The Social Economy in the European Union. EESC Publications Unit, Brussels (2009)
2. Dvarionas, D.: The role of Social Enterprises in Lithuania. UNDP/BRC Research Project Report, Kaunas (2006)
3. Statistics Lithuania. http://www.stat.gov.lt/en/news/view?id=9572

4. EUROSTAT. Comparison of GDP growth rates: Europe, Japan and United States (2011)
5. Dvarionas, D.: Socialinio verslo prielaidos Lietuvoje. Socialinė ekonomika: vietos ben-druomenių poreikiai ir galimybės, pp. 58–68. VDU, Kaunas (2004)
6. Kučikas, A.: Savivaldybių ir nevyriausybinių organizacijų partnerystė. NIPC, Vilnius (2001)
7. Laville, J.-L., Borzaga, C., Defourny, J., Evers, A., Lewis, J., Nyssens, M., Pestoff, V.: Third System: A European Definition. EMES Research Report, Paris (1999)
8. A map of social enterprises and their eco-systems in Europe: Lithuania report to European Commission (2014)

Convertion of Dual Fuel Opportunities in Indonesia Waterways

Rizqi Fitra[1], Fredhi Agung Prasetyo[1], Rudiyanto Rudiyanto[1], and Tutut Herawan[2,3(✉)]

[1] Biro Klasifikasi Indonesia, Jl. Yos Sudarso 38-40, Tg. Priok,
North Jakarta, Indonesia
{m.rizqi, fredhiagung, rudiyanto}@bki.co.id
[2] Universitas Teknologi Yogyakarta, Yogyakarta, Indonesia
tutut@uty.ac.id
[3] AMCS Research Center, Yogyakarta, Indonesia

Abstract. Indonesian shipping generally used heavy fuel oil (HFO) or Marine diesel oil (MDO) as diesel engine fuel. Exhaust gas emission produced by fossil fuel such as SO_X, NO_X, CO_2, and PM, contributed total global emission during 2007–2012 period as much as 3% of CO_2, NO_X for 12%, and SO_X for 13%. To comply Emission Control Area (ECA), and high fuel efficiency is Dual fuel, where this method is combined from fuel oil with gas such as LNG or CNG. This paper analyzes dual fuel conversion in container ship 368TEU. Payback period and Rate Of Interest (ROI) have been adopted as method adopted for analyzing investment. From payback period, the results show that if dual fuel 80:20 is faster than dual fuel 70:30, and single fuel, where in 9th years get profit 257.743, and 10th years for single fuel and dual fuel 70:30. From the ROI method, the results show that for each methhod is 18.65% for dual fuel 80:20, 17.43% for dual fuel 70:30, and 16% for single fuel.

Keywords: Convertion · Dual fuel opportunities · Indonesia waterways

1 Introduction

Indonesian shipping generally uses HFO or MDO as diesel engine fuel. Exhaust gas emission produced by fossil fuel is SO_X, NO_X, CO_2, & PM. This fossil fuel contributed total global emission during 2007–2012 period is 3% of CO_2, NO_X for 12%, and SO_X for 13%. Due to the global warming issue, International Maritime Organization (IMO) as maritime regulator makes regulation to control emission from maritime activity. The IMO TIER III is implemented on 1st January 2016, where ECA (Emission Control Area) restricted value of sulphur from ship fuel in 0.1%.

Many kinds of method to reduce emission from exhaust gas [1–3]. They usually divided into three methods i.e. the first method is prior to combustion with using fuel oil with low sulphur to reduce SOX and PM [4]. The second method is during combustion, for reduce NO_X use EGR (Exhaust Gas Recirculation) [5–7]. The third method is after combustion, if want to reduce SOX used wet scrubber, else if want to reduce NOX used SCR (Selective Catalytic Reducer) [8, 9], and if want to reduce PM used dry scrubber.

O. Gervasi et al. (Eds.): ICCSA 2017, Part VI, LNCS 10409, pp. 610–619, 2017.
DOI: 10.1007/978-3-319-62407-5_44

Method to reduce emission describe above are having limitations to comply with Emission Control Area (ECA) regulations, and low of fuel efficiency. To comply ECA, and high on fuel efficiency, another method offered is Dual fuel, where this method is combined from fuel oil, with gas (LNG or CNG). Indonesia's natural gas stockpile as of January 1st, 2015 increased by 151.33 Trillion Cubic Feet or a 1.36% increase compared to natural gas stockpile status January 1st, 2014 amounted to 149.3 Trillion Standard Cubic Feet [10]. Indonesia potentially uses natural gas as main fuel in maritime industry due to natural gas stockpile, however the owner is facing problem with high investment.

This paper analyzes the feasibility of dual fuel conversion in Indonesia with the limitations of economical view for operational cost i.e. fuel, maintenance, and insurance, initial investment due to dual fuel conversion, and revenue from shipping cargo container. Specifically, we analyze dual fuel conversion in container ship 368TEU. Payback period and Rate Of Interest (ROI) have been adopted as method adopted for analyzing investment.

The rest of this paper is organized as follow: Sect. 2 describes proposed method. Section 3 describes obtained results and following by discussion. Section 4 concludes this paper.

2 Proposed Method

Dual fuel conversion in this paper are analyzes Container ship 368 TEU. Before conducting dual fuel conversion need to assess of appropriate diesel engine. To convert from a single fuel to dual fuel need to minimalized changes engine component, easy method to convert is change intake manifold of the engine. For control amount of gas fuel to the combustion chamber, then need to be installed converter kit. Converter kit is responsible to control the composition of fuel oil, and gas fuel in the combustion chamber, because if the amount of gas fuel are bigger than fuel oil it can affect to engine knocking. Converter kit fitted to the engine are depend by the type of engine low speed, medium speed, or high speed engine. Composition maximum for dual fuel are 70:30 [11], where 70% is LNG fuel, and 30% of fuel oil. In many papers said composition maximum can reach 80:20, where 80% is LNG and 20% is fuel oil.

Retrofit existing ship must be fulfill with the *Biro Klasifikasi Indonesia* (BKI) class requirement, where already stated in rules and guidelines. Retrofit existing ship have many challenges to fulfill a class requirement, such as LNG storage, and safety devices need to be fitted in ship. Type of LNG storage fitted in ship is type C, because this is easier to install, and doesn't required integrated with existing hull construction.

Operational cost are included in maintenance cost, fuel cost, and insurance cost. Fuel cost used 2 compositions of dual fuel, 70:30 and 80:20, then compared with fuel oil. The maintenance cost is divided by 2 part hull and machinery [12]. Insurance cost depends by value of the vessel. The initial cost component is consultant engineering cost, shipyard cost, engine conversion component, automation & control, and fuel gas system including LNG storage, bunkering, process equipment. Revenue comes from voyage ship Tanjung Perak, Surabaya to Tanjung Priok, Jakarta, Indonesia. A voyage of ship are considered with periodical survey to maintain class.

Method to use for analyzing investment worthed or not is payback period, and ROI. Where payback period is depend by capital cost, and cumulative income netto. If this retrofit worthed than single fuel, then payback period for dual fuel are more faster. If ROI are more higher than interest rate of debt in Bank, then this project is worthed.

2.1   Operational Cost

Fuel consumption is depending on power of main engine, SFOC of vessel, distance of voyage, and speed of vessel. Fuel consumption between single fuel and dual fuel will be different, because the different of Low Heating Value (LHV). The LHV for MDO is 42.8 MJ/Kg, and LHV for LNG is 49.5 MJ/Kg [13]. To get fuel consumption for dual fuel are used formula below:

$$LHV_{MDO\,(MJ/Kg)} \times Fuel\ Consumption_{(Kg/day)} = LHV_{MDO\,ENGINE(MJ)} \qquad (1)$$

$$LHV_{MDO\,ENGINE} \times Fuel\ Ratio = LHV_{Dual\,Fuel\,(MJ)} \qquad (2)$$

$$LHV_{Dual\,Fuel\,(MJ)}\big/LHV_{LNG\,(MJ/Kg)} \qquad (3)$$

From formulas in (1) to (3), we obtain the result of LNG consumption is 24.07 m^3 per day. The LNG consumption depends on fuel ratio between MDO and LNG. Fuel cost is depend by fuel price today, and fuel prices rises next 20 years. To reduce LNG tank storage is optimised distances of ship voyage, in this research is used distance of voyage maximum 500 Nautical Mile [14]. In this research, we assume if fuel rises 5% each years. Fuel prices for MDO is 465 USD/ton, and LNG prices is 3.71 USD/mmBtu. Another operational cost is maintenance cost, where is assumed rises 5% each years. Maintenance cost annually are divided in 2 part, first is hull construction, and second is machinery part [12]. For hull construction cost, it is used formula below:

$$10000 \times CN^{2/3} = Hull\ Maintenance \qquad (4)$$

$$(L \times B \times D)/100 = CN \qquad (5)$$

For maintenance cost machinery is used formula below:

$$10000 \times (SHP/1000)^{2/3} = Machinery\ Maintenance \qquad (6)$$

Maintenance cost for dual fuel are more higher 20%–25% than single fuel, due to material and equipment fitted in dual fuel [15]. Another operational cost is insurances cost depend by type of insurances owners have registered. Owner can choose insurance total loss, or all risk condition [18]. In this research, the owner chooses the total loss insurance with flat premi until 20 years, because lower cost than all risk insurance [12]. To calculate insurance premi is using formula below:

$$2,551 \times 10^{-5} \times \text{Value of Ship} = \text{Premi Insurance} \tag{7}$$

The value of ship is capital cost to build this ship. Capital cost from owner is getting from bank loan, then owner need to pay installment to bank each years. Bank loan is 70% from capital cost 25 million USD, where length of loan is 10 years. Installment each years are flat with interest rate loan is 10% per years. To get installment per year is using the formula as follow:

$$\text{Int. rate} \times ((1 + \text{Int. rate})^T / (1 + \text{Int. rate})^T - 1) \times \text{Loan Debt} = \text{Installment} \tag{8}$$

2.2 Initial Cost

Initial Cost for this retrofit is divided by Cost for Surveys, consultant engineering represent 12% of inital cost total. Inital cost is spend higher in cost docking, and LNG storage, process equipment represent 62% of total initial cost. Cost for docking are included with material cost, routing of pipe, and installation cost. LNG storage are included with engine commissioning, safety equipment, and vendor trainee. Surveys cost are cost for class surveys during engine modification, and change of class notation. Consultant engineering are include pre engineering deisgn about retrofit ship, and stability after fitted with LNG storage on cargo deck. Initial cost will add capital cost in dual fuel, in 5th year. Ideally retrofit of ship from single fuel to dual fuel are in years 5th [16].

2.3 Revenue

Revenue is come from rate container form Tanjung Perak, Surabaya, East Java, Indonesia to Tanjung Priok, Jakarta, Indonesia, where per container charge is 250 USD. Revenue for single fuel and dual fuel are different, because in dual fuel space for container are reduced to 360 TEU. Each year cumulative value are different due to class maintain during periodic survey. Operational of ship in year is 360 day, where for annual survey ship operational is 353 day, intermediate survey is 345 day, and for special survey or renewal survey is 330 day. Trip per voyage are spent 4 day, then for annual ship can do 88 trip, and in intermediete can do 86 trip, and renewal survey ship can do 83 trip. Each year charge of container raised to 5%.

2.4 Payback Period and ROI

Payback period in feasibilty study are needed for information to owner when this project can reach Break Event Point or profit. If this project break event point reach outside length of investment 20 years [17], then this project are not worthed. Payback period representative which project profitable for owner. To calculate payback period can used formula below:

$$\text{Income cumulative nett} - \text{Capital Cost} = \text{Payback} \tag{9}$$

The ROI are representative of interest rate from project compare with interest rate in Bank, because if interest rate in Bank are more higher than interest rate of project then the project not worthed to executed. To calculate ROI can used formula below:

$$(\text{Inc. cumulative nett}/\text{Time Investment})/\text{Capital Cost} = \text{ROI} \tag{10}$$

3 Result and Discussion

This section presents obtained results and following by discussion.

3.1 Operational Cost

Operational Cost for single fuel are shown in Table 1. Single Fuel Operational Cost, where total cost are cumulative from maintenance cost, fuel cost, and insurances cost.

Table 1. Single fuel operational cost (in USD)

Year	Maintenance	Fuel	Insurance	Total cost
1	369.520	2.271.060	229.590	2.870.170
2	387.996	2.384.613	229.590	3.002.199
3	407.396	2.503.844	229.590	3.140.830
4	427.766	2.629.036	229.590	3.286.392
5	449.154	2.760.488	229.590	3.439.232
6	471.612	2.898.512	229.590	3.599.714
7	495.193	3.043.438	229.590	3.768.220
8	519.952	3.195.609	229.590	3.945.152
9	545.950	3.355.390	229.590	4.130.930
10	573.247	3.523.159	229.590	4.325.997
11	601.910	3.699.317	229.590	4.530.817
12	632.005	3.884.283	229.590	4.745.878
13	663.605	4.078.497	229.590	4.971.693
14	696.786	4.282.422	229.590	5.208.798
15	731.625	4.496.543	229.590	5.457.758
16	768.206	4.721.371	229.590	5.719.167
17	806.616	4.957.439	229.590	5.993.646
18	846.947	5.205.311	229.590	6.281.848
19	889.295	5.465.577	229.590	6.584.461
20	933.759	5.738.856	229.590	6.902.205

Table 2. Dual fuel operational cost (70:30)

Year	Maintenance	Fuel	Insurance	Total cost
1	461.900	1.392.551.50	229.590	2.084.042
2	484.995	1.629.905.06	229.590	2.344.490
3	509.245	1.672.504.86	229.590	2.411.340
4	534.707	1.796.970.33	229.590	2.561.268
5	561.443	1.779.613.23	229.590	2.570.646
6	589.515	1.981.159.79	229.590	2.800.265
7	618.991	2.080.217.78	229.590	2.928.798
8	649.940	2.134.587.11	229.590	3.014.117
9	682.437	2.293.440.11	229.590	3.205.467
10	716.559	2.271.287.56	229.590	3.217.437
11	752.387	2.528.517.72	229.590	3.510.495
12	790.006	2.654.943.60	229.590	3.674.540
13	829.507	2.724.334.17	229.590	3.783.431
14	870.982	2.927.075.32	229.590	4.027.647
15	914.531	2.898.802.43	229.590	4.042.924
16	960.258	3.227.100.54	229.590	4.416.948
17	1.008.271	3.388.455.57	229.590	4.626.316
18	1.058.684	3.477.017.48	229.590	4.765.292
19	1.111.618	3.735.772.26	229.590	5.076.981
20	1.167.199	3.699.688.10	229.590	5.096.477

Operational cost for dual fuel are shown in Table 2. Operational Cost dual fuel 70:30, and Table 3. Operational Cost dual fuel 80:20.

From Tables 1, 2, and 3, we draw a conclusion i.e. if fuel cost for dual fuel are lower than single fuel.

3.2   Revenue

Revenue from single fuel is more higher than dual fuel, due to number of container can be load on single fuel. Single fuel can load 368 TEU, then for dual fuel is 360 TEU (Table 4).

3.3   Payback Period

Payback period for single fuel can be seen in Fig. 1 Payback Period Single Fuel. Payback period is reach in years 10th with profit 727.646 USD, and then for investment until year 20th can get profit 56.502.470 USD. For payback period dual fuel can be seen in Fig. 2 Payback Period Dual Fuel 70:30, payback period reach in years 10th with profit 1.196.615 USD, where the end of the investment get profit 62.142.955 USD. Even payback period of single fuel and dual fuel 70:30 in the same years, but dual fuel

Table 3. Dual fuel operational cost (80:20)

Year	Maintenance	Fuel	Insurance	Total cost
1	461.900	1.339.293.75	229.590	2.030.784
2	484.995	1.406.258.44	229.590	2.120.844
3	509.245	1.443.012.92	229.590	2.181.848
4	534.707	1.550.399.93	229.590	2.314.697
5	561.443	1.535.424.48	229.590	2.326.457
6	589.515	1.709.315.92	229.590	2.528.421
7	618.991	1.794.781.72	229.590	2.643.362
8	649.940	1.841.690.79	229.590	2.721.221
9	682.437	1.978.746.85	229.590	2.890.774
10	716.559	1.959.633.95	229.590	2.905.783
11	752.387	2.181.568.40	229.590	3.163.545
12	790.006	2.290.646.82	229.590	3.310.243
13	829.507	2.350.516.00	229.590	3.409.613
14	870.982	2.525.438.12	229.590	3.626.010
15	914.531	2.501.044.68	229.590	3.645.166
16	960.258	2.784.295.52	229.590	3.974.143
17	1.008.271	2.923.510.30	229.590	4.161.371
18	1.058.684	2.999.920.23	229.590	4.288.194
19	1.111.618	3.223.170.11	229.590	4.564.378
20	1.167.199	3.192.037.21	229.590	4.588.826

Table 4. Revenue single fuel & dual fuel

Year	Cost per TEU	Single fuel	Dual fuel
1	250	8.096.000	7.568.000
2	263	8.500.800	7.946.400
3	269	8.515.290	7.959.945
4	276	8.931.153	8.348.687
5	283	8.634.294	8.071.188
6	290	9.383.293	8.771.339
7	297	9.617.875	8.990.622
8	304	9.634.269	9.005.947
9	312	10.104.780	9.445.772
10	320	9.768.911	9.131.808
11	328	10.616.334	9.923.965
12	336	10.881.743	10.172.064
13	344	10.900.291	10.189.403
14	353	11.432.631	10.687.025
15	362	11.052.626	10.331.802
16	371	12.011.408	11.228.055
17	380	12.311.693	11.508.757
18	390	12.332.679	11.528.374
19	399	12.934.973	12.091.387
20	409	12.505.032	11.689.486

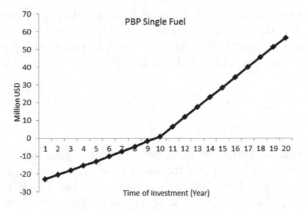

Fig. 1. Payback period single fuel

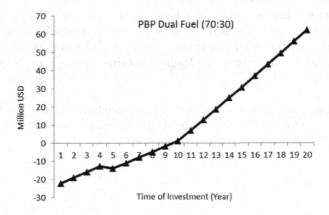

Fig. 2. Payback period dual fuel 70:30

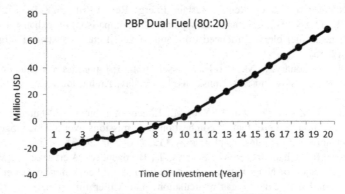

Fig. 3. Payback period dual fuel 80:20

are more profitable than single fuel, it can be seen in collecting of profit on payback period, and profit in the end of investment.

Payback period for dual fuel 80:20 can be seen in Fig. 3, where payback period reach in years 9th with profit 257.743 USD. In the end of investment can make profit 68.229.870 USD.

4   Conclusion

Payback period Dual fuel 80:20 are faster than Dual fuel 70:30, and single fuel, where reach payback period at 9th years, and the others is 10th years. From the result payback period can get conclusion Dual Fuel are worthed to execute. The ROI for Dual Fuel 80:20 is higher than Dual Fuel 70:30, and Single Fuel. Meanwhile, the ROI for Dual Fuel 80:20 is 18.65%, ROI for Dual Fuel 70:30 is 17.43%, and ROI for Single Fuel is 16%, ROI in Bank is 13%. Future of gas as ship fuel are profitable than fossil fuel or single fuel, even maintenance cost gas fuel higher than single fuel. In Indonesia waterways the challenges of the dual fuel, or gas fuel are still low because the infrastructure to support this system in Indonesia doesn't available. In the future, it is suggested to develop: (1) A feasible study on optimised bunkering system in Indonesia waterways; (2) on board bunkering or floating bunkering by using feasible study or any appropriate way.

Acknowledgment. This research is supported by Biro Klasifikasi Indonesia. The work of Tutut Herawan is supported by Universitas Teknologi Yogyakarta Research Grant no vote O7/UTY-R/SK/0/X/2013.

References

1. Schmelz, H.: U.S. Patent No. 5,628,186. Patent and Trademark Office, Washington, DC, U.S. (1997)
2. Basha, S.A., Gopal, K.R., Jebaraj, S.: A review on biodiesel production, combustion, emissions and performance. Renew. Sustain. Energy Rev. **13**(6), 1628–1634 (2009)
3. Behçet, R., Oktay, H., Çakmak, A., Aydin, H.: Comparison of exhaust emissions of biodiesel–diesel fuel blends produced from animal fats. Renew. Sustain. Energy Rev. **46**, 157–165 (2015)
4. Seddiek, I.S., Elgohary, M.M.: Eco-friendly selection of ship emissions reduction strategies with emphasis on SOx and NOx emissions. Int. J. Nav. Archit. Ocean Eng. **6**(3), 737–748 (2014)
5. Liu, H., Li, S., Zheng, Z., Xu, J., Yao, M.: Effects of n-butanol, 2-butanol, and methyl octynoate addition to diesel fuel on combustion and emissions over a wide range of exhaust gas recirculation (EGR) rates. Appl. Energy **112**, 246–256 (2013)
6. Verschaeren, R., Schaepdryver, W., Serruys, T., Bastiaen, M., Vervaeke, L., Verhelst, S.: Experimental study of NO x reduction on a medium speed heavy duty diesel engine by the application of EGR (exhaust gas recirculation) and Miller timing. Energy **76**, 614–621 (2014)

7. Asad, U., Tjong, J., Zheng, M.: Exhaust gas recirculation–zero dimensional modelling and characterization for transient diesel combustion control. Energy Convers. Manag. **86**, 309–324 (2014)
8. Paolucci, C., Verma, A.A., Bates, S.A., Kispersky, V.F., Miller, J.T., Gounder, R., Delgass, W.N., Ribeiro, F.H., Schneider, W.F.: Isolation of the copper redox steps in the standard selective catalytic reduction on Cu-SSZ-13. Angew. Chem. Int. Ed. **53**(44), 11828–11833 (2014)
9. Bates, S.A., Verma, A.A., Paolucci, C., Parekh, A.A., Anggara, T., Yezerets, A., Schneider, W.F., Miller, J.T., Delgass, W.N., Ribeiro, F.H.: Identification of the active Cu site in standard selective catalytic reduction with ammonia on Cu-SSZ-13. J. Catal. **312**, 87–97 (2014)
10. http://statistik.migas.esdm.go.id/index.php?r=pengolahanNaturalGas/index
11. Kraipat, C., Chedthawut, P., Choi, G.H.: Performance and emission of heavy duty diesel engine fuelled with diesel and LNG (Liquid Natural Gas). Energy **53**, 52–57 (2013)
12. Harry, B.: The practical application of economics merchant ship design. J. Mar. Technol. Soc. Nav. Archit. Mar. Eng. (1967)
13. Puji, D.W, Ariana, I.M., Semin: Rancang Bangun Sistem Penginjeksian Gas Pada Modifikasi Dual Fuel Diesel Engine. Propul. J. BKI 2014, pp. 1–6 (2014)
14. Daniel, D.: Feasibility of dual fuel engines in short sea shipping lines, Final Project. Naval Engineering, Universitat Politècnica de Catalunya (2012)
15. Herdzik, J.: LNG as a marine fuel-possibilities and problem. J. KONES **18**, 169–176 (2011)
16. Guðrún, J.J.: LNG as ship fuel in Iceland. Thesis of Master of Science in Construction Management (2013)
17. Watson, D.G.: Practical ship design, vol. 1. Gulf Professional Publishing, Houston (2002)
18. Jinca, M.Y.: Transportasi Laut Indonesia, cetakan ke-1 Agustus 2011

Workshop on Advances in Spatio-Temporal Analytics (ST-Analytics 2017)

Classification Method by Information Loss Minimization for Visualizing Spatial Data

Toshihiro Osaragi[✉]

School of Environment and Society, Tokyo Institute of Technology,
Tokyo, Japan
osaragi.t.aa@m.titech.ac.jp

Abstract. It is necessary to classify numerical values of spatial data when representing them on a map and so that, visually, it can be clearly understood as possible. Inevitably some loss of information from the original data occurs in the process of this classification. A gate loss of information might lead to a misunderstanding of the nature of original data. In this study, a classification method for organizing spatial data is proposed, in which any loss of information is minimized. When this method is compared with other existing classification methods, some new findings are shown.

Keywords: Information loss · Classification · Visualization · Spatial data

1 Introduction

The advantage of using a map to visually represent the complex information present in spatial data is that it is possible to use people's innate powers of discrimination and interpretation (i.e. the ability to understand colors, patterns and spatial relevance). However, when preparing thematic maps and other materials using a geographic information systems, there is a risk of overlooking characteristics of the original data, or causing misjudgment, if inadequate attention is paid to the representation method. In other words, it is necessary to consider the method of representation—i.e. "What is the best way to represent (map) data?" When visually understanding analysis results, it is necessary to consider not only the problems of uncertainty in the data, but also the uncertainty, which arises in the processes of data processing and visualization [1].

When visualizing spatial data for which attribute values are quantitatively defined, there is always a need for classification. More specifically, the general method is to classify data within a certain range into the same class, and indicate it with the same color. Existing geographic Information systems are equipped with a number of different methods for automatically performing classification [2], but the displayed thematic maps vary greatly in appearance depending on the method used. For example, the spatial distributions of the "the number of industries" (used as an illustrative example in this research, Fig. 5) appear completely different—to the extent that a viewer would not think that the same data is being used. Existing classification methods should be used selectively to suit the purpose of analysis or the properties of the spatial data to be displayed, but at the early stage of analysis, or for general end-users, it is necessary to

© Springer International Publishing AG 2017
O. Gervasi et al. (Eds.): ICCSA 2017, Part VI, LNCS 10409, pp. 623–634, 2017.
DOI: 10.1007/978-3-319-62407-5_45

have a method which is simpler, applicable to any type of spatial data, and enables display of the properties of the data without bias. That is the focus of this research.

Basically, it is possible to faithfully represent the detailed distribution characteristics in the original data if the number of classes is increased. However, if the number of classes is too large, the legend becomes cumbersome, complicated and difficult to understand. Conversely, if the range of each class is too large, there is a risk of losing visibility of the spatial distribution characteristics in the original data—such as components, which vary in small increments and information relating to peaks and pits. In other words, it is necessary to consider the following two aspects of classification problems [3]:

(a) What is the best number of classes?
(b) Where should the class boundary values be set?

For problem (a), an algorithm has been developed for effectively classifying data, which does not have information regarding the number of classes [2]. Other methods, such as Jenks' optimization [4], have been proposed for problem (b). The Jenks' method determines boundary values so as to minimize the average variance within each class. Various other methods have been devised, and incorporated into existing geographic information systems. Andrienko et al. [5] developed a set of tools for classification that facilitate looking on data from various viewpoints and thereby investigate different aspects of the data. In order to balance between these requirements in search of an acceptable compromise solution, they employed the interactive tools for classification.

Osaragi [6] has previously studied classification methods, which simultaneously take into account problems (a) and (b) by using an Akaike's Information Criterion [7, 8]. The AIC is an evaluation criterion for making an overall judgment about relevance in the trade-off between the simplicity and fitness of models. However, it has been determined that likelihood is dominant over degree of freedom when the number of observed samples is large, and this results in the optimal number of classes being extremely large. That is, it may be more realistic to examine only problem (b) described above, that is, to examine how we should set up the class boundary values with fixing the number of classes. As for problem (a), GIS users should set up a-priori the number of classes to be employed according to their demands. This paper considers problem (b)—i.e. classification methods from the perspective of "Where should the class boundary values be set?"

2 Classification Based on Minimizing Information Loss

When data whose original values are different is displayed by classifying the data into the same class, part of the information contained in the data is lost. Minimizing this information loss is thought to be effective as a method for minimizing errors in judgment. To minimize information loss, Roy et al. [9] propose a method for creating a more compact cross table, by integrating the columns or rows in that table. This method can be applied to classification problems for spatial data.

First, we consider the case where data has been defined with discrete values, as in the case of point sampling data. It is assumed that at each point i ($i = 1, 2, ..., n$), x_i objects are observed. In this spatial distribution of observed objects, all of the observed objects X ($= x_1 + x_2 + ... + x_n$) are the instantiated values of a multinomial distribution assigned to n points in accordance with the assignment probability characteristic of each point. When the original data items x_i ($i = 1, 2, ..., n$) are classified into n classes, the average amount of information I_0 is defined as follows:

$$I_0 = \sum_{i=1}^{n} p_i \log_2 p_i, \tag{1}$$

where $p_i = x_i / \sum_{j=1}^{n} x_j$. On the other hand, suppose that the number of points with the same assignment probability q_k ($k = 1, 2, ..., m$) is N_k ($N_1 + N_2 + ... + N_m = n$) and regard this as the number of points contained in the same class C_k. That is, the average information when the original data x_i ($i = 1, 2, ..., n$) is divided into m classes C_k ($k = 1, 2, ..., m$), can be written as follows:

$$I = \sum_{k=1}^{m} N_k q_k \log_2 q_k, \tag{2}$$

where $q_k = \sum_{i \in G_k} x_i / N_k \sum_{j=1}^{n} x_j$. Here, G_k is the set of subscripts of samples contained in C_k. The following statistic, found using the above Eqs. (1) and (2) is defined as the "information loss rate: L,"

$$L = \frac{I - I_0}{I_0} \times 100 \ (\%). \tag{3}$$

This is a good classification method in the sense that the smaller this value is, the smaller the loss of information contained in the original data. The information loss rate in Eq. (3) is an index, which indicates the degree to which information in the original data is lost during classification, and thus is one reference point for understanding the quality of the display results for a thematic map etc.

In many cases, spatial data is data observed as a continuous value (such as area or density). The above approach can be naturally extended to the continuous case, and the information loss rate can be found using a calculation, which is procedurally the same. That is, if we let x be a random variable for a continuous information source and let p (x) be its density, the average information of the continuous information source can be defined as follows [10, 11]:

$$I_0 = \int_{-\infty}^{\infty} p(x) \log_2 p(x) dx. \tag{4}$$

However, in many cases the actual spatial data x_i ($i = 1, 2, ..., n$) is not obtained in a continuous form, and is obtained instead as values tabulated within a certain spatial range. In the actual calculation, we adopt the following approach.

We let $p(x_i)\Delta x_i$ be the probability of data x_i being observed in the certain spatial range Δx_i. Since the interval of integration in Eq. (4) corresponds to the entire region, the average information I_0 can be defined by the following equation:

$$I_0 = \sum_{i=1}^{n} p(x_i)\Delta x_i \log_2 p(x_i)\Delta x_i, \tag{5}$$

where $p(x_i)\Delta x_i = x_i / \sum_{j=1}^{n} x_j$. Furthermore, the average information I when classified into m classes can be written as in the following equation:

$$I = \sum_{k=1}^{m} N_k q(x_k)\Delta x_i \log_2 q(x_k)\Delta x_i, \tag{6}$$

where $q(x_k)\Delta x_{i(i \in G_k)} = \sum_{i \in G_k} x_i / N_k \sum_{j=1}^{n} x_j$. As can be seen by comparing Eqs. (1) and (5), and Eqs. (2) and (6), the amount of information can be found using a calculation method exactly the same as that for discrete quantities, even in the case of continuous variables.

3   Method of Determining Classification Boundary Values

Cumulative frequency curve is one of well-known methods for graphical representation of statistical distribution of attribute values. Andrienko and Andrienko [12] introduced the idea of generalizing a cumulative frequency curve to show arbitrary cumulative counts, which can be a valuable instrument for exploratory data analysis. Cumulative frequency curve can be applied to determine classification boundary values as follows.

If data are sorted by the size of their values, the problem of determining classification boundary values is equivalent to the problem of determining the ranking of data corresponding to boundary values (hereafter called "boundary ranking") (see Fig. 1). More specifically, the boundary ranking can be determined according to the procedure shown in Fig. 2: (1) First, if there are ties in the data, they are collected into one group. (2) The boundary ranking for (m-1) classes to be the boundaries of the classes is set, and this is taken to be the initial value. (3) The boundary ranking is guided in the direction, which lowers the information loss rate by moving each boundary ranking up and down. (4) The previous step (3) is repeated, and the process ends when the value of the information loss rate can no longer be reduced. Since the information loss rate always decreases and a lower limit exists, it is evident that the iterative calculation shown in Fig. 2 will always converge. Furthermore, there are a finite number of ways to determine the boundary ranking, and thus it will always converge in a finite number of iterations.

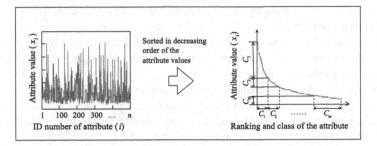

Fig. 1. Basic concept of classification for data representation. The problem of setting the boundary values of classification is equivalent to the problem of setting the boundary ranking (the ranking of data corresponding to boundary values), if the data is sorted in decreasing order of the attribute values.

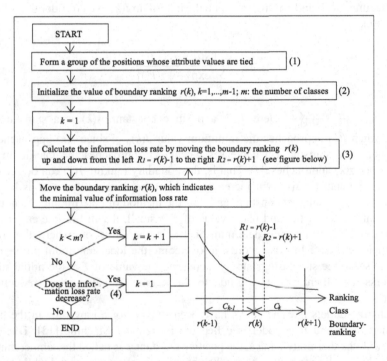

Fig. 2. Algorithm for minimizing the information loss rate. (1) Form a group of spatial units, if their values are tied. (2) Set up the boundary ranking of $(m - 1)$ classes, and consider them as initial values. (3) Calculate one value of information loss rate by moving the boundary ranking up or down, and set the boundary ranking such that the value of information loss rate becomes minimum. (4) Repeat the above operation until the value of information loss does not decrease.

Here, if it is desirable to round the classification boundary values to a value with better cut-off (downward truncation, upward truncation, rounding etc.), then the analyst can round the original data to numerical values with better cut-off in step (1), and perform grouping by configuring data in a virtual way. The above method can flexibly respond to requirements like this.

However, in the above procedure, there is a danger that the value of the information loss rate will get trapped in a local minimum. One way of avoiding this problem is the method of stochastically changing the boundary ranking in step (3). That is, the boundary ranking can be stochastically guided so that the smaller the information loss rate, the easier it is to move in that direction, and the larger the information loss rate, the harder it is to move in that direction. For example, the following is one method of stochastically changing the k-th boundary ranking $r(k)$.

As it is clear from Eq. (3), minimization of the information loss rate L is equivalent to maximization of the average information I, and thus we can shift our thinking to the problem of maximizing the average information I. As shown in Fig. 2, we let $I(k)$ be the value of the average information I calculated from Eq. (2) when the boundary ranking $r(k)$ is changed to the data ranking $R_1 = r(k)-1$ to $R_2 = r(k)+1$. Letting $I(k)_{max}$ be the maximum $I(k)$ and letting $I(k)_{min}$ be the minimum $I(k)$, we consider the following statistic $s(k)$:

$$s(k) = \frac{\exp\left[-\frac{1}{T}z(k)\right]}{\sum_{k'} \exp\left[-\frac{1}{T}z(k')\right]}, \tag{7}$$

where $z(k) = \frac{I(k)_{max}-I(k)}{I(k)_{max}-I(k)_{min}}$. Here T is a positive constant. $s(k)$ is a statistic which increase when $I(k)$ approaches the minimum value $I(k)_{min}$, since $z(k)$ approaches 1.0. Furthermore, the statistic $s(k)$ decreases when $I(k)$ approaches the maximum value $I(k)_{max}$, since $z(k)$ approaches 0.0. That is, the boundary ranking $r(k)$ can be set so that the size of the statistic $s(k)$ will increase. At this time, if the value of T is large, the value of $s(k)$ will be almost constant, and thus the method of determining the boundary ranking will be stochastic; and if the value of T is small, it will become easier for the boundary ranking to change deterministically in the direction where the average information increases. In other words, to overcome the local minimum problem, it is enough to set to the stochastic state by increasing the value of T at the initial stage of the process, and then gradually guide to the maximum value while reducing the value of T.

Stochastic models like the above have been studied for a long time in the field of neural networks, and the model is called the Boltzmann Machine [13]. Due to the analogy with the thermodynamics, the positive constant T is called the temperature, and the method of guiding to a state where the objective function is minimal while lowering the temperature T is called simulated annealing. If variation is done stochastically, the expected value for the potential energy (here this is the absolute value of the average information I) changes in the decreasing direction, and thus it is known that the system will arrive at a stochastic equilibrium state after a sufficient number of calculation steps.

The conceptual diagram in Fig. 3 shows the differences in behavior of the information loss rate in the case when the boundary value is moved deterministically, and the case where it is moved stochastically.

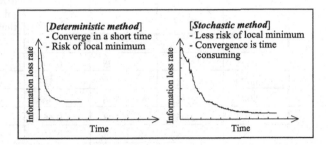

Fig. 3. Comparison of the deterministic and stochastic methods in the minimization process of information loss rate. The deterministic method allows for easy calculations; however, it may create the risk of the value of information loss rate being a local minimum. The stochastic method is more desirable to avoid the problem of the local minimum; however, the calculations are time consuming.

However, in the experience of the author, it is possible to obtain a boundary ranking very close to the minimum value, and serious practical problems do not arise, if the initial values are varied and the process in the above steps (2) through (4) in Fig. 2 is repeated a few times.

4   Data Properties and Comparison of Classification Methods

In order to conduct a comparative study of the features of each classification method, an analysis was done using actual spatial data. More specifically, for the 10 types of data shown in Table 1, the original data was classified into 9 classes using 5 existing classification methods and the Classification Method by Information Loss Minimization (CMILM). The information loss was found for each method, and the results are shown in Table 1. Moreover, the existing classification methods with the smallest and second smallest information loss rate are shown with hatching. Furthermore, in order to understand the properties of the original data, we found the distribution of the original data sorted by value size. Some of those results are shown in Fig. 4 with the information loss rate. Furthermore, Fig. 5 shows an example of the display results after classification.

First, looking at the results shown in Table 1 and the form of the graph in Fig. 4, it is evident that the information loss rate with each method varies greatly depending on the features of the original data. For example, the classification method of Natural-Breaks is effective for data with clear break points such as "the number of industries" or "area of building lots." For other data too, this is a method with comparatively little information

Table 1. The information loss rate L: comparison of existing classification methods [14] and a method based on the minimization of information loss

Classification Methods	Quantile/ Equal Area	Equal Interval	Standard Deviation	Natural Breaks	Info-loss Minimization
Industry *	0.986	2.615	0.520	0.356	0.210
	0.975	2.653	0.505	0.349	0.215
Company *	3.080	6.323	1.899	0.583	0.332
	2.945	6.322	1.866	0.584	0.335
Factory *	1.194	2.353	0.708	0.465	0.209
	1.315	2.393	0.515	0.497	0.242
Private Shop *	0.759	3.100	0.513	0.373	0.219
	0.792	3.105	0.507	0.390	0.224
Shop *	1.376	5.461	1.407	0.553	0.280
	1.323	4.427	0.915	0.589	0.283
Population A **	1.052	3.282	5.851	1.134	0.837
	1.058	3.241	5.875	1.129	0.840
Population B *	0.639	1.233	1.360	0.932	0.562
	0.645	1.275	1.415	0.922	0.563
Area of building lots (m^2) *	1.344/ 0.901	3.551	0.751	0.297	0.221
	1.320/ 0.863	3.531	0.762	0.310	0.221
Sex ratio (%) **	4.286	1.941	1.657	1.377	0.892
	4.371	1.860	1.891	1.376	0.893
Nuclear family ratio (%) **	4.197	1.297	1.714	1.407	1.015
	4.218	1.290	1.716	1.402	1.015

Upper cell: Normal classification
Lower cell: * Round down the first figure/** Round down the second figure

loss. However, it is clearly not well suited to data with unclear break points, as in the case of "population of city B." With the Equal-interval classification method, on the other hand, the boundary values of each class are equally spaced, and thus the relative relationship between each class is easy to understand. However, for almost all data other than "ratio of nuclear families," it is clear that a large amount of information is lost. With the Quantile classification method (Equal-Area classification is the same in the case of raster data), it is possible to display data with lower information loss if the data is distributed linearly, as in the example of the "population of city A" but information loss is large with data such as "ratio of nuclear families."

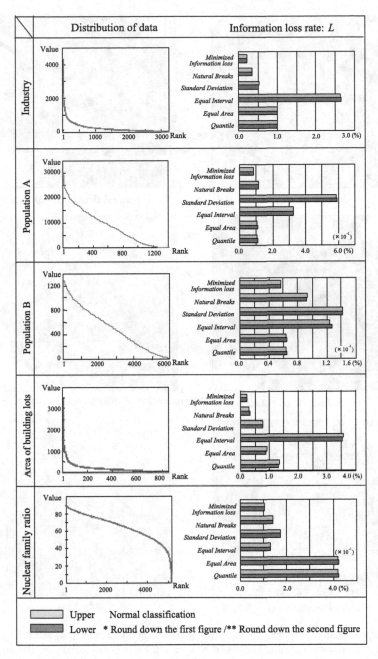

Fig. 4. Shapes of data distribution and information loss rate. The data is sorted in decreasing order of the attribute values. The information loss rate L is strongly dependent on the features of the original data.

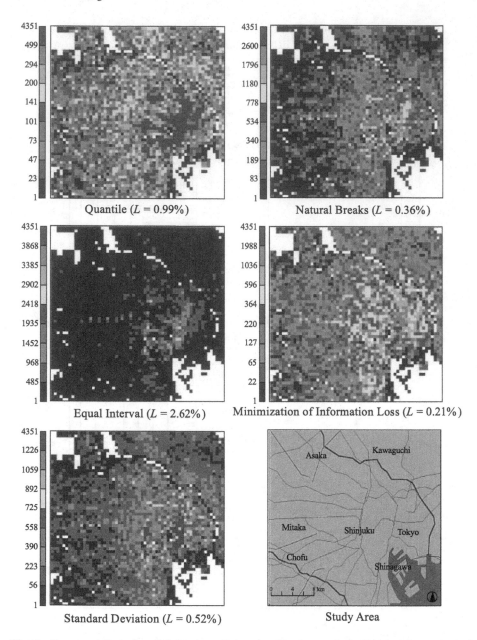

Fig. 5. Representation of spatial data by existing classification methods [14] and the method of minimizing information loss rate. The data used here is the number of industries. The number of classes is 9 and the number of cells is 3,220. The information loss rate L is also shown below each figure.

5 Summary and Conclusions

The method of classification for visually displaying spatial data depends on the purpose, i.e. "what information is to be read off (or communicated)?" However, it is necessary to carefully investigate the distribution characteristics of the spatial data to be displayed, and if adequate care is not taken, there is a risk of losing a large amount of information and overlooking data characteristics. Therefore, this paper proposes the Classification Method by Information Loss Minimization (CMILM) for visualizing spatial data. This method can flexibly cope with various kinds of spatial data consisting with numerical values (integer or decimal values), and is particularly effective at the initial stage of analysis, or for browsing of spatial data by an ordinary end user.

When actual calculation was attempted, no significant difference was seen in the time needed for convergence, or the results—even when (1) the initial boundary rankings are set randomly, (2) the number of samples in each class is set almost equal, and (3) the total of values for samples in each class is set almost equal. However, there is a danger (depending on the initial values) of falling into a local solution, and thus it is best to perform the calculation a few times while varying the initial values. Also, the "Natural-Breaks classification method (Jenks' method)" has little information loss, and thus if class boundary values found with the Natural-Breaks classification method are used as the initial values, there is a high probability that the number of calculation steps will be reduced, and the calculation will converge comparatively quickly.

One issue for the future will be the development of more efficient algorithms, including a method for setting the initial values (boundary rankings) for the information loss minimization method.

Acknowledgements. A portion of this paper was published in [15]. The author would like to give his special thanks to Mr. Hiroki Nakayama for computer-based numerical calculations. The authors would like to acknowledge the valuable comments and useful suggestions from reviewers of Scientific Program Committee of ICCSA 2017.

References

1. Goodchild, M.F., Guoqing, S., Shiren, Y.: Development and test of error model for categorical data. Int. J. Geogr. Inf. Syst. **6**(2), 87–104 (1992)
2. Umesh, R.M.: A technique for cluster formation. Pattern Recogn. **21**(4), 393–400 (1988)
3. MacEachren, A.M.: Some Truth with Maps: A Primer on Symbolization and Design. Association of American Geographers, Washington (1994)
4. Jenks, G.F.: The data model concept in statistical mapping. Int. Year book Cartography **7**, 186–190 (1967)
5. Andrienko, G., Andrienko, N., Savinov, A.: Choropleth maps: classification revisited, In: Proceedings ICA 2001, Beijing, China, vol. 2, pp. 1209–1219 (2001)
6. Osaragi, T.: Classification methods for spatial data representation. In: CASA Working Papers 40. Centre for Advanced Spatial Analysis (UCL), London (2002)

7. Akaike, H.: Information theory and an extension of the maximum likelihood principle, In: Petron, N., Csak, F. (eds.) Proceedings of the 2nd International Symposium on Information Theory, Akademiai kaido, Budapest, pp. 267–281 (1972)
8. Akaike, H.: A new look at the statistical model identification. IEEE Trans. Autom. Control **19**, 716–723 (1974)
9. Roy, J.R., Batten, D.F., Lesse, P.F.: Minimizing information loss in simple aggregation. Environ. Plann. A **14**, 973–980 (1982)
10. Batty, M.: Spatial entropy. Geogr. Anal. **6**, 1–31 (1974)
11. Batty, M.: Entropy in spatial aggregation. Geogr. Anal. **8**, 1–21 (1976)
12. Andrienko, N., Andrienko, N.: Cumulative curves for exploration of demographic data: a case study of Northwest England. Comput. Stat. **19**(1), 9–28 (2004)
13. Aarts, E., Korst, J.: Simulated annealing and Boltzmann machines, a stochastic approach to combinatorial optimization and neural computing (Wiley Series in Discrete Math. and Optimization), Wiley (1988)
14. ESRI: ArcView GIS - The Geographic Information System for Everyone. Environmental Systems Research Institute, USA (1996)
15. Osaragi, T.: Information loss minimization for spatial data representation. J. Architectural Plann. Eng, AIJ **574**, 71–76 (2003). (in Japanese)

Online Reverse Subpattern Matching for Reproducing Trajectories from Sub-paths

Marie Kiermeier[(✉)]

LMU München, Munich, Germany
marie.kiermeier@ifi.lmu.de

Abstract. In this paper, we present an online method for reverse sub-pattern matching (RSM). RSM is a special kind of subpattern matching, whereby the query trajectory is longer than the patterns for matching. This is the case, if trajectories have to be reproduced from given sub-paths. In particular, we use as a showcase an anomaly detection method for Self-Organizing Industrial Systems (SOIS) where the movements of the robots in the factory have to be reproduced online by given moving patterns. Assuming network-constrained trajectories, we introduce *edge lists* as suitable data structure for indexing the pattern dictionary. Based on them *candidate lists* are built and updated each time a new edge of the query trajectory is recorded. Patterns of the candidate lists which are completely contained in the trajectory are shifted to the *matched pattern list*. In addition, we propose the *covering rate* as a quality indicator to discover trends about the reproducibility of the trajectory as soon as possible. The work-flow of the presented method is illustratively evaluated based on a SOIS scenario where anomalies are detected by reproducing the trajectories from given patterns online.

Keywords: Pattern matching · Spatio-temporal index structures · Online trajectory computing · Network-constraint trajectories

1 Introduction

Modern tracking systems allow to record the movement of an object over time with relatively small effort. Be that smart phones or navigation systems of cars, which use GPS-signals for positioning, or special indoor tracking systems, which use cameras for positioning. Independent from that, we always end up with a vast amount of spatio-temporal data. Therefore, special techniques are required which process automatically the bulk of data and extract the requested information. One discipline of spatio-temporal analysis is *pattern matching*. Thereby, a query trajectory, representing the movement of an object, is compared to pattern trajectories and their match is evaluated.

Depending on the length ratio of the two trajectories, three kinds of pattern matching can be distinguished: whole pattern matching, subpattern matching, and reverse subpattern matching (RSM) [8,9]. In Fig. 1 the differences are illustrated, whereby the solid line is the query trajectory and the dashed line is a

© Springer International Publishing AG 2017
O. Gervasi et al. (Eds.): ICCSA 2017, Part VI, LNCS 10409, pp. 635–646, 2017.
DOI: 10.1007/978-3-319-62407-5_46

pattern trajectory. For whole pattern matching the two trajectories have similar length (see Fig. 1a). A typical application for this is finding groups of persons taking similar routes. For subpattern matching the query trajectory is shorter and the trajectories which are compared against it are longer (see Fig. 1b). This can be used for getting routes which all share a certain sub-path. In the last case the problem definition is inverted. This means, the query trajectory is longer than the patterns for matching. With it, patterns which are contained in a query trajectory can be found.

While the first two problems make up the majority of pattern matching tasks, there are, so far, few use cases which require RSM. The anomaly detection approach for Self-Organizing Industrial Systems (SOIS), which is introduced in [4], is one of it. Incoming trajectories, representing the movements of objects in the factory, are evaluated by trying to put them together from sub-trajectories stored in a dictionary. Obviously, this process is a RSM task. In addition, what is special in this case, is that the pattern matching must be performed online to detect anomalous behaviour has soon as possible. This aspect of online applicability is mostly left out, since the query trajectory is usually already available as a whole. Therefore, in this work, we will present an online RSP method for reproducing trajectories from sub-paths at run-time.

Thereby, we assume network-constrained trajectories. It turns out, that for many scenarios the routes which have to be processed are not arbitrary, but adhere to an underlying structure. For example, movement data from cars are all constrained to streets of the road network. So, if the network structure is known, trajectories can easily represented by a sequence of touched network edges instead of a sequence of imprecise x-/y-coordinates. With it, on the one hand, inaccuracies with regard to the exact position of an object can be compensated. On the other hand, the problem of different sampling rates is avoided. So, as long as the spatio-temporal data to be processed is network-constrained, this representation can be used to process the data more efficiently.

As in many other cases, the basis for the efficient processing of a query are suitable data structures for indexing the data. The objective is here to prune as early as possible as many irrelevant data as possible and so reduce subsequently the set of possible solutions. Accordingly, we present in this paper data structures which can be used for solving the online RSM problem for network-constrained trajectories efficiently. In addition, we introduce a quality which gives online information about the reproducibility of the query trajectory.

(a) Whole pattern matching.

(b) Subpattern matching.

(c) Reverse subpattern matching.

Fig. 1. Three kinds of pattern matching.

The organization of the paper is as follows. After the introduction section, existing indexing structures for network-constrained trajectories are reviewed in Sect. 2. In Sect. 3, we present our approach for online RSM in detail. The results of the evaluation which was performed on a SOIS scenario is described in-depth in Sect. 4. The paper is ended by Sect. 5 with the conclusion.

2   Related Work

As already mentioned, suitable data structures for indexing trajectories are essential for the efficient analysis of spatio-temporal data. Accordingly, there are various approaches for such index techniques in literature. A general introduction and good overview of the basic principles and methods for indexing spatio-temporal data can be found for example in [2]. In addition, there are several approaches focusing on network-constrained trajectories, e.g. [1,3,7,10]. While these methods are practicable for general range and k(NN) queries, in [6], Krogh et al. present the first index for strict path queries (SPQ). It is specialized efficiently retrieving trajectories which follow strictly a specific path. With it, these queries can be categorized as subpattern matching tasks. The index in [6] is built based on an encoding scheme for the trajectories, which allows to answer SPQ by considering only the start and end edges of the trajectories. In [5], a further index for SPQ is introduced. Thereby, the FM-Index, which has its origin in information retrieval and substring matching, is modified so that the temporal dimension is considered, too.

All of the previous mentioned approaches for indexing spatio-temporal data are evaluated for offline pattern matching and subpattern matching applications. This means, the query trajectory is always available as a whole. In contrast, there are applications where this is not the case and the query trajectory is only successively available. For processing pattern matching tasks anyway, special techniques are required. While the aforementioned indices are designed for efficiently processing individual queries, the idea of the method presented in this work is, to have data structures which allow to pass over the information already gained over time. With it, online RSP tasks can be processed much more efficient.

3   Online Reverse Subpattern Matching

Initially, for RSM a dictionary of subpatterns is given. The challenge is then to find a suitable data structure which allows an efficient access to these patterns. For it, we use in this paper *edge lists*. Based on these edge lists, subpatterns can easily be identified which are contained in the query trajectory. The design of the edge lists, and the subsequent matching process are described in detail in the following sections.

3.1   Edge Lists

One of the big advantages when working with network-constrained trajectories is, that each edge $e_i \in E$ in the network can be labelled with a unique identifier

ID(e_i) (see Fig. 2). Based on this, for each edge a list $list(\text{ID}(e_i))$ is built. For each pattern $p_i \in P$ which contains the edge e_j an entry of the form $(\text{ID}(p_i), t_{in}, t_{out})$ is added to $list(\text{ID}(e_j))$, where $\text{ID}(p_i)$ is the unique identifier of pattern p_i. t_{in} and t_{out} denote for which points in the time structure of the pattern the corresponding edge e_j occurs.

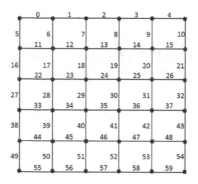

Fig. 2. Each edge is labelled with a unique identifier $\text{ID}(e_i) = i$, $i = 0, 1, ..., 59$.

Exemplary, we show based on our example network of Fig. 2 how the edge list for edge e_{46} is built. Therefore, all patterns which contain edge e_{46} are identified (see Fig. 3). As pattern $\text{ID} = 4$ enters edge $\text{ID} = 46$ at the time of 4 and leaves it after 2 time units (see Fig. 3d), the first line in Table 1 is $(\text{ID}(p_4), t_{in}, t_{out}) = (4, 4, 6)$. Next, pattern $\text{ID} = 7$ uses the edge at the time of 11 to 13, and finally, pattern $\text{ID} = 12$ from 7 to 9, resulting in entries $(7, 11, 13)$ and $(12, 7, 9)$. The final edge list for edge e_{46} is given with Table 1.

After having built this data structure for all edges by scanning all patterns, possible candidate patterns for the RSM can easily be identified. The exact procedure is described in detail in the following section.

Table 1. Edge list for edge e_{46}.

ID(p)	t_{in}	t_{out}
4	4	6
7	11	13
12	7	9

3.2 Identifying Candidates

Since our approach is an online method, the query trajectory is processed edge-wise. This means that every time a new edge is recorded, the process for RSM

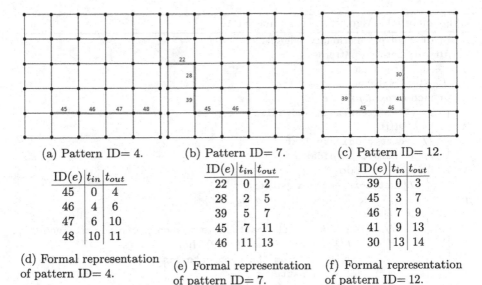

(a) Pattern ID= 4.   (b) Pattern ID= 7.   (c) Pattern ID= 12.

ID(e)	t_{in}	t_{out}
45	0	4
46	4	6
47	6	10
48	10	11

ID(e)	t_{in}	t_{out}
22	0	2
28	2	5
39	5	7
45	7	11
46	11	13

ID(e)	t_{in}	t_{out}
39	0	3
45	3	7
46	7	9
41	9	13
30	13	14

(d) Formal representation of pattern ID= 4.

(e) Formal representation of pattern ID= 7.

(f) Formal representation of pattern ID= 12.

Fig. 3. Patterns which contain edge e_{46}.

is continued. The key element of this process is the *candidate list*. It contains subpatterns which are possibly contained in the current trajectory. Therefore, the first column contains the unique identifiers of the pattern candidates. The second column is used as a counter for the number of edges which are shared so far by the pattern and the current trajectory (see Table 2a and corresponding example at the end of this section).

Because of the online mode, the candidate list has to be updated each time a new edge is recorded. Algorithm 1 gives more detail about the overall procedure of online RSM.

Basically, the update process is divided into two steps: first (line 5–16), candidates which are completely contained in the current trajectory are removed. Instead, they are consequently added to the *matched pattern list*. Alternatively, candidates are removed if they are not conform any more to the current trajectory. In the second step (line 17–21), new candidates are identified by scanning the corresponding edge list for patterns which have this edge as starting edge.

The method $next\_edge(c, x)$ (line 6) returns for a given pattern c the $(x + 1)$th row of the table representing the pattern (see Fig. 3d). This means, for the condition in line 7 both the edge IDs, and the time constraints, given by t_{in} and t_{out}, are compared.

For illustration, we show in Table 2 the candidate and matched pattern lists for a trajectory of the example network of Fig. 2 (see Fig. 4a). The time of consideration is $t = 11$. This means, so far, only the first four edges of the trajectory are apparent ($e_{33}, e_{39}, e_{45}, e_{46}$). Since pattern ID $=1$ (see Fig. 4b and d) is contained completely in the current trajectory, it is recorded as entry in the matched

---

Algorithm 1. Algorithm for online RPM

---

    input   : pattern dictionary, edge lists
    output: matched patterns

1 $C \leftarrow \emptyset$ `// candidate list`
2 $M \leftarrow \emptyset$ `// matched pattern list`
3 **foreach** *incoming edge e* **do**
      `// get corresponding edge list`
4      $l \leftarrow list(\text{ID}(e))$
      `// update candidate list:`
      `// 1. remove candidates`
5   **foreach** $c \in C$ **do**
         `// get next edge of pattern c`
6         $e_{next} \leftarrow next\_edge(c.ID, c.matchedEdges)$
7   **if** $e == e_{next}$ **then**
           `// check whether the pattern is now completely contained`
8   **if** $e_{next}.ID == lastElement(c.ID)$ **then**
              `// remove c from the candidate list and add it to the`
                 `matched patterns`
9               $C \leftarrow C.remove(c)$
10              $M \leftarrow M.add(c)$
11   **else**
              `// increase counter for matched edges`
12              $c.matchedEdges \leftarrow c.matchedEdges+1$
13   **end**
14   **else**
          `// remove c from the candidate list`
15          $C \leftarrow C.remove(c)$
16   **end**
       `// 2. check edge list for new candidates`
       `// add all patterns which starts at edge e`
17   **foreach** $p \in l$ **do**
18   **if** $p.t_{in} == 0$ **then**
19             $C \leftarrow C.add(p)$
20   **end**
21   **end**
22   **end**
23 **end**

---

pattern list (see Table 2b). Pattern ID = 4 and pattern ID = 12, meanwhile, are registered as candidates (see Table 2a), since with regard to pattern ID = 4, its first two edges, and with regard to pattern ID = 12, its first three edges coincide with parts of the current trajectory. Accordingly, the current candidate list consists of the two entries $(4, 2)$ and $(12, 3)$. Other patterns, like pattern ID = 7 (see Fig. 3b), are not considered, since they are at the moment not relevant for the reproduction of the current trajectory.

Table 2. Lists for example trajectory at time $t = 11$.

ID(p)	# matched edges
4	2
12	3

(a) Candidate list.

ID(p)
1

(b) Matched pattern list.

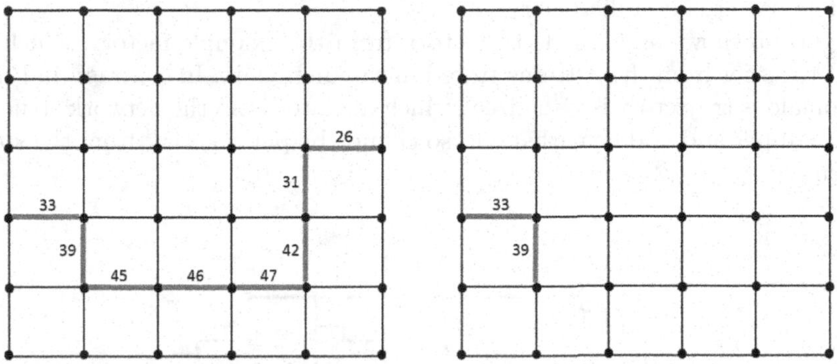

(a) Example trajectory.

(b) Pattern ID= 1.

ID(e)	t_{in}	t_{out}
33	0	3
39	3	5
45	5	9
46	9	11

(c) Formal representation of the trajectory at time $t = 11$.

ID(e)	t_{in}	t_{out}
33	0	3
39	3	5

(d) Formal representation of pattern ID= 1.

Fig. 4. Example trajectory and pattern ID $= 1$.

4 Evaluation

For demonstrating the work-flow of our algorithm, we use the application presented in [4]. For detecting anomalous behaviour in Self-Organizing Industrial Systems (SOIS), routes of work-pieces are reproduced using a pattern dictionary. Therefore, we first introduce an example scenario of such a factory. Then, the procedure for online reproduction of "normal" trajectories is demonstrated. The evaluation is concluded by introducing the covering rate as a quality, which gives online information about the reproducibility of trajectories.

4.1 Scenario

In Fig. 5a the underlying network structure of an example factory is visualized. Basically, the work-pieces start on the left hand side (three possible start points), take their way through the factory while visiting several production stations, until the desired product finally leaves the factory on the right hand side (again three possible end points). What is special about SOIS is, that the routes are not defined in advance, but scheduled at run-time. With it, the routes certainly consists of similar sub-paths (e.g., shortest paths from one production station to another), but each time in a different order. Therefore, in [4] the task of anomaly detection in SOIS is transformed into a RSM problem. Thereby, the routes are represented by network-constrained trajectories and the common sub-paths are stored as patterns in a dictionary.

Figure 5b shows a "normal" trajectory from the example factory, which can be put together from the patterns stored in the dictionary. In contrast, in Fig. 5c an anomalous trajectory is visualized, which deviates from the network structure (see the upper and middle paths) and so cannot be put together from the stored patterns.

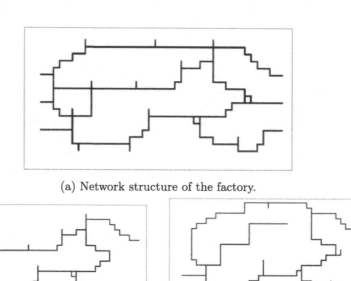

(a) Network structure of the factory.

(b) Example of a "normal" trajectory.

(c) Example of an anomalous trajectory.

Fig. 5. Evaluation scenario.

4.2 Reproducing Trajectories from Sub-paths

Based on the "normal" trajectory, we will now illustrate a run of our online RSM algorithm as an example. Therefore, in Figs. 6 and 7, the candidate and matched patterns are visualized after having received the first, respectively, second edge of the "normal" trajectory. As it can be seen in Fig. 6, there is already one pattern which completely matches the current trajectory and so is directly added to the matched pattern list (see Fig. 6b). Additionally, there are three patterns which share the first edge with the current trajectory. Therefore, they build the current candidate list (see Fig. 6c). Figure 7a shows the current trajectory after having received the second edge. Since the first and second patterns of the candidate list also share the second edge with the trajectory, they remain in the candidate list (see Fig. 7c). In contrast, the third candidate pattern is removed from the candidate list, since it differs from the current trajectory with regard to the third edge. The set of matched patterns remains the same (see Fig. 7b). By continuing this process, we are able to reproduce online the trajectory step by step from the given patterns.

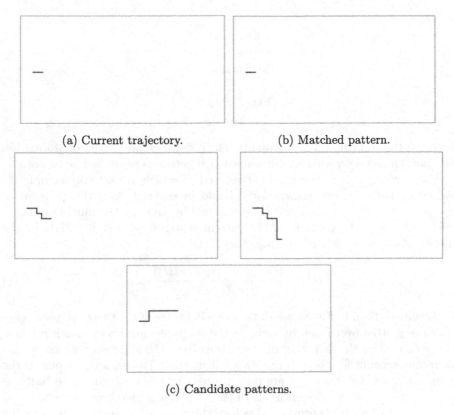

(a) Current trajectory.   (b) Matched pattern.

(c) Candidate patterns.

Fig. 6. Step 1.

4.3   Covering Rate

In case of online RMS, it is useful to have already information about the reproducibility of the trajectory even if it is not terminated yet. Therefore, we

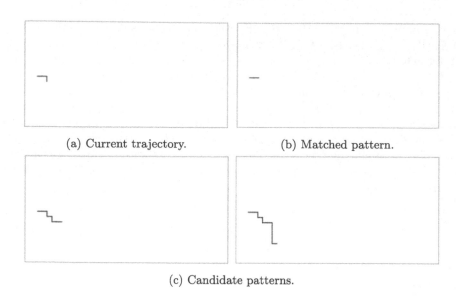

(a) Current trajectory.   (b) Matched pattern.

(c) Candidate patterns.

Fig. 7. Step 2.

introduce the covering rate as a quality which delivers intermediate information and shows a trend on whether the current trajectory is reproducible or not.

The covering rate is the ratio between edges which are actually completely covered by patterns, and those which should be covered. Accordingly, for each time step t the covering rate $c(t)$ is determined by dividing the number of edges which are covered by patterns of the current matched pattern list $M(t)$ by the number of edges of the current trajectory $T(t)$:

$$c(t) = \frac{|\{e|e \in p_i, \forall p_i \in M(t)\}|}{|\{e|e \in T(t)\}|}, \tag{1}$$

For illustration, in Fig. 8a and b the covering rates of the two example trajectories are plotted over time. In both cases, the graphs form a sawtooth pattern. This is caused by the fact, that only patterns from the matched pattern list are taken into account for the covering rate calculation. This means, as long as the recent edges of the trajectory are only covered by edges of candidate patterns the covering rate decreases until one of the candidates match completely.

With regard to the "normal" trajectory these variations level off between 0.8 and 1.0 (see Fig. 8). In contrast, the covering rates of the anomalous trajectory start collapsing around step $t = 25$ and then recover only to a maximum of 0.65 until the end. This reflects, that there are edges which can principally not be covered by patterns of the dictionary.

So, overall, this example illustrates that by monitoring the covering rates, useful information about the reproducibility of trajectories from sub-paths can already be achieved even if the trajectory is not terminated yet.

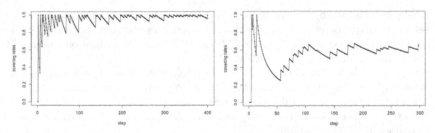

(a) Covering rates of a "normal" tra- (b) Covering rates of an anomalous tra-
jectoriy.   jectoriy.

Fig. 8. Covering rates.

5   Conclusion

In this paper, we presented an online RSM method for reproducing trajectories
from sub-paths. Assuming network-constrained trajectories, we introduced edge
lists as suitable data structure for indexing the pattern dictionary. For every
edge of the network, information about the patterns which cover it, are stored.
Based on these edge lists a candidate list is built and updated each time a new
edge of the query trajectory is recorded. Patterns of the candidate lists which are
completely contained in the trajectory are shifted to the matched pattern list.
To get intermediate information about the reproducibility of the trajectory, and
to show a trend on whether the incoming trajectories can be put together from
the patterns of the dictionary, we proposed to use the covering rate as quality.

We illustrated the work-flow and operation functionality of our approach on
the basis of an anomaly detection method for SOIS which uses online RSM.

Overall, our approach differs from the existing methods in its ability to pass
over information already gained over time. With it online RSM tasks can be
processed much more efficiently than processing all queries individually and put
the results together afterwards.

References

1. De Almeida, V.T., Güting, R.H.: Indexing the trajectories of moving objects in
   networks. GeoInformatica **9**(1), 33–60 (2005)
2. Deng, K., Xie, K., Zheng, K., Zhou, X.: Trajectory indexing and retrieval. In:
   Zheng, Y., Zhou, X. (eds.) Computing with Spatial Trajectories, pp. 35–60.
   Springer, New York (2011). doi:10.1007/978-1-4614-1629-6_2
3. Frentzos, E.: Indexing objects moving on fixed networks. In: Hadzilacos, T.,
   Manolopoulos, Y., Roddick, J., Theodoridis, Y. (eds.) SSTD 2003. LNCS, vol.
   2750, pp. 289–305. Springer, Heidelberg (2003). doi:10.1007/978-3-540-45072-6_17
4. Kiermeier, M., Werner, M., Linnhoff-Popien, C., Sauer, H., Wieghardt, J.: Anomaly detection in self-organizing industrial systems using pathlets. In: 2017 IEEE
   International Conference on Industrial Technology (ICIT) (2017, accepted)

5. Koide, S., Tadokoro, Y., Yoshimura, T.: SNT-index: spatio-temporal index for vehicular trajectories on a road network based on substring matching. In: Proceedings of the 1st International ACM SIGSPATIAL Workshop on Smart Cities and Urban Analytics, pp. 1–8. ACM (2015)
6. Krogh, B., Pelekis, N., Theodoridis, Y., Torp, K.: Path-based queries on trajectory data. In: Proceedings of the 22nd ACM SIGSPATIAL International Conference on Advances in Geographic Information Systems, pp. 341–350. ACM (2014)
7. Pfoser, D., Jensen, C.S.: Trajectory indexing using movement constraints. GeoInformatica 9(2), 93–115 (2005)
8. Roh, G.P., Hwang, S.W.: TPM: supporting pattern matching queries for road-network trajectory data. In: Proceedings of the 14th International Conference on Extending Database Technology, pp. 554–557. ACM (2011)
9. Roh, G.P., Roh, J.W., Yi, B.K., et al.: Supporting pattern-matching queries over trajectories on road networks. IEEE Trans. Knowl. Data Eng. 23(11), 1753–1758 (2011)
10. Sandu Popa, I., Zeitouni, K., Oria, V., Barth, D., Vial, S.: Indexing in-network trajectory flows. VLDB J. 20(5), 643–669 (2011)

An Algorithm to Discover Partners in Trajectories

Diego Vilela Monteiro[✉], Karine Reis Ferreira[✉], and Rafael Santos[✉]

National Institute for Space Research, Av. dos Astronautas, 1758,
São José dos Campos, SP 12227-010, Brazil
dvm1607@gmail.com, {karine.ferreira,rafael.santos}@inpe.br

Abstract. Spatiotemporal data is everywhere, being gathered from different devices such as Earth Observation and GPS satellites, sensor networks and mobile gadgets. Spatiotemporal data collected from moving objects is of particular interest for a broad range of applications. In the last years, such applications have motivated many researches on moving object trajectory data mining. In this paper, we propose an efficient method to discover partners in moving object trajectories. Such method identifies pairs of trajectories whose objects stay together during certain periods, based on distance time series analysis. We present two case studies using the proposed algorithm.

Keywords: Moving objects · Trajectory · Pattern · Data mining

1 Introduction

Recent advances on sensors and communication technologies have produced massive spatiotemporal data sets that allow scientists to observe the world in novel ways. Earth observation satellites capture changes over time in cities and forests. Environmental sensors measure the variation of air pollution, temperature and humidity in specific locations. GPS satellites and devices collect locations of animals, vehicles and people over time. Mobile gadgets, sensor networks, social media and GPS tools create useful data for planning better cities, capturing human interactions and improving life quality.

Spatiotemporal data collected from moving objects is of particular interest for a wide range of applications. Moving objects are entities whose spatial positions or extents change over time (Erwig et al. 1999). Examples of moving objects are cars, aircraft, ships, mobile phone users, polar bears, hurricanes, forest fires, and oil spills on the sea. Trajectories are countable journeys associated to moving objects (Spaccapietra et al. 2008).

Nowadays, moving object trajectories have been used in a broad range of applications, such as location-based social networks, intelligent transportation systems, and urban computing (Zheng 2011; Zheng et al. 2014b). These applications have motivated research on novel data mining techniques to discover

© Springer International Publishing AG 2017
O. Gervasi et al. (Eds.): ICCSA 2017, Part VI, LNCS 10409, pp. 647–661, 2017.
DOI: 10.1007/978-3-319-62407-5_47

patterns in trajectories, attracting attention from many areas including computer science, sociology, and geography (Zheng 2015).

Along the years many patterns have been proposed to extract information from trajectories. These patterns range from those that take into account semantically enhanced trajectories, such as CB-SMoT (Palma et al. 2008), to those that analyze trajectories based solely on their geometries, like Convergence or Flock (Laube et al. 2005).

In this work, we focus on a specific group of trajectory pattern called *Moving Together Patterns* (Zheng 2015). We propose a new approach called Partner to discover objects that move or stay together during certain periods, based on trajectory distance time series analysis. Next section presents related work, the main differences between the existing moving together patterns and the proposed Partner as well as its main advantages.

2   Trajectory Pattern Mining

Recently, the research area on trajectory data mining has grown a lot. Studies on this area consist in analyzing the mobility patterns of moving objects and in identifying groups of trajectories sharing similar patterns. In last years, many methods and techniques for trajectory pattern discovering have been proposed to meet a broad range of applications. (Zheng 2015) presents a systematic survey on the major research into trajectory data mining and classifies existing patterns in four categories: (1) *Moving together patterns*; (2) *Clustering*; (3) *Frequent sequence patterns*; and (4) *Periodic patterns*. In this work, we focus on the first category and propose a new approach to identify moving together objects.

Examples of patterns that discover a group of objects that move together for a certain period are flock (Gudmundsson and van Kreveld 2006; Vieira et al. 2009; Tanaka et al. 2015), group (Wang et al. 2006), convoy (Jeung et al. 2008), swarm (Li et al. 2010), traveling companion (Tang et al. 2012), gathering (Zheng et al. 2013, 2014a) and co-movement (Fan et al. 2016). *Moving together patterns* are useful for a high number of applications, such as monitoring of delivery trucks (Jeung et al. 2008) and identification of vessels that fish together.

The flock pattern has attracted a lot of interest from the community with many studies being published over the years regarding this pattern. A flock is a group of objects that stay together within a disk with a user-defined radius for at least K consecutive time stamps. (Vieira et al. 2009) propose a framework and polynomial-time algorithms, called *Basic Flock Evaluation* (BFE), to discover such pattern in streaming spatiotemporal data. (Tanaka et al. 2015) present variations of the BFE algorithm, employing the plane sweeping technique, binary signatures and/or an inverted index. Similar to flock, group pattern identifies moving objects that travel within a radius for certain timestamps that are possibly nonconsecutive (Wang et al. 2006). The main difference between both is that group considers relaxation of the time constraint.

According to (Zheng 2015), a major concern with flock and group patterns is the predefined circular shape, which may not well describe the shape of a group in reality. Since they use a disk with rigid limits, they miss objects that are close to a group but outside the disk limits. This drawback is called *lossy-flock* problem. The chosen disk size has a substantial effect on the results of the discovery process. The selection of a proper disk size is very difficult. Besides the *lossy-flock* problem, for some data sets, no single appropriate disc size may exist that works well for all parts of the space and time domain (Jeung et al. 2008).

The `convoy` pattern uses density-based clustering in order to capture groups of arbitrary extents and shapes. Instead of using a rigid size disk as `flock`, such pattern requires a group of objects to be density connected during k consecutive time points. While both `flock` and `convoy` have a strict requirement on consecutive time period, (Li et al. 2010) propose a more general type of trajectory pattern, called `swarm`, which captures the moving objects that move within arbitrary shape of clusters for certain timestamps that are possibly nonconsecutive. Even though `swarm` and `group` patterns consider relaxation of the time constraint, the `group` pattern definition restricts the size and shape of moving object clusters by specifying the disk radius. Moreover, redundant `group` patterns make the algorithm exponentially inefficient (Li et al. 2010).

Aiming to overcome the limitations brought by the global consecutiveness of the `convoy` and aiming to be more selective than the `swarm`, (Li et al. 2015) have proposed the `platoon`. The `platoon` is not as restrictive as the `convoy` regarding the consecutiveness of the timestamps, nor is as loose as the `swarm` on the same matter. It uses the concept of local consecutiveness, in which objects can separate for a while, as long as the periods in which they are together meet the minimum temporal requirements.

The `traveling companion` (Tang et al. 2012) proposes a data structure, called traveling buddy, to improve the efficiency of the algorithms to find moving together patterns from trajectories that are being streamed into a system. The concepts of `convoy` and `swarm` patterns are similar to `traveling companion`. The main difference is that `convoy` and `swarm` need to load entire trajectories into memory for a pattern mining. Hence it is impractical to use them in a data stream environment. The `traveling companion` pattern can be considered an online (and incremental) detection fashion of convoy and swarm (Zheng 2015). The `gathering` pattern (Zheng et al. 2013, 2014a) detects some incidents, such as celebrations and parades, in which objects join in and leave an event frequently. This pattern loses the constraints of the aforementioned patterns by allowing the membership of a group to evolve gradually (Zheng 2015).

(Fan et al. 2016) propose a more general patterns, called `co-movement`, to unify those to identify moving together patterns, such as `convoy`, `swarm`, `flock` and `group`. They argue that `co-movement` pattern can avoid the *loose-connection anomaly* and can be reduced to any of the previous pattern by customizing its parameters. *Loose-connection anomaly* refers to the problem in which clusters that are clearly too distant in time are considered part of the same pattern. Moreover, the authors propose two types of parallel and scalable frameworks to process the `co-movement` method and deploy them on MapReduce platform.

2.1 Difference Between Our Proposal and the Existing Moving Together Patterns

All methods to identify moving together objects presented in this section are based on two steps: (1) cluster the objects of each snapshot and (2) intersect the clustering results to retrieve moving-together objects. Both clustering and intersection steps involve high computational overhead. Differently from these methods, we propose a new approach, called `Partner`, to detect moving together objects, based on trajectory distance time series analysis. The method `Partner` does not use disk with predefined circular shape, or density-based clustering, or cluster of objects. The method `Partner` analyzes the distance time series of each pair of trajectories whose spatiotemporal boxes intersect.

The main advantages of our approach are: (1) The distance time series analyses are completely independent and so can be processed in parallel. (2) Users do not have to predefine either the shape and density of a group or the number of objects in a group. From the resulting pairs of partners, users can easily extract all trajectories that stayed together. (3) The method `Partner` finds objects that stayed together for certain periods, including objects that were moving as well as stationary. (4) Partners are allowed to stay apart for a maximum user-defined time. (5) The method `Partner` does not suffer the loose-connection problem, since it verifies if two trajectories are not too far for too long. (6) The method `Partner` does not suffer the lossy-flock problem, since it is not based on a rigid circular disk.

3 The `Partner` Method and Definitions

We define two trajectories as partners when they stay together within a maximum distance during a certain period. This pattern is useful for many applications that need identity objects that are performing activities jointly, such as marine vessels that are fishing together or trucks that are working in cooperation, as well as objects that are stationary close to others.

We propose a method to recognize partners based on distance time series analyzes. For each pair of trajectories selected based on *spatiotemporal boxes*, the method calculates its distance time series, that is, a time series that represents the euclidean distance variation over time of these two trajectories. Then, the method analyses such time series regarding user-defined restrictions that define partnership.

One efficient way to filter trajectories data is using their *spatiotemporal boxes* (STBox). A STBox of a trajectory is an entity defined in two domains, space and time, that surrounds all observations of this trajectory. Using STBox, we can filter trajectories considering their intersections in space and time, reducing the amount of data for further processing.

Definition 1 (Trajectory). A trajectory Trj_i is represented by a set of observations τ temporally ordered in the form $\{(x_1, y_1, t_1), ..., (x_n, y_n, t_n)\}$. Each observation (x, y, t) contains the spatial location x and y of a moving object at a certain time t.

Definition 2 (Trajectory spatiotemporal box). A *spatiotemporal box* (STBox) of a trajectory Trj_i, STBox(Trj_i), is an entity defined in two domains, space and time. Spatially, it is defined as the rectangular region that contains all spatial locations of a trajectory. Such region is represented by two points $((x_{min}, y_{min}), (x_{max}, y_{max}))$, which are the bottom-left and top-right corners, respectively, of the rectangle. Temporally, it is delimited by (t_{min}, t_{max}) that are the minimum and maximum time instants associated to a trajectory. Thus, the STBox of a trajectory Trj_i can be represented as $((x_{min}, y_{min}, t_{min}), (x_{max}, y_{max}, t_{max}))$ given that all observations of Trj_i satisfy $x_{min} < x < x_{max}$, $y_{min} < y < y_{max}$ and $t_{min} < t < t_{max}$.

Definition 3 (Trajectory distance time series). The distance time series of two trajectories, symbolized here as $\Delta\, Trj_{ij}$, represents the euclidean distance variation over time between the two trajectories Trj_i and Trj_j. Further $\Delta\, Trj_{ij}$ can be described as a set of tuples in the form $\{(\Delta d_1, t_1), ..., (\Delta d_n, t_n)\}$ in which Δd_n is the euclidean distance between the spatial locations x_n and y_n of the trajectories Trj_i and Trj_j at the instant t.

Definition 4 (Partner). Two trajectories Trj_i and Trj_j are considered partners during a certain period (t_{ini}, t_{fin}) when the distance variation between these trajectories in such period obeys two rules: (1) the distances between the two trajectories must be less than a distance threshold (d_{max}) for, at least, a minimum period (p_{min}); (2) separation periods are allowed, that is, the distance between the two trajectories can be more than a distance threshold (d_{max}) for a maximum period (p_{max}). The same pair of trajectories may be considered partner more than once on detached periods. A partner is represented by a tuple (i, j, t_{ini}, t_{fin}), where i and j are the unique identifiers of Trj_i and Trj_j and t_{ini} and t_{fin} are respectively the initial and final instants of the period when such trajectories are partners.

Figures 1 and 2 show trajectories of four objects and their corresponding STBoxes. Taking the object 1 STBox, we can notice that only the object 3 STBox intersects it. Object 2 STBox intersects spatially the object 1 STBox, but not temporally. And object 4 intersects temporally object 1 STBox, but not spatially.

After selecting trajectories based on their STBoxes, the method calculates a distance time series for each pair of trajectories and analyzes such times series regarding user-defined parameters. These parameters represent a set of restrictions that define when two trajectories are considered partners. The three parameters are: (1) a distance threshold (d_{max}), that is, a maximum distance that two objects can stay apart; (2) the minimum period (p_{min}) that two objects must stay together, that is, the distance between the two objects must be under the distance threshold (d_{max}) during, at least, the period $((p_{min}))$; (3) a maximum period (p_{max}) that two objects are allowed to stay apart, that is, over the distance threshold. The method analyses the distance time series in order to verify whether they comply with the user-defined parameters.

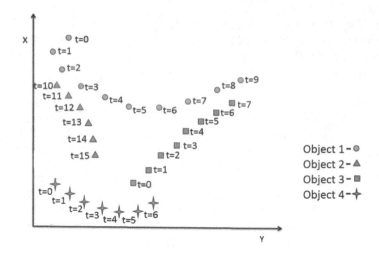

Fig. 1. Trajectories of four objects

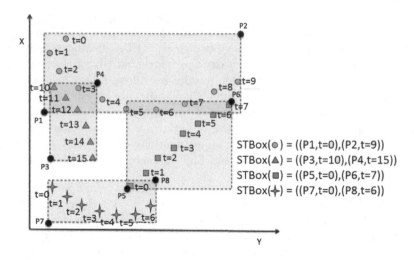

Fig. 2. Trajectories and their STBoxes

Figures 3, 4 and 5 present examples of distance time series extracted from two trajectories. Figure 3 shows an example of two trajectories that are partners because all their distances are less than the maximum distance threshold d_{max}. Figure 4 shows two trajectories that are not partners, considering the parameter values p_{max} and p_{min} showed in the picture. In this case, objects 1 and 2 go in different directions, not matching the minimum time together requirement.

Figure 5 shows two trajectories that start and end together, however they stay apart for a period during their paths. Even though they surpass the maximum distance threshold d_{max}, they might be still considered partners depending on the

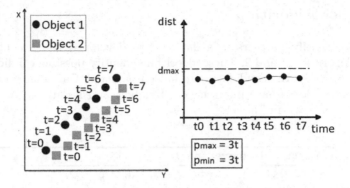

Fig. 3. Example of **partner** trajectories

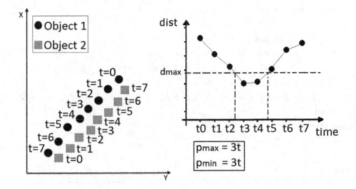

Fig. 4. Example of trajectories which are not **partners**.

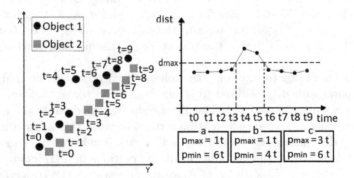

Fig. 5. Trajectories with three possible **partner** interpretations

user-defined parameter values p_{max} and p_{min} showed in the picture. Considering the parameter set (a) they are not partners. They are partners when we consider the parameter sets (b) and (c).

4   Partner Algorithm

In order to identify partners in a dataset, we developed the two algorithms shown in Algorithms 1 and 2. These algorithms identify possible candidates and then verify whether the candidates are indeed partners. Candidates are trajectories whose STBoxes have intersections. Symbols and notations used in both algorithms are shown in Table 1.

Algorithm 1. `Partner Discovering`

Require: $T = \{T_0, ..., T_n\}, d_{max} \in \mathbb{Q}, p_{max} \in \mathbb{Q}, p_{min} \in \mathbb{Q}$
1: **procedure** DISCOVERPARTNER($T, d_{max}, p_{max}, p_{min}$)
2:      $P \leftarrow \emptyset$
3:      **for all** $T_i \in T$ **do**
4:          $sb \leftarrow CreateSTBox(T_i, d_{max})$
5:          $[T] \leftarrow GetTrajectories(sb, T)$
6:          **for all** $T_j \in [T]$ **parallel do**
7:              $DTS \leftarrow CreateDistanceTimeSeries(T_i, T_j)$
8:              $PP \leftarrow PeriodAsPartners(DTS, d_{max}, p_{max}, p_{min})$
9:              **if** $PP \neq \emptyset$ **then**
10:                 $P \leftarrow P \cup \{(PP_1, i, j), ..., (PP_n, i, j)\}$
11:             **end if**
12:         **end for**
13:     **end for**
14:     Return P
15: **end procedure**

The first algorithm is responsible for the overall analysis. The user must input a set of trajectories T, and values for d_{max}, p_{max} and p_{min}. For each trajectory T_i of the set T, the algorithm repeats the following steps. The line 4 of the Algorithm 1 creates a STBox of the trajectory T_i (Definition 2) increased in all sides by d_{max}.

We create a STBox to filter only the relevant trajectories to be analyzed further. The parameter d_{max} is used to ensure that trajectories that are marginal to the STBox will be considered. This filter reduces considerably the number of trajectories being analyzed, which is important because this algorithm is computationally demanding. Its complexity is $O(n^2 m)$, in which n is the number of different trajectories and m the number of observations in a trajectory.

After creating a STBox, line 5 of Algorithm 1 retrieves the trajectories from the set T which intersect the STBox. Trajectories in T have their distance time series calculated (Definition 3) and such time series are analyzed independently by the partners verification algorithm (shown in Algorithm 2). Trajectories that are partners, according to Definition 4, are then saved in the set P.

Algorithm 2 is responsible for the fine-tuning of the partner discovery. This algorithm receives the distance time series DTS calculated between the two

---

Algorithm 2. Partner Verification

Require: $DTS = \{DTS_1, ..., DTS_n\}$, $d_{max} \in \mathbb{Q}$, $p_{max} \in \mathbb{Q}$, $p_{min} \in \mathbb{Q}$
1: **procedure** PERIODASPARTNERS(DTS,d_{max},p_{max},p_{min})
2:      $PP \leftarrow \emptyset$
3:      $begin \leftarrow NULL$
4:      $end \leftarrow NULL$
5:      $timeAway \leftarrow NULL$
6:      **for** all point $DTS_i \in DTS$ **do**
7:          **switch** DTS_i **do**
8:              **case** $DTS_i.dist \leq d_{max}$ **and** $begin = NULL$
9:                  $begin \leftarrow DTS_i.time$
10:             **case** $DTS_i.dist \leq d_{max}$ **and** $timeAway \neq NULL$
11:                 $timeAway \leftarrow NULL$
12:             **case** $DTS_i.dist \leq d_{max}$ **and** $begin \neq NULL$ **and** $DTS_i = LAST$
13:                 **if** $DTS_i.time - begin > p_{min}$ **then**
14:                     $end \leftarrow DTS_i.time$
15:                     $PP \leftarrow PP \cup \{(begin, end)\}$
16:                 **end if**
17:             **case** $DTS_i.dist > d_{max}$ **and** $begin \neq NULL$ **and** $timeAway = NULL$
18:                 $begin \leftarrow DTS_i.time$
19:             **case** $DTS_i.dist > d_{max}$ **and** $begin \neq NULL$ **and** $timeAway > p_{max}$
20:                 **if** $timeAway - begin > p_{min}$ **then**
21:                     $end \leftarrow timeAway$
22:                     $PP \leftarrow PP \cup \{(begin, end)\}$
23:                     $end \leftarrow begin \leftarrow NULL$
24:                 **end if**
25:             **case** $DTS_i.dist > d_{max}$ **and** $begin \neq NULL$ **and** $DTS_i = LAST$
26:                 **if** $timeAway - begin > p_{min}$ **then**
27:                     $end \leftarrow timeAway$
28:                     $PP \leftarrow PP \cup \{(begin, end)\}$
29:                     $end \leftarrow begin \leftarrow NULL$
30:                 **end if**
31:         **end for**
32:     Return PP
33: **end procedure**

---

trajectories T_i and T_j. It checks all tuples of DTS (Definition 3) and verifies if there are one or more periods when the trajectories T_i and T_j can be considered partners, taking into account the user-defined requirements through the input values d_{max}, p_{max} and p_{min}.

Algorithm 2 checks if the distance time series DTS verifies the two rules described in Definition 4. Finally, it returns a list with all periods when the trajectories T_i and T_j are partners.

The proposed algorithms identify pairs of trajectories that are partners, returning the periods when they are nearby. Based on these pairs and periods, we can infer groups of trajectories that are moving together. For instance, if the pair

Table 1. List of symbols and notations used in Algorithms 1 and 2.

T	A set of trajectories
$[T]$	A set of trajectories whose STBoxes have intersections
T_i	The i-th trajectory of a set (T)
d_{max}	Maximum distance that two objects can stay apart
p_{max}	Maximum time period that two objects can stay apart
p_{min}	Minimum time period that two objects must stay together
DTS	A distance time series calculated between two trajectories
DTS_i	The i-th tuple of the distance time series DTS
$DTS_i.dist$	The euclidean distance component of the i-th tuple of the distance time series DTS
$DTS_i.time$	The time component of the i-th tuple of the distance time series DTS
PP	A set of time periods when two trajectories are partners
PP_i	The i-th time period of the set PP
P	A set of partners. Each partner is represented by the identifiers of the two trajectories and the period when they are close together
CreateSTBox	Method that returns the STBox of a given trajectory increased by d_{max}
GetTrajectories	Method that returns the trajectories from a set whose STBoxes intersect a given STBox
CreateDistanceTimeSeries	Method that returns the distance variation between two objects over time as a time series
PeriodAsPartners	Method that returns the periods when two trajectories are partners

of trajectories T_i and T_j are partners from t_i to t_{i+10} and the trajectories T_i and T_k are partners from t_{i+5} to t_{i+15}, then the trio are partners from t_{i+5} to t_{i+10}.

4.1   Algorithm Parallelization

Parallel computing is one of the major techniques of High-Performance Computing (HPC). The main idea of HPC is to solve large problems faster than it would be possible using standard methods (Kumar et al. 2003), and is a very desirable feature of systems that must process high volumes of data.

Among the many parallel computing techniques, Fork/Join parallelism is one of the most well-know, easily implemented and effective design techniques of parallel computing. Fork/Join algorithms are parallel versions of familiar divide-and-conquer algorithms (Lea 2000).

An important feature of the proposed method is the capability of the discovery process being executed in parallel. Given the independent nature of time series analysis and partners verification, Fork/Join algorithm can be used to increase processing speed, since evaluation of one pair of trajectories does not have any influence on the others.

Figure 6 shows how part of the partners discovery algorithm can be executed in parallel. In this example, the distance time series for the trajectories are created and analyzed separately. The analysis results are joined by saving the set of partners.

Another possible way to parallelize this algorithm is to do the task division on the first loop of the algorithm (line 3 of Algorithm 1). In this case, the processes of selecting trajectories, creating their STBoxes (function `CreateSTBox`) and filtering the trajectories whose STBoxes have intersections (function `GetTrajectories`) are executed in parallel.

Even though both parallelization alternatives are possible and can generally improve the time of the algorithm, both have shortcomings. The adequate placement of the parallel loop should be chosen according to the user needs.

When the parallelization is done on the second loop (line 6 of Algorithm 1), it might bring no improvements in areas of low trajectory density. Such lack of improvement occurs because in low density areas the trajectories generally do not have many possible partners. Therefore, in these cases, divide and conquer techniques are irrelevant. On the other hand, if the parallelization is done on the first loop (line 3 of Algorithm 1), it might consume too much memory, specially in high density areas where there are many possible partners.

It is important to mention that the parallel version of the proposed partner method may be useful only for large amounts of data. For smaller datasets the time to divide and parallelize the tasks can make the total processing time larger and not smaller.

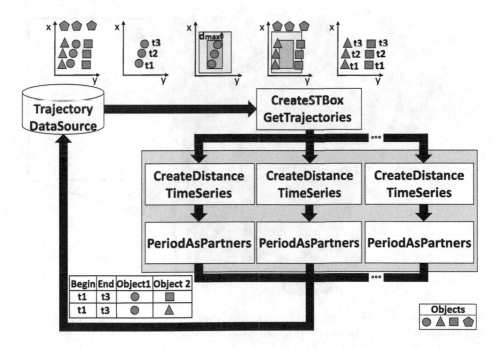

Fig. 6. Steps in `Partner` discovery

5   Case Studies

To test and validate the proposed method, we developed two case studies with different datasets. In the first case study, we used trajectories of over 1000 vessels around the Brazilian coast collected during 6 months in 2008. These trajectories are stored in a PostGIS database in over 3 million rows in a single table. On this data set, we executed the algorithm using the following parameters: $d_{max} = 1.5\,\mathrm{Km}$, $p_{max} = 1\,\mathrm{h}$, and $p_{min} = 5\,\mathrm{h}$. As a result we obtained over 50 thousand partners. Possibly, these partners are vessels that were fishing together.

Figure 7 shows, on the left side, all vessel trajectories observations and, on the right side, two of these trajectories that are partners identified by the proposed method. On the right side, the region bound by the rectangle is shown in detail, and on it we can see two objects (identified by red and green dots) that are partners. The partners shown in that figure have spent around 11 h together.

The second dataset consists of 276 trajectories of 50 delivery trucks in Athens, Greece. This dataset was obtained from `chorochronos.org`. Its structure is rather similar to the one of the previous dataset, but since each truck has a different number of trajectories we added an extra column to uniquely identify the pair object-trajectory.

On a first try, we examined the dataset for partners using the same d_{max} value used on the first case study. However, this value is too large for this dataset. Using this value, all trajectories were identified as partners, even the ones going

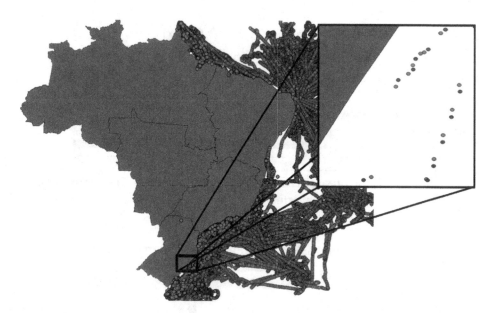

Fig. 7. Vessels around the Brazilian coast. Blue dots/lines: original trajectories. Red and Green dots: partners identified by our proposed algorithm. (Color figure online)

on different directions and different roads. Thus, we chose values that seemed more reasonable for data collected within cities.

The number of discovered partners depends on the given parameters. We looked for partners that were separated at most by 10, 50 and 100 m. It is important to reiterate that the analyzed area increases according to d_{max} (Algorithm 1 line 4), which affects the running time of the algorithm.

For $d_{max} = 10$ m, the algorithm identified only 31 partners. Since this value of d_{max} is a lot more restrictive, trajectories like the one from Fig. 8 are not completely identified as partners. For $d_{max} = 50$ m, we found 863 partners (one is shown in Fig. 8). Intuitively the number of partners declined as the acceptable separation distance was shortened, and increased as the acceptable separation distance was extended.

For $d_{max} = 100$ m, the algorithm identified 1327 partners. We consider this distance as the more meaningful for the city truck data. Nevertheless, since this

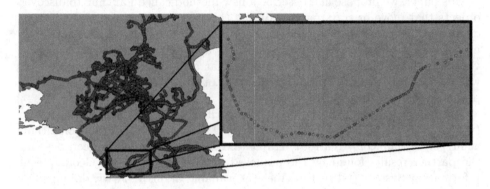

Fig. 8. Truck trajectories in Greece. Blue dots/lines: original trajectories. Red and Green dots: partners identified by our proposed algorithm. (Color figure online)

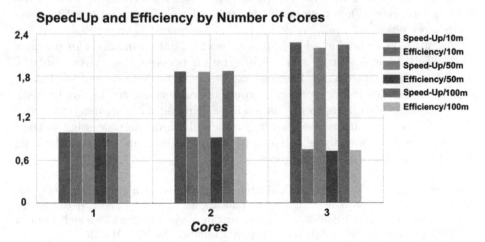

Fig. 9. Speed-Up comparison with changing parameters and number of cores

value caused an increase on the areas to be analyzed, the algorithm took longer to execute.

This dataset presents an ideal size to test the parallelization capabilities of the algorithm. It is big enough to gain processing time from parallelization, but not too large to be too much time-consuming. To measure performance gains, we executed the algorithm twice for each distance, using one, two, and three CPU cores.

Even though each increase in d_{max} made the algorithm more time consuming since a larger area was analyzed, we verified that the speed-up was practically constant for all distances, being 1.87 faster for 2 cores and 2.24 faster for 3 cores, when comparing running time using only one core. Shown on Fig. 9.

6   Conclusions and Future Work

In this paper, we propose and describe a new method called **Partner** to discover objects that move or stay together during certain periods, based on trajectory distance time series analysis. This pattern is useful for many applications that need to identify objects that are performing activities jointly, such as marine vessels that are fishing together or trucks that are working in cooperation, as well as objects that are stationary close to others. We test and validate the proposed method through two case studies.

We implemented the proposed algorithms in a new R package (www.r-project.org) called **TrajDataMining**. This package is available on github at https://github.com/dvm1607/TrajDataMining. To visualize trajectories and the partner results found by the proposed method, we used the Geographical Information System (GIS) TerraView 5 available at http://www.dpi.inpe.br/terralib5.

After experimenting the proposed method with different values for the input parameters, we could observe that the adequate values for them are vital for the discovery process. When they are inadequate, false positives and negatives may occur more often.

As future work, we intend to improve the visualization methods for partners in the TerraView 5 system and the integration between this system and the R package **TrajDataMining**. Besides that, we want to describe and implement algorithms for identification of bigger groups of partners and to improve the data retrieval of the algorithm, so less data will be required for in-memory analysis.

The algorithm can be executed in parallel, allowing the processing of large datasets, but further analyses still need to be done to evaluate when parallelizing this algorithm becomes advantageous.

Acknowledgements. Diego Vilela Monteiro is supported by a grant from the Coordination for the Improvement of Higher Education Personnel (CAPES) during his Master's studies. Rafael Santos would like to thank the Brazilian Research Council (CNPq) for support during this research (grant number 206785/2014-3).

References

Erwig, M., Gu, R.H., Schneider, M., Vazirgiannis, M., et al.: Spatio-temporal data types: an approach to modeling and querying moving objects in databases. GeoInformatica **3**(3), 269–296 (1999)

Fan, Q., Zhang, D., Wu, H., Tan, K.-L.: A general and parallel platform for mining co-movement patterns over large-scale trajectories. Proc. VLDB Endow. **10**(4), 313–324 (2016)

Gudmundsson, J., van Kreveld, M.: Computing longest duration flocks in trajectory data. In: Proceedings of the 14th Annual ACM International Symposium on Advances in Geographic Information Systems, pp. 35–42. ACM (2006)

Jeung, H., Yiu, M.L., Zhou, X., Jensen, C.S., Shen, H.T.: Discovery of convoys in trajectory databases. Proc. VLDB Endow. **1**(1), 1068–1080 (2008)

Kumar, V., Grama, A., Gupta, A., Karypis, G.: Introduction to Parallel Computing, 2nd edn. Addison Wesley, Boston (2003)

Laube, P., van Kreveld, M., Imfeld, S.: Finding REMO-detecting relative motion patterns in geospatial lifelines. In: Fisher, P.F. (ed.) Developments in Spatial Data Handling, pp. 201–215. Springer, Heidelberg (2005)

Lea, D.: A java fork/join framework. In: Proceedings of the ACM 2000 Conference on Java Grande [S.l.], pp. 36–43. ACM (2000)

Li, Y., Bailey, J., Kulik, L.: Efficient mining of platoon patterns in trajectory databases. Data Knowl. Eng. **100**, 167–187 (2015)

Li, Z., Ding, B., Han, J., Kays, R.: Swarm: mining relaxed temporal moving object clusters. Proc. VLDB Endow. **3**(1–2), 723–734 (2010)

Palma, A.T., Bogorny, V., Kuijpers, B., Alvares, L.O.: A clustering-based approach for discovering interesting places in trajectories. In: ACMSAC, pp. 863–868 (2008)

Spaccapietra, S., Parent, C., Damiani, M.L., de Macedo, J.A., Porto, F., Vangenot, C.: A conceptual view on trajectories. Data Knowl. Eng. **65**(1), 126–146 (2008)

Tanaka, P.S., Vieira, M.R., Kaster, D.S.: Efficient algorithms to discover flock patterns in trajectories (2015)

Tang, L.-A., Zheng, Y., Yuan, J., Han, J., Leung, A., Hung, C.-C., Peng, W.-C.: On discovery of traveling companions from streaming trajectories. In: 2012 IEEE 28th International Conference on Data Engineering (ICDE), pp. 186–197. IEEE (2012)

Vieira, M.R., Bakalov, P., Tsotras, V.J.: On-line discovery of flock patterns in spatio-temporal data. In: Proceedings of the 17th ACM SIGSPATIAL International Conference on Advances in Geographic Information Systems, pp. 286–295. ACM (2009)

Wang, Y., Lim, E.-P., Hwang, S.-Y.: Efficient mining of group patterns from user movement data. Data & Knowledge Engineering **57**(3), 240–282 (2006)

Zheng, Y.: Location-based social networks: users. In: Zheng, Y., Zhou, X. (eds.) Computing with spatial trajectories, pp. 243–276. Springer, Heidelberg (2011)

Zheng, K., Zheng, Y., Yuan, N.J., Shang, S.: On discovery of gathering patterns from trajectories. In: 2013 IEEE 29th International Conference on Data Engineering (ICDE), pp. 242–253. IEEE (2013)

Zheng, K., Zheng, Y., Yuan, N.J., Shang, S., Zhou, X.: Online discovery of gathering patterns over trajectories. IEEE Trans. Knowl. Data Eng. **26**(8), 1974–1988 (2014a)

Zheng, Y., Capra, L., Wolfson, O., Yang, H.: Urban computing: concepts, methodologies, and applications. ACM Trans. Intell. Syst. Technol. (TIST) **5**(3), 38 (2014b)

Zheng, Y.: Trajectory data mining: an overview. ACM Trans. Intell. Syst. Technol. (TIST) **6**(3), 29 (2015)

Clustering Methods to Asses Land Cover Samples of MODIS Vegetation Indexes Time Series

Lorena A. Santos[✉], Rolf E.O. Simoes, Karine R. Ferreira,
Gilberto R. de Queiroz, Gilberto Camara, and Rafael D.C. Santos

National Institute for Space Research - INPE, Av. dos Astronautas, 1758,
São José dos Campos, SP 12227-010, Brazil
{lorena.santos, rolf.simoes, karine.ferreira, gilberto.queiroz,
gilberto.camara, rafael.santos}@inpe.br

Abstract. MODIS vegetation indexes time series have been widely used to build land cover change maps on large scales. In this scope, to obtain good quality maps using supervised classification methods, it is crucial to select representative training samples of land cover change classes. In this paper, we evaluate two clustering methods, Hierarchical and Self-Organizing Map (SOM), to assess land cover samples of MODIS vegetation indexes time series. As we show, these techniques are suitable tools for assisting users to select representative land cover change samples from MODIS vegetation indexes time series. We present the accuracy of both methods for a case study in Ipiranga do Norte municipality in Mato Grosso state, Brazil.

Keywords: Time series clustering · MODIS vegetation indexes · Land cover change classification · Self-Organizing Map (SOM)

1 Introduction

Remote sensing images play a crucial role in land cover change classification on global and continental scales. Recently, time series of vegetation indexes, such as NDVI (Normalized Difference Vegetation Index) and EVI (Enhanced Vegetation Index), from MODIS (Moderate Resolution Imaging Spectroradiometer) products have been widely used to build land cover change maps on large scales [1–4].

Aguiar et al. [1] identify pasture land and its different levels of degradation in Mato Grosso do Sul state, Brazil, using MODIS NDVI time series and a J48 classifier with wavelet technique. Arvor et al. [2] use MODIS EVI time series to quantify the evolution of agricultural area from 2000 to 2006 in Mato Grosso state, Brazil. Bagan et al. [3] propose an approach to classify land cover from MODIS EVI time series using the Self-Organizing Map (SOM) neural network technique and present a case study in eastern China from March to December in 2002. Maus et al. [4] propose an algorithm called Time-Weighted Dynamic Time

© Springer International Publishing AG 2017
O. Gervasi et al. (Eds.): ICCSA 2017, Part VI, LNCS 10409, pp. 662–673, 2017.
DOI: 10.1007/978-3-319-62407-5_48

Warping (TW-DTW), based on the classical Dynamic Time Warping (DTW) method, for land cover and land use classification and present a case study using MODIS EVI time series in the Porto dos Gaúchos municipality in Mato Grosso state, Brazil.

In general, supervised classification methods require a training step, which consists in gathering training samples to represent the classes to be identified. The quality of such samples is crucial in the classification process. Representative samples of classes lead to good classification results. Therefore, specially in land cover change classification, there is a need for techniques that help users to select representative land cover change samples from vegetation indexes time series of remote sensing images.

Remote sensing time series usually contains spatio-temporal phenomena that implies in monitoring of land use and land cover, i.e. atmospheric effects or cloud cover. NDVI index has several limitations that affect the accuracy of classification, including atmospheric conditions. To account for these limitations EVI was proposed to remove residual atmosphere contamination [7]. However, Heute et al. [6] studied vegetation index product, they found that EVI and NDVI of the biotic formation in different regions were sensitive to seasonal change, land cover change and biophysical parameters change. They demonstrated that NDVI was highly relevant to EVI and the value of NDVI was always bigger than EVI when the soil background and the atmospheric aerosol vary less [6].

In this context, time series clustering techniques can assist in an exploratory analysis to evaluate which vegetation indexes will be used to select representative land cover change samples. These samples allow extract temporal patterns that concerns the seasonal periodicity of vegetation, can be improve the training step of the land cover change classification.

It is important to develop an appropriate and validation scheme to assess the performance and limitation of clustering algorithms. In this paper, we present a ground truth based comparative study accuracy and performance of two clustering methods, Hierarchical and Self-Organizing Map (SOM), to assess the separability of land cover samples of MODIS vegetation indexes time series [5,10,11]. In this work, we present the accuracy of these two clustering methods for a case study in Ipiranga do Norte municipality in Mato Grosso state, Brazil.

2   Background

In this section, we present concepts and algorithms used in our study: the Self-Organizing Map (SOM) Neural Network, the Dynamic Time Warping Distance (DTW) and the Hierarchical Clustering Algorithm.

2.1   SOM Neural Network

A SOM (Self-Organizing Map) is an unsupervised neural network that consists in competitive learning for providing a topology-preserving mapping from a high-dimensional input onto a low-dimensional output. The structure of a SOM is composed by input and output layers. The training data or input data are in the input layer whereas the output layer is formed by a set of neurons that are trained to extract patterns from the input data [5].

An important property of SOM is the neighborhood relationship among neurons in the output layer, i.e., vectors in the input layer with similar characteristics can be mapped into either a neuron that represents those characteristics or neighboring neurons in the output layer [3].

Each neuron j in the set of J neurons has a n-dimensional weight vector $w_j = [w_{j1}, \ldots, w_{jn}]$ associated to it. At each training step t, an input vector $x(t) = [x(t)_1, \ldots, x(t)_n]$ is randomly chosen, and then the Euclidean distance D_j is calculated between this input vector and each neuron j for all the neurons in the output layer (Eq. 1).

$$D_j = \sum_{i=1}^{N} \sqrt{(x(t)_i - w_{ji})^2}. \tag{1}$$

The next step is to determine the Best-Matching-Unit (BMU), i.e. the neuron d_b with weight vector closer to $x(t)$ (Eq. 2):

$$d_b = min\{D_1, \ldots, D_J\}. \tag{2}$$

The weight vector of the neuron chosen from the BMU is updated, i.e. adjusted to be closer to the input vector (Eq. 3). The weights of the neurons $N_b(t)$, neighbors of the BMU, are also updated with a smaller weight.

$$w_{ji}(t+1) = w_{ji}(t) + \alpha(t)[x(t)_i - w_{ji}(t)], \tag{3}$$

In Eq. 3, $\alpha(t)$ is the learning rate, set as $0 < \alpha(t) < 1$. An iteration ends when all vectors of the input layer are trained, then $\alpha(t)$ must be reduced [8]. The number of iterations must be high in order to allow the neurons to fit accurately to the data sets [9].

2.2   Dynamic Time Warping Distance

Dynamic Time Warping (DTW) is a classical algorithm that produces the most robust distance used to align two time series, allowing the alignment of similar sequences that match even if they are out of phase in the time axis [13].

Consider two time series $Q = [q_1, \ldots, q_i, \ldots, q_n]$ and $C = [c_1, \ldots, c_i, \ldots, c_m]$. The first step of DTW is to compute a cost matrix, $n \times m$, given by the squared distance between the elements of the two time series:

$$\Psi_{i,j} = (q_i - c_j)^2 \tag{4}$$

From Ψ, we can find the best matching between two time series, getting an optimal path that minimizes the cost warping. This warping path can be found using dynamic programming, transforming a complex global problem into a number of local optimization subproblems [11].

$$d_{i,j} = \Psi_{i,j} + \min \begin{cases} d_{i-1,j} \\ d_{i-1,j-1} \\ d_{i,j-1} \end{cases} \tag{5}$$

2.3 Hierarchical Clustering

Hierarchical Clustering is another well-known method used to cluster data points. There are two types of hierarchical clustering: agglomerative and divisive. In this paper we use the agglomerative type, where each sample starts in its own cluster, and the clusters are then grouped with large clusters based on linkage criteria, until all samples are contained in a single cluster.

The linkage criterion determines the distance between sets of data as a function of the pairwise distances between the data [12]. There are several linkage criteria, some of them are presented next.

Ward's criterion merges two clusters that result in the smallest increase in the value of the sum-of-squares variance. At each clustering step, all possible mergers of two clusters are tried. The sum-of-squares variance is computed for each cluster, and the one with the smallest value is selected [11]. Other popular linkage criteria are *average*, *single* and *complete*. All these are used to determine which pair of clusters are going to be merged in the next step of the hierarchical algorithm: the *average* criterion calculates, for each pair of clusters, the average distance between all data points in each cluster; the **single** criterion calculates the distance between two clusters A and B as $Dist(A, B) = \min_{a,b} d(a, b)$, and the **complete** criterion calculates the distance as $Dist(A, B) = \max_{a,b} d(a, b)$ [12]. For each criterion, the smallest distance between the two clusters is selected and the clusters are merged, and the process repeated until there is only one cluster with all data points.

A dendrogram can be used to visualize the hierarchy obtained from the hierarchical clustering method. The dendrogram helps visualization of the merging of the clusters, and can be used to evaluate the height in where the largest change in dissimilarity occurs, so it can be cut at such height for the clusters extraction. It is also possible to specify the number of clusters and then cut the dendrogram in such a way that the chosen number is obtained [10, 11].

3   Materials and Methods

3.1   Data

The data used in this study was extracted from MODIS sensor of the Terra satellite developed by NASA. This sensor monitors the state of Earth's environment. The MOD13Q1 product from MODIS provides per pixel values of vegetation indexes. These indexes are used for global monitoring of vegetation conditions and land cover classifying on large spatial scales [14]. In this product, there are two vegetation layers, the Normalized Difference Vegetation Index (NDVI) and Enhanced Vegetation Index (EVI) [15]. Time series MODIS data, used in this paper, produce vegetation indexes at each 16-day with 250-meter spatial resolution. During plant growth periods, different vegetation styles can be distinguished by vegetation indexes time series [17].

The study area comprises a region of $9.6\,km \times 8\,km$ and it is located in Ipiranga do Norte (*Mato Grosso*, Brazil) municipality as shown in Fig. 1.

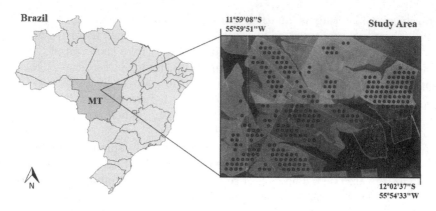

Fig. 1. The area of study corresponds to approximately $76.8\,Km^2$ in the Ipiranga do Norte municipal

The study region was chosen considering the existence of a dataset containing 603 ground truth sample points, from 2007 to 2013. This dataset is organized in five classes: 138 samples for "forest", 68 for "cotton-fallow", 79 for "soybean-cotton", 134 for "soybean-maize" and 184 for "soybean-millet". Each data sample has the spatial location (latitude and longitude), the start and end dates, with an one year interval, and the label representing the class.

From the ground truth, the MODIS vegetation indexes time series of each sample was extracted, for each spatial location and date. In total, we have collected 603 one-year-spanned time series from different years. Figure 2, shows the EVI and NDVI time series between 2011 and 2012. This time series correspond to the point $lat = -12.0385$, $lon = -55.9844$ for the "cotton-fallow" class.

3.2   Analysis

In order to evaluate the accuracy of the clustering result we use the Shannon information entropy [16] over all clusters and its capability of representing one class. If there is confusion between two or more known classes belonging to the same cluster then the entropy metric will increase up to a maximum. The entropy e can be obtained for each cluster taking into account all its classes frequencies p_i in each cluster, using Eq. 6.

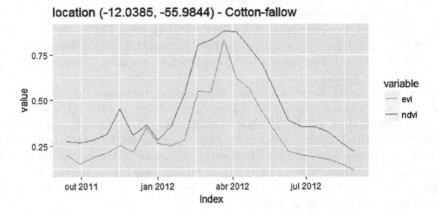

Fig. 2. EVI and NDVI time series from a random chosen sample point.

$$e = -\sum_i p_i \log p_i \qquad (6)$$

This gives us a measure of separability for each cluster and informs how much a set of samples are distinguishable. In order to get a range from 0 to 1 we just divide the Eq. (6) by the logarithm of the number of classes obtaining a relative entropy metric. To compute an overall metric, we just have taken the weighted average of all clusters with respect to the amount of samples linked to it.

Differently from internal cluster validity indexes (e.g. Silhouette) that aims to measure compactness and separation between clusters, entropy aims to measure the clusters from a ground truth and can be viewed as an external cluster validity index [18]. This choice may overcome the disadvantage of internal cluster validity indexes in a high dimensionality context, such as time series data, due to the curse of dimensionality [19].

The clusters were produced by two methods. In the first method, the hierarchical clustering (**HC**) over a dissimilarity matrix among all sampled time series using DTW distance. Hence, a dendrogram can be computed according to a linkage method and once produced, the dendrogram can provide any number of clusters, from one to the number of samples, just informing a dissimilarity parameter: those samples with dissimilarity bellow that parameter will be tied together in a same cluster and all those samples above that value will pertaining others clusters.

In the second method, which uses the SOM as a preprocessing step of the HC (**SOM+HC**), we conducted a SOM neural network before the hierarchical clustering. In SOM, a neuron, through a weight vector represents a pattern of samples. These patterns are organized into a meaningful two-dimensional order in which similar models are closer to each other in the grid than the more dissimilar ones and the neighboring models are mutually similar, in this way, a neuron contains all the samples which are mapped to it. This step allows the samples to be represented by these patterns. As a result, we have obtained a set of representative neurons that were used as the input of the hierarchical clustering step. From them, we compute a dissimilarity matrix using DTW distances, and what follows is similar with the first method. Hence, the main difference here is that the hierarchical clustering is being applied to a reduced but representative number of time series.

Our experiments were conducted using some variations of the methods described in previous sections, with different parameters. For the **HC** method, we chose the parameters *number of clusters* (n_clusters), *vegetation index combination* (bands) and *linkage method* (linkage). For the **SOM+HC** method, the parameters chosen were *number of clusters, vegetation index combination, linkage method, SOM grid size* (grid_size), *SOM learning rates* (learnr_init and learnr_fin), and *SOM iteration steps* (iterations). The parameter's ranges were defined as described in Table 1.

Table 1. Experiments parameters' range

Parameter	Range	Methods
n_clusters	{3, 5, 7, 9}	HC and HC+SOM
bands	{EVI, NDVI, EVI& NDVI}	HC and HC+SOM
linkage	{Ward, average, complete, single}	HC and HC+SOM
grid_size	{49, 81, 121, 169}	HC+SOM
learnr_init	{0.1, 0.2, 0.3, 0.4}	HC+SOM
learnr_fin	{0.02, 0.04, 0.06, 0.08}	HC+SOM
iterations	{1000, 1200, 1400, 1600}	HC+SOM

All possible parameters values were combined producing 48 different experiments for the **HC** method and 12,288 experiments for the **SOM+HC**. As output of both methods, we have got the same set of samples labeled with the corresponding computed cluster identification. This information allowed us to compute the separability level given by Eq. 6. The methods are summarized in the Figs. 3 and 4.

The experiments produced a database relating each parameter's values to the entropy dependent variable. The subsequent analysis were made on this dataset using descriptive statistics and correlation analysis in order to show the parameters-entropy behavior.

4   Results

In order to verify how the numeric variables correlate to the entropy we calculated the Pearson correlation. Both **HC** and **SOM+HC** methods showed a similar result. The correlation between the parameter n_clusters and entropy in the case of **HC** method was −0.55. For **SOM+HC** method, the n_clusters

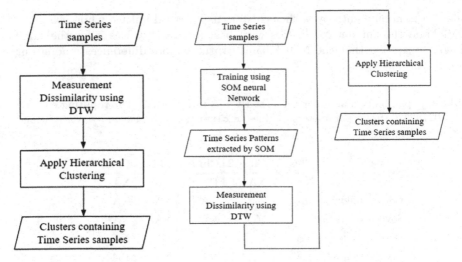

Fig. 3. Clustering time series using hierarchical clustering

Fig. 4. Clustering time series using SOM neural network and hierarchical clustering

and `grid_size` correlations against `entropy` are −0.58 and 0.15, respectively (all measures have a p-value less than 1.0×10^{-7}). The `n_clusters` correlation was expected as a fine grained clustering can capture more subtleties that the data may present by reducing the confusion and consequently the entropy for each cluster. However, the positive correlation between SOM grid size may suggest that this is not the case for SOM stage, at least in the range values used in the experiments.

The best achievement of **HC** experiments is shown in the Table 2, which parameters' values were `n_clusters` = 5, `bands` = $NDVI$, and `linkage` = $Ward$. The resulting entropy was 0.02782576 indicating a reasonably separability between classes using only the NDVI band. When considering EVI and NDVI bands, the lowest entropy (0.02872378) was achieved only by increasing the clusters to 9 with the same linkage criterion. The first non **Ward** linkage criterion with the lowest entropy comes at 7th position with 0.03011122.

All the **HC** experiments' results can be seen in Fig. 5. As the number of clusters increases, the overall entropy stabilizes, suggesting an optimal `n_cluster` value. The graphs shows that **single** linkage criterion was outperformed in terms of separability and so are inadequate to our data samples. Maybe this is the case for all land use and land cover spectral data as different classes exhibit considerably variance and are, sometimes, very similar between them. The same is also observed in **SOM+HC** experiments where **single** linkage resulted in higher entropies. The Fig. 6 depicts its entropy results for those experiments with `grid_size` = 49, `iterations` = 1000, `learnr_init` = 0.1, and `learnr_fin` = 0.02. Figures 5 and 6 show that the number of clusters depends on the data bands and the linkage criterion. For example, using only the EVI band we have reached a minimum entropy with 7 clusters using Ward linkage while using the NDVI band the amount of clusters with best separability was 5 for the same linkage, suggesting that the NDVI band captures more differences among our sample classes.

Table 2. Separability matrix for the best HC clustering result. The resulting entropy was 0.02782576 from parameters' values **n_clusters** = 5, **bands** = $NDVI$, and **linkage** = $Ward$.

Classes	Clust.1	Clust.2	Clust.3	Clus.4	Clust.5
Forest	NA	NA	138	NA	NA
Cotton-fallow	66	2	NA	NA	NA
Soybean-cotton	3	76	NA	NA	NA
Soybean-maize	NA	1	NA	133	NA
Soybean-millet	NA	NA	NA	NA	184

The lowest entropy was achieved by **average** linkage criterion that outperformed **Ward** entropies only when the number of clusters were 9. The parameters used were: **n_clusters** = 9, **grid_size** = 49, **iterations** = 1000, **learnr_init** = 0.2, **learnr_fin** = 0.04, and **bands** = $EVI\&NDVI$. The respective separability matrix is shown in Table 3.

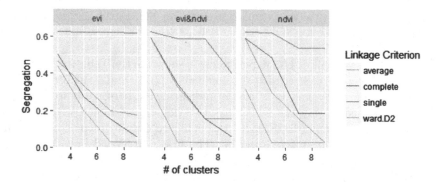

Fig. 5. HC experiments entropy results.

We can see from the the best separability results (Tables 2 and 3) that some classes are mixed up inside a same cluster. This is the case of "cotton-fallow" and "soybean-cotton" for both methods. Specific investigations may provide some understanding why this may be the case for those classes or if we may consider discard the confusions cases as outliers or probably miss classified data sample.

Despite the fact that we can obtain the lowest relative entropy by setting **n_clusters** to the sample size, our results show that the relative entropy drops

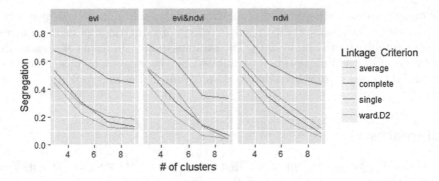

Fig. 6. SOM+HC experiments behavior for fixed parameters grid_size= 49, iterations = 1000, learnr_init = 0.2, learnr_fin = 0.04

Table 3. Separability matrix of the best SOM+HC clustering separability result. The resulting entropy was 0.02872378 from parameters' values n_clusters = 9, grid_size = 49, iterations = 1000, learnr_init = 0.1, learnr_fin = 0.04, and bands = *EVI&NDVI*

Classes	Clust.1	Clust.2	Clust.3	Clust.4	Clust.5	Clust.6	Clust.7	Clust.8	Clust.9
Forest	NA	NA	25	113	NA	NA	NA	NA	NA
Cotton-fallow	NA	NA	NA	NA	NA	NA	NA	NA	68
Soybean-cotton	NA	NA	NA	NA	2	NA	NA	74	3
Soybean-maize	NA	NA	NA	NA	NA	134	NA	NA	NA
Soybean-millet	52	4	NA	NA	94	NA	34	NA	NA

rapidly to an inflection point for a low value of n_clusters (1.5% of the sample size) after which it decreases very slowly. The existence of such inflexion point for a low value relative to the sample can give us a measure of data separability. However, an unexplored question is if such separability is stable.

5 Conclusions

Hierarchical clustering methods has disadvantage that the dissimilarity matrix is calculated for all given dataset. Then, for big dataset the process becomes expensive computationally, due to a complexity of $O(n^2)$, where n is the amount of data. Thus, hierarchical clustering may be prohibitive for large datasets.

In remote sensing context, using hierarchical clustering may become a problem due the large amount of data. A way to soften the complexity problem is to use iterative techniques such as SOM to reduce the amount of input data by creating patterns to represent those data. In this way, the dissimilarity matrix can be computed with much less data. Here we showed that the results presented by SOM method produced a good separability compared with hierarchical clustering approach.

Our experiments showed that vegetation indexes and linkage criterion inter-
fere directly in the separability result. An entropy metric was used to assess
class separability on the input data. The samples quality in terms of separability
can be given by the inflexion point exhibited by the entropy against number of
clusters. However, further investigations on stability of cluster validity index are
needed.

References

1. Aguiar, D., Adami, M., Silva, W., Rudorff, B., Mello, M., Silva, J.: MODIS time
   series to assess pasture land. In: Geoscience and Remote Sensing Symposium
   (IGARSS) (2010)
2. Arvor, D., Meirelles, M., Dubreuil, V., Begue, A., Shimabukuro, Y.: Analyzing the
   agricultural transition in Mato Grosso, Brazil, using satellite-derived indices. Appl.
   Geogr. **22**, 702–713 (2012)
3. Bagan, H., Wang, Q., Watanabe, M., Yang, Y., Ma, J.: Land cover classification
   from MODIS EVI time-series data using SOM neural network. Int. J. Remote Sens.
   26, 4999–5012 (2005)
4. Maus, V., Camara, G., Cartaxo, R., Sanchez, A., Ramos, F., Ribeiro, G.: A time-
   weighted dynamic time warping method for land cover mapping. IEEE J. Sel. Top.
   Appl. Earth Obs. Remote Sens **20**, 3729–3739 (2015)
5. Augustijn, E., Zurita-Milla, R.: Self-organizing maps as an approach to exploring
   spatiotemporal diffusion patterns. Int. J. Health Geogr. **12**, 60–74 (2013)
6. Huete, A., Didan, K., Miura, T.: Overview of the radiometric and biophysical
   performance of the MODIS vegetation indices. Remote Sens. Environ. **83**(1), 195–
   213 (2002)
7. Xiao, X., Braswell, B., Zhang, Q., Boles, S., Frolking, S., Moore, B.: Sensitivity of vegetation indices to atmospheric aerosols: continental-scale observations in
   Northern Asia. Remote Sens. Environ **84**, 385–392 (2003)
8. Natita, W., Wiboonsak, W., Dusadee, S.: Appropriate learning rate and neighborhood function of self-organizing map (SOM) for specific classification over Southern
   Thailand. Int. J. Model. Optim **6**, 61–65 (2016)
9. Silva, A.L., Peres, M.S., Boscarioli, C.: Introduo a Minerao de Dados: com Aplicaes
   em R (2016)
10. Sarda-Espinosa, A.: Comparing time-series clustering algorithms in R using the
    dtwclust package (2017). https://CRAN.R-project.org/package=dtwclust
11. Liao, T.: Clustering of time series data-a survey. J. Pattern Recogn. Soc. **8**, 1857–
    1874 (2004)
12. Tatarinova, T., Schumitzky, A.: Nonlinear Mixture Models: A Bayesian Approach.
    Imperial College Press (2014)
13. Keogh, E., Ratanamahatana, C.: Exact indexing of dynamic time warping. Knowl.
    Inform. Syst. **7**, 358–386 (2005)
14. Huete, A., Liu, H.Q., Batchily, K., Leeuwen, V.: A comparison of vegetation indices
    over a global set of TM images for EOS-MODIS. Remote Sens. Environ. **59**, 440–
    451 (1997)
15. Vegetation Indices 16-Day L3 Global 250m. http://pdaac.usgs.gov/dataset_
    discovery/modis/modis_products_table/mod13q

16. Bock, H.H.: Information and entropy in cluster analysis. In: Bozdogan, H., Sclove, S.L., Gupta, A.K., Haughton, D., Kitagawa, G., Ozaki, T., Tanabe, K. (eds.) Proceedings of the First US/Japan Conference on the Frontiers of Statistical Modeling: An Informational Approach. Springer, Netherlands (1994). doi:10.1007/978-94-011-0800-3_4

17. Boles, S.H., Xiao, X., Liu, J., Zhang, Q., Sharav, M., Chen, S., Ojima, D.: Land cover characterization of temperate East Asia using multi-temporal VEGETA-TION sensor data. Remote Sens. Environ. **90**, 477–489 (2004)

18. Arbelaitz, O., Gurrutxaga, I., Muguerza, J., Perez, J.M., Perona, I.: An extensive comparative study of cluster validity indices. Pattern Recogn. **46**(1), 243–256 (2013)

19. Tomašev, N., Radovanović, M.: Clustering evaluation in high-dimensional data. In: Celebi, M.E., Aydin, K. (eds.) Unsupervised Learning Algorithms. Springer International Publishing, Cham (2016). doi:10.1007/978-3-319-24211-8_4

Gisplay- Extensible Web API for Thematic Maps with WebGL

Diogo Cardoso, Rui Alves$^{(\boxtimes)}$, João Moura Pires, Fernando Birra,
and Ricardo Silva

NOVA Laboratory for Computer Science and Informatics,
Departamento de Informática, Faculdade de Ciências e Tecnologia,
Universidade Nova de Lisboa, 2829.516 Caparica, Portugal
{dam.cardoso,rr.alves}@campus.fct.unl.pt, {jmp,fpb}@fct.unl.pt,
ricardofcsasilva@gmail.com

Abstract. This paper analyses and shows the need of a client side web API devoted to present and explore spatial information through thematic maps. We define a set of requirements for such API, most notably the ability to process datasets with many millions of points, allowing full interactivity, providing a high level of abstraction and defining clear paths for easy extension at many levels. The Gisplay API is implemented using WebGL, enabling the required speed for full interactive thematic maps with millions of points. Such claims are experimentally demonstrated. Gisplay already provides 4 types of thematic maps and very detailed discussion is presented showing the high level of abstraction and the different mechanisms to extend it. This extensibility is based on a modular architecture which includes an intermediate API that deals with WebGL complexity.

Keywords: Thematic maps · WebGL · Spatial data visualisation

1 Introduction

The ever growing amount of data that is continually being produced everywhere is a trend of our times. People carry along a myriad of sensors embedded in their smart-phones, smart-watches and other electronic gadgets that more or less continually monitor their activity and/or the environment. Data is also being collected through a large number of vast sensor networks, such as satellite data acquisition, video surveillance or automated control systems and less automated processes such as polls, inquiries and census.

Around 80% of the data produced also contains a spatial component [1], tying the information to a specific geographical location. In this context, maps appear as the natural visualization, analysis and decision support tool [4] since our brain is capable of understanding images better than numbers or text.

Cartography distinguishes between reference maps and thematic maps [5]. Reference maps visually represent geographic features without putting special

© Springer International Publishing AG 2017
O. Gervasi et al. (Eds.): ICCSA 2017, Part VI, LNCS 10409, pp. 674–689, 2017.
DOI: 10.1007/978-3-319-62407-5_49

emphasis on any type of feature over the others, while thematic maps display characteristics or attributes of features that are varying spatially. These features are related to the specific theme addressed by the map. Common examples are ground occupation, population density and economic growth rates.

Due to the ability of visually connecting the displayed data to a geographic location, thematic maps are extremely useful as an analysis tool to understand or gain insight into what geographical features may influence the displayed data.

One of the top applications on the internet is Google Maps that started by focusing on navigation. It allows users to visualize and even build reference maps. Several other competitors emerged such as Here Maps or OpenStreet Maps. These services expose APIs that enable their integration into all kinds of applications, both Web and Mobile. Their respective Web APIs can also be used to display thematic maps on top of the offered reference maps.

The actual limitations of current APIs targeting the creation of thematic maps for the Web are twofold. One hand they fail to provide an adequate level of abstraction to facilitate the task of the programmer while, on the other hand, they also suffer from severe performance limitation that prevent their use for large datasets. Well known APIs that exhibit these limitations for thematic maps are the ArcGis Javascript API[1], MapMap.js [6] and Plotly[2].

2   Related Work

The web is a client-server environment that allows different approaches to build thematic maps. Server-centric solutions have the server treating all the data and generating thematic map visualizations as images that get sent to the client who, in turn, only need to show them on screen. Client-centric solutions involve the client in the computation efforts and not only on the display tasks. Standalone solutions represent the extreme client-centric case where every task is performed on client side and the server is probably as thin as a mere data repository. Naturally, we are leaving servers for the background maps out of this discussion as they provide essentially static images.

Nowadays, there are multiple APIs available to create general purpose or thematic maps in a web context. The first one and the most well known is Google Maps API[3], which is a general purpose map API, that can be used to create a thematic map using it's capabilities to draw polygons, lines and points. Other general purpose APIs include: Leaflet[4] and OpenLayers[5].

APIs that were built with thematic maps in mind include: ESRI ArcGIS Javascript API and CartoDB[6] which were based on their desktop application

---

[1] http://www.arcgis.com/features/.
[2] http://plot.ly/.
[3] https://developers.google.com/maps/.
[4] http://leafletjs.com/.
[5] http://openlayers.org/.
[6] https://cartodb.com/solutions/web-mobile/.

counterparts. CartoDB is the only studied API which uses a server centric architecture, which has some drawbacks in terms of responsiveness because every time the user performs any operation that causes changes on the visualization, there's the need to obtain those changes from the server. Last but not least there is MapMap.js [6] whose primary goal is to build thematic maps on the web with a standalone architecture in mind.

The comparative performance tests made in [2] show that Plotly cannot build a map with more than one hundred thousand points and ESRI ArcGIS cannot handle more than two hundred thousand points. The only API that is able to perform better is MapMap.js, but not without problems. When, for example, we use a Choropleth map with two million vertices, it takes a lot of time to build the map and the responsiveness is severely affected.

All the above mentioned APIs either use SVG[7], Canvas[8] or allow the programmer to work with both of these technologies. And that can be one of the main reasons why they do not perform well with large datasets, because the upper mentioned technologies were not built with this kind of datasets in mind.

Considering this and knowing that there's still room for improvement for standalone APIs capable of dealing with large datasets, we propose the Gisplay API[9]. This API was built with thematic maps in a web context in mind and it deals with large datasets (millions of points) with ease. To achieve this kind of performance it makes use of WebGL[10], which in short brings desktop application performance to the web, by enabling direct access to the GPU (Graphics Processing Unit).

WebGL not only gives us direct access to the graphics card silicon but it also provides a totally different paradigm to rendering. Instead of making individual calls to draw each primitive, it allows us to send buffered data up to the graphics card's memory and, with a single API call, draw thousands of primitives at once. Another interesting feature of WebGL is that it allows us to inject code, the so called shaders, that get executed for each vertex (during projection) and for each pixel (during rasterization). This is specially interesting because their execution happens in parallel for several hundreds or even thousands of vertices and pixels.

Experimental work can be found online[11,12], showing that WebGL can efficiently be used to build standalone dot maps for specific examples. Cesiumjs[13] is an API using WebGL that can be used to create thematic maps. This API was built with 3D globe visualizations in mind, but since it contains polygons, points and other objects it can also be used to create thematic maps. To create a basic Choropleth map, this API requires the programmer to write much more code than with the Gisplay API, since it does not provide the adequate

---

[7] www.w3.org/Graphics/SVG/.

[8] https://developer.mozilla.org/en-US/docs/web/API/HTMLCanvasElement.

[9] https://bitbucket.org/Gisplay_Team/gisplay_1_1.

[10] http://www.khronos.org/webgl/.

[11] http://build-failed.blogspot.pt/2013/02/displaying-webgl-data-on-google-maps.html.

[12] http://bl.ocks.org/Sumbera/c6fed35c377a46ff74c3.

[13] http://cesiumjs.org.

level of abstraction needed for thematic maps. Another example of abstraction inadequacy is the creation of the maps legend, that needs to be fully built by the programmer with no automatisms. In our comparisons[14], Cesiumjs was also slower to read and process a sample dataset in comparison with the Gisplay API.

3   Overview and Architecture

This section presents a Gisplay's overview by discussing the main goals that guided its design (Subsect. 3.1) and by detailing its overall architecture (Subsect. 3.2).

3.1   Gisplay Objectives

The goal of most APIs is to facilitate, the programmers task. To be easily adopted, the API's concepts and the provided mechanisms have to be simple to use, have to produce the expected results and the implementation has to perform well. One very important design decision was to choose a client side design approach. The main rational behind this decision is to avoid a setup of server component and enabling quick and easy experimentation. It also allows for an interactive experience at the expense of some startup delay. The Gisplay API was designed with the following goals in mind, discussed inline:

- **To provide a very high level of abstraction**, meaning that the gap between the data and the thematic maps should be as short as possible and so the use of Gisplay API should be declarative and concise.
- **To provide the best defaults** in terms of cartography and data visualization based on the current best practices. These defaults are essential to get concise API and, at the same time, to assure best data visualization results. **To provide a great level of customization**, namely by allowing overwriting the previously mentioned defaults and by offering pre-prepared choices or enabling full customization. For instance, the Gisplay offers a default for computing the classes of the thematic attribute for each thematic map type. However it also offers many different methods for computing the classes and the programmer can even implement and provide its own method.
- **To provide many types of thematic maps**. The available thematic maps include Dot Maps, Choropleth, Proportional Symbol Maps, and Change Maps. **Has to be easily extensible** by providing clear mechanisms either to create a variant of an existing thematic map or to build a new one.
- **Gisplay should be able to smoothly manage datasets with many millions of points** in order to cope with most real and actual datasets.
- **To provide the required interactivity for analytical solutions**, namely zoom, pan, and selection, across all existing thematic maps and any new thematic map built with Gisplay.

---

[14] https://bitbucket.org/Gisplay_Team/gisplay_1_1/src/master/comp/.

- To provide a geral framework to build, place, and customize the legend associated with the thematic map. Furthermore, the legend should provide mechanisms to select and filter the displayed data.
- To enable the use of more than one map projection in the same thematic map to address situations that need to present on the same map different regions, like for instance to represent the USA including the Alaska and Hawai regions, or Portugal with the islands of Azores and Madeira.
- To enable the use of more than one thematic map on the same view. For analytical purposes or just to better communicate your ideas it can be very convenient to display multiple thematic maps, typically covering the same region, where each thematic maps may show different perspectives of the some phenomenon, for instance, different years or different atributes.
- To enable the use of most common background maps providers and to offer clear mechanisms to adopt new ones. A thematic map may show some spatial patterns that can be interpreted in their geographical context which is understood through a background map. Since the thematic map overlays the background maps visual interference may occurs, and the possibility to select different background maps is important.
- To be possible to integrate Gisplay with other frameworks to build modern data visualization solutions. It is thus essencial to combine and link different visualizations, such that the user is presented with different but related visualizations of the same phenomenon. Interacting with one view should also interact with the other views. Gisplay should be able to receive from and to pass to external frameworks the user interactions.
- To accept the data in most common formats and to easily extend to other formats by providing clear extension mechanisms.
- To enable easy share of any thematic map visualization state. Modern analytical processes require the ability to share different states and points of view of studied phenomena and in most cases data visualization is a essential component to be shared.
- To be prepared to run on web platforms other than the desktop browser, namely tablets or very large displays.

These requirements/goals may look a long list but each one individually or in group are essential to enable the Gisplay usage and adoption as an interactive analytical tool.

3.2   Architecture

In Fig. 1, we present an overview diagram for the Gisplay architecture. This architecture is clearly divided in two layers, each one providing its own specific API targeting different objectives.

The Gisplay High Level API is designed to be used by programmers to produce a thematic map in a very declarative and concise way. The **Data Reader** module is an extensible module designed to expose different data readers for different data formats. Each data reader is responsible for loading and parsing

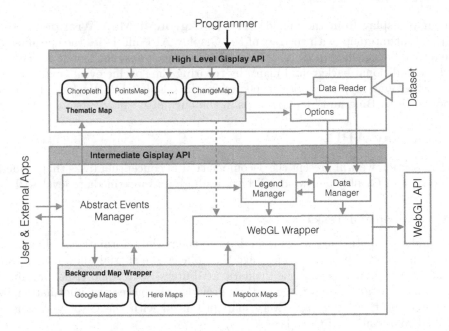

Fig. 1. Gisplay architecture

the data and delivering it in a format independent way to the **Data Manager** module of the lower layer. The current Gisplay version only implements one data reader, for the GeoJSON format. The **Options** component handles user specified options to build the thematic map, such as legend placement, number of classes or even the intervals for their computation. For options not provided in the call, appropriate defaults are considered depending on the type of map. Each specific thematic map is implemented as a specialization of an abstract **Map** object, thus minimizing the implementation efforts, mostly devoted to the areas of effective drawing and legend construction, as we will see in Sect. 4.2.

The Intermediate Level Gisplay API provides the required methods to build specific thematic map types on top of it. A very important component in this layer is the **WebGL Wrapper**. It takes care of all the aspects related to the use of this technology in this context, like compiling the shaders, linking them into a program, setting up the buffers with the data to be sent to the graphics card. One important aspect of the vertex shader is that it handles the actual map projection from geographic coordinates into the WebGL normalized coordinates. The Web Mercator projection is performed in parallel by the shader units of the graphics card and it is in fact a major advantage of using this technology, sparing the CPU from millions of intensive floating point operations. The **Abstract Events Manager** takes care of event passing to and from the outside world. External event handlers can be registered into the API receiving information about zoom, pan and click events that happen by interaction with the thematic map. On the opposite direction, we provide functions to be called externally that can control the map display from the outside. The **Background Map Wrapper** is the object that interacts with the rest of the Gisplay API and hides the specifics of the actual background map used. This design makes it easy for a programmer to include additional background maps, by rewiring the specific events exposed by the background map to the ones expected by the Gisplay API. We have included Google, Here, Bing and Mapbox maps in the current Gisplay version.

4   Gisplay API

In this section we will present the decomposition in conceptual tasks that guided the design of Gisplay and we detail both its high and intermediate level APIs.

4.1   Conceptual Tasks

To build a thematic map we go from data (with spatial information and at least one thematic atribute) up to an image that places the spatial information over a background map, encodes the thematic atribute(s) with some visual variables and builds an appropriate legend that provides the necessary information to interpret the thematic map. Every map produced with the Gisplay API is a stacked composition of three layers:

Background map layer – responsible for displaying and dealing with events such as dragging the map or zooming it.

Thematic map layer – contains all the graphical elements (lines, points, polygons), drawn using WebGL, that make up the thematic map.

Legend – an interactive legend for what is displayed on the thematic map.

In order to display these 3 layers, we addressed the problem by decomposing it into the following sequence of conceptual tasks:

Data reading — This task includes accessing the data, parsing it according to a known format (eg. CSV, GeoJSON), processing the spatial information and the selected thematic attributes to be mapped, and storing them in an internal data structure that is format independent.

Classes computation — It may be necessary to compute classes for some thematic attributes. The class computation methods used, as well the number of classes, may depend on the chosen thematic map and on the data itself.

Mappings to visual variables — For a given thematic map type, the thematic attribute classes previously computed are mapped to visual variables and its associated values.

Setting visual variable values for each spatial object — Once the mappings to visual variables are defined, it is possible to set the visual variable values for each spatial object.

Building the thematic map — During this task the thematic map is actually built based on the settings defined on previous tasks.

Building the legend — In most cases the legend can be automatically built based on the previous settings, i.e. the computed classes (if any) and the chosen mapping for the visual variables.

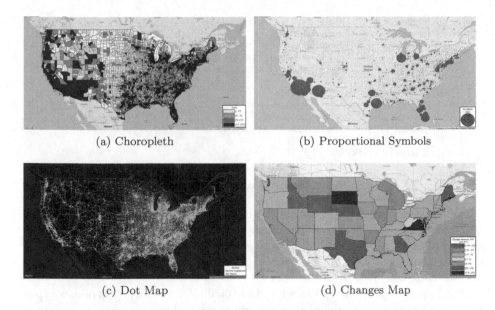

(a) Choropleth   (b) Proportional Symbols

(c) Dot Map   (d) Changes Map

Fig. 2. Examples of thematic maps created with Gisplay (Color figure online)

4.2 High Level API

Figure 2 shows four different types of thematic maps, built with the Gisplay API and focusing on the same geographical area. The datasets used[15] are all about traffic accidents in the USA, during the period from 1975 to 2015. In the first two figures (2a and b) the data is aggregated by county. Figure 2a is a Choropleth map with four different classes, built using the *k-means* algorithm. Figure 2b is a Proportional Symbols map where each point is at the center of the respective county. Figure 2c is a Dot map of accidents where each point has one of two colors, depending on whether alcohol was involved in the respective accident or not. Figure 2d is a Change map which has seven different classes and represents the change in the number of accidents between 2009 and 2013 in each state of the USA.

The High Level Gisplay API provides the required functions to create each of the implemented thematic maps. Figure 3 illustrates the use of Gisplay to build a Choropleth. The first line initializes the Gisplay API, which is immediately followed by the creation of a background map - in this case a Google Maps map, with the specified zoom level and center position. The third sentence contains the user provided options. These are provided in dictionary form and in this example only the number of classes, the selected thematic attribute and the legend title are provided by the user. The code as exemplified in Fig. 3 is mostly declarative, concise and allows for user customization via the options object, thus fulfilling some of the objectives stated in Subsect. 3.1.

---

[15] ftp://ftp.nhtsa.dot.gov/fars/.

```
1 let gisplay = new Gisplay();
2 let gm = new BGMapGoogleMaps(
3      new google.maps.Map(canvas,
4                 {zoom: 4, center: { lat: 49.36855556, lng: -81.663716667 }})));
5 let options = { numberOfClasses: 4, attr: 'f3', legendTitle: 'Fatals' };
6
7 let datareader = new FileReader();
8 datareader.onloadend = function () {
9    let data = JSON.parse(datareader.result);
10   gisplay.makeChoropleth(gm, data, options);
11 }
12 datareader.readAsText(url_to_data);
```

Fig. 3. Example code from Fig. 2a

When compared with the code required to create the map in Fig. 2c, the required changes are minimal. The changes are on the initialization of the background map, which is provided by Here Maps in this case and replacing the call to *makeChoropleth()* with a call to *makeDotMap()*, apart from the selection of the thematic attribute from the data set.

We have designed the API in a way that facilitates the inclusion of new types of maps. The specific types of thematic maps are lightweight components that derive from a base class **Map**. There are two methods that the programmer of a new thematic map has to provide. Method *draw()* implements the logic for drawing the spatial features depicted in the thematic map, while *buildLegend()* builds the interactive legend of the thematic map.

Apart from these two methods, there is also the need to program the wrapper function exposed by the High Level API that creates the specific thematic map. This is a function similar to *makeChoropleth()* and *makeDotMap()* on the two given examples, it would be possible to devise the API without the need to program it but we prefer it this way for the sake of simplicity on the end user of the Gisplay API. In fact these functions are so small that they only contain a couple lines of code, as show in Fig. 4.

```
1 makeChoropleth(bgmap, data, options)
2 {
3    const gismap = new Choropleth(bgmap, data, options);
4    gismap.makeMap();
5 }
```

Fig. 4. Implementation of *makeChoropleth()* convenience function

4.3   Intermediate API

The Gisplay intermediate API has the main goal of acting as a bridge between the High level API and WebGL, providing the abstractions to deal with this technology. The **Map** class is the basis from which all thematic maps will be built. All those functions that are basically the same for all types of thematic maps are placed in the same class and this allows one to use them in other thematic maps that are built using the Gisplay API, even new ones added in the future by means of the extensibility offered by the API.

WebGL offers a very limited set of graphics primitives, namely points, lines and triangles that are drawn in batches. The graphics rendering pipeline is fed with values – known as attributes – that are assigned to vertices and packed into data buffers. Each data buffer can have several attributes with an interleaved or packed layout or, alternatively, different attributes can be split into distinct buffers. The most important attribute is, of course, the vertex position, but others like color, transparency or texture coordinates are also common. It is also possible to feed the pipeline with constant data during primitive rendering. Constant values, known as uniforms, can only change between drawing calls.

The Gisplay intermediate API maps visual variables that vary on a per vertex basis to WebGL vertex attributes. A typical scenario is the drawing of points, each with its own color and/or alpha value.

Visual variables that have the same value for a set of graphics primitives, like several points, triangles or lines, are handled by Gisplay using uniforms. This typically occurs with visual variables that are assigned to classes. Gisplay packs data related to the primitives belonging to each class in a separate buffer. During rendering Gisplay iterates through the classes, drawing their primitives and fixing the specific value for the visual variable in a uniform parameter. In Gisplay we model those objects that share the same value for the visual variables through the **Aesthetic** class. An instance of this class stores the common values for the visual variables of its children geometry.

Load Customization Options. To create thematic maps it's necessary to load all customization options provided by the programmer. *loadOptions(userOptions, bgmap)* loads all programmer provided options and fills any remaining option with the default options provided by the Gisplay API. Only the features' attribute(s) to be mapped are mandatory, all other are optional. Examples of customization options are color scheme, number of legend classes, whether the thematic map elements are interactive or not or the location where to place the legend.

Process Geometry Data. It may be necessary to compute classes for thematic attributes. When the classes are not specified by the programmer, the method *preProcessData(geojson, numClasses, classBreakMethod, colorScheme)* computes the classes by using *calcClassBreaks(values, classBreaksMethod, numClasses)*. The available algorithms to compute class breaks are: *k-means*, *equidistant* and *quantile*. One aesthetic object is created for each of the those classes. The color used for each class is based on ColorBrewer [3].

Once the class intervals are computed and the corresponding aesthetic objects created, we have to insert data onto those aesthetic objects. There's more than one method to do this in the intermediate API, depending on the context.

Method *insertFeature()* is called for triangles and lines, while *insertGroupedFeature()* gets called for thematic maps dealing with points. Each of the above functions still has to deal with the limitation that WebGL imposes on the size of the data buffers and additional buffers are created as needed.

If we are dealing with polygons, we have to ensure they are split up into individual triangles by *processPolygon(polygon)*, which in turn uses the earcut API 16. This functionality is part of the WebGL Wrapper module in Fig. 1.

Draw Thematic Map. WebGL resources are tied to the WebGL rendering context, an object connected to a canvas in our DOM tree. Since all resources are connected to this rendering context, such as programs, shaders and buffers, the rendering context needs to exist from the very early states of map creation. This is done using *createCanvas()*. After getting the WebGL context ready, we call the method *createWebGLProgram(webgl)* that creates a flexible program that is capable of dealing with all the thematic maps that we have developed.

Now that we have the program ready to be used we can start drawing stuff according to the type of map we want to use. As a result we have a set of different methods that can be used to draw different primitives.

To draw points, one can use *drawPoints(aesthetic)*, and this method draws points applying the values present in the given aesthetic object with it's color and size, knowing that this method draws all points with the same size. If we want to draw points with different sizes one could use *drawProportionalPoints(aesthetic)* instead.

To draw polygons there is *drawTriangles(aesthetic)* that draws all triangles present in the given aesthetic object using its specified fill color. To draw the borders of any polygon there's *drawBorders(aesthetic)*, using the color stored in the aesthetic.

All previous drawing commands are invoked inside the **Data Manager** module and are programmed inside the **WebGL Wrapper** shown in Fig. 1.

Legend and Interactivity. In order to add interactivity to the API we must get the events from the background map. To add listeners for these events we use **setupEvents(id)**, which adds several events that we want to listen to. At the current state of the Gisplay API we want to take in consideration three different events: pan, zoom and click.

Each background map will have a different name for those events, so the job of connecting those events is given to the background map wrapper which acts as a link between the background map and the Gisplay API map.

To build the map legend one can take advantage of the provided **buildLegend()** method. This is suitable for polygon based map visualizations, but for other visualizations the programmer must override this method according to the type of map.

5   Evaluation and Experimentation

The evaluation of Gisplay is addressed through three main questions:

1. How does Gisplay compare with the most common web client APIs used for building thematic maps?

2. How does Gisplay fulfill the goals presented in Sect. 3?
3. How does Display perform with millions of points?

The first question was mostly discussed in Sect. 2 since there is no web client side API specifically designed for thematic maps there is able to deal with datasets with million of points.

5.1 Qualitative Evaluation

For the second question, i.e. how does Gisplay fulfill the design goals let's briefly recall the first group of four goals. The API should provide a very high level of abstraction, using the best defaults in terms of cartography and data visualization, but allowing a great level of customization. It must also be extensible in the sense that new thematic maps should be easily added. As shown in Sect. 4.2, to build one of the already provided thematic maps, requires as little as the dataset and one can mostly rely on the provided defaults (e.g. for class computation and for color mappings), but all of those defaults can be overwritten in a declarative way using the **Options** object. The ability to extend the Gisplay API with new thematic maps can, at certain level, be evaluated by the fact that all the provided thematics maps are built, in a relatively compact way, on top of Intermediate API. Table 1 shows the required lines of code by each method for implementing the four provided thematic maps. The Change Map requires modification on method *processJSON()* for processing the input data which accounts for more 10 lines do code.

Table 1. Lines of code per type of thematic map

Map	construtor	draw	buildLegend	getDefaultColors
DotMap	3	4	4	1
Choropleth	3	5	4	1
Change Map	4	5	4	1
PSymbols	4	8	10	1

What about other new thematic maps? Let's consider variants of the existing thematics maps or even some new and briefly discuss how they can be addressed to extend the Gisplay API. One issue that may arise with the Dot Maps is the overlapping of points with the same coordinates. At least three proposals can be found in the literature: (*i*) using the opacity to express the intensity of overlapping; (*ii*) avoiding the overlapping by placing the points in the nearest available coordinates [9]; (*iii*) using a 2.5D approach where the height is controlled by the density of overlapping points. For (*i*), in *processJSON* all points considered at the same coordinates will be aggregated into one point and the alpha level will be proportional to the number of represented points. For (*ii*) the extension is easily achieved by modifying the position of the point in *processJSON*. For (*iii*), part of the extension will also be done in *processJSON* but it will also require an extension to the Intermediate API to let WebGL deal with the 3D component of this approach.

A common thematic map that also deals with data points is the HeatMap [7]. For this one an extension is required at the Intermediate Level API to add new shaders and use of textures. In general new thematic maps will require an extension only at High Level API, with a few lines of code per few methods. Some may require an extension at Intermediate Level API, and consequently some knowledge on WebGL is necessary. Overall the Gisplay is not only very convenient to produce thematic maps but also an appropriate framework to implement and test new types of thematic maps.

The goals on interactivity and legend are fully achieved and the provided performance (discussed later) enables a smooth interaction in all operations on the map (zoom, pan, selection) and filtering using the legend. Three main providers of background maps are already supported and, as has been shown in Sect. 4, it is easy to add support for more. The ability to pass to external applications any event realized on the thematic map is built in as well as the possibility that external applications send events that are processed by the Gisplay API. The Gisplay is used in [8] where multiple thematic maps are presented on the same view and the events on any of those maps are communicated to other components of the application and some events originated on the other components of the application are sent to the appropriate thematic maps.

The following goals were not completely achieved with the current version: (*i*) more than one map projection in the same thematic map; (*ii*) Gisplay runs on tablets with WebGL support but the interaction was not specically designed for tablets.

The third question, i.e., how does Display performs with millions of points? is addressed in Sect. 5.2.

5.2   Experimental Performance Evaluation

The next results[16] were executed on a Core i5-4690K at 3.5 Ghz computer with 8 GB of RAM and an AMD R9/290 graphics card with 4 GB, running Windows 10. The Gisplay is compatible, at least with Firefox, Opera and Chrome web browsers, but these tests were executed with Chrome.

All times are presented in milliseconds (ms) and are averages of 15 consecutive runs. All times were gathered using Chrome's profiling tool. The experiments were carried out with each one of the already provided thematic maps, varying the size of the used dataset and measuring the time to: (*i*) process the data, (*ii*) draw the thematic map; (*iii*) execute zoom, pan and filtering operations.

We will present details of performance tests for the Dot Maps and for the Choropleth, since the performance tests for the other thematic maps showed

---

[16] Results and datasets can be found at https://bitbucket.org/Gisplay_Team/ gisplay_1_1.

similar conclusions. Furthermore the spatial features of Dot Maps are points
and those of Choropleth are multi-polygons.

Figure 2c shows a Dot Map of 446,578 traffic accidents in USA. Each accident is represented by a dot in red color if alcohol was involved, and green
otherwise. The background map used was Map Box. Based on this dataset, four
more datasets were built by randomly selecting 100 K, 200 K, 300 K and 400 K
accidents. Figure 5a shows the average time necessary to process each 1 K points,
which remains around 12 ms for all tested datasets. Figure 5b shows the time
(also per 1K points) needed for the draw, zoom, pan, and filtering operations.
Notice that the initial data processing is done by the CPU, requiring around
12 ms per 1K points while the operations are executed by the GPU (through
WebGL) requiring less than 0.01 ms per 1K points.

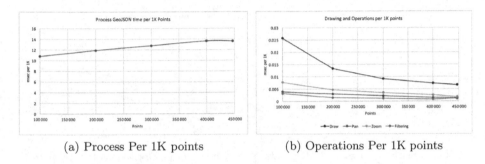

(a) Process Per 1K points   (b) Operations Per 1K points

Fig. 5. Dot Map performance experiments

For evaluating the Choropleth, the same original data about traffic accidents
was used, now aggregated at county level as shown in Fig. 2a where 4 classes were
computed using the percentile method. In this example, Google was the background map provider. The number of spatial features for the whole dataset, i.e.
the number of counties is 3141 which corresponds to 11,789 polygons requiring a
total of 2,460,570 points. Based on this dataset five more datasets were produced
by randomly selecting, 100, 500, 1200, 1700 and 2000 counties. In Fig. 6 we show
the relationship between Features, Polygons and Points, for the different used
datasets. Since the relation between features and points is strongly linear, we will
use the number of points to compare the performance on the different datasets.

Figure 7a shows the time in ms needed to process 1 K points, which remains
around 1.5 ms. Figure 7b shows the time in ms per each 1 K points, spent for the
draw, zoom, pan, and filtering operations. Initial data processing done by the
CPU requires around 1.5 ms per 1 K points while the operations are executed at
GPU level (through WebGL) requiring less than 0.06 ms per 1 K points. Similar
patterns were observed for Change Map and Proportional Symbols Map.

The use of WebGl clearly payoff both in terms of memory, speed and
interactivity.

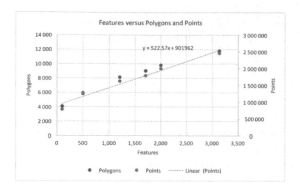

Fig. 6. Spatial features vs polygons vs points at Choropleths

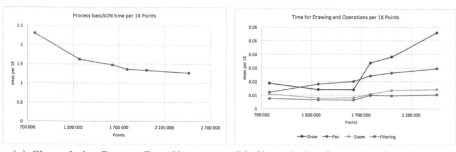

(a) Choropleth - Process Per 1K points   (b) Choropleth - Operations Per 1K points

Fig. 7. Choropleth performance experiments

6   Conclusions and Future Work

This paper proposes the Gisplay API, a client side web API to build thematic maps using webGL. The performance experimentation proved that the produced thematic maps are fully interactive with datasets with millions of points. The proposed architecture enabled the achievement of two very convenient and important objectives: a high level of abstraction allowing a declarative and concise use; and, at the same time, an extensible API.

The future work can be considered at two tracks: (*i*) Gisplay incremental improvements; (*ii*) Gisplay for spatio-temporal data.

The first and more immediate improvements will include: support for other data formats such as CSV, Shapefile, etc.; to build new types of thematic maps, like for instance, heatmaps; to introduce a mechanism to enable more than one map projection in the same thematic map. At the Intermediate Level API, some improvements are also being considered such as the possibility of easily adding custom shaders.

The other track, a Gisplay for data spatio-temporal, will benefit from the previous mentioned improvements, but will address the use of time in thematic maps, which is an unresolved issue. The Gisplay underlying technology (WebGL) will provide a solid base to deal with larger datasets and the increase requirement of interactivity to deal with spatio-temporal data on thematic maps.

Acknowledgments. This work has been supported by FCT - Fundação para a Ciência e Tecnologia MCTES, UID/CEC/04516/2013 (NOVA LINCS).

References

1. Bédard, Y., Rivest, S., Proulx, M.J.: Spatial Online Analytical Processing (SOLAP): concepts, architectures, and solutions. In: Data Warehouses and OLAP: Concepts, Architectures, and Solutions, pp. 298–319. Idea Group Inc. (2007)
2. Cardoso, D.: Gisplay: A High-Level Client-Side API for Interactive Thematic Maps using WebGL. Master's thesis, Faculdade de Ciências e Tecnologia da Universidade Nova de Lisboa (2016)
3. Harrower, M., Brewer, C.A.: ColorBrewer.org: an online tool for selecting colour schemes for maps. Cartogr. J. **40**(1), 27–37 (2003)
4. Jankowski, P., Andrienko, N., Andrienko, G.: Map-centred exploratory approach to multiple criteria spatial decision making. Int. J. Geogr. Inf. Sci. **15**(2), 101–127 (2001)
5. Kraak, M.J., Ormeling, F.: Cartography: Visualization of Spatial Data. Guilford Press, New York City (2011)
6. Ledermann, F., Gartner, G.: Mapmap. js: a data-driven web mapping API for thematic cartography. In: Proceedings of the 27th International Cartographic Conference (ICC2015) (2015)
7. Perrot, A., Bourqui, R., Hanusse, N., Lalanne, F., Auber, D.: Large interactive visualization of density functions on big data infrastructure. In: 2015 IEEE 5th Symposium on Large Data Analysis and Visualization (LDAV), pp. 99–106. IEEE, October 2015. doi:10.1109/ldav.2015.7348077
8. Silva, R.A., Pires, J.M., Santos, M.Y., Datia, N.: Enhancing exploratory analysis by summarizing spatiotemporal events across multiple levels of detail. In: Sarjakoski, T., Santos, M.Y., Sarjakoski, L.T. (eds.) Geospatial Data in a Changing World. LNGC, pp. 219–238. Springer, Cham (2016). doi:10.1007/978-3-319-33783-8_13
9. Ward, M., Grinstein, G., Keim, D.: Interactive Data Visualization: Foundations, Techniques, and Applications, 2nd edn. Taylor & Francis, Abingdon (2015). https://books.google.pt/books?id=BonfoAEACAAJ

Short Papers

Theoretical Model of the Total Mass of Ejecta from Unmelted Metals

Anmin He, Jun Liu, JianLi Shao, and Pei Wang[✉]

Institute of Applied Physics and Computational Mathematics,
Beijing 100094, China
wangpei@iapcm.ac.cn

Abstract. This work is dedicated to establishing a theoretical model to describe the mass of ejecta from unmelted metals, based on the Richtmyer-Meshkov theory in solids. Hydrodynamic simulations are performed, and the effect of material strength on the perturbation growth and material ejection are analyzed. After revealing a relationship between the mass of ejected material and the saturated bubble amplitude, a theoretical model describing the mass of ejecta from unmelted metals is developed.

Keywords: Ejecta metal · Richtmyer-Meshkov instability

1 Introduction

The machining of metals usually produces micron-scale surface perturbations. Under shock wave loading, the interaction between shock waves and surface perturbations usually results in the ejection of a great amount of small particles called ejecta. Due to its great importance in many fields such as inertial confinement fusion (ICF) and implosion dynamics, extensively experimental and theoretical studies have been conducted on this phenomenon [1–6]. The chief goal of this research is to develop a physical-based ejection model that describes the total amount and velocity of ejecta over a wide range of shock, material, and surface conditions.

Theoretically, regarding ejection process as a special case of a Richtmyer-Meshkov instability (RMI) where the Atwood number is $A_t = -1$, Buttler *et al.* and Dimonte *et al.* develop mathematical models of ejecta based on the RMI theory [7,8]. In these models, the total amount of ejecta is equal to the mass excavated by the RMI bubble and the maximum ejecta velocity is determined by the RMI spikes. The models reproduce well the numerical results from both molecular dynamics and hydrodynamics simulations [9]. However, the previous studies mainly concentrate on the conditions where metals have melted during shock wave loading or unloading, and the RMI theories on ideal fluids are often adopted. Material ejection from unmelted metals where the effect of material strength can not be ignored remains virtually unexplored. Here, we focus on the ejection of metal Pb that remain in solid state under shock loading, and try

© Springer International Publishing AG 2017
O. Gervasi et al. (Eds.): ICCSA 2017, Part VI, LNCS 10409, pp. 693–700, 2017.
DOI: 10.1007/978-3-319-62407-5_50

to develop a theoretical model to predict the mass of ejecta from the unmelted metals with the help of hydrodynamic simulations and RMI theories in solids.

1.1   Methodology

In this paper, we perform hydrodynamic simulations using the 2D multi-component elastic-plastic hydrodynamic Eulerian code (MEPH). This code has been successfully used to study the hypervelocity impact and interface instability problems [10]. Pb is chosen as the simulation sample and is modeled using Mie-Gruneisen equation of state and Steinberg-Guinan constitutive relation (SG model). Detailed formula and parameters of the models can be found in Ref. [11].

A sinusoidal perturbation of wavelength λ and amplitude η_0 is set on the upside surface along the x-axis, as shown in Fig. 1. Shock waves are generated using momentum mirror method, where the sample is thrown into a wall of infinite mass with a velocity u_p along the x-direction. The length of the sample along the shock direction is long enough to delay the effects of release fans originated from the downstream surface. Periodic boundary condition is applied along the y-direction during shock wave loading. Finally, in order to obtain the displacement and velocity of bubble and spike relative to the free surface, separate simulations with $\eta_0 = 0$ are performed.

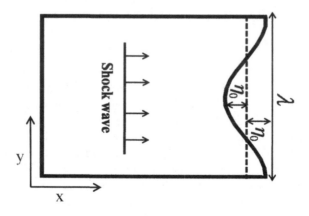

Fig. 1. Initial configuration of Pb sample.

1.2   Results and Discussion

It is well-known that the RMI occurs when shock waves pass through interfacial perturbations between materials of different densities. Figure 2 shows the representative pictures of the RMI growth for metals before (Fig. 2a and b) and after release melting (Fig. 2c). It can be seen that when the shock wave passes through the metal/vacuum interface, the perturbation inverts and grows asymmetrically, leading to the formation of typical RMI spike and bubble both for the solid and liquid metals.

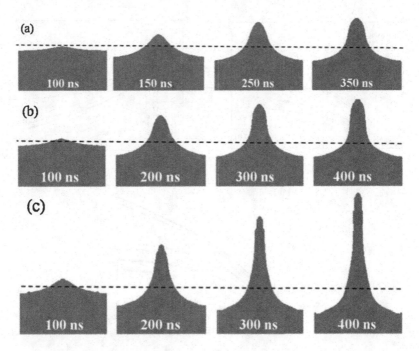

Fig. 2. Snapshots of RMI growth for $k\eta_0 = 0.25$ under various shock pressures. (a) 14.3 GPa (b) 17.8 GPa (c) 25.3 GPa. Release melting of Pb occurs at shock pressures higher than $\sim$23 GPa. The dashed lines indicate the location of the unperturbed interface

Figure 3 gives the temporal evolution of spike and bubble amplitudes at various shock pressures. For the unmelted Pb that we are concerned, the spikes and bubbles evolve asymmetrically. The bubble always saturates but the spike growth only arrests at relatively lower pressures. In the cases where the spike grows uninhibitedly, the solid state spike will eventually break up into ejecta particles, indicating the formation of ejecta as a result of RM-unstable solid flow. With respect to the total mass, or areal density, of ejecta σ_{ej}, its growth rate is arrested and σ_{ej} approaches to nearly a constant value for the solids but increases continuously for the liquids, indicating a significant enhancement in σ_{ej} for melted metals at a much later time. This phenomenon can also help us to understand the experimental observations where the ejected mass increases abruptly upon release melting. Moreover, for the unmelted cases, it can also be seen from Fig. 3 that σ_{ej} stops increasing as the bubble reaches a saturation amplitude, suggesting an underlying relationship between them.

A typical snapshot of RMI of the unmelted metal at the instant when the bubble have saturated is shown in Fig. 4. In this figure, the saturated bubble amplitude is denoted as η_b^{max}; L_b and L_s are the half-width of the bubble and spike, respectively, and $L_b + L_s = \lambda/2$. According to the basic hypothesis of ejecta model developed by Dimonte $et\ al.$ [8], the mass of ejecta is equal to that of matters excavated by the RMI bubble, and suppose that the bubble has a

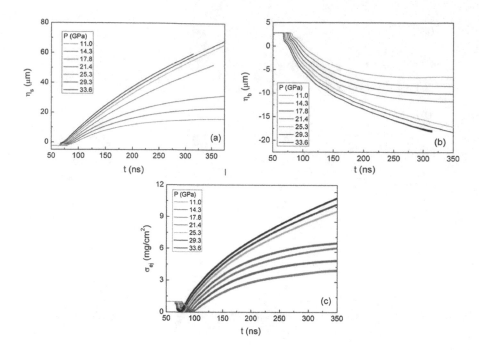

Fig. 3. Temporal evolution of spike (a) and bubble (b) amplitude, and total amount of ejected mass per unit area for Pb with $k\eta_0 = 0.25$ under various shock pressures.

parabolic shape, then the maximum ejecta areal density σ_{ej}^{max} from an unmelted metal is given by

$$\sigma_{ej}^{max} = \frac{4}{3\lambda}\rho\eta_b^{max}L_b. \tag{1}$$

Given the fact that L_b is usually smaller than $\lambda/2$, denoting $L_b = \alpha\lambda/2$, Eq. (1) becomes

$$\sigma_{ej}^{max} = \frac{2}{3}\alpha\rho\eta_b^{max}, \tag{2}$$

where $0 < \alpha \leq 1$. According to Eq. (2), besides the saturated bubble amplitude, σ_{ej}^{max} is also proportional to the ratio of the width of the saturated bubble and the initial perturbation wavelength.

Variation of the scaled ejecta areal density $\sigma_{ej}^{max}/\sigma_{def}$ with saturated bubble amplitude $k\eta_b^{max}$ is shown in Fig. 5. As can be seen, the ejecta areal density increases linearly with $k\eta_b^{max}$ for a given initial perturbations $k\eta_0$, consistent with the theoretical results from Eq. (2). Furthermore, the simulation results show that the coefficient α in Eq. (2) varies from 0.6 to 0.75 as $k\eta_0$ is changed from 0.125 to 0.25.

To complete the ejecta model for unmelted metals, we still need an analytic expression for the saturated bubble amplitude. Theoretically, Mikaelian [12] finds that the saturated bubble amplitude η_b^{max} of material with strength can be described by the following formula

$$\eta_b^{max} = \eta_b^0 + \frac{2}{k+ck} ln[1 + (1+c)\frac{\rho_0|\dot{\eta}_b^0|^2}{6Y}] \tag{3}$$

where η_b^0 and $\dot{\eta}_b^0$ are the initial amplitude and peak growth rate of bubbles, Y is yield stress, and $c = 2$ for 2D problems. For $\dot{\eta}_b^0$, we find that the analytic expression developed by Buttler et $al.$ [7] agree quite well with our numerical simulations (Fig. 6), which is written as

$$\dot{\eta}_b^{0,Mod} = F^l F_b^{nl}|k\eta_0 u_{fs}|. \tag{4}$$

In Eq. (4), $F^l = 1 - \frac{u_{fs}}{2D}$ is a linear compression factor; $F_b^{nl} = \frac{1}{(1+k\eta_0/6)}$ is a nonlinear bubble factor; D and u_{fs} are the shock and free surface velocities, respectively, in the frame of reference where the interface is initially at rest.

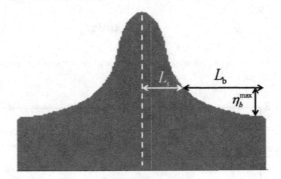

Fig. 4. Snapshot of RM instability for unmelted Pb at the instant when bubble have saturated.

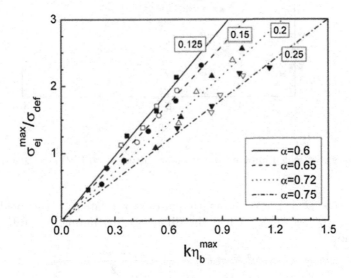

Fig. 5. Scaled ejecta areal density vs. bubble saturation amplitude for different initial perturbations. $\sigma_{def} = \rho_0\eta_0$ is the initial defect mass per unit area.

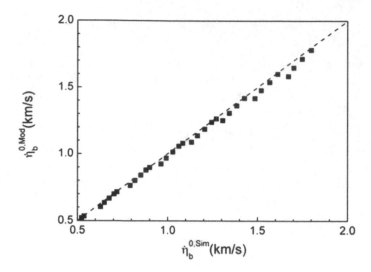

Fig. 6. Comparison between numerical and analytical results of the peak bubble growth rates. Dashed line indicates $\dot{\eta}_b^{0,Mod} = \dot{\eta}_b^{0,Sim}$.

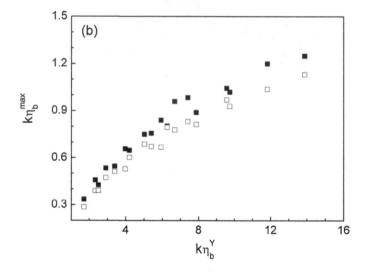

Fig. 7. Variation of scaled bubble saturation amplitude with RMI strength parameter.

If we define scaled RMI strength parameter related to the peak bubble growth rate, $k\eta_b^Y \equiv \frac{\rho_0}{Y}|\dot{\eta}_b^0|^2$, then the dimensionless bubble saturation amplitude becomes

$$k\eta_b^{max} = k\eta_b^0 + \frac{2}{3}ln[1 + \frac{k\eta_b^Y}{2}] \qquad (5)$$

The variation of scaled bubble saturation amplitude with RMI strength parameter is displayed in Fig. 7. The theoretical results (open squares) are basically consistent with that of simulations (solid squares). The maximum and minimum relative deviations are 15% and 5%, respectively. It should be pointed out that because Y is not a constant in SG constitutive relation, an effective yield strength is adopted in calculations of the RMI strength parameter. In doing this, we perform separated simulations of samples modeled by ideal elastic-perfectly plastic constitutive relation with different parameters of Y. The effective yield strength is then defined as the one that best reproduce the results of SG models. Therefore, the total mass of ejecta from unmelted metals can be calculated theoretically by combining Eqs. 2, 3, and 4.

2 Conclusion

We performs hydrodynamic simulations to investigate the Richtmyer-Meshkov instability and material ejection from unmelted metals. It is found that the RMI bubbles in the unmelted metals always saturate due to the effect of material strength. At the same time. the total mass of ejecta also stops increasing and approaches a constant value. Based on the numerical results, an analytic expression describing the relationship between the maximum ejecta mass and the bubble saturation amplitude is established. By utilizing the existing RMI theories for materials with strength, a theoretical model of mass of ejecta from unmelted metals is established. Consequently, combining with the previous ejecta model for liquids, the total mass of ejecta at the entire range of shock pressure can be theoretically calculated. Finally, it should be pointed out that because previous experimental studies focus mainly on the material ejection from melted metals, our theoretical model needs further experimental verification.

Acknowledgment. This work is supported by the National Natural Science Foundation of China (Grant Nos. 11402032 and U1530261), the Science and Technology Development Foundation of CAEP (Grant No. 2015B0201039).

References

1. Asay, J.R., Mix, L.P., Perry, F.C.: Appl. Phys. Lett. **29**, 284 (1976)
2. Zellner, M.B., Grover, M., Hammerberg, J.E., Hixson, R.S., Iverson, A., Macrum, G., Morley, K., Obst, A., Olson, R.T., Payton, J.: J. Appl. Phys. **102**, 013522 (2007)
3. Buttler, W.T., Zellner, M.B., Olson, R.T., Rigg, P.A., Hixson, R.S., Hammerberg, J.E., Obst, A.W., Payton, J.R., Iverson, A., Young, J.: J. Appl. Phys. **101**, 063547 (2007)
4. Monfared, S.K., Grover, D.M., Hammerberg, J.E., Lalone, B.M., Pack, C.L., Schauer, M.M., Stevens, G.D., Stone, J.B., Turley, W.D., Buttler, W.T.: J. Appl. Phys. **116**, 063504 (2014)
5. Wang, P., Shao, J.L., Qin, C.S.: Acta Phys. Sin. **58**, 1064 (2009). (in Chinese)
6. He, A.M., Wang, P., Shao, J.L.: Model. Simul. Mater. Sci. Eng. **24**, 025002 (2016)
7. Buttler, W.T., Or, D.M., Preston, D.L., Mikaelian, K.O., Cherne, F.J., Hixson, R.S., Mariam, F.G., Morris, C., Stone, J.B., Terrones, G., Tupa, D.: J. Fluid. Mech. **703**, 60 (2012)

8. Dimonte, G., Terrones, G., Cherne, F.J., Ramaprabhu, P.: J. Appl. Phys. **113**, 024905 (2013)
9. Cherne, F.J., Hammerberg, J.E., Andrews, M.J., Karkhanis, V., Ramaprabhu, P.: J. Appl. Phys. **118**, 185901 (2015)
10. Liu, J., Feng, Q.J., Zhou, H.B.: Acta Phys. Sin. **63**, 155201 (2014). (in Chinese)
11. Li, M.S., Chen, D.Q.: Chin. J. High Press. Phys. **29**, 63 (2001). (in Chinese)
12. Mikaelian, K.O.: Phys. Rev. B **87**, 031003 (2013)

Influence of Void Coalescence by Direct Impingement on Spall Response of Polycrystalline Metal

Feng-guo Zhang[✉], Jian-li Shao, Pei Wang, and Qi-jing Feng

Institute of Applied Physics and Computational Mathematics,
Beijing 100094, China
zhang_fengguo@iapcm.ac.cn

Abstract. It is well known that the spall response of ductile metal is sensitive to its initial microstructures. The present work is focused on the grain size effect and void coalescence during the damage evolution. According to the geometric relationship between voids derived from the porosity and voids size, a criterion for the onset of void coalescence was proposed. Then, we propose an extended damage formulation combining void coalescence by direct impingement, which can describe the increase of porosity rate in view of the reduction of void number and increase of void size after the void coalescence. Our simulations show that the pull-back velocity increases with grain size. Besides, the void coalescence will enhance the magnitude of the spall peak and the acceleration rate after the pull-back minimum. These numerical results are in qualitative agreement with the corresponding experiment data.

Keywords: Spall · Grain size · Void coalescence · Ductile metal

1 Introduction

Damage in materials subjected to dynamic tension remains a very important study field. Extensive investigation shave shown that the spall response of materials is significantly dependent on their microstructures [1–5]. Recently, Chang et al. analyzed size effects on void growth. Their simulations showed the development of a size effect on the void growth rate of type "the larger, the faster" [6]. Alinaghain et al. studied the effect of pre-strain and work hardening rate on void growth and coalescence and proposed a simple model based on the local state of hardening between voids which is able to capture the change in failure strains [7]. The influence of grain size on the spall response of polycrystalline aluminum was studied by Trivedi et al. in experiments, where they observed that the pull-back velocity of ultrapure aluminum increases with the grain size at peak compressive stress of 21 GPa [8]. Escobedo et al. studied the spall of polycrystalline copper by considering different grain sizes (30 μm, 60 μm, 100 μm and 200 μm). Their results showed that the void volume fraction and the average void size increase with increasing grain size. And in the 30 μm and 200 μm samples, void coalescence is observed to dominate the void growth behavior [9].

© Springer International Publishing AG 2017
O. Gervasi et al. (Eds.): ICCSA 2017, Part VI, LNCS 10409, pp. 701–709, 2017.
DOI: 10.1007/978-3-319-62407-5_51

In fact, void coalescence is especially important and yet the least understood. In some cases, void may even begin to coalesce at the onset of nucleation, preventing any dilatational void growth [10]. Therefore, void coalescence in damage modeling should be considered to better describe the fracture of ductile materials [11–13]. Several mechanisms of void coalescence have been explored in the literature [14, 15]. The basic mechanism is focused on the void linking through an instability in the inter void ligament [10, 16]. But under dynamic loading, void coalescence may also occur through a direct impingement mechanism. In this case, voids may grow in spherical shape, without the occurrence of localized phenomena. The complete fracture of the material appears at high porosity levels when voids impinge each other [17, 18].

In the present work, we mainly consider the porosity rate derived from voids direct impinge, ignoring the inter void ligament snecking. Besides, the effects of gain size on free surface velocity, void sizes and void number are also discussed based on our previous work. Based on the proposed work, we simulate the corresponding spall experiment results on polycrystalline copper with a 1-D finite element code.

2   A Criterion for the Onset of Void Coalescence

As suggested by Jacques et al. [13] and Zhang et al. [19], the porous ductile material is modelled as an assemblage of hollow spheres with the internal radius ai and external radius bi, and the ratio of bi to ai is assumed to be identical for all hollow spheres, i.e. $\alpha = b_i^3/(b_i^3 - a_i^3)$. So, the macroscopic damage D can be described by the porosity α according to the formulation $D = 1 - 1/\alpha$.

Considering the two conterminous hollow spheres (see Fig. 1), the distance between the two voids is defined as:

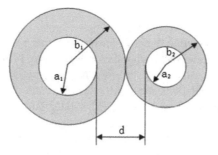

Fig. 1. Distance between void sin porous material

$$d = \left[\left(\frac{\alpha}{\alpha - 1}\right)^{1/3} - 1\right](a_1 + a_2) \tag{1}$$

Some criteria for the onset of void coalescence have been developed based on the distance between the voids, such as the works by Horstemeyer et al. [20], Seppala et al.

[21]. In fact, the distance between voids includes effects of porosity α and void size a_i. Therefore, we use Eq. (1) in describing the criterion for the onset of void coalescence. In other words, if intervoid distance $d \leq d_{cr}$, void coalescence by direct impingement occurs, ignoring the intervoid ligament snecking. There d_{cr} is critical value for the onset of void coalescence, and the value of dcr is confirmed by the experiment data (here dcr = 4 * min(a_1,a_2)).

3   Influence of Void Coalescence on Spall Response

For sufficiently high stress triaxiality and loading rate, void coalescence is found to occur by direct impingement, instead of ligament necking [13]. In this work, we neglected the intervoid ligament snecking, supposing that voids remain almost spherical and total volume of voids will not change shortly after coalescence. Voids coalescence satisfies energy conservation:

$$E_i + E_k = \bar{E}_i + \bar{E}_k \tag{2}$$

where E_i and E_k are internal energy and kinetic energy of hollow spheres, $\bar{E}_i$ and $\bar{E}_k$ are internal energy and kinetic energy after coalescence.

Under adiabatic conditions in process of void coalenscence, the internal energy change rate is given by

$$\dot{e} = -p\frac{\partial v}{\partial t} \tag{3}$$

where e, p are internal energy per unit volume, pressure and $v = \rho_0/\rho$. ρ_0 and ρ are initial density and current density.

During the voids coalescence, the voids volume will keep a constant (see Fig. 2),

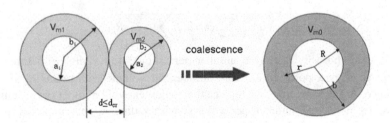

Fig. 2. Void coalescence

$$\frac{4}{3}\pi\left(a_1^3 + a_2^3\right) = \frac{4}{3}\pi R^3 \tag{4}$$

if we consider that the matrix is imcompressible,

$$V_{m1} + V_{m2} = V_{m0} \tag{5}$$

Furthermore, For the hollow sphere, let r_0 be the Lagrangian radial position coordinate in the impressible solid surrounding the void and let r be the corresponding Eulerian coordinate, Thus:

$$\dot{r} = \frac{a_0^3}{3r^2(\alpha_0 - 1)} \cdot \dot{\alpha} \tag{6}$$

So, the kinetic energy of a hollow sphere:

$$E_k = \frac{1}{2} \int_a^b 4\pi \rho r^2 \dot{r}^2 dr = \frac{4\pi \rho}{9} \cdot \frac{a_0^6}{(\alpha_0 - 1)^2} \cdot \left[1 - \left(\frac{\alpha - 1}{\alpha} \right)^{1/3} \right] \cdot \dot{\alpha}^2 \cdot \frac{1}{a} \tag{7}$$

During damage evolution, there is a population of different-sized voids. The total kinetic energy of all hollow spheres before void coalescence:

$$E_k = \sum n_i E_k^i = \frac{4\pi \rho}{9} \cdot \frac{a_0^6}{(\alpha_0 - 1)^2} \cdot \left[1 - \left(\frac{\alpha - 1}{\alpha} \right)^{1/3} \right] \cdot \dot{\alpha}^2 \cdot \sum \frac{n_i}{a_i} \tag{8}$$

and the total kinetic energy of all hollow spheres after void coalescence:

$$\bar{E}_k = \sum n_i \bar{E}_k^j = \frac{4\pi \rho}{9} \cdot \frac{a_0^6}{(\alpha_0 - 1)^2} \cdot \left[1 - \left(\frac{\alpha - 1}{\alpha} \right)^{1/3} \right] \cdot \dot{\bar{\alpha}}^2 \cdot \sum \frac{N_j}{R_j} \tag{9}$$

According to the kinetic energy conservation $E_k = \bar{E}_k$, we can obtain the following equation

$$\dot{\bar{\alpha}} = \dot{\alpha} \cdot \sqrt{\sum (n_i/a_i) / \sum (N_j/R_j)} \tag{10}$$

where $\dot{\alpha}$, n_i are porosity rate, void number per volume with radius a_i before coalescence, $\dot{\bar{\alpha}}$, N_j are new porosityrate, void number per volume with radius R_j after void coalescence. At the same time, we replace the coalescent void number per volume with radius a_i by new void number per volume with radius R_j in computing process. Therefore, Eq. (10) includes the effect of void number and void size changes on the porosity rate after coalescence.

4 Void Nucleation and Growth

In the previous work [22], void nucleation rate model by Seaman et al. is extended to account for effect of grain size [23]. The rate of voids nucleation is given as following

$$\dot{N}_t = \dot{N}_0 \cdot b^{-3} \cdot EXP\left(\frac{p - p_0}{p_1}\right) p > p_0 \qquad (11)$$

where $\dot{N}_0$ and p_1 are material constants, p_0 is the threshold for all voids nucleation, p is the mean stress in the porous region containing voids, and b is the average grain size.

So, new total number of nucleation void per volume is $N_t = \dot{N}_t \bullet \Delta t$, and the number N_t^i of nucleation void was recorded at the different time.

The growing rate of porosity $\dot{\alpha}$ with the inertial effects is given according to Johnson's work [24]

$$\tau Q(\ddot{\alpha}, \dot{\alpha}, \alpha) = \alpha p - \frac{2}{3}Y_0 \ln\left(\frac{\alpha}{\alpha - 1}\right) - \frac{2}{3}\eta\dot{\alpha}\left[(\alpha - 1)^{-1/3}(\alpha_0 - 1)^{-2/3} - \alpha^{-1/3}\alpha_0^{-2/3}\right]$$

$$(12)$$

Here $\tau = \rho\frac{a_0^2}{3(\alpha_0 - 1)^{2/3}}$, a_0 is nucleation void radius, α_0 provides the initial distension to get the void-growth process started, computed from initial nucleated voids, Y_0 is the yield stress, η is a material constant, $Q(\ddot{\alpha}, \dot{\alpha}, \alpha) = \ddot{\alpha}[(\alpha - 1)^{-1/3} - \alpha^{-1/3}] - \frac{\dot{\alpha}^2}{6}[(\alpha - 1)^{-4/3} - \alpha^{-4/3}]$. So, in the n + 1 time the void radius ai change as $a_i^{n+1} = a_i^n \cdot ((\alpha^{n+1} - 1)/(\alpha^n - 1))^{1/3}$.

5 Numerical Analysis

As an application of the foregoing work, plate-impact experiments on OFHC copper (initial void radii $a_0 = 1.82\,\mu m$, $\dot{N}_0 = 2.8 \times 106/s$) are simulated with a 1-D lagrangian elasto-plastic code with the Zerilli-Armstrong dynamic strength model [25].

There, the targets (4 mm) are the OFHC copper with the average grain sizes 30, 60, 100, and 200 μm. The impact velocity of quartz impactor (2 mm) is ~131 m/s [9]. Results from the experiments show that void coalescence was enhanced in samples with relatively small grain sizes (30 μm) in which voids were nucleated on different grain boundaries and are close together, or on samples with large grain sizes (200 μm) in which several voids were nucleated along the same grain boundary and are close together. At same time, there was a correlation between higher acceleration rates after the minima and large sized voids that resulted from coalescence. On the contrary, in samples with intermediate grain sizes (60 and 100 μm), the voids tend to grow but remain isolated for the most part. However, in numerical simulation, the total number of voids decreases with increasing grain size. Since voids distribute uniformly inside the sample, the distance between voids increases with increasing grain size. Therefore, void coalescence occurs only in the sample with 30 μm grain size.

The numerical results (Fig. 3) show that the effect of grain size and void coalescence (30 μm sample) on the spallation response. First, the increase of grain size can significantly affect the pull-back of surface velocity by lowing its valley value, but the grain size has small effect on the time for the rebound of the surface velocity. Second, the void coalescence does not change pull-back velocity and lead to increase of the spall peak magnitude and the acceleration rate after the pull-back minimum. At the same time, the optical analysis [9] and numerical results (in Table 1) show that the void number decreases, void size increases with increasing grain size and voids coalescence lead to decrease of void number and to increase of void size. Because very small voids could not be observed in experiments, we have not compared directly the size and number of voids in simulations to that of experiments. However, the variation of size and number of voids with grain sizes in numerical results agrees qualitatively with experiments.

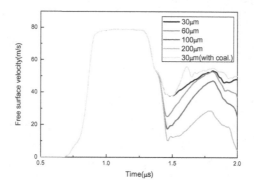

Fig. 3. Influences of void coalescence and grain size on the free surface velocities

Table 1. Damage statistics from optical analysis and numerical result

Grain size (μm)	Number of voids		Void diameter (μm)	
	Exp.	Calcu. (*10^7number/cm^3)	Exp. (average)	Calcu. (maximum)
30	236	11.460	38.1	12.17
		0.044 (with coal.)		37.12 (with coal.)
60	363	3.2358	22.7	22.36
100	267	1.4214	33.0	34.51
200	111	0.5663	55.1	42.60

Figure 4 shows that influence of void coalescence on the spallation response in spall plane. By comparing the simulations with void coalescence and without void coalescence, we find the porosity evolution can be roughly divided two stages. First, void coalescence is the dominant factor leading to the void growth behavior, but the rate of porosity does not change obviously at the early period of coalescence (1.28–1.31 μs). Second, we can see the increase of both the porosity rate and size of voids after 1.31 μs.

Furthermore, the numerical results clearly show that the number of small voids decreases rapidly due to the continuous void coalescence in the late stage of damage evolution. Figure 5 shows the relation between the presence of a large void and the

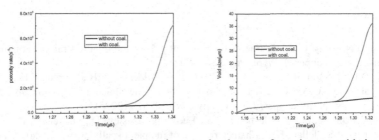

a. change of porosity rate   b. change of maximum void size

Fig. 4. Influences of void coalescence on the spall response

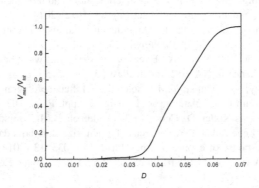

Fig. 5. Relation betweenthe presence of a large void and the damage of the material

damage of the material, and there is a critical void concentration damage(D ≈ 0.035) at which the strength of the largest void Vmax/Vtot starts to grow steeply (Vmax is the volume of the largest void and Vtot is the total volume contained in voids). And at the late stage, there mainly exit large voids in materials because of continuous coalescence of voids. Our results are qualitatively consistent to the MD simulations by Alejandro et al. [26].

6   Conclusions

A micromechanical constitutive model is proposed to describe the spallation response of ductile polycrystalline metals under dynamic loading. An important feature of this model is that the effects of grain and void coalescence by direct impingement are incorporated into the formulation. Our numerical results show that the void number decrease, pull-back velocity and void size increase with increasinggrain size; void coalescence lead to void size and porosity rate rapid growth and there is a critical void concentration damage at whichthe strength of the largest void starts to grow steeply. There is a qualitative agreement between numerical results and corresponding observations [9, 10].

References

1. Johnson, J.N., Gray, G.T., Bourne, N.K.: Effect of pulse duration and strain rate on incipient spall fracture in copper. J. Appl. Phys. **86**, 4892–4901 (1999)
2. Koller, D.D., Hixson, R.S., Gray, G.T., Rigg, P.A., Addessio, L.B., Cerreta, E.K., Maestas, J.D., Yablinsky, C.A.: Influence of shock-wave profile shape on dynamically induced damage in high purity copper. J. Appl. Phys. **98**, 103518 (2005)
3. Peralta, P., DiGiacomo, S., Hashemian, S., Luo, S.N., Paisley, D., Dickerson, R., Loomis, E., Byler, D., McClellan, K.J.: Characterization of incipient spall damage in shocked copper multicrystals. Int. J. Damage Mech **18**, 393 (2009)
4. Wilkerson, J.W., Ramesh, K.T.: A dynamic void growth model governed by dislocation kinetics. J. Mech. Phys. Solids **70**, 262–280 (2014)
5. Glam, B., Strauss, M., Eliezer, S., Moreno, D.: Shock compression and spall formation in aluminum containing helium bubbles at room temperature and near the melting temperature: Experiments and simulations. Int. J. Impact Eng **65**, 1–12 (2014)
6. Chang, H.-J., Segurado, J., LLorca, J.: Three-dimensional dislocation dynamics analysis of size effects on void growth. ScriptaMaterialia **95**, 11–14 (2015)
7. Alinaghian, Y., Asadi, M., Weck, A.: Effect of pre-strain and work hardening rate on void growth and coalescence in AA5052. Int. J. Plast **53**, 193–205 (2014)
8. Trivedi, P.B., Asay, J.R., Gupta, Y.M., Field, D.P.: Influence of grain size on the tensile response of aluminum under plate-impact loading. J. Appl. Phys. **102**, 08351 (2007)
9. Escobedo, J.P., Dennis-Koller, D., Cerreta, E.K., Patterson, B.M., Bronkhorst, C.A., Hansen, B.L., Tonks, D., Lebensohn, R.A.: Effects of grain size and boundary structure on the dynamic tensile response of copper. J. Appl. Phys. **110**, 033513 (2011)
10. Thomason, P.F.: A view on ductile-fracture modelling. Fatigue Fract. Eng. Mater. Struct. **21**, 1105–1122 (1998)
11. Benzerga, A.A.: Micromechanics of coalescence in ductile fracture. J. Mech. Phys. Solids **50**, 1331–1362 (2002)
12. Gao, X., Kim, J.: Modelling of ductile fracture: significance of void coalescence. Int. J. Solids Struct. **43**, 6277–6293 (2006)
13. Jacques, N., Mercier, S., Molinari, A.: Void coalescence in a porous solid under dynamic loading conditions. Int. J. Fract. **173**, 203–213 (2012)
14. Benzerga, A.A., Leblond, J.-B.: Ductile fracture by void growth to coalescence. Adv. Appl. Mech. **44**, 169–305 (2010)
15. Besson, J.: Continuum models of ductile fracture: a review. Int. J. Damage Mech **19**, 3–52 (2010)
16. Tonks, D.L., Zurek, A.K., Thissell, W.R.: Coalescence rate model for ductile damage in metals. Journal de Physique IV France **110**, 893–898 (2003)
17. Llorca, F., Roy, G.: Metallurgical investigation of dynamic damage in tantalum. In: Furnish, M.D., Gupta, Y.M., Forbes, J.W. (eds.) Shock Compression of Condensed Matter 2003, AIP Conference Proceedings, Portland, vol. 706, pp. 589–592 (2003)
18. Gray III, G.T., Bourne, N.K., Vecchio, K.S., Millett, J.C.F.: Influence of anisotropy (crystallographic and microstructural) on spallation in Zr, Ta, HY-100 steel, and 1080 eutectoid steel. Int. J. Fract. **163**, 243–258 (2010)
19. Zhang, F.G., Zhou, H.Q., Hu, J., Shao, J.L., Zhang, G.C., Hong, T., He, B.: Modelling of spall damage in ductile materials and its application to the simulation of the plate impact on copper. Chin. Phys. B **21**(9), 094601 (2012)

20. Horstemeyer, M.F., Matalanis, M.M., Sieber, A.M., Botos, M.L.: Micromechanical finite element calculations of temperature and void configuration effects on void growth and coalescence. Int. J. Plast **16**, 979–1015 (2000)
21. Seppala, E.T., Belak, J., Rudd, R.E.: Three-dimensional molecular dynamic simulation of void coalescence during dynamic fracture of ductile metals. Phys. Rev. B **71**, 064112 (2005)
22. Zhang, F.G., Zhou, H.Q.: Effects of grain size on the dynamic tensile damage of ductile polycrystalline metall. Acta Phys. Sin. **62**(16), 164601 (2013). (in Chinese)
23. Seaman, L., Curran, D.R., Shockey, D.A.: Computational models for ductile and brittle fracture. J. Appl. Phys. **47**, 4814–4826 (1976)
24. Johnson, J.N.: Dynamic fracture and spallation in ductile solids. J. Appl. Phys. **52**(4), 2812–2825 (1981)
25. Zerilli, F.J., Armstrong, R.W.: Dislocation-mechanics-based constitutive relations for material dynamics calculations. J. Appl. Phys. **61**(5), 1816–1825 (1987)
26. Alejandro, S., Tahir, C., William III, A.G.: Critical behavior in spallation failure of metals. Phys. Rev. B **63**, 060103 (2001)

Sentiment Analysis Applied to Hotels Evaluation

Gustavo Soares Martins, Alcione de Paiva Oliveira$^{(\boxtimes)}$, and Alexandra Moreira

Departament of Informatics, Universidade Federal de Viçosa,
Viçosa, Minas Gerais 36570-900, Brazil
gustavo.mar71ns@gmail.com, alcione@gmail.com, xandramoreira@gmail.com
http://www.dpi.ufv.br

Abstract. Websites evaluating products and services are becoming quite common. The large number of evaluations form a substantial corpus that can be used to train and test sentiment analysis tools. The analyzes produced by these tools allow companies and institutions in general to make important decisions that may be vital to the institution's future. This paper describes an implementation of the Naïve Bayes algorithm for the polarity analysis of the reviews from Rio de Janeiro hotel services, reporting the development and difficulties of the data extraction, processing and analysis methods of a corpus with 69076 comments. The results show that the tool is suitable for detecting feelings of positive and negative polarity, but does not present satisfactory results for neutral polarity.

Keywords: Sentiment analysis · Naïve Bayes · Hotels evaluation

1 Introduction

Websites evaluating products and services are becoming quite common. Some of these well-known sites are TripAdvisor (www.tripadvisor.com) that receives evaluations related to services for tourism, the site Booking (booking.com) that receives evaluations on hotels and the Amazon (www.amazon.com) itself that receives evaluations on the products that it commercializes. The large number of evaluations form a substantial corpus that can be used to train and test sentiment analysis tools. According to [4] sentiment analysis is the task that seeks to determine whether a particular sentence or document expresses a positive or negative opinion. This task has gained importance because it can enable companies and institutions to assess how much products or people are viewed by the general public. This consolidated information enables valuable decision making for both individuals and institutions, making it a crucial competitive advantage. It is still a very active area of research, with challenges due to the diversity of contexts and the inherent difficulty of natural language processing. This paper describes an implementation of the Naïve Bayes algorithm for the polarity analysis of the reviews from Rio de Janeiro hotel services voluntarily posted by users on TripAdvisor. The results showed good accuracy for the positive and negative

© Springer International Publishing AG 2017
O. Gervasi et al. (Eds.): ICCSA 2017, Part VI, LNCS 10409, pp. 710–716, 2017.
DOI: 10.1007/978-3-319-62407-5_52

polarities, however, there is a greater difficulty to obtain good results on neutral polarity evaluations. The choice of Naïve Bayes technique is due to its ease of implementation, resistance to noise and overfitting, combined with good precision. Other techniques, such as maximum entropy tend to produce results more accurately, but are more difficult to train and may present overfitting.

This paper is organized as follows: the next section presents the work previously developed that are related to this research; Sect. 3 describes how the corpus was built; Sect. 4 describes the classifier technique; Sect. 5 presents the results obtained; and Sect. 6 presents the final remarks.

2   Related Work

The number of works related to sentiment analysis is immense, therefore, we will focus on related works that address the area of tourism.

Kozak and Arslan [3] have conducted a study to determine the cause of the customers online complaints of the employee at the TripAdvisor site but they did not use any tool for sentiment analysis.

Ye et al. [9] performed a sentiment analysis using as input travel blogs texts extracted from seven popular travel destinations in the US and Europe. They have applied three supervised machine learning algorithms, namely, Naïve Bayes, SVM and the character based N-gram model. According to the authors all three approaches have reached accuracies of at least 80%.

Kasper and Vela [2] developed a system that collects German reviews about hotels from the web and creates classified and structured overviews of theses reviews. They used statistical polarity classifier based on character *n-grams* to assign to each text segment a polarity value. The best F value obtained was 81%, which was less than what was obtained in our experiments.

Shimada et al. [8] extracted tweets relating to the target locations and tourism events and analyzed the polarity of the extracted tweets. They employed an unsupervised machine learning approach based on a Naïve Bayes classifier that used two seed words. They have used a non-tagged corpus of over 10 million tweets for the pseudo training data acquisition. They obtained an accuracy of 0.89. The main differences of the work presented in this article is the use of a distinct corpus, the fact that we didn't use seeds words and that our training method was supervised.

Vincent and Winterstein [6] built a corpus for sentiment analysis for the French language. They used three domains: the analysis of movies, the amazon analysis romance books and the review of TripAdvisor hotels. After the construction of the corpus, two techniques were used for sentiment analysis: logistic regression and classification based on Support Vector Machines (SVM). The best F value obtained was 90.16% with logistic regression. The F value obtained in our study was 87.2%, but it was achieved with a much simpler technique to implement.

3   The Corpus Development

To compose the *corpus* were used 69,075 reviews, in Portuguese language, of hotels of Rio de Janeiro available on a site for evaluation of tourist attractions. The date of the last review was October 31, 2016. Each user's rating has a title of up to 10 words and the rating can have an average of 60 words. The extracted corpus consists of a total of words: 3,994,201. The user also indicates the degree of satisfaction with the hotel through the attribution of a number of stars with the following meaning: 1: very negative; 2: negative; 3: neutral; 4: positive; and 5: very positive. In order simplify our analysis, we restricted the number of classes from 5 to 3, with only the following classes remaining: 0: negative; 1: neutral; and 2: positive. The mapping of the original classes to the new classes was done as follows: 1 and 2 stars (4860 reviews; 7,03%) mapped to class 0; 3 stars (12117 reviews; 17,55%) mapped to class 1; and 4 and 5 stars (52098 reviews; 75,42%) mapped to class 3. The imbalance in evaluations can be attributed to the tendency of persons to make complimentary evaluations rather than negative evaluations.

For training and testing the corpus was divided as follows: 69% (47,702 reviews) for training and 31% (21,374 reviews) for testing.

Figure 1 shows the steps of sentiment analysis process. After the creation of the corpus is applied a normalization process and then assigned probability of occurrence to words that are stored in tables, one for each class. The normalization step is described in the next section.

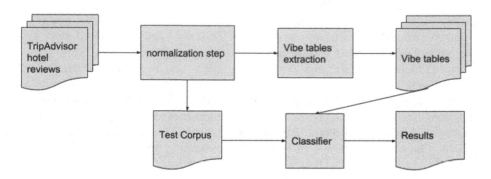

Fig. 1. Steps of the sentiment analysis process over the hotel reviews.

3.1   Text Normalization

User reviews went through a normalization process, with words being converted to lowercase, accents were taken, as well as punctuation marks. A list of stopwords was also drawn up to be deleted from the reviews.

Some normalizations were made for some tests, in order to verify if these changes resulted in some gain. These normalizations were as follows.

Lematization. Inflected words have been replaced by your dictionary entry (for example: gosta → gostar; gostado → gostar; gostam → gostar).

Inversion of Polarity. As suggested by [5], all words that occur after the word "no" have received a reverse polarity represented by "!". This was done to ensure that the inversion effect provided by the "no" was computed. In this way the excerpt "não era ruim" (was not bad) becomes "não !era !ruim".

Vibe Tables. In order to identify the words that have the highest probability of occurrence within a given document d, a 500-word vibe (vibration) dictionary Fig. 2 was created for each class c. The vibration of a word x is a function of the probability of the word belonging to a class c_i expressed by the following equation.

$$vibe(x, c) = \frac{P(x|c)}{\prod_{i=1}^{n} P(x|c_i)} \tag{1}$$

vibeNegativa

#	word	value
1	lamentavel	88.6349040414782649577
2	pessimo	87.7087738122062603451922
3	!ninguem	87.249377441079445816285
4	pior	86.9943578140045588042994495
5	decepcao	86.7427759182373421253394
6	decadente	85.0991409935101330 5749
7	sujo	84.8289830824553376942276372
8	horrivel	83.5898124907255308828190
9	pessima	82.8588587584921327788833
10	sujeira	81.7269690724025467917 3792
11	decepcionante	80.7396853239123686
12	terrivel	80.4932812556260302017 10
13	ma	80.1471232462987615008387 3537
14	mal	75.96644935823418620657321298
15	nunca	75.8699506945850714600965 14
16	ninguem	75.3480591175269580617168
17	gerente	74.4576201391991361333566
18	!neste	73.9060410966985870118151 0
19	telefone	73.5396212385751170472758
20	!reserva	73.4939605681304755080041
21	!dormir	72.4948263132041574483652
22	chao	72.3396094545306880263524362
23	informar	71.6471631374248829615 68
24	vaso	71.5949543376854506959716673
25	quebrar	70.9134643333881768967330
26	mofo	70.7549541733964844070214894
27	mau	70.5697499913059118625824 3396
28	cheirar	70.3481929938802892365856
29	seguinte	70.0219484291912124 26303
30	velho	69.8170399316340279938201710
31	perguntar	69.3636481362312622422 9
32	ruim	68.9899356962616110422459314
33	ligar	68.5878806364577826570 89410
34	total	68.4030815518179053924541 22
35	!pena	68.3060782698278626014250248
36	fugir	68.2552300265017772920697
37	tentar	68.1036037393195670119894 2
38	cartao	67.6087622472200564516242 5
39	!agua	66.459868349927830877277301
40	fotos	66.357973920058924477416439
41	absurdo	66.35671365850564100912237891
42	!conseguir	66.1318932058680815089
43	recomendar	66.1029230676364960572
44	impossivel	66.0286085060539988944
45	!hospedar	65.7251978957519753 5214

vibeNeutra

#	word	value
1	razoavel	91.6724860787745271868516
2	mediano	70.3433528606328621890497
3	normal	63.9280814607894539070552
4	basico	62.6727252916254187198 3194
5	simples	62.0046644029535443110034
6	curto	60.41937031009668855969653 1
7	regular	60.3479054141110893283439
8	honesto	58.4427895999583242314656
9	relativamente	57.9276208620836285
10	adequado	57.6563345345820863 48564
11	entanto	57.0432365990593289193384
12	melhorar	56.7697515538557553 32733
13	entretanto	53.6553267214230658055
14	dumont	53.5558423315235074824158 8
15	aeroporto	52.94661145621331854727
16	atendimento	52.301987191325096
17	pouca	51.8721456803783027567 07853
18	ok	51.70079330329740230354 2185109
19	negocios	51.1355571641554433 88153
20	porem	50.6220802381885803811 20118
21	negativo	50.0941755864449973 50099
22	!muito	49.5370496271989111392 0311
23	!bom	49.3355486499453732562869 845
24	vantagem	49.2618623397029082 86653
25	santo	48.9556138778443994397 09377
26	poder	48.820225349082569721304025
27	antigo	48.7371053237322513496 1103
28	relacao	48.7216759991193413270593
29	desejar	48.5995582322810548703 273
30	centro	48.3260924502014432846 4084
31	considerar	48.284481729163850 6162
32	precisar	48.131840449279850 48361
33	pouco	47.6846587995998021369 81452
34	lapa	47.6567271368277758369913499
35	mas	47.63473811567669002897673635
36	cozinhar	47.5897240228617803 83664
37	padrao	47.5840748302599223507 0589
38	geral	47.3300125956220227863 01386
39	itens	47.2910136316864679884 02000
40	pequeno	47.2563965034970721 035278
41	wi	47.1352321800065610091 22378891
42	ideal	47.1295929071443708835 43160
43	central	46.5786369814625373 919625
44	peco	46.5254204397782658020 332746
45	quem	46.3373192462914218481 273564

vibePositiva

#	word	value
1	parabens	91.5538435563986646796878
2	amar	91.5041905975724461086429073
3	fantastico	91.0935920798749805271
4	impecavel	90.1278332623627704 8337
5	adorar	89.7234309626973214355 5213
6	sensacional	87.6037343152745 64892
7	delicioso	85.94461593496805562608
8	maravilhoso	85.20983095424985 2652
9	top	84.8412289540765272022 23268261
10	deslumbrante	84.566067685977 51511
11	espetacular	83.4987526624562 40614
12	perfeito	83.40806248320680538 21996
13	incrivel	79.7186816305790415 52664
14	aconchegante	79.4959362190 8135787
15	recomendi	78.7702673965823 5499064
16	luxuoso	76.917369985783906 82072369
17	otimas	75.91050706286015575 491910
18	indico	75.679958078718073011 25954
19	otimo	75.413396068677881430 154957
20	excelente	75.0247897891432 2738091
21	certeza	73.8868119443403372 770
22	satisfeito	73.8868119443403 372770
23	lindo	73.6195263303373366170 49004
24	otimos	72.9049144731631599 8250502
25	moderno	72.203651458906804 8053377
26	fantastica	72.135890590489 9837756
27	excepcional	70.713374016854 857018
28	saboroso	69.046981383568834 1896645
29	super	69.023950687783042212 686268
30	completo	68.476522543074963 778053
31	organizado	68.3885240773034 297490
32	belo	68.3176218484889687942 995806
33	eficiente	67.910358553753653 15467
34	aniversario	67.741464130848 001445
35	farto	67.5261251566716964 58943188
36	cristo	66.9140425713141269 1705918
37	terraco	66.398201407522080 3306846
38	privilegiado	66.26129716136 270531
39	cobertura	66.260426998211 41468419
40	orlar	66.2147747218810422 96445230
41	catete	65.7167876062837166 27313115
42	variado	65.265402191865206 3050103
43	vista	64.748424140898379164 354992
44	expectativas	64.381918778 63232548
45	pertinho	63.8772091210025 50773482

Fig. 2. A sample of the Vibe table.

4   The Classifier

The classification technique used in the classifier was Naïve Bayes [1] with Laplace smoothing [7]. The probability of a review r belonging to a class c_i is given by the following formula.

$$p(r|c_i) = P(x_1, x_2, ..., x_n|c_i) \tag{2}$$

where x are the words in the review.

The Laplace smoothing is done adding 1 to the occurrence number of each word in the dictionary, and and thus, prevent a particular word has zero probability of occurring in a class. The Laplace smoothing expressed by the following equation.

$$\hat{P}(x_i, c) = \frac{count(X_i, c) + 1}{count(c) + |V|} \tag{3}$$

where V is the number of individual words.

5   Results

The lemmatization of the vocabulary allowed a considerable reduction in the number of tokens. The application resulted in the following reductions set out in Table 1.

Table 1. Reduction of tokens after the lemmatization

Polarity	Number of tokens	Number of tokens after lemmatization	Reduction rate (%)
Positive	30,703	20,014	34.81
Neutral	18,001	11,528	35.5
Negative	14,785	9,587	35.5

Several versions of the algorithm were implemented, with the purpose of evaluating the gains of each variation. The versions are shown in Table 2. As can be seen in the table the version with more precision was version number 6.

There are scenarios where we only want to analyze whether the review was negative or positive, and for those cases we apply algorithm version 6 to evaluate its performance.

It is noted that the result for the negative evaluations was very poor and it was suspected that the reason for this is the fact that there is a much greater number of positive reviews. In order to verify this hypothesis, another test was performed, and this time the same number of reviews (1499) was used in the training. In this case the performance for the negative ratings improved noticeably, as can be seen in the Table 6. This last test obtained a measure F with a value of 87.2%, which is very close to values obtained with more sophisticated techniques described in the related work (Tables 3, 4 and 5).

Table 2. Characteristics of each version of the algorithm and the accuracy of the tests.

Version	Stopwords	Inversion	Lemmatization	Add-1	Accuracy (%)
1	X				52.22
2	X	X			66.14
3	X		X		60.77
4	X			X	68.32
4	X	X		X	74.36
6	X	X	X	X	74.48

Table 3. Confusion matrix for version 6.

	Real		
	Positive	Neutral	Negative
Positive	**12810**	1207	86
Neutral	2963	**2073**	385
Negative	376	436	**1038**

Table 4. Precision and recall for version 6.

	Precision	Recall
Positive	90.83%	79.32%
Neutral	38.24%	55.79%
Negative	56.11%	65.79%

Table 5. Confusion matrix for version 6 only for positive and negative classes.

	Positive	Negative
Positive	**12810**	86
Negative	3339	**1423**

Table 6. Precision and recall with the same number of evaluations in training classes.

	Precision	Recall
Positive	97.90%	98.38%
Negative	78.69%	73.89%

6 Conclusions

Although it is a simple classification technique, the Naïve Bayes algorithm is able to classify the evaluations of hotels written in Brazilian Portuguese with a satisfactory precision and with a low computational cost.

The main difficulty is to classify a review as neutral. This is because it contains terms that occur in both the positive and negative classes and being a class with whose assignment has a certain subjectivity. An example of a review rated as neutral is *"banheiro horrivel, o hotel e bem localizado, restaurante excelente"* (bathroom horrible, the hotel is well located, restaurant excellent), but a person reading this comment would have trouble sorting it out.

Acknowledgments. This research is supported in part by the funding agencies FAPEMIG, CNPq, and CAPES.

References

1. Friedman, N., Geiger, D., Goldszmidt, M.: Bayesian network classifiers. Mach. Learn. **29**, 131–163 (1997)
2. Kasper, W., Vela, M.: Sentiment analysis for hotel reviews. In: Computational Linguistics-Applications Conference, vol. 231527 (2011)
3. Kozak, M.A., Arslan, E.: Evaluation of customer complaints of employees: the case of tripadvisor. In: Proceedings of The 2015 ICBTS International Academic Research Conference in Europe & America (2015)
4. Jurafsky, D., Martin, J.H.: Speech and Language Processing of the An Introduction to Natural Language Processing, Computational Linguistics, and Speech Recognition, 2nd edn. MIT Press, Cambridge (2008)
5. Pang, B., Lee, L., Vaithyanathan, S.: Thumbs up?: sentiment classification using machine learning techniques. In: Proceedings of the ACL 2002 Conference on Empirical Methods in Natural Language Processing, vol. 10, pp. 79–86. Association for Computational Linguistics (2002)
6. Vincent, M., Winterstein, G.: Construction et exploitation d'un corpus franais pour l'analyse de sentiment. In: TALN-RÉCITAL, pp. 764–771 (2013)
7. Yuan, Q., Cong, G., Thalmann, N.M.: Enhancing naive bayes with various smoothing methods for short text classification. In: Proceedings of the 21st International Conference on World Wide Web, pp. 645–646. ACM (2012)
8. Shimada, K., Inoue, S., Maeda, H., Endo, T.: Analyzing tourism information on twitter for a local city. In: 2011 First ACIS International Symposium on Software and Network Engineering (SSNE). IEEE (2011)
9. Ye, Q., Zhang, Z., Law, R.: Sentiment classification of online reviews to travel destinations by supervised machine learning approaches. Expert Systems with Applications **36**(3), 6527–6535 (2009)

Analyzing and Inferring Distance Metrics on the Particle Competition and Cooperation Algorithm

Lucas Guerreiro[✉] and Fabricio Breve

São Paulo State University (UNESP), Rio Claro, SP, Brazil
guerreiroluc@gmail.com, fabricio@rc.unesp.br

Abstract. Machine Learning is an increasing area over the last few years and it is one of the highlights in Artificial Intelligence area. Nowadays, one of the most studied areas is Semi-supervised learning, mainly due to its characteristic of lower cost in labeling sample data. The most active category in this subarea is that of graph-based models. The Particle Competition and Cooperation in Networks algorithm is one of the techniques in this field, which has always used the Euclidean distance to measure the similarity between data and to build the graph. This project aims to implement the algorithm and apply other distance metrics in it, over different datasets. Thus, the results on these metrics are compared to analyze if there is such a metric that produces better results, or if different datasets require a different metric in order to obtain a better correct classification rate. We also expand this gained knowledge, proposing how to identify the best metric for the algorithm based on its initial graph structure, with no need to run the algorithm for each metric we want to evaluate.

Keywords: Artificial intelligence · Semi-supervised learning · Distance metrics · Graphs

1 Introduction

Machine Learning (ML) is a field in Artificial Intelligence which can be applied to many areas. ML aims at developing algorithms and techniques based on previous knowledge, in an analog way to how the human brain works when acquiring some new knowledge [1, 2]. In ML we can highlight the semi-supervised learning class of algorithms. In this group of techniques, we use a small portion of labeled data, and the rest in unlabeled data, which we want to label or classify. The goal here is to use a combination of these data in order to achieve an efficient classification or labeling, taking into account that the labeling of training data is a task with high-costs and is slow. In this class of algorithms, we can talk about the group of graph-based algorithms, which has been the area with more studies in this type of algorithms [3]. We will work on an algorithm in this area, which is the Particle Competition and Cooperation algorithm (PCC) [4]. The premise of the algorithm is to construct a graph based on the dataset, then label a small portion of the graph nodes, and insert "particles" on these labeled points. These particles will walk through the graph and broadcast their labels on the unlabeled data; their labels are based on their original source node.

© Springer International Publishing AG 2017
O. Gervasi et al. (Eds.): ICCSA 2017, Part VI, LNCS 10409, pp. 717–725, 2017.
DOI: 10.1007/978-3-319-62407-5_53

Particles of the same class will cooperate to gain more territory when walking on the graph, making their class stronger. Particles of different classes will compete to gain territory, trying to strength nodes of its class and weakening nodes of different classes. When a particle visits one node of the same class as it, the particle gets stronger and raises the dominance of the node. When the particle visits a node of a different class, it gets weaker and the dominance level of the node is decreased. On the end of the execution, we attribute the class with the higher dominance on each node as its class and the algorithm is finished. Such as in many different graph-based algorithms, the initial structure of the graph is very important in determining a good or bad classification rate. In this algorithm, we create a graph at first, based on the distance between the parameters of each pair of nodes, and then we use the k-nearest-neighbors to determine if there is an edge between two nodes or not. The algorithm since its conception has always used the Euclidean Distance to construct the graph. In this work, we implement other 6 different metric distances to evaluate if we can get better results. We also propose a method to determine the best possible distance metric without running the algorithm for all the seven different metrics we are using.

2   Distance Metrics

For our experiments we are using seven of the most recommended distance metrics in the literature, these metrics are seen in Table 1.

Table 1. Distance Metrics

Distance	Equation
Euclidean [5]	$d(x,y) = \sum_i \sqrt{(x_i - y_i)^2}$
Mahalanobis [5]	$d(\vec{x}, \vec{y}) = \sqrt{(\vec{x} - \vec{y})^T S^{-1} (\vec{x} - \vec{y})}$
City Block [5]	$d(x,y) = \sum_{i=1}^{n} |x_i - y_i|.$
Chebyshev [6]	$d(x,y) = max|x_i - y_i|$
Minkowski [5]	$d(x,y) = \sqrt[m]{\sum_{i=1}^{n} |x_i - y_i|^m}$
Bray-Curtis [7]	$d(x,y) = \frac{\sum |x_i - y_i|}{\sum (x_i + y_i)}$
Canberra [8]	$d(x,y) = \sum_{i=1}^{n} \frac{|x_i - y_i|}{|x_i| + |y_i|}$

3   Methodology

In this work we propose the use of different distance metrics on the PCC algorithm, building different graphs. Considering the "walking" property, we want to evaluate whether we can achieve better results with these changes or not. In order to analyze such suppositions, we are implementing and testing these concepts on three UCI [9] datasets: Iris, Wine and Banknote Authentication. Iris dataset has 150 instances, with three parameters and 3 classes. Wine dataset has 178 instances, with 13 parameters and 3 classes. Banknote Authentication dataset has 1372 instances, with 4 parameters and 2 classes. After that, we want to find an equation which can describe a method to find the best metric based on the graph structure.

Therefore, we divide our work to two experiments: (i) Evaluating the influence of different distance metrics on PCC algorithm; (ii) Finding the best distance metric based on the graph structure.

For the experiment (i) we firstly implement the original algorithm [4], then we apply the modifications, by providing all the seven distance metrics, and then we compare the classification results of the Euclidean distance and the others in order to evaluate whether we achieve some gains. We run the algorithm 100 times and take the average since it is stochastic. At the end of the experiment for each dataset, we will have the results for each distance metric and the best metric for that dataset. For the experiment (ii) we work in an analog way to the first one, but we are just concerned with the assertive of best metric for the datasets and not the results themselves.

We apply the *z-score* normalization for the Wine and Banknote Authentication datasets, in order to get more consistent results. The Iris dataset already comes normalized so we did not need to apply any normalizations. To note this classification difference we applied the experiment for normalized and non-normalized data. Also, since Minkowski is a generalized equation, sensitive to the parameter m, we want to use a value that can generate interesting results. Considering we are already using $m = 1$ (City-Block), $m = 2$ (Euclidean) and $m = \infty$ (Chebyshev), we decided to use $m = 4$ in order to use a different distance metric. Therefore, when we say Minkowski in this work, we are referring to its form with $m = 4$.

Parameters
The PCC algorithm is sensitive to its initial parameters [4]. The aim of this word is to achieve good classification results, evaluating the influence of different distance metrics; therefore, it is important to optimize such parameters. To analyze and identify the best parameters we used the Iris dataset and applied all the definitions on it, fixing 2 parameters and varying the others. The main parameters in the PCC algorithm are: learning rate (Δv), the number of particles (p), the number of neighbors for each node (k), the number of iterations (z) and greedy walk rate (p_{grd}). The least sensitive among these five parameters is the number of iterations, considering a large number can produce good results, so we fixed it on 100,000 iterations. The number of particles has an influence on the final results as well; however, since we are using a semi-supervised algorithm and the labeled data on the datasets are random, we are using 10% of nodes as particles for all the cases. We have left the parameters we are investigating: Δv, k e p_{grd}.

Firstly, we have fixed k and p_{grd} to find Δv, we have varied its value between 0.05 and 0.5, in 0.05 intervals, we did not use bigger values to avoid overfitting the model and also because tests have shown it did not improve the results. We fixed the other parameters as recommended by the original algorithm, with $k = 10$ and $p_{grd} = 0.5$. These results are seen in Fig. 1.

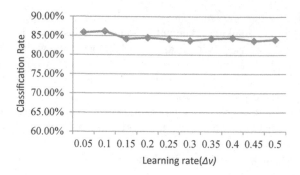

Fig. 1. Graph of classification rate versus learning rate, Δv

As we can observe the best learning rate was 0.1. We can also notice that learning rates larger than 0.1 produce worse results due to some overfitting. Therefore, we have chosen learning rate as 0.1.

The next parameters to be explored was the number of neighbors. We have varied k from 5 to 20, or proportionally to n the variation was from 3% to 13%, in intervals of 3. The other parameters were fixed as before. These test results are shown in Fig. 2.

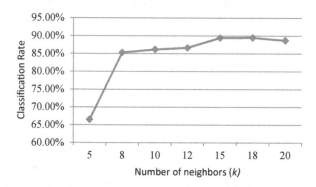

Fig. 2. Graph of classification rate vs number of neighbors, k

We see that the best results have been around $k = n/10$. Greater values have not been tested to avoid too many connections on large datasets. Therefore we have chosen $k = 10\%$, such as the number of particles p.

The last parameter to be found was p_{grd}, which represents the probability between 0 and 1 of having a greedy walk during the algorithm's execution. The original algorithm recommends this value near to 0.5, so we varied it from 0.3 to 0.7. The results are seen below (Fig. 3).

We observe that the value recommended on the reference of 0.5 proved to be among the best. Although some values have generated results close to this one, we kept it as 0.5.

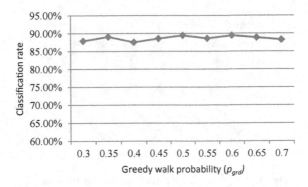

Fig. 3. Graph of classification rate versus greedy walk probability, p_{grd}

4 Results

Firstly, we will discuss the results we have found for the first experiment – executing the algorithm for different distance metrics and not only the Euclidean distance as proposed originally.

Using the parameters and definitions exposed on the previous section, we execute the algorithm, and the results are shown in this section. We show all the results after the average of 100 executions for the first experiment in Table 2.

Table 2. Classification rate for experiment I. Standard deviation represented beween brackets.

Distance metric	Classification rate		
	Iris	Wine	Banknote Authentication
Euclidean	89.44% (± 4.69%)	63.58% (± 7.02%)	95.48% (± 2.80%)
Normalized Euclidean	–	91.50% (± 7.19%)	**96.71% (± 0.86%)**
Mahalanobis	**91.17% (± 4.06%)**	64.73% (± 5.54%	68.63% (± 4.99%)
Normalized Mahalanobis	–	90.14% (± 12.01%)	79.69% (± 1.94%)
City Block	90.37% (± 5.51%)	65.40% (± 7.30%)	95.84% (± 2.46%)
Normalized City Block	–	**93.68% (± 8.02%)**	96.49% (± 0.66%)
Chebyshev	85.28% (± 8.87%)	63.31% (± 7.28%)	94.01% (± 2.97%)
Normalized Chebyshev	–	84.53% (± 15.28%)	94.81% (± 1.00%)
Minkowski	87.16% (± 5.64%)	64.40% (± 6.08%)	94.85% (± 2.18%)
Normalized Minkowski	–	86.53% (± 10.18%)	95.95% (± 0.81%)
Bray-Curtis	89.32% (± 8.27%)	65.67% (± 6.53%)	64.82% (± 19.48%)
Normalized Bray-Curtis	–	34.56% (± 6.30%)	51.76% (± 3.90%)
Canberra	81.87% (± 5.20%)	80.09% (± 19.06%)	52.04% (± 4.85%)
Normalized Canberra	–	33.72% (± 3.12%)	51.47% (± 3.41%)

We observe that the best metric for the Iris dataset was the Mahalanobis distance with 91.17% of correct classification. We already can prove that a different metric really improves the results. We also see that City-Block, Euclidean and Bray-Curtis distances did achieve good results as well. This result can also be seen in Fig. 4.

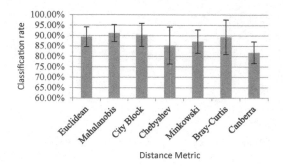

Fig. 4. Graph of classification rate on the Iris dataset

For the Wine dataset, we have run the algorithm for normalized and non-normalized data. For most cases, the normalized data have produced the best results. The best distance metric for this dataset was City-Block normalized, which had 93.68% of correct classification. We can also highlight the Euclidean and Mahalanobis distances, which also got good results. Details are shown in Fig. 5.

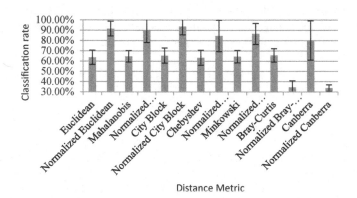

Fig. 5. Graph of classification rate on the Wine dataset

In our experiments, we also tested the Banknote Authentication dataset with normalized and non-normalized data. For this dataset, we can see that the Euclidean distance with normalized data achieved the best results with 96.71% of successful classification. City-Block distance achieved good results as well. These results are exposed in Fig. 6.

Therefore, for the three datasets we tested, we had three different metric distances as the best one. We can affirm that there is not one absolute best metric to use in every case. One possible solution is to run the algorithm for every distance metric we have, as we did in this section. However, it is a high-cost operation. We propose, then, a method to evaluate which metric is the best, based on the graph structure; this method and the results are explained in the further sections.

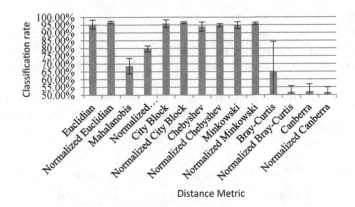

Distance Metric

Fig. 6. Graph of the classification rate on the Banknote Authentication dataset

5 Defining the Best Distance Metric

Based on tests and analyzing the algorithm and graph structure, we have got to the Eq. 1 below.

$$C_m = \frac{\sum_{i=1}^{n}(\max_c A_{p_i} - \sum A_{p_{i-k}})}{\sum_{i=1}^{n} A_{p_i}} \tag{1}$$

With A_{p_i} representing an edge connecting an unlabeled node i to a labeled node p; $\max_c A_{p_i}$ represents de amount of edges A_{p_i} which have the most classes in common with relation to the node i; $A_{p_{i-k}}$ represents edges which are not those with the most common class connected to the node i. Therefore, the coefficient C_m of a metric m, is the relation for all the unlabeled data, with the difference between the number of edges which link the most recurrent class to a given node and the others classes, for all the edges which connect labeled and unlabeled data. A coefficient C_m 1, then, represents a "cohesive" graph, with a good split between nodes of different classes, as seen in the example on Fig. 7a; on Fig. 7b we can observe the opposite example with a low value of coefficient C_m. Hypothetically, the distance metric with greater value of C_m. would provide a greater classification rate over a metric with low C_m due to the best split on the graph.

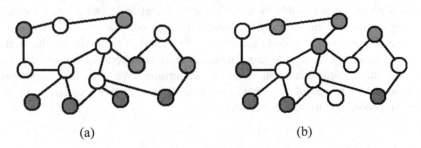

(a)   (b)

Fig. 7. Hypothetical examples of graphs with: (a) high C_m; (b) low C_m

6   Results

For this experiment, we have used the same parameters defined on the first experiment. However, in order to have more consistency, for each of the randomly generated group of particles, we repeated the experiment 10 times. The experiment for each dataset was repeated 100 times, and taken the average. The results were annotated as 1st and 2nd guess. Being 1st guess the greatest value of C_m and 2nd guess the second greatest value of C_m. The results are shown on Table 3 below.

Table 3. Correct guesses for experiment II

Dataset	1st guess	2nd guess	Total
Iris	91.2%	4%	95.2%
Wine	60%	34%	94%
Banknote Authentication	77.6%	21.2%	98.8%

As seen in the table above, for all the three datasets we have tested, summing up the 1st and 2nd guess we obtained to the best metrics on over 90% of the cases. Therefore, we can assume that using the Eq. 1 we are most of the time certain that we will get to the best distance metric, running the algorithm two times (for the 1st and 2nd guesses) with no need to execute the algorithm for all the metrics we want to evaluate.

7   Conclusions

In this work we had two goals: (i) evaluate the influence of different distance metrics on the PCC algorithm and achieve better results with such implementations; (ii) apply a method to identify the best metric previously to the execution of the algorithm in order to find the best distance metric without executing the algorithm several times.

We could see that the first objective was successfully reached, proving we can achieve better results than the original algorithm by applying other distance metrics. We can also notice that there is not a best metric for all the cases, needing to evaluate the metric that is the best for each dataset.

On the second goal, we also can call it successful, based on the results shown. We did infer a method to evaluate the best distance metric for each case. Our results demonstrate that we were able to identify the best metric in two guesses, which is promising considering the time we can save with such premise.

As future work, we propose to use more datasets and evaluate the results on those ones. It is interesting as well to use a more accurate parameter definition, maybe by optimizing those parameters with a genetic algorithm. We also find important to analyze deeply the results of the experiments to understand how the algorithm works to maybe infer those results in a better way.

Acknowledgment. The authors would like to thank the São Paulo Research Foundation - FAPESP (grant #2016/05669-4) and the National Counsel of Technological and Scientific Development - CNPq (grant #475717/2013-9) for the financial support.

References

1. Alpaydin, E.: Introduction to Machine Learning. MIT Press, Cambridge (2004)
2. Mitchell, T.: Machine Learning. McGraw Hill, New York (1997)
3. Chapelle, O., Schölkopf, B., Zien, A.: Semi-Supervised Learning, Adaptive Computation and Machine Learning. MIT Press, Cambridge (2006)
4. Breve, F.A., Zhao, L., Quiles, M.G., Pedrycz, W., Liu, J.: Particle competition and cooperation in networks for semi-supervised learning. IEEE Trans. Knowl. Data Eng. **24**, 1686–1698 (2012)
5. Liu, Q., Chu, X., Xiao, J., Zhu, H.: Optmizing non-orthogonal space distance using PSO in software cost estimation. In: IEEE Computer Software Applications Conference (COMPSAC), pp. 21–26 (2014)
6. Yang, Z., Shufan, Y., Yang, X., Liqun, G.: High-dimensional statistical distance for object tracking. In: International Conference on Measuring Technology and Mechatronics Automation (ICMTMA), vol. 2, pp. 386–389 (2010)
7. Shyam, R., Singh, Y.N.: Evaluation of eigenface and fisherfaces using Bray Curtis dissimilarity metric. In: 9th International Conference on Industrial and Information Systems (ICIIS), pp. 1–6 (2014)
8. Kokare, M., Chatterji, B.N., Biswas, P.K.: Comparison of similarity metrics for texture image retrieval. In: Conference on Convergent Technologies for the Asia-Pacific Region, vol. 2, pp. 571–575 (2003)
9. Bache, K., Lichman, M.: UCI Machine Learning Repository http://archive.ics.uci.edu/ml. Irvine, CA. University of California, School of Information and Computer Science (2013)

The D'Alembert Type Solution of the Cauchy Problem for the Homogeneous with Respect to Fourth Order Derivatives for Hyperbolic Equation

Mahir Rasulov[1,2], Zafer Aslan[3,6]($\boxtimes$), Bahaddin Sinsoysal[4], and Hakan Bal[5]

[1] Department of Mathematics and Computing, Beykent University, Istanbul, Turkey
[2] Faculty of Engineering and Architecture,
Istanbul Esenyurt University, Istanbul, Turkey
[3] Faculty of Engineering, Department of Computer Engineering,
Istanbul Aydın University, Istanbul, Turkey
zaferaslan@aydin.edu.tr
[4] Department of Management Information Systems,
Beykent University, Istanbul, Turkey
[5] Faculty of Administrative Sciences, Beykent University, Istanbul, Turkey
[6] International Centre for Theoretical Physics (ICTP), Trieste, Italy

Abstract. In this paper the D'Alembert type solution for the following

$$\sum_{i=0}^{4} a_i \frac{\partial^4 u(x,t)}{\partial t^{4-i} \partial x^i} = 0, \quad \frac{\partial^k u(x,t)}{\partial t^k}\bigg|_{t=0} = \phi_k(x), \quad (k = 0, 1, 2, 3)$$

Cauchy problem is constructed. Here, a_i, $(i = 1, 2, 3, 4)$ and $\phi_k(x)$, $(k = 1, 2, 3, 4)$ are given constants and functions, respectively. The cases where the roots of the characteristic equation are folded are examined and compact expressions for the solutions are obtained. The obtained solutions allow proving the uniqueness and existence of the solutions of the considered problem.

Keywords: D'Alembert type solution · Fourth order hyperbolic equation

1 Introduction

The theoretical investigation of many important problems in practice is often reduced to the solution of the Cauchy problem for partial differential equations. The solution of the Cauchy problem for hyperbolic type (usually linear) equations is studied by [1–11].

Formally, we divide the applied methods to solve the Cauchy problem, one of which is the variation methods [5,11] and the second is the methods using classical Fourier or other transformations [4,7–10]. In such studies, the formulas obtained which express the solution of the problem permit us both to verify the

© Springer International Publishing AG 2017
O. Gervasi et al. (Eds.): ICCSA 2017, Part VI, LNCS 10409, pp. 726–734, 2017.
DOI: 10.1007/978-3-319-62407-5_54

uniqueness and the existence of the solution of the Cauchy problem, and the continuous dependence of the solution of the initial data. As a matter of fact, certain difficulties are encountered when these solutions are used for practical purposes. It is clear that, obtaining easy-to-use solution formulas for high-order partial differential equations is important both from the practical point of view and theoretical point of view. With the help of such solutions, the existence and the uniqueness of the solution of the problem, as well as the problem of the solution being well defined can be easily examined. In addition, such solutions also allow evaluation of the solutions obtained using numerical methods applicable to differential equations. In this article, D'Alembert type solutions have been obtained for the solution of the Cauchy problem written for the homogeneous hyperbolic type equation according to the fourth-order derivative.

In upper half plane of $R^2(x,t)$ Euclid space we consider for unknown function $u = u(x,t)$ the following

$$a_0\frac{\partial^4 u}{\partial t^4} + a_1\frac{\partial^4 u}{\partial t^3 \partial x} + a_2\frac{\partial^4 u}{\partial t^2 \partial x^2} + a_3\frac{\partial^4 u}{\partial t x^3} + a_4\frac{\partial^4 u}{\partial x^4} = 0 \qquad (1)$$

$$\left.\frac{\partial^k u(x,t)}{\partial t^x}\right|_{t=0} = \phi_k(x), \quad (k = 0,1,2,3) \qquad (2)$$

Cauchy problem. Here, $\phi_k(x)$, $(k = 0,1,2,3)$ and a_i, $(i = 0,1,2,3,4)$ are certain functions and real constants, respectively. To solve the problem (1), (2) we will search a solution of the form $u(x,t) = \varphi(x + \lambda t)$ with four times differentiable function $\varphi(x)$. Here, number λ which described speed of traveling wave will be find from

$$a_0\lambda^4 + a_1\lambda^3 + a_2\lambda^2 + a_3\lambda + a_4 = 0, \qquad (3)$$

which is called characteristic equation. Since the Eq. (3) is a fourth-order algebraic equation, it may have up to 4 roots.

We will investigate in following cases:

1. The simple roots case,
2. $\lambda = \lambda_1$ is a four-multiple root,
3. $\lambda_1 = \lambda_3$, $\lambda_2 = \lambda_4$ i.e. λ_1 and λ_2 are 2-multiple roots,
4. $\lambda_1 = \lambda_3 = \lambda_4 \neq \lambda_2$, i.e. $\lambda = \lambda_1$ is a three-multiple root and $\lambda = \lambda_2$ is a simple root.

1.1   The Simple Roots Case

It is clear that, the function

$$u(x,t) = \varphi_1(x + \lambda_1 t) + \varphi_2(x + \lambda_2 t) + \varphi_3(x + \lambda_3 t) + \varphi_4(x + \lambda_4 t) \qquad (4)$$

is a general solution of (1). In order to find these unknown functions $\varphi_1, \varphi_2, \varphi_3$ and φ_4 by substituting (4) into (2), we get following algebraic equation as

$$\begin{cases} \varphi_1\left(x\right) + \varphi_2\left(x\right) + \varphi_3\left(x\right) + \varphi_4\left(x\right) = \phi_0\left(x\right), \\ \lambda_1\varphi_1'\left(x\right) + \lambda_2\varphi_2'\left(x\right) + \lambda_3\varphi_3'\left(x\right) + \lambda_4\varphi_4'\left(x\right) = \phi_1\left(x\right), \\ \lambda_1^{2}\varphi_1''\left(x\right) + \lambda_2^{2}\varphi_2''\left(x\right) + \lambda_3^{2}\varphi_3''\left(x\right) + \lambda_4^{2}\varphi_4''\left(x\right) = \phi_2\left(x\right), \\ \lambda_1^{3}\varphi_1'''\left(x\right) + \lambda_2^{3}\varphi_2'''\left(x\right) + \lambda_3^{3}\varphi_3'''\left(x\right) + \lambda_4^{3}\varphi_4'''\left(x\right) = \phi_3\left(x\right). \end{cases}$$

By integrating the second equation of this system once, third equation twice and fourth equation three times, we get

$$\begin{cases} \varphi_1\left(x\right) + \varphi_2\left(x\right) + \varphi_3\left(x\right) + \varphi_4\left(x\right) = \phi_0\left(x\right), \\ \lambda_1\varphi_1\left(x\right) + \lambda_2\varphi_2\left(x\right) + \lambda_3\varphi_3\left(x\right) + \lambda_4\varphi_4\left(x\right) = \int_0^x \phi_1(\xi)d\xi + C_{20}, \\ \lambda_1^2\varphi_1\left(x\right) + \lambda_2^2\varphi_2\left(x\right) + \lambda_3^2\varphi_3\left(x\right) + \lambda_4^2\varphi_4\left(x\right) = \int_0^x (x-\xi)\,\phi_2\left(\xi\right)d\xi + C_{30} + C_{31} \\ \lambda_1^3\varphi_1\left(x\right) + \lambda_2^3\varphi_2\left(x\right) + \lambda_3^3\varphi_3\left(x\right) + \lambda_4^3\varphi_4\left(x\right) = \frac{1}{2}\int_0^x (x-\xi)^2\phi_3\left(\xi\right)d\xi \\ \qquad\qquad\qquad\qquad\qquad\qquad\qquad\qquad + C_{40}\xi^2 + C_{41}\xi + C_{40}. \end{cases}$$

It is clear that, for existence of a single solution of this system, the determinant Δ, which is composed of the coefficients of unknown functions, is a Vandermonde's determinant and it must be nonzero. Since this determinant is nonzero, then the solution of the system is

$$\varphi_i\left(x\right) = \frac{1}{\Delta}\left(I_{i,1} + I_{i,0}\right), \quad (i = 1, 2, 3, 4). \tag{5}$$

Here,

$$I_{1,1} = \begin{vmatrix} \phi_0(x) & 1 & 1 & 1 \\ \int_0^x \phi_1(\xi)d\xi & \lambda_2 & \lambda_3 & \lambda_4 \\ \int_0^x (x-\xi)\,\phi_2\left(\xi\right)d\xi & \lambda_2^2 & \lambda_3^2 & \lambda_4^2 \\ \frac{1}{2}\int_0^x (x-\xi)^2\phi_3\left(\xi\right)d\xi & \lambda_2^3 & \lambda_3^3 & \lambda_4^3 \end{vmatrix}, I_{1,0} = \begin{vmatrix} 0 & 1 & 1 & 1 \\ C_{20} & \lambda_2 & \lambda_3 & \lambda_4 \\ C_{30}x + C_{31} & \lambda_2^2 & \lambda_3^2 & \lambda_4^2 \\ C_{40}x^2 + C_{41}x + C_{42} & \lambda_2^3 & \lambda_3^3 & \lambda_4^3 \end{vmatrix},$$

$$I_{2,1} = \begin{vmatrix} 1 & \phi_0(x) & 1 & 1 \\ \lambda_1 & \int_0^x \phi_1(\xi)d\xi & \lambda_3 & \lambda_4 \\ \lambda_1^2 & \int_0^x (x-\xi)\,\phi_2\left(\xi\right)d\xi & \lambda_3^2 & \lambda_4^2 \\ \lambda_1^3 & \frac{1}{2}\int_0^x (x-\xi)^2\phi_3\left(\xi\right)d\xi & \lambda_3^3 & \lambda_4^3 \end{vmatrix}, I_{2,0} = \begin{vmatrix} 1 & 0 & 1 & 1 \\ \lambda_1 & C_{20} & \lambda_3 & \lambda_4 \\ \lambda_1^2 & C_{30}x + C_{31} & \lambda_3^2 & \lambda_4^2 \\ \lambda_1^3 & C_{40}x^2 + C_{41}x + C_{42} & \lambda_3^3 & \lambda_4^3 \end{vmatrix},$$

$$I_{3,1} = \begin{vmatrix} 1 & 1 & \phi_0(x) & 1 \\ \lambda_1 & \lambda_2 & \int_0^x \phi_1(\xi)d\xi & \lambda_4 \\ \lambda_1^2 & \lambda_2^2 & \int_0^x (x-\xi)\,\phi_2\left(\xi\right)d\xi & \lambda_4^2 \\ \lambda_1^3 & \lambda_2^3 & \frac{1}{2}\int_0^x (x-\xi)^2\phi_3\left(\xi\right)d\xi & \lambda_4^3 \end{vmatrix}, I_{3,0} = \begin{vmatrix} 1 & 1 & 0 & 1 \\ \lambda_1 & \lambda_2 & C_{20} & \lambda_4 \\ \lambda_1^2 & \lambda_2^2 & C_{30}x + C_{31} & \lambda_4^2 \\ \lambda_1^3 & \lambda_2^3 & C_{40}x^2 + C_{41}x + C_{42} & \lambda_4^3 \end{vmatrix},$$

$$I_{4,1} = \begin{vmatrix} 1 & 1 & 1 & \phi_0(x) \\ \lambda_1 & \lambda_2 & \lambda_3 & \int_0^x \phi_1(\xi)d\xi \\ \lambda_1^2 & \lambda_2^2 & \lambda_3^2 & \int_0^x (x-\xi)\,\phi_2\left(\xi\right)d\xi \\ \lambda_1^3 & \lambda_2^3 & \lambda_3^3 & \frac{1}{2}\int_0^x (x-\xi)^2\phi_3\left(\xi\right)d\xi \end{vmatrix}, I_{4,0} = \begin{vmatrix} 1 & 1 & 1 & 0 \\ \lambda_1 & \lambda_2 & \lambda_3 & C_{20} \\ \lambda_1^2 & \lambda_2^2 & \lambda_3^2 & C_{30}x + C_{31} \\ \lambda_1^3 & \lambda_2^3 & \lambda_3^3 & C_{40}x^2 + C_{41}x + C_{42} \end{vmatrix}.$$

Lemma 1. Let $f_{ij}(y) = f_{ij}\left(y_1, y_2, y_3, y_4\right), (i = 2, 3, 4), (j = 0, 1, 2)$ be single valued functions in domain of definition D, then for any $y \in D$,

$$A_4 = \begin{vmatrix} 0 & 1 & 1 & 1 \\ f_{20}(y) & \lambda_2 & \lambda_3 & \lambda_4 \\ f_{30}\left(y\right)\lambda_1 + f_{31}(y) & \lambda_2^2 & \lambda_3^2 & \lambda_4^2 \\ f_{40}\left(y\right)\lambda_1^2 + f_{41}\left(y\right)\lambda_1 + f_{42}\left(y\right) & \lambda_2^3 & \lambda_3^3 & \lambda_4^3 \end{vmatrix}$$

$$+ \begin{vmatrix} 1 & 0 & 1 & 1 \\ \lambda_1 & f_{20}(y) & \lambda_3 & \lambda_4 \\ \lambda_1^2 & f_{30}(y)\lambda_1 + f_{31}(y) & \lambda_3^2 & \lambda_4^2 \\ \lambda_1^3 & f_{40}(y)\lambda_1^2 + f_{41}(y)\lambda_1 + f_{42}(y) & \lambda_3^3 & \lambda_4^3 \end{vmatrix}$$

$$+ \begin{vmatrix} 1 & 1 & 0 & 1 \\ \lambda_1 & \lambda_2 & f_{20}(y) & \lambda_4 \\ \lambda_1^2 & \lambda_2^2 & f_{30}(y)\lambda_1 + f_{31}(y) & \lambda_4^2 \\ \lambda_1^3 & \lambda_2^3 & f_{40}(y)\lambda_1^2 + f_{41}(y)\lambda_1 + f_{42}(y) & \lambda_4^3 \end{vmatrix}$$

$$+ \begin{vmatrix} 1 & 1 & 1 & 0 \\ \lambda_1 & \lambda_2 & \lambda_3 & f_{20}(y) \\ \lambda_1^2 & \lambda_2^2 & \lambda_3^2 & f_{30}(y)\lambda_1 + f_{31}(y) \\ \lambda_1^3 & \lambda_2^3 & \lambda_3^3 & f_{40}(y)\lambda_1^2 + f_{41}(y)\lambda_1 + f_{42}(y) \end{vmatrix} = 0.$$

We can easily prove this Lemma 1 using mathematical induction method. Using the Lemma 1, it is clear to show that $\sum_{i=1}^{4} I_{i,0}(x) = 0$. According into consideration of (4) for the solution of problem (1), (2) we have

$$u(x,t) = \frac{1}{\Delta}\{I_{1,1}(x + \lambda_1 t) + I_{2,1}(x + \lambda_2 t) + I_{3,1}(x + \lambda_3 t) + I_{4,1}(x + \lambda_4 t)\}. \tag{6}$$

(6) is the D'Alembert's type solution of the problem (1), (2).

Now let's get the D'Alembert's formula for the classic vibration of string equation from Eq. (6). In this case, the roots of the characteristic equation corresponding to Eq. (3) are $\lambda_1 = a$, $\lambda_2 = -a$ and $\Delta = \begin{vmatrix} 1 & 1 \\ a & -a \end{vmatrix} = -2a \neq 0$. From (6) we have

$$u(x,t) = -\frac{1}{2a}\left\{ \begin{vmatrix} \phi_0(x+at) & 1 \\ \int_0^{x+at}\phi_1(\xi)\,d\xi & -a \end{vmatrix} + \begin{vmatrix} 1 & \phi_0(x-at) \\ a & \int_0^{x-at}\phi_1(\xi)\,d\xi \end{vmatrix} \right\}$$

$$= \frac{\phi_0(x+at) + \phi_0(x-at)}{2} + \frac{1}{2a}\int_{x-at}^{x+at}\phi_1(\xi)\,d\xi.$$

This is the classical D'Alembert formula, [6].

1.2 Multiple Roots ($\lambda_1 = \lambda_2 = \lambda_3 = \lambda_4 = \lambda$)

In this section, we will consider the case when all roots of the characteristic equation are equal. In this case the general solution is written as

$$u(x,t) = \varphi_1(x + \lambda t) + t\varphi_2(x + \lambda t) + t^2\varphi_3(x + \lambda t) + t^3\varphi_4(x + \lambda t). \tag{7}$$

In order to obtain the unknown functions $\varphi_1(x), \varphi_2(x), \varphi_3(x)$ and $\varphi_4(x)$ we will use the initial conditions (2). Solving this system of algebraic equations, we get

$$\varphi_1(x) = \phi_0(x), \quad \varphi_2(x) = \phi_1(x) - \lambda\phi_0'(x),$$

$$\varphi_3\left(x\right) = \frac{1}{2}\left[\phi_2\left(x\right) - 2\lambda\phi_1'\left(x\right) + \lambda^2\phi_0''\left(x\right)\right],$$

$$\varphi_4\left(x\right) = \frac{1}{6}\left[\phi_3\left(x\right) - 3\lambda\phi_2'\left(x\right) + 3\lambda^2\phi_1''\left(x\right) - \lambda^3\phi_0'''\left(x\right)\right].$$

Substituting these obtaining functions in (7) for the solution we have

$$u\left(x,t\right) = \phi_0\left(x + \lambda t\right) - \lambda t\phi_0'\left(x + \lambda t\right) + \frac{\lambda^2 t^2}{2}\phi_0''\left(x + \lambda t\right)$$

$$-\frac{\lambda^3 t^3}{6}\phi_0'''\left(x + \lambda t\right) + \left[\phi_1\left(x + \lambda t\right) - \lambda t\phi_1'\left(x + \lambda t\right) + \frac{\lambda^2 t^2}{2}\phi_1''\left(x + \lambda t\right)\right]$$

$$+\frac{t^2}{2}\left[\phi_2\left(x + \lambda t\right) - \lambda t\phi_2'\left(x + \lambda t\right)\right] + \frac{t^3}{6}\phi_3\left(x + \lambda t\right) \tag{8}$$

$$= \sum_{j=0}^{3}\sum_{k=0}^{3-j}\left(-1\right)^k\frac{t^j}{j!}\phi_j^{(k)}\left(x + \lambda t\right)\left(x - \xi\right)^k, \quad \left(\xi = x - \lambda t\right).$$

If we assume that functions $\varphi_j\left(x\right)$, $\left(j = 0, 1, 2, 3\right)$ are $\left(7\text{-}j\right)$ times continuously differentiable functions, then it is easy to show that the function (8) is solution of problem (1), (2). As you can see, this solution shows the superposition of simple waves scattered at $-\lambda$ speed.

1.3 $\lambda_1 = \lambda_3, \lambda_2 = \lambda_4$, i.e. λ_1 and λ_2 are 2-Multiple Roots

It is clear that in this case the general solution of Eq. (1) is

$$u\left(x,t\right) = \varphi_{11}\left(x + \lambda_1 t\right) + t\varphi_{12}\left(x + \lambda_1 t\right) + \varphi_{21}\left(x + \lambda_2 t\right) + t\varphi_{22}\left(x + \lambda_2 t\right). \tag{9}$$

Unknown functions φ_{ij}, $\left(i, j = 1, 2\right)$ are founded from

$$\varphi_{11}\left(x\right) + \varphi_{21}\left(x\right) = \phi_0\left(x\right),$$

$$\lambda_1\varphi_{11}'\left(x\right) + \lambda_2\varphi_{21}'\left(x\right) + \varphi_{12}\left(x\right) + \varphi_{22}\left(x\right) = \phi_1\left(x\right),$$

$$\lambda_1^2\varphi_{11}''\left(x\right) + \lambda_2^2\varphi_{21}''\left(x\right) + 2\lambda_1\varphi_{12}'\left(x\right) + 2\lambda_2\varphi_{22}'\left(x\right) = \phi_2\left(x\right),$$

$$\lambda_1^3\varphi_{11}'''\left(x\right) + \lambda_2^3\varphi_{21}'''\left(x\right) + 3\lambda_1^2\varphi_{21}''\left(x\right) + 3\lambda_2^2\varphi_{22}''\left(x\right) = \phi_3\left(x\right)$$

system of algebraic equations.

To find the solution of this system, we first differentiate the first equation once, then integrate of the fourth equation twice, and the third equation once with respect to the variable x from zero to x. After this process, for $\varphi_{11}'\left(x\right)$, $\varphi_{21}'\left(x\right)$, $\varphi_{12}\left(x\right)$, $\varphi_{22}\left(x\right)$ we have the following system of equations as

$$\varphi_{11}'\left(x\right) + \varphi_{21}'\left(x\right) = \phi_0'\left(x\right),$$

$$\lambda_1\varphi_{11}'\left(x\right) + \lambda_2\varphi_{21}'\left(x\right) + \varphi_{12}\left(x\right) + \varphi_{22}\left(x\right) = \phi_1\left(x\right),$$

$$\lambda_1^2 \varphi_{11}'(x) + \lambda_2^2 \varphi_{21}'(x) + 2\lambda_1 \varphi_{12}(x) + 2\lambda_2 \varphi_{21}(x) = \int_0^x \phi_2(\xi)\, d\xi + C_{20},$$

$$\lambda_1^3 \varphi_{11}'(x) + \lambda_2^3 \varphi_{21}'(x) + 3\lambda_1^2 \varphi_{12}(x) + 2\lambda_2^2 \varphi_{21}(x) = \int_0^x d\xi \int_0^x \phi_3 d\xi + C_{30}x + C_{31}.$$

The determinant which is composed of coefficients of unknown functions must be nonzero, that is

$$\Delta_1 = \begin{vmatrix} 1 & 1 & 0 & 0 \\ \lambda_1 & \lambda_2 & 1 & 1 \\ \lambda_1^2 & \lambda_2^2 & 2\lambda_1 & 2\lambda_2 \\ \lambda_1^3 & \lambda_2^3 & 3\lambda_1^2 & 3\lambda_2^2 \end{vmatrix} \neq 0.$$

According the Cramer rule we have

$$\varphi_{11}'(x) = \frac{1}{\Delta_1}\left[I_{111}^{(2)}(x) + I_{110}^{(2)}(x) \right], \quad \varphi_{21}'(x) = \frac{1}{\Delta_1}\left[I_{211}^{(2)}(x) + I_{210}^{(2)}(x) \right], \quad (10)$$

$$\varphi_{12}(x) = \frac{1}{\Delta_1}\left[I_{121}^{(2)}(x) + I_{120}^{(2)}(x) \right], \quad \varphi_{22}(x) = \frac{1}{\Delta_1}\left[I_{221}^{(2)}(x) + I_{220}^{(2)}(x) \right], \quad (11)$$

where

$$I_{111}^{(2)}(x) = \begin{vmatrix} \phi_0 & 1 & 0 & 0 \\ \phi_1 & \lambda_2 & 1 & 1 \\ \int_0^x \phi_2(\xi)d\xi & \lambda_2^2 & 2\lambda_1 & 2\lambda_2 \\ \int_0^x d\xi \int_0^x \phi_3(\xi)d\xi & \lambda_2^3 & 3\lambda_1^2 & 3\lambda_2^2 \end{vmatrix}, \quad I_{110}^{(2)}(x) = \begin{vmatrix} 0 & 1 & 0 & 0 \\ 0 & \lambda_2 & 1 & 1 \\ C_{20} & \lambda_2^2 & 2\lambda_1 & 2\lambda_2 \\ C_{30}x + C_{31} & \lambda_2^3 & 3\lambda_1^2 & 3\lambda_2^2 \end{vmatrix},$$

$$I_{211}^{(2)}(x) = \begin{vmatrix} 1 & \phi_0 & 0 & 0 \\ \lambda_1 & \phi_1 & 1 & 1 \\ \lambda_1^2 & \int_0^x \phi_2(\xi)d\xi & 2\lambda_1 & 2\lambda_2 \\ \lambda_1^3 & \int_0^x d\xi \int_0^x \phi_3(\xi)d\xi & 3\lambda_1^2 & 3\lambda_2^2 \end{vmatrix}, \quad I_{210}^{(2)}(x) = \begin{vmatrix} 1 & 0 & 0 & 0 \\ \lambda_1 & 0 & 1 & 1 \\ \lambda_1^2 & C_{20} & 2\lambda_1 & 2\lambda_2 \\ \lambda_1^3 & C_{30}x + C_{31} & 3\lambda_1^2 & 3\lambda_2^2 \end{vmatrix},$$

$$I_{121}^{(2)}(x) = \begin{vmatrix} 1 & 1 & \phi_0 & 0 \\ \lambda_1 & \lambda_2 & \phi_1 & 1 \\ \lambda_1^2 & \lambda_2^2 & \int_0^x \phi_2(\xi)d\xi & 2\lambda_2 \\ \lambda_1^3 & \lambda_2^3 & \int_0^x d\xi \int_0^x \phi_3(\xi)d\xi & 3\lambda_2^2 \end{vmatrix}, \quad I_{120}^{(2)}(x) = \begin{vmatrix} 1 & 1 & 0 & 0 \\ \lambda_1 & \lambda_2 & 0 & 1 \\ \lambda_1^2 & \lambda_2^2 & C_{20} & 2\lambda_2 \\ \lambda_1^3 & \lambda_2^3 & C_{30}x + C_{31} & 3\lambda_2^2 \end{vmatrix},$$

$$I_{221}^{(2)}(x) = \begin{vmatrix} 1 & 1 & 0 & \phi_0 \\ \lambda_1 & \lambda_2 & 1 & \phi_1 \\ \lambda_1^2 & \lambda_2^2 & 2\lambda_1 & \int_0^x \phi_2(\xi)d\xi \\ \lambda_1^3 & \lambda_2^3 & 3\lambda_1^2 & \int_0^x d\xi \int_0^x \phi_3(\xi)d\xi \end{vmatrix}, \quad I_{220}^{(2)}(x) = \begin{vmatrix} 1 & 1 & 0 & 0 \\ \lambda_1 & \lambda_2 & 1 & 0 \\ \lambda_1^2 & \lambda_2^2 & 2\lambda_1 & C_{20} \\ \lambda_1^3 & \lambda_2^3 & 3\lambda_1^2 & C_{30}x + C_{31} \end{vmatrix}.$$

Using the Lemma 1, we get

$$I_{110}^{(2)}(x) + I_{210}^{(2)}(x) = 0, \quad I_{120}^{(2)}(x) + I_{220}^{(2)}(x) = 0. \quad (12)$$

By integrating (10) with respect to the variable x from 0 to x for $\varphi_{11}(x)$ and $\varphi_{21}(x)$ we have

$$\varphi_{11}(x) = \frac{1}{\Delta_1}\int_0^x [I_{111}(\xi) + I_{110}]\, d\xi = \frac{1}{\Delta_1}\int_0^x I_{111}(\xi)d\xi + \frac{1}{\Delta_1}\int_0^x I_{110}d\xi, \quad (13)$$

$$\varphi_{21}(x) = \frac{1}{\Delta_1} \int_0^x [I_{211}(\xi) + I_{210}]d\xi = \frac{1}{\Delta_1} \int_0^x I_{211}(\xi) \, d\xi + \frac{1}{\Delta_1} \int_0^x I_{210}d\xi. \quad (14)$$

As it is from $\frac{1}{\Delta_1} \int_0^x (I_{110} + I_{210}) \, d\xi = 0$. According into consideration (11), (13) and (14), from (9) for the solution of problem (1), (2) we get

$$u(x,t) = \frac{1}{\Delta_1} \left[\int_0^{x+\lambda_1 t} I_{111}(\xi) \, d\xi + t \int_0^{x+\lambda_1 t} I_{121}(\xi) \, d\xi \right]$$

$$+ \left[\int_0^{x+\lambda_2 t} I_{211}(\xi) \, d\xi + t \int_0^{x+\lambda_2 t} I_{221}(\xi) \, d\xi \right]$$

$$= \frac{1}{\Delta_1} \left\{ \int_0^{x+\lambda_1 t} \left[I_{111}(\xi) + t I_{121}(\xi) \right] d\xi + \int_0^{x+\lambda_2 t} \left[I_{211}(\xi) + t I_{210}(\xi) \right] d\xi \right\}.$$

1.4 $\lambda_1 = \lambda_2 = \lambda_3 \neq \lambda_4$, i.e. $\lambda = \lambda_1$ is a Three-Multiple Root and $\lambda = \lambda_2$ is a Simple Root

In this section, let us assume that λ_1 is a three-multiple root and λ_2 is a simple root. In this case, it is clear that the solution of Eq. (1) has in the following form

$$u(x,t) = \varphi_{11}(x + \lambda_1 t) + t\varphi_{12}(x + \lambda_1 t) + t^2\varphi_{13}(x + \lambda_1 t) + \varphi_{21}(x + \lambda_2 t) \quad (15)$$

where $\varphi_{11}(x)$, $\varphi_{12}(x)$, $\varphi_{13}(x)$ and $\varphi_{21}(x)$ are unknown functions. Using the initial conditions (2) to find these unknown functions, the following system of algebraic equations are obtained

$$\varphi_{11}(x) + \varphi_{21}(x) = \phi_0(x),$$

$$\lambda_1\varphi'_{11} + \varphi_{12}(x) + \lambda_2\varphi'_{21}(x) = \phi_1(x),$$

$$\lambda_1^2\varphi''_{11} + 2\lambda_1\varphi'_{12}(x) + \lambda_2^2\varphi''_{21}(x) = \phi_2(x),$$

$$\lambda_1^3\varphi'''_{11} + 3\lambda_1^2\varphi''_{12}(x) + 6\lambda_1\varphi'_{13}(x) + \lambda_2^3\varphi'''_{21}(x) = \phi_3(x).$$

By integrating the first equation of this system twice, also the fourth equation of this system once and assuming $\varphi''_{11}(x)$, $\varphi'_{12}(x)$, $\varphi_{13}(x)$, $\varphi''_{21}(x)$ are unknowns, then we get

$$\varphi''_{11}(x) + \varphi''_{21}(x) = \phi''_0(x),$$

$$\lambda_1\varphi''_{11}(x) + \varphi'_{12}(x) + \lambda_2\varphi''_{21}(x) = \phi'_1(x),$$

$$\lambda_1^2\varphi''_{11}(x) + 2\lambda_1\varphi'_{12}(x) + \lambda_2^2\varphi''_{21}(x) = \phi_2(x),$$

$$\lambda_1^3\varphi''_{11}(x) + 3\lambda_1^2\varphi'_{12}(x) + 6\lambda_1\varphi_{13}(x) + \lambda_2^3\varphi''_{21}(x) = \int_0^x \phi_3(\xi)d\xi + C_{40}$$

for unknown functions. Since the main determinant of the system which consists of the coefficients of unknown functions,

$$\Delta_2 = \begin{vmatrix} 1 & 0 & 0 & 1 \\ \lambda_1 & 1 & 0 & \lambda_2 \\ \lambda_1^2 & 2\lambda_1 & 0 & \lambda_2^2 \\ \lambda_1^3 & 3\lambda_1^2 & 6\lambda_1 & \lambda_2^3 \end{vmatrix} \neq 0$$

the solution of the system is

$$\varphi_{11}''(x) = \frac{1}{\Delta_2}\left[I_{111}^{(3)}(x) + I_{110}^{(3)}(x)\right], \quad \varphi_{12}'(x) = \frac{1}{\Delta_2}\left[I_{121}^{(3)}(x) + I_{120}^{(3)}(x)\right],$$

$$\varphi_{13}(x) = \frac{1}{\Delta_2}\left[I_{131}^{(3)}(x) + I_{130}^{(3)}(x)\right], \quad \varphi_{21}''(x) = \frac{1}{\Delta_2}\left[I_{211}^{(3)}(x) + I_{210}^{(3)}(x)\right],$$

respectively. Here,

$$I_{111}^{(3)}(x) = \begin{vmatrix} \phi_0''(x) & 0 & 0 & 1 \\ \phi_1'(x) & 1 & 0 & \lambda_2 \\ \phi_2(x) & 2\lambda_1 & 0 & \lambda_2^2 \\ \int_0^x \phi_3(\xi)d\xi & 3\lambda_1^2 & 6\lambda_1 & \lambda_2^3 \end{vmatrix}, \quad I_{110}^{(3)}(x) = \begin{vmatrix} 0 & 0 & 0 & 1 \\ 0 & 1 & 0 & \lambda_2 \\ 0 & 2\lambda_1 & 0 & \lambda_2^2 \\ C_{40} & 3\lambda_1^2 & 6\lambda_1 & \lambda_2^3 \end{vmatrix},$$

$$I_{121}^{(3)}(x) = \begin{vmatrix} 1 & \phi_0''(x) & 0 & 1 \\ \lambda_1 & \phi_1'(x) & 0 & \lambda_2 \\ \lambda_1^2 & \phi_2(x) & 0 & \lambda_2^2 \\ \lambda_1^3 & \int_0^x \phi_3(\xi)d\xi & 6\lambda_1 & \lambda_2^3 \end{vmatrix}, \quad I_{120}^{(3)}(x) = \begin{vmatrix} 1 & 0 & 0 & 1 \\ \lambda_1 & 0 & 0 & \lambda_2 \\ \lambda_1^2 & 0 & 0 & \lambda_2^2 \\ \lambda_1^3 & C_{40} & 6\lambda_1 & \lambda_2^3 \end{vmatrix},$$

$$I_{131}^{(3)}(x) = \begin{vmatrix} 1 & 0 & \phi_0''(x) & 1 \\ \lambda_1 & 1 & \phi_1'(x) & \lambda_2 \\ \lambda_1^2 & 2\lambda_1 & \phi_2(x) & \lambda_2^2 \\ \lambda_1^3 & 3\lambda_1^2 & \int_0^x \phi_3(\xi)d\xi & \lambda_2^3 \end{vmatrix}, \quad I_{130}^{(3)}(x) = \begin{vmatrix} 1 & 0 & 0 & 1 \\ \lambda_1 & 1 & 0 & \lambda_2 \\ \lambda_1^2 & 2\lambda_1 & 0 & \lambda_2^2 \\ \lambda_1^3 & 3\lambda_1^2 & C_{40} & \lambda_2^3 \end{vmatrix},$$

$$I_{211}^{(3)}(x) = \begin{vmatrix} 1 & 0 & 0 & \phi_0''(x) \\ \lambda_1 & 1 & 0 & \phi_1'(x) \\ \lambda_1^2 & 2\lambda_1 & 0 & \phi_2(x) \\ \lambda_1^3 & 3\lambda_1^2 & 6\lambda_1 & \int_0^x \phi_3(\xi)d\xi \end{vmatrix}, \quad I_{210}^{(3)}(x) = \begin{vmatrix} 1 & 0 & 0 & 0 \\ \lambda_1 & 1 & 0 & 0 \\ \lambda_1^2 & 2\lambda_1 & 0 & 0 \\ \lambda_1^3 & 3\lambda_1^2 & 6\lambda_1 & C_{40} \end{vmatrix}.$$

Since

$$I_{110}^{(3)}(x) + I_{120}^{(3)}(x) + I_{130}^{(3)}(x) + I_{210}^{(3)}(x) = 0, \tag{16}$$

for unknown functions $\varphi_{11}(x)$, $\varphi_{12}(x)$, $\varphi_{13}(x)$ and $\varphi_{21}(x)$ finally, we get

$$\varphi_{11}(x) = \frac{1}{\Delta_2}\int_0^x d\xi \int_0^x \left[I_{111}^{(3)}(\xi) + I_{110}^{(3)}(\xi)\right]d\xi,$$

$$\varphi_{12}(x) = \int_0^x \left[I_{121}^{(3)}(\xi) + I_{120}^{(3)}(\xi)\right]d\xi,$$

$$\varphi_{13}(x) = I^{(3)}_{131}(x) + I^{(3)}_{130}(x),$$

$$\varphi_{21}(x) = \int_0^x d\xi \int_0^x \left[I^{(3)}_{211}(\xi) + I^{(3)}_{211}(\xi) \right] d\xi.$$

To take these expressions and (16) into consideration, we find the following representation for solution of (1), (2)

$$u(x,t) = \int_0^{x+\lambda_1 t} (x - \xi - \lambda_1 t) I^{(3)}_{111}(\xi)\, d\xi + t \int_0^{x+\lambda_1 t} I^{(3)}_{121}(\xi)\, d\xi$$

$$+t^2 I^{(3)}_{131}(x)(x + \lambda_1 t) + \int_0^{x+\lambda_2 t} (x - \xi - \lambda_2 t) I^{(3)}_{211}(\xi)\, d\xi.$$

2   Conclusion

In this article, representations that express the exact solution of the Cauchy problem written for the hyperbolic equation which is homogeneous with respect to the fourth order derivative are obtained. With the help of the solutions obtained here, it is possible to investigate the propagation dynamics of waves in certain mediums.

Acknowledgment. The authors gratefully appreciate and acknowledge the support of International Centre for Theoretical Physics (ICTP)- Associateship Program.

References

1. Garding, L.: Linear hyperbolic partial differential equations with constant coefficients. Acta Math. **85**, 1–62 (1950)
2. Jost, J.: Partial Differential Equations. Springer, New York (2002)
3. Lax, A.: On cauchy's problem for partial differential equations with multiple characteristics. Commun. Pure Appl. Math. **9**, 135–169 (1956)
4. Leray, J.: Hyperbolic Differential Equations. Institute for Advanced Study, Princeton (1953)
5. Mizohata, S.: Lectures on Cauchy Problem. Tata Institute of Fundamental Research, Bombay (1965)
6. Petrowski, I.G.: On cauchy problem for system of linear partial differential equations in domain of non analytic functions. Bul. Mosk. Gos. Univ. Mat. Mekh. **7**, 1–74 (1938)
7. Pinchover, Y., Rubinstein, J.: An Introduction to Partial Differential Equations. Cambridge University Press, Cambridge (2005)
8. Polyanin, A.D.: Handbook of Linear Partial Differential Equations for Engineers and Scientists. Chapman and Hall/CRC Press, Boca Raton (2002)
9. Polyanin, A.D., Schiesser, W.E., Zhurov, A.I.: Partial differential equations. Scholarpedia **3**(10), 4605 (2008)
10. Rasulov, M.L.: Methods of Contour Integration. North Holland, Amsterdam (1967)
11. Sobolev, S.L.: Some Applications of Functional Analysis in Mathematical Physics. American Mathematical Society, Providence (1963)

A Nonlinear Finite Element Formulation Based on Multiscale Approach to Solve Compressible Euler Equations

Sérgio Souza Bento[1,2]([✉]), Paulo Wander Barbosa[1,2], Isaac P. Santos[2],
Leonardo Muniz de Lima[1], and Lucia Catabriga[1]

[1] High Performance Computing Lab, Federal University of Espírito Santo,
Vitória, ES, Brazil
{sergio.bento,paulo.barbosa}@ufes.br, lmuniz@ifes.edu.br,
luciac@inf.ufes.br
[2] Department of Applied Mathematics, Federal University of Espírito Santo,
São Mateus, ES, Brazil
isaac.santos@ufes.br

Abstract. This work presents a nonlinear finite element method for solving compressible Euler equations. The formulation is based on the strategy of separating scales – the core of the variational multiscale (finite element) methodology. The proposed method adds a nonlinear artificial viscosity operator that acts only on the unresolved mesh scales. The numerical model is completed by adding the YZβ shock-capturing operator to the resolved scale, taking into account the Mach number. We evaluate the efficiency of the new formulation through numerical studies, comparing it with other methodologies such as the SUPG combined with a shock-capturing operator.

Keywords: Finite element method · Multiscale formulation · Bubble function · Compressible flow problems

1 Introduction

Coming up with numerical methods to solve compressible Euler equations can be complicated by the presence of shocks and boundary layers in the computational domain. To solve this problem, researchers [4,7] have developed numerical formulations based on stabilized finite element method, since the beginning of the 1980s. The stabilization techniques prevent numerical oscillations and other instabilities when solving problems with high Reynolds and/or Mach numbers and shocks or strong boundary layers [7].

One well-known stabilized method for obtaining an accurate solution to the compressible Euler equations is the Streamline Upwind Petrov-Galerkin (SUPG) method coupled with the YZβ shock-capturing operator proposed by Tezduyar and Senga [7]. The stabilization parameter of the YZβ shock-capturing operator is calculated in an adaptive way, taking into account the directions of high

© Springer International Publishing AG 2017
O. Gervasi et al. (Eds.): ICCSA 2017, Part VI, LNCS 10409, pp. 735–743, 2017.
DOI: 10.1007/978-3-319-62407-5_55

gradients and the spatial discretization domain. The resulting stabilization parameter acts adaptively and is useful in avoiding excessive viscosity; this helps to maintain smaller numerical dissipations.

In order to solve the compressible Euler equations, Bento et al. [3] modified the Dynamic Diffusion (DD) method [2] by substituting its stabilization parameter by the YZβ parameter, yielding a self-adaptive numerical formulation, named Nonlinear Multiscale Viscosity (NMV) method. The NMV method was compared with the CAU/PG and SUPG + YZβ methods, providing good results. One drawback of the NMV method is that the $L^2(\Omega)$-norm of the density residue remains approximately constant throughout the time evolution.

In this paper, we propose a numerical formulation to solve inviscid compressible flow problems in conservative variables. The method, that is a modification of the method proposed in [3], adds different nonlinear artificial viscosity operators acting on the unresolved and the resolved scales. On the unresolved (or fine) scale of the discretization is added a nonlinear dissipative operator, similar to the Nonlinear Subgrid Scale (NSGS) method presented in [5]. This goes along with a special fine-scale stabilization parameter and the YZβ shock-capturing operator on the resolved scale. The stabilization parameter of the YZβ method is modified, taking into account the Mach number of the problem. The fine space is constructed by bubble functions defined into elements. We compare our method with the SUPG + YZβ and NMV methods. It is worth pointing out that this formulation is self-adaptive, a property inherited from the NSGS and YZβ methods.

The rest of this paper is organized as follows. Section 2 presents a discussion of the numerical formulation. Section 3 reports on numerical experiments conducted to show how our numerical formulation behaves in a variety of supersonic flow problems. Finally, Sect. 4 presents our conclusions.

2   Governing Equations and Numerical Formulation

We consider the two-dimensional compressible Euler equations for an ideal gas, written in conservative variables without source terms in its quasi-linear form,

$$\mathcal{L}\mathbf{U} := \frac{\partial \mathbf{U}}{\partial t} + \mathbf{A_x}\frac{\partial \mathbf{U}}{\partial x} + \mathbf{A_y}\frac{\partial \mathbf{U}}{\partial y} = \mathbf{0}, \text{on } \Omega \times (0, T_f], \tag{1}$$

where $\mathbf{A}_x = \frac{\partial \mathbf{F}_x}{\partial \mathbf{U}}$ and $\mathbf{A}_y = \frac{\partial \mathbf{F}_y}{\partial \mathbf{U}}$ are the Jacobian matrices, where $\mathbf{U} \in \mathbb{R}^4$ is the vector of conservative variables and $\mathbf{F}(\mathbf{U}) \in \mathbb{R}^{4 \times 2}$, the Euler flux vector. Associated with Eq. (1) we have a proper set of boundary and initial conditions. For simplicity, in the variational formulation we consider homogeneous Dirichlet boundary conditions.

The numerical formulation proposed for the Euler equation consists of finding $\mathbf{U}_E = \mathbf{U}_h + \mathbf{U}_B \in \mathcal{V}_E$ with $\mathbf{U}_h \in \mathcal{V}_h$, $\mathbf{U}_B \in \mathcal{V}_B$ such that

$$\underbrace{\int_{\Omega} \mathbf{W}_E \cdot \mathcal{L}\mathbf{U}_E d\Omega}_{\text{Galerkin term on } \mathcal{V}_E} + \underbrace{\sum_{e=1}^{nel} \int_{\Omega_e} \delta_B \left(\frac{\partial \mathbf{W}_B}{\partial x} \cdot \frac{\partial \mathbf{U}_B}{\partial x} + \frac{\partial \mathbf{W}_B}{\partial y} \cdot \frac{\partial \mathbf{U}_B}{\partial y} \right) d\Omega}_{\text{Nonlinear subgrid stabilization term}} +$$

$$\underbrace{\sum_{e=1}^{nel} \int_{\Omega_e} \delta_h \left(\frac{\partial \mathbf{W}_h}{\partial x} \cdot \frac{\partial \mathbf{U}_h}{\partial x} + \frac{\partial \mathbf{W}_h}{\partial y} \cdot \frac{\partial \mathbf{U}_h}{\partial y} \right) d\Omega}_{\text{Shock-capturing term on } \mathcal{V}_h} = \mathbf{0}, \; \forall \mathbf{W}_E \in \mathcal{V}_E, \tag{2}$$

where $\mathbf{W}_E = \mathbf{W}_h + \mathbf{W}_B \in \mathcal{V}_E$ with $\mathbf{W}_h \in \mathcal{V}_h$, $\mathbf{W}_B \in \mathcal{V}_B$. Here, $\mathcal{V}_h$ is the standard test space for first order approximation in a triangular mesh and $\mathcal{V}_B$ is a bubble function space, both are defined in details in [3]. The amount of fine-scale artificial viscosity is defined on the element-level by

$$\delta_B = \frac{h}{2} \frac{\|\mathbf{Y}^{-1} R(\mathbf{U}_h)\|}{\|\mathbf{Y}^{-1}(\nabla \mathbf{U}_h)^T\|}, \tag{3}$$

where $R(\mathbf{U}_h) = \frac{\partial \mathbf{U}_h}{\partial t} + \mathbf{A}_x^h \frac{\partial \mathbf{U}_h}{\partial x} + \mathbf{A}_y^h \frac{\partial \mathbf{U}_h}{\partial y}$ is the residue of the problem on Ω_e and $\mathbf{Y} = diag((U_1)_{\text{ref}}, (U_2)_{\text{ref}}, (U_3)_{\text{ref}}, (U_4)_{\text{ref}})$ is a diagonal matrix constructed from the reference values of the components of $\mathbf{U}$. The local length scale h is defined as in [6],

$$h = \left(\sum_a |\mathbf{j} \cdot \nabla N_a| \right)^{-1}, \tag{4}$$

with $\mathbf{j} = \nabla \rho / \|\nabla \rho\|$ a unit vector, and N_a is the interpolation function associated with node a. An expression similar to (3) is proposed in [5], based on the concept of minimum kinetic energy in order to measure the quantity of fine-scale artificial dissipation needed in scalar advection-diffusion-reaction problems. But the parameter introduced in [5] takes no account of information about the reference values of the problem as in (3) via matrix $\mathbf{Y}$ or of the subgrid mesh parameter h defined by Eq. (4).

The stabilization parameter of the shock-capturing operator is based on YZβ shock-capturing [7], and is written as

$$\delta_h = \frac{M}{4} \left(\frac{h}{2} \frac{\|\mathbf{Y}^{-1} R(\mathbf{U}_h)\|}{\|\mathbf{Y}^{-1}(\nabla \mathbf{U}_h)^T\|} + \frac{h^2}{4} \frac{\|\mathbf{Y}^{-1} R(\mathbf{U}_h)\|}{\|\mathbf{Y}^{-1} \mathbf{U}_h\|} \right), \tag{5}$$

where we introduce the Mach number dependence.

For both parameters, the local length h is defined automatically, accounting for the directions of high gradients and the spatial discretization domain. It is important to note that the stabilized terms in Eq. (2) are meant to be really considered in regions of the domain where the residual of the equation is relevant.

The numerical solution is obtained using iterative procedures for space and time. The iterative procedure for space is defined as follows: given $\mathbf{U}_E^i$ at iteration i, we find $\mathbf{U}_E^{i+1}$ satisfying the formulation (2). To improve convergence, the following convex combination to determine $\delta_B = \delta_B(\mathbf{U}_h)$ is used in the experiments [5]:

$$\delta_B(\mathbf{U}_h^i) = \nu^{i+1};$$
$$\nu^{i+1} = \omega\tilde{\nu}^{i+1} + (1-\omega)\nu^i, \text{ with } \omega \in [0,1] \text{ suitably chosen};$$
$$\tilde{\nu}^{i+1} = \frac{h}{2}\frac{\|\mathbf{Y}^{-1}R(\mathbf{U}_h)\|}{\|\mathbf{Y}^{-1}(\nabla\mathbf{U}_h)^T\|}, \text{ if } \|\mathbf{Y}^{-1}(\nabla\mathbf{U}_h)^T\| > tol_\delta$$

and $\tilde{\nu}^{i+1} = 0$, otherwise, for $i = 0, 1, \cdots, i_{MAX}$. The same strategy is applied to determine $\delta_h = \delta_h(\mathbf{U_h})$.

The formulation (2) can be partitioned in two subproblems, one related to the resolved scale, given by

$$\int_\Omega \mathbf{W}_h \cdot \mathcal{L}\mathbf{U}_h^{i+1} d\Omega + \int_\Omega \mathbf{W}_h \cdot \left(\frac{\partial \mathbf{U}_B^{i+1}}{\partial t} + \mathbf{A}_x^h \frac{\partial \mathbf{U}_B^{i+1}}{\partial x} + \mathbf{A}_y^h \frac{\partial \mathbf{U}_B^{i+1}}{\partial y}\right)d\Omega +$$

$$\sum_{e=1}^{nel} \int_{\Omega_e} \delta_h^i \left(\frac{\partial \mathbf{W}_h}{\partial x} \cdot \frac{\partial \mathbf{U}_h^{i+1}}{\partial x} + \frac{\partial \mathbf{W}_h}{\partial y} \cdot \frac{\partial \mathbf{U}_h^{i+1}}{\partial y}\right)d\Omega = \mathbf{0}, \quad \forall \mathbf{W}_h \in \mathcal{V}_h, \qquad (6)$$

and the other, representing the subgrid scale, is written as

$$\int_\Omega \mathbf{W}_B \cdot \frac{\partial \mathbf{U}_B^{i+1}}{\partial t} d\Omega + \int_\Omega \mathbf{W}_B \cdot \left(\frac{\partial \mathbf{U}_h^{i+1}}{\partial t} + \mathbf{A}_x^h \frac{\partial \mathbf{U}_h^{i+1}}{\partial x} + \mathbf{A}_y^h \frac{\partial \mathbf{U}_h^{i+1}}{\partial y}\right)d\Omega +$$

$$\sum_{e=1}^{nel} \int_{\Omega_e} \delta_B^i \left(\frac{\partial \mathbf{W}_B}{\partial x} \cdot \frac{\partial \mathbf{U}_B^{i+1}}{\partial x} + \frac{\partial \mathbf{W}_B}{\partial y} \cdot \frac{\partial \mathbf{U}_B^{i+1}}{\partial y}\right)d\Omega = \mathbf{0}, \quad \forall \mathbf{W}_B \in \mathcal{V}_B. \qquad (7)$$

The numerical solution is advanced in time by the implicit predictor-multi-corrector algorithm, adapted for the multiscale framework in [3,6] for the Euler equations.

3   Numerical Experiments

In this section we present numerical experiments considering three well-known 2D benchmark supersonic problems: "oblique shock", "reflected shock", and "blast wave/explosion", discretized by unstructured triangular meshes using Delaunay triangulation. The first and second problems used the GMRES solver with 5 vectors to restart. The third problem used the GMRES solver with 30 vectors to restart. In all problems, GMRES tolerance equal to 10^{-5}, the time integration tolerance is 10^{-5}, the non-linear tolerance – correction phase – is 10^{-2}, the maximum number of multi-corrections is fixed to 10, and the time-step size is 10^{-3}. We compare the new method with the NMV method and the SUPG formulation with the YZβ shock-capturing operator. The tests were performed on a machine with an Intel Core i7-4770 3.4 GHz processor with 16 GB of RAM and Ubuntu 14.04 operating system. The applications were compiled with the GNU gcc-4.8.4 using optimization flags -Ofast -march=native.

3.1 2D Oblique Shock Problem

The first problem is a Mach 2 uniform flow over a wedge, at an angle of $-10°$ with respect to a horizontal wall. The solution involves an oblique shock at an angle of $29.3°$ emanating from the leading edge of the wedge, as shown in Fig. 4(a). The computational domain is a square with $0 \leq x \leq 1$ and $0 \leq y \leq 1$. The inflow conditions are given at the left, up and bottom boundaries of the domain as described in [3] and no boundary condition is imposed at the outflow (right) boundary.

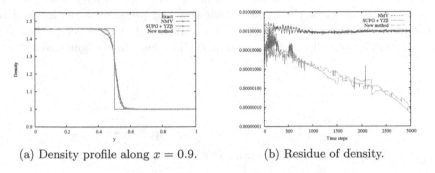

(a) Density profile along $x = 0.9$.   (b) Residue of density.

Fig. 1. Oblique shock problem.

For all simulations we consider an unstructured mesh consisting of 1,676 nodes and 3,202 elements. For all three methods, we plot in Fig. 1(a), the computed densities along line $x = 0.9$. We can observe that the new method and the SUPG + YZβ solutions are similar and more accurately represented than the NMV solution. Figure 1(b) shows the time evolution of the $L^2(\Omega)$-norm of the density residue. As we can see, the new method and SUPG + YZβ residual sequence converges to approximately 10^{-7} in less than 3,000 steps. For the NMV method, the residue remains approximately constant at the order of 10^{-3} throughout the time evolution. The computational performance of the new method is shown in Table 1 in terms of GMRES iterations and CPU time. The new method needs far fewer GMRES iterations and CPU time than the others. That is because, for the new method the nonlinear process converges before reaching the maximum number of non-linear iterations, whereas in the other two methods 10 iterations are performed at almost all time steps. The new method computational effort requires approximately 64% and 67% of the CPU times required by the NMV and the SUPG + YZβ methods.

3.2 2D Reflected Shock Problem

This problem consists of three regions (R1, R2, and R3) separated by an oblique shock and its reflection from a wall, as shown in Fig. 2. The inflow data in the first region R1 are known. Specifying those conditions and requiring the incident

Table 1. Oblique shock problem: computational Performance

Method	GMRES iterations	CPU time (s)
NMV	415,251	465.314
SUPG + YZβ	501,757	441.919
New method	234,160	298.665

shock to be at an angle of 29° results in an exact solution for regions R2 and R3 as described in [3]. The computational domain is a rectangle with $0 \leq x \leq 4.1$ and $0 \leq y \leq 1$. We prescribe the density, velocities and pressure at the left and top boundaries, the slip condition with $v = 0.0$ is imposed at the bottom boundary, and no boundary condition is imposed at the outflow (right) boundary.

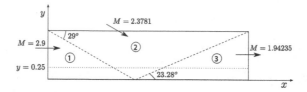

Fig. 2. Reflected shock problem description.

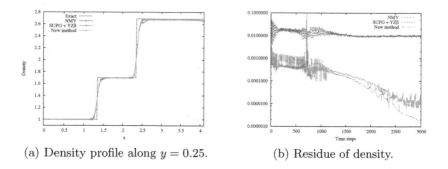

(a) Density profile along $y = 0.25$.   (b) Residue of density.

Fig. 3. Reflected shock problem.

For this example we consider an unstructured mesh consisting of 4,976 nodes and 9,626 elements. The density profile along the line $y = 0.25$ is shown in Fig. 3(a). We observe good agreement between the SUPG + YZβ and the new method solutions, and both are more accurate than the solution obtained by the NMV method. Figure 3(b) shows the time evolution of the $L^2(\Omega)$-norm of the density residue. The residual sequence of the SUPG + YZβ method converges to approximately 10^{-6} in less than 3,000 steps, whereas the new method converge to approximately 10^{-5}. The behavior of the NMV residual sequence is similar to the previous example, remaining approximately constant in the order of 10^{-2}. The computational performance for the reflected shock of the new method is shown in Table 2 in terms of GMRES iterations and CPU time. Once again, the new method needs far fewer GMRES iterations and CPU time than the others. The new method needs approximately half the number of GMRES iterations required by the SUPG + YZβ and NMV methods. Additionally, it requires approximately 65% and 56% of the CPU times required by the SUPG + YZβ and NMV methods.

Table 2. Reflected shock problem: computational Performance

Methods	GMRES iterations	CPU time (s)
NMV	459,364	2,125.755
SUPG + YZβ	476,337	1,836.932
New method	232,577	1,190.327

3.3 2D Explosion Problem

We consider the explosion problem as described by [1]. The 2D Euler equations are solved on a 2.0×2.0 square domain in the xy−plane. The initial condition consists of the region inside of a circle with radius $R = 0.4$ centered at $(1,1)$ and the region outside the circle, see Fig. 4(b). The flow variables are constant in each of these regions and are separated by a circular discontinuity at time $t = 0$. Details can be found in [3].

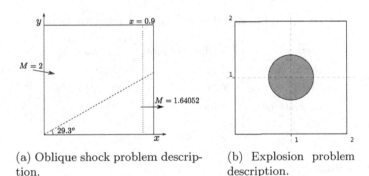

(a) Oblique shock problem description.

(b) Explosion problem description.

Fig. 4. Problems description.

In our simulation, we consider an unstructured mesh with 24,003 nodes and 47,468 elements. The radial variations of the density profile are shown in Fig. 5(a). Visually we have good agreement among the methods. Figure 5(b) shows the time evolution of the $L^2(\Omega)$-norm of the density residue for all three

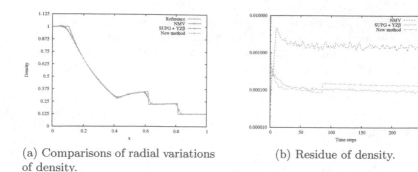

(a) Comparisons of radial variations of density.

(b) Residue of density.

Fig. 5. Explosion problem.

methods. The residual sequence of the SUPG + YZβ and the new method decrease until approximately 50 steps, remaining in the vicinity of the value 10^{-4} after this point. The NMV residual sequence presents an accentuated increase initially, followed by a small decrease, and remaining in the vicinity of the value 10^{-3}. The computational performance for the explosion problem of the new method is shown in Table 3 in terms of GMRES iterations and CPU time. Again, the new method needs fewer GMRES iterations and less CPU time than the others, requiring approximately 78% and 16% of the CPU times required by the SUPG+YZβ and NMV methods. However, for this problem the nonlinear SUPG+YZβ phase also converges before reaching the maximum number of nonlinear iterations, as the new method, while for the NMV method 10 iterations are performed at almost all time steps.

Table 3. Explosion problem: computational performance

Methods	GMRES iterations	CPU time (s)
NMV	141,824	3,072.606
SUPG + YZβ	26,760	637.239
New method	18,083	498.643

4   Conclusions

We have presented a numerical formulation to solve inviscid compressible flow problems in conservative variables. The method is a modification of the NMV method (proposed in [3]) adding different nonlinear artificial viscosity operators acting on the unresolved and the resolved scales. We have evaluated the performance of the new methodology by comparing it with the NMV and the SUPG + YZβ methods in the solution of three problems: oblique shock, reflected shock, and explosion problems. It is worth pointing out that the numerical experiments in [3] was obtained with the number of nonlinear multicorrections fixed equals 3, whereas in this work the nonlinear process goes on until the convergence criteria is satisfied or the maximum number of 10 iteration is reached. This strategy modified the quality of the solutions obtained with the NMV method, whereas the new numerical formulation provides similar solutions to those obtained by the SUPG + YZβ method, requiring fewer GMRES iterations, once the nonlinear procedure converges faster resulting in less CPU time. Furthemore, the $L^2(\Omega)$-norm of the density residue decays during the temporal evolution.

Acknowledgments. This work has been supported in part by CNPq, CAPES, and FAPES.

References

1. Abbassi, H., Mashayek, F., Jacobs, G.B.: Shock capturing with entropy-based artificial viscosity for staggered grid discontinuous spectral element method. Comput. Fluids **98**, 152–163 (2014). http://dx.doi.org/10.1016/j.compfluid.2014.01.022
2. Arruda, N.C.B., Almeida, R.C., Carmo, E.G.D.: Dynamic diffusion formulations for advection dominated transport problems. Mec. Comput. **29**, 2011–2025 (2010)
3. Bento, S.S., de Lima, L.M., Sedano, R.Z., Catabriga, L., Santos, I.P.: A nonlinear multiscale viscosity method to solve compressible flow problems. In: Gervasi, O., et al. (eds.) ICCSA 2016. LNCS, vol. 9786, pp. 3–17. Springer, Cham (2016). doi:10. 1007/978-3-319-42085-1_1
4. Hughes, T.J.R.: Multiscale phenomena: Green's functions, the Dirichlet-to-Neumann formulation subgrid scale models bubbles and the origins of stabilized methods. Comput. Methods Appl. Mech. Eng. **127**(1–4), 387–401 (1995). http://dx.doi.org/10.1016/0045-7825(95)00844-9
5. Santos, I.P., Almeida, R.C.: A nonlinear subgrid method for advection-diffusion problems. Comput. Methods Appl. Mech. Eng. **196**, 4771–4778 (2007). http://dx.doi.org/10.1016/j.cma.2007.06.009
6. Sedano, R.Z., Bento, S.S., Lima, L.M., Catabriga, L.: Predictor-multicorrector schemes for the multiscale dynamic diffusion method to solve compressible flow problems. In: CILAMCE2015 - XXXVI Ibero-Latin American Congress on Computational Methods in Engineering, November 2015
7. Tezduyar, T.E., Senga, M.: Stabilization and shock-capturing parameters in SUPG formulation of compressible flows. Comput. Methods Appl. Mech. Eng. **195**(13–16), 1621–1632 (2006). http://dx.doi.org/10.1016/j.cma.2005.05.032

Pump Efficiency Analysis of Waste Water Treatment Plants: A Data Mining Approach Using Signal Decomposition for Decision Making

Dario Torregrossa[1]([⊠]), Joachim Hansen[2], Francesc Hernández-Sancho[3],
Alex Cornelissen[4], Georges Schutz[4], and Ulrich Leopold[1]

[1] Department of Environmental Research and Innovation (ERIN),
Luxembourg Institute of Science and Technology (LIST),
Avenue des Hauts-Fourneaux, 4362 Esch-sur-Alzette, Luxembourg, Luxembourg
dario.torregrossa@list.lu, dariotorregrossa@gmail.com
[2] Université du Luxembourg, 6 Rue Richard Coudenhove-Kalergi,
1359 Luxembourg City, Luxembourg
[3] Water Economic Group, Universitat de València,
Avda dels Tarongers, s/n, 46022 Valencia, Spain
[4] RTC4Water s.a.r.l, 9, av. des Hauts-Fournaux,
4362 Esch-sur-alzette (belval), Luxembourg

Abstract. In Waste Water Treatment Plants (WWTPs), the pump systems are one of the most energy intensive processes. An efficient energy management of pumps should produce environmental and economic benefits. In this paper, we propose a daily data-driven approach for a detailed pump efficiency analysis that reduces the time gap between an inefficiency and its detection, provides detailed information for decision making by using new Key Performance Indicators (KPIs), and detects inefficient pump set-ups and designs. The proposed approach based on signal decomposition relies on sensors generally available in WWTPs, e.g. daily pump inflow and energy consumption. Moreover, it allows decomposing the data signal in an automatic way into a long-term trend and short-term fluctuations. This information can then be used to support plant managers more effectively.

Keywords: Waste Water Treatment Plant · Pump energy efficiency · Signal analysis · Decision support

1 Introduction: WWTP Pump Efficiency and Its Most Recurring Problems

In the last years, the economic and environmental benefits associated with a proper energy management in Waste Water Treatment Plants (WWTPs) have received considerable critical attention by several authors [1–5]. The detailed analysis of WWTP energy consumption showed the importance of pumps (12% of total consumption) in energy balance and an attractive global energy saving potential (up to 25%) [2,6]. A broad expert knowledge base - described in

© Springer International Publishing AG 2017
O. Gervasi et al. (Eds.): ICCSA 2017, Part VI, LNCS 10409, pp. 744–752, 2017.
DOI: 10.1007/978-3-319-62407-5_56

literature - has been developed to increase the energy efficiency of the pump systems. Some approaches are based on the comparison of efficiency against reference values, such as [7,8], while other authors worked on predictive maintenance ([9–11]). From our experience, the use of these approaches has some limitations in standard WWTP management conditions: regularity of analysis, quality of the information, sensor availability. For example, the benchmark analysis is generally done on a yearly basis and using average values with the consequence that the results are not accurate and inefficiencies are not detected at their early stage. On the other side, the main limitation of predictive maintenance is that some important pump parameters (such as vibration and temperature) are usually not measured in WWTPs.

The aim of this paper is to propose a plant-generic and user-friendly approach to support the daily performance analysis in the WWTP centrifugal pump systems. This method can investigate two important typologies of pump management issues: pump wear and set-up issues (like pump over-sizing and pump system controller inefficiencies [12]). The main idea is that pump efficiency is influenced by "slow" or long-term processes, such as typical pump ageing, and "fast" or short-term processes, such as changes in pump inflow. In this paper, we classify "slow" processes that can be observed at a monthly resolution and "fast" processes that can be observed at a daily resolution.

The novelty of the approach consists in applying algorithms typically developed for econometric time-series to analyse the energy efficiency of the pump and to calculate new indicators. With these algorithms, we can distinguish slow and fast processes applying a data signal decomposition. This approach shows promising results and can be easily replicated as it is based on data typically available in WWTPs (pump energy consumption, water flow and static head). We can apply this approach to single pumps and pump systems regardless their configuration and their capacity.

The approach has been tested in the WWTP of Solingen-Burg (GER) that is equipped with an intermediate pumping station with six pumps of 80 kW, driven by a controller. The data are collected and stored in the Energy Online System (EOS) which was designed in the framework of the INNERS project [13] and is currently further developed in the EdWARDS project [5]. Table 1 reports the statical description of the input variables.

In the methodological section, the reader will find the description of the algorithms used for the detrending of the η signal, the equations to calculate the new indicators and the approach to identify the flow-related issues. In Sect. 3, the reader can find the results and the operative conclusion of the methodology applied to the Burg WWTP. Discussion and conclusion report a general view of the advantages of this approach, potential limitations and further lines of research.

2   Methodology: Data-Driven Approach

The main index to estimate the pump performance is η, calculated with Eq. 1, in which m [kg] is the mass, g is the gravity acceleration [m/s^2], h is the static head

Table 1. Dataset description

	Energy [kWh/day]	Inflow [m3/day]	eta
Min	368	8589	0.17
1st quartile	1359	11752	0.23
Median	1537	14586	0.25
Mean	1647	16253	0.28
3rd quartile	1876	19260	0.27
Max	2995	37862	0.94

$[m]$ and E $[J]$ is the energy consumed by the pumps [10]. For high efficient pump $\eta = 0.85$ and for normal pump $\eta = 0.32$ [7]. The preliminary operation consists in calculating η with Eq. 1 and removing the out-layers values $(0 < \eta < 1)$.

$$\eta = \frac{m * g * h}{E} \qquad (1)$$

Multiple elements influence the value of η by altering the value of E (such as pump characteristic, operational conditions, system controller, wear rate) [12]. The operational conditions (such as rotation per minute, volume, current, temperature) can vary instantaneously while other phenomena (such as wear effects, or changes in dry weather flow profile), in normal condition, can be observed over several years [14–16].

In EOS, the η values of the WWTP pumps are calculated for each day and recorded in a database. The result is a time-series that can be decomposed into signals to distinguish the long-term trend and short-term fluctuations. This approach works under 2 assumptions: (1) the trend (η_t) depends on long-term phenomena, such as wear effects; (2) short term fluctuations (ϕ) depend on operational conditions. For each day we assume that the efficiency η (cf. Eq. 1) can be decomposed according to Eq. 2.

$$\eta = \eta_t + \phi \qquad (2)$$

We decompose the pump efficiency signal in η_t and ϕ by using four different signal analysis algorithms: rolling median, Kalman, Hodrick-Prescott (HP) and Baxter-king (BK) filters. These algorithms were implemented using R packages [17].

After decomposing η in trend and fluctuations, we propose a key performance indicator (KPI) to identify whether the pump is ageing as expected or irregularly. The pump efficiency is supposed to decrease slowly except for the last years of pump life in which the degradation rate is higher [11]. Normally a pump lifespan is between 15 and 20 years [16] in which the efficiency is reduced in a range of 10–25% [15]. Consequently, the standard yearly degradation rate of η is between 0.4% and 1.6% [18]. In [11] the wear rate is a factor depending on age and material composition that varies in a range between 0.8%/year and 2.4%/year. We propose to compare this information with a new KPI (τ) calculated by Eq. 3 in which η_{t_d} is the trend calculated at the reference day 'd', $\eta_{t_{d-180}}$ is the trend calculated at 180 days before 'd'. Consequently, τ represents the average slope of the trend. In Eq. 3, we have chosen to calculate the average trend slope for 180 days. According to the benchmark definition, we should have calculated the average slope of 1 year but a shorter period of investigation can reduce the detection time of irregular ageing phenomena. The decision about the dimension of this window of analysis can be customized. However, we suggest taking into account a period large enough to be representative of the most recurring operational conditions of the plants.

$$\tau = 365 * \frac{\eta_{t_d} - \eta_{t_{d-180}}}{180 * \eta_{t_{d-180}}} \qquad (3)$$

Typically, τ is expected to increase after maintenance and decrease because of ageing phenomena. The expected τ is then compared with four reference values $\pm 0.4\%/year$ and $\pm 1.6\%/year$. If $\tau < |0.4\%/year|$ no action is required, if $\tau > |0.4\%/year|$ and $\tau < |1.6\%/year|$ we suggest to plan a pump inspection, if $\tau > |1.6\%/year|$ we suggest an urgent maintenance.

In parallel, the analysis of fluctuations can provide interesting information for the early detection of the problems and the identification of inefficiency patterns. The ϕ sequence is a time-series and its trend can be used for the early detection of failures. In fact, we expect that ϕ fluctuates between positive and negative values, while a failure produces series of negative ϕ. A simple algorithm can alert the plant manager if a threshold number of consecutive negative ϕ is registered. The authors aim to improve this analysis in further works to produce more performing results.

The parameter ϕ can also be used to investigate issues related to an over-estimation of the design flow. These kind of issues are frequent [12,20] above all in a context as WWTPs in which the accurate estimation of design parameters is complicate.

The proposed approach can consequently address this research question: *Are pump inefficiencies depending on water flow conditions?* In order to provide an answer, we investigate if there are patterns between inflow and efficiency fluctuation (Table 2). We can identify four patterns in the database: (A) positive fluctuation and dry weather flow, (B) negative fluctuation and dry weather flow, (C) positive fluctuation and rain weather flow, (D) negative fluctuation and rain weather flow. The different data points should be distributed homogeneously inside the subsets composed by the dry weather flow and the rain weather flow. If this does not happen, it is probable that the undesirable efficiency patterns depend on the volume. This behaviour can be explained for example with a sub-optimal set-up of the pump controller, with a mismatching between pumps and variable speed driver (VSD) or with a pump over-sizing (in the case of a single pump). In fact, each pump works at its optimal efficiency when the flow correspond to the best efficient point (BEP) [14]. A common design error is generally to size a single pump for the maximum volume [20], with a consequent frequent low efficiency due to part-flow condition [14]. In multi-pump systems, the electronic control should ensure that the pump system performance does not decrease in the expected range of daily volumes.

The analysis of indicators about trend η_t, trend speed τ and ϕ patterns provide the plant manager with robust support for decision-making. For example, a well-performing trend with a negative trend speed (i.e. the pump is working well but is ageing fast) should be treated differently than a low stationary trend performance.

3   Results: Data-Driven Signal Decomposition

We have tested different algorithms on Burg WWTP intermediate pumping station. Kalman filter is based on a state transition, the rolling moving median is the median of previous 180 observed values, Baxter-King and Hodrick-Prescott are band-pass filters, originally used for business cycles analysis [19].

Even if the algorithms work in a different way, they produce a similar result (Fig. 1): a trend line mostly below the reference benchmark line ($\eta = 0.32$) with an increasing trend in the last part of the time-series. The first conclusion is that

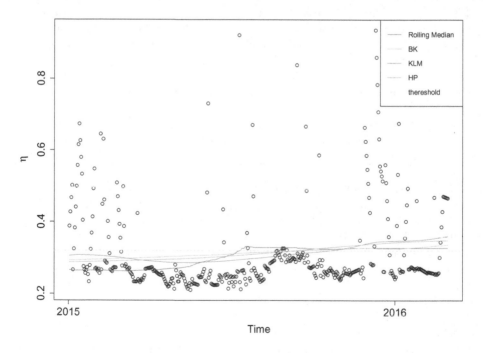

Fig. 1. Comparison between trend algorithms. The circled points are the observed points, the lines represent the trend obtained by different algorithms in accordance with the legend. The dotted blue line corresponds to the threshold value of $\eta = 0.32$ (Color figure online)

the energy performance of the pump system is not optimal. Consequently, we suggest an additional investigation to identify the sources of this inefficiency.

The benchmark analysis is performed also taking into consideration the values of τ. With all the decomposition algorithms, the most frequent condition is that $\tau < |0.4\%/year|$ (Fig. 2). In order to make a balanced decision when the algorithms produce diverging results, we take into account the output of each method. For example, the rolling median shows a peak in the period in the first part of 2016 not observed with other algorithms (Fig. 2). In this case, an urgent maintenance is not required.

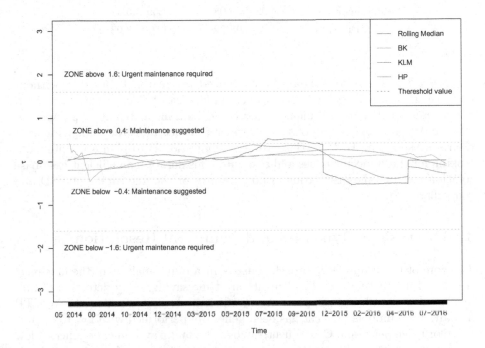

Fig. 2. Each line reports the τ control chart for different algorithms. When τ is between the horizontal grey line no action is required, when τ is between a dotted grey and a dotted blue lines a maintenance is required and when τ is not between the dotted blue lines an urgent action is required. (Color figure online)

Table 2 reports the distribution of the points depending on fluctuation and volume for Burg WWTP intermediate pumping station. The analyses performed with distinct decomposition algorithms produce similar results. In particular, we can observe that during the dry weather flow, the points are concentrated in the zone of negative fluctuations (B zone has more points than A one). During the rain weather flow, the point density is higher in the D zone (negative fluctuations) than in C zone (positive fluctuations). The comparative analysis between dry and rain conditions shows that the pump system produces better performance during the rain weather flow.

Table 2. Table: conditional distribution of points depending on fluctuation and volume. We can observe 4 zones corresponding to different operational conditions: (A) positive fluctuation and dry weather flow, (B) negative fluctuation and dry weather flow, (C) positive fluctuation and rain weather flow, (D) negative fluctuation and rain weather flow.

	Dry weather flow		Rain weather fow	
Operational conditions	A[%]	B[%]	C[%]	D[%]
Kalman filter	1.78	98.22	22.46	77.54
Baxter-king	1.78	98.22	24.05	75.95
Rolling median	4.92	95.08	54.59	45.41
Hodrick-Prescott	1.78	98.22	21.92	78.08

A multi-pump system should be optimized to treat different flow without decreasing the efficiency and such a performance can be considered inefficient. The sources of flow-related inefficiencies could be a mismatching of pumps and VSD and/or a wrong set-up of the controller that activates the pumps far from their corresponding BEP. In this case, the suggestion is to plan a maintenance looking for these specific issues and take into consideration an efficient case-base solution (such as the installation of pony pumps or the fine tuning of VSD and controller [12]).

4 Discussion: Dynamics and Temporal Resolution

The core of the proposed approach consists in a daily analysis of the efficiency trend and its fluctuations. The benefits are time saving, early detection of failures, and a data-based support for an efficient decision making. In WWTP domain, the pump assessment has been generally performed with low frequency (once or twice per year). Consequently, when the pump system experiences a low energy efficiency condition, the problem and its potential solutions are identified with a time gap in which environmental and economic costs increase. Since SCADA systems start to be a common equipment in WWTPs [5], this time gap can be decreased by increasing the regularity of analysis. In this paper, we suggest a daily analysis because the flow and energy sensors have distinct measurement time (the flow sensor reports one value each 15 min, the energy sensor reports one value each 2 h) and the values are aggregated to daily resolution.

The same approach can be performed with a higher resolution with the only limitation of the sensor measurement interval. At this time resolution, the value of η, strictly depending on operational condition, is not a satisfactory indicator because it produces the not realistic conclusion that the pump efficiency characteristic changes day by day. The signal decomposition (trend and fluctuations) enables the plant managers to analyse the performance more in detail distinguishing between long-term phenomena and short-term fluctuations. This approach shows a great added value when compared with the current practice;

in fact, with a classical approach, the only information for the plant manager should be the average η for the period, equal to $\eta = 0.28$.

5 Conclusion

Pumps are one of the most important energy consumers in the WWTP domain. Nowadays the efficiency analysis is performed with low regularity. The availability of high-frequency information (second, hours, day) enables the plant manager to analyse the pump system more in detail and at a higher resolution.

In this paper, an effective approach is presented to support the daily performance analysis of pump system. This approach enables plant managers to distinguish long-term and short-term phenomena, detect problems related to pump degradation and set-up issues. Consequently, it produces economic and environmental benefits because the problems can be detected at their early stage and the solutions are based on a detailed analysis.

Moreover, this approach relies on information generally available in WWTPs and it is free of budget costs because it can be implemented with open source software.

In the perspective of the installation of a wider sensor-set (such as vibration and temperature), this approach could be extended to produce a more detailed analysis. Moreover, future developments of this approach will take into account additional managerial, economic and environmental parameters in order to provide the plant managers with additional information. Last but not least, the authors are improving this approach to include the analysis of other specific pump problems such as cavitation, the effect of water quality parameters (such as suspended solids, pH, temperature) or fluid leakages.

Acknowledgements. The authors gratefully acknowledge the financial support of Luxembourg Institute of Science and Technology (LIST) and the National Research Fund (FNR) in Luxembourg (7871388-EdWARDS). The methodology and the results presented in this paper are part of the PhD project of Dario Torregrossa (EdWARDS).

References

1. Gude, V.G.: Energy and water autarky of wastewater treatment and power generation systems. Renew. Sustain. Energy Rev. **45**, 52–68 (2015)
2. Castellet, L., Molinos-Senante, M.: Efficiency assessment of wastewater treatment plants: a data envelopment analysis approach integrating technical, economic, and environmental issues. J. Environ. Manage. **167**, 160–166 (2016)
3. Becker, M., Hansen, J.: Is the energy-independency already state-of-art at NW-European wastewater treatment plants. In: Proceedings of the IWAconference Asset Management for enhancing energy efficiency in water and wastewater systems', pp. 19–26 (2013)
4. Hernández-Sancho, F., Molinos-Senante, M., Sala-Garrido, R.: Energy efficiency in Spanish wastewater treatment plants: a non-radial DEA approach. Sci. Total Environ. **409**(14), 2693–2699 (2011)

5. Torregrossa, D., Schutz, G., Cornelissen, A., Hernandez-Sancho, F., Hansen, J.: Energy saving in WWTP: daily benchmarking under uncertainty and data availability limitations. Environ. Res. **148**, 330–337 (2016)
6. Shi, C.Y.: Mass Flow and Energy Efficiency of Municipal Wastewater Treatment Plants. IWA Publishing, London (2011)
7. Spellman, F.R.: Handbook of Water and Wastewater Treatment Plant Operations. CRC Press, Boca Raton (2003). doi:10.1016/B978-0-12-369520-8.50043-7
8. WEF. Operation of Municipal Wastewater Treatment Plants, vol. 1542. WEF Press, New York (2008). doi:10.1017/CBO9781107415324.004
9. Berge, S.P., Lund, B.F., Ugarelli, R.: Condition monitoring for early failure detection. Frognerparken pumping station as case study. Procedia Eng. **70**(1877), 162–171 (2014)
10. Marchi, A., Simpson, A.R., Ertugrul, N.: Assessing variable speed pump efficiency in water distribution systems. Drinking Water Eng. Sci. **5**(1), 15–21 (2012)
11. Beebe, R.S.: 2 - Pump performance and the effect of wear. In: Predictive Maintenance of Pumps Using Condition Monitoring (2004)
12. Department of Environment (UK): Improving Pumping System Performance: A Sourcebook for Industry. Technical report (2006)
13. INNERS: Innovative energy recovery strategies in the urban water cycle. Technical report (2015)
14. Gülich, J.F.: Centrifugal Pumps (1998)
15. Hydraulic Institute and Energy, U.D. of. Energy Tips - Pumping Systems (2005)
16. Hydraulic Institute, Europump, Office of Industrial Technologies - US Department of Energy: Pump Life Cycle Costs: A Guide to LCC Analysis for Pumping Systems - Executive Summary. Renewable Energy (2001)
17. R Core Team: R: A Language and Environment for Statistical Computing. R Foundation for Statistical Computing, vol. 1 (2011)
18. Torregrossa, D., Leopold, U., Hernández-Sancho, F., Hansen, J., Cornelissen, A., Schutz, G.: A tool for energy management and cost assessment of pumps in waste water treatment plants. In: Linden, I., Liu, S., Colot, C. (eds.) ICDSST 2017. LNBIP, vol. 282, pp. 148–161. Springer, Cham (2017). doi:10.1007/978-3-319-57487-5_11
19. Balcilar, M.: mFilter (2007). https://cran.rproject.org/web/packages/mFilter/mFilter.pdf
20. European Association of Pump Manufacturers, and DOE: Variable Speed Pumping, A guide to successful applications. Elsevier Advanced Technology, Amsterdam (2004)

"It Sounds Wrong…" Does Music Affect Moral Judgement?

Alessandro Ansani$^{(\boxtimes)}$, Francesca D'Errico, and Isabella Poggi

University of Rome 3, Rome, Italy
alessandro.ansani@gmail.com, {francesca.derrico,
isabella.poggi}@uniroma3.it

Abstract. Can simply listening to a music piece affect the harshness of a moral judgement? A priming experiment was run to answer this question. Participants gave moral judgements, after listening to musical pieces inducing certain emotions (Joy, Relax, Sadness, Annoyance). After reading some vignettes about moral transgressions and rating them, they were asked to fill in a self-report affect questionnaire concerning the emotions experienced during their listening, and a test assessing musical sensitivity. In accordance with Greene's dual-process moral theory, classic moral vignettes fell into two categories: "high emotional involvement" (HEI) vs. "low emotional involvement" (LEI). Results show that the two emotions with a negative valence (Sadness, Annoyance) worsened the overall harshness of participants' moral judgements while the positive emotions (Joy, Relax) weakened it; in the most arousing ones (Joy, Annoyance) the effect was increased, and the annoyance condition determined the highest moral harshness. This effect was stronger in the HEI moral questions, as predicted by the dual-process moral theory.

Keywords: Affective priming · Musical priming · Moral judgement · Music & emotions · Emotion induction

1 Introduction

In the late moral psychology literature, several studies have investigated the influence that semantic and non-semantic stimuli can exert over moral evaluations:

- **Semantic**: reading or listening to a tale, retrieving some memory, viewing some picture [1], watching a film, a parody sketch [2] or a short clip [3] may all be regarded as semantic stimuli since they require high-level complex processing.
- **Non-semantic**: a smell [4], a taste [5], a color or even the subliminal presentation of a facial expression [6, 7] are all contingencies that have a deep influence on thought, and therefore decision-making, subliminally, since they require a lower and more peripheral processing.

Starting from such previous research, our goal in this work is to test the influence of music on moral judgement. We chose this kind of stimuli precisely because a typical (though not totally pervasive) feature of music is its a-semanticity [8].

© Springer International Publishing AG 2017
O. Gervasi et al. (Eds.): ICCSA 2017, Part VI, LNCS 10409, pp. 753–760, 2017.
DOI: 10.1007/978-3-319-62407-5_57

This is not the place to enter into the debate about musical meanings (interesting introductive treatises may be found in [9–11]); nevertheless a clarification is needed: here we refer to "a-semantic" in the sense of "devoid of referential content".

Actually, certain music pieces, such as "descriptive" music (e.g., Rimskij Korsakov's *Flight of the bumblebee*) or culturally loaded music (e.g. Chopin's *Funeral march*, or Joplin's *Entertainer*), thanks to their auditory-visual iconicity, or to cultural connections, may remind or evoke specific memories, past experience, people or places[1]. Yet, for the majority of music pieces emotions are the only content conveyed (or induced). Using this kind of a-semantic music was the best way to be certain that influence over moral judgment was only due to the music itself and not to other "referential" contents.

2 Musical Primes and Moral Judgements

Our hypothesis is that music influences the harshness of moral judgements much like other non-semantic stimuli do. Since the influence of music on moral judgements is presumably mediated by the emotions elicited by music [12], we deal with a double-step process: the first step is the relationship between music and emotions (music alters the affective state) [13, 14]; the second concerns emotions and moral judgement (the ongoing affective state influences the moral choice) [15–17].

Therefore, to work as primes for the subsequent moral judgements, we looked for music fragments apt to induce four emotions varying in terms of valence and arousal (Table 1), like those elicited in previous studies. Yet, whereas studies with visual and taste stimuli had elicited *disgust* (as a –Valence and + Arousal emotion), given the difficulty of finding "disgusting" music, we looked for one eliciting *annoyance*: these two emotions beside the same valence-arousal pattern also share a "mental ingredient" of "distance": the subject who experiences them feels the pressing need to distance himself from the eliciting object.

Table 1. Experimental design.

	Valence	Arousal
Joy	+	+
Sadness	–	–
Annoyance	–	+
Relax	+	–

---

[1] To be more accurate, according to Scherer and Zentner [13]: "an emotion that is actually experienced by a listener while listening to music is determined by a multiplicative function consisting of several factors: *Experienced emotion = Structural features × Performance features × Listener features × Contextual features.*

Priming Items Selection Criteria

Not to expose the priming items to inter-subject variability, we opted for unknown pieces. While to properly induce a mood 8-minutes long pieces are generally used [18], since in our study the experimenter was not physically present with participants, in order to avoid boredom or distraction effects, we shortened the listening to no more than 3.40 min long.

In accordance with the framework by Juslin and Laukka [19] (Fig. 1), for the joy condition we chose a fast piece, major tonality, fluent rhythm, pronounced dynamics. Criteria for the sadness condition were above all slowness and minor tonality. For the annoyance condition, we chose a contemporary music piece to elicit an estrangement-like feeling thanks to the complexity of the rhythmic structure, the absence of any tonal center and the incessant variation of tempo and dynamics[2]. For the

Emotions	Musical features
Happiness	Fast tempo, small tempo variability, major mode, simple and consonant harmony, medium-high sound level, small sound level variability, high pitch, much pitch variability, wide pitch range, ascending pitch, perfect 4th and 5th intervals, rising micro intonation, raised singer's formant, staccato articulation, large articulation variability, smooth and fluent rhythm, bright timbre, fast tone attacks, small timing variability, sharp contrasts between "long" and "short" notes, medium-fast vibrato rate, medium vibrato extent, micro-structural regularity
Sadness	slow tempo, minor mode, dissonance, low sound level, moderate sound level variability, low pitch, narrow pitch range, descending pitch, "flat" (or falling) intonation, small intervals (e.g., minor 2nd), lowered singer's formant, legato articulation, small articulation variability, dull timbre, slow tone attacks, large timing variability (e.g., rubato), soft contrasts between "long" and "short" notes, pauses, slow vibrato, small vibrato extent, ritardando, micro-structural irregularity
Fear	Fast tempo, large tempo variability, minor mode, dissonance, low sound level, large sound level variability, rapid changes in sound level, high pitch, ascending pitch, wide pitch range, large pitch contrasts, staccato articulation, large articulation variability, jerky rhythms, soft timbre, very large timing variability, pauses, soft tone attacks, fast vibrato rate, small vibrato extent, micro-structural irregularity
Tenderness	Slow tempo, major mode, consonance, medium-low sound level, small sound level variability, low pitch, fairly narrow pitch range, lowered singer's formant, legato articulation, small articulation variability, slow tone attacks, soft timbre, moderate timing variability, soft contrasts between long and short notes, accents on tonally stable notes, medium fast vibrato, small vibrato extent, micro-structural regularity

Fig. 1. Musical features correlated with discrete emotions by Juslin and Laukka [19] [reduced].

---

[2] We hope not to hurt the feelings of those who like atonal music, but we do believe it would have been hard to elicit this kind of feeling otherwise.

relax condition (similar to tenderness in Juslin and Laukka [19]) we opted for a quite slow piece, with large and classic intervals, tranquil tendency and quiet dynamics.

3 Experiment

Method
The experiment was realized and distributed on www.qualtrics.com. Participants were licensed to open the correct link of the survey.

Experimental Design and Participants
The study was based on monofactorial design between subjects with affective priming manipulation as main independent variable with four levels (Joy, Relax, Sadness, Annoyance). Gender, age and nationality were checked as control variables. Dependent variables are represented by basic manipulation checks (experienced emotion on an 11-point Likert scale), evaluation of moral harshness (on a 9-point Likert scale). 222 participants[3] (95 males, mean age: 30,6 – 126 females, mean age: 31,2): 130 by way of microworkers.com (native English-speakers), 91 through Facebook (Italian speakers). The first ones received 0,26$ each as a reward.

Materials
- 4 musical priming items (1 per condition):
  Joy: Wolfgang Amadeus Mozart – Sonata K448 in D major
  Relax: Nils Frahm – Do
  Sadness: Fryderyk Chopin – Nocturne op. 15 n.6 in G minor
  Annoyance: György Ligeti – Étude pour piano n.1
- Moral questionnaire with 6 moral vignettes
  - 2 High Emotional Involvement (HEI):

Example: *Frank's dog was killed by a car in front of his house. Frank had heard that in China people occasionally eat dog meat, and he was curious what it tasted like. So he cut up the body and cooked it and ate it for dinner. How wrong is it for Frank to eat his dead dog for dinner?-*

- 2 Low Emotional Involvement (LEI)

Example :*You have a friend who has been trying to find a job lately without much success. He figured that he would be more likely to get hired if he had a more impressive resume. He decided to put some false information on his resume in order to make it more impressive. By doing this he ultimately managed to get hired, beating out several candidates who were actually more qualified than he. How wrong was it for your friend to put false information on his resume in order to help him find employment?*

---

[3] The total number of the participants was 291; 69 of them have been excluded since they had not listened to the whole piece.

- 1 Trolley dilemma
- 1 Distractor
- Self-report affect questionnaire
- Self-made test we called Music Feelings Index (MFI) assessing participants' sensitivity to music.

3.1 Procedure

Each participant was randomly assigned to one condition; immediately after the listening session, six moral vignettes were given in a randomized order. The participant had to answer to each of them through a 9 points Likert scale by rating "To what extent do you think it is morally unacceptable to do what the character of the vignette does?" (the higher the score, the more unacceptable the action). Finally the participant was submitted the affect questionnaire, the MFI and the demographics.

3.2 Results

Affective Priming. From ANOVA analysis on manipulation checks a main effect of the affective priming emerged on Joy [$F(3,163) = 18,98$; $p < ,000$], Relax [$F(3,163) = 38,32$; $p < ,000$], Sadness [$F(3,163) = 28,64$; $p < ,000$] and Annoyance [$F(3,163) = 31,65$; $p < ,000$]. The results in Table 2 showed that the highest experienced emotion in each condition was coherent with the affective priming condition.

Table 2. Descriptive statistics of emotions felt * affective priming

		Mean experienced emotions							
		Joy	SD	Relax	SD	Sadness	SD	Annoyance	SD
AF	Joy	7,57	(2,3)	4,17	(2,7)	4,10	(2,6)	3,65	(2,7)
	Relax	6,92	(2,1)	**8,31**	(2,1)	6,95	(2,5)	3,15	(2,7)
	Sadness	2,41	(1,9)	5,88	(2,7)	**6,15**	(2,3)	2,96	(2,2)
	Annoyance	2,24	(1,8)	1,88	(1,6)	2,26	(2,2)	**6,09**	(3,2)

Moral Harshness. We used three composite scores: one for the LEI vignettes, one for the HEI vignettes and one for the overall moral harshness.

ANOVA analysis showed that the overall participants' moral harshness was significantly affected by the priming [$F(3,163) = 3,46$; $p < ,01$] in the sense that annoyance was the emotion that most of all worsened the participants' moral harshness ($M_{annoyance} = 7,78$). At first glance, describing descriptive means we noticed a strong polarization in the outcomes: the positive valence emotions were opposed to the negative valence ones with the most arousing at the ends (+ Valence emotions = 6,88; − Valence emotions = 7,60).

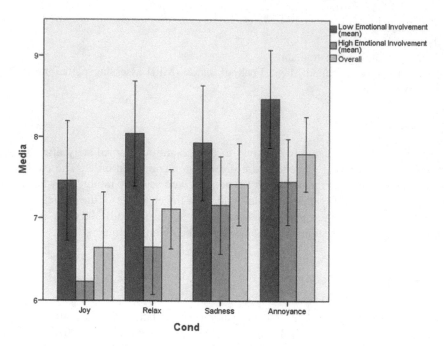

Fig. 2. Affective priming/moral harshness.

In particular, the moral questions that have been affected the most by our manipulation were those with a stronger emotional involvement [F(3,163) = 3,06; p < ,03]; differently, LEI moral questions were not affected [F(3,163) = 1,56; p = NS] (see Fig. 2). Apparently, our data point in the same direction as Greene's dual-process moral theory [20] that highlights that there would be at least two different ways in which moral information may be processed: one deals with emotional involvement, the other is more similar to a cost-benefit analysis.

The highest moral harshness mean were found in the annoyance ($M_{annoyance}$ = 7,44) and in the sadness condition ($M_{sadness}$ = 7.16) coherently with previous studies on affective priming that showed that negative emotions elicited a more analytic - and then severe - evaluations [21] while "inducing positive emotion might lead to more utilitarian approval" [20].

The vignette that mainly contributed to the HEI significance was the one called "Dog" [F(3,163) = 4,54; p < ,004], where the character ate his dog after it had been accidentally ran over and killed by a car. In particular, it elicited a harsher judgement in both sadness and annoyance conditions; in this case, we might account for the high sadness mean ($M_{sadness}$ = 8,59) by assuming that this vignette induced a sense of compassion for the dead dog. Yet, further studies have to be made in this direction.

Concerning MFI, nothing relevant emerged from the analysis of moral harshness, the only important correlation being that the higher the MFI score (i.e., responsiveness to music), the higher the perceived sadness (r = ,192; p < ,014).

Other Control Variables: Nationality, Gender and Age

United States vs. Italy. The participants in our study were Italian, mostly southerner and native-English speakers especially from Canada and United States.

The popular saying according to which Italians are more permissive than other cultures concerning morality seems to be confirmed by our data: $M_{ita} = 6,81 - M_{eng} = 7,42$; $p < ,008$).

Females vs. Males. The analysis on the correlation between males' and females' moral judgements appears to confirm well-known data from the literature on gender morality [22–24]: women are in general more uncompromising than men: $M_F = 7,58 - M_M = 6,61$; $p < ,00003$).

The older, the wiser? We also noticed a tendency concerning age and overall moral harshness: older participants tended to be harsher than younger ones, especially in the LEI vignettes. However, those data are not statistically significant ($p = .10$), partly because of the lack of homogeneity of our sample. We hope to better investigate these and other issues in our forthcoming studies.

4 Conclusion, Limits, and Future Work

The study presented demonstrates that negative emotions (i.e., annoyance and sadness) induced through music can worsen moral judgment.

Some refinement is necessary in the choice of the moral vignettes, since the stimuli we used, drawn from literature, mix up different criteria of moral evaluation, to be disentangled instead. For example, degrees of severity of the moral transgression cut across different fields of human experience (e.g. theft vs. sexual transgression). We will improve and better differentiate vignettes taking into account clear moral intuitions [25] such as those of Haidt's Moral Foundations Theory [26].

The general results suggest that music has a strong framing effect, thus involving relevant implications for marketing, advertising, political communication, persuasive computing. Therefore, in future studies, the relationship between music, induced emotions and moral judgment will be investigated in other more controlled scenarios.

The influence of emotions over decision and judgment, and the possibility they can be induced in ways that make awareness less likely (such as with subliminal perception or music) poses an urgent necessity: to have – and to provide people with – a clear consciousness of the human's possibility of non-awareness. This would be a task of research in order to make people keener to free (and informed) choice.

References

1. Baumgartner, T., Esslen, M., Jäncke, L.: From emotion perception to emotion experience: emotions evoked by pictures and classical music. Int. J. Psychophysiol. **60**(1), 34–43 (2006)
2. D'Errico, F., Poggi, I.: "The Bitter laughter". When parody is a moral and affective priming in political persuasion. Front. Psychol. **7**, 1–16 (2016)
3. Valdesolo, P., DeSteno, D.: Manipulations of emotional context shape moral judgement. Psychol. Sci. **17**(6), 476–477 (2006)

4. Inbar, Y., Pizarro, D.A., Bloom, P.: Disgusting smells cause decreased liking of gay men. Emotion **12**(1), 23 (2012)

5. Eskine, K.J., Kacinik, N.A., Prinz, J.J.: A bad taste in the mouth gustatory disgust influences moral judgement. Psychol. Sci. **22**(3), 295–299 (2011)

6. Winkielman, P., Berridge, K.C., Wilbarger, J.L.: Unconscious affective reactions to masked happy versus angry faces influence consumption behavior and judgements of value. Personal. Soc. Psychol. Bull. **31**(1), 121–135 (2005)

7. Ong, H.H., Mullette-Gillman, O., Kwok, K., Lim, J.: Moral judgement modulation by disgust is bi-directionally moderated by individual sensitivity. Front. Psychol. **5**, 194 (2014)

8. Mila, M.: L'esperienza musicale e l'estetica. Einaudi, Torino (1950)

9. Leman, M.: An embodied approach to music semantics. Musicae Sci. **14**(1_suppl), 43–67 (2010)

10. Besson, M., Schön, D.: Comparison between language and music. Ann. New York Acad. Sci. **930**(1), 232–258 (2001)

11. Cross, I., Tolbert, E.: Music and meaning. In: Hallam, S., Cross, I., Thaut, M. (eds.) Oxford Handbook of Music Psychology. Oxford University Press (2016)

12. Zentner, M., Grandjean, D., Scherer, K.R.: Emotions evoked by the sound of music: characterization, classification, and measurement. Emotion **8**(4), 494 (2008)

13. Scherer, K.R., Zentner, M.R.: Emotional effects of music: production rules. In: Music and Emotion: Theory and Research, pp. 361–392. Oxford University Press, Oxford, New York (2001)

14. Kreutz, G., Ott, U., Teichmann, D., Osawa, P., Vaitl, D.: Using music to induce emotions: influences of musical preference and absorption. Psychol. Music **36**(1), 101–126 (2008)

15. Prinz, J.: The emotional basis of moral judgements. Philos. Explor. **9**(1), 29–43 (2006)

16. Lerner, J.S., Li, Y., Valdesolo, P., Kassam, K.S.: Emotion and decision making. Psychology **66** (2015)

17. Naqvi, N., Shiv, B., Bechara, A.: The role of emotion in decision making a cognitive neuroscience perspective. Curr. Dir. Psychol. Sci. **15**(5), 260–264 (2006)

18. van der Zwaag, M.D., Dijksterhuis, C., de Waard, D., Mulder, B.L., Westerink, J.H., Brookhuis, K.A.: The influence of music on mood and performance while driving. Ergonomics **55**(1), 12–22 (2012)

19. Juslin, P.N., Laukka, P.: Expression, perception, and induction of musical emotions: a review and a questionnaire study of everyday listening. J. New Music Res. **33**(3), 217–238 (2004)

20. Greene, J.D.: Why are VMPFC patients more utilitarian? A dual-process theory of moral judgment explains. Trends Cogn. Sci. **11**(8), 322–323 (2007)

21. Forgas, J.P.: When sad is better than happy: negative affect can improve the quality and effectiveness of persuasive messages and social influence strategies. J. Exp. Soc. Psychol. **43** (4), 513–528 (2007)

22. Shaub, M.: An analysis of the association of traditional demographic variables with the moral reasoning of auditing students and auditors. J. Account. Educ. **12**(1), 1–26 (1994)

23. Cohen, J., Pant, L., Sharp, D.: The effect of gender and academic discipline diversity on the ethical evaluations, ethical intentions and ethical orientation of potential public accounting recruits. Account. Horiz. **12**(3), 250–270 (1998)

24. Davidson, R.A., Casey Douglas, P., Schwartz, B.N.: Differences in ethical judgements between male and female accountants (2000). Unpublished

25. Graham, J., Haidt, J., Nosek, B.A.: Liberals and conservatives rely on different sets of moral foundations. J. Personal. Soc. Psychol. **96**(5), 1029 (2009)

26. Graham, J., Haidt, J., Koleva, S., Motyl, M., Iyer, R., Wojcik, S.P., Ditto, P.H.: Moral foundations theory: the pragmatic validity of moral pluralism (2012)

Using Accelerometers to Improve Real Time Railway Monitoring Systems Based on WSN

Suzane G. dos Santos[1], Iury R.S. de Araújo[1], Eudisley G. Anjos[1,3(✉)],
Romulo C.C. Araújo[2], and Franscisco A. Belo[3]

[1] Informatics Center, Federal University Paraíba, João Pessoa, Brazil
suzane.gomes@eng.ci.ufpb.br, iuryrogerio@gmail.com, eudisley@ci.ufpb.br
[2] Federal Institute of Pernambuco, Recife, Brazil
romuloaraujo@recife.ifpe.edu.br
[3] Mechanical Engineering Post-Graduate Programme, UFPB, João Pessoa, Brazil
belo@pq.cnpq.br

Abstract. Rail transport systems require constant monitoring for better control and safety of passengers. Due to this, there is a great growth in the use of real time monitoring systems for trains. The use of WSN has been configured as one of the technologies that most allow such evolution. The sensors help in obtaining and verifying existing data as well as estimating and deducing new information. This paper proposes the use of WSN, accelerometers and gyroscopes as complementary technologies for monitoring several railway system variables. As validation, we present a prototype and initial experiments performed in the Railway system of Recife - Brazil (METROREC).

Keywords: Accelerometer · Real time monitoring · Railway monitoring · Wireless sensor network

1 Introduction

Nowadays it is noticeable the large increasement of transports, causing a saturation in vehicular infrastructure. This situation affects the live of users, leading to traffic jams, accidents, delays and many other problems. As consequence, a growing number of researches are directing more attention to the field of Intelligent Transportation Systems (ITS). These researches are creating solutions to improve the transportation monitoring, security, efficiency and services [9]. In this field, monitoring systems are one of the largely used. The systems are concerned to implement real time monitoring to get different objects data and use them to provide information to an efficient system management. The accuracy of these information has to be high, principally when the focus are railway monitoring systems which can impact the life of several people.

In terms of urban transportation monitoring systems, there is a variety of technologies being used in many different situations. Most of them need tracking technologies and the use of GPS (Global Positioning System) is the most common

© Springer International Publishing AG 2017
O. Gervasi et al. (Eds.): ICCSA 2017, Part VI, LNCS 10409, pp. 761–769, 2017.
DOI: 10.1007/978-3-319-62407-5_58

one. However, when it is about railway systems the GPS has a series of problems (e.g. low signal coverage because of obstacles), which make it not the best suitable solution in these systems for tracking [12]. Therefore, due to the low cost and easy management and implementation, several researches as [1, 2, 4, 8] have used accelerometers and gyroscopes as devices to obtain data about the train. These data can be then analyzed to get information and detect patterns. Some of these information are: train location, permitting the train and passengers localization; detection of rail problems, from small cracks to ruptures between the rails; train vibration analysis, with the intent on control it to avoid health problems of passengers; and others.

One interesting system was developed and is being studied in the subway of the city of Recife, Brazil. In this context, an innovative system is in constant evolution to provide state-of-the-art technologies to improve the operation of the metro systems of Recife, Brazil. The RailBee Project developed a railway monitoring system which uses Wireless Sensor Network (WSN) to send data from the trains to the Center of Railway Control [14]. The initial system gets data directly from the train and sends it to the monitoring center. In way to improve the data acquisition process and subsequently provide more accurate information and high precision, an expansion of the inicial system was proposed, using accelerometers and gyroscopes and integrating with the pre-existed WSN infrastructure. These new system allows a better management and more reliable decisions regarding various aspects of train operation, as well as improves the services provided.

This paper presents the proposed expansion, detailing the architecture and the operation of the system. The system can provide the previous information acquired from the train plus new data from accelerometers, gyrscopes and other sensors, while allowing the comparison between measures obtained through different ways. The paper is divided in three main sections: Sect. 2 presents the state-of-art relating researches that use accelerometers in railway systems. Section 3 detail the proposed system, its characteristics, architecture and the initial tests performed. Finally, the conclusion of the paper is presented in Sect. 4.

2   State of the Art

In the literature accelerometers are used in a larger number of researches and fields, as healthcare, transportation, industry, military, etc. In the field of railway transportation a considerable number of papers presents the use of accelerometers to train monitoring. The work presented in [1] proposes the AARA (Accelerometer-Assisted Rate Adaptation Mechanism) to improve the Wi-fi connection in trains. Because of the signal range limit in vehicle environments, the signal quality can vary along the way. To ensure a better Wifi connection and so the data transfer, the authors used the moving acceleration, calculated through the three coordinates valors from the accelerometer. The AARA can estimate four phases of the train motion: arrival, stop, departure and cruising, as presented in Fig. 1. Each phase has peculiar characteristics which help the system to localize the train, define the next station and connect the inside devices to it.

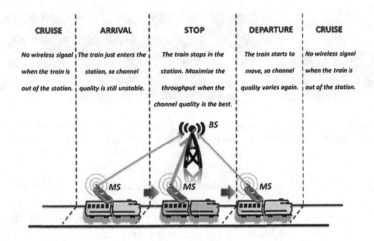

Fig. 1. AARA train motion phases. (Source: [1])

System security is another relevant topic on the field of railways. As mentioned in [10], security may be directly related to the quality of the railroad. Thus, detecting railroad flaws in advance helps to prevent derailments, reducing the cost of maintenance on trains and rails.

Another work proposed by [2] presents an interaction model between the train wheels and the railroad, which generates training data to a rail flaws detection algorithm. This algorithm was embedded in a circuit board installed on the side of the train. The circuit obtains and analyse the vertical acceleration, identifying since small gaps in the surface to ruptures between rails. In a similar matter, [4] proposes the use of accelerometer and gyroscope within the board, which installed on a train gather vertical and lateral acceleration data. The data are used to find irregularities on the railroad.

The results in [10] also present the use of accelerometers to reduce the problems on rail monitoring systems. During the work the authors studied different techniques to analyze the data of a triple-axis accelerometer. Among the analyzed techniques, the use of wavelets seems to increase the autocorrelation to detect singularities. The tests presented promising results to identify rail fractures, chipped rails, and broken concrete foundations as presented in Fig. 2. These results are similar to the ones found in [11]. In this research the methods adopted to process the data users continuous and discrete derivative wavelets filters to find problems in the railroad, as cracks, ruptures and etc.

The rise of smartphones has made localization systems very common in all types of applications, mainly railway systems. These systems allow the user locate he/she position in an indoor environment and also having access to a variety of services such as guidance of navigation to a determined location, not wasting time [3]. To achieve this goal, a variety of existing systems are using accelerometers, gyroscopes and other similar sensors to obtain the user localization. However, researches on rail users localization are still rare.

Fig. 2. Rail defects found by the solution used in [10]: (a) Low battered weld in rail; (b) Broken concrete ties; (c) Chipped rail at a joint (Source: [10])

The work presented in [5] proposes a localization system of mobile users in subway lines called M-Ioc, which operates in two main phases. In the first phase the system collects data from the accelerometer, barometer and magnetometer present in smartphones. Thus, M-Ioc analyses the information and creates a pattern map to each subway line and according each user. In the second phase, to find their position, the users can download the pattern map to their smartphones. Thus, the train position it is discovered according to real-time information from monitoring sensors and the comparison to pattern maps.

The research in [13] presents the StationSense, a mobile solution based on smartphones' accelerometer and magnetometer which permits acquire the subway passenger location. In this solution the magnetometer is used to detect the train magnetic field, which is bigger when the train is speeding up or breaking. This phenomenon permits to identify if the train is stopped or moving. The results were twice as accurate as existing solutions, which use only accelerometer.

In the Wireless Sensor Network (WSN) the use of accelerometers has shown importance in research. Since these networks use batteries and in most cases the difficulty and cost for change is high, the main problem approached on the researches is related to energy consumption. The work proposed by [7] sought to improve energy efficiency of WSN nodes installed in railroad cars used for real time monitoring. The authors developed a protocol called Time Adaptive-Bit Map Assisted (TA-BMA), which permits the communication between the nodes. Using this protocol, the node wakes up to send data and after that enter in sleep mode, saving 39,7% of energy. The important point in this protocol is because it is only possible with the use of accelerometers to get the vertical and lateral speed of the train.

Almost all solutions found in literature have very specific uses, restricting the systems and making it impossible the use on real time monitoring. It happens because in most of them the data gathering is done by local storage to further analysis. Even solutions that use accelerometers have very restricted focuses. The need for a more complete system for integration with our project led us to the development of a prototype that would allow the use of WSN and could be easily installed and flexible to change, when necessary.

3   Real Time Monitoring System Using Accelerometers

As mentioned earlier, the common railways monitoring systems provide to the operators the same information visualized by the trains drivers. The addition of new sensors technologies (e.g. accelerometers) allows the achievement of new data and sources of information. In the case study performed for this work a new prototype was added to RailBee Project in a train of Recife Subway System (METROREC). The experiments aimed to evaluate the improvements provided to the system by the use of accelerometers. The circuit board developed was installed in a railroad car and the data is being collected since then. In the next subsection the RailBee Project and the new module architecture are explained.

3.1   RailBee System

RailBee is a telemetry system created with the intention of providing more services and the better use of resources found in ITS for railroad systems. A prototype of this system is being tested in the subway of Recife, Pernambuco, Brazil. The RailBee system is composed by three subsystems: a telemetry system, to gather the signals in the trains traffic routes; a receiving system, receiving and decodifying data to send to the supervision subsystem; and a supervision system, to store and visualize the data [15].

The RailBee is capable of obtaining data, from preexistent sensors or new ones installed in the trains, process and send them to the control and monitoring center. The system uses a wireless sensor network to data acquisition, communication and distribution. The WSN is based on ZigBee Protocol and allows the monitoring of performance of the actives trains on their routes as presented in [15]. The use of ZigBee provides to send data in a fast, cheap and efficient way [16]. The ZigBee modules can be used in a variety of projects since they can offer a excellent immunity against interferences and are capable to host a large number of devices in one network. The Fig. 3 shows a model of operation of the RailBee project.

Obtaining real time data, as speed and railroad cars airbags pressure, are of great value to the central of command, which can take faster and accurate decisions [15]. Also, the data analysis allows identifying the need for real-time service improvements, such as the addition of new cars. Some examples are: speed analysis, making possible to identify in real time the train performance; airbags pressure measurement, allowing to estimate the number of passengers in the train and the density of them inside the railroad cars. In addition many indirect measures can be deduced, such as the comfort level of users through the measures mentioned before.

3.2   System Prototype Architecture

The initial architecture implemented in the RailBee Project is based in the principle of getting data directly from the engine through wires connected to the

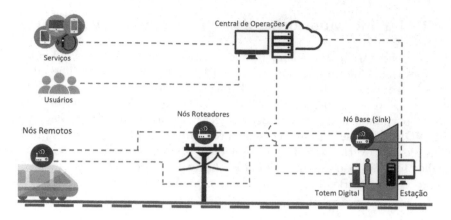

Fig. 3. RailBee system operation diagram.

ZigBee Module channels. However, depending on the communication technology, a data preprocessing is necessary before transmission by ZigBee modules. The use of microcontrollers allows the accomplishment of this data conversion and extends the possibility of information manipulation *in-loco*, allowing that alert messages or data compression are made before to be sent. This way, the accelerometer and gyroscope used in our prototype will work as a wireless sensor node, mostly because the trains used on the Recife subway does not have them installed.

To test the use of a accelerometer was created a board containing all necessary components. In the first prototype this second board is isolated from the one used for RailBee Project, containing another ZigBee module and the other components used for our experiments. To process information about train acceleration and orientation was used the module GY-521/MPU-6050. The integrated circuit MPU-6050 contains an accelerometer and gyroscope communicating through the I2C protocol. The conversion between I2C protocol to ZigBee is executed by a script code located on the microcontroller. Besides the use of WSN a local SD memory card was also installed through SPI interface. It makes possible the comparison of network received data with the locally stored data and assists in analyzing network traffic, packet loss, and data consistency. Due to that, all stored data must have a timestamp value, which are used to compare the acquired information, including the ones obtained by accelerometer data transformation and similar data obtained directly from the train. The timestamp was obtained by a DS3231 module, a Real Time Clock (RTC) which provides information of seconds, minutes, days, months and years and operates on I2C protocol. For integration of all components, the microcontroller used was the Atmel Mega328p of the Arduino Nano. The Arduino acted as the processing center as the protocol converter and storing the data on the local memory card (Fig. 4).

Fig. 4. Prototype board and installed box in the trains for tests.

During the tests, the data is collected from the SD card weekly to be compared with the WSN data. The data are then separated into three axes: X, Y and Z, both for gyroscope and accelerometer. From this point, mathematical calculations using already known methods popular in literature are applied to the information [10, 11]. For velocity and distance data, the values are compared with the values obtained directly from the train. This makes it possible to calibrate the system more accurately. Therefore, the initial tests confirmed the accuracy of using accelerometers and corroborated the potential of its working on a wireless sensor network. Thus, the final prototype can be integrated to the RailBee system board to a more complete test.

4   Conclusion

This paper presents a improvement to real time railway monitoring systems based in WSN, accelerometers and gyroscope. The use of these technologies together allows a variety of monitoring possibilities. When coupled with the use of WSN, the power of real-time monitoring is even greater, leading to an integration of services and features that may be of great interest to current research in the area. In addition to allowing the verification of existing parameters, the proposed technology supports the control center operators with relevant information for constant decisions to improve the railways activities.

The results of the initial tests carried out in the dependencies of Recife Subway System (METROREC) show the relevance of joining the potentialities of the accelerometer and gyroscope to wireless sensor networks. Through the analyzes of the data it was possible to infer new information that led our team to propose several future works, many of them in progress at the moment. In addition, the use of data mining and neural networks has allowed the discovery of knowledge that was not previously expected by the research or technical teams. These additional information can be used to provide better services for daily passengers.

Despite the promising results, there are still some issues to be addressed in future work. Further analysis and comparison is needed on data conversion algorithms as well as device calibration. This analysis is beyond the scope of our initial proposal. In addition, it is important to verify how inconsistencies in the operation of accelerometers can be detected and to estimate the frequency that such inconsistencies may occur, avoiding erroneous decisions about the control of the system. Finally, it is important that the circuit board on which the accelerometer is coupled can be developed in accordance with possible adaptations to the train models to which they will be adapted.

References

1. Lai, Y.J., Kuo, W.H., Chiu, W.T., Chang, S.T., Wei, H.Y.: Accelerometer-assisted 802.11 rate adaptation on mass rapid transit system. In: ACM SIGCOMM Computer Communication Review, vol. 40, no. 4, pp. 421–422. ACM, August 2010
2. Hopkins, B.M., Taheri, S.: A dynamic wheel/rail interaction model based on an Euler-Bernoulli beam rail model. In: ASME 2011 Rail Transportation Division Fall Technical Conference, pp. 17–22. American Society of Mechanical Engineers, January 2011
3. Google: Indoor Maps - about - Google Maps. https://www.google.com/maps/about/partners/indoormaps/
4. Weston, P.F., Roberts, C., Goodman, C.J., Ling, C.S.: Condition monitoring of railway track using in-service trains (2006)
5. Ye, H., Gu, T., Tao, X., Lu, J.: Crowdsourced smartphone sensing for localization in metro trains. In: 2014 IEEE 15th International Symposium on A World of Wireless, Mobile and Multimedia Networks (WoWMoM), pp. 1–9. IEEE, June 2014
6. Hua, J., Shen, Z., Zhong, S.: We can track you if you take the metro: tracking metro riders using accelerometers on smartphones. IEEE Trans. Inf. Forensics Secur. 12(2), 286–297 (2017)
7. Philipose, A., Rajesh, A.: Investigation on energy efficient sensor node placement in railway systems. Eng. Sci. Technol. Int. J. 19(2), 754–768 (2016)
8. DiFiore, A., Zaouk, A., Durrani, S., Mansfield, N., Punwani, J.: Long-haul whole-body vibration assessment of locomotive cabs. In: 2012 Joint Rail Conference, pp. 635–641. American Society of Mechanical Engineers, April 2012
9. Figueiredo, L., Jesus, I., Machado, J.T., Ferreira, J.R., De Carvalho, J.M.: Towards the development of intelligent transportation systems. In: 2001 IEEE Proceedings of Intelligent Transportation Systems, pp. 1206–1211. IEEE (2001)
10. Abuhamdia, T., Taheri, S., Meddah, A., Davis, D.: Rail defect detection using data from tri-axial accelerometers. In: 2014 Joint Rail Conference, pp. V001T06A001-V001T06A001. American Society of Mechanical Engineers, April 2014
11. Taheri, S., Taheri, S.: Rail track defect detection using derivative wavelet transform. In: ASME 2012 Rail Transportation Division Fall Technical Conference, pp. 51–56. American Society of Mechanical Engineers, October 2012

12. Stockx, T., Hecht, B., Schöning, J.: SubwayPS: towards smartphone positioning in underground public transportation systems. In: Proceedings of the 22nd ACM SIGSPATIAL International Conference on Advances in Geographic Information Systems, pp. 93–102. ACM, November 2014

13. Higuchi, T., Yamaguchi, H., Higashino, T.: Tracking motion context of railway passengers by fusion of low-power sensors in mobile devices. In: Proceedings of the 2015 ACM International Symposium on Wearable Computers, pp. 163–170. ACM, September 2015

14. Araújo, R.C., Santos, J.L.A., Anjos, E.G., Lima Filho, A.C., Belo, F.A., Lima, J.A.: "RailBee" Sistema de instrumenta virtual de veículos em malhas metroferroviárias. In: Seminário Negócios nos Trilhos. Prêio Alston de Tecnologia, Revista Ferroviária (2009)

15. Araújo, R.: Sistema telemétrico dinâico sem fio aplicado aos veículos rodoferroviários em malhas metroferroviárias. 147 f. Theses, Doctorate Degree in Mechanic Engineering, Federal University of Paraíba, João Pessoa (2009)

16. Araujo, I., Castro, J., Matos, F., Anjos, E.: MUV-Bee: Using WSN to monitoring urban vehicles. In: 2015 Latin American Network Operations and Management Symposium (LANOMS), pp. 99–102. IEEE, October 2015

Comparative Analysis of Two Variants of the Knox Test: Inferences from Space-Time Crime Pattern Analysis

Monsuru Adepeju$^{(\boxtimes)}$ ⓘ and Andy Evans ⓘ

School of Geography, University of Leeds, Leeds LS2 9JB, UK
M.O.Adepeju@leeds.ac.uk

Abstract. This paper compares two variants of the Knox test in relation to space-time crime pattern analysis. A case study of burglary and 'stolen-vehicle' crime data sets of San Francisco city is presented. The comparative analysis shows that while one variant is designed to detect the sizes of the spatio-temporal neighbourhoods at which clustering (hotspots) is prominent within a data set, the other variant is able to reveal the spatial and temporal windows/bands at which crime events are frequently repeated to form clusters (hotspots) across an area.

Keywords: Knox test · Space-time · Crime · Repeat pattern · Critical distances

1 Introduction

The Knox test [1] is ubiquitous in the analysis of spatio-temporal patterns of crime data. The Knox test offers an interesting means of determining whether there are more instances of observed pairs of events within a defined spatio-temporal neighbourhood than would be expected on the basis of chance. In the literature, two variants of the Knox test can be identified [2]. They are fundamentally distinguishable by the way in which the spatio-temporal neighbourhoods are defined. We have, one, an enclosed neighbourhood, in which the event count is conducted within a space, formed by a reference point r and the critical distances (CDs), measured along the spatial and the temporal dimensions (Fig. 1a). Two, we have a binned neighbourhood, in which the event count is conducted within the space formed from subtracting two CDs measured along each dimension (Fig. 1b), relative to a reference point r. As a revised version, the binned neighbourhood is intended to filter out anomalous highs which are less likely to result from interactions due to r.

The Knox test has been widely used for the purpose of understanding the space-time pattern of crime data sets, particularly the repeat and near-repeat (RNR) pattern [3]. The RNR is the concept that if a location is the target of a particular crime type, such as burglary, the houses within a relatively short distance of it have an increased risk of being burgled over a period of a limited number of weeks. While both variants of the Knox test have been used to describe the RNR patterns, there is yet to be any empirical studies that compare such results, and further, to highlight their relationship with regards to the space-time patterns displayed by the data

© Springer International Publishing AG 2017
O. Gervasi et al. (Eds.): ICCSA 2017, Part VI, LNCS 10409, pp. 770–778, 2017.
DOI: 10.1007/978-3-319-62407-5_59

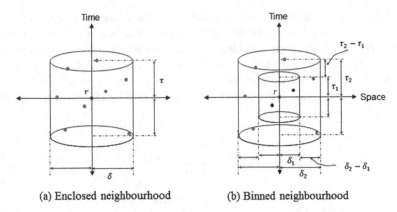

(a) Enclosed neighbourhood   (b) Binned neighbourhood

Fig. 1. Two types of neighbourhood definitions for the Knox test. (a) Enclosed neighbourhood, (b) Binned neighbourhood: each dimension requires two of δ and τ to form the region (i.e. $\{\delta_1$ and $\delta_2\}$ and $\{\tau_1$ and $\tau_2\}$, respectively). The cases count in (a) is 8, while the cases count in (b) is 6 (Diagram from: [8]).

set. Therefore, the purpose of this study is to address this research gap. For a comprehensive comparison, each variant of the Knox test will be used to examine the space-time pattern in relation to two different crime types, using a list of neighbourhood values.

2 The Knox Test and the Space-Time Neighbourhood Definition

The Knox test looks at the relationships between all pairs of events in a spatio-temporal dataset, generating a test statistic $(n_{\delta,\tau})$ that is larger when, for example, short space-time distances appear more frequently than would be expected by chance (indicating RNR behaviour). The test statistic is generated from a table where spatial and temporal distances are binned and pairings allocated to the cells. Rather than testing all $n(n-1)/2$ possible combinations, it is usual to pass a moving window repeatedly over the data point-by-point, varying the critical spatial (δ) and temporal (τ) widths, measured from the moving reference point, r; the variation size corresponding to the bins in the table. This allows only events within a reasonable space/time distance of r to be taken, improving calculation speeds. Structuring the window to remove sections allows for more complicated distance filters, as is the case in the "binned" windows discussed here.

As the full test is still computationally laborious, a close analogue of the full test can be implemented in a programming environment by constructing two 'closeness matrices'. A closeness matrix describes the closeness of all pairs of events, either in space or in time. Thus, one matrix X is created for closeness in space and the other Y for closeness in time. For the first matrix, an X_{ij} will have an entry 1 for the cell $[i,j]$ if event i occurred within some spatial distance δ of the event j and 0 otherwise. The spatial closeness is calculated by the 2D Euclidean distance measurement. Similarly for the second matrix, a Y_{ij} will have an entry 1 for the cell $[i,j]$ if the event i occurred within some temporal distance τ of the event j and 0 otherwise. The temporal closeness is calculated by the difference between the times of occurrences of any pair of events. Both matrices would have a dimension $n \times n$, where n is the total number of point events. For both matrices, if $i = j$, then the entry is 0. The test statistic $n_{\delta,\tau}$, is then formulated by the cross-product:

$$n_{\delta,\tau} = \frac{1}{2} \sum_{i=1}^{n} \sum_{j=1}^{n-1} X_{ij} Y_{ij}$$

$$X_{ij} = \begin{cases} 1, & \text{if the distance between cases i and j} < \delta \\ 0, & \text{otherwise} \end{cases}$$

$$Y_{ij} = \begin{cases} 1, & \text{if the distance between cases i and j} < \tau \\ 0, & \text{otherwise} \end{cases}$$

If $n_{\delta,\tau}$ is large enough, the null hypothesis of no space-time interaction may be rejected. A number of approaches have been used to derive the point distribution under the null hypothesis. These include the chi-square approach [4], standardised residual analysis [5], and the Monte Carlo (MC) simulation approach [6]. The MC simulation approach has become very popular due to the greater processing power of modern computers. Moreover, it has the advantage of minimising the impacts of the edge effect of the Knox test, compared to the other available approaches. The MC involves randomising the time attributes of the cases, while the spatial attributes are kept constant. The process is repeated multiple times (usually 999), so that a pseudo p-value can be calculated as $P = 1 - n_e/(n_s + 1)$, where n_e is the number of times that the expected count, based on MC, was exceeded by the observed count and n_s is the number of simulations. Here we utilise this MC process. The MC process is extremely computationally intensive, thus, a parallel computing approach, in which multiple analyses are run simultaneously, was used.

2.1   Types of Space-Time Neighbourhoods

The parameters δ and τ described above can be referred to as the spatial and temporal neighbourhoods, respectively, within which events that fall inside are considered to be 'close'. This is the original version of the Knox test, as proposed by [1]. In this study, this was referred to as an enclosed neighbourhood variant. In other words, an enclosed neighbourhood involves counting events within a region defined by critical spatial (δ) and a critical temporal (τ) distances, measured from a reference point r (Fig. 1a). A revised neighbourhood definition was later proposed [7], in which event counts are conducted within a region, which are created by the subtraction of the two distances measured along each dimension. In this version of the Knox test, the resulting neighbourhood is referred to as the binned neighbourhood (Fig. 1b).

Thus, for the binned neighbourhood, two critical distances $\{\delta_1$ and $\delta_2\}$ and $\{\tau_1$ and $\tau_2\}$ are defined for the δ and τ, respectively.

The above two neighbourhood definitions, that is, the enclosed neighbourhood and the binned neighbourhood, underlie the two variants of the Knox test and will be used in this study.

In the crime domain, a contingency table is usually created in which the Knox test is repeatedly run using different values of δ and τ, and the contingency table is populated accordingly. The values of δ and τ are usually set arbitrarily. Although most studies often justify the value they use from a theoretical and/or practical point of view. In this study, the primary focus is to compare the two variants as regards to long-term local neighbourhood crime prevention. Therefore, the values of δ and τ will be made to vary at 30 day interval and 200 m in the temporal and spatial dimensions, respectively. In the case of the binned variant, the inner bin is set such that the neighbourhood created with the outer bin is of 30 days and 200 m in the temporal and spatial dimensions, respectively.

3   Case Study of San Francisco Crime Data Sets

3.1   Data Sets

The data for this study are the burglary and 'stolen-vehicle' crimes in San Francisco, CA, U.S.A, for the year 2015. During this year, the city had some of the highest rates of burglary and 'theft-of-motor-vehicles', at approximately 20% and 150%, respectively, higher than the national average. The burglary and 'stolen-vehicles' data contains 4,267 and 4,970 records, respectively. The maps of spatial distribution of both crime events are shown in Fig. 2. The burglary crime reveals some easily identifiable local aggregations, especially within the most highly populated residential regions (i.e. the north-eastern parts of San Francisco). In contrast, the various local aggregations of stolen-vehicle have wider coverages, resulting into more evenly distributed events in comparison with burglary events. The temporal distribution of both crime types (based on monthly aggregations) are also different (Fig. 2). From January to September, the monthly count of burglary crime looks very similar, with the exception of May, which is $\sim 15\%$ higher than the monthly average (Fig. 2c). On the other hand, 'stolen-vehicle' crime shows a continuous rise for monthly counts and peaks in May, yet immediately falls to its lowest frequency in September. The remaining three months have roughly the same crime count, while in the case of burglary, the crime level rose above the average level.

3.2   Profiles of Events' Repeat Patterns

In crime pattern analysis, a common approach for visualising the repeat and near repeat (RNR) pattern in crime data sets is to draw the repeat pattern profile [9]. The repeat pattern profile usually provides a picture of how many events are repeated within a time window for a defined neighbourhood size, across the entire area. This same idea

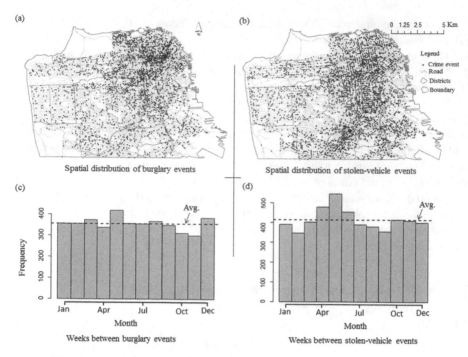

Fig. 2. Spatial and temporal distributions of San Francisco crime data sets

underlies the Knox test, except that the Knox test incorporates the statistical significance evaluation. Examples of repeat pattern profiles are shown in Fig. 3.

Figure 3 shows the repeat pattern of the data set, measured within a neighbourhood of 400 m in radius of an initial event. A relatively high level of repeats is indicated by the undulations in the profiles. A reference line is created in order to make the undulations (repeats) more visible by joining the first and last event. In Fig. 3a, it can be observed that the repeat pattern is more prominent in the first five months of burglary crime, while the repeat extends up to the sixth month in the 'stolen-vehicle'

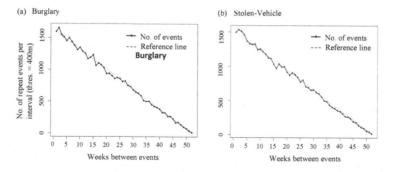

Fig. 3. Repeat pattern profile. (a) and (b) are the repeat pattern profiles of burglary and 'stolen-vehicle' profiles, respectively, within a radius of 400 m of every event.

data. Thus, it can be easily be inferred that the repeat of crimes is prominent in the first five months for both crime types. However, the statistical quantification is missing in this type of analysis. This is where the Knox test plays an important role.

4 Results and Discussion

Tables 1 and 2 shows the results of the two variants of the Knox test for the burglary and 'stolen-vehicle' crimes respectively. The results for the enclosed neighbourhood variant and binned neighbourhood variant are presented in table labelled (a) and (b). Highlighted values indicate where the pseudo p-value is smaller than 0.05, meaning that the likelihood of it occurring by chance is less than one in 20.

Table 1. The *enclosed* neighbourhood variant Knox test: (a) Burglary crime. (b) Stolen-vehicle

(a)

days / mts	0 - 30	0 - 60	0 - 90	0 - 120	0 - 150	0 - 180	0-210	0-240
0 - 200	0.005	0.004	0.003	0.043	0.069	0.103	0.081	0.197
0 - 400	0.006	0.013	0.002	0.042	0.089	0.245	0.250	0.201
0 - 600	0.018	0.008	0.003	0.086	0.188	0.162	0.266	0.324
0 - 800	0.004	0.002	0.002	0.065	0.094	0.102	0.131	0.204
0 - 1000	0.002	0.005	0.014	0.104	0.145	0.109	0.112	0.179
0 - 1200	0.010	0.018	0.510	0.123	0.223	0.194	0.166	0.205
0 - 1400	0.081	0.110	0.079	0.194	0.310	0.304	0.214	0.278
0 - 1600	0.721	0.074	0.073	0.188	0.280	0.253	0.209	0.220
0 - 1800	0.060	0.081	0.081	0.163	0.253	0.229	0.150	0.152
0 - 2000	0.051	0.082	0.060	0.142	0.234	0.173	0.134	0.211

(b)

days / mts	0 - 30	0 - 60	0 - 90	0 - 120	0 - 150	0 - 180	0-210	0-240
0 - 200	0.031	0.035	0.202	0.448	0.466	0.515	0.663	0.85
0 - 400	0.002	0.001	0.003	0.026	0.615	0.762	0.65	0.744
0 - 600	0.001	0.001	0.002	0.039	0.397	0.638	0.488	0.503
0 - 800	0.001	0.001	0.003	0.059	0.394	0.621	0.55	0.631
0 - 1000	0.001	0.021	0.005	0.117	0.465	0.625	0.556	0.593
0 - 1200	0.009	0.042	0.128	0.436	0.834	0.801	0.678	0.72
0 - 1400	0.036	0.061	0.249	0.574	0.829	0.798	0.664	0.727
0 - 1600	0.054	0.054	0.371	0.558	0.845	0.803	0.635	0.684
0 - 1800	0.059	0.071	0.438	0.677	0.859	0.832	0.643	0.712
0 - 2000	0.097	0.136	0.543	0.771	0.911	0.827	0.665	0.736

In Tables 1a and b, representing the results of the enclosed neighbourhood variant of Knox test, all the cells with statistically significant values are arranged close to one another. In both cases, as the statistically significant cells extend in temporal sizes from 30 to 120 days, the corresponding spatial sizes decreases. This pattern indicates that significant levels of events are much closer in space at much smaller temporal sizes than at larger temporal sizes, larger distances and times being associated with events which blend with background noise. The relative closeness of events, simultaneously in both space and time, is generally referred to as space-time clustering. Considering the construct of the enclosed neighbourhood, as shown in Fig. 1a, it is argued that this variant of the Knox test is suited to the detection of events' concentration and its

Table 2. The *Binned* neighbourhood variant Knox test: (a) Burglary crime. (b) 'Stolen-vehicle' crime

(a)

mts \ days	0 - 30	31 - 60	61 - 90	91 - 120	121 - 150	151 - 180	181 - 210	> 210
0 - 200	0.005	0.040	0.917	0.418	0.081	0.425	0.967	0.968
201 - 400	0.108	0.038	0.567	0.691	0.966	0.720	0.066	0.723
401 - 600	0.107	0.248	0.736	0.528	0.017	0.973	0.919	0.643
601 - 800	0.017	0.117	0.402	0.201	0.212	0.342	0.992	0.997
801 - 1000	0.252	0.081	0.854	0.351	0.057	0.123	0.805	0.976
1001 - 1200	0.294	0.304	0.476	0.886	0.727	0.004	0.583	0.852
1201 - 1400	0.938	0.693	0.165	0.617	0.567	0.042	0.643	0.573
1401 - 1600	0.126	0.313	0.278	0.673	0.337	0.054	0.774	0.634
1601 - 1800	0.341	0.293	0.232	0.234	0.004	0.017	0.811	0.999
1801 - 2000	0.363	0.042	0.416	0.481	0.030	0.720	0.402	0.342

(b)

mts \ days	0 - 30	31 - 60	61 - 90	91 - 120	121 - 150	151 - 180	181 - 210	> 210
0 - 200	0.034	0.766	0.809	0.498	0.590	0.699	0.882	0.201
201 - 400	0.001	0.577	0.607	1.000	0.863	0.097	0.432	0.653
401 - 600	0.022	0.169	0.906	0.684	0.935	0.125	0.426	0.710
601 - 800	0.014	0.579	0.671	0.866	0.837	0.724	0.713	0.951
801 - 1000	0.129	0.375	0.733	0.931	0.407	0.225	0.770	0.832
1001 - 1200	0.479	0.999	0.924	0.978	0.031	0.016	0.200	0.208
1201 - 1400	0.560	0.856	0.745	0.664	0.385	0.053	0.815	0.671
1401 - 1600	0.577	0.693	0.247	0.844	0.464	0.065	0.624	0.985
1601 - 1800	0.459	0.884	0.982	0.702	0.387	0.001	0.837	0.502
1801 - 2000	0.921	0.380	0.975	0.943	0.029	0.021	0.841	0.533

extensions in both space and time. In other words, we can say that space-time clustering of events is being detected. Specifically, it detects the critical spatial and temporal distances at which events' concentrations are statistically significant, and thus, form space-time clusters (hotspots). These critical spatial and temporal distances can then be described as the spatial and temporal scales of the most prominent clusters within the data set. Since the results in Table 1 are product of spatial and temporal aggregations of the point events, the resulting patterns may not necessarily reflect the visible purely spatial or purely temporal patterns of the data in question. For example, Fig. 2a shows that the local aggregations in the burglary crime are of few hundred metres in sizes. However, the results in Table 1a shows significant values at very large spatial neighbourhood sizes (e.g. [0–1000 m] & [0–1200 m]), which are not reflective of the patterns in Fig. 2a.

In Tables 2a and b, representing the results of the binned neighbourhood variant of the Knox test, all the cells with statistically significant values are disjointed and distributed within four temporal bands, namely: {0–30 days}, {31–60 days}, {121–150}, and {151–180}. The critical spatial distances of the cells against their corresponding critical temporal distances can be described as the spatial and temporal windows at which there are more interactions or repeats in the data set. Two events interact when they are at a distance from one another in space and time different from that which would be expected on aggregate on the basis of chance. In both crime types, events are clustered within the first 2 months and also within 5 to 6 months of one another, which may be translated as the short-term and long-term repeat patterns. The ability to this variant of the Knox test to clearly isolate repeat patterns in terms of their temporality constitutes its major improvement over the events' repeat profile of Fig. 3. Besides, the lack of statistical baseline (such as in Knox test) for the repeat profile, it does not allow the magnitude of interactions, represented by various undulations along the profile, to be quantified statistically. This represents the major advantage of the Knox test over the conventional repeat profile method. For example, out of all the undulations in the repeat profile of stolen-vehicle for the spatial interval of 400 m (as shown in Fig. 3), only those incidents (undulations) within 0–4 weeks interval which is equivalent to 0–30 days for Knox test are truly significant.

Comparing the general distribution of the significant cells in all the tables, it can be observed that the patterns at smaller spatial and temporal bands/neighbourhoods (i.e. between 0 m and 800 m) are much stronger, and thus persist across the two variants. On the other hand, the significant cells in the distant bands for the binned variant were not reflected in the enclosed variant, and are thus tagged as weak patterns.

Crime clusters (hotspots) are formed as a result of many crimes interacting within a common neighbourhood. In practice, while the results of the binned neighbourhood variant of the Knox test can be used to identify the spatial and the temporal lags at which the next set of crimes are likely to occur, the results of the enclosed neighbourhood variant of the Knox test can be used to investigate the sizes of the spatial and temporal windows (neighbourhoods) to target during crime intervention over a coherent period after the event. This has advantages in terms of both public fear of crime and utilisation of householders fresh commitment to target hardening. While the Knox test is relatively simple and efficient to calculate, a more locally detailed answer as to the structure of short-term spatio-temporal clustering may be obtainable by employing a local clustering test, such as space-time scan statistics (STSS) [10]. An STSS is able to detect and isolate the specific space-time regions that contribute to the overall clustering derived by a global test such as the Knox test.

5   Conclusions

This study presents a comparative analysis of two variants of the Knox test, distinguished by the manner in which the space-time neighbourhoods (i.e. critical distances) are defined. These two neighbourhood definitions were referred to here as the enclosed and binned neighbourhoods. The goal was to compare the results generated by these two variants in relation to the same data set, as well as to describe their relationships from both theoretical and practical perspectives. The data set used for the analysis was the burglary and 'stolen-vehicles' crime dataset for the city of San Francisco.

The results reveal that the enclosed variant helped to detect the sizes of the spatio-temporal neighbourhoods at which clustering (hotspots) was prominent within the dataset. The binned variant, on the other hand, revealed the spatial and temporal windows at which crime events were repeated more broadly to form clusters (hotspots) across an area.

The Knox test has been widely applied in many different geographical domains, such as epidemiology, ecology and forestry, in order to study phenomena that are peculiar to each domain. In most of these applications, usually only one variant of the Knox test is used at a time. It is proposed in this study that the utility of the two variants of the Knox test may facilitate the revelation of new meanings and inferences in relation to the space-time pattern of the data under study.

References

1. Knox, G.: Epidemiology of childhood leukaemia in Northumberland and Durham. Br. J. of Prev. Soc. Med. **18**, 17–24 (1964)
2. Townsley, M., Homel, R., Chaseling, J.: Infectious burglaries: a test of the near repeat hypothesis. Br. J. Criminol. **43**, 615–633 (2003)
3. Farrell, G., Pease, K.: Once bitten, twice bitten: repeat victimisation and its implications for crime prevention. Great Britain Home Office, UK (1993)
4. Ohno, Y., Aoki, K., Aoki, N.: A test of significance for geographic clusters of disease. Int. J. Epidemiol. **8**(3), 273–281 (1979)
5. Agresti, A., Finlay, B.: Statistical Methods for the Social Sciences, 3rd edn. Prentice Hall, New Jersey (1997)
6. Mantel, N.: The detection of disease clustering and a generalized regression approach. Cancer Res. **27**, 209–220 (1967)
7. Knox, G.: Detection of low intensity epidemicity application to cleft lip and palate. Br. J., Prev. Soc. Med. **17**(3), 121–127 (1963)
8. Adepeju, M.: Modelling of sparse spatio-temporal point process (STPP) – an application in predictive policing. A Ph.D. thesis submitted to the University College London (2017)
9. Johnson, S.D., Bowers, K., Hirschfield, A.: New insights into the spatial and temporal distribution of repeat victimization. Br. J. Criminol. **37**, 224–241 (1997)
10. Kulldorff, M., Heffernan, R., Hartman, J., Assunção, R., Mostashari, F.: A space–time permutation scan statistic for disease outbreak detection. PLoS Med. **2**(3), e59 (2005)

Testing the Adequacy of a Single-Value Monte Carlo Simulation for Space-Time Interaction Analysis of Crime

Monsuru Adepeju[(✉)]

School of Geography, University of Leeds, Leeds LS2 9JB, UK
M.O.Adepeju@leeds.ac.uk

Abstract. The goal of this study is to determine the number of iterations (r) required in a Monte Carlo based space-time interaction analysis of crime data sets, in order to test the adequacy of using a single value of 999 iterations. A case study of burglary crime data sets is presented in which Knox test is used for the analysis of space-time interactions. The outcomes of this analysis demonstrate that the use of a single value, such as 999, does not always represent the most appropriate number of iterations especially when multiple ST neighbourhood sizes are involved. This analysis opens further research opportunities into determining the best strategy to defining the expected distribution in a space-time interaction analysis of crime.

Keywords: Space-time neighbourhoods · Monte Carlo simulation · Crime · Knox test

1 Introduction

The use of a Monte Carlo (MC) simulation in space-time interaction analysis using the Knox test [1] usually involves 999 iterations in order to generate the expected distribution under the assumption of no space-time interactions. Despite the potentials of varying reliabilities relating to the underlying normal distribution for different pairs of spatial and temporal thresholds, the same number of iterations is usually employed in crime applications [2]. This study therefore aims to test the adequacy of the generally adopted 999 iterations at a chosen reliability level for different spatial and temporal thresholds.

One way to test the reliability of a MC simulation for a normal distribution is to specify a desired percentage error for the computed mean value of the random variables (in this case, the Knox statistic), while the iterations are continuously repeated [3]. Thus, the number of iterations needed (r) to attain the specified error can be determined by monitoring the convergence of r in relation to the actual number of iterations being performed. To the best of the author's knowledge, this type of analysis has not been carried out for space-time interaction analysis in relation to a crime data set. Therefore, the major goal of this study is to address this research gap by determining the number of iterations (r) required for different spatio-temporal (ST) neighbourhoods of crime,

© Springer International Publishing AG 2017
O. Gervasi et al. (Eds.): ICCSA 2017, Part VI, LNCS 10409, pp. 779–786, 2017.
DOI: 10.1007/978-3-319-62407-5_60

and subsequently, examine the adequacy of using a single value of 999 iterations in the context of the generated results.

2 Space-Time Interaction Analysis with the Knox Test

The Knox test is the most commonly used technique for the analysis of the spatio-temporal interactions of crime data sets [2]. The Knox test measures whether there are disproportionate instances of observed pairs of events within a defined spatio-temporal neighbourhood than would be expected if the events had occurred randomly. Therefore, the hypothetical random occurrences represent the expected distribution, which is generally modelled as a normal distribution.

Mathematically, the Knox statistic is a product of two 'closeness matrices'. The first matrix (X_{ij}) describes the closeness of all pairs of events in space, while the second matrix (Y_{ij}) describes the closeness of all pairs of events in time. The closeness is defined by specifying a spatial neighbourhood (δ) and a temporal neighbourhood (τ), within which event j is considered close to event i in space and time dimensions, respectively. Technically, each neighbourhood is the intersection of two distance thresholds; $[\delta_1, \delta_2]$ and $[\tau_1, \tau_2]$, where $\delta_2 > \delta_1$, and $\tau_2 > \tau_1$(see Fig. 1 for an illustration).

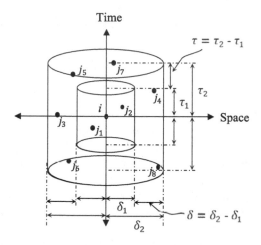

Fig. 1. An illustration of the spatio-temporal neighbourhood around a point i (in a Knox test). The event i is the reference, while events $j_1, j_2, \ldots, j_8$ are examined for 'closeness' to i. Events $j_3, j_4, \ldots j_8$ fall within the spatial neighbourhood $\delta = \delta_2 - \delta_1$ and temporal neighbourhood $\tau = \tau_2 - \tau_1$, and are therefore considered close to i in space and time (Diagram from [4]).

For each pair of spatial and temporal neighbourhood, the closeness is evaluated for every point (i) across the entire study area and finally added together in order to derive the Knox statistic as follows:

$$n_{\delta,\tau} = \frac{1}{2} \sum_{i=1}^{n} \sum_{j=1}^{n-1} X_{ij} Y_{ij} \qquad (1)$$

$$X_{ij} = \begin{cases} 1, & \text{if event } j \text{ is within } \delta \text{ of } i \\ 0, & \text{otherwise} \end{cases}$$

$$Y_{ij} = \begin{cases} 1, & \text{if event } j \text{ is within } \tau \text{ of } i \\ 0, & \text{otherwise} \end{cases}$$

The $n_{\delta,\tau}$ is referred to as the *observed*, with which the expected statistics $e_{\delta,\tau}$ are compared to estimate the critical value (p) through using the formula:

$$p = \frac{1 + \sum_{v=1}^{r} I(n_{\delta,\tau} \geq e_{\delta,\tau})}{r+1} \qquad (2)$$

Where r is the number of iterations generated, $e_{\delta,\tau}$ is the equivalent list of expected statistics, and $I(.)$ is the indication function.

An expected statistic is calculated via a new replica of the original data set, which is generated by randomising the time attribute of the events, while the spatial locations are kept constant. This process is the MC simulation (also called the iteration process). The iteration process is usually repeated in a specified number of times, to derive a list of expected test statistics ($e_{\delta,\tau}$). If each $e_{\delta,\tau}$ is considered a random variable, a plot of all $e_{\delta,\tau}$'s should assume a normal distribution defined by a mean value, and standard errors which can be evaluated at varying confidence levels. Theoretically, the mean value of the 'obtained' normal distribution is close to the mean value of the hypothetical normal distribution within an error bound. Hence, it is assumed that this hypothetical normal distribution can only be attained if the iteration is run infinitely. Both distributions can be compared in order to examine the reliability of the obtained normal distribution.

3 Determining the Number of Iterations (r) Needed to Attain a Specified Error Bound

It is possible to determine the minimum number of iterations (r) needed to provide a desired degree of reliability for the expected distribution ($e_{\delta,\tau}$), as described in Sect. 2. This process was used in [3] so as to estimate the r required in a precision analysis of military weapon effectiveness. Firstly, it is argued that r needs to be large enough to obtain sufficient granularity in the cumulative density function of the $e_{\delta,\tau}$. For example, if $r = 30$, it would not be possible to obtain a 1% rank. This is because an analyst needs at least 100 iterations. Using the [3] approach therefore requires a continuous generation of large number of replicas in which a plot of r against the number of replicas, at a specified maximum acceptable percentage error of the mean value can be used to monitor the convergence of r.

Table 1 shows the values of the confidence coefficient z_c for different confidence levels of a normally distributed random variable. The ranges for a given z_c are usually expressed in the form of an upper (U) and lower bound (L), whereby:

$$U = \mu_x + z_c \sigma_x \tag{3}$$

$$L = \mu_x - z_c \sigma_x \tag{4}$$

Where μ_x is the population mean and σ_x, the population standard deviation of the random variables x.

Table 1. Values of z_c for different confidence levels for a normally distributed random variable.

Confidence level (C.L) %	99.75	99	98	96	95.5	95	90	80	68	50
z_c		3	2.58	2.33	2.05	2	1.96	1.65	1.28	1	0.6745

Given a confidence level of 95% for example, the confidence interval of the mean is therefore as follows:

$$(L, U)_{0.95} = \mu_x \pm 1.96\sigma_x \tag{5}$$

This is stated as: we are 95% confident that the true mean is within (L, U) of a sample of the mean of x. The general form of Eq. 5 can then be written as:

$$(L, U)_{C.L} = \mu_x \pm z_c \sigma_x \tag{6}$$

If the simulation is run for a finite number of iterations (r), the sample mean $\bar{x}$ and the standard error S_x are thus estimates of the population statistics.

$$(L, U) = \bar{x} \pm z_c \left(S_x/\sqrt{r}\right) \tag{7}$$

By considering the confidence interval as representing twice this maximum error, we have:

$$error_{max} = \bar{x} \pm z_c \left(S_x/\sqrt{r}\right) \tag{8}$$

Hence, the percentage error of the mean becomes:

$$E = \frac{100 \times z_c S_x}{\bar{x}\sqrt{r}} \tag{9}$$

By solving the Eq. 9 for r, we have:

$$r = \left[\frac{100 \times z_c S_x}{\bar{x}E}\right]^2 \tag{10}$$

r in Eq. 10 is the number of iterations needed to be carried out for a given error bound E and a confidence interval whose coefficient is z_c. Let's imagine we want a confidence level of 95% at an error percentage of 1% of the mean, z_c and E will be 1.96 and 1, respectively, $\bar{x}$ and S_x can be estimated from the generated random samples. The value of r can thus be calculated continuously as the replicas increase. Furthermore, the required value of r can be taken as the point where r stabilises or converges in a plot of r against number of replicas.

4   Case Study: Data Sets and Experimental Parameters

This test is demonstrated using the burglary crime data set of the San Francisco area of the United States for the year 2015. The two most prominent sub-categories of burglary are used. They are, 'burglary-in-residence' (2,990 records) and 'burglary-of-shops' (1,166 records). The choice of these data sets is based on a previous finding that demonstrates that sub-categories of burglary crimes possess distinct spatio-temporal interactions [5]. Besides, the 3-D visual exploration of both data sets illustrates that 'burglary-in-residence' is denser than 'burglary-of-shops' spatio-temporally; indicating a potential for two distinct spatial and temporal interactions (Fig. 2).

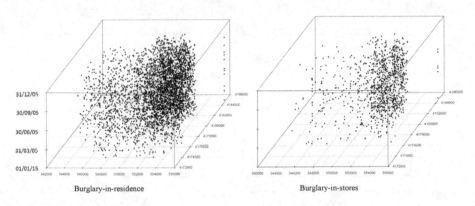

Fig. 2. A 3-D scatterplot of the case study data sets

In order to ensure a robust analysis, three common levels (sizes) of spatial and temporal neighbourhoods in both dimensions are considered. These include small, medium and large levels. Thus, the following lists are defined:

- Spatial neighbourhoods, δ = [0–200 m], [301–500 m], [701–900 m]
- Temporal neighbourhoods, τ = [0–2days], [7–21days], [30–60days]

The three levels are as demarcated with the brackets. Based on the two lists, corresponding spatio-temporal neighbourhood are formed by pairing each spatial neighbourhood with a temporal neighbourhood. Hence, a total of nine ST neighbourhoods are formed. This covers a range of levels commonly used in crime analysis.

In this study, the percentage error of 1% is chosen for the mean value, and a confidence interval of 95%. Both the mean value ($\bar{x}$) and the standard deviation (S_x) are calculated after each iteration step. The values, $\bar{x}$ and S_x are then substituted into Eq. 10, where $E = 1$ and $z_c = 1.96$.

5   Results and Discussions

Figure 3 shows the plot of the number of iterations needed (r) for the specified error parameters against the actual number of iterations (replicas) carried out. These plots enable the convergence of r to be monitored.

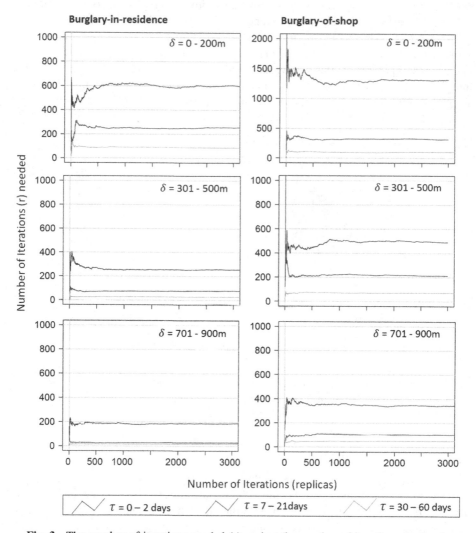

Fig. 3. The number of iterations needed (r) against the number of iterations (replicas).

Each plot shows the results generated for the selected temporal thresholds at each spatial threshold. For example, the top-most left and the top-most right plots are the results of the three temporal thresholds at the spatial threshold of 200 m, for the 'burglary-in-residence' and 'burglary-in-shops' crimes, respectively. The general pattern across all the plots is that the number of iterations needed (r) reduces as the sizes of the spatial and temporal thresholds increase. That is, in each crime sub-category, the highest value of r is obtained at the smallest spatiotemporal neighbourhood (i.e. intersection of $\delta = 0$–200 m and $\tau = 0$–2 days), while the lowest r is obtained at the largest ST neighbourhood (i.e. $\delta = 701$–900 m and $\tau = 30$–60 days). Technically, at large ST neighbourhoods, the mean values become relatively large in comparison to their respective standard errors, thereby allowing r values to converge faster. Whereas, there is higher variabilities at smaller ST neighbourhoods because of the relatively small values of mean in comparison the standard error, thereby requiring larger replicas to converge.

The results of 'burglary-in-residence' shows that the number of iterations (r) needed at the $\tau = 2$ days across all spatial thresholds is multiple times larger than all of the other temporal thresholds. The largest value of $r = 590$ is obtained at the spatial threshold of 0–200 m; a value which stabilises after only around 500 iterations. At the other spatial thresholds, r stabilises much faster; converging even before 250 iterations. In this case, the values of r are generally between 20 and 300. These are relatively small numbers compared to the commonly used 999 iterations. A similar result is obtained for 'burglary-of-shops', except with a slightly higher value of r for each corresponding ST neighbourhoods. Thus, this indicates that the 'burglary-of-shops' crime possesses a relatively higher variability in the ST distribution compared to the 'burglary-in-residence' crime. This is apparent in the 3D scatterplots (Fig. 2), in which 'burglary-of-shops' appears sparser compared to the findings for the 'burglary-in-residence' crime. The result of 'burglary-of-shops' at the smallest spatial and temporal neighbourhoods also shows that r could exceed 999. Additionally, this demonstrates that if the percentage error E value is reduced or the confidence level increased, the value of r can be much greater than 999. In summary, it is observed that the ST distribution of a data set, as well as the ST neighbourhood sizes used, influence the reliability of the expected distribution in a MC based space-time interaction analysis of crime data sets. Thus, the use of a single value, such as 999, may not adequately represent the most appropriate number of iterations in the case of multiple ST neighbourhood sizes.

6 Conclusion and Recommendations

This study has examined the number of iterations required for different sizes of ST neighbourhoods in a space-time interaction analysis of crime data sets. The aim was to test the reliability of the practice of using a single value, such as 999, in an MC simulation process involving many different ST neighbourhood sizes. The results obtained show that given some specified errors, different ST neighbourhood sizes require different numbers of iterations in order to generate reasonable expected (normal) distribution. This is in contrary to the general practice in which a uniform (single) value, particularly 999, is often used.

This study presents a demonstration involving the use of Knox test. For a given ST neighbourhood size, the convergence of the computed test statistics based on the generated expected distributions, is monitored. It was found that small valued ST neighborhoods require a relatively large number of iterations (r) compared to large valued of ST neighbourhoods. This is because the test statistics in the former is characterized by a much higher variation as compared with the latter. Additionally, the relatively sparse density in the case of small ST neighbourhood values strongly varies the number of events that will fall within any defined ST neighborhood size. For example, the 'burglary-of-shop' crime is sparser than 'burlgary-of-residence' and therefore shows higher variations in the test statistics. Overall, the results show that a smaller number of iterations (often less than 999) is required for most of the ST neighbourhood sizes experimented; and that the value is not uniform for different values of ST neighbourhood sizes.

Although the use of 999 iterations in the analysis space-time interaction of crime is the most common practice for reasons that may include convenience and ensuring uniformity in the precision (e.g. to 3 d.p) of the reported critical values, it is however shown in this study that the practice may produce bias expected distributions for different values of ST neighbourhoods. In the future, the author would like to investigate how the findings of this study could be employed to achieve a more reliable result using the Knox test.

Based on the results in this study, it is recommended that this type of analysis should first be carried out in any spatial and temporal point pattern analysis. It will help to establish the reliability of the MC simulation process.

References

1. Knox, G.: Epidemiology of childhood leukaemia in Northumberland and Durham. Br. J. of Prev. Soc. Med. **18**, 17–24 (1964)
2. Johnson, S.D., Bernasco, W., Bowers, K.J., Elffers, H., Ratcliffe, J., Rengert, G., Townsley, M.: Space-time patterns of risk: a cross national assessment of residential burglary victimization. J. Quant. Criminol. **23**, 201–219 (2007)
3. Driels, M.R., Shin, Y.S.: Determining the Number of Iterations for Monte Carlo Simulations of Weapon Effectiveness. Naval Postgraduate School, Monterey (2004)
4. Adepeju, M.: Modelling of sparse spatio-temporal point process (STPP) – an application in predictive policing. A Ph.D. thesis submitted to the University College London (2017)
5. Adepeju, M.: Investigating the repeat and near-repeat patterns in sub-categories of burglary crime. In: An Abstract Submitted to the 2017 International Conference on GeoComputation (2017)

Transport Properties of Liquid Aluminum at High Pressure from Quantum Molecular Dynamics Simulations

Shuaichuang Wang[(⊠)] and Haifeng Liu

Institute of Applied Physics and Computational Mathematics,
Beijing 100094, China
scwang@iapcm.ac.cn

Abstract. Increasing demands to subsequent design and engineering of new high performance materials are boosting the precise knowledge of the transport properties of liquid metals. Here we report a quantum molecular dynamics study of diffusion coefficients and viscosity of liquid aluminum under high pressure. The diffusion coefficients and viscosity are calculated up to 140 GPa and 10000 K. The results deviate from the Rosenfeld scaling law, but the exponential relationships still exist between the dimensionless diffusion coefficients and viscosity with the pair correlation entropy.

Keywords: Transport property · Liquid aluminum · Quantum molecular dynamics · High pressure

1 Introduction

Knowledge of the transport properties in liquid states is of fundamental importance for the subsequent design and engineering of new high performance materials, as well as for understanding the phenomena occurring in the Earth's core. Under ambient pressure, the transport properties of liquid aluminum have been investigated by molecular dynamics using empirical potential [1] or by means of the density functional theory (DFT) method [2, 3], and in experiments [4–6]. Under high pressure, experimental data for the transport properties of liquid aluminum, to our knowledge, are scarce, since the experimental determination currently seems to be extremely expensive and difficult.

Molecular dynamics (MD) is a powerful simulation technique to investigate transport properties of liquid phase. Equilibrium MD and non-equilibrium MD can both be used to obtain viscosity. Cherne et al. [7] calculated viscosity of liquid plutonium using the two methods and found their results agree well with each other. The accuracy of MD results is usually dependent on interatomic potential. Alfe and Gillan [2] implemented DFT calculations and found the self-diffusion coefficient and shear viscosity are sensitive to the choice of the density. Their calculations gave the diffusion coefficient at ambient melting point within the range 0.52–0.68 A^2/ps while shear viscosity are in the range 1.4–2.2 mPa.s. Jakse and Pasturel [3] also calculated the diffusion and viscosity by *ab initio* molecular dynamics. Their viscosity results are obtained from the transverse current correlation function, for the temperatures from

O. Gervasi et al. (Eds.): ICCSA 2017, Part VI, LNCS 10409, pp. 787–795, 2017.
DOI: 10.1007/978-3-319-62407-5_61

850 K to 1250 K. The liquid phase has remained relatively unexplored except around the melt temperature at the ambient pressure.

More recently, the diffusion coefficients and viscosity of many liquid metals at the high temperature and pressure have been investigated using MD simulations. [8, 9] It is demonstrated that the entropy-scaling laws for transport coefficients, proposed by Rosenfeld [10] at ambient pressure, hold well for several liquid metals under high pressure conditions. [8, 9] According to the entropy-scaling laws, the self-diffusion coefficient and viscosity of liquid metals can be expressed as single-valued functions of the pair correlation entropy. But the precise of the atomic interaction potentials is unknown at the high temperature and pressure, so the above knowledge needs to be further checked in the frame of quantum theory.

The main purpose of this paper is to further check the entropy-scaling law of liquid aluminum at a wider range of temperature and pressure, using quantum molecular dynamics simulations.

2 Methodology

The calculations are based on density functional theory in the finite-temperature formulation with the generalized gradient approximation (GGA) exchange-correlation functional of Perdew, Burke, and Ernzerhof (PBE) [11], as implemented in the Vienna ab initio Simulation Package (VASP) [12]. The projector-augmented-wave (PAW) method [13] was adopted with an atomic dataset involving 3 valence electrons with 3p and 3 s states. The plane-wave cut off energy of 500 eV was employed, and the precision was 0.1 meV/atom.

The computation supercells were obtained as $4 \times 4 \times 4$ cells corresponding to 256 atoms, which guaranteed that a single k point is sufficient to sample the Brillouin zone. We conducted our QMD simulations in the NVT ensemble (constant number of particles, constant volume and temperature). The ionic temperature was adjusted by Nose-Hoover thermostat, and the electron temperature was equal to the ionic temperature to achieve local thermodynamic equilibrium. The calculations were done at two fixed densities of 2.45 g/cc and 4.27 g/cc, with the temperature spanning from 900 K to 10000 K. We set the timestep of 1 fs with a negligible energy drift during tens of thousands of timesteps. The first 3000 timesteps were used for equilibrium, and subsequently atomic positions, velocities, stresses and other physical data were analyzed. The simulations run totally for from 30,000 to 100,000 timesteps.

The self-diffusion coefficient is computed from the trajectory by the mean-square displacement

$$D = \lim_{t \to \infty} \frac{1}{6t} \langle |\vec{R}_i(t) - \vec{R}_i(0)|^2 \rangle \tag{1}$$

or from the velocity autocorrelation function by the Green-Kubo formula

$$D = \frac{1}{3} \int_0^\infty \langle \vec{v}_i(t) - \vec{v}_i(0) \rangle dt, \qquad (2)$$

where $\vec{R}_i(t)$ and $\vec{v}_i(t)$ is the vectors of position and velocity of ith atom at the time t, respectively.

The viscosity is computed from the autocorrelation function of the off-diagonal components of the stress tensor by the Green-Kubo formula

$$\eta = \frac{V}{k_B T} \int_0^\infty \langle \sigma_{ij}(0) \sigma_{ij}(t) \rangle dt, \qquad (3)$$

where V, k_B, T and $\sigma_{ij}(t)$ are the atomic volume, Boltzmann constant, temperature and the off-diagonal components of the stress tensor, which includes P_{xy}, P_{xz}, P_{yz}, $(P_{xx} - P_{yy})/2$ and $(P_{xx} - P_{zz})/2$.

3 Results and Discussions

QMD simulations of aluminum were performed with the densities of 2.45 g/cc and 4.27 g/cc and the temperatures ranging from 900 K to 10000 K. The pair distribution functions (PDF) were used to determine the phase states. It is found that the solid aluminum above 1100 K changes into its liquid phase for the density 2.45 g/cc. The transition temperature is higher than the ambient melting temperature, because of the overheating effects. When the simulations started from liquid, the states were still liquid with the temperatures 1100 K and 900 K. For the density of 4.27 g/cc, the transition temperature from solid to liquid is above 4000 K. The results of the PDFs are shown in Fig. 1.

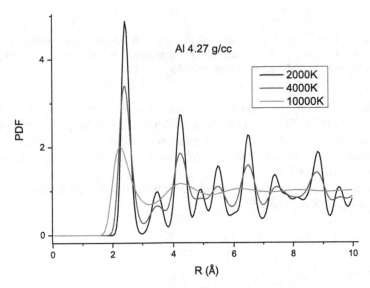

Fig. 1. The pair distribution functions of aluminum at the density of 4.27 g/cc.

The diffusion coefficients calculated are shown in Fig. 2. To our best knowledge, no experimental results exist for the diffusion constants of aluminum. Here we compared our results with the previous QMD results [3] (diamond), and found that the agreement was achieved between them. The results of the density 2.45 g/cc from Eq. (1) (square) are in good accordance with those from Eq. (2) (circle), and the similar case was found for the density 4.27 g/cc (up-triangle and down-triangle). Our calculated results also illustrate that the diffusion coefficient increases with the temperature, and a linear relationship exists between them in a wide range of temperature. This is different with the Arrhenius-type behavior for the diffusion process in the liquid state slightly above the melting temperature.

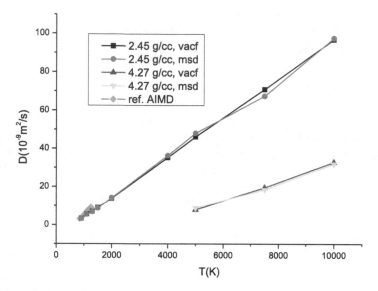

Fig. 2. The diffusion coefficients of aluminum as a function of temperature, with the densities of 2.45 g/cc and 4.27 g/cc: the present QMD calculations (square, circle, uptriangle and down-triangle) and the QMD results (diamond) [3].

The shear viscosity for liquid aluminum with the densities of 2.45 g/cc and 4.27 g/cc is shown in Fig. 3, where the experimental results [5] are also plotted for comparison. Our results are in good agreement with the experimental results [5] (blue circle). Obviously, the shear viscosity of 2.45 g/cc decreases with the increasing temperature up to 10000 K.

The Stokes-Einstein relation gives a connection between the diffusion and shear viscosity though the expression

$$F_{SE} = \frac{D\eta}{k_B T n^{1/3}} \qquad (4)$$

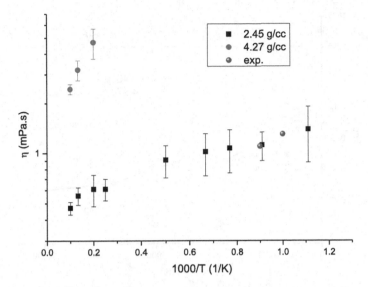

Fig. 3. The shear viscosity of aluminum as a function of temperature, with the densities of 2.45 g/cc and 4.27 g/cc: the present QMD calculations (black square, red circle) and the experimental results (blue sphere) [5]. (Color figure online)

where n is the number density, and F_{SE} is a constant. F_{SE} is about $\frac{1}{4\pi}$ when the slip coefficient limits to zero, based on the motion of a test particle through a solvent. Chisolm and Wallace [15] determined and empirical value of 0.18 ± 0.02 in a global fit to 21 metal species from a theory of liquids near melting. Here we check the Stokes-Einstein expression as a function of temperature, as shown in Fig. 4. Our QMD results are bounded by the classical slip values from below and the Chisolm-Wallace value of liquid metal [14] from above. For the density of 2.45 g/cc, F_{SF} has a slight decline with the temperature increasing. When the temperature is up to 10000 K, F_{SE} is near to the classical slip values.

To understand the relationship between the transport coefficients and structural properties of dense liquids, Rosenfeld [10] established a connection between the transport coefficients and the internal entropy, which is a universal scaling law between a dimensionless form of the diffusion coefficient $D^{\star}$, or of the viscosity $\eta^{\star}$, and the excess entropy S_{ex} of a liquid. The relationships are as below:

$$D^{\star} \approx 0.6e^{0.8S_{ex}} \tag{5}$$

$$\eta^{\star} \approx 0.2e^{-0.8S_{ex}} \tag{6}$$

In Eqs. (5) and (6), where $D^{\star}$ and $\eta^{\star}$ are given by:

$$D^{\star} = D\frac{\rho^{1/3}}{(k_B T/m)^{1/2}} \tag{7}$$

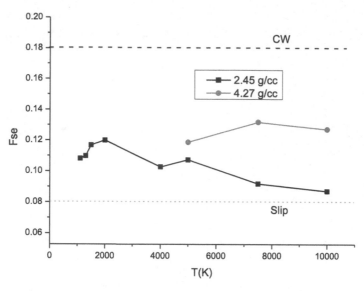

Fig. 4. The Stokes-Einstein relation as a function of temperature, with the densities of 2.45 g/cc and 4.27 g/cc: the present QMD calculations (black square, red circle). The flat lines show the slip boundary conditions (dashed) and the empirical result of Chisolm and Wallace [14] (dotted). (Color figure online)

$$\eta^{\star} = \eta \frac{\rho^{-2/3}}{(k_B T/m)^{1/2}} \tag{8}$$

In Eqs. (7) and (8), ρ and m are the density and atomic mass, respectively. The excess entropy can be approximated by the two-body expression denoted:

$$S_2 = -2\pi n \int_0^{\infty} \{g(r)\ln g(r) - [g(r) - 1]\} r^2 dr \tag{9}$$

Korkmaz and Korkmaz [15] have shown that it is less accurate for liquid iron, cobalt and nickel with a pseudopotential perturbation theory. Here we check the scaling law using the calculated diffusion coefficients and pair distribution functions.

As shown in Fig. 5, our QMD data of the density 2.45 g/cc (black square) give a linear relationship between the dimensionless diffusion coefficients and the pair correlation entropy S_2. And the line is also supported by those of the higher density 4.27 g/cc (red square), which illustrates the linear relationship is independent of the density. Our results deviate obviously from the original scaling law by Rosenfeld [10] (red line), because the two-body correlation entropy is inadequate to express the excess entropy. The similar phenomenon is also found in the previous MD simulations performed by Cao et al. [1]. Our QMD results are in agreement with the line proposed by Cao et al. [1].

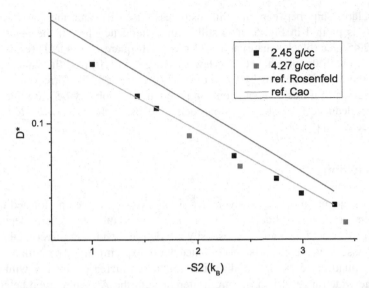

Fig. 5. The dimensionless diffusion coefficient as a function of the entropy S_2, with the densities of 2.45 g/cc and 4.27 g/cc: the present QMD calculations (black square, red square). The scaling law by Rosenfeld [10] (red line) and the line from the MD simulations by Cao et al. [1] (green line) are also plotted for comparison. (Color figure online)

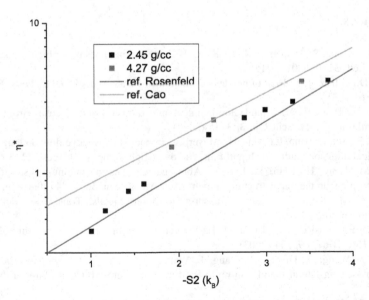

Fig. 6. The dimensionless viscosity as a function of the entropy S_2, with the densities of 2.45 g/cc and 4.27 g/cc: the present QMD calculations (black square, red square). The scaling law by Rosenfeld [10] (red line) and the line from the MD simulations by Cao et al. [1] (green line) are also plotted for comparison. (Color figure online)

The relationship between the dimensionless viscosity and the pair correlation entropy S_2 is plotted in Fig. 6. It is still a linear tendency for all the results of the densities 2.45 g/cc (black square) and 4.27 g/cc (red square). Our QMD results deviate slightly from both the original scaling law by Rosenfeld [10] (red line) and the fitting line from the MD simulations by Cao et al. [1] (green line). The former is because of the same reason of diffusion coefficient, and the latter is probably because that the MD viscosity is deduced from the diffusion coefficient using the Stokes-Einstein relation with a fixed S_{se} [1].

4   Conclusions

In this paper, quantum molecular dynamics simulations have been performed to obtain the transport properties of liquid aluminum. The temperature and density dependences of the diffusion coefficient and the viscosity are determined using direct methods. Our results of viscosity are in good accordance with experiments. The Stokes-Einstein relation is found to be valid. The diffusion coefficient increases linearly with temperature in the wide range of temperature, different with the Arrhenius-type behavior for the diffusion process in the liquid state slightly above the melting. The viscosity declines with temperature. The dimension diffusion coefficient or viscosity has a linear relationship with the pair correlation entropy S_2, independent of the density, which is slightly deviated of the scaling law by Rosenfeld.

References

1. Cao, Q., Wang, P.: Properties of liquid metals along melting lines under high pressure. Chin. Phys. Lett. **32**, 086201 (2015)
2. Alfe, D., Gillan, M.J.: First-principles calculation of transport coefficients. Phys. Rev. Lett. **81**, 5161 (1998)
3. Jakse, N., Pasturel, A.: Liquid aluminum: atomic diffusion and viscosity from ab initio molecular dynamics. Sci. Rep. **3**, 3135 (2013)
4. Kargl, F., Sondermann, E., Weis, H., Meyer, A.: Impact of convective flow on long-capillary chemical diffusion studies of liquid binary alloys. High Temp. High Press **42**, 3 (2013)
5. Sun, M., Geng, H., Bian, X., Liu, Y.: Abnormal changes in aluminum viscosity and its relationship with the microstructure of melts. Acta Metall. Sin. **36**, 1134 (2000). (in Chinese)
6. Shimoji, M., Itami, T.: Atomic Transport in Liquid Metals. Trans Tech Publications, Aedermannsdorf (1986)
7. Cherne, F.J., Baskes, M.I., Holian, B.L.: Predicted transport properties of liquid plutonium. Phys. Rev. B **67**, 092104 (2003)
8. Cao, Q., Wang, P., Huang, D., Yang, J., Wan, M., Wang, F.: Transport coefficients and entropy-scaling law in liquid iron up to Earth-core pressures. J. Chem. Phys. **140**, 114505 (2014)
9. Cao, Q., Huang, D., Yang, J., Wan, M., Wang, F.: Transport properties and the entropy-scaling law for liquid tantalum and molybdenum under high pressure. Chin. Phys. Lett. **31**, 066202 (2014)

10. Rosenfeld, Y.: A quasi-universal scaling law for atomic transport in simple fluids. J. Phys.: Condens. Matter **11**, 5415 (1999)
11. Perdew, J.P., Burke, K., Ernzerhof, M.: Generalized gradient approximation made simple. Phys. Rev. Lett. **77**, 3865 (1996)
12. Kresse, G., Furthmuuller, J.: Efficient iterative schemes for ab initio total-energy calculations using a plane-wave basis set. Phys. Rev. B **54**, 11169 (1996)
13. Blochl, P.E.: Projector augmented-wave method. Phys. Rev. B **50**, 17953 (1994)
14. Chisolm, E., Wallace, D.: Shock compression of condensed matter—2005. In: Furnish, M. D., Elert, M., Russell, T.P., White, C.T. (eds.) AIP Conference on Proceedings No. 845, AIP, New York (2006)
15. Korkmaz, S.D., Korkmaz, S.: Investigation of atomic transport and surface properties of liquid transition metals using scaling laws. J. Mol. Liq. **150**, 81 (2009)

Author Index

Printed in the United States
By Bookmasters